Radiation Protection Dosimetry

ISSN 0144-8420

Editor in Chief:
Mr. T. F. Johns U.K.

Executive Editor:
Mr. E. P. Goldfinch U.K.

Staff Editor:
Mrs. M. E. Calcraft

Consultant Editors:
Dr. G. Dietze, F.R.G.
Dr. Y. Horowitz, Israel.
Dr. J. Rundo, U.S.A.

Members of the Editorial Board:

Dr. R. M. Alexakhin, Russia
Prof. Dr. K. Becker, F.R.G.
Dr. M. A. Bender, U.S.A.
Dr. L. Bötter-Jenson, Denmark
Dr. G. Busuoli, Italy
Mr. M. W. Carter, Australia
Dr. M. W. Charles, U.K.
Mr. G. Cowper, Canada
Dr. Li Deping, Peoples Republic of China
Prof. Dr. B. Dörschel, F.R.G.
Dr. K. Duftschmid, Austria
Dr. V. J. Fominych, Russia
Miss F. A. Fry, U.K.
Mr. R. V. Griffith, U.S.A.
Dr. K. Harrison, U.K.
Mr. J. R. Harvey, U.K.
Dr. H. Ing, Canada
Dr. K. Irlweck, Austria
Prof. Dr. W. Jacobi, F.R.G.
Dr. R. L. Kathren, U.S.A.
Dr. E. Kunz, Czechoslovakia

Dr. A. R. Lakshmanan, India
Dr. D. C. Lloyd, U.K.
Mr. K. O'Brien, U.S.A.
Dr. H. F. Macdonald, U.K.
Mr. T. O. Marshall, U.K.
Dr. S. W. S. McKeever, U.S.A.
Dr. A. Moghissi, U.S.A.
Dr. M. Moscovitch, U.S.A.
Prof. Y. Nishiwaki, Japan
Mr. E. Piesch, F.R.G.
Dr. G. Portal, France
Dr. A. S. Pradhan, India
Dr. D. Ramsden, U.K.
Prof. Dr. A. Scharmann, F.R.G.
Mr. J. A. Selby, U.S.A.
Dr. F. Spurný, Czechoslovakia
Dr. R. H. Thomas, U.S.A.
Mr. I. M. G. Thompson, U.K.
Dr. L. Tommasino, Italy
Mr. J. W. N. Tuyn, Switzerland
Dr. M. E. Wrenn, U.S.A.

Published by Nuclear Technology Publishing, P.O. Box No. 7, Ashford, Kent, England.

Advertising Office – Mrs. L. Richmond, Nuclear Technology Publishing, P.O. Box No. 7, Ashford, Kent, England.

Subscription rates 1992: Volumes 40, 41, 42, 43 and 44, UK £430 p.a.; outside UK: US $950 p.a. Orders accompanying remittance should be sent to:

Subscription Department, Nuclear Technology Publishing, P.O. Box No. 7, Ashford, Kent, England. Tel. 0233 641683, Telex 966119 NTP UK G, Facsimile (0233) 610021

COPYRIGHT © 1992 Nuclear Technology Publishing

Legal disclaimer. The publisher, the editors or the editorial board accept no responsibility for the content of papers or the views expressed by authors.

Typeset by Lin-Art, West Station Yard, Godinton Road, Ashford, Kent.
Printed by Geerings of Ashford Ltd., Cobbs Wood House, Chart Road, Ashford, Kent, England

NTP.J.O.1

Who's got the world's widest range of Geiger Müller Tubes?

Centronic's range now includes The ZP Range (formerly Philips), plus The 'C' Range, plus The TGM Range making us the world's leading manufacturer and supplier of Geiger Müller tubes.

So, whatever your need; from the traditional glass-walled tube with halogen or organic vapour quenching, to thin-walled miniature metal GM tubes; Centronic Nuclear can supply it. Quickly and competitively. Added to which, our experience, know-how and high standards of production and service are second to none.

We couldn't make our Geiger Müller tubes any better; we just added the capability to make more of them.

Write for full technical information, prices and delivery, or ask for a visit from one of our Technical Sales Engineers.

CENTRONIC

The Nuclear Division of Centronic Ltd
Centronic House, King Henry's Drive, New Addington, Croydon CR9 0BG. Surrey, England.
Tel: 0689-842121. Fax: 0689-843053.
Telex: 896474 Centro G.

For North America, contact:
TGM Detectors Inc, 160 Bear Hill Road, Waltham, Massachusetts, 02154
Tel: (617) 890-2090. Fax: (617) 890-4711.

Morgan
NUCLEAR SAFETY PRODUCTS

NEUTRON DOSIMETRY

Proceedings of the
Seventh Symposium on Neutron Dosimetry
held at
Berlin, Federal Republic of Germany
October 14-18, 1991

Organised by:
Commission of the European Communities
Physikalisch – Technische Bundesanstalt
US-Department of Energy

Co-sponsored by:
GSF-Forschungszentrum für Umwelt und Gesundheit
Helmholtz-Fonds e.V.

Proceedings Editors:
R. Jahr, PTB
W. G. Alberts, PTB
H. Menzel, CEC
H. Schraube, GSF

CONF 911029
EUR 14547 EN
ISBN 1 870965 16 7
RADIATION PROTECTION DOSIMETRY Vol 44 Nos 1–4, 1992
Published by Nuclear Technology Publishing

SEVENTH SYMPOSIUM ON NEUTRON DOSIMETRY
Berlin, October 14-18, 1991

Organisation

CEC Commission of the European Communities
Directorate General for Science, Research and Development
Radiation Protection Programme

PTB Physikalisch-Technische Bundesanstalt
Braunschweig und Berlin

DOE United States Department of Energy
Office of Health and Environmental Research

Co-sponsors
GSF-Forschungszentrum für Umwelt und Gesundheit Helmholtz-Fonds e.V.

Programme Committee:

W. G. Alberts	PTB Braunschweig
D. Blanc	Univ. Toulouse
M. Coppola	ENEA Casaccia
P. DeLuca	Univ. Madison
J. A. B. Gibson	AEA Harwell
J. R. Harvey	BNL Berkeley
E. Piesch	KfK Karlsruhe
G. Portal	CEA Fontenay-aux-Roses
H. Schraube	GSF Neuherberg
F. Spurný	CSAV Praha
M. N. Varma	DOE Washington
A. Wambersie	UCL Bruxelles
M. Zielczinski	IAE Swierk
J. Zoetelief	TNO Rijswijk

Scientific Secretaries:

R. Jahr	PTB Braunschweig
H. G. Menzel	CEC Bruxelles

Local Organisation:

W. G. Alberts	PTB Braunschweig
A. Rimpler	BfS Berlin

British Library Cataloguing in Publication Data

A catalgoue record for this book is available at the British Library

LEGAL NOTICE
Neither the Commission of the European Communities nor any person acting on behalf of the Commission is responsible for the use which might be made of the following information.

Publication No EUR 14547 EN of the European Communities,
Dissemination of Scientific and Technical Knowledge Unit,
Directorate-General Telecommunications,
Information Industries and Innovation, Luxembourg

© ECSC-EEC-EAEC, Brussels and Luxembourg, 1992
ISBN 1 870965 16 7

Radiation Protection Dosimetry

INSTRUCTIONS TO AUTHORS

SCOPE: The scope of this journal covers all aspects of personnel and environmental dosimetry and monitoring for ionising and non-ionising radiations, including biological aspects, physical concepts, external personnel dosimetry and monitoring, internal dosimetry and monitoring, environment and work place monitoring and dosimetry related to protection of patients. Animal experiments and ecological sample measurements are not included.

Scientific papers should be of a theoretical or discursive nature as opposed to covering practical matters.

Technical papers should be of a practical nature relating to dosimetry or monitoring i.e. experimental or measurement techniques, instruments, etc.

Scientific or technical notes should be of six pages or less including figures or tables. (1 page contains about 900 words).

Review articles may be proposed by authors or commissioned by the Editor-in-Chief. A fee may be payable for their preparation.

Letters to the Editor should be written as letters with the authors' names and addresses at the end and should be marked 'For publication'.

LANGUAGE: All contributions should be in **English**. Spelling should be in accordance with the Concise Oxford Dictionary. However please use dosemeter and not dosimeter. Authors whose mother tongue is not English are requested to ask someone with a good command of English to review their contribution *before* submission.

TITLES should be brief and informative as possible. A short title of not more than 50 characters for a running head should be supplied.

AUTHORS' names and addresses (with full postal address) should appear immediately below the title.

ABSTRACTS containing up to 150 words should be provided. The abstract should appear on a separate page, headed by the title and authors' names.

SCRIPTS must be typewritten and **double** spaced. One copy must be directly typed and **three** additional (photocopies) should be provided for refereeing purposes to minimise the time required for refereeing. Headings should be given to main sections and sub-sections which should not be numbered. The title page should contain just the title, authors' names and addresses and a short running title. If your manuscript is prepared using a computer or word processor it would be helpful if you could also send a copy of the computer disc (3½″ or 5¼″ – please specify software).

FIGURES AND TABLES should not be inserted in the pages of manuscript but should be supplied on separate sheets. One high quality set of illustrations and figures, suitable for direct reproduction, e.g. indian ink or good quality black and white prints of line drawings and graphs, should be provided with the original typed manuscript. These should be approximately twice the final printed size (full page printed area = 19 cm × 15 cm). The lettering should be of such a size that the letters and symbols will remain legible after reduction to fit the printed area available. Tables should be typed using carbon ribbon. Tables should be lightly lined in pencil. All figures and tables should be numbered, using Arabic numerals, on the reverse side of each copy. Numbered captions or titles should be typed on a separate sheet. Figures and tables should be kept to the minimum consistent with clear presentation of the work reported. Half-tone photographs should only be included if absolutely necessary. Figures generated by computer graphics are generally NOT suitable for direct reproduction. Photocopies of all figures and tables should accompany each copy of the manuscript for refereeing purposes. Colour figures can be reproduced at cost.

UNITS, SYMBOLS AND EQUATIONS: SI units should be used throughout but other established units may be included in brackets. Any Greek letters or special symbols used in the text should be identified in the margin on each occasion they are used. Isotope mass numbers should appear at the upper left of the element symbol e.g. ^{90}Sr. Equations should be fully typed or lettered carefully in indian ink in the proper size.

FOOTNOTES should only be included if absolutely necessary. They should be typed on a separate sheet and the author should give a clear indication in the text by inserting (see footnote) so that they may appear on the correct page.

ABBREVIATIONS which are not in *common* usage should be defined when they first appear in the text.

REFERENCES should be indicated in the text by superior numbers in parenthesis and the full reference should be given in a list at the end of the paper in the following form, in the order in which they appear in the text:—

1. Crase, K. W. and Gammage, R. B. *Improvements in the Use of Ceramic BeO in TLD.* Health Phys. **29**(5) 739-746 (1975).
2. Clarke, R. H. and Webb, G. A. M. *Methods for Estimating Population Detriment and their Application in Setting Environmental Discharge Limits.* IAEM-SM-237/6. Proceedings of Symposium — Biological implications of radionuclides released from nuclear industries, Vienna, March 1979.
3. Aird, E. G. A. *An Introduction to Medical Physics.* William Heineman Medical Books Ltd. ISBN 0 433 003502.
4. Duftschmid, K. E. *TLD Personnel Monitoring Systems — The Present Situation.* Radiat. Prot. Dosim. **2**(1) 2-12 (1982).

All the authors' names and initials, the title of the paper, the abbreviated title of the journal, volume number, page numbers and year should be given. Abbreviated journal titles should be in accordance with the current World List of Scientific Periodicals.

PROOFS will be sent to any nominated author for final proof reading and must be returned within 3 days of receipt using the addressed label which will be provided. Corrections to proofs should be restricted to typographical or printer's errors etc. Any other changes may be charged to the authors. The Editor reserves the right to make editorial corrections to manuscripts. An order form for additional reprints will accompany proofs.

SUBMISSION: All manuscripts (original and three copies) and correspondence should be addressed to **Mr. E. P. Goldfinch, Nuclear Technology Publishing, PO Box No. 7, Ashford, Kent TN23 1YW England.** It is *essential* that they are accompanied by *six* fully addressed adhesive labels addressed to the author nominated to receive proofs and correspondence. These will be used for acknowledgement of receipt of the manuscript, notification of acceptance, return of proofs to authors and supply of reprints. Papers will be considered only on the understanding that they are not currently being submitted to other journals. The Publishers, the Editor and the Editorial Board do not accept responsibility for the technical content or the views expressed by authors.

CORRESPONDENCE: Please ensure that you provide telephone, fax and telex numbers if available. Please quote the manuscript number in any correspondence once receipt of your manuscript has been acknowledged.

COMPUTER MANUSCRIPTS: If your manuscript is prepared using a computer or wordprocessor, publication may be quicker if you submit a copy of the disc with the manuscript copies. The following programmes can be readily accommodated: 5¼″ disc (double density). – Multimate, Wordstar, MS Word, Word Perfect, Displaywrite and ASCII files. 3½″ – Wordstar, Multimate, MS Word, ASCII and IBM MS DOS Pro Dos files.

COPYRIGHT: Authors submitting manuscripts do so on the understanding that if it is accepted for publication, copyright of the article shall be assigned to Nuclear Technology Publishing.

In order to ensure rapid publication it is most important that **all** of the above instructions are complied with in **full**. Failure to comply may result in considerable delay in publication or the **return** of manuscripts to the author. In case of difficulty with illustrations and figures please consult the photo-reprographic section of your establishment. If illustrations of a quality high enough for direct off-set photographic reproduction cannot be supplied they may be redrawn by the publishers at the request of authors if all relevant details are provided. A charge will be made if requirements are extensive.

NTP.J.L.2

RADIATION PROTECTION DOSIMETRY
Previous Proceedings published by Nuclear Technology Publishing on behalf of the Commission of the European Communities

INDOOR EXPOSURE TO NATURAL RADIATION AND ASSOCIATED RISK ASSESSMENT – Proceedings of an International Seminar held at Anacapri, October 3-5, 1983 – Edited by G. F. Clemente, H. Eriskat, M. C. O'Riordan and J. Sinnaeve.
Radiat. Prot. Dosim. Vol. 7, Nos 1-4, 1984   440pp   £70

MICRODOSIMETRIC COUNTERS IN RADIATION PROTECTION – Proceedings of a Workshop on the Practical Aspects of Microdosimetric Counters in Radiation Protection held at Homburg/Saar May 15-17, 1984 – Edited by J. Booz, A. A. Edwards and K. G. Harrison.
Radiat. Prot. Dosim. Vol. 9, No. 3, 1984   120pp   £25

RADIATION PROTECTION QUANTITIES FOR EXTERNAL EXPOSURE – Proceedings of a Seminar held in Braunschweig, March 19-21, 1985 – Edited by J. Booz and G. Dietze.
Radiat. Prot. Dosim. Vol. 12, No. 2, 1985   166pp   £35

MICRODOSIMETRY – Proceedings of the Ninth Symposium on Microdosimetry held in Toulouse, May 20-24, 1985 – Edited by J. A. Dennis, J. Booz and B. Bauer.
Radiat. Prot. Dosim. Vol. 13, Nos. 1–4, 1985   400pp   £70

DOSIMETRY OF BETA PARTICLES AND LOW ENERGY X RAYS – Proceedings of a Workshop held at Saclay (France), October 7-9, 1985 – Edited by J. Booz, W. A. Jennings and G. Portal.
Radiat. Prot. Dosim. Vol. 14, No. 2, 1986   140pp   £30

ENVIRONMENTAL AND HUMAN RISKS OF TRITIUM – Proceedings of a Workshop held at Karlsruhe (Germany), February 17-19, 1986 – Edited by G. Gerber, C. Myttenaere and H. Smith.
Radiat. Prot. Dosim. Vol. 16, Nos. 1–2, 1986   192pp   £40

ETCHED TRACK NEUTRON DOSIMETRY – Proceedings of a Workshop held at Harwell, U.K., May 12-14, 1987 – Edited by D. T. Bartlett, J. Booz and K. G. Harrison.
Radiat. Prot. Dosim. Vol. 20, Nos. 1-2, 1987   130pp   £25

ACCIDENTAL URBAN CONTAMINATION – Proceedings of a Workshop held at Roskilde, Denmark, June 9-12, 1987 – Edited by H. L. Gjørup, F. Heikel Vinther, M. Olast and J. Sinnaeve.
Radiat. Prot. Dosim. Vol. 21, Nos. 1-3, 1987   192pp   £40

NEUTRON DOSIMETRY – Proceedings of the Sixth Symposium on Neutron Dosimetry held at Neuherberg, Germany, October 12-16, 1987 – Edited by H. Schraube, G. Burger and J. Booz.
Radiat. Prot. Dosim. Vol. 23, Nos. 1-4, 1988   498pp   £75

NATURAL RADIOACTIVITY – Proceedings of the Fourth International Symposium on the Natural Radiation Environment held at Lisbon, Portugal, December 7-11, 1987 – Edited by A. O. de Bettencourt, J. P. Galvao, W. Lowder, M. Olast and J. Sinnaeve.
Radiat. Prot. Dosim. Vol. 24, Nos. 1-4, 1988   560pp   £85

BIOLOGICAL ASSESSMENT OF OCCUPATIONAL EXPOSURE TO ACTINIDES – Proceedings of a Workshop held at Versailles, (France,) May 30-June 2, 1989 – Edited by G. B. Gerber, M. Métivier and J. Stather.
Radiat. Prot. Dosim. Vol. 26, Nos. 1-4, 1989   400pp   £65

IMPLEMENTATION OF DOSE-EQUIVALENT OPERATIONAL QUANTITIES INTO RADIATION PROTECTION PRACTICE – Proceedings of a Seminar, Braunschweig (Germany), June 7-9, 1988 – Edited by G. Dietze, H. G. Paretzke and J. Booz.
Radiat. Prot. Dosim. Vol. 28, No. 1/2, 1989   166pp   £35

IMPLEMENTATION OF DOSE-EQUIVALENT METERS BASED ON MICRODOSIMETRIC TECHNIQUES – Proceedings of a Seminar, Schloss Elmau (Germany), October 18-20, 1988 – Edited by H. Menzel and J. Booz.
Radiat. Prot. Dosim. Vol. 29, Nos. 1/2, 1989   156pp   £30

MICRODOSIMETRY – Proceedings of the Tenth Symposium on Microdosimetry, Rome, May 22-26, 1989 – Edited by J. Booz, J. A. Dennis and H. Menzel.
Radiat. Prot. Dosim. Vol. 31, Nos. 1-4, 1990   450pp   £80

STATISTICS OF HUMAN EXPOSURE TO IONISING RADIATION – Proceedings of a Workshop held in Oxford, April 2-4, 1990 – Edited by M. C. O'Riordan and J. Sinnaeve.
Radiat. Prot. Dosim. Vol. 36, Nos. 2-4, 1991   280pp   £60

SKIN DOSIMETRY – Radiological Protection Aspects of Skin Irradiation – Proceedings of a Workshop held in Dublin, May 13-15, 1991 – Edited by H. G. Menzel, P. Christensen and J. A. Dennis
Radiat. Prot. Dosim. Vol. 39, Nos. 2-4, 1991   220pp   £60

AGE-DEPENDENT FACTORS IN THE BIOKINETICS AND DOSIMETRY OF RADIONUCLIDES – Proceedings of a Workshop held in Schloss Elmau, November 5-8, 1991 – Edited by D. M. Taylor, G. B. Gerber and J. W. Stather.
Radiat. Prot. Dosim. Vol. 41, Nos. 2-4, 1992   252pp   £60

DOSIMETRY IN DIAGNOSTIC RADIOLOGY – Proceedings of a Seminar held in Luxembourg, March 19-21, 1991 – Edited by H. M. Kramer and K. Schnuer.
Radiat. Prot. Dosim. Vol. 43, Nos. 1-4, 1992   320pp   £80

The above Proceedings are available from Nuclear Technology Publishing, PO Box No. 7, Ashford, Kent, England.

NTP.J.K.3

Natural Radiation

Proceedings of the
Fifth International Symposium
held in Salzburg, Austria
September 22–28 1991

Published as a Supplement to Radiation Protection Dosimetry
(Vol 45 Nos 1–4, 1992)

Topics: Measurement Techniques and Metrology

Exposure to Natural Radiation in Non-domestic Environments

Natural Radionuclides and Transfer Pathways

Radioactivity and Radiation in the Human Environment

Surveys of Natural Radiation

Characteristics and Behaviour of Rn Progeny

Industrially Modified Levels of Radiation Exposure

Radon Mitigation Methods

Health Effects of Natural Radiation

Radon Control Policies and Recommendations

This supplement extends to some 1000 pages and is NOT automatically included in the 1992 subscription to Radiation Protection Dosimetry

ISBN 1 870965 14 0 (hardback) Dec 1992 £150 (UK) US$300 (Outside UK)

Also available:
INDOOR EXPOSURE TO NATURAL RADIATION – Proceedings of a Seminar held in Anacapri, Oct 1983 – published as Radiat. Prot. Dosim, Vol. 7 Nos 1–4 (1984) 440pp £70 (UK) (Outside UK US$140).

NATURAL RADIOACTIVITY – Proceedings of the Fourth International Symposium on the Natural Radiation Environment held in Lisbon, December 1987 – Published as Radiat. Prot. Dosim. Vol 24 Nos. 1–4 (1988) –560pp £85 (UK) (Outside UK US$170).

SPECIAL DISCOUNTS: Prepayment 5%
2 Volumes 15%
3 Volumes 25%   VISA accepted.

Orders to
Nuclear Technology Publishing
PO Box 7
Ashford
Kent TN23 1YW
England

NTP.J.P.1

Contents

	page
Contents	vii

Editorial: 7th Symposium on Neutron Dosimetry
– R. Jahr, W. G. Alberts, H. G. Menzel and H. Schraube xiii

Keynote Address

The Biological Effectiveness of Neutrons: Implications for Radiation Protection
E. J. Hall and D. J. Brenner 1

Physical Data

Status of Nuclear Data for Use in Neutron Therapy
R. M. White, J. J. Broerse, P. M. DeLuca Jr., G. Dietze, R. C. Haight, K. Kawashima, H. G. Menzel, N. Olsson and A. Wambersie (INVITED PAPER) 11

Determination of Kerma Factors of A-150 Plastic and Carbon at Neutron Energies Between 45 and 66 MeV
U. J. Schrewe, H. J. O'Brede, S. Gerdung, R. Nolte, P. Pihet, P. Schmelzbach and H. Schuhmacher 21

Measurement of Neutron Kerma Factors in C, O, and Si at 18, 23, and 25 MeV
C. L. Hartmann, P. M. DeLuca, Jr and D. W. Pearson 25

Secondary Alpha Particle Spectra and Partial Kerma Factors of the Reaction $n + {}^{12}C \rightarrow n + 3\alpha$
B. Antolkovic, G. Dietze and H. Klein 31

Radiobiological Effectiveness

RBE Modifying Factors
M. Coppola, V. Di Majo, S. Rebessi and V. Covelli 35

Variation of Neutron RBE as a Function of Energy for Different Biological Systems: A Review
J. Van Dam, M. Beauduin, V. Grégoire, J. Gueulette, G. Laublin and A. Wambersie 41

Inverse Dose Rate Effects for Neutrons: General Features and Biophysical Consequences
D. J. Brenner, R. C. Miller, S. A. Marino, C. R. Geard, G. Randers-Pehrson and E. J. Hall 45

Radiation Quality and Energy Deposition

Neutron Energy Deposition on the Nanometer Scale
J. J. Coyne and R. S. Caswell 49

A Microdosimetric Monte Carlo Code for Neutrons up to 60 MeV
G. C. Taylor and M. C. Scott 53

Determination of Quality Factor in Mixed Radiation Fields Using a Recombination Chamber
N. Golnik and M. Zielczynski 57

Cavity Detectors

Energy Deposition in a Spherical Cavity of Arbitrary Size and Composition
E. Kearsley 61

Wall Component of $r_{m,g}$ Versus Cavity Size: A Summary of the Experiments with Alpha Particles
S. Pszona 65

CONTENTS

Time-Resolved Microdosimetry in a Quasi-Monoenergetic Neutron Beam
P. J. Binns, J. H. Hough and B. R. S. Simpson 67

Simulation of the Response of an Ultraminiature Microdosimetric Counter for Fast Neutrons
P. Olko, K. Morstin and T. Schmitz 73

Measurement and Prediction of Real Tissue Microdosimetric Responses for Neutrons up to 60 MeV
A. C. A. Aro, S. Green, R. Koohi-Fayegh, M. C. Scott, T. Shahid and G. C. Taylor 77

Spectrometry and Fluence Measurements

Comparison of Response Function Calculations for Multispheres
C. A. Perks, D. J. Thomas, B. R. L. Siebert, S. Jetzke, G. Hehn and H. Schraube 85

Comparison of Measured and Calculated Bonner Sphere Responses for 24 and 144 keV Incident Neutron Energies
B. R. L. Siebert, E. Dietz and S. Jetzke 89

Determination of a Photon Response Matrix for Simultaneous Measurement of Neutron and Photon Spectra in Mixed Fields
T. Novotný 93

Absolute Neutron Fluence Determination with a Spherical Proton Recoil Proportional Counter
M. Weyrauch and K. Knauf 97

Measurement of High Energy Neutron Fluence with Scintillation Detector and Proton Recoil Telescope
R. Nolte, H. Schuhmacher, H. J. Brede and U. J. Schrewe 101

Calibration Fields and Techniques

Neutron Measurement Intercomparisons Sponsored by CCEMRI, Section III (Neutron Measurements)
R. S. Caswell and V. E. Lewis 105

BIPM Neutron Dosimetry Comparison Based on the Circulation of a Set of Transfer Instruments
V. D. Huynh 111

Critical Assessment of Calibration Techniques for Low Pressure Proportional Counters Used in Radiation Dosimetry
P. Pihet, S. Gerdung, R. E. Grillmaier, A. Kunz and H. G. Menzel 115

The Application of a New Geometry Correction Function for the Calibration of Neutron Spherical Measuring Devices Using Large Volume Neutron Sources
M. Khoshnoodi and M. Sohrabi 121

Experimental Assembly for the Simulation of Realistic Neutron Spectra
J. L. Chartier, F. Posny and M. Buxerolle 125

Extended Use of a D_2O-Moderated ^{252}Cf Source for the Calibration of Neutron Dosemeters
S. Jetzke, H. Kluge, R. Hollnagel and B. R. L. Siebert 131

A Computer Library of Neutron Spectra for Radiation Protection Environments
B. R. L. Siebert, H. Schraube and D. J. Thomas 135

Dosimetric Parameters of Simple Neutron + Gamma Fields for Calibration of Radiation Protection Instruments
K. Józefowicz, N. Golnik and M. Zielczyński 139

Establishment of a Procedure for Calibrating Neutron Monitors at the Physics Institute of the University of São Paulo, Brazil
M. T. Cruz and L. Fratin 143

CONTENTS

Development of a Neutron Calibration Facility at NRPB
T. M. Francis 147

Quality Factors and Dose Quantities in Radiation Protection

New Quantities for Use in Radiation Protection
C. B. Meinhold (INVITED PAPER) 151

Calculated Effective Doses in Anthropoid Phantoms for Broad Neutron Neutron Beams with Energies from Thermal to 19 MeV
R. A. Hollnagel 155

Equivalent Dose Versus Dose Equivalent for Neutrons Based on New ICRP Recommendations
K. Morstin, M. Kopec and Th. Schmitz 159

Implications of New ICRP and ICRU Recommendations for Neutron Dosimetry
G. Portal and G. Dietze (INVITED PAPER) 165

Verification of an Effective Dose Equivalent Model for Neutrons
J. E. Tanner, R. K. Piper, J. A. Leonowich and L. G. Faust 171

Moderator Type Area Monitors

A Moderator Type Dose Equivalent Monitor for Environmental Neutron Dosimetry
A. Esposito, C. Manfredotti, M. Pelliccioni, C. Ongaro and A. Zanini 175

Neutron Dose Equivalent Rate Meter on the Basis of the Single Sphere Albedo Technique
B. Burgkhardt, E. Piesch and M. I. Al-Jarallah 179

Study of the Response of Two Neutron Monitors in Different Neutron Fields
A. Aroua, M. Boschung, F. Cartier, K. Gmür, M. Grecescu, S. Prêtre, J.-F. Valley and Ch. Wernli 183

Dose Equivalent Response of Neutron Survey Meters for Several Neutron Fields
A. Rimpler 189

A Neutron Survey Meter with Sensitivity Extended up to 400 MeV
C. Birattari, A. Esposito, A. Ferrari, M. Pelliccioni and M. Silari 193

Tissue Equivalent Proportional Counters

Tissue-Equivalent Proportional Counters in Radiation Protection Dosimetry: Expectations and Present State
H. Schuhmacher (INVITED PAPER) 199

Influence of Photon Radiation on Neutron Dose Equivalent Measurement in Radiation Protection with CIRCEG
A. Marchetto and Y. Herbaut 207

The Homburg Area Neutron Dosemeter Handi: Characteristics and Optimisation of the Operational Instrument
A. Kunz, E. Arend, E. Dietz, S. Gerdung, R. E. Grillmaier, T. Lim and P. Pihet 213

Spectrometry and Dosimetry in Special Environments

An Intercomparison of Neutron Field Dosimetry Systems
D. J. Thomas, A. J. Waker, J. B. Hunt, A. G. Bardell and B. R. More 219

Neutron Field Spectrometry for Radiation Protection Dosimetry Purposes
A. V. Alevra, H. Klein, K. Knauf and J. Wittstock 223

CONTENTS

Neutron Spectrometry and Dosimetry Measurements Made at Nuclear Power Stations with Derived Dosemeter Responses
H. J. Delafield and C. A. Perks 227

Neutron Spectra, Radiological Quantities and Instrument and Dosemeter Responses at a Magnox Reactor and a Fuel Reprocessing Installation
D. T. Bartlett, A. R. Britcher, A. G. Bardell, D. J. Thomas and I. F. Hudson 233

Neutron Spectrometry System for Radiation Protection: Measurements at Work Places and in Calibration Fields
F. Posny, J. L. Chartier and M. Buxerolle 239

Comparison of Different Neutron Area Monitors as Routine Radiation Protection Devices Around a High Energy Accelerator
M. Boschung, C. Wernli and A. Kunz 243

Measurement of Neutron Dose Equivalent and Penetration in Concrete for 230 MeV Proton Bombardment of Al, Fe, and Pb Targets
J. V. Siebers, P. M. DeLuca, Jr., D. W. Pearson and G. Coutrakon 247

HZE Cosmic Rays in Space. Is it Possible that They are Not the Major Radiation Hazard?
J. F. Dicello (INVITED PAPER) 253

Individual Dosimetry

Individual Neutron Monitoring – Needs for the Nineties
R. V. Griffith (INVITED PAPER) 259

Properties of Personnel Neutron Dosemeters on the Basis of Intercomparison Results
E. Piesch, B. Burgkhardt and M. Vilgis 267

Type Testing and Routine Calibration of Neutron Personal Dosemeters: Phantoms and Phantom Backscatter
R. J. Tanner, D. T. Bartlett, T. M. Francis and J. D. Steele 273

Calibration Method for Personnel Neutron Dosemeters in Stray Radiation Fields
A. V. Sannikov 277

Albedo Dosemeters

The Role of Phantom Parameters on the Response of the AEOI Neutriran Albedo Neutron Personnel Dosemeter
M. Sohrabi and M. Katrrouzi 281

Comparison of Two Types of Albedo Dosemeters in Several Mixed Neutron-Gamma Fields
A. Aroua, M. Grecescu, P. Lerch, S. Prêtre and J.-F. Valley 287

Personal Albedo Neutron Dosemeter Using Highly Sensitive LiF TL Chips
D. Nikodemová, A. Hrabovcová, M. Vičanová and S. Kaclík 291

Calibration of the Brazilian Albedo Dosemeter at a CV-28 Cyclotron
P. W. Fajardo and C. L. P. Maurício 293

Thermoluminescent Detectors

Neutron Response of LiF TL Detectors
T. Hahn, J. Fellinger, J. Henniger, K. Hübner and P. Schmidt 297

TLD-300 Dosimetry at Chiang Mai 14 MeV Neutron Beam
W. Hoffmann and P. Songsiriritthigul 301

CONTENTS

Thermoluminescence Dosimetry in Mixed (n,γ) Radiation Fields Using Glow Curve Superposition
T. M. Piters, A. J. J. Bos and J. Zoetelief 305

Calibration Methods of TLD-300 Dosemeters in a Clinical 14 MeV Neutron Beam
M. Kriens, R. Schmidt, A. Hess and W. Scobel 309

Solid State Nuclear Track Detectors

A Simple Personal Dosemeter for Thermal, Intermediate and Fast Neutrons Based on CR-39 Etched Track Detectors
M. Luszik-Bhadra, W. G. Alberts, E. Dietz, S. Guldbakke and H. Kluge 313

A Comparison of the Neutron Response of CR-39 Made by Different Manufacturers
N. E. Ipe, J. C. Liu, B. R. Buddemeier, C. J. Miles and R. C. Yoder 317

International Study of CR-39 Etched Track Neutron Dosemeters (Eurados-Cendos 1990)
W. G. Alberts 323

A Three Element Etched Track Neutron Dosemeter with Good Angular and Energy Response Characteristics
J. R. Harvey 325

Recent Developments on the CRS PADC Fast Neutron Personal Dosemeter
S. Djeffal, Z. Lounis and M. Allab 329

Angle of Energy Response to Fast Neutrons of CR-39 Covered with a Radiator
E. Pitt, A. Scharmann and R. Simmer 333

Experimental and Theoretical Determination of the Fast Neutron Response Using CR-39 Plastic Detectors and Polyethylene Radiators
F. Fernández, C. Domingo, E. Luguera and C. Baixeras 337

A CR-39 Fast Neutron Dosemeter Based on an (n,α) Converter
E. Savvidis, D. Sampsonidis and M. Zamani 341

Bubble Detectors

Characterisation of New Passive Superheated Drop (Bubble) Dosemeters
R. E. Apfel 343

Bubble Detectors in Fusion Dosimetry
N. Smirnova, N. Semaschko and Y. Martinuk 347

Measurements of Fast Neutrons by Bubble Detectors
J. Schulze, W. Rosenstock and H. L. Kronholz 351

New Dosimetric Methods and Techniques

Some New Techniques for Neutron Radiation Protection Measurements
B. Dörschel, H. Seifert and G. Streubel (INVITED PAPER) 355

Principles of an Electronic Neutron Dosemeter Using a Pips Detector
B. Barelaud, D. Paul, B. Dubarry, L. Makovicka, J. L. Decossas and J. C. Vareille 363

Electronic Sensor Response in Neutron Beams
B. Dubarry, B. Barelaud, J. L. Decossas, L. Makovicka, D. Paul and J. C. Vareille 367

Gamma Interference on an Electronic Dosemeter Response in a Neutron Field
D. Paul, B. Barelaud, B. Dubarry, L. Makovicka, J. C. Vareille and J. L. Decossas 371

CONTENTS

Detection of Neutron-Induced Heavy Charged Particle Tracks in RPL Glasses
B. Lommier, E. Pitt and A. Scharmann 375

Fast Neutron and Photon Therapy

Neutron Therapy: From Radiobiological Expectation to Clinical Reality
A. Wambersie (INVITED PAPER) 379

Dosimetric Characteristics of Proton, Neutron and Negative Pion Beams at the Phasotron in Dubna
F. Spurný (INVITED PAPER) 397

Studies Relating to 62 MeV Proton Cancer Therapy of the Eye
V. P. Cosgrove, A. C. A. Aro, S. Green, M. C. Scott, G. C. Taylor, D. E. Bonnett and A. Kacperek 405

The Primary Attenuation Coefficient of a p(66)+Be(40) Neutron Therapy Beam
A. N. Schreuder, D. T. L. Jones, S. Pistorius and W. A. Groenewald 411

Tissue-Maximum Ratios for a p(66)+Be(40) Neutron Therapy Beam
M. Yudelev, A. N. Schreuder and D. T. L. Jones 417

Monte Carlo Calculations of the Effect of Air Cavities on the Dose Distribution of d(14)+Be Neutrons
P. Meissner 421

Boron Neutron Capture Therapy

Neutron Spectrometry and Dosimetry for Boron Neutron Capture Therapy
C. A. Perks and J. A. B. Gibson 425

Determination of Dose Enhancement by Neutron Capture of ^{10}B in a d(14)+Be Neutron Beam
F. Pöller, W. Sauerwein and J. Rassow 429

In-Phantom ^{10}B Capture Rates for Medical Applications at a Reactor Therapy Facility
H. Schraube, F. M. Wagner, V. Mares and G. Pfister 433

A Ferrous Sulphate Gel Dosimetry System for NCT Studies: Response to Slow Neutrons
M. C. Cantone, C. Canzi, U. Cerchiari, D. de Bartolo, L. Facchielli, G. Gambarini, N. Molho, L. Pirola and A. E. Sichirollo 437

Treatment Planning of Boron Neutron Capture Therapy: Measurements and Calculations
M. W. Konijnenberg, C. P. J. Raaijmakers, L. Dewit, B. J. Mijnheer, R. L. Moss, F. Stecher-Rasmussen and P. R. D. Watkins 443

Neutron and Gamma Introduction Facilities

Review on the Physical and Technical Status of Fast Neutron Therapy in Germany
J. Rassow, U. Haverkamp, A. Hess, K. H. Höver, U. Jahn, H. Kronholz, P. Meissner, K. Regel and R. Schmidt 447

Neutron Capture Therapy Beam on the LVR-15 Reactor
M. Marek, J. Burian, Z. Prouza, J. Rataj and F. Spurný 453

Evaluation of the Undesired Neutron Dose Equivalent to Critical Organs in Patients Treated by Linear Accelerator Gamma Ray Therapy
C. Manfredotti, U. Nastasi, E. Ornato and A. Zanini 457

List of Participants 463

Author Index 469

Radiation Protection Dosimetry is abstracted or indexed in APPLIED HEALTH PHYSICS ABSTRACTS AND NOTES, Chemical Abstracts, CURRENT CONTENTS, Energy Information Abstracts (Cambridge), EXCERPTA MEDICA (EMBASE), FLASH (Belgium), Health & Safety Science Abstracts (Cambridge), INIS ATOMINDEX (hard copy and CD-ROM), INSPEC, Nuclear Energy (Czechoslovakia), QUEST, Referativja Zhurnal (MOCKBA), RUSH ABSTRACTS (Cambridge).

Editorial

7th Symposium on Neutron Dosimetry

The Symposium in Berlin continued the series of six successful Symposia on Neutron Dosimetry previously held at GSF-Forschungszentrum für Umwelt und Gesundheit (formerly: Gesellschaft für Strahlen und Umweltforschung), in München-Neuherberg (Germany). It was jointly organised by the Commission of the European Communities, Directorate-General for Science, Research and Development, and the Physikalisch-Technische Bundesanstalt (PTB), Braunschweig and Berlin. The Symposium was co-sponsored by the US Department of Energy, Office of Health and Environmental Research, GSF and Helmholtz Fonds e.V. The Bundesamt für Strahlenschutz (BFS, German Federal Office for Radiation Protection) provided the excellent conference facilities in Berlin. The geographical location of Berlin and the recent fundamental political changes contributed to the fact that more scientists from East European countries participated in the conference than in previous Symposia.

This Symposium was again concerned with the research and development of neutron dosimetry in radiation protection, radiation biology and radiation therapy. The common research problems specific to neutron dosimetry in these areas are obviously of more relevance than the problems relating to the differences in application. In all applications, dosimetry has to provide a quantitative measure of the radiation interaction with the tissue considered so that this measure can be related to the observed or, in radiation protection, to the expected biological effect. For physical and biological reasons, neutron dosimetry poses more complex and more difficult technical and conceptual problems than photon dosimetry. From the physical point of view, the complexity of nuclide specific interactions of neutrons with matter for a large range of energy is a challenging task when quantities such as absorbed dose or dose equivalent have to be determined. From the biological point of view, the marked dependence of biological effectiveness on neutron energy has to be accounted for. Specific aspects of neutron radiation biology such as the inverse dose rate effect, observed in several investigations, were presented and discussed and they document the necessity to continue research in this field.

In radiation protection, the concept of using quality factors and dose equivalent was introduced in order to account for radiation quality and to have single quantities for all types of radiation. More recently, the International Commission on Radiological Protection (ICRP), in its publication 60, has put forward an alternative concept for risk related quantities, based on radiation weighting factors, and uses the quantity effective dose, E, instead of effective dose equivalent, H_E. The consequence of these new recommendations was discussed extensively and controversially during the Symposium.

The ICRP recommendations to reduce exposure limits and to increase the quality factor for neutrons also have implications on practical neutron monitoring. One of the important tasks for immediate neutron dosimetry research is therefore to develop dosemeters with considerably lower detection limits.

The Symposium revealed a continued increase in attention given to neutrons of high energies. In radiation protection, high energy neutron exposures occur near accelerators, during jet aircraft flights and during space missions. In tumour therapy with fast neutrons higher neutron energies (up to 70 MeV) are used to improve the dose distribution depth. At these energies the implications of the new ICRP recommendations for neutron and mixed field exposures including protons in addition to gamma rays require further detailed studies.

In the field of 'classic' radiation protection, progress was achieved in decreasing the dosemeters' energy dependence and in reducing the uncertainties of dose determination by applying more than one single measuring method. Apart from the use of conventional area and individual dosemeters, spectrometric and microdosimetric methods were emphasised. An increasing number of tasks in the fields of neutron dosimetry and spectrometry are also expected to arise at tokamaks and other nuclear fusion plants. Another aim is the design of electronic and direct reading personal dosemeters. Computational simulation has meanwhile become an indispensable aid in solving many such problems. Particularly in view of the aforementioned large energy range, development in all these areas needs to be continued in the future.

The continued interest in tumour radiation therapy using fast neutrons and charged particles was documented at the Symposium, as well as the renewed activities in boron neutron capture therapy. The complexity of the related dosimetric problems and the high accuracy requirements in clinical dosimetry are reasons for the need for further dosimetric research.

EDITORIAL

The Symposium addressed a wide range of specific topics and provided a forum for multi-disciplinary exchanges of information. The large number of participants documented the continued importance of this research field.

We would like to thank the Bundesamt für Strahlenschutz, which not only made available its conference facilities, but also cooperated in the local organisation of the Symposium. Special thanks are due to the sponsors, the CEC for the financial support granted to young scientists, to the members of the programme committee, and to those scientists who cooperated in the preparation and holding of the Symposium and in the evaluation of the papers published in these proceedings. We also express our thanks to Mr E. P. Goldfinch for his patience in the preparations for publication.

With respect to the form of the manuscripts, it has turned out to be something of a drawback that the international standards institutes have not yet succeeded in agreeing on a uniform nomenclature for certain detector quantities: there has been some confusion regarding the terms 'response' (defined as detector indication divided by the quantity to be measured, according to ISO standard 8529, which is recommended by those chairing the Symposium) and 'sensitivity' (according to the 'International Vocabulary of Basic and General Terms in Metrology' published by ISO in 1984 and also by DIN and as BSI PD 6461, part 1) which has also been used.

R. Jahr
W. G. Alberts
H. G. Menzel
H. Schraube

THE BIOLOGICAL EFFECTIVENESS OF NEUTRONS; IMPLICATIONS FOR RADIATION PROTECTION

E. J. Hall and D. J. Brenner
Center for Radiological Research
College of Physicians & Surgeons of Columbia University
New York, NY 10032, USA

INVITED PAPER

Abstract — The radiobiology of neutrons has important societal implications. In nuclear facilities worldwide, several hundred thousand individuals are monitored as potentially receiving doses of neutrons, while there is increasing concern about the dose to which airline crew members are exposed, which includes a substantial neutron component. This represents an even larger number of individuals. The relative biological effectiveness (RBE) for neutrons varies with neutron energy. With cell killing as an endpoint, the maximum RBE occurs at about 300 to 400 keV, and falls off above and below this energy. For oncogenic transformation, the variation of effectiveness with energy is less marked. However, for some neutron energies and at some doses, there is a clear inverse dose-rate effect — i.e. an increased rate of transformation with dose protraction, the magnitude of which depends on neutron energy, dose and dose rate. There is considerable uncertainty at the present time concerning the RBE of low energy 'soft' neutrons in the range 10 to 100 keV.

INTRODUCTION

A significant number of people have the potential to be occupationally exposed over a protracted period to low doses of neutrons. In the United States about 92,000 individuals are monitored as potentially receiving neutron doses in Department of Energy facilities and about 7000 individuals receive measurable neutron doses[1]. In addition, of the approximately 600,000 monitored workers in the United States under Nuclear Regulatory Commission regulation, about 6000 per year, primarily research workers, well loggers and reactor workers receive measurable neutron doses[2]. There must be at least an equal number of workers similarly exposed in Europe.

There is also increasing concern about the neutron dose to which airline crew members (300,000 in US airlines, about one million worldwide) are exposed. Calculations[3,4] indicate that in some cases crew members will receive more than the maximum permissible dose for non-radiation workers, about half the dose equivalent coming from neutrons.

It is thus of considerable interest to understand the biological hazards posed by these radiations, and to be able to relate them to the better understood hazards of X rays.

A major problem in understanding such hazards is the lack of human data. The recent reassessment of radiation doses delivered during the A bomb attacks on Hiroshima and Nagasaki have resulted in a major reduction in the estimated neutron doses at Hiroshima compared with that suggested by the old 'T65D' dosimetry. Previously, the generally observed increased risk per unit dose at Hiroshima compared with Nagasaki was attributed to neutrons, which were considered to be a significant component of the dose at Hiroshima, but not at Nagasaki. By comparing observed risks at Hiroshima and Nagasaki, the relative biological effectiveness (RBE) of the neutrons compared to gamma rays could be estimated[5]. However, since the reassessed neutron doses at Hiroshima are much smaller, such RBE estimates have become extremely uncertain[6,7]. Thus, any estimates of the biological hazards posed by neutron exposures must be primarily based on *in vitro* radiobiological experiments, and on experiments with animals.

In the present system of radiation protection the relative risks associated with exposure to different ionising radiations at low doses and low dose rates are compared quantitatively by multiplying the average dose to an organ or tissue D, by a number, w_r, previously termed the quality factor and now referred to as the radiation weighting factor:

$$H = w_r D$$

where the resulting quantity, H, is called equivalent dose. H is, by definition, independent of radiation quality.

The problem, then, is to generate appropriate values of w_r for the various neutron fields (at appropriate dose rates) to which workers are exposed. For many years the quality factor 10 was recommended for all neutrons, but in recent years the recommended value, both from the ICRP and the NCRP, has increased to 20[8]. Currently, the ICRP[8] has proposed neutron weighting (quality) factors of 5 for neutron energies less than 10 keV, 10 for energies between 10 and 100 keV, 20 for energies between 100 and 2000 keV, 10 between 2 and 20 MeV, and 5 at higher energies.

The data on which Q estimates have been based have been summarised by Sinclair[10]. The endpoints are primarily *in vitro* oncogenic transformation, mutation, and chromosomal aberration in a variety of (mostly) mammalian cells (primarily of rodent origin), and tumour induction and life shortening in rats and mice. There is considerable spread in the results from available data, from a low of around 15 to a high of around 100.

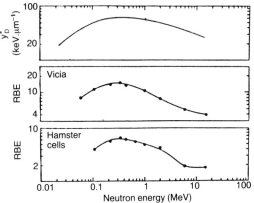

Figure 1. RBE as a function of neutron energy for inhibition of root growth in *Vicia faba* and for lethality in V79 Chinese hamster cells. (From ICRU 40[14]. Data from References 11 and 12. Also shown for comparison is y* as a function of neutron energy[13]).

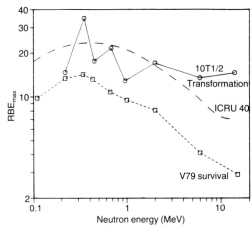

Figure 2. Relative biological effectiveness (RBE) as a function of neutron energy as measured by different biological endpoints. The transformation data, obtained with C3H 10T1/2 cells, are from Miller et al 1989[50]. The cell lethality data, obtained with V79 Chinese hamster cells are from Ref 12. The dotted line represents values of Q suggested in ICRU report 40[14].

RBE RELATIVE TO NEUTRON ENERGY

There are many data available for the biological effects of neutrons from a few hundred keV to 14 MeV. Figure 1 shows RBE as a function of neutron energy from 0.1 to 15 MeV, based on survival of V79 Chinese hamster cells and seedlings of *Vicia faba*[11,12]. The biological data are in accord with biophysical theory based on microdosimetry[13], with the RBE reaching a maximum value for neutron energies around 300 keV. At this energy most of the recoil protons released in elastic scattering processes of neutrons with hydrogen would have an LET close to 100 keV.μm^{-1}, which is biologically most effective.

In more recent years, neutrons covering the same energy range have been used for experiments involving oncogenic transformation using C3H 10T1/2 cells as an *in vitro* assay. These data are also shown in Figure 2 where they are compared with the earlier V79 data for cell survival. There is broad agreement, particularly in shape, between the various data sets, within the limits of experimental error, which is considerable, especially in the case of the transformation data. It is of interest to note that RBE_M values for transformation are in close accord with the Q values suggested by the ICRU 40 report[14].

TRANSFORMATION FREQUENCY PER INITIAL CELL AT RISK

Transformation data are usually plotted in terms of transformation incidence per surviving cell. However, for extrapolating *in vitro* data to an *in vivo* situation where a whole organ or organism

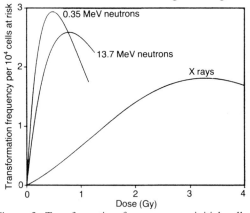

Figure 3. Transformation frequency per initial cell at risk as a function of dose for C3H 10T1/2 cells exposed to X rays and to neutrons of 0.35 and 13.7 MeV. Transformation frequency per initial cell at risk rises rapidly with dose at low doses, reaches a peak, and then falls as cell killing takes over. (Fitted curves based on data in Miller et al[50]).

is exposed, a more relevant quantity may be transformation incidence per initial cell at risk. This quantity is shown as a function of dose for X rays and for two neutron energies in Figure 3. The transformation incidence per initial cell at risk rises rapidly with dose at low doses, reaches a peak and then falls off rapidly with dose at higher doses, as transformation saturates and cell killing takes over. There are several points of interest:

(i) The maximum transformation incidence per initial cell at risk (sometimes called the maximum oncogenic potential, MOP) occurs at a lower dose for high LET radiation (<1 Gy) than for X rays.
(ii) The maximum transformation potential (MOP) does not vary greatly between high LET radiations.

In fact it has been shown elsewhere[15] on theoretical grounds that the maximum oncogenic potential (MOP) should not vary strongly between high LET radiations, but would be expected to be different from that produced by low LET radiation. In principle the high LET MOP could be greater or less than that from low LET — in this case it is greater.

SOFT NEUTRONS

The energy spectrum to which occupationally exposed workers will be subject varies widely, even within a given reactor facility, depending on the neutron source, and the degree of shielding, and thus the moderation to which the neutrons are subject. In addition, of course, the neutrons are moderated by the body of the exposed individual. Whether this is significant in terms of the biological effectiveness, and thus the appropriate Q factor, depends on whether the neutron biological effectiveness varies over the neutron energy range of significance for occupational exposure.

The significant neutron energy range, in terms of dose deposited, varies according to the fluence spectrum to which the individual is exposed. A typical example is shown in Figure 4 showing kerma weighted fluences, as a function of neutron energy, for one position in a particular PWR nuclear reactor. For these situations, on average, about half the dose comes from neutrons with energies below 140 keV.

The neutron energy range from 30 to 500 keV is, however, the energy range where there may be significant variations in biological response. In the neutron energy range below ~ 100 keV, there are basically two data sets available, both based on filtered reactor beams; these are from Sevankaev et al[16] in the Soviet Union (nominal 40, 90 keV) and Lloyd and colleagues (nominal 24 keV), in the UK (e.g. Refs. 17, 18). The yield (per unit dose at low doses) of dicentric chromosomal aberrations in human lymphocytes, as measured by Sevankaev et al[16], is considerably decreased compared with the yield at a neutron energy of a few hundred keV. This is in accord with earlier results for cellular survival (shown in Figure 1) and is also in accord with biophysical expectations[13,14]. On the other hand, the results of the Harwell group[17,18], both for dicentric chromosomal aberration yields in human lymphocytes, and for several other end points in rodent cells suggested comparable yields to those at a few hundred keV. This disagreement is significant on two levels. First, in the terms of radiation protection issues, since a significant decrease in the biological effectiveness of neutrons from the hundreds of keV to the tens of keV range would result in a decrease in the quality factor appropriate for most occupational exposure situations. Second, in terms of biological mechanisms, since radiobiological models based on energy deposition in cellular or nucleus sized targets unequivocally predict a decrease in biological effect as the neutron energy decreases; if this decrease were not to be confirmed, then such models would be substantially falsified.

To summarise, neutron quality factors currently suffer from several significant uncertainties:

1. There are no appropriate human data available, and limited *in vivo* data, for rodents.

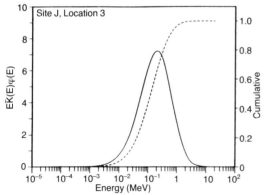

Figure 4. Distribution of neutron energies responsible for the deposition of neutron dose for various positions in different commercial PWR reactors where personnel might be exposed. The representation of the spectra is such that equal areas under the full curve correspond to equal depositions of dose. The dashed curve is a cumulative representation of the same data, such that, for example, 50% of the dose is deposited at energies below 100 keV. The graph corresponds to a particular PWR reactor, as measured by Endres et al[51].

2. The *in vitro* radiobiological data available are primarily for cells of mammalian, but non-human origin.
3. It is unclear whether the neutron quality factor should vary significantly with neutron energy in the energy range appropriate for radiation protection, i.e. below 100 keV.
4. It is unclear whether the neutron quality factor should vary with dose rate, in the dose and energy range of interest.

It is evident, therefore, that much needs to be done in radiobiological research with neutrons. Soft neutron facilities need to be designed and constructed and radiobiological experiments performed to address issues of both pragmatic and basic importance.

THE INVERSE DOSE RATE EFFECT FOR RADIATION OF INTERMEDIATE LET

In the practical case of radiation protection, doses to personnel are usually accumulated over a protracted period of time at low dose rate, whereas cancer risk estimates, based on the experience of the survivors of Hiroshima and Nagasaki, relate to a single acute exposure. In the case of low LET radiation, such as X and γ rays, it is generally agreed that the biological effects of a given total dose are reduced when the dose is delivered over a protracted period of time at low dose rate. This is the conventional dose rate effect, and appears to apply to cell killing, mutation or cancer induction. The same is not true for more densely ionising radiation, such as neutrons, where a greater biological effect may result from a protracted exposure than from the same total dose delivered in an acute exposure. This has come to be known as the inverse dose rate effect and may be of special importance in radiation protection. Some evidence for this inverse dose rate effect for neutrons or other high LET radiation comes from life-shortening, tumour induction and cataract induction in laboratory animals[19–24]. There are also some epidemiological studies that are supportive of the animal studies, though none are conclusive[25–28]. However, the most comprehensive set of data for the inverse dose rate effect relate to studies with *in vitro* assays for oncogenic transformation.

Elkind and his colleagues first reported an inverse dose rate effect for fission spectrum neutrons in the induction of oncogenic transformation in C3H 10T1/2 cells[29]. They found that for a given low dose of neutrons (50–300 mGy), the transformation incidence was much higher if the radiation were delivered at low dose rate, or in a series of fractions over five days, than if delivered in a single high dose rate exposure.

Since the initial report by Hill *et al*[29] at least 15 further publications have appeared reporting experimental results that address the possible enhancements of transformation by dose protraction (i.e. by low dose rate or fractionation). A review has been published elsewhere[30]. These measurements cover a variety of radiations of intermediate to high LET including fission neutrons[29,31,32], high energy therapy neutrons[33], monoenergetic neutrons[34,35], as well as charged particles[35–38]. A variety of doses, dose rates, and fractionation schemes were utilised. Although, at first sight, the data appear to be confusing and conflicting, in fact a pattern does emerge. The largest enhancement was observed for fission neutrons at dose rates below about 5 mGy.min^{-1}, while little or no enhancement is apparent for all radiations at dose rates above 5 mGy.min^{-1}. It is apparent, too, that charged particles having LET above about 140 keV.μm^{-1} produce little or no enhancement. In addition, the inverse dose rate effect appears most prominent at doses around 0.2 Gy, with less evidence of enhancement at doses much above or below this region.

Over the years a number of suggestions have been published to account for the enhancement of transformation by dose protraction[31,39–46]. The simplest and most credible model is based on an idea by Rossi and Kellerer[42]. This model postulates that, for the biological endpoint of interest, cells at some period of their cell cycle are more sensitive to radiation than at other times, i.e. there is a 'window' of increased sensitivity. A further postulate is that an acute high LET exposure of cycling cells results in some fraction of these sensitive cells receiving very large depositions of energy — much greater than that required to produce oncogenic transformation. On the other hand, if the exposure is protracted or fractionated, a larger proportion of sensitive cells will be exposed, though to smaller numbers of energy depositions; however, since higher LET radiations are involved it is postulated that the total specific energy deposited in the sensitive cells by this smaller number of energy depositions will still be large enough to produce transformation. To the extent that this latter postulate will not hold at low LET, the inverse dose rate effect would not be expected to apply to such radiation.

The development of this model, and the methods used to fit the experimental data, have been described in detail by Brenner and Hall[45,46], only the general conclusions will be discussed here. Figure 5 shows the transformation enhancement ratio as a function of dose for several different neutron beams. The predictions of the model are also shown, assuming a window of sensitivity of 1 h.

It is apparent that the predictions are in reasonable accord with the data for the various different radiations. In particular, although 6 MeV neutrons and fission neutrons have comparable values of y_D, their markedly differing values of y_F result in a much larger enhancement from fission neutrons at low dose rates. For all the radiations, the predicted enhancements at dose rates above 5 mGy.min^{-1} are small enough to be experimentally undetectable, again in agreement with experiment. Figure 6 shows the experimental data together with the corresponding predictions from the model, for higher LET charged particle radiation (140-200 keV.µm^{-1}); the predicted enhancements are very small, and are consistent with the lack of a significant effect seen experimentally.

The model outlined here, and described in detail by Brenner and Hall[45,46] appears to be consistent with all the data currently available in the C3H 10T1/2 system for the enhancement of transformation by dose protraction of intermediate and high LET radiation. Whether or not enhancement of transformation is seen as a consequence of dose protraction is a complex function resulting from the interplay of several factors, including dose, dose rate and the LET of the radiation. One unequivocal prediction is that no enhancement would be expected for high LET alpha particles from domestic radon-produced exposure. It is, however, possible that the higher exposures to which uranium miners were exposed may produce an inverse dose rate effect[25,27], which could bias domestic radon risk estimates. Another possible area of concern involves the fast trapped protons to which astronauts on a space station would be exposed, since they produce a significant fraction of the total dose from medium LET heavy fragments. For intermediate LET radiation such as fission neutrons the effect would be confined to intermediate doses, as the model predicts that the acute and continuous transformation rates will have the same initial slope at extremely low doses.

To assess the possible implications of the inverse dose rate effect to the field of radiation protection in general, the model, with parameters derived from the C3H 10T1/2 data, has been extrapolated to the wide range of very low dose rates encountered in radiation protection involving fission spectrum neutrons, or other radiation of intermediate LET. For doses less than 1 mGy (corresponding to non-occupational annual effective dose limits), the enhancement would be

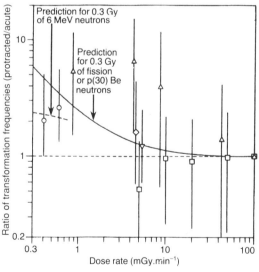

Figure 5. Published data with 95% confidence limits for enhancement of oncogenic transformation of C3H 10T1/2 cells by low doses (0.2–0.3 mGy) of 'medium' LET radiation ($y_D = 60$–70 keV.µm^{-1} in a 1µm site). (o) 6 MeV neutrons, (∇,◊,□) fission neutrons or p(30)–Be and p(46) Be neutrons (0.5 Gy). Predictions of model: Full curve; $y_F = 1$ keV.µm^{-1}, 0.3 Gy fission neutrons, refers to (∇,◊,□). Dash: $y_F = 15.7$ keV.µm^{-1}; 0.3 Gy of 6 MeV neutrons, refers to (o). References to data as in Ref. 45.

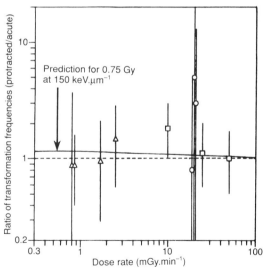

Figure 6. Published data with 95% confidence limits for enhancement of oncogenic transformation of C3H 10T1/2 cells by low doses (0.5–1 Gy) of high LET charged particles. (Δ) 150 keV.µm^{-1}; (□) 140 keV.µm^{-1}; (o) 200 keV.µm^{-1}. (O) predictions of model. The predicted enhancement for 140, 150 and 200 keV.µm^{-1} are virtually identical, and cannot be resolved experimentally from zero enhancement (redrawn from Brenner and Hall[45]).

insignificant at any dose rate. The annual occupational effective dose equivalent limit is 50 mSv, corresponding (assuming a quality factor of 10) to 5 mGy of fission neutrons; if this dose were delivered continuously over a year at a dose rate of 10^{-4} mGy.min^{-1}, then an enhancement of about 2 might be expected. Larger enhancements could theoretically be possible for astronauts receiving bigger doses of protons at very low dose rates. For example, NCRP Report 98 recommends that the effective dose equivalent for space missions in the shuttle be 0.5 Sv.y^{-1} or 1–1.5 Sv in a lifetime; if the model can be believed, the enhancement ratio may then be as high as 10 to 20.

PHOTONEUTRONS

The production of photoneutrons in high energy medical linacs is a subject of considerable topical interest, and a real challenge to the neutron dosimetrist. More and more high energy machines of around 18 to 20 MeV or higher are in routine clinical use for radiation therapy, particularly in treatment plans designed with curative intent; this is because of the superior dose distributions that can be obtained for deep seated tumours. Photoneutrons are generated, primarily, in the high Z material of the collimators, with a threshold of about 6 to 8 MeV for lead, tungsten or uranium. There is a giant resonance (i.e. a peak) in the photoneutron production cross section in these heavy materials at a photon energy around 14 MeV.

Neutrons with a broad energy range are generated, the mean being around 1–3 MeV. Various attempts have been made to measure the quantity and quality of these neutrons, against the background of high energy X rays. The most elegant attempt to date utilised an ultra-miniature tissue-equivalent proportional counter (Kliauga, personal communication, 1991). Neutrons are emitted isotropically and contribute about 0.1 to 0.3% by dose to the primary X ray field; because of their range, this represents a whole-body neutron dose to the patient.

Medium energy neutrons are highly effective biologically, and so it is of interest to make an assessment of the impact of photoneutrons on radiotherapy in terms of the possible induction of second malignancies:

Assuming a treatment dose of 6×10^4 mGy over 6 weeks, a neutron dose of 0.1 to 0.3%, with a quality factor 10 to 20 for the mixture of neutron energies (some of which would qualify for 20 and some for 10 according to ICRP), then the whole body dose equivalent to the patient would amount to:

$$6 \times 10^4 \times \frac{0.1 \text{ to } 0.3}{100} \times (10 \text{ to } 20) \text{ mSv}$$
$$= 600 \text{ to } 3600 \text{ mSv}$$

Using a risk estimate of about 3% per Sv for protracted whole-body irradiation (based on BEIR V and UNSCEAR), this whole-body dose would produce an incidence of second malignancies of about 2 to 10%. This is, of course, in addition to the risk from the photon dose, which would not be significantly different from that characteristic of lower energy linacs or cobalt teletherapy units.

In the United States, about one million new cases of cancer are reported annually, about half of which are treated with radiotherapy. Half of these are treated with intent to cure, and about half of these are cured. This amounts to about $1/8 \times 10^6$ long-term survivors per year. If about 1/3 of these patients were treated with high energy linacs, the number of second malignancies that might be induced by photoneutrons would be:

$$(2 \text{ to } 10\%) \times 1/3 \times (1/8 \times 10^6)$$
$$= 800 \text{ to } 4000$$

This estimate is for the United States, and there would be a comparable number in Europe, as more high energy linacs come into routine clinical use. These secondary malignancies are the price paid for an improved dose distribution which, presumably, leads to better tumour control and/or reduced normal tissue morbidity. It might seem to be a high price that needs to be considered further.

ONCOGENE ACTIVATION

Neutrons, as well as X rays, have been studied in the context of oncogene expression in radiation induced tumours, though in no instance has a specific oncogene been identified as the causal step either for a tumour *in vivo*, or even with transformation *in vitro*. Pellicer and colleagues[47] showed that in X ray induced lymphomas in mice, activated ras could be identified in 9/37 or 24% of the tumours, and that seven of these involved the same mutation. In further studies with neutrons[48], activated ras was observed in only 4/25 or 16% of the tumours and each instance involved a different mutation. It was concluded from these studies that while the activated ras oncogene may well be involved with the development of the tumour, at least in some cases it was not the causal event. This interesting study *in vivo* finds its parallel in the work of Trutschler *et al*[49], who reported a study by Northern analysis of the expression of cellular oncogenes in a number of transformed cell lines from Syrian hamster embryos (SHE cells). All cell lines showed increased expression of the c-Ha-ras gene by factors of two or three.

This enhanced Ha-ras expression was also found in spontaneously immortalised cells even in low passage number, before they became tumorigenic. They interpreted these data as evidence that the enhanced expression of the c-Ha-ras gene is an early event during the process of neoplastic transformation of SHE cells, and that it is not radiation-specific. In none of the cell lines were transcripts of c-Ki-ras or N-ras found.

In contrast to the c-Ha-ras gene the expression of the c-myc gene was increased (by factors of two to four) only in SHE cells transformed by carbon ions, in related tumour cell lines and in spontaneously immortalised cells at higher passages. The other transformants expressed the c-myc gene to the same extent as primary SHE cells. Trutschler et al[49] also sequenced the c-Ha-ras gene lines and normal SHE cells, but could find no changes in the nucleotide sequence; thus radiation induced transformation of SHE cells does not appear to be caused by, or accompanied by, activation of the Ha-ras gene via a point mutation.

ACKNOWLEDGEMENT

This paper is based on research supported by grant number DE-FG02-88ER60631 from the United States Department of Energy and by grants CA 12536, CA 49062 and CA 24232 from the National Cancer Institute.

REFERENCES

1. Merwin, S. E., Millete, W. H. and Traub, R. J. *Twenty-first Annual Report on Radiation Exposures for DOE and DOE Contractor Employees — 1988.* DOE Report DOE/EH-0171P (1990).
2. Brooks, B. G. *Occupational Radiation Exposure at Commercial Nuclear Power Reactors and Other Facilities, 1985: Eighteenth Annual Report.* NUREG-0713, Vol. 7 (US Nuclear Regulatory Commission, Washington, DC) (1988).
3. Friedberg, W., Faulkner, D. N., Snyder, L., Darden, E. B. and O'Brien, K. *Galactic Cosmic Radiation Exposure and Associated Health Risks for Air Carrier Crew Members.* Aviat. Space Environ. Med. **60**, 1104-1108 (1989).
4. Wilson, J. W. and Townsend, L. W. *Radiation Safety in Commercial Air Traffic: A Need for Further Study.* Health Phys. **55**, 1001-1003 (1988) with erratum, Health Phys. **56**, 973-974 (1989).
5. National Research Council: Committee on the Biological Effects of Ionizing Radiation. *The Effects on Populations of Exposure to Low Levels of Ionizing Radiation: 1980. (Washington, DC: National Academy Press) (1980).*
6. Preston, D. L. and Pierce, D. A. *The Effect of Changes in Dosimetry on Cancer Mortality Risk Estimates in the Atomic Bomb Survivors.* Radiat. Res. **114**, 437-466 (1988).
7. Brenner, D. J. *Significance of Neutrons from the Atomic Bomb at Hiroshima for Revised Radiation Risk Estimates.* Health Phys. **60**, 439-442 (1991).
8. ICRP. *The 1990–1991 Recommendations of the International Commission on Radiological Protection.* Publication 60 (Oxford: Pergamon) Ann. ICRP **21**(1-3) (1991).
9. Sinclair, W. K. *Trends in Radiation Protection — A View from the National Council on Radiation Protection and Measurement (NCRP).* Health Phys. **55**, 149-157 (1988).
10. Sinclair, W. K. *Fifty Years of Neutrons in Biology and Medicine.* In: Proc. Eighth Symp. on Microdosimetry (London: Harwood for CEC) EUR 8395. Eds J. Booz and H. G. Ebert, pp. 1-37 (1983).
11. Hall, E. J., Rossi, H. H., Kellerer, A. M., Goodman, L. and Marino, S. *Radiobiological Studies with Monoenergetic Neutrons.* Radiat. Res. **54**, 431-443 (1973).
12. Hall, E. J., Novak, J. K., Kellerer, A. M., Rossi, H. H., Marino, S. and Goodman, L. J. *RBE as a Function of Neutron Energy 1. Experimental Observations.* Radiat. Res. **64**, 245-255 (1975).
13. Kellerer, A. M. and Rossi, H. H. *The Theory of Dual Radiation Action.* Curr. Topics Radiat. Res. Q. **8**, 85 (1972).
14. International Commission on Radiation Units and Measurements. *The Quality Factor in Radiation Protection.* Report 40 (Bethesda, MD: ICRU Publications) (1986).
15. Brenner, D. J. *High LET Radiation Risk Assessment at Medium Doses.* Rad. Res. (submitted).
16. Sevankaev, A. V., Zherbin, E. A., Obaturov, G. M., Kozlov, V. M., Tjatte, E. G. and Kapchigashev, S. P. *Cytogenic Effects Produced by Neutrons in Lymphocytes of Human Peripheral Blood In Vitro. II Relative Biological Efficiency of Neutrons of Different Energies.* Genetika **15**, 1228-1234 (1979).
17. Lloyd, D. C., Edwards, A. A., Prosser, J. S., Finnon, P. and Moquet, J. E. *In Vitro Induction of Chromosomal Aberrations in Human Lymphocytes, with and without Boron 10, by Radiation Concerned in Boron Capture Therapy.* Br. J. Radiol. **61**, 1136-1141 (1988).
18. Morgan, G. R., Mill, A. J., Roberts, C. J., Newman, S. and Holt, P. D. *The Radiobiology of 24 keV Neutrons: Measurements of the RBE free-in-air, Survival and Cytogenic Analysis of the Biological Effect at Various Depths in a Polyethylene Phantom and Modification of the Depth-Dose Profile by Boron 10 for V79 Chinese Hamster and HeLa Cells.* Br. J. Radiol. **61**, 1127-1135 (1988).

19. Ullrich, R. L. *Tumour Induction in BALB/c Mice after Fractionated or Protracted Exposures to Fission-Spectrum Neutrons.* Radiat. Res. **97**, 587-597 (1984).
20. Vogel, H. H. and Dickson, H. W. *Mammary Neoplasia in Sprague-Dawley Rats following Acute and Protracted Irradiation.* In: Proc. European Seminar on Neutron Cardiogenesis. Ed. J. J. Broerse and G. W. Gerber. (Luxembourg: CEC) pp. 84-90 (1982).
21. Cross, F. T. *Evidence of Lung Cancer from Animal Studies in Radon and its Decay Products in Indoor Air.* In: Radon and its Decay Products in Indoor Air. Eds. W. W. Nazaroff and A. V. Nero (New York: John Wiley) pp. 373-404 (1988).
22. Little, J. B., Kennedy, A. R. and McGandy, R. B. *Effect of Dose Rate on the Induction of Experimental Lung Cancer in Hamsters by γ–radiation.* Radiat. Res. **103**, 293-299 (1985).
23. Muller, W. A., Linzer, U. and Luz, A. *Early Induction of Leukaemia (Malignant Lyphoma) in Mice by Protracted Low γ-doses.* Health Phys. **54**, 461-463 (1988).
24. Worgul, B. V., Merriam, Jr, G. R., Medevdovsky, C. and Brenner, D. J. *Accelerated Heavy Particles and the Lens. III. Cataract Enhancement by Dose Fractionation.* Radiat. Res. **118**, 93-100 (1989).
25. Hornung, R. W. and Meinhardt, T. J. *Quantitative Risk Assessment of Lung Cancer in US Uranium Miners.* Health Phys. **52**, 417-430 (1987).
26. Lubin, J. H., Qiao, Y. L., Taylor, P. R., Yao, S. X., Schatzin, A., Mao, B. L., Rao, J. Y., Xuan, X. Z. and Li, J. Y. *A Quantitative Evaluation of the Radon and Lung Cancer Association in a Case Control Study of Chinese Tin Miners.* Cancer Res. **50**, 174-180 (1990).
27. Darby, S. C. and Doll, R. *Radiation and Exposure Rate* (Letter), Nature **344**, 824 (1990).
28. Chemelevsky, D., Spiess, H., Mays, D. W. and Kellerer, A. M. *The Reverse Protraction Factor in the Induction of Bone Sarcomas in Radium-224 Patients.* Radiat. Res. **124**, 569 (1990).
29. Hill, C. K., Buonoguro, F. J., Myers, C. P., Han, A. and Elkind, M. M. *Fission-Spectrum Neutrons at Reduced Dose Rate Enhance Neoplastic Transformation.* Nature **298**, 67-68 (1982).
30. Brenner, D. J. and Hall, E. J. *Radiation-induced Oncogenic Transformation: The Interplay Between Dose, Dose Protraction and Radiation Quality.* Adv. Radiat. Biol. (in press) (1992).
31. Hill, C. K., Han, A. and Elkind, M. M. *Fission-Spectrum Neutrons at Low Dose Rate Enhance Neoplastic Transformation in the Linear, Low-dose Region (0-10 cGy).* Int. J. Radiat. Biol. **46**, 11-15 (1984).
32. Hill, C. K., Carnes, B. A., Han, A. and Elkind, M. M. *Neoplastic Transformation is Enhanced by Multiple Low Doses of Fission-Spectrum Neutrons.* Radiat. Res. **102**, 404-410 (1985).
33. Hill, C. K. *Is the Induction of Neoplastic Transformation by Radiation Dependent upon the Quality and Dose Rate?* Sci. Pap. Inst. Phys. Chem. Res. **83**, 31-35 (1989).
34. Miller, R. C., Brenner, D. J., Geard, C. R., Komatsu, K., Marino, S. A. and Hall, E. J. *Oncogenic Transformation by Fractionated Doses of Neutrons.* Radiat. Res. **114**, 589-598 (1988).
35. Miller, R. C., Brenner, D. J., Randers-Pehrson, G., Marino, S. A. and Hall, E. J. *The Effects of the Temporal Distribution of Dose on Oncogenic Transformation by Neutrons and Charged Particles of Intermediate LET.* Radiat. Res. **124**, 562-568 (1990).
36. Hieber, L., Ponsel, G., Roos, H., Feen, S., Fromke, E. and Kellerer, A. M. *Absence of a Dose-rate Effect in the Transformation of C3H 10T1/2 Cells by α-particles.* Int. J. Radiat. Biol. **52**, 859-869 (1989).
37. Yang, T. C., Craise, L. M., Mei, M. T. and Tobias, C. A. *Dose Protraction Studies with Low and High-LET Radiations on Neoplastic Cell Transformation In Vitro.* Adv. Space Res. **6**, 137-147 (1987).
38. Miller, M. C., Randers-Pehrson, G., Hieber, L., Marino, S. A., Richards, M. and Hall, E. J. H. *The Inverse Dose-Rate Effect for Oncogenic Transformation by Charged Particles is LET Dependent.* Radiat. Res (in press) (1992).
39. Barendsen, G. W. *Do Fast Neutrons at Low Dose Rate Enhance Cell Transformation in Vitro? A Basic Problem of Microdosimetry and Interpretation.* Int. J. Radiat. Biol. **47**, 431-434 (1985).
40. Burch, P. R. J. and Chesters, M. S. *Neoplastic Transformation of Cells In Vitro at Low and High Dose-rates of Fission Neutrons: an Interpretation.* Int. J. Radiat. Biol. **49**, 495-500 (1986).
41. Elkind, M. M. and Hill, C. K. *Age-Dependent Variations in Cellular Susceptibility to Neoplastic Transformation.* Reply to Letter to the Editor by H. H. Rossi and A. M. Kellerer (1986). Int. J. Radiat. Biol. **50**, 1117-1122 (1986).
42. Rossi, H. H. and Kellerer, A. M. *The Dose Rate Dependence of Oncogenic Transformation by Neutrons may be due to Variation of Response During the Cell Cycle.* Int. J. Radiat. Biol. **50**, 353-361 (1986).
43. Dennis, J. A. and Dennis, L. A. *Neutron Dose Effect Relationships at Low Doses.* Radiat. Environ. Biophys. **27**, 91-101 (1988).
44. Sykes, C. E. and Watt, D. E. *Interpretation of the Increase in Efficiency of Neoplastic Transformation Observed for some Ionizing Radiations at Low Dose Rate.* Int. J. Radiat. Biol. **55**, 925-942 (1989).

45. Brenner, D. J. and Hall, E. J. *The Inverse Dose Rate Effect for Oncogenic Transformation by Neutrons and Charged Particles. A Plausible Interpretation Consistent with Published Data.* Int. J. Radiat. Biol. **58**, 745-758 (1990).
46. Brenner, D. J. *The Influence of LET on the Inverse Dose-rate Effect.* In: Biophysical Modelling of Radiation Effects. Eds K. H. Chadwick, G. Moschini and M. N. Verma. (Bristol: Adam Hilger) pp. 145-153 (1992).
47. Sloan, S., Newcomb, E. W. and Pellicer, A. *Ionizing Radiation and Ras Oncogene Activation.* Cancer Res. Clin. Oncol. **116**, Suppl. 808 (1990).
48. Guerrero, J., Viollasonte, A., Corces, V. and Pellicer, A. *Activation of a cK-ras Oncogene by Somatic Mutation in Mouse Lymphomas Induced by Gamma Radiation.* Science **225**, 1159-1162 (1984).
49. Trutschler, K., Hieber, L. and Kellerer, A. M. *Cytological and Oncogene Alterations in Radiation-Transformed Syrian Hamster Embryo Cells.* In: New Developments in Fundamental and Applied Radiobiology. Eds. C. B. Seymour and C. Mothersill. (London: Taylor and Francis) (1991).
50. Miller, R. C., Geard, C. R., Brenner, D. J., Komatsu, K., Marino, S. A. and Hall, E. J. *Neutron-Energy Dependent Oncogenic Transformation of C_3H 10T1/2 mouse cells.* Radiat. Res. **117**, 114-127 (1989).
51. Endres, G. W. R., Aldrich, J. M., Brackenbush, L. W., Faust, L. G., Griffith, R. V. and Hankins, D. E. *Neutron Dosimetry at Commercial Nuclear Plants.* Pacific Northwest Laboratory Report PNL-3585 (1981).

STATUS OF NUCLEAR DATA FOR USE IN NEUTRON THERAPY

R. M. White*, J. J. Broerse, P. M. DeLuca Jr., G. Dietze, R. C. Haight, K. Kawashima, H. G. Menzel, N. Olsson and A. Wambersie
Coordinated Research Programme
Nuclear Data Section
International Atomic Energy Agency
Vienna, Austria

INVITED PAPER

Abstract — Optimisation of neutron therapy requires nuclear cross section data for: (1) the source reactions for neutron production, (2) the design of collimators and shields, (3) the calculation of absorbed dose in the irradiated tissues, (4) microdosimetry, and (5) studies of the influence of radiation quality on biological effects. Under the auspices of the International Atomic Energy Agency, a Coordinated Research Programme (CRP) has been underway since 1987 to assess the status of these nuclear data, to coordinate research efforts, to report recent progress, and to recommend acceptance of appropriate data and further research where necessary. In this paper, we outline the results of the CRP's final report (to be published) and evaluate the status of the most critical nuclear data needs for therapy, i.e. neutron kerma calculations and measurements from 10 to 70 MeV. Recommended values for the hydrogen kerma factor and the carbon-to-oxygen kerma factor ratio from 10 to 70 MeV are given with estimates of the current uncertainties.

INTRODUCTION

In 1987, the International Atomic Energy Agency's Nuclear Data Section initiated a Coordinated Research Programme (CRP) to deal with nuclear data needed for neutron therapy. Three Research Coordination Meetings (RCM) were held to consider the nuclear data needs for optimum neutron therapy. The objective of the first RCM was to assess the deficiencies in the required nuclear data. Improved nuclear data could be produced and used in the dosimetry protocols for radiotherapy applications using neutron beams. Accomplishments of the first RCM included presentation of the most recent measurements, exchange of views of participants concerning the status of nuclear data needed to optimise neutron therapy, and specification of research to be carried out under this CRP.

The second RCM produced the outline of a technical document[1], to be an IAEA-TECDOC, that will be published in the future to provide up-to-date information on the status and needs of nuclear data for neutron therapy. The report is intended for both the radiation therapists who need the data and the nuclear physicists who must produce them and thereby to help bridge the gap between the two.

In the third and final RCM, the outline of the IAEA-TECDOC was completed. The document will contain the following sections: present status of fast neutron therapy — survey of clinical data and of the clinical research programmes; protocols for the determination of absorbed dose in mixed neutron–photon beams; neutron source reactions; collimation and shielding; microscopic data and kerma factors; absorbed dose and radiation quality; conclusions and recommendations; and an appendix containing calculated kerma factors based on the 1991 version of the Lawrence Livermore National Laboratory's Evaluated Neutron Data Library[2]. The CRP intends that this IAEA-TECDOC provide the necessary background to prompt proposals for the funding of new measurements to improve neutron therapy.

Below, a brief summary of the success of neutron therapy is presented along with an overview of the nuclear data needed to optimise therapy. The status of the most critical nuclear data needs for therapy, i.e. kerma calculations and measurements, from low neutron energies to ~70 MeV, the upper limit of neutron energies in use today is addressed. Recommended values for hydrogen, carbon, and oxygen kerma factors and the carbon-to-oxygen neutron kerma factor ratio for neutron energies from 10 to 70 MeV are given with estimates of the uncertainties.

OPTIMISATION OF NEUTRON THERAPY

Success of neutron therapy

The motivation to improve the nuclear data

---

* Chairman and to whom correspondence should be addressed: Nuclear Data Group, L-298, Lawrence Livermore National Laboratory, PO Box 808, Livermore, CA 94550, USA.

from which neutron therapy can be optimised comes principally from the present success of neutrons to treat specific kinds of tumours. Fast neutron therapy is routinely applied today at 18 centres throughout the world. Over 15,000 patients have been treated with fast neutrons, either as the sole irradiation modality or in combination with other radiotherapy techniques. For some centres the follow-up period of patients now exceeds 15 years. A discussion of the overall local control rates for specific kinds of tumours when irradiated with fast neutrons or neutrons plus photons relative to the control rates for photons alone is given in a following paper[3]. These data will be reviewed in comprehensive detail in Reference 1.

Roughly half of all cancer patients are referred to some type of radiotherapy. Given the present uncertainties in most of the nuclear data needed for therapy, the patients treated with neutrons thus far have been treated in suboptimal technical conditions. However, results to date suggest that fast neutrons are superior to photons for up to 15% of the patients currently referred to radiotherapy[3]. Clinical results indicate that the dose response for tumour control and normal tissue complications are as steep for neutrons as for photons. Therefore the same accuracy in dose delivery and physical selectivity are necessary. The differential between the dose necessary to control the tumour and that which is unacceptable to the patient is less than 5%. Since we do not know the basic nuclear data to this accuracy, particularly at the higher neutron energies currently being used (10–70 MeV), neutron therapy has been demonstrated successfully in less than optimal conditions. Quite likely neutron therapy can be more successful for a larger fraction of patients referred to radiotherapy if improvements are made in the nuclear database.

Neutron source reactions for therapy

Since the relative biological effectiveness (RBE) depends on the neutron spectrum bombarding the tumour, knowledge of the neutron source energy spectrum and the effects of collimation and shielding upon this spectrum are essential. All modern neutron therapy facilities use hospital-based proton cyclotrons to produce neutrons via the (p,n) nuclear reaction. The ideal nuclear reaction would yield the maximum high energy neutrons only in the forward direction. Reactions best meeting this requirement are: ^{7}Li(p,n)^{7}Be, ^{9}Be(p,n)^{9}B, ^{11}B(p,n)^{11}C, ^{13}C(p,n)^{13}N, and ^{15}N(p,n)^{15}O, i.e. reactions which have large Fermi (charge-exchange) and Gamow–Teller (charge-exchange plus spin-flip) transitions in light nuclei. Of these reactions, targets of ^{13}C and ^{15}N are ruled out because they represent only 1.1% and 0.37%, respectively, of naturally occurring carbon and nitrogen and would be prohibitively expensive to prepare. Lithium (92.5% ^{7}Li) is a difficult material to work with so ^{9}Be and ^{11}B are the remaining choices.

Most high energy neutron therapy facilities currently use the ^{9}Be(p,n)^{9}B reaction. Figure 1 shows a plot of the neutron emission spectra from this reaction for an incident proton energy of 135 MeV[4] and for neutron emission angles of 0, 9, and 12 degrees. This figure shows that Fermi and Gamow–Teller interactions play crucial roles in high energy neutron production through both the dramatic forward peaking of the high energy part of the spectrum and how greatly it decreases in magnitude with angle. What have not been compared are the equivalent spectra for the ^{11}B(p,n)^{11}C reaction which, from a nuclear structure viewpoint, may be superior to the ^{9}Be(p,n)^{9}B reaction for the purpose of high energy neutron production for therapy. Specific recommendations for the cross section measurements needed for both reactions as well as charged particle Monte Carlo techniques for accurate calculation of thick target yields and optimisation of target design are given in Reference 1.

Collimation and shielding

High energy neutrons from (thick target) source-producing reactions are broadly distributed in energy and to some degree in emission angle

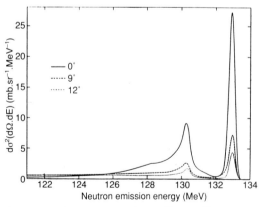

Figure 1. Measurement of the ^{9}Be(p,n)^{9}B neutron emission spectra at 0, 9, and 12 degrees for 135 MeV incident protons[4]. The data in this figure show that both the Fermi and Gamow–Teller interactions play crucial roles in the high energy neutron production from this reaction through both the dramatic forward peaking of the high energy part of the spectrum and how greatly it decreases in magnitude with angle. For clarity, the energy scales of the 9 and 12 degree spectra have been shifted to correspond to the 0 degree peak.

relative to the initial charged particle beam. To localise these neutrons to the therapy volume, a collimator is placed between the source and the patient. Additional shielding is used to protect other parts of the patient and to confine radiation to the therapy room. Collimators must be adjustable and, because of the high penetrating power of neutrons, must present a thickness of several mean free paths — often nearly one metre of material. The mechanical complexity and mass of the collimator lead to a cost of the order of $500,000 (US). Shielding of irradiation facilities is accomplished by building materials such as concrete, the composition of which must be carefully chosen for shielding characteristics, e.g. concrete that is limestone-based is preferred to silicaceous concrete[1].

Nuclear data and radiation transport codes are needed to calculate the performance of collimators and shields. For neutron data, the energy region between 10 and 70 MeV is an intermediate region that is difficult to describe with simple theories or with complete databases. In this energy region, compound nuclear processes, direct processes, and intermediate or 'pre-compound' processes are important. Most evaluated databases were intended for use in the development of nuclear fission and fusion energy sources and have a 20 MeV upper energy limit. Recent measurements and the modification of transport codes to reproduce new benchmark tests better are given in Reference 1.

High energy neutrons also interact with materials to produce activation (radioactive nuclides). Depending on the materials selected for the collimator and shielding, the activated isotopes have a variety of half-lives and can emit energetic radiation. Although the added dose to patients from this induced activity is small when compared with the therapeutic dose, the medical staff and accelerator technicians will receive additional dose. Hence, there exist safety requirements and economic benefits for developing databases and codes that lead to optimised collimation and shielding. Reference 1 discusses new sources of experimental information from which these databases can be assembled.

Neutron kerma

Effects produced by nuclear reactions in matter are due to the energetic charged particles produced, including residual nuclei, and the subsequent energy loss of these particles by ionisation and excitation. Kerma is defined as the average kinetic energy released in matter (per unit mass) and is the sum of all energy transferred to light charged particles and residual nuclei in a reaction. The neutron kerma factor is the kerma produced per unit neutron fluence. The absorbed dose is the energy deposited per unit mass.

For biological purposes, accurate measurement of neutron absorbed dose is best accomplished in wet tissue. Such a medium is not amenable to the construction of detectors, hence tissue substitute materials, more suitable for measurement purposes, are used. Ideally, a tissue-equivalent material would exactly duplicate the atomic composition of tissue and a measurement would yield the absorbed dose in tissue. As this is never the case, the tissue absorbed dose determination depends upon knowledge of the relative rate of charged particle energy production per unit mass for the tissue and tissue substitute material, i.e. the kerma or kerma factor ratio. As the kerma ratio is a function of the neutron energy, information about the neutron energy spectrum is needed. Hence, neutron dose determinations using tissue substitute materials require accurate kerma factor values for all constituent materials for both the tissue and tissue substitute over the entire neutron energy range.

Standard man[5] consists of hydrogen (10%), carbon (18%), nitrogen (3%), oxygen (65%), and various trace elements (4%). A typical tissue substitute material such as A-150 plastic has the corresponding percentages: 10.1%, 77.6%, 3.5%, 5.3%, and 3.5%[6]. Hydrogen, which contributes most significantly to the kerma and absorbed dose, is well matched for these mixtures. For A-150 plastic, and most other substitute materials, carbon is exchanged for oxygen. Thus, information about the hydrogen kerma factor and the carbon-to-oxygen kerma factor ratio are essential for accurate fast neutron absorbed dose determinations.

EVALUATED KERMA FACTORS

There are two ways kerma factors are determined: (1) from direct measurement of kerma, and (2) from calculation of kerma factors from basic cross section information contained in nuclear data libraries[2,7] where all significant reaction channels are explicitly represented, including angular and/or energy distributions of secondary reaction particles, etc. While some experimental kerma factor data exist, the direct measurement of kerma is difficult and values are available for only a few elements and neutron energies. Some new measurements are discussed in a following paper[8]. The explicit representation of all significant reaction channels and secondary reaction product properties in nuclear data libraries comes either from evaluation of experimental microscopic cross section data or from nuclear model predictions. The degree to which an evaluation can be based on experimental data determines the uncertainty of the kerma factors calculated from evaluated

Table 1. Recommended neutron kerma factor for hydrogen. The table is designed for linear-linear interpolation. The uncertainty (1 standard deviation) is ±1% at all energies.

E_n (MeV)	K_f (fGy.m^2)
10.00	45.69
12.38	46.59
15.62	46.93
20.46	46.50
31.42	44.21
46.19	41.11
59.79	38.92
70.00	37.80

nuclear data libraries. For purposes of therapy, the uncertainties in kerma factor values are as important as the values themselves. In practice, realistic uncertainties are difficult to assess. The following sections discuss neutron kerma factors for hydrogen, carbon, and oxygen, and the associated uncertainties in those kerma factors, given the databases currently available.

Evaluation of the neutron kerma factor for hydrogen

The database of experimental cross section values for the ^{1}H(n,n)^{1}H reaction is extensive since this reaction is fundamental to the understanding of the nucleon–nucleon interaction. These cross sections are some of the best determined of all nuclear reactions. The calculation of the neutron kerma factor for hydrogen depends only upon the integrated elastic cross section and the average energy given to the recoil proton — which in turn is directly related to the a_1 coefficient of the Legrendre polynomial expansion of the differential elastic scattering cross section. (The capture cross section is insignificant.)

Recommended hydrogen kerma factors for neutron energies from 10 to 70 MeV calculated from cross sections obtained from the 'VL40' phase shift analysis of Arndt[9] are given in Table 1. The values in Table 1 can be interpolated linearly (as can the rest of the tables in this paper) and have an assigned uncertainty of ±1.0% (one standard deviation). This uncertainty is based on comparing kerma factor values calculated with the VL40 phase shifts to those calculated from the cross sections in the ENDF/B-VI evaluation[7].

The ENDF/B-VI evaluation of the integrated ^{1}H(n,n)^{1}H reaction cross section is based on an R-matrix analysis[10] at energies below 26 MeV, a phase shift analysis[9] above 30 MeV, and a sliding

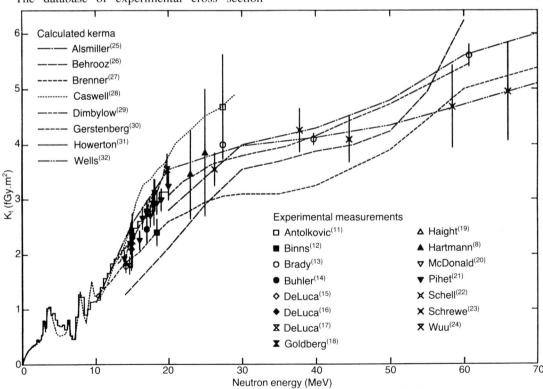

Figure 2. Measured and calculated neutron kerma factors for carbon as a function of incident neutron energy from 0 to 70 MeV.

average of the two between 26 and 30 MeV. The differential elastic cross section values below 26 MeV are taken from the R-matrix analysis and those at and above 26 MeV are taken from the phase shift analysis. The kerma factors calculated from ENDF/B-VI and VL40, based on 1988 and 1992 phase shift analyses by Arndt, disagree by less than 0.5% over the 30 to 70 MeV region. In the 10 to 30 MeV region, the cross sections for the two kerma factor calculations were based on two completely different analyses (phase shift rather than R-matrix) of the cross sections and the resulting kerma factors disagreed by up to 1.0%.

Evaluation of the neutron kerma factor for carbon

Figures 2 and 3 show the information currently known to us for both measured[8,11–25] and calculated[26–33] neutron kerma factors for carbon as a function of incident neutron energy. The works of Caswell et al[29] and Howerton[32] were calculated from evaluated nuclear databases[7,2] and the other calculated kerma factors were obtained using nuclear models based on various theoretical approaches. The nuclear models may or may not be normalised to one or more experimental measurements. The work of Caswell has been recently revised above ~13 MeV[34] and is now in substantial agreement with the work of Howerton[32] at the higher neutron energies.

Until quite recently, the only experimental data available at high energies was the work of Brady and Romero[13]. Based upon the thresholds of reactions on carbon, their probable cross sections, and numerous modelling calculations, the kerma factor measured by Brady at the highest energy point (60.7 MeV) appears too high. As this paper was being prepared, new measurements from Schrewe et al[23,24] were received which provided four experimental points above 30 MeV. These new data are in excellent agreement with our preliminary evaluation. As shown in Figure 4, ±8% was assigned as the minimum uncertainty in

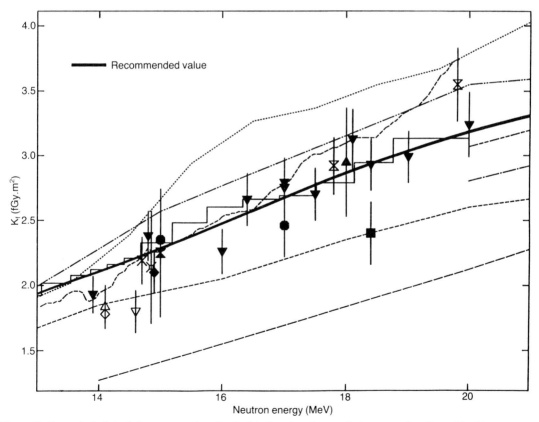

Figure 3. Expanded plot of the measured and calculated carbon kerma factors as a function of incident neutron energy from 13 to 21 MeV. Symbols and lines keyed as in Figure 2. Also shown is the recommended carbon kerma factor. See Figure 4, Table 2, and text for further explanation of the recommended values and associated minimum uncertainties.

the recommended kerma factor up to neutron energies of ~50 MeV. That is, given the current database, no realistic uncertainty of less than ±8% can be assigned to the recommended values. A greater uncertainty could be assigned. Above ~50 MeV, a minimum uncertainty of ±16% was assigned based on the only two sets of experimental data available. While the errors reported by Schrewe are significantly larger than those of Brady, we have given more weight to the shape and normalisation of the Schrewe data based partially on guidance from the change in slope of several nuclear model calculations. Table 2 gives the recommended values of the carbon kerma factor as a function of neutron energy from 10 to 70 MeV.

Evaluation of the neutron kerma factor for oxygen

Figure 5 shows a summary of measured and calculated neutron kerma factors for oxygen as a function of incident neutron energy. As in the case

Table 2. Recommended neutron kerma factor for carbon. The table is designed for linear-linear interpolation. The minimum uncertainty (see text) is ±8% below 50 MeV and ±16% above 50 MeV.

E_n (MeV)	K_f (fGy.m^2)
10.0	1.20
12.1	1.75
19.5	3.12
23.1	3.51
30.1	3.86
70.0	5.14

for carbon, the values of Caswell et al[29] and Howerton[32] are calculated from evaluated nuclear databases[2,7] and the other calculations employ various nuclear models. As seen in Figure 5, the experimental database for oxygen is limited. There exists reasonable agreement in shape and normalisation between most of the measurements and calculations up to 40 MeV. With this limited database, however, the uncertainties are much greater than in the case of carbon. Figure 6 shows our recommended values for oxygen kerma

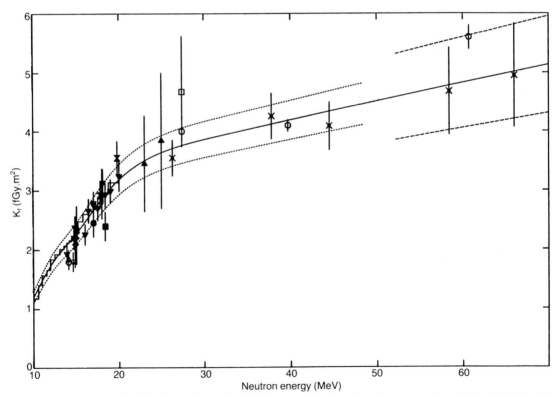

Figure 4. Recommended carbon kerma factor (——) values as a function of neutron energy from 10 to 70 MeV with experimental measurements included. Figure is keyed the same as Figure 2. Included is the calculation of Howerton[32] at the lower neutron energies. Also included are the minimum uncertainties on the recommended values as discussed in the text: (·····) ±8% at lower neutron energies, (-----) ±16% at energies above 50 MeV.

factors from 10 to 70 MeV. The assignment of ±15% from 10 to 30 MeV for the minimum uncertainty in the recommended kerma factors is dictated by the spread in the experimental data.

Above 30 MeV, a minimum uncertainty of ±20% has been assigned based on our belief that the value of the 60.7 MeV data point of Brady is too large. We based this conclusion upon knowledge of thresholds of reactions, the systematics of cross sections for those reactions, several model calculations, and the fact that the same energy point for Brady's carbon value is higher than the latest measurements[23,24] would indicate. However, this value is the only experimental point available at high energy and, until further measurements are carried out, must fall within the minimum uncertainty estimate on our recommended values. Values of the recommended kerma factor for oxygen are given in Table 3.

The carbon-to-oxygen kerma ratio

In Figure 7 recommended carbon to oxygen kerma factor ratios are plotted for neutrons from 10 to 70 MeV. Corresponding values are given in Table 4. Because the recommended carbon and oxygen kerma values each had two (non-overlapping) regions where the minimum assigned uncertainty changed, the recommended ratio has three regions of differing uncertainties as indicated in Figure 7. Because Brady and Romero[13] and Hartmann et al[8] each measured carbon and oxygen kerma factors at the same energies, Figure 7 includes their ratios with their corresponding uncertainties plotted. Also plotted is the ratio of DeLuca's carbon measurement at 14.9 MeV[16] to his measurement of oxygen at 15 MeV[35]. DeLuca's measurements of oxygen at 17.5 and 18.1 MeV[35] were interpolated linearly and used with his carbon measurement at 17.8 MeV[17] to obtain the ratio point at 17.8 MeV. DeLuca's measurement of carbon at 19.8 MeV[17] was used with his oxygen measurement at 19.1 MeV[35] and plotted as 19.45 MeV. These ratios are plotted in Figure 7 for informational purposes only and were not used to determine the recommended carbon-to-oxygen kerma ratio. The recommended ratio was determined from the separate evaluations of carbon and oxygen.

The purpose of this carbon-to-oxygen ratio evaluation is: (1) to establish a recommended value for this ratio based upon all available direct

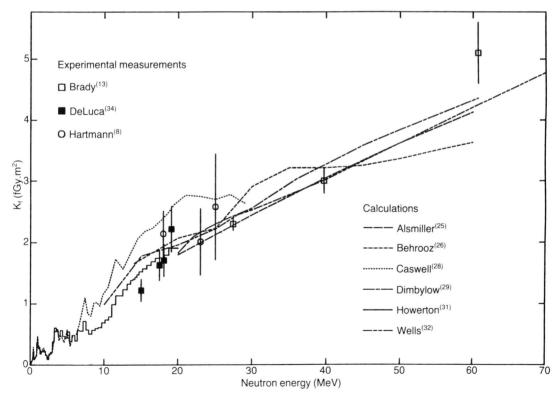

Figure 5. Measured and calculated neutron kerma factors for oxygen as a function of incident neutron energy from 0 to 70 MeV.

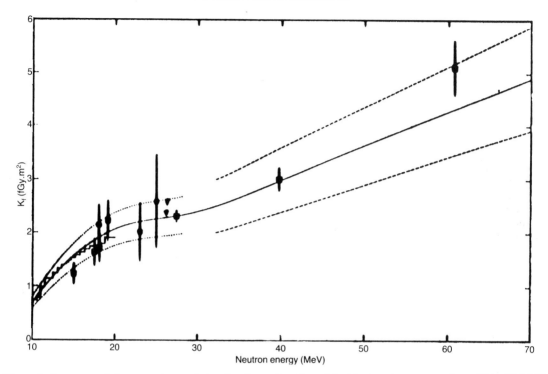

Figure 6. Recommended oxygen kerma factor (——) as a function of incident neutron energy from 10 to 70 MeV with experimental measurements included. Figure is keyed the same as Figure 5. Included is the calculation of Howerton[32] at the lower neutron energies. Also included are the minimum uncertainties on the recommended values as discussed in the text: (·····) ±15% at energies up to 30 MeV, (-----) ±20% at higher neutron energies.

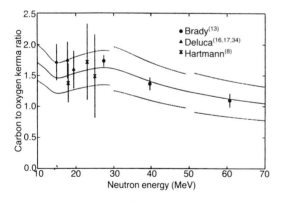

Figure 7. Recommended carbon-to-oxygen kerma factor ratio (——) as a function of neutron energy from 10 to 70 MeV with the ratios of specific data included. Also included are the minimum uncertainties on the recommended values of the ratio as described in the text: (·····) ±17% for 10 to 30 MeV, (-----) ±22% from 30 to 50 MeV, (– – –) ±26% for 50 to 70 MeV.

measurements and model calculations currently known to us, and (2) to give an uncertainty estimate to this ratio. A minimum uncertainty which is only meant to indicate that no lesser uncertainty could reasonably be placed on the values has been chosen. It does not mean that the uncertainty is not greater. It is not a precise definition, but only meant to serve as a guide to the current state-of-the-art and to indicate clearly where further measurements are needed.

CONCLUSIONS

A brief description of a Coordinated Research Programme sponsored by the Nuclear Data Section of the International Atomic Energy Agency to assess the status of nuclear data needed for neutron therapy has been provided. An outline of a comprehensive technical document has been given that will be published in the future by the IAEA. Recommended values, along with our estimate of the current uncertainties, for the most important quantities which depend upon nuclear measurements that are needed to help optimise neutron therapy have also been provided. From this work it is clear that, with the exception of the hydrogen neutron kerma factor, we do not yet have the nuclear data necessary to allow neutron therapy to reach its full potential.

Table 3. Recommended neutron kerma factor for oxygen. The table is designed for linear-linear interpolation. The minimum uncertainty (see text) is ±15% below 30 MeV and ±20% above 30 MeV.

E_n (MeV)	K_f (fGy.m^2)
10.0	0.70
13.2	1.28
15.8	1.67
18.9	1.99
22.5	2.19
30.5	2.42
40.7	3.03
70.0	4.91

Table 4. Recommended ratio of carbon to oxygen kerma factors. The table is designed for linear-linear interpolation. The minimum uncertainty (see text) is ±17% below 30 MeV, ±22% between 30 and 50 MeV, and ±26% above 50 MeV.

E_n (MeV)	Ratio
10.0	1.72
14.6	1.47
17.3	1.49
26.1	1.62
30.9	1.60
46.0	1.30
61.5	1.12
70.0	1.05

REFERENCES

1. White, R. M., Broerse, J. J., DeLuca, Jr, P. M., Dietze, G., Haight, R. C., Kawashima, K., Menzel, H. G., Olsson, N. and Wambersie, A. *Nuclear Data Needed for Neutron Therapy.* IAEA Technical Document, Coordinated Research Programme, Nuclear Data Section (Vienna: International Atomic Energy Agency) To be published.
2. Howerton, R. J. *The LLL Evaluated Nuclear Data Library (ENDL).* UCRL-50400 **15**, revised (University of California, Lawrence Livermore National Laboratory, Livermore, CA, USA) (1978).
3. Wambersie, A. *Neutron Therapy: From Radiobiological Expectation to Clinical Reality.* Radiat. Prot. Dosim. **44**(1-4) 379-395 (1992) (This issue).
4. Pugh Jr, B. G. *The (p,n) Reaction on ^9Be and ^{17}O at 135 MeV.* Ph D Thesis, Massachusetts Institute of Technology, Cambridge, MA (1985).
5. ICRP *Report of Committee II on Permissible Dose for Internal Radiation.* Publication 2 (Oxford: Pergamon Press) (1959).
6. Smathers, J. B., Otte, V. A., Smith, A. R., Almond, P. R., Attix, F. H., Spokas, J. J., Quam, W. M. and Goodman, L. J. *Composition of A-150 Tissue Equivalent Plastic.* Med. Phys. **4**, 74-77 (1975).
7. *ENDF/B-VI Summary Document.* Compiled and edited by P. F. Rose, ENDF-201 (National Nuclear Data Center, Brookhaven National Laboratory, Upton, NY) (1991).
8. Hartmann, C. L., DeLuca Jr, P. M. and Pearson, D. W. *C, O, Al, and Si Kerma Factors for 18-25 MeV Neutrons.* Radiat. Prot. Dosim. **44**(1-4) 25-30 (1992) (This issue).
9. Arndt, R. A. Private communication.
10. Dodder, D. C. and Hale, G. M. (in preparation); see also Reference 7.
11. Antolkovic, B., Slaus, I. and Plenkovic, D. *Determination of the Kerma Factors for the Reaction $^{12}C(n,n'3\alpha)$ at E_n = 10-35 MeV.* Radiat. Res. **97**, 253-261 (1984).
12. Binns, P. J. and Hough, J. H. *NAC Annual Report.* Technical Report NAC/AR/90-01, (National Accelerator Center, Faure, RSA) (June 1990).
13. Brady, F. P. and Romero, J. L. *Neutron Induced Reactions in Tissue Resident Elements.* Final Report to the National Cancer Institute, Grant No. 1R01 CA16261 Technical Report (University of California-Davis, Davis, CA) (1979).
14. Bühler, G., Menzel, H., Schuhmacher, H. and Guldbakke, S. *Dosimetric Studies with Non-hydrogenous Proportional Counters in Well-defined High-energy Neutron Fields.* In: Proc. 5th Symp. on Neutron Dosimetry, EUR-9762 (Luxembourg: CEC) pp. 309-320 (1985).
15. DeLuca, Jr, P. M., Barschall, H. H., Haight, R. C. and McDonald, J. C. *Kerma Factor of Carbon for 14.1 MeV Neutrons.* Radiat. Res. **100**, 78-86 (1984).
16. DeLuca, Jr, P. M., Barschall, H. H., Haight, R. C. and McDonald, J. C. *Measured Neutron Carbon Kerma Factors from 14.1 MeV to 18 MeV.* In: Proc. 5th Symp. on Neutron Dosimetry, EUR-9762 (Luxembourg: CEC) pp. 193-200 (1985).
17. DeLuca, Jr, P. M., Barschall, H. H., Burhoe, M. and Haight, R. C. *Carbon Kerma Factor for 18- and 20-MeV Neutrons.* Nucl. Sci. Eng. **94**, 192-198 (1986).
18. Goldberg, E., Slaughter, D. R. and Howell, R. H. *Experimental Determination of Kerma Factors at E_n = 15 MeV.* Technical Report UCID-17789 (Lawrence Livermore National Laboratory, Livermore, CA) (1978).
19. Haight, R. C., Grimes, S. M., Johnson, R. G. and Barschall, H. H. *The $^{12}C(n,\alpha)$ Reaction and the Kerma Factor of Carbon at E_n = 14.1 MeV.* Nucl. Sci. Eng. **87**, 41-47 (1984).

20. McDonald, J. C. *Calorimetric Measurements for the Carbon Kerma Factor for 14.6-MeV Neutrons*. Radiat. Res. **109**, 28-35 (1987).
21. Pihet, P., Guldbakke, S., Menzel, H. G. and Schuhmacher, H. *Measurement of Kerma Factors for Carbon and A-150 Plastic: Neutron Energies from 13.9 to 20.0 MeV*. Phys. Med. Biol. submitted.
22. Schell, M. C., Pearson, D. W., DeLuca, Jr, P. M. and Haight, R. C. *Measurement of Dose Distributions of LET in Matter Irradiated by Fast Neutrons*. Med. Phys. **17**(1), 1-9 (1990).
23. Schrewe, U. J., Brede, H. J., Henneck, R., Gerdung, S., Kunz, A., Menzel, H. G., Meulders, J. P., Pihet, P., Schuhmacher, H. and Slypen, I. *Determination of Kerma Factors for A-150 Plastic and Carbon for Neutron Energies above 20 MeV*. In: Proc. Int. Conf. on Nuclear Data for Science and Technology, Jülich, Germany, (1991) Springer-Verlag ed, S. m. Qaim pp. 586-588 (1992).
24. Schrewe, U. J., Brede, H. J., Gerdung, S., Nolte, R., Pihet, P., Schmelzbach, P. and Schuhmacher, H. *Determination of Kerma Factors of A-150 Plastic and Carbon at Neutron Energies between 45 and 66 MeV*. Radiat. Prot. Dosim. **44**(1-4) 21-24 (1992) (This issue).
25. Wuu, C. S. and Milavickas, L. R. *Determination of the Kerma Factors in Tissue-equivalent Plastic, C, Mg, and Fe for 14.7 MeV Neutrons*. Med. Phys. **14**(6), 1007-1014 (1987).
26. Alsmiller Jr, R. G. and Barish, J. *Neutron Kerma Factors for H, C, N, O, and Tissue in the Energy Range of 20–70 MeV*. Health Phys. **33**, 98-100 (1977).
27. Behrooz, M. A. and Watt, D. E. *An Appraisal of Partial Kerma Factors for Neutrons Up to 60 MeV*. In: Proc. 4th Symp. on Neutron Dosimetry, EUR-7448 (Luxembourg: CEC) pp. 353-360 (1981).
28. Brenner, D. J. *Neutron Kerma Values Above 15 MeV Calculated with a Nuclear Model Applicable to Light Nuclei*. Phys. Med. Biol. **29**, 437-441 (1984).
29. Caswell, R. L., Coyne, J. J. and Randolph, M. L. *Kerma Factors for Neutron Energies Below 30 MeV*. Radiat. Res. **83**, 217-254 (1980).
30. Dimbylow, P. J. *Neutron Cross Section and Kerma Value Calculations for C, N, O, Mg, Al, P, S, Ar, and Ca from 20 to 50 MeV*. Phys. Med. Biol. **27**(8), 989-1001 (1982).
31. Gerstenberg, H. M., Caswell, R. L. and Coyne, J. J. *Initial Spectra of Neutron-induced Secondary Charged Particles*. Radiat. Prot. Dosim. **23**(1), 41-44 (1988).
32. Howerton, R. J. *Calculated Neutron KERMA Factors Based on the LLNL ENDL Data File*. UCRL-50400 **27**: revised (University of California, Lawrence Livermore National Laboratory, Livermore, CA) (1986).
33. Wells, A. H. *A Consistent Set of Kerma Values for H, C, N, and O for Neutrons of Energies from 10 to 80 MeV*. Radiat. Res. **80**, 1-9 (1979).
34. Caswell, R. S., Coyne, J. J., Gerstenberg, H. M. and Axton, E. J. *Basic Data Necessary for Neutron Dosimetry*. Radiat. Prot. Dosim. **23**(1), 11-17 (1988).
35. DeLuca Jr, P. M., Barschall, H. H., Sun, Y. and Haight, R. C. *Kerma Factor of Oxygen, Aluminium, and Silicon for 15- and 20-MeV Neutrons*. Radiat. Prot. Dosim. **23**, 27-30 (1988).

DETERMINATION OF KERMA FACTORS OF A-150 PLASTIC AND CARBON AT NEUTRON ENERGIES BETWEEN 45 AND 66 MeV

U. J. Schrewe†, H. J. Brede†, S. Gerdung‡, R. Nolte†, P. Pihet‡, P. Schmelzbach§ and H. Schuhmacher†
†Physikalisch-Technische Bundesanstalt
W-3300 Braunschweig, Germany
‡Fachrichtung Biophysik und Physikalische Grundlagen der Medizin
Universität des Saarlandes, W-6650 Homburg, Germany
§Paul Scherrer Institut, CH-5232 Villigen PSI, Switzerland

Abstract — Kerma factors of carbon and A-150 tissue-equivalent plastic have been measured in nearly monoenergetic neutron beams at energies between 45 and 66 MeV. The kerma was measured with low pressure proportional counters (PC), the walls consisting either of graphite or A-150 plastic material, and the neutron beam fluence was measured with a proton recoil telescope. The kerma factor corrections were determined from time-of-flight measurements with NE213 scintillation detectors and PCs. At neutron energies of (44.5 ± 2.6) MeV and (66.0 ± 3.0) MeV, the kerma factors were (40.9 ± 4.1) pGy.cm² and (49.5 ± 8.7) pGy.cm² for carbon and (80.4 ± 6.3) pGy.cm² and (80.1 ± 8.2) pGy.cm² for A-150 plastic.

INTRODUCTION

The kerma K (initial kinetic energy of charged secondaries per unit mass) produced by monoenergetic neutrons of energy E and fluence Φ may be expressed as $K(E) = \Phi\, k_f(E)$, where k_f is the kerma factor[1]. Kerma factors can be calculated from the cross sections $\sigma_j(E)$ as the sum of the interactions of type j which produce charged particles of an average energy $\bar{E}_j(E)$, multiplied by the number of target atoms per unit mass n:

$$k_f = K/\Phi = n \sum_j \bar{E}_j(E)\sigma_j(E) \quad (1)$$

The kerma factors of hydrogen, carbon, oxygen and nitrogen are of great importance in neutron dosimetry, since these elements are the main constituents of tissue and of the detector materials which are used to substitute tissue in measuring devices. Kerma factors are required, in particular, to convert the dose measured with dosemeters made of A-150 plastic, a tissue substitute, into the tissue dose.

For neutron energies of less than 20 MeV, the evaluated data files of neutron cross sections are often used to calculate the kerma factors according to Equation 1. Above 20 MeV, however, calculations are much more complex since the number of reaction channels rises drastically with increasing neutron energy, there is a lack of precise cross section data, and the reaction kinematics are scarcely known, particularly for the multiparticle break-up reactions.

The energy range above 20 MeV is of interest in neutron therapy if broad neutron energy distributions between 20 and 70 MeV are used. Because of the lack of microscopic data, we have determined the kerma factors of A-150 plastic and carbon in the neutron energy range of 20 to 70 MeV directly by measuring both the kerma and the corresponding fluence in approximately monoenergetic neutron fields. Other kerma factor measurements of this type have recently been reported[2,3]. Schuhmacher et al[3] describe the experimental methods which have also been used in the present work. Special aspects of the calibration and corrections of kerma measurements with cavity chambers are treated by Pihet et al[4]. For detailed information we refer the reader to previous works[3,4]. The present paper reports on a new kerma factor measurement at neutron energies of 45 MeV and 66 MeV.

EXPERIMENTAL METHODS

The experiments were performed at the monoenergetic neutron beam facility of the Paul Scherrer Institute, Switzerland. Intensive neutron beams were produced by bombarding a 2 mm thick beryllium target with protons of energies of 50.0 and 71.2 MeV. Neutron beams with a small solid angle of about 5.10^{-5} sr were produced by means of a collimator.

The neutron beams were not ideally monoenergetic. Various effects influenced the actual spectral distribution[3]. The spectral fluences exhibited strong peaks at 44.5 and 66.0 MeV, denoted as the nominal neutron energy, with standard deviations of 2.6 and 3.0 MeV, respectively, and also significant low energy tails.

Since kerma and absorbed dose are numerically almost equal if the charged secondary particle achieves equilibrium, the kerma was determined by measuring the absorbed dose with a cavity chamber. Low pressure proportional counters (PC) with walls of A-150 plastic or graphite were used (types: LET1/2, Far West Technology, Goleta, USA, cavity diameters 12.7 mm, wall thickness 2.5 mm). Measurements with different build-up caps of the respective materials with thicknesses between 5 and 15 mm served to ensure the equilibrium of the secondaries. The PCs were operated and calibrated as described previously[3,4].

The neutron fluence was measured with a proton recoil telescope (PRT)[5]. Recoil protons from elastic scattering in a polyethylene radiator containing hydrogen (PE) were detected and separated from other reaction products with two proportional counters and two solid-state silicon detectors.

A significant fraction of the kerma in the measurements caused by low energy neutrons had to be subtracted. Time-of-flight (TOF) measurements carried out with both a PC and an NE213 scintillation detector were combined to determine the disturbing influences over the entire energy range: (i) TOF techniques used with the PCs enabled the kerma to be directly measured as a function of neutron energy and hence the kerma fractions below fixed neutron energy thresholds to be subtracted. However, the applicability of this method was limited to the neutron energy range of 0.1 to about 7 MeV, since the timing resolution capacity of the PCs, the flight paths and the neutron intensities were limited. (ii) The main fraction of disturbing kerma was determined from the spectral fluence measured with an NE213 scintillation detector (51 mm in diameter and 102 mm in length) which was applied using pulse-shape techniques for photon–neutron separation, and TOF techniques for neutron energy determination.

In order to cover a wide range of neutron energy, the repetition rate of the proton beam pulses had to be smaller than that normally produced by the cyclotron. By means of an electrostatic deflector in the beam line (which deflected 16 out of 15 pulses for beams of proton energies of 50 MeV, and 19 out of 20 for beams of proton energies of 71 MeV), it was possible to achieve pulse repetition times of ~ 1 μs on the Be target. The total background intensity relative to the intensity of the non-deflected pulse was less than 2%.

The number of neutron-induced events detected in the NE213 detector was recorded as a function of TOF, setting various thresholds on the scintillator pulse height by software. The detector efficiencies of the respective thresholds were calculated with two different Monte Carlo codes, (SCINFUL[6] and KENT01[7]). A comparison between absolute fluence measurements with the scintillation detector and the PRT is reported in a further contribution to this conference[8].

Various measurements in a neutron beam were performed under various conditions. For normalisation, each measurement was related to two different monitors: (i) the proton beam charge on the Be target and (ii) the number of events produced in a thin NE102 plastic scintillator (0.5 mm thick) which was positioned behind the neutron collimator. The ratio of the two monitors varied significantly according to the conditions and had to be corrected for a time-dependent drift of the leakage current of the charge integrator device and for pile-up effects in the NE102 detector. However, the kerma and the fluence measurements were carried out with almost identical proton beam currents and thus the monitor ratio was stable within 1–2% during these measuring periods.

Table 1. Experimental results for A-150 plastic (A) and carbon (C). The numbers in brackets denote the uncertainty (one standard deviation) of the last digits of the respective values.

E_p MeV	E_n MeV	M	K_{tot}/Q pGy.nC^{-1}	k_c	E_2 MeV	K_o/Q pGy.nC^{-1}	Φ_o/Q cm^{-2}.nC^{-1}	k_f pGy.cm^2	k_f^C/k_f^A	Ref.
31.9	26.3(29)	A	1539(58)	0.562(11)	20	865(37)	11.44(64)	75.6(54)	0.469(36)	Schuhmacher et al[3]
		C	563(35)	0.721(11)	20	406(26)	11.44(64)	35.5(30)		
43.4	37.8(25)	A	1573(59)	0.493(11)	32	775(34)	9.81(55)	79.0(56)	0.539(45)	Schuhmacher et al[3]
		C	691(48)	0.605(11)	32	418(30)	9.81(55)	42.6(39)		
50.0	44.5(26)	A	1799(75)	0.470(10)	38	845(40)	10.52(70)	80.4(65)	0.506(45)	Present work
		C	768(59)	0.560(10)	38	430(32)	10.52(70)	40.9(41)		
63.8	58.4(29)[a]	A	1899(195)	0.427(12)[a]	(50)[a]	811(86)[b]	10.50(8)[a]	77.0(10)[c]	0.61(11)[c]	Menzel et al[11] and
		C	976(134)	0.504(11)[a]	(50)[a]	492(68)[b]	10.50(8)[a]	46.8(74)[c]		Present work
71.2	66.0(30)	A	2030(128)	0.405(10)	58	822(55)	10.26(79)	80.1(82)	0.62(11)	Present work
		C	1058(169)	0.480(10)	58	508(82)	10.26(79)	49.5(88)		

[a]Interpolated value, see text.
[b]Obtained from measured K_{tot}/Q and interpolated k_c.
[c]Obtained from K_o/Q and interpolated Φ_o/Q.

DATA ANALYSIS AND RESULTS

The total kerma, K_{tot}, obtained from the PC measurement can be written as the sum of three portions[3]: (i) the desired kerma of the nominal energy, K_0, which refers to the neutron energy range E_2 to E_{max}, and two background contributions, (ii) K_1 and (iii) K_2, which are made accessible by the two TOF measurements. K_1 is produced by neutrons of energy $0 - E_1$, and K_2 by $E_1 - E_2$.

The correction factor k_c, defined as the ratio K_0/K_{tot}, can be expressed as:

$$k_c = K_0/K_{tot} = (1 - K_1/K_{tot})/(1 + K_2/K_0) \qquad (2)$$

where K_1/K_{tot} is directly obtained from the PC-TOF measurements, and K_2/K_0 is calculated from the spectral neutron fluences and kerma factors extrapolated from recent measurements[3,4] and theoretical predictions[9,10]. Although the kerma fraction of low energy neutrons, $(1 - k_c)$, was quite substantial, k_c was determined with a relatively low uncertainty (see Table 1). The uncertainty of k_c includes the contributions due to (i) the influence of variations in the spectral fluences predicted by different scintillator efficiency codes[6,7], (ii) the uncertainty of extrapolating the kerma factors and (iii) the uncertainty of the PC-TOF calibration. The k_c factors have a smooth dependence on the neutron energy which also allowed k_c to be interpolated for another kerma measurement carried out under similar experimental conditions but without TOF measurements.

The experimental results of the present investigation obtained for kerma and fluence normalised to the proton beam charge on the Be target and the resulting kerma factors are shown in Table 1. For a better identification of systematic trends, the data from our earlier investigation[3] are included. The data points at $E_n = 58.4$ MeV represent the results of a kerma measurement in a neutron field produced by protons with an energy of 63.75 MeV on the Be target. In this experiment no TOF or fluence measurements were performed. The mean neutron energy and the correction factor k_c were interpolated. As can be seen in Table 1, the fluence per unit proton beam charge is approximately constant for all neutron fields. The kerma factor was determined by also adopting the mean fluence per unit beam charge $\Phi_0/Q = 10.5 \pm 0.8$ cm^{-2}.nC^{-1}, for the $\bar{E}_n = 58.4$ MeV field.

The kerma factors are shown in Figure 1 together with calculated and other experimental data. In the energy range of 40 to 70 MeV, the calculations of Brenner[10] and Wells[13] agree best with the experimental data. A detailed discussion of the present status of kerma factors will be given by White et al[20] in another contribution to this conference.

ACKNOWLEDGEMENT

The authors would like to express their gratitude to H. G. Menzel who initiated and promoted this investigation. The work was partly supported by the Commission of the European Communities, Contract No. BI7-030.

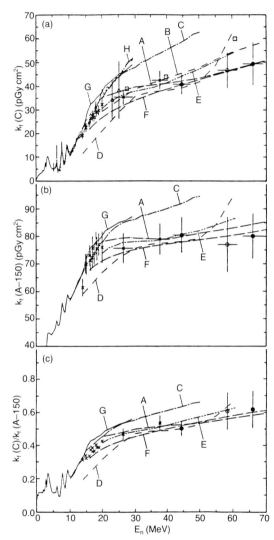

Figure 1. Kerma factor, k_f, for carbon (a) and A-150 plastic (b) and the ratio k_f^C/k_f^{A-150} (c) plotted against neutron energy. The figure shows experimental values of this work (filled circles), the results of Schuhmacher et al[3] and Pihet et al[4] (crosses), Hartmann et al[2] (triangles) and Romero et al[12] (squares) together with calculated data of (A) Wells[13], (B) Alsmiller and Barish[14], (C) Dimbylow[15], (D) Behrooz et al[16], (E) Dimbylow[17], (F) Brenner[10], (G) Caswell et al[18] and a recent evaluation of (H) Axton[13]. The data points obtained from measured kerma values and interpolated data of k_c and (Φ_0/Q) (see text) are also shown (open circles).

ACKNOWLEDGEMENT

The authors would like to express their gratitude to H. G. Menzel who initiated and promoted this investigation. The work was partly supported by the Commission of the European Communities, Contract No. BI7-030.

REFERENCES

1. ICRU. *Radiation Quantities and Units*. Report 33 (Bethesda, MD: International Commission on Radiation Units and Measurements) (1980).
2. Hartmann, C. L., DeLuca Jr, P. M. and Pearson, D. W. *Measurement of C, Mg, and Fe Kerma Factors and the $^{19}F(n,2n)^{18}F$ Cross Section for 18 to 27 MeV Neutrons*. In: Proc. Int. Conf. on Nuclear Data for Science and Technology (1991) Ed.: S. M. Qaim, Springer Verlag, Berlin – Heidelberg – New York, p.583-591 (1992).
3. Schuhmacher, H., Brede, H. J., Henneck, R., Kunz, A., Menzel, H. G., Meulders, J. P., Pihet, P. and Schrewe, U. J. *Measurement of Neutron Kerma Factors for Carbon and A-150 Plastic at Neutron Energies of 26.3 MeV and 37.8 MeV*. Phys. Med. Biol. **37**, 1265-1281 (1992). Also Schrewe, U. J., Brede, H. J., Henneck, R., Gerdung, S., Kunz, A., Menzel, H. G., Meulders, J. P., Pihet, P., Schuhmacher, H. and Slypen, I. *Determination of Kerma Factors for A-150 Plastic and Carbon for Neutron Energies above 20 MeV*. In: Proc. Int. Conf. on Nuclear Data for Science and Technology (1991) Ed.: S. M. Qaim, Springer Verlag, Berlin – Heidelberg – New York, p. 586-588 (1992).
4. Pihet, P., Schuhmacher, H. and Guldbakke, S. *Measurement of Kerma Factors for Carbon and A-150 Plastic: Neutron Energies from 13.9 to 20.0 MeV*. Phys. Med. Biol. (in press).
5. Schuhmacher, H., Siebert, B. R. L. and Brede, H. J. *Measurement of Neutron Fluence for Energies between 20 MeV and 65 MeV Using a Proton Recoil Telescope*. In: Proc. of a Specialists' Meeting on Neutron Cross Section Standards for the Energy Region above 20 MeV, Uppsala 1991, OECD: NEANDC-305 'u' Paris 1991, p. 123-134.
6. Cecil, R. A., Anderson, B. D. and Madey, R. *Improved Predictions of Neutron Detection Efficiency for Hydrocarbon Scintillators from 1 MeV about 492 MeV*. Nucl. Instrum. Methods **161**, 439 (1989). Also Sailor, W. C., Byrd, R. C. and Yariv, Y. *Calculation of the Pulse-Height Response of Organic Scintillators for Neutron Energies $28 < E_n < 492$ MeV*. Nucl. Instrum. Methods **A277**, 599-607 (1989).
7. Dickens, J. K. *SCINFUL: A Monte Carlo Based Computer Program to Determine a Scintillator Full Energy Response to Neutron Detection for E_n between 0.1 and 80 MeV: Program Development and Comparison of Program Predictions with Experimental Data*. Oak Ridge National Laboratory Report ORNL 6463 (1988).
8. Nolte, R., Schuhmacher, H., Brede, H. J. and Schrewe, U. J. *Measurements of High-Energy Neutron Fluences with Scintillation Detectors and Proton Recoil Telescopes*. Radiat. Prot. Dosim. **44**(1-4) 101-104 (1992) (This issue).
9. Caswell, R. S., Coyne, J. J. and Randolph, M. L. *Basic Data Necessary for Neutron Dosimetry*. Radiat. Prot. Dosim. **23**, 11-17 (1988).
10. Brenner, D. J. *Neutron Kerma Values above 15 MeV Calculated with a Nuclear Model Applicable to Light Nuclei*. Phys. Med. Biol. **29**, 437-441 (1983).
11. Menzel, H. G., Pihet, P., Folkerts, K. H., Dahmen, P. and Grillmeier, R. E. *Dosimetry Research Using Low Pressure Proportional Counters for Neutrons with Energies up to 60 MeV*. Radiat. Prot. Dosim. **23**, 389-392 (1988).
12. Romero, J. L., Brady, F. P. and Subramanian, T. S. *Neutron Induced Charged Particle Spectra and Kerma from 25 to 60 MeV*. In: Proc. Int. Conf. on Nuclear Data for Basic and Applied Science, Santa Fe, New Mexico (New York: Gordon and Breach Science Publishers) p. 687-699 (1985).
13. Wells, A. H. *A Consistent Set of Kerma Values for H, C, N, and O for Neutrons of Energies from 10 to 80 MeV*. Radiat. Res. **80**, 1-9 (1979).
14. Alsmiller, R. G. and Barish, J. *Neutron Kerma Factors for H, C, N, O and Tissue in the Energy Range 20 – 70 MeV*. Health Phys. **33**, 98-100 (1977).
15. Dimbylow, P. J. *Neutron Cross-Sections and Kerma Values for Carbon, Nitrogen and Oxygen from 20 to 50 MeV*. Phys. Med. Biol. **25**, 637-649 (1980).
16. Behrooz, M. A., Gillespie, E. J. and Watt, D. E. *Kerma Factors for Neutrons with Energies up to 60 MeV*. Phys. Med. Biol. **26**, 507-515 (1990).
17. Dimbylow, P. J. *Neutron Cross-Section and Kerma Value Calculations for C, N, O, Mg, Al, P, S, Ar and Ca from 20 to 50 MeV*. Phys. Med. Biol. **27**, 989-1001 (1982).
18. Caswell, R. S., Coyne, J. J. and Randolph, M. L. *Kerma Factors for Neutron Energies below 30 MeV*. Radiat. Res. **83**, 217-254 (1980).
19. Axton, E. J. private communication (1988).
20. White, R. M., Broerse, J. J., DeLuca, P. M., Dietze, G., Haight, R. C., Kawashima, K., Menzel, H. G., Olsson, N. and Wambersie, A. *Status of Nuclear Data for Use in Neutron Therapy*. Radiat. Prot. Dosim. **44**(1-4) 11-20 (1992) (This issue).

MEASUREMENT OF NEUTRON KERMA FACTORS IN C, O, AND Si AT 18, 23, AND 25 MeV

C. L. Hartmann, P. M. DeLuca, Jr and D. W. Pearson
Department of Medical Physics
University of Wisconsin - Madison
Madison, WI 53706, USA

Abstract — Carbon, oxygen, and silicon kerma factor values were measured at neutron energies of 18–25 MeV using microdosimetric techniques. Small spherical and cylindrical proportional counters served to measure neutron kerma, while detection of the activity induced in Teflon $(C_2F_4)_n$ samples by the $^{19}F(n,2n)^{18}F$ reaction determined the fast neutron fluence. Deuteron bombardment of a tritium gas target provided a monoenergetic neutron source plus a secondary flux of 0 to 6 MeV neutrons. At each energy, a liquid scintillator-based time-of-flight spectrometer measured the neutron spectrum. In order to correct our measurements for kerma contributed by off-energy neutrons, we combined the secondary neutron spectrum with tabulated energy-dependent kerma factors and subtracted the result from the total measured kerma.

INTRODUCTION

Neutron kerma is the average initial energy transferred from neutrons to charged particles per unit mass of material. The kerma factor is the kerma per unit fluence. Kerma factors provide necessary data for clinical neutron dosimetry as well as for radiation damage estimation. Carbon and oxygen kerma factors provide the principal data required to convert the dose measured in A-150 plastic ionisation chambers to dose in human tissue, a relationship that is of importance to clinical neutron dosimetry. Silicon kerma factors provide a tool to predict neutron radiation damage to electronic components. Below 20 MeV neutron energy, kerma factors can be calculated from measured microscopic cross section data. Above 20 MeV, however, measured cross section data are sparse, and neutron kerma factors are determined primarily from nuclear model calculations. The kerma factors calculated by nuclear models vary by as much as a factor of two in the 20 to 60 MeV neutron energy range for carbon and oxygen and do not exist for silicon.

Kerma factor measurements for carbon, oxygen, and silicon at 18, 23, and 25 MeV neutron energy are reported. The carbon and silicon kerma were measured directly by proportional counters with walls constructed of each of the respective materials. A matched pair of instruments constructed of zirconium and zirconium oxide (Zr and ZrO_2) yielded the oxygen kerma factor by subtraction. During proportional counter irradiation, a liquid scintillator monitored the neutron fluence. For a fraction of the time, the scintillator was calibrated by exposing it to neutrons simultaneously with two Teflon (poly-tetrafluoromethane or $(C_2F_4)_n$) samples. Detection of fluorine activation radiation determined the fast neutron fluence.

Deuterons bombarding a small tritium gas target provided the neutron source, which consisted of monoenergetic $^3H(d,n)^4He$ neutrons accompanied by a contaminating flux of lower energy neutrons. The lower energy neutrons contributed significantly to the 23 and 25 MeV measurements, but did not affect measurements at 18 MeV. At each deuteron bombarding energy, we determined the neutron spectrum by time-of-flight spectrometry. The lower energy neutron spectrum, convoluted with tabulated energy-dependent kerma factors, provided the secondary neutron kerma required to correct the total kerma measurement.

EXPERIMENTAL PROCEDURE

Neutrons were produced by bombarding a tritium gas target with deuterons from the University of Wisconsin tandem accelerator. A 101.6 mm long and 6.4 mm inner diameter stainless steel cylinder, with a 2.5 μm Mo entrance foil and 0.25 mm Au beam stop, contained the tritium gas at about 0.1 MPa pressure. The neutron spectrum consisted of a monoenergetic group of neutrons at the energy of interest, and a distribution of lower energy neutrons with a maximum energy about 19 MeV lower than the monoenergetic group.

The proportional counters were positioned symmetrically about the deuteron beam axis at 50° at distances of 40–60 cm from the centre of the tritium target. Activation samples were also positioned at 50°, about 40 cm from the tritium target.

Proportional counters with walls of graphite (CPC), zirconium (ZrPC), zirconium oxide (ZrO_2PC), and silicon (SiPC) measured the neutron dose. The carbon-walled proportional counter was a 12.7 mm inner diameter, 1.3 mm thick spherical shell of the standard Far West Technology 1/2 in LET counter design, while the Si, Zr, and ZrO_2 counters were 15.9 mm inner diameter, 52.3 (Zr and ZrO_2) mm or 55.9 (Si) mm long cylindrical shells with wall thicknesses of 2.5 to 4.3 mm. Although the cylindrical shells of the Si and Zr/ZrO_2 detectors are 55.9 and 52.3 mm long, their collection wires are surrounded by field tubes that extend 7.0 mm into the cylinder on each end, making their 'collection lengths' 41.9 mm and 38.3 mm. The cylindrical proportional counters are described in more detail elsewhere[1]. Each counter was filled to a pressure of 8 kPa with a mixture of Ar and CO_2 gas (10% CO_2 by weight).

Proportional counter response was calibrated in terms of absorbed dose primarily with an internal ^{244}Cm alpha particle source. The average ^{244}Cm alpha particle stopping power used to calibrate each proportional counter was 488 MeV.cm^2.g^{-1}. To determine the average alpha particle stopping power, we accounted for energy loss in the 0.2 mg.cm^{-2} of Au (manufacturer specified) source encapsulation as well as energy loss in the gas cavity.

Proportional counter calibration using the internal alpha particle source requires the ratio of the cavity diameter to the cavity volume, (d/V). The spherical (CPC) detector diameter and volume are determined by the inner diameter of the graphite wall. The effective detector diameter-to-volume ratio for the cylindrical (Si, Zr, and ZrO_2) proportional counters, $(d/V)_{cyl}$, by simultaneously irradiating spherical and cylindrical A-150 plastic proportional counters with 1.85 MeV neutrons from the ^{3}H(p,n)^{3}He reaction was measured. The A-150 counters were filled with 8 kPa of tissue-equivalent propane gas. Each detector was positioned at a distance of 61 cm from the centre of the neutron source at an angle of 50°. The spherical A-150 plastic-walled detector has the same dimensions as the CPC, while the cylindrical A-150 plastic-walled detector is 15.9 mm in diameter (the same as all the cylindrical counters) and 55.9 mm long (the same length of the Si detector, but 3.5 mm longer than the Zr and ZrO_2 counters). By comparing the ionisation collected in the spherical and cylindrical A-150-walled proportional counters, $(d/V)_{cyl}$ to be 2.65±0.04×10^{-3} mm^{-2} was measured. This value was corrected for the ZrPC and ZrO_2PC to account for their slightly shorter collection lengths.

Another method of calibrating the proportional counter data is to fix the position of a known feature of the final event size spectrum, such as the maximum stopping power of a particular charged particle reaction product. For the carbon proportional counter, a distinct signal fall-off occurs at the maximum stopping power of recoil carbon particles in Ar+(10%)CO_2 gas, which is 4300 MeV.cm^2.g^{-1}. This appears as an abrupt edge in the event size spectrum, the Bragg edge, and was used as an alternative to the internal alpha particle source to calibrate the detector. Although precise edge calibration by the position of proton, alpha particle, and heavy recoil Bragg edges was not always possible, their identification provides substantial supporting evidence for the accuracy of the internal alpha particle calibration.

The dose to the carbon and silicon proportional counter gas was determined directly from proportional counter data. The proportional counter gas dose due to charged particles resulting from n-O reactions in the ZrO_2 detector was determined by subtracting the dose to the ZrO_2 gas caused by neutron interactions in zirconium, determined from the Zr proportional counter dose measurements.

The energy-dependent neutron fluence per deuteron charge collected on the tritium target was determined for the 18, 23 and 25 MeV measurements. The neutron spectrum was measured with a time-of-flight spectrometer that consisted of a 5.1 × 5.1 cm^2 NE 213 liquid scintillator positioned 2 to 2.5 m from the pulsed neutron source (a bunched, chopped deuteron beam incident on the tritium gas target). The Monte Carlo scintillator response code SCINFUL[2] was used to calculate the energy-dependent scintillator efficiency. Accuracy of the absolute neutron spectrum measurement was verified at 1.8, 23, and 25 MeV neutron energies. In the 1.8 MeV test, the number of 1.8 MeV ^{3}H(p,n)^{3}He neutrons per proton determined from scintillator measurements was compared with the value expected from activation measurements of the number of 18 MeV ^{3}H(d,n)^{4}He neutrons per deuteron, combined with the ^{3}H(p,n)^{3}He and ^{3}H(d,n)^{4}He cross sections. The scintillator result was within about 5% of the value expected from activation, which is well within the uncertainty of the activation measurement. At 23 and 25 MeV, the number of ^{3}H(d,n)^{4}He neutrons per deuteron determined from scintillator measurements was compared to the results of activation measurements carried out with the same tritium target. The scintillator results were within about 15% of activation values.

During the kerma factor measurements, the ^{3}H(d,n)^{4}He neutron fluence was monitored by periodically recording the integrated response of another liquid scintillator and by recording the deuteron current on a chart recorder while periodically recording the integrated deuteron charge. The neutron fluence monitors were calibrated during a portion of each irradiation by exposing them simultaneously with two 2.54 cm diameter, 0.63 cm thick Teflon discs. By comparing the response of the fluence monitors to activation induced in the Teflon, the absolute neutron fluence for the full irradiation period was determined. A high purity germanium spectrometer detected annihilation quanta resulting from fluorine activation. For the 25 MeV measurement activation results were corrected for a small amount of ^{11}C activity due to the ^{12}C(n,2n)^{11}C reaction. Neutron fluence was determined by combining activation measurements with ^{19}F(n,2n)^{18}F cross section values of 83.0±3.0, 72.7±4.2, and 57.2±2.9 mb for 18, 23, and 25 MeV, respectively, taken from Hartmann and De Luca[3].

The proportional counter dose measurements were corrected for the presence of gamma rays as well as for neutrons due to deuteron interactions in the tritium containment cell by subtracting the signal acquired with an empty target. By subtracting the empty-target measurement from the tritium target-in dose, the dose to the proportional counter gas due to neutrons from deuteron interactions in the tritium gas target was determined.

For the 23 and 25 MeV measurements, the kerma contributed to carbon, oxygen, and silicon by d-T breakup neutrons was determined by folding the breakup neutron spectra with energy-dependent kerma factors tabulated by Caswell et al[4]. The breakup dose in the wall was converted to breakup dose in the gas for each detector by multiplying by the average ratio of mass stopping powers calculated for recoil nuclei produced by the lower energy neutron interactions in the wall. Dose due to d-T breakup neutrons was subtracted from the proportional counter gas dose measurements to obtain the net dose in the proportional counter gas due to monoenergetic ^{3}H(d,n)^{4}He neutrons.

The net dose to the proportional counter gas due to 18, 23, and 25 MeV neutrons was converted to dose in the wall (equivalent to wall kerma), by multiplying the gas dose by the average ratio of mass stopping powers in the wall to the gas. The ratios of stopping powers for charged particles resulting from 18, 23, and 25 MeV neutron interactions in the proportional counter walls were determined using Brenner and Prael's double-differential secondary-particle production cross section calculations[5] for the CPC and the ENDF-B/VI cross section tabulations[6] for Si, Zr, and ZrO$_2$. For 23 and 25 MeV neutron energies, we assumed that cross sections for neutron interactions in oxygen, zirconium, and silicon have the same relative magnitude as in the ENDF-B/VI evaluation at 20 MeV. Table 1 summarises the average ratios of mass stopping powers calculated for 18 to 25 MeV neutrons, as well as the values determined for the off-energy breakup neutrons.

RESULTS

For neutron energies below 20 MeV, kerma factors have been calculated from measured microscopic cross sections. Above 20 MeV, the sparseness of cross section information has resulted in kerma factors derived almost completely from nuclear models, with concomitant large uncertainties. It is difficult to measure accurately all the cross sections necessary to make kerma calculations similar to those made by Caswell et al below 20 MeV[4]. However, direct measurement of kerma factors can be used to normalise and verify model-based calculations.

The uncertainties in our measurements are caused primarily by subtraction of the dose due to d-T breakup neutrons. Subtraction of this background resulted in a 40–60% adjustment to the measured proportional counter dose. The dose due to lower energy breakup neutrons was assumed to be uncertain by about 9%, due primarily to the 7% uncertainty in the calculated scintillator efficiency, and 5% uncertainty in the kerma factors used to convert neutron fluence to dose. The measured dose to the proportional counter gas is uncertain by about 7% due to error caused by assuming that W/e is constant for all charged particle types and energies (3%), uncertainty related to the alpha particle or Bragg edge calibration (5%), and statistical error in the dose measurement (1–2%). Errors stated for final kerma factor values also include the uncertainty in the average ratio of stopping powers (11%), as well as the uncertainty in the absolute fluence determination (6–7%).

The kerma factor results in carbon, oxygen, and silicon are summarised in Table 2. Figure 1 compares the measured carbon kerma factors with several measurements[7–17] and evaluations[4,18,19], and one calculation[20]. Kerma factors determined from Brady's charged particle production measurements[8] have been corrected for the effect of high

Table 1. Ratio of average wall to gas stopping powers calculated for charged particles produced by monoenergetic ^{3}H(d,n)^{4}He and lower energy breakup neutrons incident on C, Zr, ZrO$_2$, and Si walls enclosing Ar + (10%) CO$_2$ gas.

Element	$_m\overline{S}_g^w$			
	d-T breakup neutrons	18 MeV	23 MeV	25 MeV
C	1.50	1.49	1.48	1.47
Zr	0.507	0.510	0.510	0.510
ZrO$_2$ (Zr(n,X) only)	–	0.673	0.674	0.674
ZrO$_2$ (O(n,X) only)	0.675	0.733	0.748	0.756
Si	1.01	1.05	1.05	1.05

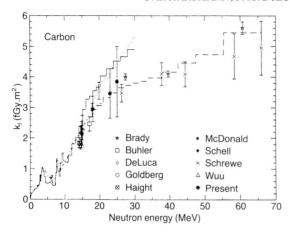

Figure 1. Carbon kerma factor measurements and calculations. Key to lines: (———) Caswell *et al* (1980); (.......) Gerstenberg (1988); (— - —) Howerton (1991); (– – –) Dimbylow (1982).

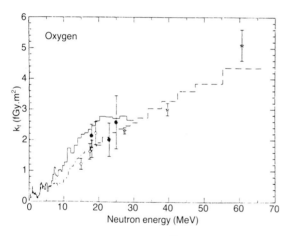

Figure 2. Oxygen kerma factor measurements and calculations. Key to lines as Figure 1.

Table 2. Measured kerma factors for C, O, and Si at 18, 23, and 25 MeV neutron energy.

Element	Energy (MeV)	k_f (fGy.m^2)
C	18	2.95 ± 0.42
C	23	3.46 ± 0.81
C	25	3.85 ± 1.15
O	18	2.14 ± 0.37
O	23	2.01 ± 0.54
O	25	2.58 ± 0.86
Si	18	1.63 ± 0.27
Si	25	1.28 ± 0.29

detector thresholds, as discussed by Dimbylow[20]. Our 18 MeV kerma factor is in substantial agreement with previous measurements. Although the values measured at 23 and 25 MeV are consistent with those based on integral dose[16] and charged particle production[8] measurements as well as Dimbylow's calculation[20], the uncertainty introduced by the correction for lower energy neutrons only allows the rejection of the evaluation based on limited microscopic data. Gerstenberg's kerma factor values use a recent evaluation of the $^{12}C(n,n')3\alpha$ and $^{12}C(n,\alpha)$ reaction cross sections combined with values from ENDF/B-V, which provides a full cross section evaluation for carbon up to 30 MeV neutron energy. Although Gerstenberg's kerma factors agree with measurements at neutron energies below 18 MeV, they appear to diverge from measured data at higher neutron energies[18].

For the oxygen kerma determination, the measured dose to the ZrPC gas was converted to dose to the Zr wall by multiplying by the average ratio of mass stopping powers, $_m\overline{S}^{Zr}{}_{Ar+(10\%)CO_2}$. The dose to zirconium was scaled by the weight fraction of Zr in ZrO$_2$ (0.7403) and then converted to the dose that the n-Zr charged particles would produce in the ZrO$_2$ proportional counter gas by dividing the Zr dose by $_m\overline{S}^{ZrO_2}{}_{Ar+(10\%)CO_2}|_{(Zr(n,X))}$. The gas dose due to Zr in the ZrO$_2$ wall was subtracted from the gas dose measured by the ZrO$_2$ proportional counter to obtain the dose to the ZrO$_2$ counter gas from the oxygen component in the counter wall at 18 and 23 MeV[4]. Figure 2 compares our oxygen results with other measured kerma factors[8,1], as well as values calculated from measured and evaluated microscopic cross sections[8,4,19] and one nuclear model calculation[20]. Our 18 MeV result agrees with the previous measurement as well as with Caswell's evaluation, but is somewhat higher than the Livermore evaluation. At higher neutron energies the trend indicated by our results agrees with DeLuca's and Brady's measurements as well as with Caswell's and Dimbylow's calculated values.

Figure 3 compares the silicon kerma factor values with the only other integral kerma factor measurement[1], as well as with neutron kerma factor determinations from the ENDF/B-IV[4], ENDF/B-VI[21] and Livermore[19] cross section evaluations. Our 18 MeV result agrees with the previous measurement as well as kerma factors determined from ENDF/B-IV, but is somewhat higher than the ENDF/B-VI value and significantly higher than the Livermore estimate. Our 25 MeV measurement, however, is lower than Caswell's evaluation, and more in line with the latest kerma factor determination from the ENDF/B-VI data. The 25 MeV value was measured on two separate occasions. The kerma per neutron fluence from each measurement agree to within 1%, which is well within the statistical and fluence calibration uncertainty.

The results reported here provide the only integral kerma factor measurements for oxygen and silicon and one of only two integral measurements of the carbon kerma factor above 20 MeV neutron energy. A more complete discussion of the kerma factor measurements is given elsewhere[22].

ACKNOWLEDGEMENT

This work was supported in part by the US Department of Energy through grant DE-FG02-86-ER60417 and the National Institutes of Health through grant 5-T32-CA-09206 (Hartmann).

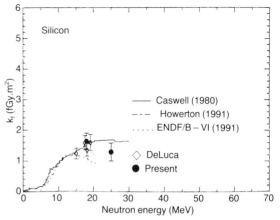

Figure 3. Silicon kerma factor measurements and calculations.

REFERENCES

1. DeLuca, P. M. Jr, Barschall, H. H., Sun, Y. and Haight, R. C. *Kerma Factor of Oxygen, Aluminium, and Silicon for 15 and 20 MeV Neutrons.* Radiat. Prot. Dosim. **23**, 27-30 (1988).
2. Dickens, J. K. *Scintillator Full Response to Neutron Detection.* Volume PSR-267 ORNL-6462 of RSIC Peripheral Shielding Routine Collection (Radiation Shielding Information Center, Oak Ridge, TN) (1988).
3. Hartmann, C. H. and DeLuca, P. M. Jr *Measurement of the $^{19}F(n,2n)^{18}F$ Cross Section from 18 to 27 MeV.* Nucl. Sci. Eng. **109**, 319-323 (1991).
4. Caswell, R. L., Coyne, J. J. and Randolph, M. L. *Kerma Factors for Neutron Energies below 30 MeV.* Radiat. Res. **83**, 217-254 (1980).
5. Brenner, D. J. and Prael, R. E. *Calculated Differential Secondary-particle Production Cross Sections after Nonelastic Neutron Interactions with Carbon and Oxygen between 15 and 60 MeV.* Atomic Data and Nuclear Data Tables **41**, 71-130 (1989).
6. ENDF. *ENDF/B-VI: Evaluated Nuclear Data File.* Technical Report (Brookhaven National Laboratory, National Nuclear Data Center, Upton, NY) (1990).
7. Binns, P. J. and Hough, J. H. *NAC Annual Report.* Technical Report NAC/AR/90-01, (National Accelerator Center, Faure, RSA) (June 1990).
8. Brady, F. P. and Romero, J. L. *Neutron Induced Reactions in Tissue Resident Elements, Final Report to the National Cancer Institute, Grant No. 1R01 CA16261.* Technical Report (University of California-Davis, Davis, CA) (1979).
9. Bühler, G., Menzel, H., Schuhmacher, H. and Guldbakke, S. *Dosimetric Studies with Non-hydrogenous Proportional Counters in Well Defined High Energy Neutron Fields.* In: Proc. 5th Symp. on Neutron Dosimetry, Munich/Neuherberg, FRG, EUR-7448, pp. 309-320 (Luxemburg: Commission of the European Communities) (1985).
10. DeLuca, P. M. Jr, Barschall, H. H., Haight, R. C. and McDonald, J. C. *Kerma Factor of Carbon for 14.1-MeV Neutrons.* Radiat. Res **100**, 78-86 (1984).
11. DeLuca, P. M. Jr, Barschall, H. H., Burhoe, M. and Haight, R. C. *Carbon Kerma Factor for 18- and 20-MeV Neutrons.* Nucl. Sci. Eng. **94**, 192-198 (1986).
12. Goldberg, E., Slaughter, D. R. and Howell, R. H. *Experimental Determination of Kerma Factors at E_n=15 MeV.* Technical Report UCID-17789 (Lawrence Livermore Laboratory, Livermore, CA) (1978).
13. Haight, R. C., Grimes, S. M., Johnson, R. G. and Barschall, H. H. *The $^{12}C(n,\alpha)$ Reaction and the Kerma Factor of Carbon at $E_n = 14.1$ MeV.* Nucl. Sci. Eng. **87**, 41-47 (1984).
14. McDonald, J. C. *Calorimetric Measurements of the Carbon Kerma Factor for 14.6 MeV Neutrons.* Radiat. Res. **109**, 28-35 (1987).
15. Schell, M. C., Pearson, D. W., DeLuca, P. M. Jr and Haight, R. C. *Measurement of Dose Distributions of Linear Energy Transfer in Matter Irradiated by Fast Neutrons.* Med. Phys. **17** (1), 1-9 (1990).
16. Schrewe, U. J., Brede, H. J., Henneck, R., Gerdung, S., Kunz, A., Menzel, H. G., Meulders, J. P., Pihet, P., Schuhmacher, H. and Slypen, I. *Determination of Kerma Factors for A-150 Plastic and Carbon for Neutron Energies above 20 MeV.* In: Proc. Int. Conf. on Nuclear Data for Science and Technology, Jülich, Germany (in press).
17. Wuu, C. S. and Milavikas, L. R. *Determination of the Kerma Factors in Tissue-equivalent Plastic, C, Mg, and Fe for 14.7-MeV Neutrons.* Med. Phys. **14** (6), 1007-1014 (1987).

18. Caswell, R. L., Coyne, J. J., Gerstenberg, H. M. and Axton, E. J. *Basic Data Necessary for Neutron Dosimetry.* Radiat. Prot. Dosim. **23** (1), 11-17 (1988).
19. Howerton, R. J. Private communication (1990).
20. Dimbylow, P. J. *Neutron Cross-section and Kerma Value Calculations for C, N, O, Mg, Al, P, S, Ar, and Ca from 20 to 50 MeV.* Phys. Med. Biol. **27** (8), 989-1001 (1982).
21. Hetrick, D. M. Private communication (1991).
22. Hartmann, C. L. *Measurement of Neutron Kerma Factors at 18, 23, and 25 MeV.* PhD Thesis, University of Wisconsin-Madison, Madison, WI (1991).

SECONDARY ALPHA PARTICLE SPECTRA AND PARTIAL KERMA FACTORS OF THE REACTION $n+{}^{12}C \rightarrow n+3\alpha$

B. Antolkovic†, G. Dietze‡ and H. Klein‡
†Ruder Boskovic Institute
POB 1016, 41001 Zagreb, Croatia
‡Physikalisch-Technische Bundesanstalt
POB 3345, W-3300 Braunschweig, Germany

Abstract — The alpha particle spectra from the $n+{}^{12}C \rightarrow n+3\alpha$ break-up at E_n = 11.9, 12.9, 14.0, 14.8, 17.0 and 19.0 MeV have been deduced by measuring the three correlated alpha particles in nuclear emulsion. The alpha particle spectra have a low energy cut-off at E_{coff} = 0.45 MeV. To account for the loss of 3α events due to the low energy cut-off correction factors have been calculated using the Monte Carlo method. The mean alpha particle energies and kerma factors for the reaction ${}^{12}C(n,n3\alpha)$ have been deduced from the corrected alpha particle spectra.

INTRODUCTION

Secondary charged particle spectra play an important role in neutron metrology and various biomedical applications (kerma factors, linear energy transfer, detector response, etc.). In neutron interaction with carbon at energies above 12 MeV, α particle spectra are almost entirely due to α particles emitted in the $n+{}^{12}C \rightarrow n+3\alpha$ reaction ($Q \leq -7.2$ MeV). So far, only a few experimental results have been obtained on α particle spectra from the $n3\alpha$ break-up[1-3], and several calculations[4-8] based on various theoretical assumptions show considerable discrepancies. The aim of this work was to determine experimentally the α particle spectra of the ${}^{12}C(n,n3\alpha)$ reaction in the neutron energy range from 12 to 19 MeV, and to evaluate partial kerma factors for this reaction using partial cross section data published earlier[9]. The present data are compared with values obtained by α particle spectrometry[1,2] and by measurements of ionisation yield spectra[10].

DATA REDUCTION

In a recent kinematically complete measurement of the ${}^{12}C(n,n3\alpha)$ reaction[9], nuclear emulsions acting as a target and a 4π detector were exposed to monoenergetic neutrons produced via the $D(d,n){}^3He$ and $T(d,n){}^4He$ reactions at the PTB's fast neutron facility. The cross section data and the ${}^{12}C$ and ${}^8Be$ excitation energy spectra were determined by detecting the three correlated α particles. In the present work the same raw data have been re-analysed in order to deduce the α particle spectra at E_n = 11.9, 12.9, 14.0, 14.8, 17.0 and 19.0 MeV. An extensive description of the processing and analysis of data, including corrections due to the loss of events during the scanning procedure, is given elsewhere (Reference 11 and references therein). The largest correction is due to the loss of events in which at least one of the alpha particle tracks is too short, i.e. is of too low energy to be detected or reliably measured. This correction influences the shape of the alpha energy spectrum and it will therefore be briefly reviewed after the main features of the Monte Carlo method, as used for the calculation of the correction, are discussed.

MONTE CARLO CALCULATION

As shown recently[11], a sample of events defining the final state of the $n+{}^{12}C \rightarrow n+3\alpha$ break-up can be generated by a Monte Carlo calculation provided the input data, i.e. the ${}^{12}C$ excitation energy spectra for the decay via the ground and the first excited state of ${}^8Be$, are known. The experimentally extracted ${}^{12}C$ excitation energy spectra[9] implicitly include the 3α and 2α correlations from all possible reaction chains, but in an *a priori* unknown ratio. The detailed analysis[11] has also shown that using these data as input for a Monte Carlo calculation, the distributions and correlations of kinematical variables, i.e. the energies and the emission angles of the α particles, can be calculated by assuming only a sequential decay mechanism, ${}^{12}C(n,n){}^{12}C(\alpha){}^8Be(2\alpha)$, regardless of the presence of other reaction mechanisms contributing to the $n+{}^{12}C \rightarrow n+3\alpha$ break-up.

The input data for the Monte Carlo calculation consist of a set of experimentally obtained cross section distributions with respect to the ${}^{12}C$ excitation energies[9], bin averaged over 1 MeV. Furthermore, the transitions via the ${}^8Be$ ground and first excited state were experimentally separated. The data are presented in the form directly applicable for the calculation of the reaction sequence via the ground and first excited state of ${}^8Be$ (Tables 1 and 2, respectively).

In treating the history of each event the Monte Carlo program assumes the Breit-Wigner shape of the intermediate states and takes into account the angular distribution of inelastic neutron scattering. The program can also accommodate various experimental conditions, e.g. low energy cut-off, various preset criteria and energy and angular spread.

LOW ENERGY CUT-OFF CORRECTION

In order to assess the expectation energy spectra of alpha particles, the Monte Carlo program, fed by the input data from Tables 1 and 2, is used. The three alpha particle energies of the generated events were sorted into energy bins of two separate alpha particle spectra, one containing the three alpha particle energies of all generated events (total spectrum), and the other spectrum including only those events in which all three alpha particle energies are above a certain cut-off energy, E_{coff} (cut-off spectrum). The correction factor is then expressed as the ratio of the number of events in the bin of the total spectrum and the corresponding number of events in the cut-off spectrum. Since the events omitted from the cut-off spectrum contain alpha particles with energies below as well as above the cut-off energy, it is obvious that the whole energy range of the alpha particle spectrum is affected by the loss. The largest corrections are necessary at the extreme parts of the spectra. The corrections for the low energy cut-off are considerably higher at E_n = 11.9 MeV, which is 3.8 MeV above the $^{12}C \to 3\alpha$ threshold, than at E_n = 19 MeV, which is 10.3 MeV above the threshold. For these two extreme neutron energies the total correction factors for low energy cut-off amount to 1.88 and 1.21, respectively[9].

In the present work the sharp cut-off at E_{coff} = 0.45 MeV (see References 9, 11) is applied to the experimental data and to the Monte Carlo calculation of the cut-off spectra.

RESULTS AND DISCUSSION

The measured alpha particle spectra at E_n = 11.9, 12.9, 14.0, 14.8, 17.0 and 19.0 MeV are shown in Figure 1 by open triangles, and the corresponding corrected alpha particle spectra are presented by full points. In the same figure the Monte Carlo calculated total and cut-off spectra are shown by full and dashed lines, respectively. All diagrams represent the cross section $d\sigma/dE_\alpha$ for the production of alpha particles in the n+^{12}C→n+3α break-up (note: $\int dE_\alpha (d\sigma/dE_\alpha)$ = $3\sigma_{n,n3\alpha}$). The agreement between experimental and theoretical spectra is very good, confirming that

Table 1. Partial cross sections for the transition via $^8Be_{gs}$ in the n+^{12}C→n+3α break-up reaction.

^{12}C exitation energy		Neutron energy (MeV)					
		11.9	12.9	14.0	14.8	17.0	19.0
Boundaries (MeV)	Mean energy (MeV)	Cross section (mb)					
< 10.25	9.6	134.8 ± 4.6	71.7 ± 8.5	76.5 ± 8.1	69.0 ± 8.4	44.5 ± 4.9	45.9 ± 5.3
10.25 – 11.25	10.75	21.0 ± 3.0	51.6 ± 4.7	30.8 ± 2.6	22.5 ± 2.9	14.7 ± 1.8	10.7 ± 1.9
11.25 – 12.25	11.75	1.5 ± 0.7	10.3 ± 1.7	23.1 ± 1.8	20.6 ± 2.6	11.9 ± 1.5	6.0 ± 1.3
12.25 – 13.25	12.75		0.7 ± 0.5	4.5 ± 0.6	15.1 ± 1.9	14.6 ± 1.8	4.7 ± 1.0
13.25 – 14.25	13.75			0.5 ± 0.2	2.7 ± 0.7	11.2 ± 1.5	5.7 ± 1.1
14.25 – 15.25	14.75					8.9 ± 1.4	5.5 ± 1.1
15.25 – 16.25	15.75					1.9 ± 0.9	4.7 ± 1.0
16.25 – 17.25	16.75						3.9 ± 1.0
17.25 – 18.25	17.75						0.5 ± 0.4

Table 2. Partial cross sections for the transition via $^8Be_{1.ex.st.}$ in the n+^{12}C→n+3α break-up reaction.

^{12}C exitation energy		Neutron energy (MeV)					
		11.9	12.9	14.0	14.8	17.0	19.0
Boundaries (MeV)	Mean energy (MeV)	Cross section (mb)					
10.25 – 11.25	10.75	19.3 ± 2.1	23.7 ± 2.4	16.1 ± 1.8	13.6 ± 1.7	10.3 ± 1.6	10.7 ± 1.7
11.25 – 12.25	11.75	0.1 ± 0.1	10.6 ± 1.7	35.2 ± 2.6	32.6 ± 2.5	23.8 ± 1.8	27.2 ± 2.4
12.25 – 13.25	12.75		0.5 ± 0.3	14.3 ± 1.9	41.1 ± 4.0	44.2 ± 3.8	26.1 ± 2.4
13.25 – 14.25	13.75			0.7 ± 0.3	10.1 ± 1.7	39.7 ± 3.5	31.6 ± 2.8
14.25 – 15.25	14.75					31.8 ± 3.3	39.6 ± 3.6
15.25 – 16.25	15.75					14.2 ± 2.0	43.4 ± 3.8
16.25 – 17.25	16.75						28.2 ± 2.6
17.25 – 18.25	17.75						10.2 ± 1.9

the set of given input data forms a valuable basis for the calculation of any kinematical variable of the reaction $^{12}C(n,3\alpha)$. The slight difference in the lowest part of the alpha particle energy spectrum is chiefly due to the bin width (1 MeV) of the input data which is too broad to describe the fine structure of this part of the spectrum. The calculated total alpha particle energy spectra have been used to extrapolate the corrected experimental alpha particle spectra to the extreme low and high energy range, for which, due to the low energy cut-off, there are no data.

The kerma factor for the reaction $^{12}C(n,n3\alpha)$ is defined by the product

$$KF_{n,n3\alpha} = N_C \varepsilon_\alpha \sigma_{n,n3\alpha} \qquad (1)$$

where N_C is the number of ^{12}C nuclides per unit mass, ε_α is the mean energy transferred to the kinetic energy of alpha particles and $\sigma_{n,n3\alpha}$ is the cross section for the $^{12}C(n,n3\alpha)$ reaction. The mean energy transfer is deduced from the mean alpha energy of the corrected experimental alpha particle spectra. Since there are three indistinguishable alpha particles in the final state, the mean alpha particle energy must be multiplied by a factor of 3 to obtain the mean energy transfer ε_α to all three alpha particles in the $n+^{12}C \rightarrow n+3\alpha$ break-up. The cross sections for the $^{12}C(n,n3\alpha)$ reaction are taken from Reference 9.

The extracted mean alpha energies and the kerma factors deduced from Relation 1 are listed in Table 3, second and third row, respectively.

In order to evaluate the uncertainties of the extracted data, it is worthwhile examining the distribution $f = f(E_{\alpha i} \cdot \sigma_i)$ obtained by introducing the relation for the mean alpha energy into Relation 1. One obtains

$$KF_{n,n3\alpha} = 3N_C \Sigma E_{\alpha i} \cdot \sigma_i \qquad (2)$$

where $E_{\alpha i}$ is the mean α particle energy of the bin i and σ_i the corresponding cross section. The shape of the experimental distribution $E_{\alpha i} \cdot \sigma_i$ is shown in Figure 2 for $E_n = 12.9$ and 17 MeV. The

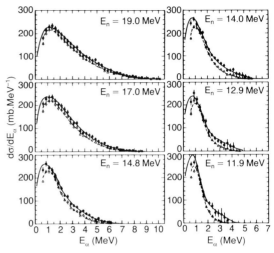

Figure 1. Alpha particle spectra of the reaction $n+^{12}C \rightarrow n+3\alpha$. The measured and the corresponding corrected α particle spectra are indicated by triangles and full points, respectively. The Monte Carlo calculated cut-off and total α particle distributions are indicated by dashed and full lines, respectively.

error bars of the experimental points are obtained by summing up the statistical errors and the correction factor uncertainties. The latter amount to 10% of the correction itself (see Reference 9). As can be seen, only the upper part of the distribution is due to larger uncertainties. However, this represents only a small part of the whole distribution and contributes considerably less to the total uncertainty.

The mean alpha energies and the kerma factors deduced in another experiment performed using a similar technique[2] are given in Table 3, rows 4 and 5. In the latter experiment a continuous energy neutron source was used, and the cut-off energy was $E_{coff} = 0.6$ MeV (appropriately accounted for in the cut-off correction calculation). The mean alpha particle energies are in excellent agreement with the present results, though there are some

Table 3. Mean alpha energies and kerma factors of the reaction $^{12}C(n,n3\alpha)$.

E_n	Present data		Data of Ref. 2[a]		Other data	
	$\overline{E}_\alpha$ (MeV)	KF (fGy.m²)	$\overline{E}_\alpha$ (MeV)	KF (fGy.m²)	KF (fGy.m²)	Ref.
11.9	1.17	0.49±0.07	1.18[b]	0.21±11[b]		
12.9	1.37	0.55±0.07	1.38±[c]	0.43±0.16		
14.0	1.55	0.75±0.07	1.60	1.16±0.34	0.56±0.08	1
14.8	1.74	0.95±0.08	1.89[d]	1.41±0.38[d]	0.96±0.18[d]	10
17.0	2.24	1.46±0.13	2.22	1.86±0.41	0.80±0.18	10
19.0	2.63	1.89±0.16	2.67	2.35±0.45		

[a] Data are averaged over 1 MeV of neutron energy.
[b,c,d] Data refer to $E_n = 12.0$, 13.0 and 15 MeV, respectively.

differences in the kerma factors which are mainly due to larger uncertainties in extracting the cross section data, and ambiguities in the determination of the white neutron spectrum fluence in Reference 2.

The kerma factors extracted from the angle-dependent α particle spectra at 14.1 MeV[1] and from the ionisation yield spectra at E_n = 15.0 and 17.0 MeV[10] are also included in Table 3 and Figure 3.

CONCLUSION

For neutron energies above 10 MeV the ^{12}C(n,n3α) reaction contributes strongly to the carbon kerma factors but only few experimentally based partial kerma factor data for this reaction exist. Theoretical calculations have shown considerable discrepancies due to their different assumptions on the reaction mechanism of the n3α break-up.

In the present work the α particle spectra at E_n = 11.9, 12.9, 14.0, 14.8, 17.0 and 19.0 MeV have been deduced by analysing three-prong stars in irradiated nuclear emulsions. Corrections for the loss of events due to the low energy cut-off (E_{coff} = 0.45 MeV) and the theoretical α particle energy distributions have been calculated on the basis of the Monte Carlo calculation, fed with the input data which completely describe the n+^{12}C→n+3α break-up. A good fit of the theoretically calculated spectra to the experiment data has enabled us to extrapolate the α particle spectra to extremely low and high energy regions, for which, due to the detection cut-off conditions, there are no experimental data. The kerma factors extracted in this work are shown in Table 3 and Figure 3 together with the other published data. The uncertainties of the present data are significantly lower than those of the data published previously.

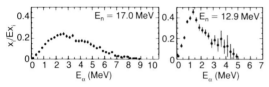

Figure 2. Typical $E_{αi} \cdot σ_i$ (=x_i) distributions of the alpha particle spectra in the n+^{12}C→n+3α break-up. Distributions are normalised to unity.

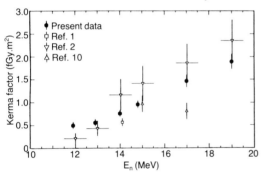

Figure 3. Kerma factors of the reaction n+^{12}C→n+3α.

REFERENCES

1. Haight, R. C., Grimes, S. M., Johnson, R. G. and Barschall, H. H. *The ^{12}C(n,α) Reaction and the Kerma Factor for Carbon at E_n = 14.1 MeV.* Nucl. Sci. Eng. **87**, 41-47 (1984).
2. Antolkovic, B., Slaus, I. and Plenkovic, D. *Experimental Determination of the Kerma Factor for the Reaction ^{12}C(n,n3α) at E_n = 10–35 MeV.* Radiat. Res. **97**, 253-261 (1984).
3. Subramanian, T. S., Romero, J. L., Brady, F. P., Watson, J. W., Fitzgerald, D. H., Garrett, R., Needham, G. A., Ullmann, J. L. and Zanelli, C. I. *Double Differential Inclusive Hydrogen and Helium Spectra from Neutron-Induced Reactions on Carbon at 27.4, 39.7 and 60.7 MeV.* Phys. Rev. **C28**, 521-528 (1983).
4. Alsmiller Jr, R. G. and Barish, J. *Neutron Kerma Factors for H, C, N, O and Tissue in the Energy Range of 20–70 MeV.* Health Phys. **33**, 98-100 (1977).
5. Wells, A. H. *A Consistent Set of Kerma Values for H, C, N, and O for Neutrons of Energies from 10 to 80 MeV.* Radiat. Res. **80**, 1-9 (1979).
6. Dimbylow, P. J. *Neutron Cross-Sections and Kerma Values for Carbon, Nitrogen and Oxygen from 20 to 50 MeV.* Phys. Med. Biol. **25**, 637-649 (1980).
7. Caswell, R. S., Coyne, J. J. and Randolph, M. L. *Kerma Factors for Neutron Energies below 30 MeV.* Radiat. Res. **83**, 217-254 (1980).
8. Gerstenberg, H. M., Caswell, R. S. and Coyne, J. J. *Initial Spectra of Neutron-Induced Secondary Charged Particles.* Radiat. Prot. Dosim. **23**, 41-44 (1988).
9. Antolkovic, B., Dietze, G. and Klein, H. *Reaction Cross-Sections on Carbon for Neutron Energies from 11.5 to 19 MeV.* Nucl. Sci. Eng. **107**, 1-21 (1991).
10. Bühler, G., Menzel, H. G., Schumacher, H. and Dietze, G. *Neutron Interaction Data in Carbon Derived from Measured and Calculated Ionisation Yield Spectra.* Radiat. Prot. Dosim. **13**, 13-17 (1985).
11. Antolkovic, B. and Turk, M. *Monte Carlo Analysis of the Four-Body Final State in the Neutron-Induced ^{12}C Breakup.* Nucl. Phys. **A524**, 285-305 (1991).

RBE MODIFYING FACTORS

M. Coppola, V. Di Majo, S. Rebessi and V. Covelli
Department of Health Effects, ENEA, CRE Casaccia
C.P. 2400, 00100 Rome, Italy

Abstract — Experimental studies, when carried out under well controlled experimental conditions and planned to investigate specific endpoints, can provide reliable information on many aspects of the biological action of radiation. In particular, animal studies are suitable to help in determining the shapes of the dose–response relationships for tumour induction, and the influence of dose rate, and in understanding the species and strain dependence, as well as the effect of factors such as sex, age and hormonal status. All these variables can diversely influence the biological response to radiations of different qualities, and thus the RBEs. These aspects are examined and discussed in the paper, special attention being addressed to the tumour induction response of the haemolymphopoietic tissue and of epithelial tissues, such as liver and ovary, which have shown an appreciable degree of susceptibility to radiation carcinogenesis.

INTRODUCTION

The determination of radiation RBE values at low doses and dose-rates is difficult to obtain from epidemiological human data, as these data are incomplete, often do not provide statistically significant evidence for radiation effects at low doses and, in general, are obtained from uncontrolled situations where the possibility of deriving incontrovertible information relevant to radiological protection is questionable. In addition, radiation is just one among the large number of agents present in the environment to which the biological tissues are sensitive. Thus, a number of confounding factors mostly related to socio-economic, dietary and smoking habits may substantially alter the effects of radiation, and even the basic form of the dose response may be dependent on the interplay of more factors. On the contrary, experimental studies carried out under well controlled laboratory conditions and designed to investigate specific endpoints in the whole-body system, or with specialised systems *in vivo* and/or *in vitro*, are suitable to provide reliable information on many aspects of the biological action of radiation.

In particular, animal studies have proved to be an irreplaceable source of data to define the shape of the dose–response relationships for the induction of a variety of tumour types, and to assess the influence of factors pertaining to the radiation, such as radiation quality, dose rate, local irradiation and non-uniformity of dose distribution, as well as the effect of so-called 'host factors', which are of a biological nature, such as species and strain, sex, age and hormonal status.

The complexity of these factors and the uniqueness of each tissue can influence to a variable extent the biological response to radiation of different qualities, and therefore, can also variably affect the RBE values for the induction of specific effects, in particular the various tumour types.

In the following we shall briefly analyse some of the above mentioned aspects, and we shall limit ourselves to consideration of tumour induction.

FACTORS INFLUENCING THE RELATIVE RESPONSE

Species

One of the most debated aspects of risk estimation concerns the possibility of extrapolating the data on similar biological effects from one species to another, as the hope would be to extrapolate this information eventually to man. This might represent a severe problem especially when dealing with a particular tumour type, as the susceptibility to tumour induction by radiation for a certain tissue or organ may depend very much on the species. Some indication in the direction of a confirmation of such a species dependence comes from studies on the induction of malignant neoplasms by radiation in burros or in monkeys in comparison to rodents; however, the information reported in the literature about different species is, in general, too scarce for any definite conclusion to be made[1].

For this reason the possibility of inter-species comparison offered by the data on mammary carcinomas in mice and rats has been considered with much interest, since different sets of results obtained from different species are available for similar neutron sources (fission neutrons) and similar dose rates. In particular, we have compared the data of mammary adenocarcinomas induced by fission neutrons in BALB/C mice[2] and in Sprague-Dawley rats[3] (Figure 1). It has been assumed that the number of malignant tumours in the rats is 40% of the total number of tumours as shown by Vogel and Dickson[4]. The two induction curves appear to run parallel, as a function of the dose, the differences being essentially due to the contribution of the spontaneous incidence in the two species. Thus, one may expect to obtain similar neutron RBE values in the two cases. In fact, Vogel quotes the following fission neutron RBE values relative to 250 keV X rays for the induction of mammary tumours in Sprague-Dawley rats: 4 at high doses, 20 to 60 at 50 mGy, and 50 at 20 mGy[5]. In the case of BALB/c mice, Ullrich calculates a fission neutron RBE of 33, if the initial slope of the dose–response curves is assumed to be linear for both neutrons and ^{137}Cs γ rays. Therefore, the estimates of RBE are similar in the two species. On the other hand, Broerse et al[6] report an RBE of 15 at 10 mGy of monoenergetic neutrons of 0.5 MeV, referred to X rays, for induction of mammary carcinomas in WAG/Rij rats, estimated from linear extrapolations to low doses of the experimental results for both neutrons and X rays. Deviations from the results of Vogel[3] may be due at least partially to differences in radiation quality as well as in the choice of the dose–response relationships utilised at low doses.

The influence of species can also be investigated with respect to the life shortening consequent to irradiation. The results of an analysis carried out by UNSCEAR[7] indicate that the dose dependence of the percentage life-span shortening following single exposures to low LET radiation, if expressed as the percentage of the life span of controls, is fairly similar, in a wide dose range, in rats, chinese hamsters and dogs. It has also been shown that the life shortening at low and moderate radiation doses is essentially related to the induction of tumours[8]. Therefore, it can be assumed that the variability of the overall tumour response to a specific radiation, relative to the incidence in the unirradiated control, is seemingly rather small among these species. In conclusion, from this point of view, data on life shortening appear to be well suitable for RBE determinations, which should not be much influenced by species differences.

Strain

To investigate the influence of strain on the susceptibility to tumour induction, a good set of data is that for myeloid leukaemia induced by radiation in CBA[9–11] and in BC3F1 mice[12]. In both cases the dose–response curves after X rays appear to be bell shaped, with maximum incidence between 2 and 3 Gy. This shape is interpreted as the result of competition between induction of malignant transformation and inactivation of the haemopoietic stem cells. However, the frequencies of myeloid leukaemia observed at the same doses are very different in the two mouse populations, indicating a higher sensitivity of CBA mice for induction of this lesion by a factor of 5. Still, similar values of the fission neutron RBE are found for the two strains, e.g. about 4 at neutron doses around 0.2 Gy. This would imply that a variation between the two strains in the susceptibility to malignant transformation of the haemopoietic stem cells by the same factor also affects the neutron dose response.

Other interesting information on the influence of strain on tumour susceptibility comes from consideration of the spontaneous incidence of two other tumour types observed in CBA and BC3F1 mice, namely malignant lymphoma[9,12] and hepatocellular tumours[13]. In fact, malignant lymphoma is absent in unirradiated CBA mice, while its spontaneous incidence is around 60% in hybrids BC3F1 mice; on the contrary, the incidence of liver tumours is about 70% in untreated CBA mice in contrast to a 10% in BC3F1.

However, it is difficult to anticipate how such differences, observed in the unirradiated two mouse strains, may impact on RBE for malignant lymphoma or hepatocellular tumours, as the available information is insufficient for a comparison

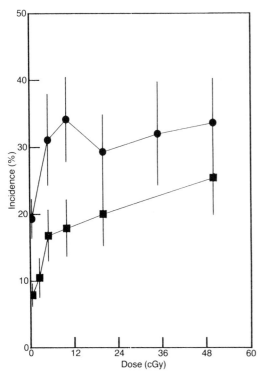

Figure 1. Mammary adenocarcinomas induced by single acute doses of fission neutrons in BALB/C mice[2] (squares) and in Sprague-Dawley rats[3] (circles).

among the two strains with respect to the relative effect of different radiation qualities. This fact stresses the need of continuing the studies of the influence of strain on tumour induction by radiation.

Age

The information on the dependency of sensitivity to radiation on age is also limited. However, experiments carried out at our laboratory on the induction of liver tumours by either X rays or fission neutrons in BC3F1 mice[14] provide a complete set of data suitable to allow us to evaluate whether age at irradiation affects neutron RBE for the induction of these tumours. These experiments have shown a marked dependence of the frequency of hepatocellular tumours induced by radiation on age, with a much higher susceptibility in young than in old animals. In addition, in young animals this age dependence appears to be more significant in the case of neutron irradiation than for X ray exposure (Figure 2). Thus, the neutron RBEs also prove to be age dependent with a value of 28 at a neutron dose of 90 mGy for prenatal irradiation and 13 at a dose of 170 mGy in adult animals.

Sex

An interesting example of sex dependence of the tumour response to irradiation is provided by data on thymic lymphoma and myeloid leukaemia induced by γ rays[15], showing a much higher incidence of thymic lymphoma and a much lower incidence of myeloid leukaemia in female RFM mice in comparison to male of the same strain. Unfortunately, no inference regarding RBE can be derived from this information at the moment.

Hormonal status

As far as the influence of the hormonal status on the development of tumours is concerned, an example is offered by the results of experiments on ovarian tumour induction in BALB/C female mice, irradiated with fission neutrons and γ rays[16], and in BC3F1 irradiated with 1.5 MeV neutrons and X rays[17]. In both cases the data suggest the presence of a threshold dose somewhere around 50 mGy, followed by a rapid increase in the induction (Figure 3).

This shape of the dose–response curves, as well as the absence of a clear radiation quality dependence has been related to the effect of hormonal

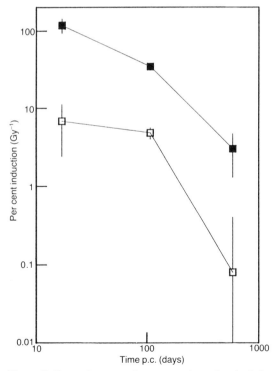

Figure 2. Dependence on time post-coitum (p.c.) of the per cent increment per unit dose of the frequency of hepatocellular tumours induced by fission neutrons at all doses (full symbols) and X rays at 2 Gy (open symbols)[13].

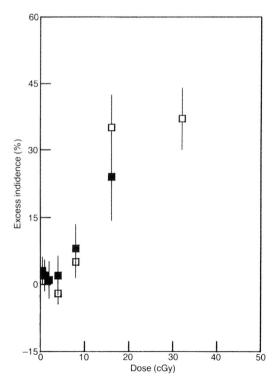

Figure 3. Ovarian tumour induction in BC3F1 female mice, irradiated with 1.5 MeV neutrons (full symbols) and X rays (open symbols)[17].

unbalance following irradiation. The hypothesis that is put forth is that, when a certain number of oocytes are killed, this provokes a reaction from the hormonal system that might correspond to an excess of gonadotropic hormone, which then becomes the important causative factor for the development of ovarian tumours. In this sense this process may loose its character of stochasticity.

Due to the shape of the dose–response relationships no RBE value can be obtained in the region below 0.1 Gy, while the rapid increase in the frequency of induced neoplasms in the region up to the plateau has a similar slope for neutrons and X rays, suggesting an RBE of about 1. Since a similar RBE value was obtained for oocyte killing in the mouse at doses higher than 0.1 Gy[18], this finding supports the hypothesis previously put forward.

Mammary carcinogenesis is another endpoint with a large dependence on the hormonal status of the animal. Shellabarger et al[19] reported an appreciable increase in neutron RBE relative to X rays for tumour induction in ACI rats after DES administration. Conversely, an increased response in mammary carcinogenesis was observed by Broerse et al[20] in different rat strains after hormone administration for both X rays and neutrons without a significant increase in RBE.

Dose rate and fractionation

Coming now to the factors pertaining to the irradiation conditions, the one which appears to be the most important is the temporal distribution of the dose delivery. The possible existence of an inverse dose rate effect for neutron irradiations has stimulated great interest in this matter by a considerable number of laboratories. The results of in vitro experiments on neoplastic transformation of C3H10T1/2 cells are at the moment fairly discordant, in that they have shown either no effect from fractionated rather than single doses[21] or only a small effect by a factor of around two at some neutron energies[22]. As the existence of such an effect would modify the neutron RBE values, when passing from single acute to fractionated or protracted exposures, we have also felt it worthwhile to consider this aspect here.

For this purpose we examine first the life span shortening of mice irradiated with fission neutrons at our laboratory, applying a dose fractionation protocol consisting in 5 fractions at 24 h intervals.

The results indicate no increase in mortality of our BC3F1 mice irradiated with this fractionation scheme in respect of single acute exposures[23]. Also raw data of tumour incidence at death, now being analysed, do not reveal any marked variation in the effectiveness of fission neutrons when delivered acutely or in the fractionated mode.

This is consistent with previous data on lung adenocarcinoma induced by fission neutrons in BALB/c mice, where a 24 h separation of dose fractions did not significantly affect the dose–effect curve, and a separation of 30 days increased the induction at 0.5 Gy only by a factor 2. Furthermore, no effect of dose fractionation was seen in the induction of mammary adenocarcinoma[2]. However, it cannot be excluded from Ullrich's data that dose protraction at a neutron dose rate lower than 100 mGy per 20 h day (i.e. 0.083 mGy.min^{-1}) might increase the low dose response for mammary tumours by a factor of two. This is not in contradiction with a recent review of the effect of dose fractionation or reduction in dose rate on tumour induction or longevity in experimental animals after high LET radiation[24] that has indicated the possibility of an increased effectiveness in mammary tumours, pulmonary tumours and longevity.

CONCLUSIONS

In conclusion, several factors seem to be able to affect, to a varied extent, the RBE values for stochastic effects. In particular, age at irradiation and dose protraction appear to have a more variable impact on the effect of different radiation qualities, and so to affect the determination of RBE. However, the influence of each of these factors does not appear to exceed a factor two. Conversely, differences in species do not appear to play a major role in determining a different effect of radiation, at least in the cases examined here. This might have important implications in the framework of radiation protection, and stresses the importance of continuing animal studies for the assessment of radiological risks.

ACKNOWLEDGMENTS

This paper has been prepared in the course of contract Bi6-004-I of the Radiation Protection Programme of CEC. Contribution n. 2590.

REFERENCES

1. Fry, R. J. M. and Storer, J. B. *External Radiation Carcinogenesis.* In: Advances in Radiation Biology, pp. 31-90. (London: Academic Press) (1987).
2. Ullrich, R. L. *Tumor Induction in BALB/c Mice after Fractionated or Protracted Exposures to Fission-spectrum Neutrons.* Radiat. Res. **97**, 587-597 (1984).
3. Vogel, H. H. *Mammary Gland Neoplasms after Fission Neutron Irradiation.* Nature **222**, 1279-1281 (1969).

4. Vogel, H. H. and Dickson, H. W. *Mammary Neoplasia following Neutron and Gamma Irradiation.* In: Proc. Int. Colloquium on Neutron Radiation Biology, 5-7 November 1990, Rockville, MD (abstract).
5. Vogel, H. H. *High LET Irradiation of Sprague-Dawley Female Rats and Mammary Neoplasm Induction.* In: Late Biological Effects of Ionizing Radiation, STI/PUB/489, Vol. II, pp. 147-163 (Vienna: IAEA) (1978).
6. Broerse, J. J., Hennen, L. A. and Van Zwieten, M. J. *Radiation Carcinogenesis in Experimental Animals and its Implications for Radiation Protection.* Int. J. Radiat. Biol. **85**(2), 167-187 (1985).
7. United Nations Scientific Committee on the Effects of Atomic Radiation. *Ionizing Radiation: Sources and Biological Effects.* Report to the General Assembly, with annexes, pp. 655-725, (New York: United Nations) (1982).
8. Coppola, M., Covelli, V. and Di Majo, V. *Analysis of Mortality from Different Causes of Death in Mice Exposed to Fast Neutrons.* In: Proc. of Fifth Symp. on Neutron Dosimetry, EUR 9762, pp. 35-43, (Luxembourg: CEC) (1984).
9. Di Majo, V., Coppola, M., Rebessi, S., Bassani, B., Alati, T., Saran, A., Bangrazi, C. and Covelli, V. *Dose-Response Relationship of Radiation-induced Harderian Gland Tumors and Myeloid Leukemia of the CBA/Cne Mouse.* J. Natl Cancer Inst. **76**(5), 955-966 (1986).
10. Mole, R. H., Papworth, D. G. and Corp, M. J. *The Dose-response for X-ray Induction of Myeloid Leukaemia in Male CBA/H Mice.* Br. J. Cancer **47**, 285-291 (1983).
11. Mole, R. H. *Dose-response Relationships.* In: Radiation Carcinogenesis: Epidemiology and Biological Significance, pp. 403-420 (New York: Raven Press) (1984).
12. Covelli, V., Di Majo, V., Coppola, M. and Rebessi S. *The dose-response Relationships for Myeloid Leukemia and Malignant Lymphoma in BC3F1 Mice.* Radiat. Res. **119**, 553-561 (1989).
13. Di Majo, V., Coppola, M., Rebessi, S., Bassani, B., Alati, T., Saran, A., Bangrazi, C. and Covelli, V. *Radiation Induced Mouse Liver Neoplasms and Hepatocyte Survival.* J. Natl Cancer Inst. **77**(4), 933-939 (1986).
14. Di Majo, V., Coppola, M., Rebessi, S. and Covelli, V. *Age-related Susceptibility of Mouse Liver to Induction of Tumors by Neutrons.* Radiat. Res. **124**, 227-234 (1990).
15. Ullrich, R. L. and Storer, J. B. *Influence of γ Irradiation on the Development of Neoplastic Disease in Mice.* Radiat. Res. **80**, 317-324 (1979).
16. Ullrich, R. L. *Tumor Induction in BALB/c Female Mice after Fission Neutron or γ Irradiation.* Radiat. Res. **93**, 506-515 (1983).
17. Covelli, V., Coppola, M., Di Majo, V., Rebessi, S. and Bassani, B. *Tumor Induction and Life Shortening in BC3F1 Female Mice at Low Doses of Fast Neutrons and X rays.* Radiat. Res. **113**, 362-374 (1988).
18. Dobson, R. L. and Straume, T. *Cancer Risk and Neutron RBE's from Hiroshima and Nagasaki.* In: Proc. Seminar on Neutron Carcinogenesis, EUR 8084. pp. 279-300 (Luxembourg: CEC) (1982).
19. Shellabarger, C. J., Chmelevsky, D., Kellerer, A. M., Stone, J. P. and Holtzman, S. *Induction of Mammary Neoplasms in the ACI Rat by 430 keV Neutrons X rays and di-ethil-stilbestrol.* J. Natl Cancer Inst. **69**(5), 1135-1146 (1982).
20. Broerse, J. J., Hennen, L. A., Klapwijk, W. M. and Solleveld, H. A. *Mammary Carcinogenesis in Different Rat Strains after Irradiation and Hormone Administration.* Int. J. Radiat. Biol. **51**(6), 1091-1100 (1987).
21. Saran, A., Pazzaglia, S., Coppola, M., Rebessi, S., Di Majo, V., Garavini, M. and Covelli, V. *Absence of a Dose-Fractionation Effect on Neoplastic Transformation Induced by Fission-spectrum Neutrons in C3H 10T1/2 Cells.* Radiat. Res. **126**, 343-348 (1991).
22. Miller, R. C., Geard, C. R., Brenner, D. J., Komatsu, K., Marino, S. A. and Hall, E. J. *Neutron-energy-Dependent Oncogenic Transformation of C3H 10T1/2 Mouse Cells.* Radiat. Res. **117**, 114-127 (1989).
23. Di Majo, V., Coppola, M., Rebessi, S., Saran, A., Pazzaglia, S. and Covelli, V. *Do Multifractionated Neutron Doses Enhance Mortality in Mice?* In: 9th Int. Congress of Radiation Research, 6-12 July 1991, Toronto, Canada (abstract).
24. Broerse, J. J. *Influence of Physical Factors on Radiation Carcinogenesis in Experimental Animals.* In: Low Dose Radiation: Biological Bases of Risk Assessment, pp. 181-194, (London: Taylor and Francis) (1989).

VARIATION OF NEUTRON RBE AS A FUNCTION OF ENERGY FOR DIFFERENT BIOLOGICAL SYSTEMS: A REVIEW

J. Van Dam†, M. Beauduin‡, V. Grégoire§, J. Gueulette§, G. Laublin§ and A. Wambersie§
†Gezwelziekten, U.Z. St. Rafaël, 3000 Leuven, Belgium
‡Hôpital de Jolimont, 7100-Haine St-Paul, Belgium
§Université Catholique de Louvain, Cliniques Universitaires St-Luc, 1200 Bruxelles, Belgium

Abstract — RBE determinations have been performed in all neutron therapy centres prior to any clinical application. In some centres, the minimum checks were done in order to start the treatments in safe conditions. In other centres, extensive investigations were performed comparing several biological systems. A large amount of data was thus accumulated. However, only a few investigations were performed in order to systematically assess the RBE variation as a function of neutron energy for the same biological system, which could allow an accurate exchange of information between centres. From the available radiobiological data, it can be concluded that RBE increases with decreasing neutron energy, but the slope of the RBE/neutron energy relationship depends on the biological system and endpoint. RBE values as high as 1.53 were observed for d(20) + Be relative to p(65) + Be neutrons. Even when comparing the modern neutron therapy facilities (neutrons produced by protons with energy higher than 40 MeV), significant RBE variations are still observed, which needs to be taken into account when designing multicentre therapeutic protocols.

INTRODUCTION

In the field of low LET radiations (^{60}Co, photons and electrons from a few MeV to about 50 MeV), RBE variations, if any, do not exceed 1%, and the absorbed dose alone (and fractionation scheme) can be used to predict the biological/clinical effect.

The RBE of fast neutrons relative to photons is significantly different from unity. It depends on dose, biological system and endpoint, dose rate, experimental conditions (such as oxygenation), etc. In addition, the RBE varies with neutron energy[1]. Therefore, in clinical neutron therapy, two problems related to RBE can be identified:

(1) the RBE of a given neutron beam relative to photons; and
(2) the RBE of a given neutron beam relative to another neutron beam.

The second point will be mainly studied in this paper. When comparing different neutron beams, one has to specify the neutron energy spectrum (or beam quality) and the RBE with its uncertainty, and the system and dose for which it has been obtained.

Many radiobiological experiments were performed in the neutron beams in clinical use. These beams vary within a large energy range. Different biological systems were used by different teams of physicists and radiobiologists, which may introduce some variation between the results, related to the technique and environmental conditions.

SPECIFICATION OF THE NEUTRON ENERGY

In order to compare the RBE data obtained at different neutron energies, it is convenient to scale the neutron beam energy according to a single parameter. The mean energy or several parameters related to the beam penetration have been proposed[2], but none of them is fully adequate. The parameter proposed here to specify the effective neutron beam energy is the half value thickness (HVT) measured under reference conditions, i.e. infinite SSD, 10 cm × 10 cm field size, water phantom. The HVT measured between 5 cm and 15 cm was chosen as being adequate for most of the practical situations.

If the dose decreases exponentially with added thickness (beyond the build-up region), the depths between which the HVT is measured are not important and a single HVT value can be adopted. This is in fact the current situation. However, if the dose does not decrease exponentially with added thickness, HVT will depend on the depths at which it is determined. For example, if the neutron beam has been contaminated by low energy neutrons produced in, for example, the target area or in the collimation system, the depth dose curve could then consist of two components, and two RBE values could then be expected.

Accurate and complete specification of the beam quality, at a reference point, is provided by microdosimetric spectra[3]. A microdosimetric intercomparison programme has been carried out in Europe under the auspices of the EORTC-Heavy Particle Therapy Group (European Organization for Research and Treatment of Cancer) and systematic measurements were performed at all neutron facilities.

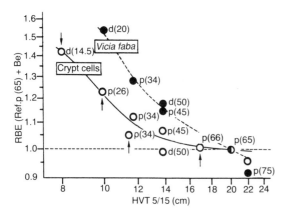

Figure 1. RBE variation of different neutron beams as a function of their energy. Summary of the data obtained at Louvain-la-Neuve. The energy of the beams, expressed by the parameter HVT (see text), is in abscissa (logarithmic scale); it covers the whole range actually used for therapy. Two biological systems were used: *Vicia faba* (closed circles) and intestinal crypt cells in mice (open circles). The different beams which were compared, were produced at the Louvain-la-Neuve cyclotron by varying the energy and the particle type. In addition, the intestinal crypt cell system was used during four site visits (results indicated by arrows, see text).

Table 1. Neutron beams intercomparison at different facilities. Biological system: jejunum crypt regeneration in mice.

Neutron facility	Nuclear reaction and energy	RBE relative to d(50)+Be*
Ghent (Belgium)	d(14.5) + Be	1.44
Essen (Germany)	d(14.5) + Be	1.52
Edinburgh (UK)	d(15) + Be	1.31
Hammersmith (UK)	d(16) + Be	1.34
Seattle (USA)	d(22) + Be	1.18
NIRS-Chiba (Japan)	d(30) + Be	1.03
Washington, DC (USA)	d(35) + Be	1.15
Houston (USA)	d(50) + Be	1.000 (ref)
Louvain-la-Neuve (B)	d(50) + Be	1.000 (ref)
Riyadh (Saudi Arabia)	p(26) + Be	1.19
Orléans (F)	p(34) + Be	1.06
Louvain-la-Neuve (B)	p(34) + Be	1.13
Louvain-la-Neuve (B)	p(45) + Be	1.08
Seattle (USA)	p(50.5) + Be	1.06-1.14
Clatterbridge (UK)	p(62) + Be	1.11-1.18
Louvain-la-Neuve (B)	p(65) + Be	0.97
Cape Town (South Africa)	p(62) + Be	0.99
Louvain-la-Neuve (B)	p(75) + Be	0.96
Hamburg (Germany)	d + T	1.27

* RBE values expressed relative to d(50) + Be, assuming that the effectiveness of the d(50) + Be neutron beams at Houston and at Louvain-la-Neuve are equal.

RADIOBIOLOGICAL RESULTS

RBE determinations at different energies by the same radiobiological team, at the same facility, at Louvain-la-Neuve

At the isochronous, variable energy cyclotron of Louvain-la-Neuve, a systematic study of neutron RBE as a function of energy has been performed. The characteristics of the machine allowed the simulation of all beams used in neutron therapy worldwide. Two biological systems were used extensively – growth inhibition in *Vicia faba*[4] and intestinal crypt colony assay in mice[5]. The results are summarised in Figure 1. In addition, experiments were made directly at other neutron therapy facilities (Ghent, Orléans, Riyadh, NAC-Faure) using the crypt colony system. Direct measurements at different facilities are indeed necessary since the neutron spectra depend not only on the nuclear reaction and the energy of the incident particles but are also influenced by other factors such as the design of the target and of the collimation system.

For these comparisons, neutrons produced by bombarding a beryllium target with 65 MeV protons were chosen as reference (Figure 1). This is the highest energy used in therapy, and it is used in at least four centres: Clatterbridge (UK), Fermilab (USA), NAC-Faure (South Africa) and Louvain-la-Neuve (Belgium). A direct comparison was performed between NAC-Faure and Louvain-la-Neuve on intestinal crypt cells, and a RBE of 1.01 was found for the NAC-Faure beam compared to the Louvain-la-Neuve beam.

Additional experiments were performed at the Louvain-la-Neuve facility, with *Allium cepa*[6], V79 cell survival *in vitro*[7], and late lung tolerance in mice[8]. For chromosome aberrations in *Allium cepa* onion roots[6], and V79 cell survival[7], RBE values obtained from p(75) + Be neutrons compared to d(50) + Be neutrons were equal to 0.85 and 0.90 respectively. For late lung tolerance in mice[8], RBE values of 1.02 and 1.26 were obtained for p(45) + Be neutrons relative to p(65) + Be neutrons for a single fraction irradiation and an 'infinitely fractionated' irradiation (RBE_{max}) respectively.

Intestinal crypt system used by different teams travelling at different facilities

The intestinal crypt system has been used by several radiobiological teams travelling at different facilities. The large majority of the neutron facilities used for cancer therapy have been compared using this system (Table 1)[1].

Survey of recent RBE studies

Recently, studies were performed on mouse intestine by Rasey et al[9] comparing p(50.5) + Be neutrons from the Radiation Oncology Dept to d(22) + Be neutrons produced at the Nuclear Physics Laboratory of the University of Washington. RBE values between 0.96 and 0.90 were obtained.

Acute skin reactions on mouse feet were used by Joiner and Field[10] to compare the effect of p(62) + Be neutrons from the cyclotron at Clatterbridge to d(16) + Be neutrons from the cyclotron at Hammersmith. Fractionated irradiation (up to 16 fractions, or 16 fractions followed by a top-up dose) allowed the authors to study the effects of neutron doses < 1 Gy per fraction. RBE of p(62) + Be neutrons, relative to d(16) + Be neutrons, was found equal to 0.92–0.86 for a fraction number ranging between 1 and 16. The top-up studies indicated that this figure might decrease to 0.88–0.76 at very low neutron doses.

Joiner[11] used renal damage in the mouse to compare the same two neutron beams. Different fraction numbers (1, 2, 4 and 8) were given, or 8 fractions, followed by a top-up dose in order to explore a range of fraction sizes between 0.2 and 9.6 Gy. The RBE of p(62) + Be neutrons was found equal to 0.72 over the complete range of dose per fraction. On the other hand, Hornsey et al[12] found RBE values ranging from 0.83 to 0.89 for intestinal crypt system in the dose range of 1.8–9 Gy for p(62) + Be relative to d(16) + Be neutrons.

Bewley et al[13] studied the change in biological effectiveness of the p(62) + Be neutron beam at Clatterbridge as a function of depth and filtration. For cell survival *in vitro* the RBE at a depth of 12 cm compared to 2 cm was 0.95–0.98. When using a 4.5 cm thick polythene filter, the RBE at 2 cm depth was 0.97 compared to the unfiltered beam.

In order to complete this review, some recent data on neutron RBE relative to photons are summarised. The RBE of the d(42) + Be neutron beam of the variable energy clyclotron at AERE was determined for three different biological systems. For effects on the epidermis and dermal vascular/connective tissues[14], the neutron RBE was found equal to 2.75 for photon fraction sizes between 2 and 5 Gy. For ischaemic dermal necrosis, the RBE exceeded 3.0 for fraction sizes smaller than 3 Gy and an upper limit value of 4.3 was calculated. For late effects on cutaneous and subcutaneous tissues[15], RBE ranges between 1.5 (single dose) and 3.4 (30 fractions in 39 days). For early lung damage[16], RBE ranges between 1.2 (single dose) and 4.8 (30 fractions in 39 days). The corresponding figures for late damage are 1.5 and 4.3 respectively. At Hammersmith, for the same lung system and a single fraction irradiation, Law and Ahier[17] found a RBE for d(16) + Be neutrons relative to photons of 1.3–1.5 and of 1.4 for early and late damage respectively.

For the p(66)+Be beam of the NAC-Faure, Blekkenhorst et al[18] used response of pig skin to compare the clinical fractionation schemes: 12 fractions in 26 days for neutrons and 24 fractions in 39 days for ^{60}Co gamma rays. For moist desquamation and dermal necrosis, the dose ratios corresponding to the ED_{50}, for ^{60}Co and neutrons, were found equal to 4.0 and 3.0 respectively. When using a flattening filter and a 2.5 cm polyethylene hardening filter, the corresponding figures were 3.0 and 2.9 respectively.

Feola et al[19] using mouse testis weight loss, obtained an average RBE of 3.5 for the p(26) + Be neutron beam at the King Faisal Specialist Hospital and Research Centre in Riyadh.

At the University of Essen, Budach et al[20] compared the responses to d(14) + Be neutrons and photons of 10 high grade human soft tissue sarcoma xenografts. At a specific growth delay (SGD) of 0.5, the neutron RBE varied between 1.6 (±0.3) and 12.7 (±7.0). RBE values for clamped tumours exceeded those of the normal tissue tolerance (RBE ≈ 3) in 6 out of 10 tumour lines.

CONCLUSIONS

This review confirms that RBE increases with decreasing neutron energy, but the slope of the RBE/neutron energy relationship depends on the biological system and endpoint.

RBE differences of up to 40% were observed between the lowest and highest neutron energies actually used in therapy. Even between the modern neutron therapy facilities, RBE differences were observed which are too large to be neglected, taking into account the required accuracy in dose delivery; any error in the selection of the clinical RBE will directly influence the doses delivered to the patients. Exchange of clinical information between different centres is becoming meaningless if differences in radiation quality are not taken into account and they should not be neglected when designing multicentre therapeutic protocols.

Due to the strong dependence of RBE on neutron energy, there is a need for an accurate and complete specification of the beam quality. Microdosimetric spectra can provide such information. However, microdosimetry could reach a real predictive value only to the extent that radiobiological data and microdosimetric data, obtained in identical conditions, could be correlated (e.g. unfolding method). Limited results have already been obtained, but acquisition of relevant radiobiological data systematically obtained, for the same system, remains the weak arm of the comparison.

REFERENCES

1. Beauduin, M. Gueulette, J., Grégoire, V. De Coster, B. Vynckier, S. and Wambersie, A. *Practical Problems raised by the Comparison of Clinical Results of Neutron Therapy. Clinical RBE and Clinical Neutron Potency Factor (CNPF), as a Function of Neutron Energy.* Survey of the Literature In: Proc. EULIMA Workshop on the Potential Value of Light Ion Beam Therapy, Nice, France (1988).
2. International Commission on Radiation Units and Measurements (ICRU). *Clinical Neutron Dosimetry, Part 1: Determination of Absorbed Dose in a Patient Treated by External Beams of Fast Neutrons.* Report 45 (Bethesda, MD: ICRU Publications) (1989).
3. Pihet, P., Gueulette, J. Menzel, H. G., Grillmaier, R. E. and Wambersie, A. *Use of Microdosimetric Data of Clinical Relevance in Neutron Therapy Planning.* 6th Symposium on Neutron Dosimetry, Neuherberg, October, 1987. Radiat. Prot. Dosim. **23** (1-4), 471-474 (1988).
4. Van Dam, J. and Wambersie, A. *OER and RBE Variation between p(75)+Be and d(50)+Be Neutron Beams.* Br. J. Radiol. **54**, 921-922 (1981).
5. Gueulette, J. *Efficacité Biologique Relative (EBR) de Neutrons Rapides pour la Tolérance de la Muqueuse Intestinale chez la Souris.* Thèse. Université Paul Sabatier de Toulouse, No. **362** (1982).
6. Laublin, G. *Efficacité Biologique Relative des Neutrons Rapides et des Hélions pour la Production d'Aberrations Chromosomiques chez Allium cepa.* Thèse de Doctorat, Université Catholique de Louvain, Louvain-la-Neuve (1981).
7. Guichard, M., Gueullette, J., Meulders, J. P., Wambersie, A. and Malaise, E. P. *Biological Intercomparison of d(50)+Be and p(75)+Be Neutrons.* Br. J. Radiol. **53**, 991-995 (1980).
8. Grégoire, V., Beauduin, M., Gueulette, J., De Coster, B., Denis, J. M., Octave-Prignot, M. Vynckier, S. and Wambersie, A. *Comparison of the α/β Ratio for Intestinal and Lung Tolerance in Mice after Fractionated Irradiations with Photons and High Energy Clinical Neutron Beams.* In: ESTRO Congress, Montecatini Terme, Italy, 12-15 September 1990. Abstract Book, p. 94.
9. Rasey, J. S., Magee, S., Nelson, N., Chin, L. and Krohn, K. A. *Response of Mouse Tissues to Neutron and Gamma Radiation : Protection by WR-3689 and WR-77913.* Radiother. Oncol. **17**, 167-173 (1990).
10. Joiner, M. C. and Field, S. B. *The Response of Mouse Skin to Irradiation with Neutrons from the 62 MeV Cyclotron at Clatterbridge, UK.* Radiother. Oncol. **12**, 153-166 (1988).
11. Joiner, M. C. *A Comparison of the Effects of p(62) + Be and d(16) + Be Neutrons in the Mouse Kidney.* Radiother. Oncol. **13**, 211-224 (1989).
12. Hornsey, S., Myers, R., Parnell, J. C., Bonnett, D. E., Blake, S. W. and Bewley, D. K. *Changes in Relative Biological Effectiveness with Depth of the Clatterbridge Neutron Therapy Beam.* Br. J. Radiol. **61**, 1058-1062 (1988).
13. Bewley, D. K., Cullen, B. M., Astor, M., Hall, E. J., Blake, S. W., Bonnett, D. E. and Zaider, M. *Changes in Biological Effectiveness of the Neutron Beam at Clatterbridge (62 MeV p on Be) measured with Cells in vitro.* Br. J. Radiol. **62**, 344-347 (1989).
14. Hopewell, J. W., Barnes, D. W. H., Robbins, M. E. C., Samson, J. M., Knowles, J. F. and van den Aardweg, G. J. M. J. *The Relative Biological Effectiveness of Fractionated Doses of Fast Neutrons (42 $MeV_{d \to Be}$) for Normal Tissues in the Pig. I. Effects on the Epidermis and Dermal Vascular/Connective Tissues.* Br. J. Radiol. **61**, 928-938 (1988).
15. Hopewell, J. W., Barnes, D. W. H., Robbins, M. E. C., Corp, M., Samson, J. M., Young, C. M. A., Wiernik, G. *The Relative Biological Effectiveness of Fractionated Doses of Fast Neutrons (42 $MeV_{d \to Be}$) for Normal Tissues in the Pig. II. Late Effects on Cutaneous and Subcutaneous Tissues.* Br. J. Radiol. **63**, 760-770 (1990).
16. Rezvani, M., Barnes, D. W. H., Hopewell, J. W., Robbins, M. E. C., Sansom, J. M., Adams, P. J. V. and Hamlet, R. *The Relative Biological Effectiveness of Fractionated Doses of Fast Neutrons (42 $MeV_{d \to Be}$) for Normal Tissues. III. Effects on Lung Function.* Br. J. Radiol. **63**, 875-881 (1990).
17. Law, M. P. and Ahier, R. G. *Vascular and Epithelial Damage in the Lung of the Mouse after X Rays or Neutrons.* Radiat. Res. **117**, 128-144 (1989).
18. Blekkenhorst, G., Hendrikse, A., Kent, C., Jones, D. and van den Aardweg G. J. M. J. *Preclinical Studies with the Faure High Energy Neutron Facility: Response of Pig Skin to Fractionated Doses of Fast Neutrons. (66 $MeV_{d \to Be}$).* Radiother. Oncol. **18**, 147-154 (1990).
19. Feola, J. M., Aissi, A., Greer, W., El-Sayed, R., Clubb, B. and El-Akkad, S. *Relative Biological Effectiveness (RBE) of p(26)+Be Neutrons from the King Faisal Specialist Hospital and Research Centre CS-30 Cyclotron Measured by Testis Weight Loss.* Strahlenther Onkol. **165**, 817-823 (1989).
20. Budach, V., Stuschke, M., Budach, W., Molls, M. and Sack, H. *Radiation Response in 10 High-Grade Human Soft Tissue Sarcoma Xenografts to Photons and Fast Neutrons.* Int. J. Radiat. Oncol. Biol. Phys. **19**, 941-943 (1990).

INVERSE DOSE RATE EFFECTS FOR NEUTRONS: GENERAL FEATURES AND BIOPHYSICAL CONSEQUENCES

D. J. Brenner, R. C. Miller, S. A. Marino, C. R. Geard, G. Randers-Pehrson and E. J. Hall
Center for Radiological Research, Columbia University
630 West 168th Street, New York, NY 10032, USA

Abstract — Evidence has accumulated to suggest that when a neutron dose is protracted, its effectiveness is enhanced for transformational endpoints. A pattern has emerged as to the dependence of the effect on dose, dose rate and radiation type. An explanation of the effect is that cells in part of their cycle are more sensitive to radiation than in the rest of the cycle. An acute high LET exposure of cycling cells will thus result in some fraction of the sensitive cells receiving large depositions of energy — greater than required for the effect. In a protracted exposure a larger proportion of sensitive cells will be exposed to smaller numbers of energy depositions — though still adequate to induce the effect. The model produces results consistent with all available data. Upper limit predictions for the field of radiation protection are made: at the non-occupational annual effective dose limit the enhancement would be insignificant; at the annual occupational limit an enhancement of ≤1.5 might be expected. For radon, environmentally, there should be little dose rate effect; for miners on whom risk estimates are primarily based, there might be an effect, resulting in misleading extrapolations to environmental risk.

INTRODUCTION

Virtually all significant human exposures to neutrons will be protracted over long periods of time. This is the case for the three most important exposed groups in the US namely DOE workers (~7000 individuals receiving measurable doses in 1988[1]), NRC workers (~6000 individuals receiving significant doses annually[2]) and airline crew personnel (~300,000 in US airlines[3]).

Until a decade ago, conventional wisdom was that protracting a dose of neutrons would make little difference to the biological outcome, because of the lack of sub-lethal damage repair in neutron-irradiated biological systems. However, in the early 1980s, *in vivo* and *in vitro* results appeared which seemed to conflict with this notion, and suggested that the frequency of oncogenic-related endpoints (in which we include life shortening) increased as the dose rate decreased — a phenomenon which has come to be known as the 'inverse dose rate' effect.

In vitro, where data can be quantified more readily, the early experiments of Hill and colleagues[4–6] with fission neutrons suggested that almost an order of magnitude increase in transformation rate was achieved by sufficient protraction of the dose; clearly, if this is true, there are major consequences in the field of radiation protection, because most of the results on which neutron risks are estimated were for acute exposures.

In the past decade, much experimental work, both *in vivo* and *in vitro*, has been devoted to investigating this effect. The effect and its underlying mechanism have certainly not yet been fully elucidated, but certain patterns and conclusions have emerged and are described here.

EMERGING PATTERNS

Summaries of the pertinent experiments in which the inverse dose rate effect is or is not seen may be found in References 7-10. Perhaps the most significant finding is that the effect, when seen, is not as large as was first suggested by the early work of Hill *et al*[4–6]. This is due to a variety of reasons: for example, it has since emerged that the effect is radiation quality dependent[9–11], and the use of fission neutrons turned out to produce about the maximum possible effect. Another complication resulted from the use[4–6] of confidence limits for the estimated transformation rates calculated with a method not applicable[12] to the small numbers of observed foci — the quoted confidence limits suggested somewhat more precise results than was actually the case. For example, Hill *et al*[4,5] observed an enhancement in transformation yield of about a factor of 6.5 when a dose of 0.1 Gy of fission neutrons was delivered in five fractions. The recent experiments by Saran *et al*[13], who duplicated the original experiments as closely as possible, gave a corresponding enhancement ratio of 1.1. However, calculation of realistic 95% confidence bounds[12] for the estimated enhancement gives 1.3 to 24 (Hill *et al*[4,5]), and 0.09 to 5 (Saran *et al*[13]).

The published reports on the inverse dose rate effect leave little doubt that the effect is real, but the available evidence indicates that the magnitude of the effect is due to a complex interplay between dose, dose rate and radiation quality:

With regard to dose, the largest effects appear at doses around a few tenths of a gray, with smaller enhancements at much larger or much smaller doses.

With regard to dose rate, *in vitro* experiments with neutrons indicate little measurable effect at dose rates above about 0.3 Gy.h^{-1}, but a significant effect at lower dose rates.

With regard to radiation quality, a series of experiments with monoenergetic charged particles by Miller *et al*[14] suggests that the effect is limited to an LET range from about 40 to 120 keV.µm^{-1}. A narrow LET region had also previously been indicated from experiments with photons[15] where no inverse dose rate effect is seen, and with monoenergetic neutrons[16] having a much higher average LET than fission neutrons, where only a small effect was seen.

CAN A MODEL MAKE SENSE OF THESE PATTERNS?

A consistent model of the inverse dose rate effect as a function of dose, dose rate, and radiation type was first proposed by Rossi and Kellerer[17] and further developed by Brenner and Hall[9] and by Elkind[18]. The basic hypothesis is that cycling cells are the targets for oncogenic transformation and that cells in part of their cycle are more sensitive to radiation (for transformation) than in the rest of the cycle. An acute medium LET exposure of cycling cells will result in some fraction of the sensitive cells receiving large depositions of energy — greater than required for the effect. If the exposure is protracted, a larger proportion of sensitive cells will be exposed, but to smaller numbers of energy depositions — though still sufficient to produce the effect.

The mathematical development of the model is described elsewhere[9,11,17]; in these papers it is shown that the model produces results that are consistent with all available data, including those results where no enhancement was observed, on dose rate effects for transformation in the $C_3H10T_{1/2}$ system by protraction or fractionation at medium or high LET. Direct experimental verification of the model, in terms of differential sensitivity to neutrons through the cell cycle of synchronised $C_3H10T_{1/2}$ cells, has also been reported[19]. Further direct evidence comes from the observation that $C_3H10T_{1/2}$ cells in plateau phase do not show an inverse dose rate effect when exposed to 40 keV.µm^{-1} deuterons, whereas, with all other conditions held the same, exponentially growing cells do show a significant effect[20]. Here we investigate qualitatively what the model would predict with regards to dose, dose rate and radiation quality:

With regard to dose, at very low doses, where the chances of a cell being hit more than once are very small, no dose rate effect would be predicted[21]. At high doses, when all or most cells exposed in their sensitive phase are hit, the effect would 'plateau' and the effects of damage in other parts of the cell cycle could dominate. Thus the inverse dose rate effect would be expected to be confined to some intermediate dose range, which is in fact the case.

With regard to dose rate, with increasing protraction (i.e. longer times or more fractions), the proportion of cells in their sensitive phase at some time during irradiation would increase, and thus the inverse dose rate effect should increase, which is in agreement with experiment.

With regard to radiation quality, for a given dose, as the LET (or its microdosimetric correlate, Y_F) increases, the mean number of energy depositions will decrease, and when this mean number becomes significantly less than one, as discussed above, no inverse dose rate effect would be expected. Again this is in agreement with the observed lack of effect at high LET[14,22]. At low LET, it is likely[23,24] that the probability that a sensitive cell will show an effect will be dependent on the specific energy (z) deposited in it. This probability might be expected[11,23,24] to decrease rapidly with decreasing z, but saturate to some maximum value with increasing z. Specifically, a response function of the form[11] $1-\exp(-z/z_c)$, where z_c is a constant (~0.5 Gy in the cellular nucleus) gives agreement with the results of the LET studies of Miller *et al*[14].

The inverse dose rate effect depends not only on dose, dose rate, and radiation quality, but on the interrelationships between these variables. For there to be a significant effect (of importance in radiation protection), an exposure would need to involve an intermediate dose of highly fractionated (or prolonged) medium LET radiation. The numerical values will, of course, be highly endpoint dependent, but it is a plausible and testable hypothesis that these trends will persist for oncogenic endpoints other than the $C_3H10T_{1/2}$ system in which they were elucidated.

SIGNIFICANCE

It is certainly tempting to draw broad conclusions from the data discussed here, and the associated models. However most of the quantitative data on which the modelling is based are for the $C_3H10T_{1/2}$ *in vitro* system, in which cells are growing exponentially. For tissues *in vivo*, the growth fraction varies considerably, from almost zero in the brain, to almost 1 in the crypt cells of the intestinal mucosa. Intermediate values apply to most human tissues, such as the breast (~0.25[25]) or the epidermis (0.05 – 0.1[26]). Thus predictions based on data from exponentially growing $C_3H10T_{1/2}$ cells must be upper limits to a situation *in vivo* where only a fraction of target cells are cycling at any given time.

Based on the model and the parameters described here, at a dose less than 1 mGy (corresponding to the annual effective dose limit for the general public), the enhancement would be insignificant at any dose rate[9]. The annual occupational effective dose equivalent limit is 50 mSv, corresponding (assuming a quality factor of 20) to 2.5 mGy of fission neutrons; if this dose were delivered continuously over a year at a dose rate of 5 nGy.min^{-1}, then a maximum enhancement of about a factor of 1.5 might be expected[9].

The most important area where the effects of protracted low doses of high LET radiation are of concern is exposure to alpha particles from radon daughters, in both the general public and uranium miners. For the environmental situation to which the general public is exposed, there should be no detectable dose rate effect. This is because, for an average house containing a radon level of about 40 Bq.m^{-3}, the average number of alpha particles traversing a cell nucleus in the bronchial epithelium would be around one per lifetime. As discussed above, no dose rate effects are possible unless significant numbers of cell nuclei are exposed to multiple tracks of radiation.

By contrast, for the uranium miners on whom risk estimates are primarily based, the doses, and thus the numbers of alpha particles traversing each cellular nucleus at risk, are much larger. For example, in the four cohorts (Colorado, Malmberget, Ontario and Beaverlodge) that were analysed for the BEIR IV[27] report on the effects of radon, the average numbers of traversals of basal cell nuclei in the segmental bronchus were, respectively, about 20, 5, 3 and 1. Thus there could be an inverse dose rate effect in some of these cohorts, especially Colorado. This suggestion is supported by recent analyses by Hornung[28] of the Colorado data who demonstrated a significant inverse dose rate effect, by Lubin *et al*[29] of Chinese tin miners exposed to comparable doses to those in Colorado, and by Darby and Doll[30] who compared results from different cohorts exposed at differing average dose rates.

The presence of an inverse dose rate effect amongst some of the miner cohorts on which environmental radon risk estimates are based would have the effect of overestimating the radon risk to the general public, when risks are extrapolated from high to low exposures. This phenomenon, if true, would be of considerable social significance.

ACKNOWLEDGEMENTS

This work was supported by grants CA-12536 and CA-49062 from the NCI, DE-FG02-88ER60631 from the US DOE, and PDT-438 from the American Cancer Society.

REFERENCES

1. Merwin, S. E., Millete, W. H. and Traub, R. J. *Twenty-first Annual Report on Radiation Exposures for DOE and DOE Contractor Employees - 1988.* DOE Report DOE/EH-0171P (1990).
2. Radditz, S. Nuclear Regulatory Commission, Private communication (1991).
3. Wilson, J. W. and Townsend, L. W. *Radiation Safety in Commercial Air Traffic: a Need for Further Study.* Health Phys. **55**, 1001-1003 (1988). With erratum, Health Phys. **56**, 973-974 (1989).
4. Hill, C. K., Buonoguro, F. J., Myers, C. P., Han, A. and Elkind, M. M. *Fission-spectrum Neutrons at Reduced Dose Rate Enhance Neoplastic Transformation.* Nature **298**, 67-68 (1982).
5. Hill, C. K., Carnes, B. A., Han, A. and Elkind, M. M. *Neoplastic Transformation is Enhanced by Multiple Low Doses of Fission-spectrum Neutrons.* Radiat. Res. **102**, 404-410 (1985).
6. Hill, C. K., Han, A. and Elkind, M. M. *Fission-spectrum Neutrons at Low Dose Rate Enhance Neoplastic Transformation in the Linear, Low Dose Region (0–10 cGy).* Int. J. Radiat. Biol. **46**, 11-15 (1984).
7. Charles, M., Cox, R., Goodhead, D. T. and Wilson, A. *CEIR Forum on the Effects of High-LET Radiation at Low Doses/Dose Rates.* Int. J. Radiat. Biol. **58**, 859-885 (1990).
8. National Council on Radiation Protection and Measurements. *The Relative Biological Effectiveness of Radiations of Different Quality.* Report 104 (Bethesda, MD: NCRP) (1990).
9. Brenner, D. J. and Hall, E. J. *The Inverse Dose Rate Effect for Oncogenic Transformation by Neutrons and Charged Particles. A Plausible Interpretation Consistent with Published Data.* Int. J. Radiat. Biol. **58**, 745-758 (1990).
10. Brenner, D. J. and Hall, E. J. *Radiation-induced Oncogenic Transformation: The Interplay between Dose, Dose Protraction and Radiation Quality.* Adv. Radiat. Biol. in press (1991).
11. Brenner, D. J. *The Influence of LET on the Inverse Dose-rate Effect.* In: Proc. CEC/DOE Workshop on Biophysical Modelling of Radiation Effects. In press (1991).
12. Brenner, D. J. and Quan, H. *Confidence Limits for Low Induced Frequencies of Oncogenic Transformation in the Presence of a Background.* Int. J. Radiat. Biol. **57**, 1031-1045 (1990).

13. Saran, A., Pazzaglia, S., Coppola, M., Rebessi, S., Di Majo, V., Garavini, M. and Covelli, V. *Absence of a Dose-fractionation Effect on Neoplastic Transformation Induced by Fission Spectrum Neutrons in $C_3H10T_{1/2}$ Cells.* Radiat. Res. **126**, 343-348 (1991).
14. Miller, R. C., Randers-Pehrson, G., Hieber, L., Marino, S. A., Kellerer, A. M. and Hall, E. J. *Influence of Dose Protraction of Intermediate and High LET Radiation on Oncogenic Transformation.* In: New Developments in Fundamental and Applied Radiobiology. Eds C. B. Seymour and C. Mothersill (London, England: Taylor and Francis) pp. 177-182 (1991).
15. Hill, C. K., Han, A., Buonaguro, F. and Elkind, M. M. *Multifractionation of ^{60}Co Gamma Rays reduces Neoplastic Transformation in vitro.* Carcinogenesis **5**, 193-197 (1984).
16. Miller, R. C., Brenner, D. J., Geard, C. R., Komatsu, K., Marino, S. A. and Hall, E. J. *Oncogenic Transformation by Fractionated Doses of Neutrons.* Radiat. Res. **114**, 589-598 (1988).
17. Rossi, H. H. and Kellerer, A. M. *The Dose Rate Dependence of Oncogenic Transformation by Neutrons may be due to Variation of Response during the Cell Cycle.* Int. J. Radiat. Biol. **50**, 353-361 (1986).
18. Elkind, M. M. *Enhanced Neoplastic Transformation due to Protracted Exposures of Fission-spectrum Neutrons: a Biophysical Model.* Int. J. Radiat. Biol. **59**, 1467-1475 (1991).
19. Hill, C. K., Renan, M. and Buess, E. *Is Neoplastic Transformation by High-LET Radiations Dose Rate Dependent or Cell Cycle Dependent?* In: Proc. Int. Congr. on Radiation Research. Eds J. D. Chapman, W. C. Dewey and G. F. Whitmore (San Diego: Academic Press) p. 344 (1991).
20. Miller, R. C., Randers-Pehrson, G., Hieber, L., Marino, S. A., Richards, M. and Hall, E. J. *The Inverse Dose Rate Effect for Oncogenic Transformation by Charged Particles is LET Dependent.* Radiat. Res. in press (1992).
21. Barendsen, G. W. *Do Fast Neutrons at Low Dose Rate Enhance Cell Transformation in vitro? A Basic Problem of Microdosimetry and Interpretation.* Int. J. Radiat. Biol. **47**, 731-734 (1985).
22. Hieber, L., Ponsel, G., Roos, H., Fenn, S., Fromke, E. and Kellerer, A. M. *Absence of a Dose-rate Effect in the Transformation of $C_3H10T_{1/2}$ Cells by α-particles.* Int. J. Radiat. Biol. **52**, 859-869 (1989).
23. Zaider, M. and Brenner, D. J. *On the Microdosimetric Definition of Quality Factors.* Radiat. Res. **103**, 302-316 (1985).
24. International Commission on Radiation Units and Measurements. *The Quality Factor in Radiation Protection.* Report 40 (Bethesda, MD: ICRU Publications) (1986).
25. Russo, J., Calaf, G., Roi, L. and Russo, I. H. *Influence of Age and Gland Topography on Cell Kinetics of Normal Human Breast Tissue.* J. Natl. Cancer Inst. **78**, 413-418 (1987).
26. Van Erp, P. E., De Mare, S., Rijzewijk, J. J., Van de Kerkhof, P. C. and Bauer, F. W. *A Sequential Double Immunoenzymic Staining Procedure to Obtain Cell Kinetic Information in Normal and Hyperproliferative Epidermis.* Histochem. J. **21**, 343-347 (1989).
27. National Research Council. *Health Risks of RADON and Other Internally Deposited Emitters.* BEIR IV (Wasington, DC: National Academy Press) (1988).
28. Hornung, R. W. and Meinhardt, T. J. *Quantitative Risk Assessment of Lung Cancer in U.S. Uranium Miners.* Health Phys. **52**, 417-430 (1987).
29. Lubin, J. H., Qiao, Y. L., Taylor, P. R., Yao, S. X., Schatzkin, A., Mao, B. L., Rao, J. Y., Xuan, X. Z. and Li, J. Y. *Quantitative Evaluation of the Radon and Lung Cancer Association in a Case Control Study of Chinese Tin Miners.* Cancer Res. **50**, 174-180 (1990).
30. Darby, S. C. and Doll, R. *Radiation and Exposure Rate.* Nature **344**, 824 (1990).

NEUTRON ENERGY DEPOSITION ON THE NANOMETER SCALE

J. J. Coyne* and R. S. Caswell
National Institute of Standards and Technology†
Gaithersburg, Maryland 20899, USA

Abstract — We have incorporated the synthesis of Monte Carlo results for proton tracks of Wilson, Metting, and Paretzke into our analytic method neutron microdosimetry code. Both 'crossers' and 'passers' are included. Results are calculated for neutrons in the energy range 1.05 MeV to 14.5 MeV for site diameters 2 nm to 1 µm.

INTRODUCTION

It has been clear for some time that although the cell nucleus is of the scale of microns, say 8 µm, smaller structures in the cell such as DNA (2 nm) or nucleosomes (5–10 nm) are important for the understanding of the biological effects of low and high LET radiations. The distribution of energy depositions in nanometer volumes, or 'nanodosimetry', is therefore of interest, but has been quite difficult to study experimentally.

CALCULATIONAL METHOD

The present approach is to incorporate the synthesis of Monte Carlo results for proton tracks interacting with small spherical sites of Wilson, Metting and Paretzke (WMP)[1] and Wilson and Paretzke (WP)[2] into our analytic method neutron microdosimetry code[3,4]. Consistent with this approach, a spherical site in tissue is assumed, although as the site size becomes very small, shape becomes unimportant. WMP have generated event size distributions for monoenergetic protons with a given chord length within a sphere of fixed diameter. The particles in WMP are assumed to be 'crossers', which is not unreasonable for tracks which intersect a nanometer scale site since the chance of a particle starting or stopping in a tiny site is small. The distributions WMP obtained with the Monte Carlo code have been fitted with log–normal distribution functions as a function of proton energy, sphere diameter, and chord length. The energy deposition distribution for crossers is then taken as log–normal with two parameters specified by WMP.

In their more recent work, WP have studied the ionisation distributions for 'touchers' or 'passers', that is, particles whose tracks pass outside the site and deposit energy in the site via delta ray transport. The event size distribution of passers is approximated by an exponential function which depends on site size but is independent of radial distance from the path of the particle. The probability of the site receiving an energy deposition at large distances is proportional to the solid angle of the site as seen from the path of the particle. In our calculation we fix the coefficient of the exponential so that the energy not deposited by charged particles in the site is compensated by the energy deposited by passers. In the calculations thus far straggling and delta ray effects have been considered only for protons. Alpha particles and heavier recoil nuclei have been treated in the continuous-slowing-down-approximation (CSDA).

RESULTS

Using the approach discussed above, lineal energy distributions for neutrons in the energy range 1.05 MeV to 14.5 MeV for site diameters from 1 µm to 2 nm have been calculated. The inclusion of straggling and delta ray effects, while visible, is relatively small near 1 MeV and 1 µm site diameter. For example, in Figure 1 y spectra for 1.05 MeV neutron energy for a site of 1 µm diameter, a typical microdosimetry site size are shown. Note that the energy deposition spectra are very similar for CSDA and for the calculation with inclusion of straggling, the main peak being slightly higher in energy and a little broader at the top for the calculation with straggling. The microdosimetric parameters are about the same, $\bar{y}_F$ = 54.5 (CSDA) and 56.6 (straggling), all in keV.µm^{-1}. Similarly, $\bar{y}_D$ = 83.3 (CSDA) and 85.5 (straggling). At a smaller site size, 100 nm, for the same energy neutrons and include straggling for both the crossers and passers, (see Figure 2) the CSDA and straggling distributions are found to be considerably different. The straggling distribution is broader than the CSDA, extends to lower y, and the peak height is lowered. The heavy particle peaks are the same since both are CSDA. $\bar{y}_F$ decreases from

---

* Present address: Superconducting Super Collider Laboratory, Dallas, Texas, USA.
† Ionizing Radiation Division, Physics Laboratory, Technology Administration, US Department of Commerce. Contribution of the National Institute of Standards and Technology. Not subject to copyright.

59.4 keV.µm^{-1} (CSDA) to 30.1 (straggling). $\bar{y}_D$, being more dependent upon the unchanged high energy depositions, that is high y values, is less changed, 98.2 as against 91.8. Results for the 2 nm site diameter are shown in Figure 3. Note that the distributions are considerably changed from CSDA, which is not appropriate to the small site size, with straggling producing considerable broadening. In this case 22% of the energy is deposited by passers. $\bar{y}_F$ values are 59.1 (CSDA) and 42.0 (straggling); $\bar{y}_D$ values are 99.8 and 104.4, respectively.

At 4.9 MeV neutron energy deposition distributions for the same three site sizes have been calculated (Figures 4–6). The results are qualitatively similar to the 1.05 MeV case, with the CSDA and straggling results similar for the 1 µm site diameter (Figure 4), and very dissimilar for 2 nm (Figure 6). At 2 nm the $\bar{y}_F$ values are 21.9 (CSDA) and 18.9 (straggling), all in keV.µm^{-1}; $\bar{y}_D$ values are 64.1 (CSDA) and 86.3 (straggling). Of the energy deposited in the 2 nm site 23% is deposited by passers.

At 14.5 MeV neutron energy, proton ranges are long compared to all three site sizes, and alpha particles and heavy recoil nuclei of C, N, and O play a larger role. The y spectra for the 1 µm site diameter show the usual broadening and increase

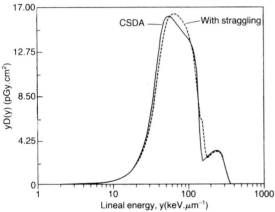

Figure 1. Calculated lineal energy, y, spectra for 1.05 MeV neutron energy, 1 µm site diameter in 4-element tissue. The solid curve shows the spectrum in the continuous-slowing-down approximation (CSDA), the dashed curve includes proton straggling. The contribution from passers is very small under these conditions.

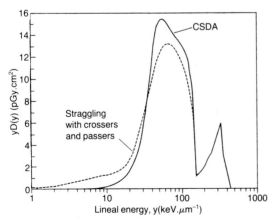

Figure 2. Calculated y spectra for 1.05 MeV neutron energy, 100 nm site diameter for tissue. Solid curve is CSDA, dashed curve includes contributions of both crossers and passers. The peak at around 300 keV.µm^{-1} is due to heavy particle recoils, mostly C and O, ejected by the neutrons.

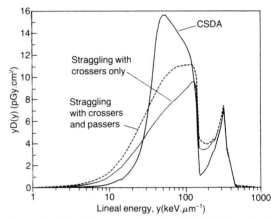

Figure 3. Calculated y spectra for 1.05 MeV neutrons 2 nm site diameter in tissue. Solid curve is CSDA, dotted curve includes straggling by crossers, and the dashed curve includes proton energy depositions by both crossers and passers.

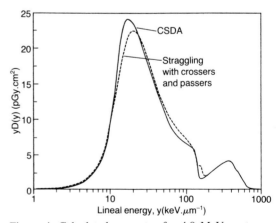

Figure 4. Calculated y spectra for 4.9 MeV neutrons, 1 µm site diameter in tissue. Solid curve is for CSDA, dashed curve includes contributions of proton crossers and passers. The peak at around 400 keV.µm^{-1} includes both alpha particles and heavy neutron recoils.

in energy of the proton peak with straggling (Figure 7). A more dramatic change is shown at the 0.1 μm site diameter (Figure 8), and at 2 nm the CSDA and straggling proton distributions bear little resemblance to each other (Figure 9). The values for $\bar{y}_F$ are not very different: 11.7 keV.μm^{-1} (CSDA) and 11.2 keV.μm^{-1} (straggling with crossers and passers). This probably means that $\bar{y}_F$ is not a good parameter to describe these distributions. On the other hand, the $\bar{y}_D$ values are considerably different: 108.5 (CSDA) as against 132.8 (straggling).

CONCLUSIONS AND FUTURE PLANS

The results presented thus far indicate that this calculational method is capable of yielding interesting and reasonable results, which should be comparable to experimental results using tissue-equivalent proportional counters or the variance–covariance method. The calculation is very fast, a few minutes on a modern personal computer. Comparisons with experimental data made thus far seem reasonable qualitatively, but are not quantitatively precise. Several points of discussion.

(1) It is planned to replace the present stopping powers for protons and alpha particles (based on fits to TE proportional counter data) with the stopping powers from a forthcoming ICRU report[5]. The absolute values of $\bar{y}_F$ and $\bar{y}_D$ given here should not be considered final.

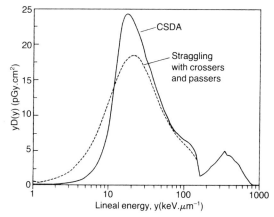

Figure 5. Calculated y spectra for 4.9 MeV neutrons, 100 nm site diameter for tissue. Solid curve is CSDA, dashed curve includes energy depositions by protons, both crossers and passers.

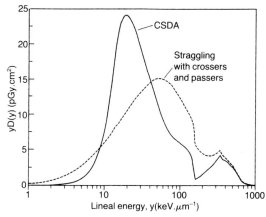

Figure 6. Calculated y spectra for 4.9 MeV neutrons, 2 nm site diameter for tissue. Solid curve is CSDA, dashed curve includes proton straggling, both crossers and passers included.

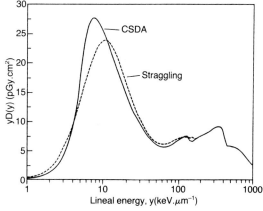

Figure 7. Calculated y spectra for 14.5 MeV neutrons, 1 μm site diameter for tissue. Solid curve is CSDA, dashed curve includes proton straggling (crossers only).

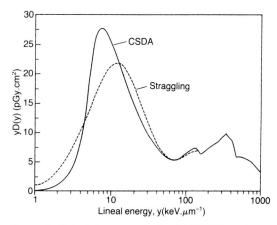

Figure 8. Calculated y spectra for 14.5 MeV neutrons, 100 nm site diameter for tissue. Solid curve is CSDA, dashed curve includes proton straggling (crossers only). Proton edge is shown at around 150 keV.μm^{-1}. The peak at around 300 keV.μm^{-1} is for alpha particles. The spectrum above about 400 keV.μm^{-1} is heavy particle neutron recoils.

(2) The calculations of Wilson and Paretzke are for water vapour and the results are ionisation yields. The calculations reported here are for energy deposition in tissue. Thus an effective W value is required to convert from WMP and WP ionisation yields to energy deposition. In this paper it has been determined by considering monoenergetic protons from 300 keV to 15 MeV and choosing an energy-dependent W value to get the same energy deposition in the CSDA and straggling calculations for large site sizes (where passers are very small). At small site sizes the energy deposited by the crossers in the straggling calculation is compared to the expected total energy deposited (Fano theorem), the remainder being attributed to passers. This provides a normalisation for the passer distribution. We do not see a way to improve this general approach which has the virtue of providing 'energy balance' in the calculation; that it, in the end the energy deposited is what is should be, even though the WMP and WP calculations are not specifically normalised to give the correct energy deposition.

(3) Straggling, and possibly passer contributions, should be included for the alpha particles and heavy ions. These effects are probably not large, but belong in a proper calculation.

(4) Comparisons should be made to the experimental data available[6–8] at small site size.

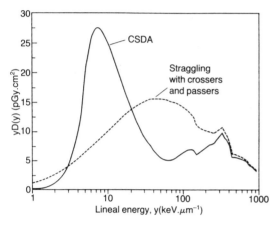

Figure 9. Calculated y spectra for 14.5 MeV neutron energy, 2 nm site diameter for tissue. Solid curve is CSDA, dashed curve includes proton straggling with contributions from both crossers and passers.

ACKNOWLEDGEMENT

This work has been supported in part by the Office of Health and Environment Research, US Department of Energy.

REFERENCES

1. Wilson, W. E., Metting, N. F. and Paretzke, H. G. *Microdosimetric Aspects of 0.3– to 20–MeV Proton Tracks: I. Crossers.* Radiat. Res. **115**, 389 (1988).
2. Wilson, W. E. and Paretzke, H. G. *A Stochastic Model of Ion Track Structure.* In: Proc. Cong. of Radiation Research, Toronto, 7-12 July 1991, Poster paper.
3. Caswell, R. S. and Coyne, J. J. *Microdosimetric Spectra and Parameters of Fast Neutrons.* In: Proc. Fifth Symp. on Microdosimetry, Verbania-Pallanza, Italy (Luxembourg: CEC) EUR 5452, p. 97 (1976).
4. Caswell, R. S. and Coyne, J. J. *Effects of Track Structure on Neutron Microdosimetry and Nanodosimetry.* Nucl. Tracks Radiat. Meas. **16**, 187 (1989).
5. International Commission on Radiation Units and Measurements. *Stopping Power for Protons and Alpha Particles.* (Bethesda, MD: ICRU Publications) (in press).
6. Lindborg, L., Kliauga, P., Marino, S. and Rossi, H. *Variance–Covariance Measurements of the Dose Mean Lineal Energy in a Neutron Beam.* Radiat. Prot. Dosim. **13**, 347 (1985).
7. Goldhagen, P., Randers-Pehrson, G., Marino, S. A. and Kliauga, P. *Variance–Covariance Measurements of $\bar{y}_D$ for 15 MeV Neutrons in a Wide Range of Site Sizes.* Radiat. Prot. Dosim. **31**, 167 (1990).
8. Kliauga, P. *Development of an Ultraminiature Proportional Counter for Nanodosimetry.* In: Center for Radiological Research, Columbia University, Annual Report, p. 71 (1990).

A MICRODOSIMETRIC MONTE CARLO CODE FOR NEUTRONS UP TO 60 MeV

G. C. Taylor* and M. C. Scott
Medical Physics Group
School of Physics and Space Research, University of Birmingham
Edgbaston, Birmingham B15 2TT, UK

Abstract — A Monte Carlo code has been developed for modelling spherical microdosimetric counters constructed from hydrogen, carbon, nitrogen, oxygen and aluminium below 20 MeV, and from hydrogen, carbon and oxygen between 20 and 60 MeV. A description of the particle transport geometry, from the location of the neutron interaction to the deposition of energy in the detector, is given. The sources of input data and an overview of the reaction kinematics are presented. Verification of the code by comparison with the analytical code NESLES is outlined and kerma values for carbon compared with those of other authors. Comparisons with experiments carried out using the Clatterbridge neutron cancer therapy cyclotron are presented.

INTRODUCTION

Following the use in recent years of high energy neutron therapy, computational microdosimetry has been unable to predict the relevant spectra due to a paucity of neutron data above 20 MeV. However, with the recent publication of double differential particle production cross sections for carbon and oxygen up to 60 MeV[1], this area can now be investigated.

PARTICLE TRANSPORT IN THE PROGRAM MITE60

MITE60 (microdosimetry in important tissue elements below 60 MeV) models the action of a parallel neutron beam incident on a sphere of one material with a spherical volume of a second material at its centre. Because the inner volume is usually a gas-filled cavity, the following discussion will refer to it in those terms.

The location of neutron interactions

Because of the cylindrical symmetry of the modelled system, the incident neutrons are treated as coming from a line source. Consequently, the only geometric variable is the radial distance, r, between a given neutron's projected path and the system's axis of symmetry. The number of neutrons, $N(r)$, is simply given by

$$N(r)\, dr = \Phi\, 2\pi r\, dr \quad (1)$$

where Φ is the neutron fluence.

The neutron's path length, $d(r)$, in a counter of external radius R_{ext} is subsequently defined by

$$d(r) = 2\sqrt{(R_{ext}^2 - r^2)} \quad (2)$$

For values of r greater than the internal radius R_{int}, this path is entirely within the wall material. However, for values of r less than R_{int}, a segment of this path, $c(r)$, falls within the cavity, where

$$c(r) = 2\sqrt{(R_{int}^2 - r^2)} \quad (3)$$

Having evaluated the path lengths $d(r)$ and $c(r)$ (taken as 0 for $r > R_{int}$), the neutron interaction probability I is calculated as

$$I = 1 - \exp[-\Sigma_w(d(r) - c(r))]\exp(-\Sigma_g c(r)) \quad (4)$$

where Σ_w and Σ_g represent the macroscopic cross sections for the wall and gas materials respectively at that neutron energy. However, because the neutron undergoes a forced first collision, this interaction probability is used as a weighting factor for all contributions to the Monte Carlo data resulting from that enforced reaction.

Figure 1 shows the relationship between the interaction point P and the other geometric parameters discussed above.

Determination of track lengths

Having selected an interaction position P, a reaction mechanism is chosen according to their macroscopic cross sections at the current neutron energy. The resulting daughter particles can each be characterised by four parameters: their species i, energy at birth E_{born}, angle of scatter in the lab frame θ and their azimuthal scatter angle ϕ. Using both the daughter particle and the interaction location parameters it can be shown that the intersection of a particle's projected track T (either forwards or backwards) with the cavity is:

$$T = 2\sqrt{\{R_{int}^2 - l^2[1-(\cos\theta\cos\rho+\sin\theta\sin\rho\cos\phi)^2]\}} \quad (5)$$

---

* Current address: Division of Radiation Science and Acoustics, The National Physical Laboratory, Teddington, Middlesex TW11 0LW, UK.

For particles whose trajectories do not intersect the cavity, the square-rooted part of this expression is negative. The portion of the expression bounded by round brackets represents the cosine of the angle between the projected particle's true (i.e. forward) path and the line joining the counter centre O to the point of interaction P. If this portion is negative (i.e. the angle is larger than 90°), the particle in question is moving away from the centre of the detector. Particles born in the wall under these circumstances will, therefore, never enter the cavity; hence they need not be tracked further.

For particles born in the wall whose true path is toward the centre of the counter, the wall thickness W to be traversed in reaching the cavity is given by:

$$W = l \, |(\cos\theta \cos\rho + \sin\theta \sin\rho \cos\phi)| - \frac{T}{2} \quad (6)$$

For those born in the gas, the track length T_{gas} is given by:

$$T_{gas} = l \, (\cos\theta \cos\rho + \sin\theta \sin\rho \cos\phi) + \frac{T}{2} \quad (7)$$

Note the similarity between the expressions for T_{gas} and W. This is due to both being expressions for the track length from the point of interaction to the cavity/wall boundary. Full derivations of these expressions are given by Taylor[2].

Determination of energy deposition

If a daughter particle i reaches the cavity, its energy (E_{in}) at the boundary is given by:

$$E_{in} = \varepsilon(R_{(w,i,E_{born})} - W) = \varepsilon(L) \quad (8)$$

where $\varepsilon(L)$ represents the energy of a particle with a range L and $R_{(w,i,E_{born})}$ the particle's range in the wall material.

In order to determine the energy deposited in the counter ($\in$), it is necessary to calculate the energy of the particle as it leaves the cavity, E_{out}, given by:

$$E_{out} = \varepsilon(R_{(g,i,E_{in})} - T), \quad R_{(g,i,E_{in})} > T \quad (9)$$

Clearly, for the case where the particle range in the gas is less than the track length, $E_{out} = 0$.

The energy deposited, $\in$, is therefore given by:

$$\in = E_{in} - E_{out} \quad (10)$$

Similarly, for particles born in the gas,

$$\in = E_{born} - E_{out} \quad (11)$$

INPUT DATA FOR MITE60

The input data required for the code MITE60 fall into two categories: neutron cross section and charged particle range-energy data. The cross section data have been compiled from a variety of sources. Below 20 MeV, data for reactions on hydrogen, carbon, nitrogen, oxygen and aluminium have been taken from ENDFB4, together with the latest branching ratios for the reaction $^{12}C(n,3\alpha)n^{(3)}$. Between 20 and 60 MeV, cross sections for reaction products from carbon and oxygen were taken from Brenner and Prael[1]. Elastic scattering cross sections for hydrogen, carbon and oxygen were taken from Dimbylow[4] and, at 60 MeV, from ENDFB4, Meigooni et al[5] and Islam et al[6] respectively.

The range–energy data for the charged particle products were calculated using the data of Ziegler et al[7].

OVERVIEW OF REACTION MECHANISMS

Secondary particle production in MITE60 is divided into four categories:

(1) Anisotropic scatter: angular distributions for both elastic and inelastic scattering have been taken from ENDF-B4 and Dimbylow[4], in the form of Legendre polynomials. All are solved using the Newton–Rhapson method.
(2) 2-body break-up: this is assumed to be isotropic in the centre of mass. Both daughters may be left in an excited state. Includes isotropic inelastic scatter.
(3) Multibody break-up: 3+-body break-up is assumed to occur via an instantaneous

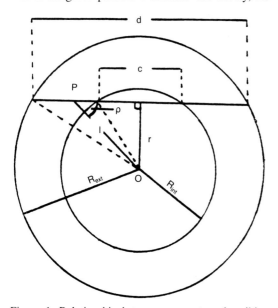

Figure 1. Relationship between parameters describing position of neutron interaction.

cascade. If branching ratio data for specific intermediate particle excited levels is known (i.e. $^{12}C(n,3\alpha)n^{(3)}$), this data may be used. If such data are unavailable, an evaporation spectrum is assumed for each particle.

(4) Secondary-particle production: the data above 20 MeV take the form of double differential (energy/angle) cross sections for daughter particle production rather than the kinematically complete reaction cross sections used below 20 MeV.

VERIFICATION OF THE CODE

The algorithms used for the various reaction mechanisms and energy deposition calculations were tested by running a series of intercomparisons with the analytical microdosimetric code NESLES[8], for a range of energies between 1 and 20 MeV. The agreement was found to be very good considering differences in the treatment of multibody break-up reactions. Details of this are given in Taylor[2].

Further verification can be seen in Figure 2, which compares carbon kerma values generated by MITE60 with Caswell et al[9] below 20 MeV and with Dimbylow[10] up to 50 MeV. Below 10 MeV, agreement is excellent, which is to be expected as this represents the most accurate data. Between 10 and 20 MeV, the MITE60 predictions are lower than those of Caswell et al, but are closer to experimental values[11]. Above 20 MeV, MITE60 predicts kerma values roughly 10% higher than those of Dimbylow.

PREDICTIONS USING MITE60

Comparisons with experiment at Clatterbridge

Figure 3 shows a comparison of yd(y) spectra between MITE60 (using the spectrum of Crout et al[12]) and experiment[13] for a 2.5 cm carbon-walled detector measured in air at the isocentre for a 10 × 10 cm² field size and a 2 μm simulated diameter. Clearly there is an underprediction in the proton contribution (~ 10 keV.μm⁻¹) and the alpha peak (~ 300 keV.μm⁻¹) is too large. Although the former may be an undervaluation of the proton production cross section, the latter is caused, at least in part, by the lack of wall effect in the modelled response. Because the cross sections above 20 MeV are for particle production rather than for specific reaction mechanisms, it is impossible to model the simultaneous detection of particles from the same reaction, e.g. two alphas from the $^{12}C(n,3\alpha)n$ reaction. The consequence of this is that two smaller events are recorded rather than one double event.

Figure 4 shows a comparison between a predicted yd(y) response for a 3 cm A-150 detector with a 2 μm simulated diameter and an isocentric measurement for a 10 × 10 cm² field size. The prediction was generated using the spectrum of Crout et al, with an empirically estimated (flat) component between 0 and 7 MeV, comprising roughly 20% of the total number of neutrons, as this energy range is absent from the above spectrum[14]. The agreement is reasonably good, but the empirical nature of the low energy component limits the conclusions that may be drawn.

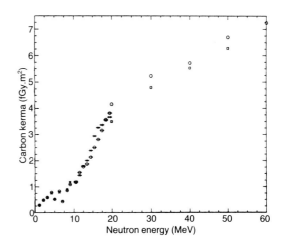

Figure 2. Comparison of carbon kerma values predicted by Monte Carlo with other authors. (○) Monte Carlo, (×) Caswell et al[8], (□) Dimbylow[10].

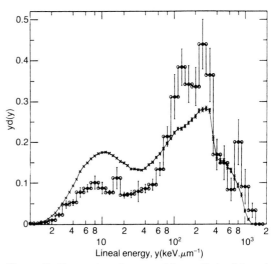

Figure 3. Comparison between Monte Carlo (○) and experiment (×) for a carbon-walled detector at Clatterbridge.

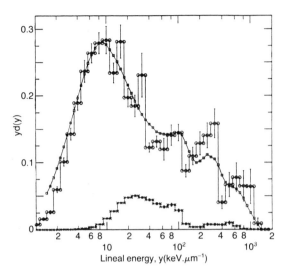

Figure 4. Comparison between Monte Carlo (o) and experiment (□) for an A-150 plastic walled detector at Clatterbridge, including low energy, 0–7 MeV, component (x).

CONCLUSIONS

A Monte Carlo code capable of predicting microdosimetric spectra for neutrons up to 60 MeV has been developed. Verification of its calculations have been demonstrated by comparing predicted spectra with NESLES and predicted kerma values with established data up to 50 MeV. It is believed that this code may help shed light on the debate concerning the appropriateness of using carbon as an oxygen substitute at these higher neutron energies.

ACKNOWLEDGEMENTS

Personal support for G.C.T. was provided by the Medical Research Council, London, during the period 1984-87, and by the Clatterbridge Trust, Clatterbridge, during 1991. Both organisations are acknowledged with gratitude. The help of the Medical Physics Group at the University of Birmingham, sadly now disbanded, is also gratefully acknowledged.

REFERENCES

1. Brenner, D. J. and Prael, R. E. *Calculated Differential Secondary-Particle Production Cross Sections after Nonelastic Neutron Interactions with Carbon and Oxygen between 15 and 60 MeV.* Atomic Data and Nuclear Data Tables **41**(1), 71-130 (1989).
2. Taylor, G. C. *The Prediction and Measurement of Microdosimetric Spectra Relating to Neutron Cancer Therapy.* PhD thesis, University of Birmingham (1990).
3. Antolkovic, B., Dietze, G. and Klein, H. *Reaction Cross Sections on Carbon for Neutron Energies from 11.5 MeV to 19 MeV.* Nucl. Sci. Eng. **107**(1), 1-21 (1991).
4. Dimbylow, P. J. *A Calculation of Neutron Cross-Sections for the Elements Hydrogen, Carbon, Nitrogen and Oxygen in the Energy Range 20–50 MeV.* NRPB-R78 (National Radiological Protection Board, Didcot, Oxon) (1978).
5. Meigooni, A. S., Petler, J. S. and Finlay, R. W. *Scattering Cross Sections and Partial Kerma Factors for Neutron Interactions with Carbon at $20 < E_n < 65$ MeV.* Phys. Med. Biol. **29**(6), 643-659 (1984).
6. Islam, M. S., Finlay, R. W., Petler, J. S., Rapaport, J., Alarcon, R. and Wierzbicki, J. *Neutron Scattering Cross Sections and Partial Kerma Values for Oxygen, Nitrogen and Calcium at $18 < E_n < 60$ MeV.* Phys. Med. Biol. **33**(3), 315-328 (1988).
7. Ziegler, J. F., Biersack, J. P. and Littmark, U. *The Stopping and Range of Ions in Solids* (Oxford: Pergamon Press) (1985).
8. Edwards, A. A. and Dennis, J. A. *NESLES — a Computer Program which Calculates Charged Particle Spectra in Materials Irradiated by Neutrons.* NRPB M-43 (National Radiological Protection Board, Didcot, Oxon) (1979).
9. Caswell, R. S., Coyne, J. J. and Randolph, M. L. *Kerma Factors for Neutron Energies below 30 MeV.* Radiat. Res. **83**, 217-254 (1980).
10. Dimbylow, P. J. *Neutron Cross-Sections and Kerma Values for Carbon, Nitrogen and Oxygen from 20 to 50 MeV.* Phys. Med. Biol. **25**(4), 637-649 (1980).
11. Gerstenberg, H. M., Caswell, R. S. and Coyne, J. J. *Initial Spectra of Neutron-Induced Secondary Charged Particles.* Radiat. Prot. Dosim. **23**(1-4), 41-44 (1988).
12. Crout, N. M. J., Fletcher, J. G., Scott, M. C. and Taylor, G. C. *Neutron Physics Studies Relating to Cancer Therapy on the Clatterbridge Cyclotron.* Radiat. Prot. Dosim. **23**(1-4), 381-384 (1988).
13. Green, S., Aro, A. C. A., Taylor, G. C. and Scott, M. C. *The Development of Microdosimetric Detectors for Investigating LET Distributions in Different Body Tissues.* Radiat. Prot. Dosim. **31**(1-4), 137-141 (1990).
14. Crout, N. M. J., Fletcher, J. G., Green, S., Scott, M. C. and Taylor, G. C. *In Situ Neutron Spectrometry to 60 MeV in a Water Phantom Exposed to a Cancer Therapy Beam.* Phys. Med. Biol. **36**(4), 507-519 (1991).

DETERMINATION OF QUALITY FACTOR IN MIXED RADIATION FIELDS USING A RECOMBINATION CHAMBER

N. Golnik and M. Zielczynski
Institute of Atomic Energy
05-400 Otwock-Swierk, Poland

Abstract — A recombination chamber was used for determination of the recombination index of radiation quality (Q_4). The possibility of calculation of the quality factor defined by ICRP Report 60 from the measured Q_4 values is discussed. The calculations were performed for the experimental results, obtained in radiation fields of fast neutrons (from 0.9 MeV up to 14 MeV), high energy (350 MeV) neutrons, high energy (200 MeV) protons and for alpha particles. It was found that the calculation procedure can be used for most existing mixed radiation fields with accuracy of about 25%, if the photon component of the absorbed dose is independently determined.

INTRODUCTION

The serious shortcoming of the quality factor concept is the fact that this factor has not been defined in terms of physically measurable quantities, and its direct determination is difficult.

It was shown previously[1,2] that the radiation quality factor in mixed radiation fields, even in those with poorly known composition and energy spectrum, can be determined using the recombination method. The experimentally determined quantity is the recombination index of radiation quality (Q_4), which is measured by a tissue-equivalent ionisation chamber. Since Q_4 depends on initial recombination of ions in the chamber, this dependence was used for determination of radiation quality.

The Q_4 approximates with good accuracy the quality factor, defined by ICRP in Publication 21[3]. A recently proposed change in the definition of quality factor requires some modification of the recombination method.

The aim of this work was to find a procedure for calculation of the quality factor from the measured values of Q_4. The resulting values should approximate the new international recommendations (ICRP Report 60[4]) and should be in agreement with Polish state recommendations, based mainly on ICRU Report 40[5].

CALCULATION PROCEDURE

Q_4 is determined by measuring the ion collection efficiency in a high pressure tissue equivalent ionisation chamber both for investigated (f_R) and for reference gamma radiation (f_γ), at a suitable collecting voltage, U_R. This voltage should be determined by a calibration procedure in such a way that $f_\gamma(U_R)=0.96$.

Q_4 is defined as:

$Q_4 = (1 - f_R)/0.04$

It was shown[1], that both Q_4 and the ICRP 21 quality factor had a similar dependence on LET. Q_4 is plotted against LET in Figure 1 as a dashed line.

Since ICRP 60 introduces a new functional relationship between the quality factor and LET (solid line in Figure 1), it is not possible to use Q_4 as a direct approximation of the ICRP 60 quality factor.

From the dependences on LET (given in Figure 1) a relationship was found between the Q_4 and the quality factor as defined in ICRP 60 ($Q_{ICRP\ 60}$). This relationship is shown by the continuous line in Figure 2, however it is valid only for hypothetical radiation with a single value of LET.

It is proposed that the same curve is used as the definition of a new quantity Q_{r1}, which will be considered in this work as a first approximation to the quality factor. It means that Q_{r1} will be calculated from the values of Q_4 using the solid curve in Figure 2, even for radiations with a LET distribution.

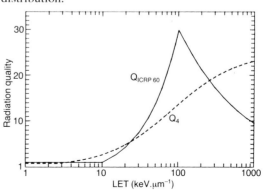

Figure 1. Dependence of Q_4 and quality factor (ICRP Report 60) on LET.

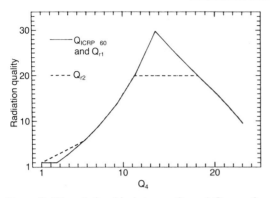

Figure 2. The relationship between Q_4 and $Q_{ICRP\,60}$ for radiations with a single value of LET. The same curve is used for the dependence of Q_{r1} on Q_4. Dashed lines represent the differences between Q_{r1} and Q_{r2}.

The difference between the quality factor $Q_{ICRP\,60}$ and our approximation Q_{r1} for a realistic radiation depends mainly on the width of the microdosimetric spectrum of the investigated radiation. In order to estimate this difference we simulated a number of LET distributions (some of them based on published data), for which we calculated the following quantities:

(i) the quality factor resulting from the definition given in ICRP Report 60 ($Q_{ICRP\,60}$).

$$Q_{ICRP\,60} = \int Q_{ICRP\,60}(L)D(L)dL / \int D(L)dL$$

where L=LET and D(L) is the dose distribution relative to LET.

(ii) Q_{r1} obtained from Q_4 according to the solid curve in the Figure 2, where Q_4 was calculated as:

$$Q_4 = \int Q_4(L)D(L)dL / \int D(L)dL$$

The results for some narrow distributions[6], chosen as examples are presented in Table 1. For these distributions and for other realistic distributions ranging from 20 keV.µm^{-1} up to 200 keV.µm^{-1} the differences between the $Q_{ICRP\,60}$ and Q_{r1} do not exceed 25%.

The values of Q_{r1} might be much too high when the LET distribution contains a significant component of radiation with LET higher than 300 keV.µm^{-1}. In the most critical case, 60% radiation with LET of 500 keV.µm^{-1} and 40% gamma radiation, the resulting Q_4=13.5 and Q_{r1} is equal to 30, while for the same LET distribution $Q_{ICRP\,60}$=8.4.

In order to improve the agreement between the Q_{r1} and the ICRP 60 recommendations, we propose to use the another quantity, denoted Q_{r2}, which depends on Q_4 according to the relationship given by the dashed line in the Figure 2.

Q_{r2} differs from Q_{r1} for Q_4<5 and in the range of Q_4 between 11 and 17.

For Q_4 ranging from 1 up to 5 the discontinuity of the recommended Q(L) dependence at 10 keV.µm^{-1} may result in too low values of Q_{r1} compared with $Q_{ICRP\,60}$ (e.g. one obtains Q_{r1}=1 from calculations based on the D(L) distribution for 8.7 MeV protons[6], whereas, for the same distribution, $Q_{ICRP\,60}$=1.9 and Q_{r2}=Q_4=2.5 was calculated).

The correction in the range of Q_4 from 11 up to 17 is proposed because the radiation weighting factors recommended by ICRP do not exceed the value of 20. The correction may influence the value of Q_{r2} only in very rare cases (for neutrons we never obtained the values of Q_4 from this range).

Table 1. Comparison of the quality factors calculated according to ICRP 60 definition ($Q_{ICRP\,60}$) and those resulting from our procedure (Q_{r1}) for simulated LET distributions. The values of better approximation by Q_{r2} and corresponding relative errors are shown in parentheses for the cases where Q_{r2} differs from Q_{r1}.

Radiation	$Q_{ICRP\,60}$	Q_{r1} (Q_{r2})	Relative error (%)
Neutrons			
0.7 MeV	16.2	17.6	9
0.9 MeV	17.1	18.5	8
^{252}Cf	13	14.2	9
7.6 MeV	7.7	8.7	13
14.9 MeV	7.5	9.0	20
Alpha ^{239}Pu	22	24.(20)	10(−10)
Protons 8.7 MeV	1.9	1.(2.3)	−50(20)
^{3}He 23 MeV	7.5	7.3	−3

Table 2. Dependence of the relative error (Q_{r2}-$Q_{ICRP\,60}$)/$Q_{ICRP\,60}$ (in per cent) on the photon component of the absorbed dose, when the calculation of Q_{r2} was not performed separately for photon and neutron components of the dose.

Photon component (%)	Relative error (%) for neutrons		
	0.9 MeV	^{252}Cf	14.9 MeV
0	2.9	9.2	20
5	3.2	5.5	18
10	−4.5	1.7	13.2
20	−13	−5.8	6.7
30	−20.3	−16	2.3
40	−36	−31	−10.2
50	−52.5	−43	−31

INFLUENCE OF THE PHOTON COMPONENT

In mixed radiation fields, the photon component broadens the LET distribution and this may influence the accuracy of our approximation.

Relative differences between the Q_{r2} and $Q_{ICRP\,60}$ are presented in Table 2 for three simulated neutron radiations, mixed with the gamma component of the dose ranging from zero to 50%. The differences depend strongly on the magnitude of the gamma component of the mixed radiation, which introduces an additional error to our method.

This problem can be avoided by the independent determination of the photon component of the dose. If the photon component is known Q_4 can be expressed as:

$$Q_4 = D_\gamma Q_4(\gamma) + D_n Q_4(n)$$

where D_γ and D_n are the photon and neutron dose fractions, $Q_4(\gamma)=1$ and $Q_4(n)$ is the value of Q_4 for neutrons alone.

In the same way the Q_{r2} can be calculated separately for neutrons and photons as:

$$Q_{r2} = D_\gamma \cdot 1 + D_n Q_{r2}(Q_4(n))$$

If this method of calculation is used the relative error of Q_{r2} does not depend on the photon component.

EXPERIMENTAL RESULTS

The values of Q_{r2} were calculated from the experimental values of Q_4 obtained earlier[2]. Results are shown in the Table 3.

Special precise measurements of Q_4 with accuracy better than 5% were performed in standard fields[7] of ^{252}Cf, ^{241}Am-Be and ^{239}Pu-Be neutron sources (for bare sources and the sources placed in paraffin and iron filters). For the same fields the photon component of the absorbed dose was also determined. Q_{r2} was calculated separately for neutrons and for the photon component. The values obtained are shown in Table 4.

CONCLUSION

The proposed calculation procedure enables the use of Q_4 for determination of the quality factor recommended by ICRP 60. The error resulting from calculation usually does not exceed 25%; however, determination of the gamma component is recommended and fields with components of very high LET should be considered with special caution.

ACKNOWLEDGMENTS

The partial financial support of the International Atomic Energy Agency under the research contract No. 6353/RB is appreciated.

Table 3. Q_{r2} calculated from the former experimental values of Q_4.

Radiation source (reaction)	E(MeV)	Q_4	Q_{r2}
Neutrons			
Accelerator (p 1.7 MeV+T)	0.9	11	19
(p 2.9 MeV+T)	2	8.9	12.6
(d 0.2 MeV +T)	5.5	7.0	8.2
(d 12 MeV+Be)	14	7.0	8.2
JINR Dubna	350	3.2	3.2
Protons in TE phantom			
Synchrocyclotron JINR Dubna	200	1.4	1.4
Alphas			
^{222}Rn	5	13	20
Natural uranium	4	18	20

Table 4. Experimental values of Q_4 obtained for standard neutron fields and the values of Q_{r2} calculated from them (photon and neutron component calculated separately).

Neutron source and filter		Photon component	Q_4	Q_{r2}
^{252}Cf	bare	0.35	6.8	10.7
	10 cm iron	0.10	10.1	17
	10 cm paraffin	0.70	4.0	6.4
^{241}Am-Be	bare	0.24	6.5	8.6
	10 cm iron	0.14	8.8	13.6
	10 cm paraffin	0.48	5.3	7.6
^{239}Pu-Be	bare	0.24	6.5	8.7
	10 cm iron	0.14	9.2	15
	10 cm paraffin	0.50	5.2	7.5

REFERENCES

1. Zielczynski, M., Golnik, N., Makarewicz, M. and Sullivan, A. H. *Definition of Radiation Quality by Initial Recombination of Ions.* In: 7th Symp. on Microdosimetry, Oxford (Luxembourg: CEC) EUR 7147, Vol. 2, 853-862 (1980).
2. Golnik, N., Wilczynska, T. and Zielczynski, M. *Determination of the Recombination Index of Quality for Neutrons and Charged Particles Employing High Pressure Ionisation Chambers.* Radiat. Prot. Dosim., 23,(1/4) 273-276, (1988).
3. ICRP. *Data for Protection against Ionizing Radiation from External Sources.* Supplement to ICRP 15, Publication 21 (Oxford: Pergamon) (1973).

4. ICRP. *Radiation Protection — Recommendations of the International Commission on Radiological Protection.* Publication 60 (Oxford: Pergamon) (1991).
5. ICRU. *The Quality Factor in Radiation Protection.* Report 40 (Bethesda, MD: ICRU Publications) (1986).
6. Edwards, A. A., Lloyd, D. C. and Prosser, J. S. *The Induction of Chromosome Aberrations in Human Lymphocytes by Accelerated Charged Particles.* Radiat. Prot. Dosim. **13**(1-4) 205-209 (1985).
7. Józefowicz, K., Golnik, N. and Zielczyński, M. *Dosimetric Parameters of Simple Neutron + Gamma Fields for Calibration of Radiation Protection Instruments.* Radiat. Prot. Dosim. **44**(1-4) 139-142 (1992) (This issue).

ENERGY DEPOSITION IN A SPHERICAL CAVITY OF ARBITRARY SIZE AND COMPOSITION

E. Kearsley
Radiation Biophysics Department
Armed Forces Radiobiology Research Institute
Bethesda, MD 20889-5145, USA

Abstract — The dose distribution inside a spherical cavity is calculated using analytical expressions for both the stopping power and the starting energy distributions for elastically scattered secondary charged particles. Cavity-generated secondaries are treated separately from secondaries generated in the surrounding medium. The result is an analytical expression for the ratio of the dose to the cavity to the dose to the surrounding medium. This expression, sometimes referred to as the 'effective stopping power' is in a form similar to the Burlin general cavity theory for photons.

INTRODUCTION

The energy deposited in a spherical cavity irradiated in a neutron field depends on the composition of the cavity and surrounding medium, the size of the cavity, and the energy of the neutrons. Caswell[1] analysed this problem in terms of 'insiders, starters, stoppers, and crossers', referring to the trajectories of secondary charged particles relative to the volume of the cavity. His objective was to provide a detailed understanding of the pulse height distributions obtained from measurements using tissue-equivalent proportional counters. Rubach and Bichsel[2-4] applied these same techniques to the study of the response of ionisation chambers with a variety of wall–gas combinations and cavity volumes. This approach provides insight into the total energy deposition in a volume but little information about the spatial distribution of the deposited energy, which may be important to our understanding of the biological response of certain tissues of the body after neutron irradiation. This paper describes a calculation of both the dose distribution and the average dose within a spherical cavity of arbitrary size and composition from neutron interactions with both the cavity material and the surrounding medium.

THE CALCULATION

The origin of a spherical coordinate system is placed at a distance, x, from the centre of a sphere of radius, a (Figure 1). The coordinate r may extend to any point inside or outside the cavity volume. The dose at x is the product of the fluence of secondary charged particles and their stopping power. Assuming an isotropic source of secondaries and neglecting any scattering effects at the interface for secondaries generated outside the cavity, the charged particle fluence at x, generated by neutron interactions in a differential volume element, dV, located at a distance, r, from the origin can be written as

$$d\Phi = \frac{N \, dV}{4\pi r^2} \quad (1)$$

where N is the number of secondary charged particles per unit volume.

If it is assumed that the range of a secondary can be written as $R = A E^m$, where A and m are constants that depend on the particle type, then the stopping power for a secondary generated in dV with an initial range, R, after travelling a distance, r, can be written

$$\frac{dE}{dx} = \frac{1}{m} \left(\frac{1}{A}\right)^{1/m} (R - r)^{1/m - 1} \quad (2)$$

The contribution to the dose at the origin of the coordinate system is determined by considering the separate contributions from neutron interactions in the cavity material (i.e. the cavity contribution) and in the surrounding wall (i.e. the wall contribution). The complete calculation

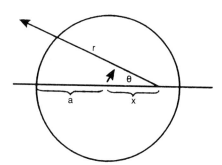

Figure 1. Geometry for the calculation of the response of a spherical cavity in a neutron field. A spherical co-ordinate system is centred at a distance x from the centre of a sphere of radius a.

considers a series of cases that depend on the range of the secondaries and the boundaries of the cavity. Only the case in which the maximum range of the secondary, R_m, is less than the cavity radius, a, will be illustrated in detail.

The cavity contribution

Every point in the region, $O < x < (a - R_m)$, is surrounded by a thickness of cavity material greater than R_m. The dose in this region can be written as

$$D_{c,1}(x) = \int_0^{2\pi} d\phi \int_0^{E_m} dE \int_0^{\pi} d\theta \int_0^{R(E)} dr \frac{N_c}{4\pi\rho_c E_m} \sin\theta \frac{1}{m}\left(\frac{1}{A}\right)^{1/m}$$

$$(R-r)^{1/m-1} \quad (3)$$

Both ϕ and θ have their usual meanings in a spherical coordinate system. The integrand is the product of the secondary fluence and stopping power, divided by the density of the cavity material. The radial integration is limited by the range of the secondary which depends on its energy. The energy integral is over the starting energy distribution, which is assumed to be a simple step function. The subscript, c, refers to the contributions to the dose to the cavity from neutron interactions within the cavity; the numerical subscript is an index to distinguish between different components of the dose.

For the region $(a-R_m) < x < a$, secondaries generated within the cavity with a range less than $(a-x)$ will contribute

$$D_{c,2}(x) = \int_0^{2\pi} d\Phi \int_0^{E1} dE \int_0^{\pi} d\theta \int_0^{R(E)} dr\, I(r,E,\theta) \quad (4)$$

where $I(r,E,\theta)$ is the same integrand used in Equation 3. E_1 is the energy for a secondary with a range equal to $(a-x)$:

$$E_1(x) = \left(\frac{a-x}{A}\right)^{1/m} \quad (5)$$

For a secondary with a range greater than $(a-x)$, the integral becomes

$$D_{c,3}(x) = \int_0^{2\pi} d\Phi \int_{E_1}^{E_m} dE \int_0^{\theta_m} d\theta \int_0^{R(E)} dr\, I(r,E,\theta)$$

$$+ \int_0^{2\pi} d\Phi \int_{E_1}^{E_m} dE \int_{\theta_m}^{\pi} d\theta \int_0^{r_s} dr\, I(r,E,\theta) \quad (6)$$

where r_s is the distance between the origin and the boundary of the cavity at a given angle θ; θ_m is the azimuthal angle at which the range of a secondary is equal to r_s. The sum of Equations 3, 4, and 6 represents the total dose to the cavity from interactions with the cavity material producing secondaries with ranges less than the cavity radius.

The wall contribution

To account for the fact that some fraction of the path of the secondary is in the wall, the residual range appearing in the stopping power becomes

$$(R-r) \rightarrow R - \tilde{n}(r-r_s) - r_s \quad (7)$$

where ñ is the ratio of the range in the cavity to the range in the wall material. The dose at x from secondaries produced from the wall can then be written

$$D_{w,1}(x) = \int_0^{2\pi} d\Phi \int_{E_1}^{E_m} dE \int_{\theta_m}^{\pi} d\theta \int_{r_s}^{r_{max}} dr\, \frac{N_w}{4\pi\rho_c E_m}$$

$$\sin\theta \frac{1}{m}\left(\frac{1}{A}\right)^{1/m} [R-\tilde{n}(r-r_s)-r_s]^{1/m-1} \quad (8)$$

where r_{max} is the maximum range of a secondary starting in the wall at an angle θ, correcting for the range differences in the two materials. That is,

$$r_{max} = r_s + \frac{R_c - r_s}{\tilde{n}} \quad (9)$$

The Brass simplification

For the special case in which the cavity and wall material are identical, the sum of the cavity contribution and the wall contribution at any point x within the cavity must be equal to the equilibrium dose to the material:

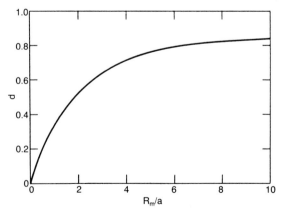

Figure 2. A plot of d as a function of R_m/a for a value of m = 1.75 (appropriate for protons).

$$D_c(x) + D_w(x)|_{w \to c} = \frac{N_c E_m}{2\rho_c} \quad (10)$$

Therefore, the cavity contribution can be determined from the wall contribution and the total dose at any point x within the cavity can be written:

$$D_T(x) = D_c(x) + D_w(x) = \frac{N_c E_m}{2\rho_c} - D_w(x)|_{w \to c} + D_w(x) \quad (11)$$

This procedure, first suggested by Bragg[5], is useful because the expression for the wall contribution, $D_w(x)$, is always much simpler to derive than the cavity contribution, $D_c(x)$. The average dose to the cavity can then be determined by integrating over all values of x. The result can be put into the simple form:

$$\langle D_T \rangle = \frac{N_w E_m}{2\rho c \tilde{n}} d + \frac{N_c E_m}{2\rho_c} (1-d) \quad (12)$$

where $d = d(R_m, a, m)$

Dividing both sides of this expression by the equilibrium dose to the wall of the cavity, we obtain an expression for what is sometimes referred to as the 'effective stopping power'

$$f_w^c = \frac{\rho_w}{\rho_c \tilde{n}} d + \frac{N_c E_m \rho_w}{N_w E_m \rho_c} (1-d) \quad (13)$$

Note that this is in the form of the Burlin general cavity theory for photons[6]. The first term is effectively the mass stopping power ratio for the secondaries multiplied by $d(R_m, a, m)$. The second term is effectively the neutron kerma factor ratio (equivalent to the ratio of the mass energy absorption coefficients in the Burlin theory) multiplied by (1–d).

The function $d(R_m, a, m)$ is illustrated in Figure 2 for recoil protons (m=1.75). The exact expression for $d(R_m, a, m)$ is a complicated, multi-termed function. Space limitations do not permit a full listing of the analytical expression for $d(R_m, a, m)$. However, an approximate expression that can be used to determine values of $d(R_m, a, m)$ to within 1% of the exact value is

$$d(R_m, a) = 1 - \left[\alpha \left(\frac{R_m}{a}\right)^2 + \beta \left(\frac{R_m}{a}\right) + \gamma\right]^{-1} \quad (14)$$

Recommended values for α, β, and γ are provided in Table 1. For a practical calculation involving complex materials such as muscle tissue or A–150 plastic, the contribution of each secondary must be calculated and summed to determine either the dose at x as in Equation 11 or the average dose to the cavity, Equation 12.

DISCUSSION

The calculation rests on several assumptions. First, it is assumed that the secondaries generated by neutron interactions either inside or outside the cavity are produced isotropically. This assumption restricts the use of the model to obtain dose distribution information to cases in which the neutron fields can be considered to be isotropic. However, relations involving the average dose to the cavity such as Equation 12 or Equation 13, are general for any neutron field because of the spherical symmetry of the cavity. Second, the simple parameterised form for the range–energy relationship and therefore the stopping power, Equation 2, is not strictly correct at the end of the track of the secondary. As before, this simplification has little effect on the total energy deposited in the cavity as long as very little of the total energy of the secondary is involved. This simplification will have a greater impact on the dose distribution, because the stopping power at the end of the track will be underestimated. Third, it is assumed that the secondary energy distribution is a simple step function appropriate for secondaries generated via elastic scattering interactions. At high neutron energies, this can introduce substantial errors in the calculation. Fourth, it is assumed that the ratio of ranges for a particular secondary in two different media is independent of the energy. This is approximately correct over a wide range of energies.

The results of this calculation have been compared with the calculations by Rubach and Bichsel[3,4] for a wide range of neutron energies (0.760 – 14 MeV), three cavity–wall combinations

Table 1. Recommended parameters for Equation 14.

$R_m/a < 2$			
m	α	β	γ
0.868 (ions)	0.07724	0.2359	1
1.500 (alphas)	0.08511	0.2511	1
1.750 (protons)	0.08117	0.2501	1
$R_m/a > 2$			
m	α	β	γ
0.868 (ions)	–0.000902	0.6257	0.5298
1.500 (alphas)	–0.001315	0.6151	0.6202
1.750 (protons)	–0.002566	0.5727	0.6908

(TE–TE, TE–air, and C–CO$_2$) and four decades of gas–filled cavity volumes (0.01 – 10 cm^3). At 2 MeV and below, the maximum difference between the ratio of the dose to the cavity to the equilibrium wall dose was less than 3%. At higher energies, the differences between the two calculations are much larger, probably as a result of the assumed shape of the secondary starting energy distribution (i.e. neutron interactions are no longer dominated by elastic scattering).

CONCLUSION

An expression has been derived for the dose distribution within a spherical cavity of arbitrary size and composition surrounded by a medium of arbitrary composition. The expression was averaged over the cavity volume to determine the ratio of the total dose to the cavity to the equilibrium dose to the surrounding medium. The form for the latter expression is identical to the form of the Burlin general cavity theory for photons.

ACKNOWLEDGEMENT

This work was supported by the Armed Forces Radiobiology Research Institute, Defense Nuclear Agency, under Work Unit 4610.

REFERENCES

1. Caswell, R. S. *Deposition of Energy by Neutrons in Spherical Cavities* Radiat. Res. **27,** 92–107 (1966).
2. Rubach, A. and Bichsel, H. *Neutron Dosimetry with Spherical Ionization Chambers I. Theory of the Dose Conversion Factors r and W_n.* Phys. Med. Biol. **27,** 893–904 (1982).
3. Rubach, A. and Bichsel, H. *Neutron Dosimetry with Spherical Ionization Chambers III. Calculated Results for Tissue–equivalent Chambers.* Phys. Med. Biol. **27,** 1231–1243 (1982).
4. Rubach, A. and Bichsel, H. *Neutron Dosimetry with Spherical Ionization Chambers IV. Neutron Sensitivities for C/CO$_2$ and Tissue–equivalent Chambers.* Phys. Med. Biol. **27,** 1455–1463 (1982).
5. Bragg, W. H. *Studies in Radioactivity* (London: Macmillan) (1912).
6. Burlin, T. E. *A General Theory of Cavity Ionization.* Br. J. Radiol. **39,** 727–734 (1966).

WALL COMPONENT OF $r_{m,g}$ VERSUS CAVITY SIZE: A SUMMARY OF THE EXPERIMENTS WITH ALPHA PARTICLES

S. Pszona
Institute for Nuclear Studies
05-400 Swierk, Poland

Abstract — The results of the experiments for determining the wall component of $r_{m,g}$, carried out with extrapolation and parallel plate ionisation chambers, are reviewed. The results are compared with calculations for protons and alpha particles from N(n,p)C and C(n,n)3α nuclear reactions. The concept of a 'cavity response function' is presented.

INTRODUCTION

The absorbed dose to the wall material of an ionisation chamber can be derived from the absorbed dose to its gas filled cavity using the gas-to-wall conversion factor[1].

The absorbed dose to the gas in a cavity of an ionisation chamber or proportional counter irradiated by neutrons is the sum of two components: the wall component and gas component. This means that the gas-to-wall conversion factor, $r_{m,g}$, is the sum of two partial conversion factors. Each of these is a function of cavity size[2,3]. For cavities approaching the case of the infinitesimal small cavity (most of the ionisation chambers used in radiotherapy beams) the wall component contributes mainly to $r_{m,g}$. Therefore the dependence of this component on the cavity size is of importance for the theory as well for practical aspects of neutron dosimetry.

The approach to the experimental evaluation of the wall component of $r_{m,g}$ has been reported in previous symposia[4,5]. The results obtained in these experiments are reviewed here and a concept of cavity response function is presented.

EXPERIMENTAL TECHNIQUES

The method has been devised for the evaluation of only one single component of $r_{m,g}$. This method is based on the use of a specially prepared electrode of an ionisation chamber. The electrode has been prepared as a mixture of electrically conductive material based on epoxy resin mixed with the alpha emitting radionuclide ^{238}Pu. In this way a uniform distribution of alpha emitter with respect to the depth of the electrode has been achieved which gives rise to a slowing down spectrum of alpha particles emerging into the gas cavity. The alpha particles, emitted isotropically from the electrode, simulate the interaction of charged particles with the chamber wall and gas, especially alpha particles emitted from the C(n,n')3α reaction. Two experimental set-ups have been used: first, one in which an extrapolation chamber was used for changing the cavity size, and a second in which a parallel plate ionisation chamber with constant distance between electrodes but variable air pressure was used.

The following expressions linking the measured quantities with the evaluated wall component of $r_{m,g}$ were used:

for the extrapolation chamber:

$$(r/S_{g,m}) W(0)/W(d) = (Q(d)/d) / (Q(0)/d(0)) \quad (1)$$

for the parallel plate ionisation chamber:

$$(r/S_{g,m}) W(0)/W(p) = (Q(p)/p) / (Q(0)/p(0)) \quad (2)$$

where $S_{g,m}$ is the mass stopping power ratio of gas to wall, $1/r$ is the partial gas-to-wall conversion factor for charged particles from the wall, Q(d), Q(0), Q(p) are ion charges measured for given distance d, for d→0, i.e. for an infinitesimally small cavity and for given pressure p, respectively. W(0), W(d), W(p) are mean energies expended in gas per ion pair for the defined cavities.

On the left side of these expressions are the evaluated quantities and on the right side the measured ones.

SUMMARY OF THE RESULTS

The results obtained with both techniques have been analysed in the light of the accuracies which can be derived, based on the Equations 1 and 2, and are shown in Figure 1. The fitting function to the results can be expressed as follows:

$$(r/S_{g,m})W(0)/W(d) = 1 - \exp(-0.4\, R/d) \quad (3)$$

within the maximum relative error of 6%, for R/d >1, where R is the range of alpha particles of ^{238}Pu in air and d is the distance between the electrodes.

It is evident that the relative error of Q(0)/d(0) is much higher than Q(0)/p(0). This is due to difficulties of getting the true value of d(0) owing to lack of parallelism of the electrodes, porosity of the alpha emitting electrode, etc. Therefore the results for R/d <2 were acceptable from the point of view of the accuracy of an extrapolation chamber measurement. On the other hand, the accuracy of pressure measurements is always less than 1%. This provides a convincing argument for the assumption that the parallel plate ionisation chamber with regulated pressure is the proper method for investigating the wall component of $r_{m,g}$ for the cavities close to infinitesimally small sizes.

COMPARISON WITH CALCULATIONS

The measurements were compared with calculations as seen in Figure 2. Two different calculations were performed assuming protons generated in the wall by the N(n,p) reaction of 1 MeV neutrons[6] and alpha particles from the C(n,n')3α reaction of 15 MeV neutrons[7]. It can be seen that for protons an excellent matching has been achieved. Obviously this is due to the fact that protons from the N(n,p) reaction are almost monoenergetic with a range close to the range of alpha particles used in the experiment. The comparison of r/S with the results of calculations for the C(n,n')3α reaction of 15 MeV neutrons also shows a reasonable agreement despite the fact that these calculations were made for tissue-equivalent gas and that the spectrum of alpha particles generated in this reaction is not monoenergetic.

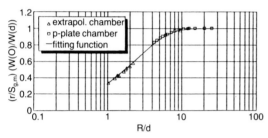

Figure 1. Cavity response relative to cavity size: experimental results.

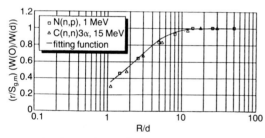

Figure 2. Comparison of the experiments with the calculations.

CONCLUSIONS

The results of the experiments for evaluating the wall component of $r_{m,g}$ were reviewed and the final dependence on R/d has been found to be given by the following fitting function:

$$(r/S_{g,m})\, W(0)/W(d) = 1 - \exp(-0.4\, R/d)$$

which is valid in the range of R/d >1, with a maximum error of 6%.

This dependence, according to experiments and comparison with the calculations, seems to be valid for all monoenergetic charged particles generated in the wall of an ionisation chamber and can be termed as 'cavity response function'. It should be useful for calculating $r/S_{g,m}$ for a particular cavity of an ionisation chamber and a given neutron spectrum.

REFERENCES

1. Broerse, J. J., Mijnheer, B. J. and Williams, J. R. *European Protocol for Neutron Dosimetry for External Beam Therapy.* Br. J. Radiol. **54**, 882-898 (1981).
2. Makarewicz, M. and Pszona, S. *Specification of Cavity Size Effect in the Ionisation Chamber used for Neutron Dosimetry.* In: Proc. Fourth Symp. on Neutron Dosimetry, Vol. II, pp. 307-314. EUR 7448 (Luxembourg: CEC) (1981).
3. Pszona, S. and Makarewicz, M. *Effect of Cavity Size on the Sensitivity of a TE-walled, TE-gas-filled Ionisation Chamber for Fast Neutrons.* Phys. Med. Biol. **27**, 1015-1022 (1982).
4. Pszona, S. *Experimental Examination of the Wall-to-dose Component and its Dependence on Cavity Size for a Parallel Plate Ionisation Chamber.* Radiat. Prot. Dosim. **23**, 449-450 (1988).
5. Pszona, S. *Wall Component of Absorbed Dose to Gas Cavity: Experiment for Microdosimetric Cavities for Alpha Particles.* Radiat. Prot. Dosim. **31**, 97-99 (1990).
6. Makarewicz, M. and Pszona, S. unpublished.
7. Makarewicz, M. and Pszona, S. *Demonstration of a Fast Method for Evaluating the Gas-to-Wall Absorbed Dose Conversion Factor vs Cavity Size.* In: Advances in Dosimetry for Fast Neutrons and Heavy Charged Particles for Therapy Applications. STI/PUB/643 (Vienna: IAEA) (1984).

(a) cross sections for water vapour are applied in generating particle tracks,
(b) slowing down of charged particles within the sensitive volume is neglected, so only crossers and touchers could have been included in present calculations,
(c) the lower energy limit for track simulations is at present 0.3 MeV.amu^{-1}, i.e. some densely ionising secondaries have not been taken into account,
(d) the track structure calculations for heavier ions are performed by scaling the proton tracks with the effective charge from Barkas[12],
(e) calculations are performed for spherical sites only.

As to the last item, Kliauga's spectra were registered with a right-cylindrical UMC (cylinder diameter equal to the height). The mean chord length in such a cylinder is equal to that of the sphere of the same diameter. Although the maximum chord length of such a cylinder exceeds by a factor of √2 the sphere diameter, no spectral edges (useful for calibration purposes) are likely to be observed in the volume of few hundred cubic nanometers. Therefore, it does not play an important role.

RESULTS AND DISCUSSION

Results of such calculations performed for 15 MeV neutrons for spherical sites of 5 and 10 nm diameter in A-150 plastic are displayed in Figure 1 in the form of step functions which reflect the discrete (in fact) scale of ionisation events. The widths of the bins have been assigned to one ionisation in the y scale (e.g. the first bin spreads over the values corresponding to 0.5–1.5 ionisations). The dashed lines display the corresponding results of UMC measurements as published[2]. The comparison indicates substantial differences between the measured and calculated spectra. The calculated distributions are sharp and reflect the quantal (discontinuous) process of energy deposition. In the real UMC only a few primary ionisations are produced in the sensitive volume. Under such conditions, gas amplification leads to strong variations in the number of electrons reaching the anode, which finally results in broad d(y) spectra.

The maxima of the measured spectra are shifted to lower values in comparison with the calculated ones. This shift may possibly result from the inconsistency in the applied calibration procedure. At such small diameters it is not possible to calibrate spectra by finding the alpha or proton edges. It was assumed in the original calibration procedure[2] that the position of the peak corresponds to the single ionisation event in the counter. To justify this, the distribution of number of electrons arriving at the anode wire after gas amplification of a single ionisation event was assumed to be described by the exponential function ~exp(–ax), which after conversion to the dose distribution in lineal energy, d(y), gives the maximum at 1/a. This has been (somewhat arbitrarily) attributed to the position of a single-ionisation peak in the registered dose distribution. This may be well justified for simulated diameters as small as 5 nm. However, if the logarithmic dose distribution, yd(y), is chosen to be displayed, the single-ionisation maximum shifts to 2/a, which can be easily shown arithmetically. Thus, the calibration procedure would have been more consistent with its rationale if the peak of the displayed yd(y) spectrum for 5 nm was positioned at two ionisations. After such a recalibration (solid line in Figure 1) the maxima of both calculated and measured distributions do overlap. In fact, a

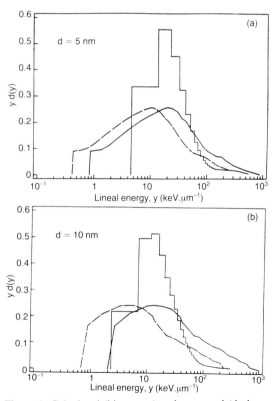

Figure 1. Calculated (histogram) and measured (dashed line) logarithmic dose distributions in lineal energy for 15 MeV neutrons in (a) 5 nm, (b) 10 nm, sites. Solid lines reflect the proposed calibration (see text for more explanation).

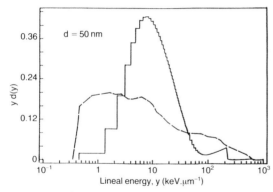

Figure 2. Calculated (histogram) and measured (dashed line) logarithmic dose distributions in lineal energy for 15 MeV neutrons in 50 nm site.

small bump at 7–8 keV.μm^{-1} seems to correspond to the single ionisation event. The similar situation is observed for 10 nm (Figure 1(b)). The broad peak at 5 keV.μm^{-1} (dashed line) was considered in calibration as a single ionisation event. The calculations indicate, however, that the maximum of yd(y) occurs most likely at a triple-ionisation event and therefore the original spectrum has been shifted by a factor of three (solid line). Also in this case the small bump at 1.5 keV.μm^{-1}, when shifted to 4.5 keV.μm^{-1}, can be interpreted as a result of single ionisation produced in a 10 nm volume. Figure 2 (for 50 nm), cannot be interpreted so readily on the basis of present calculations. The original spectrum (dashed line) is again peaked at the value corresponding to single ionisation. Present calculations suggest that for 50 nm sites the most probable (in a yd(y) scale) are 9–ionisation events. Rescaling the measured spectrum creates, however, a significant discrepancy at large y values. Possibly, this may be due to the above-mentioned limit of 0.3 MeV.amu^{-1} in the calculational approach, which results in the neglect of some densely ionising charged secondaries.

CONCLUSIONS

Comparison of the results of calculations of energy deposition spectra induced by neutrons in nanometric sites to the pulse-height distributions registered with the prototype of an ultraminiature microdosimetric counter seems positively to validate the proposed simple analytical approach based on parameterisation of the results of track structure calculations. Obviously, such calculations do not take into account all the physical processes which occur in the avalanche region of a counter and are responsible for broadening the intrinsic ion yield distribution. The influence of these processes clearly increases with decreasing the simulated diameter. The comparison seems to reveal some inconsistencies in the calibration of the originally registered pulse height distributions into the lineal energy scale. In fact, the approach presented may be useful in calibration of ultraminiature counters by indicating localisation of the maxima of measured spectra in the selected scale.

REFERENCES

1. Feinendegen, L. E., Booz, J., Bond, V. P. and Sondhaus, C. A. *Microdosimetric Approach to the Analysis of Cell Responses at Low Dose and Low Dose Rate.* Radiat. Prot. Dosim. **13**, 259-265 (1985).
2. Kliauga, P. *Measurement of Single Event Energy Deposition Spectra at 5 nm to 250 nm Simulated Site Size.* Radiat. Prot. Dosim. **31**, 119-123 (1990).
3. Kliauga, P. *Microdosimetry at Middle Age: Some Old Experimental Problems and New Aspirations.* Radiat. Res. **124**, S5-S15 (1990).
4. Wilson, W. E. and Paretzke, H. G. *Calculation of Distribution of Energy Imparted and Ionization by Fast Protons in Nanometer Sites.* Radiat. Res. **95**, 521-527 (1981).
5. Booz, J., Paretzke, H. G., Pomplun, E. and Olko, P. *Auger-electron Cascades, Charge Potential and Microdosimetry of Iodine-125.* Radiat. Environ. Biophys. **26**, 151-162 (1987).
6. Olko, P., Schmitz, Th., Morstin, K., Dydejczyk, A. and Booz, J. *Microdosimetric Distributions for Photons.* Radiat. Prot. Dosim. **29**, 105-108 (1989).
7. Wilson, W. E., Miller, J. H. and Paretzke, H. G. *Microdosimetric Aspects of 0.3 to 20 MeV Proton Tracks. I Crossers.* Radiat. Res. **115**, 339-352 (1988).
8. Olko, P. and Booz, J. *Energy Deposition in Spherical Sites due to Protons and Alpha Particles.* Radiat. Environ. Biophys. **29**, 1-17 (1990).
9. Kellerer, A. M. and Chmelevsky, D. *Criteria for the Applicability of LET.* Radiat. Res. **63**, 226-234 (1975).
10. Morstin, K., Kawecka, B. and Booz, J. *Combined Primary and Secondary Particle Transport Calculations of Microdosimetric Distributions in Tissues and Tissue Substitutes.* Radiat. Prot. Dosim. **13**, 103-110 (1985).
11. Edwards, A. A. and Dennis, J. A. *NESLES. A Computer Program which Calculates Charged Particle Spectra in Materials Irradiated by Neutrons.* Report NRPB-M43 (London: HMSO) (1979).
12. Barkas, W. H. In: *Nuclear Research Emulsion,* Vol. I, p. 371 (London, New York: Academic Press) (1963).

MEASUREMENT AND PREDICTION OF REAL TISSUE MICRODOSIMETRIC RESPONSES FOR NEUTRONS UP TO 60 MeV

A. C. A. Aro†, S. Green†, R. Koohi-Fayegh‡, M. C. Scott†, T. Shahid† and G. C. Taylor§
†Medical Physics Group, School of Physics and Space Research
University of Birmingham, Birmingham B15 2TT, UK
‡School of Sciences, Ferdowsi University of Mashhad, Mashhad, Iran
§Division of Radiation Sciences and Acoustics
National Physical Laboratory, Teddington, Middlesex TW11 0LW, UK

Abstract — Two approaches are described for determining the microdosimetric responses of real tissue exposed to neutrons. One is to measure the response for individual elements by a difference technique, so that the response for any body tissue can be synthesised. The second is to develop methods for building counters with walls of real tissue, the greatest problem to overcome being the presence of water.

INTRODUCTION

In order to make neutron dosimetry and microdosimetry measurements using proportional counters, conducting tissue-equivalent wall materials have been developed, such as A-150 plastic. However, due to limitations in the composition of such materials, it has been necessary to substitute carbon for the oxygen present in real tissues. As is well known, their elemental compositions therefore differ considerably from that of standard tissue[1]. In addition, there are several different body tissues (e.g. muscle, fat, brain and bone), each of which has a different elemental composition, so that measurements made using a single material will not reflect the behaviour in all tissue types.

At energies below 20 MeV or so, where most of the neutron dose arises from interactions with hydrogen, the energy deposition in tissue-equivalent materials is very similar to that of standard tissue. However, at higher energies, where the alpha producing reactions assume increasing importance, there are differences between the carbon and oxygen cross sections, and the resulting uncertainties in measured dosimetric quantities have not yet been evaluated.

In the present work, two approaches have been adopted to allow microdosimetric information to be obtained for real body tissues over the range of energies of interest in neutron cancer therapy. The first was to determine the microdosimetric response for individual elements, e.g. C, H, N and O, and hence to allow synthesis of the response for any body tissue. The second was to attempt to build a microdosemeter with real tissue walls.

DETERMINATION OF ELEMENTAL MICRODOSIMETRIC RESPONSE

In order to determine the elemental microdosimetric responses a difference technique was used[2]. For example, the microdosimetric response for oxygen was found by subtracting the response of an aluminium counter from that of one made from aluminium oxide. However, the use of this difference technique depends upon a number of assumptions and corrections[3]. For example, for this approach to be valid, the perturbation of the neutron field produced by each of the counters has to be small, and to be similar for all the counters. One has then to scale one of the responses in each pair of measurements, to allow for differences in the stopping powers and number densities of the two different wall materials. Finally, since the use of a scaling factor in the subtraction technique will mean that gas events are not properly allowed for, their importance has to be investigated.

Neutron perturbation

Because of the limited ranges of alpha particles and heavy ions, the neutron spectrum in the immediate vicinity of the cavity will determine their contributions to the microdosimetric response, whereas proton induced events will arise over a much larger volume. Two approaches have been used to investigate the importance of neutron perturbation on the detector responses. At energies below 20 MeV the Monte Carlo code MCNP[4] was used to determine the flux at the central cavity for different monoenergetic neutron energies, and the perturbed spectrum at the cavity was then used as input to the analytical code NESLES[5], to determine the changes in microdosimetric response.

The second approach, which was applied over the full energy range of interest, was to use the one-dimensional neutron transport code ANISN[6] in conjunction with the high energy neutron data set HILO[7] to calculate the spatial dependence of the angular (i.e. vector) flux in different regions of the detector shell. This was then combined with scalar flux calculations using the Monte Carlo code MORSE[8] to provide the input data to a microdosimetric Monte Carlo program[9]. This allowed the effects of detector perturbation on the proton response to be investigated as well.

The calculated neutron fluxes at the central cavity of detectors having walls of polythene, carbon, A150 plastic and muscle are shown in Figure 1 for a 15.5 MeV neutron beam, where it can be seen that the low energy portion of the perturbed flux is both small and similar for all the counters. Although the neutron cross section data used for aluminium was not the latest available, this does not affect comparisons between perturbed and unperturbed neutron spectra, and similar results were obtained for the Al, Al_2O_3 and AlN counters. We note that in all cases the assumed wall thickness (30 mm) was much greater than that used in the actual detectors (13 mm).

For an aluminium counter exposed to 15.5 MeV neutrons the effect of this perturbation on the microdosimetric spectrum was negligible (Figure 2); the greatest effect at 15.5 MeV was observed in the alpha particle response of the Al_2O_3 counter (Figure 3), but it was still very small.

At higher neutron energies, the use of a Monte Carlo code to predict the microdosimetric response meant that comparisons were subject to considerable statistical uncertainties. Nevertheless, the results up to a neutron energy of 55 MeV were consistent with the effects of perturbation being small. It was concluded, therefore, that neutron perturbation did not invalidate the use of a difference technique to determine the elemental response.

However, as can be seen from Figure 1 and Table 1, the main difference between the neutron spectrum for different detectors was in the attenuation of the primary neutron beam at the central cavity, and it was necessary to allow for this in the scaling factors used. Although, in principle, the attenuation factor is neutron spectrum dependant, in practice the change in neutron total cross section with energy for each of the elements is sufficiently small that the factors determined for 15.5 MeV were applied to measurements at higher energies.

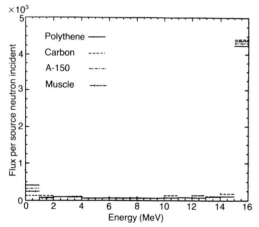

Figure 1. Calculated neutron flux at a central cavity (per source neutron incident) in a detector having 3 cm thick walls of polythene, carbon, A-150, and muscle materials, irradiated with 15.5 MeV neutrons.

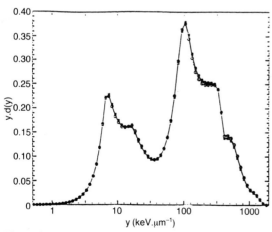

Figure 2. Comparison of the calculated microdosimetric response of a 3 cm thick aluminium detector exposed to 100% (monoenergetic) 15.5 MeV neutrons (o), with the perturbed response produced by 2 mm (x), and 3 cm (□) wall thicknesses.

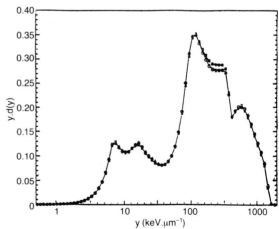

Figure 3. Comparison of the calculated microdosimetric response of a 3 cm thick aluminium oxide detector exposed to 100% (monoenergetic) 15.5 MeV neutrons (o), with the perturbed response produced by 2 mm (x), and 3 cm (□) wall thicknesses.

Stopping power effects

Differences in the stopping power of the wall materials used meant that the effective volume over which the detectors were sampling particle tracks was not the same. This difference therefore had to be allowed for in the scaling factors used. As an example, the ratios of the stopping powers for alpha particles and protons in Al and Al_2O_3 are shown in Figure 4. It can be seen that, although there is a variation with particle energy, the differences are not large, and an average stopping power can be used without introducing significant errors. Using this assumption, Table 2 gives the mean stopping power ratios which were used from 100 eV to 60 MeV. We note, however, that where the charged particle spectra are known, or can be estimated, the use of spectrum weighted mean stopping powers would reduce any uncertainties involved[10].

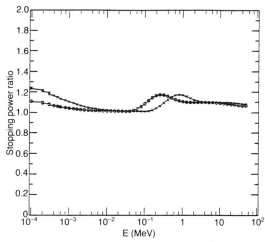

Figure 4. Stopping power ratios between Al_2O_3 and Al for protons (o) and alpha particles (x).

Elemental number density corrections

Since the elemental number density in the wall material will determine its overall contribution to the detector response, it is necessary to correct for this, and Table 3 gives the data for the five counters involved.

Gas events

The importance of gas events was discussed in an earlier paper[3], where it was shown that, at 15 MeV for example, they accounted for about 20% of the events above 100 $keV.\mu m^{-1}$, but account for less than 5% of the total measured dose. For a given neutron fluence the gas event contribution in the different detectors will be the same (neglecting any neutron perturbation effects), but the introduction of a scaling factor in the difference technique means that a small systematic difference will be introduced in the total micro-

Table 1. Fraction of 15 MeV neutrons transmitted through different wall materials.

Wall material	Carbon	Polythene	Aluminium	Al_2O_3	AlN
Density ($g.cm^{-3}$)	1.7	0.95	2.71	3.9	2.93
Wall thickness (cm)	2.5	3.0	1.37	1.37	1.37
Transmitted fluence (%)	74.2	71.9	86.5	77.0	82.1

Table 2. Average stopping power ratios for mixed wall materials and single elements for protons and alpha particles between 100 eV and 60 MeV.

Wall materials	$(CH_2)_n$ / C	Al_2O_3 / Al	AlN / Al	Al_2O_3 / O	AlN / N
Stopping power ratio	1.36	1.10	1.10	0.95	0.85

Table 3. Number density corrections.

	Carbon	C in $(CH_2)_n$	Al	Al in Al_2O_3	Al in AlN
Atoms per g ($\times 10^{22}$)	5.02	4.29	2.23	1.18	1.47
Number density ratio	–	1.17	–	1.89	1.52

dosimetric response. Although it is possible to correct for gas events if the neutron spectrum is known, this is generally not the case, and we have therefore not made any correction for this effect in the work so far.

THE ELEMENTAL RESPONSES

The elemental response for oxygen was determined from that of the Al and Al_2O_3 counters using

(Response of O) = (Response of Al_2O_3) − (Response of Al)/FAC   (1)

F, A, and C are the correction factors for differences in the number density, stopping power and neutron attenuation respectively in the two detectors. For example, F is the ratio (Al number density in an aluminium detector/Al number density in an Al_2O_3 detector). The other correction factors are defined similarly, and the values used are given in the earlier tables. The product of the correction factors was 2.33. A similar expression was used for the nitrogen response, the product of the scaling factors being 1.76.

The individual detector responses for Al and Al_2O_3 to 15 MeV neutrons from the Dynamitron accelerator, normalised using a fission chamber to the same neutron fluence, are shown in Figure 5, whilst the derived oxygen response is compared to that of carbon in Figure 6. The main difference in this case lies in the proton contribution to the response, that for oxygen being significantly higher. The difference in the proton contribution at 15 MeV is even more pronounced when carbon is compared to nitrogen (Figure 7), since nitrogen has a significant (n,p) cross section over the neutron energy range of interest.

Finally, Figure 8 shows a comparison of the oxygen and carbon elemental responses measured in air in the Clatterbridge neutron therapy beam; they are normalised to equal neutron fluences using a fission chamber. Here we see that the responses are similar in the alpha particle and heavy ion regions, but that the oxygen proton response is about 15% higher than that for carbon at its peak. Noting that this response is per atom, it can be seen that the carbon kerma factor will, nevertheless, be higher than that for oxygen over this energy spectrum. Our result is therefore consistent with the kerma factor data of other workers (e.g. Reference 11).

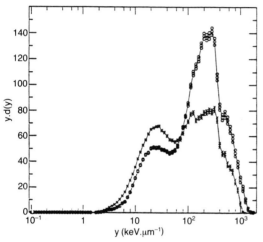

Figure 5. Microdosimetric responses of (o) Al_2O_3 and (x) Al, normalised to fission chamber measurements with the gamma ray component subtracted, following irradiation by 15 MeV neutrons.

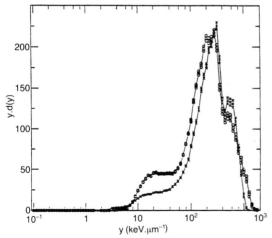

Figure 6. Comparison of microdosimetric responses of oxygen (o) and carbon (x) for 15 MeV neutrons.

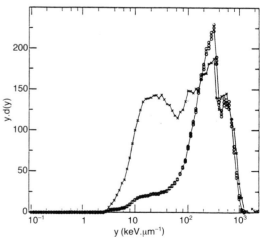

Figure 7. Comparison of microdosimetric responses of carbon (o) and nitrogen (x) for 15 MeV neutrons.

DEVELOPMENT OF REAL TISSUE COUNTERS

The main problems in the development of real tissue counters centred on (i) preventing the outgassing of water and fats into the counting gas, and (ii) the provision of a conducting surface at the tissue walls. The initial development centred on making a counter with ice walls. This is because water constitutes about 75% of normal tissues and the fact that it is strongly electronegative means that it presented the greatest challenge.

The effects of outgassing were minimised by running the detector at liquid nitrogen temperature (77 K) using a liquid nitrogen cooling jacket (Figure 9). The resulting detector could be run for several hours before the resolution started to deteriorate as a result of the build-up of water vapour. At this temperature, normal tissue-equivalent filling gases liquify, and it was necessary to use a filling gas based on hydrogen (purified using a palladium barrier) with 17% neon atoms added, to ensure that the stopping power was similar to that of tissue. As a check on this stopping power equivalence, the responses of a polythene counter exposed to 15 MeV neutrons and filled with either tissue-equivalent gas or a hydrogen–neon mixture were compared (Figure 10), and there were only minor differences.

The use of a graphite grid coating, which was developed for studies of non-tissue materials[2], was precluded by the uneven nature of the real tissue or ice surface. What was required was a cylindrical grid which was self supporting, had a high transparency, produced a uniform electric field in the multiplying region of the detector and could be rapidly inserted into a real tissue or ice annulus. Using the resolution of the proton edge as a criterion for the multiplying field uniformity, the problem was solved using a free-standing, cylindrical, copper wire grid; several configurations were tried, the one finally used having a transparency of 90%. The detector used field tubes to define the measurement volume.

The detector was assembled very rapidly at room temperature. The grid was inserted into a previously prepared ice annulus, and the whole detector was then cooled to liquid nitrogen temperature before being evacuated and filled.

Two sets of measurements have been made with the ice counter so far. The first was at 15 MeV, where the results were compared with those from

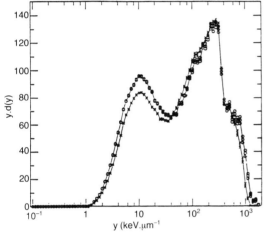

Figure 8. Comparison of microdosimetric responses of oxygen (o) and carbon (x) for p(62)Be neutrons produced at Clatterbridge cyclotron.

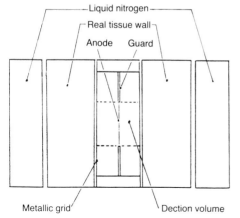

Figure 9. Schematic design of a real tissue microdosimetric counter.

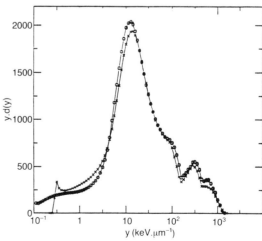

Figure 10. Comparison between the microdosimetric responses obtained with the metallic grid surrounded by polythene wall using (o) methane-based tissue-equivalent gas, and (x) 83% H_2 and 17% neon gas, following irradiation by 15 MeV neutrons (fission chamber normalised).

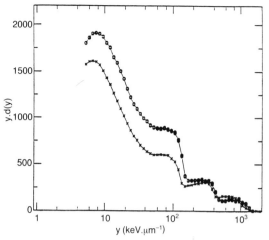

Figure 11. Microdosimetric responses obtained with the metallic grid surrounded by ice (o) and $(CH_2)_n$ (x) walls to 15 MeV neutrons (fission chamber normalised).

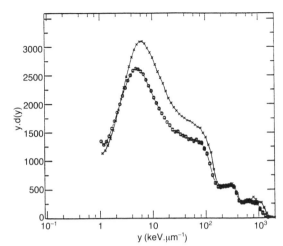

Figure 12. Comparison of microdosimetric responses of the A-150 TEP counter (o), with the ice counter (x) when exposed to p(62)Be neutrons produced at Clatterbridge cyclotron (fission chamber normalised).

a polythene walled counter, normalised to the same neutron fluence (Figure 11). The results showed that the proton and alpha particle edges were equally well resolved in the two counters, and that the dose in the ice counter was significantly higher than that in the polythene one.

The second set of measurements were made in air on the Clatterbridge therapy beam, and are shown in Figure 12, where they were compared with that from a counter having tissue-equivalent walls. We again see that the charged particle edges are extremely well resolved — indicating that recombination from outgassed water vapour was not a problem. We also see that the dose in the ice counter is significantly higher than that in the tissue-equivalent one when normalised to neutron fluence.

A detector with real bone walls has also been constructed, but is still under test.

CONCLUSIONS

It has been shown that the elemental microdosimetric responses can be determined by a difference technique using scaling factors to allow for differences in neutron attenuation, stopping power and elemental number density between the detectors used. From the results obtained for O, N, C and H it is possible to synthesise the results which would be obtained from any body tissue. We have also shown that it is possible to make a detector having ice walls; since the presence of water presents the greatest instrumental problems, the extension to real tissue walls should be straightforward.

ACKNOWLEGEMENTS

This work forms part of a programme of neutron physics relating to neutron cancer therapy which has received support from the Medical Research Council, London, and the Clatterbridge Cancer Trust. This support is gratefully acknowledged, as is the personal support for A.C.A.A. from the Brazilian Government (CNPq), for S.G. from the National Physical Laboratory, and for T.S. from the Pakistan Government. The invaluable assistance of Don Grose and Mike Smith, and of the accelerator operators and scientific staff at Birmingham and Clatterbridge, is also recorded with pleasure.

REFERENCES

1. International Commission on Radiation Units and Measurements. *Neutron Dosimetry for Biology and Medicine.* ICRU Report 26 (Bethesda, MD: ICRU Publications) (1977).
2. Green, S., Aro, A. C. A., Taylor, G. C. and Scott, M. C. *The Development of Microdosimetric Detectors for Investigating LET Distribution in Different Body Tissues.* Radiat. Prot. Dosim. **31**, 137-141 (1990).
3. Scott, M. C., Aro, A. C. A., Green, S. and Taylor, G. C. *Elemental Synthesis of Real Tissue Microdosimetric Responses to High Energy Neutrons: Principles and Limitations.* In: Proc. 2nd European Particle Accelerator Conference, pp, 1799-1801 (1990).
4. *MCNP — A General Purpose Monte Carlo Code for Neutron and Photon Transport.* LA-7396-M (Los Alamos National Laboratory) (1981).

5. Edwards, A. A. and Dennis, J. A. *NESLES — a Computer Programme which Calculates Charged Particle Spectra in Materials Irradiated with Neutrons.* NRPB/M43 (National Radiological Protection Board) (1979).
6. Engle Jr, W. W. *ANISN, a One Dimensional Discrete Ordinates Transport Code with Anisotropic Scattering.* K-1693 RSIC (1967).
7. Alsmiller, R. G. and Barish, J. *Neutron Photon Multigroup Cross-sections for Neutron Energies up to 400 MeV.* ORNLITM-7818 (Oak Ridge National Laboratory) (1983).
8. Taylor, N. P. and Needham, J. *MORSE-H — a Revised Version of the Monte Carlo Code MORSE.* AERE-R10432 (AERE Harwell) (1983).
9. Taylor, G. C. *The Prediction and Measurement of Microdosimetric Spectra Relating to Neutron Cancer Therapy.* PhD thesis, University of Birmingham (1990).
10. Bühler, G., Menzel, H. G., Schuhmacher, H., Dietze, G. and Guldbakke, S. *Neutron Kerma Factors for Magnesium and Aluminium Measured with Low-pressure Proportional Counters.* Phys. Med. Biol. **31**, 601-611 (1986).
11. Brady, F. P. and Romero, J. L. *Neutron Induced Reactions in Tissue Resident Elements.* Final Report of the National Cancer Institute (1980).

COMPARISON OF RESPONSE FUNCTION CALCULATIONS FOR MULTISPHERES#

C. A. Perks†, D. J. Thomas‡, B. R. L. Siebert§, S. Jetzke§, G. Hehn|| and H. Schraube*
†AEA Environment and Energy, B. 364, Harwell Laboratory, Oxfordshire OX11 0RA, UK
‡Division of Radiation Science and Acoustics, National Physical Laboratory
Teddington, Middlesex TW11 0LW, UK
§Physikalisch-Technische Bundesanstalt
Bundesallee 100, D-3300 Braunschweig, Germany
||IKE, Universität Stuttgart, Pfaffenwaldring 31, D-7000 Stuttgart 80, Germany
*GSF-München, Institut für Strahlenschutz, D-8042 Neuherberg, Germany

Abstract — An intercomparison of computer calculations to determine the response function of a multisphere neutron spectrometry system has been undertaken by EURADOS-CENDOS Working Committee IV. For this purpose, a standard multisphere spectrometer was adopted. This consisted of four polyethylene spheres, radii 38.1, 63.5, 101.6 and 152.4 mm with a central ^{3}He detector, diameter 32 mm, having an atomic density of 4.25×10^{19} atoms.cm^{-3} (pressure, 172 kPa). The calculations were made for two polyethylene densities (0.92 and 0.95 g.cm^{-3}). The Monte Carlo computer codes MCNP and BOKU (developed at PTB specifically for these calculations) and the discrete ordinates code ANISN were used to calculate the response of the detector for a variety of monoenergetic incident neutrons. Preliminary results for the 38.1 and 63.5 mm radii spheres are presented to compare these codes and their application at different laboratories. Discrepancies are discussed. In addition, further calculations were undertaken to compare the effects of various factors on the response functions, including: the effect of using cross sections without taking into account the binding of the carbon and hydrogen in the polyethylene; the effect of the steel walls of the ^{3}He detector; and the effect of the stem of the counter.

INTRODUCTION

A multisphere neutron energy spectrometer[1] consists of a set of polyethylene spheres of different diameters. A review of applications of multispheres has been given by Awschalom and Sanna[2]. At the centre of each sphere a thermal neutron detector (for example a ^{3}He proportional counter) is located. Fast neutrons impinging on a sphere are moderated in the polyethylene and detected by the central counter. The combination of the response of the detector as a function of neutron energy and the moderating characteristics of the polyethylene sphere give rise to an overall response function for each polyethylene sphere/detector combination. These response functions are used as input, together with the detector count rates, to an unfolding program which derives the incident neutron energy spectrum. Ideally, these response functions would be determined experimentally using monoenergetic neutrons. However, this is only practical for monoenergetic neutrons with energies greater than about a few keV and for thermal neutrons. Consequently, it is necessary to perform computer calculations to obtain the response functions over the complete range of energies, for example from thermal to 14 MeV.

Recently, EURADOS-CENDOS Working Committee IV completed an intercomparison of the unfolding codes used to determine neutron spectra from given response functions and detector count rates[3]. This committee has now turned its attention to the calculation of the response functions. For this purpose, a standard multisphere spectrometer was adopted. The Monte Carlo computer codes MCNP and BOKU (developed at PTB specifically for these calculations) and the discrete ordinates code ANISN were used to calculate the response of the detector as a function of neutron energy in the range from thermal to 14 MeV. In addition, some further calculations were made to compare the effects of various factors on the response functions, for example: the effect of using cross sections without taking into account the binding of the carbon and hydrogen in the polyethylene; the effect of the steel walls of the ^{3}He detector; and the effect of the stem of the counter.

Details of the standard multisphere spectrometer are given, followed by brief descriptions of the programs used to calculate the response functions. Preliminary results are then given. Finally there are some concluding remarks.

THE STANDARD MULTISPHERE

The standard multisphere spectrometer selected for this work consisted of four polyethylene spheres, radii 38.1, 63.5, 101.6 and 152.4 mm with a central ^{3}He detector, diameter 32 mm,

© UKAEA.

having an atomic density of 4.25×10^{19} atoms.cm^{-3} (corresponding to a pressure of 172 kPa). Two polyethylene densities were used: 0.92 and 0.95 g.cm^{-3}. In this paper, only preliminary results are given for the 38.1 and 63.5 mm radii spheres.

COMPUTER CODES AND INPUT DATA

Laboratories involved in this work used three computers codes for calculating the response functions. The discrete ordinates code, ANISN[4], run in adjoint mode, has been widely used for multisphere response function calculations in the past. This enables the complete response function to be determined in one calculation. The adjoint source is taken as the macroscopic cross section uniformly distributed throughout the ^{3}He detector volume. The response function is equivalent to the resulting adjoint neutron fluence at the surface. The cross sections for carbon and hydrogen used by NPL were extracted from the UKCTRI library[5]. In this data, the hydrogen is taken to be bound to oxygen as in water. IKE used JEF2.1 data, both taking into account the binding effects of the carbon and hydrogen in the polyethylene in the energy range up to 8.76 eV, and without taking the binding into account.

Calculations at the Harwell Laboratory, GSF, PTB and NPL used the widely adopted Monte Carlo type program called MCNP[6]. The particular advantage of this program is that it allows the use of pointwise cross section data. Consequently, in regions where the cross section

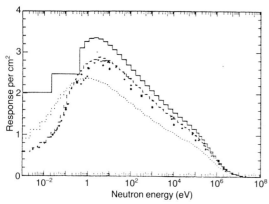

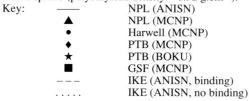

Figure 1. Response function calculations for the 38.1 mm radius sphere (polyethylene density = 0.92 g.cm^{-3}).
Key:
— NPL (ANISN)
▲ NPL (MCNP)
● Harwell (MCNP)
♦ PTB (MCNP)
★ PTB (BOKU)
■ GSF (MCNP)
--- IKE (ANISN, binding)
..... IKE (ANISN, no binding)

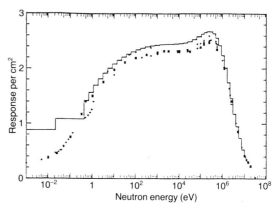

Figure 3. Response function calculations for the 63.5 mm radius sphere (polyethylene density = 0.92 g.cm^{-3}). Key as Figure 1.

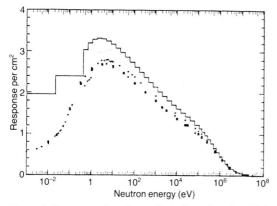

Figure 2. Response function calculations for the 38.1 mm radius sphere (polyethylene density = 0.95 g.cm^{-3}). Key as Figure 1.

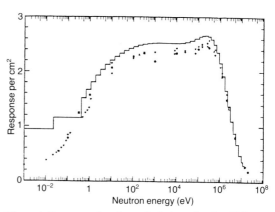

Figure 4. Response function calculations for the 63.5 mm radius sphere (polyethylene density = 0.95 g.cm^{-3}). Key as Figure 1.

is changing rapidly, it can be modelled more accurately. However, this means that the code cannot be used in the adjoint mode. Therefore, the multisphere response function has to be calculated separately for a range of incident monoenergetic neutrons. For these calculations, a broad, parallel, beam of monoenergetic neutrons impinged on the multisphere. Cross sections used by MCNP are derived from the ENDFB-IV and ENDFB-V libraries. Full account was taken of the binding of carbon and hydrogen in the polyethylene. The NPL calculations included a steel shell to model the walls of the ^{3}He counter.

Finally, the PTB made calculations using the Monte Carlo computer code called BOKU, developed by Siebert[7]. This code is an analogue Monte Carlo code with pointwise cross sections obtained from ENDFB-V. As for MCNP this program cannot be used in adjoint mode and calculations were made at a number of discrete incident neutron energies.

RESULTS

Only preliminary results are presented in this paper, full results will be published in a later publication. Figures 1 and 2 show the response function calculations for the 38.1 mm radius sphere with a polyethylene density of 0.92 and 0.95 g.cm^{-3} respectively. Similar data for the 63.5 mm radius sphere are given in Figures 3 and 4. For the Monte Carlo calculations (MCNP and BOKU), the data is in the form of discrete points corresponding to the incident neutron energy, whereas the ANISN calculations give the response function in pre-selected ranges of energy.

The effect of taking the binding of carbon and hydrogen into account is shown in Figure 1 for the 38.1 mm radius sphere with a polyethylene density of 0.92 g.cm^{-3} and using the ANISN method of calculation. The effect of taking the steel walls of the ^{3}He counter into account for the 63.5 mm radius sphere is shown in Table 1. For these calculations the MCNP code was used and the steel walls were modelled by 0.5 mm iron. Finally, the effect of the stem of the ^{3}He counter (modelled as a cylindrical void) was examined for thermal and 2.0 MeV neutrons impinging on the 38.1 mm sphere with a polyethylene density of 0.92 g.cm^{-3} (Table 2). This sphere was chosen since the relative proportion of polyethylene removed for the stem is larger than for the other spheres.

DISCUSSION

The response function calculations using the Monte Carlo codes give good agreement over the

Table 1. Comparison of the response function for the 63.5 mm radius sphere (calculated using MCNP, Harwell Laboratory) with and without the inclusion of an 0.5 mm thick iron shell to model the wall of the ^{3}He counter.

Neutron energy (MeV)	Detector response (counts per unit fluence)/cm^2		Ratio of responses (no iron shell / iron shell)
	No iron shell	0.5 mm iron shell	
Thermal	0.618 ± 0.014	0.626 ± 0.012	0.987
0.1×10^{-6}	0.802 ± 0.009	0.788 ± 0.013	1.018
1.0×10^{-6}	1.633 ± 0.023	1.573 ± 0.039	1.038
10.0×10^{-6}	2.045 ± 0.031	2.073 ± 0.047	0.986
100.0×10^{-6}	2.285 ± 0.028	2.280 ± 0.028	1.002
1.0×10^{-3}	2.389 ± 0.036	2.192 ± 0.047	1.090
10.0×10^{-3}	2.371 ± 0.030	2.341 ± 0.048	1.013
100.0×10^{-3}	2.418 ± 0.031	2.415 ± 0.028	1.001
300.0×10^{-3}	2.534 ± 0.054	2.441 ± 0.048	1.038
2.0	1.556 ± 0.035	1.495 ± 0.020	1.041
20.0	0.197 ± 0.009	0.189 ± 0.001	1.042

Table 2. Comparison of the response function for the 38.1 mm radius sphere (calculated using MCNP, Harwell Laboratory) with and without inclusion of the stem of the ^{3}He counter.

Neutron energy (MeV)	Multisphere response (counts per unit fluence)/cm^2		Ratio of responses (no stem/stem)
	No stem	Stem	
Thermal	1.284 ± 0.021	1.317 ± 0.021	0.975
2.0	0.205 ± 0.006	0.198 ± 0.006	1.035

complete energy range. To date, ANISN has been used by most people who have calculated multi-sphere response functions and many measurements made with multisphere spectrometers have used these response functions. Therefore, it is important to know how accurate they are. Using cross section data which take into account the binding of the carbon and hydrogen, the ANISN calculations performed by IKE for the 38.1 mm radius sphere are in good agreement with the Monte Carlo calculations. For the NPL calculations using ANISN (with cross section data for hydrogen bound in water), there are significant discrepancies from the Monte Carlo calculations. However, provided some normalisation is applied, for example by comparing the calculated response function with a number of measurements made with monoenergetic neutrons, the response function is good for energies > 10 keV. However, towards thermal neutron energies there are large discrepancies. For dosimetric purposes, the fact that the response function is incorrect for thermal neutrons is unimportant since many multisphere spectrometers are used under cadmium and the thermal response is removed. The thermal component is then measured independently (for example by using a calibrated thermal neutron detector with and without a cadmium cover). The effect on the response functions of modelling the walls and stem of the ^{3}He counter are clearly of little importance compared with the large effect due to the binding of the carbon and hydrogen in the polyethylene.

CONCLUSIONS

Preliminary results of a comparison of multi-sphere response functions as calculated at different laboratories and using a number of different computer codes and cross section data were briefly summarised in this paper. In general, the results for the response functions calculated at various laboratories using MCNP and BOKU are in good agreement. It is clearly necessary to use cross section data that take into account the binding of the carbon and hydrogen in the polyethylene, particularly for neutron energies approaching thermal. Detailed modelling of the structure of the ^{3}He counter, by including its steel wall and stem, leads to small variations (up to about 5%) in the calculated response function, and may be necessary for the smaller radii spheres. For the larger radii spheres, simple correction factors may be sufficient. A complete description of the techniques used and the results, including those for the 101.6 and 152.4 mm radius spheres, will be published in a future paper.

ACKNOWLEDGEMENT

This work is being performed under the auspices of EURADOS-CENDOS Working Group IV ('Numerical Dosimetry').

REFERENCES

1. Bramblett, R. L., Ewing, R. I. and Bonner, T. W. *New Type of Neutron Spectrometer*. Nucl. Instrum. Methods **9**, 1-12 (1960).
2. Awschalom, M. and Sanna, R. S. *Applications of Bonner Sphere Detectors in Neutron Field Dosimetry*. Radiat Prot. Dosim. **10**, 89-101 (1985).
3. Alevra, A. V., Siebert, B. R. L., Aroua, A., Buxerolle, M., Grecescu, M., Matzke, M., Mourgues, M., Perks, C. A., Schraube, H., Thomas, D. J. and Zaborowski, H. L. *Unfolding Bonner Sphere Data: A European Intercomparison of Computer Codes*. PTB report number PTB-7.22-90-1 (January 1990).
4. Engle, W. W. *A Users Manual for ANISN*, ORNL-K1692 (Oak Ridge National Laboratory, TN, USA) (1967).
5. Beynon, T. D. and Taylor, N. P. *The UKCTRI Data Library: 46-group Neutron Cross-sections for Fusion Reactor Calculations*. University of Birmingham, Paper No. 79-02.
6. Breismeister, J. F. (ed.) *MCNP - A General Monte Carlo Code for Neutron and Photon Transport, Version 3A*. (Los Alamos National Laboratory) Report No. LA-7396-M Rev. 2 (1986).
7. Siebert, B. R. L., Alberts, W. G. and Bauer, B. W. *Computational Study of Phantoms for Individual Neutron Dosimetry*. PTB report no. PTB-N-6 (1990).

COMPARISON OF MEASURED AND CALCULATED BONNER SPHERE RESPONSES FOR 24 AND 144 keV INCIDENT NEUTRON ENERGIES

B. R. L. Siebert, E. Dietz and S. Jetzke
Physikalisch-Technische Bundesanstalt, D-3300 Braunschweig, Germany

Abstract — Calculations and experimental calibrations in standard neutron fields are needed in order to describe adequately the response function of a Bonner sphere. Measurements of response functions in reactor filtered neutron beams are intercompared with calculations, and the influence of the moderator's mass density is studied in calculations. The ^{3}He content of the proportional counter used as the central detector has been experimentally determined and roughly confirmed by calculations. The calculational model used was found to be too simple for simulating small Bonner spheres.

INTRODUCTION

Bonner spheres (BS)[1,2] are widely used for determining spectral neutron fluences and derived dosimetric quantities such as absorbed dose and dose equivalent. A BS consists of a central detector, most sensitive to thermal neutrons, and a moderator sphere. A ^{3}He proportional counter is often used as the central detector. A knowledge of BS response as a function of incident neutron energy is required to interpret BS measurements. The response of a BS is defined as the ratio of yield over neutron fluence.

To describe the response function adequately, calculations and experimental calibrations in standard neutron fields are needed. The mass density of the moderator material and the partial ^{3}He pressure must be known as input for the calculations.

The measured and calculated fluence responses of Bonner spheres to iron and silicon filtered reactor neutrons with nominal energies of 24 keV and 144 keV, respectively, are presented here. Four spheres with diameters ranging from 7.62 to 20.32 cm (i.e. 3"–8") were used.

Discrepancies were observed and will be discussed. A method to determine the partial ^{3}He pressure from an intercomparison of measured and calculated data is demonstrated.

EXPERIMENTAL AND CALCULATIONAL METHOD

The central detector is a spherical ^{3}He proportional counter, Centronic Type SP 90, 3.2 cm in inner diameter and with a stainless steel wall 0.5 mm thick. It is affixed to a cylindrical stem 1.27 cm in diameter and extended opposite to the stem by a small cylinder of the same diameter and 0.68 cm in height (the 'nose'). The stem contains cables and perhaps ^{3}He. The specified partial ^{3}He pressure, p_{He}, for the routinely used detector is 200 kPa. For some measurements an additional detector of the same type was used but with a specified p_{He} of 20 kPa. The moderator spheres are made of polyethylene $(CH_2)_n$, and are 7.62 cm, 11.43 cm, 15.24 cm and 20.32 cm in diameter with a mass density, ρ, of 0.95 ± 0.005 g.cm^{-3}. The fittings around the detector also consist of $(CH_2)_n$, their mass density being 0.944 ± 0.003 g.cm^{-3}. Reactor filtered beams with nominal energies 24 keV and 144 keV served as the neutron source. The fluence measurements were performed with a calibrated De Pangher Precision Long Counter manufactured by 20th Century Electronic[3,4]. The overall experimental uncertainty is 4.5% for 24 keV and 4% for 144 keV measurements.

For all measurements, analogue Monte Carlo simulations were performed. A 'home-made' code described elsewhere[5] has been used. The ^{3}He proportional counter was simulated as an ideal ^{3}He sphere. Partial pressures of 172 kPa or 200 kPa — and in some cases also 228 kPa, 20 kPa and 22.8 kPa — and homogeneous moderators with mass densities of 0.95 g.cm^{-3} and 0.92 g.cm^{-3} were assumed. The statistical errors are insignificant. Systematic errors due to the database are estimated as 5%.

RESULTS

Figure 1 shows smoothed Bonner spheres (BS) responses as calculated for four sets of parameters as a function of moderator diameter. The moderator mass density is 0.95 g.cm^{-3}. The response increases with higher pressure. The peak position for the higher energy is shifted to larger diameters.

Figure 2 shows the ratio of BS responses as calculated for four sets of parameters as a function of moderator diameter for moderator densities of 0.95 g.cm^{-3} and 0.92 g.cm^{-3}, respectively. The influence of moderator density on the fluence response depends at given energies clearly on the diameter of the BS. In the case of the 11.43 cm BS

(4.5") this influence is quite small at the energies considered.

In Figure 3, the ratio of BS responses as calculated for four sets of parameters as a function of moderator diameter for p_{He} of 172 and 200 kPa is given. The data do not indicate a strong dependence on the varied parameters, i.e. on moderator diameter,

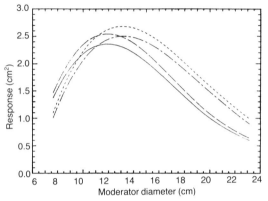

Figure 1. Smoothed responses of Bonner spheres as a function of the moderator diameter for two partial ^{3}He pressures and incident neutron energies. The moderators consists of polyethylene with a specific mass density of 0.95 g.cm^{-3}. The diameter of the central detector is 3.2 cm.

P_{He} = 200 kPa, E_n = 144 keV ------
P_{He} = 172 kPa, E_n = 144 keV – – –
P_{He} = 200 kPa, E_n = 24 keV - - -
P_{He} = 172 kPa, E_n = 24 keV ———

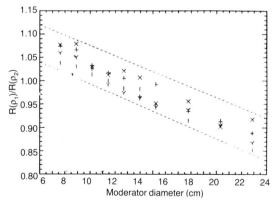

Figure 2. Influences of moderator mass density on the response of Bonner spheres as a function of the moderator diameter for two partial ^{3}He pressures and incident neutron energies. The ratios of responses for moderators with specific mass densities ρ_1 = 0.95 and ρ_2 = 0.92 g.cm^{-3} are shown. The dashed lines serve as eye guides.

+ P = 172 kPa, E_n = 24 keV
X P = 200 kPa, E_n = 24 keV
Y P = 172 kPa, E_n = 144 keV
I P = 200 kPa, E_n = 144 keV

mass density or incident energy.

Figure 4 demonstrates the method used to estimate p_{He} by linear interpolation between responses calculated for two different p_{He}. Numerical results are given in Table 1. The partial pressures estimated, p_e, by comparing measured and calculated responses are quite consistent for larger spheres. However, there is a clear indication that the calculation for the 7.62 cm sphere is inadequate, and questionable for the 11.43 cm sphere.

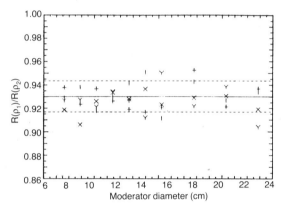

Figure 3. Influence of partial ^{3}He pressures on the response of Bonner spheres as a function of the moderator diameter for two moderator mass densities and incident neutron energies. The ratios of responses for partial ^{3}He pressures p_1 = 172 and p_2 = 200 kPa in the central detector are shown. The solid line marks the mean and the dashed lines the standard deviation.

+ ρ = 0.95 g.cm^{-3}, E_n = 24 keV
x ρ = 0.92 g.cm^{-3}, E_n = 24 keV
Y ρ = 0.95 g.cm^{-3}, E_n = 144 keV
I ρ = 0.92 g.cm^{-3}, E_n = 144 keV

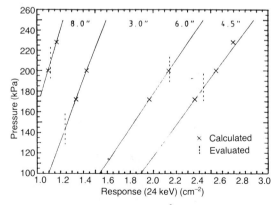

Figure 4. Evaluation of the partial ^{3}He pressure by linear interpolation between the calculated responses of Bonner spheres. The lines are determined by calculational results (crosses). The vertical dashed lines indicate the evaluated pressure at the measured response (intersection with solid lines) and its standard deviation.

Stem and 'nose' are not taken into account in the model used here for the calculations, therefore the effective moderator mass in the calculation for small spheres is larger than in actual spheres. Indeed, for small spheres (cf. Figure 2) lower mass densities would result in a decrease in the response, which qualitatively explains the overestimation of the response of the 7.62 cm sphere in calculations.

The count rates of the bare counter in the two neutron fields were used to evaluate the partial pressure using capture cross sections as specified by ENDF/B versions V or VI. The results are given in Table 2.

An unweighted mean value of 213 ± 4 kPa is found for p_e using the results for the 15.24 cm and 20.32 cm spheres for both moderator densities as shown in Table 1(a) and the bare counter results for ENDF/B V and VI as shown in Table 2. Rejecting the results for ENDF/B VI one obtains 209 ± 4.8 kPa. This rejection is arbitrary, it yields however more consistent data. The data for the 11.43 cm BS (4.5") are not used, as deviations to the simplified model are felt to be important. A more realistic model needs to be used in future calculations in order to include smaller spheres in this analysis.

A value of 10.7 for the ratio p_{low}/p_{high} was determined from bare counter measurements in the 144 keV neutron field. For the low pressure a value of about 19.5 ± 2 kPa is then found using the above determined value of 209 kPa. Again, the calculation, as shown in Table 1(b), for the smaller sphere is inadequate. However, the result for the 20.32 cm sphere agrees within somewhat more than one standard deviation.

CONCLUSIONS

From the results obtained a partial ^{3}He pressure of 209 kPa with a standard deviation of 4.8 kPa is inferred. This is somewhat higher than the specified value of 200 kPa. The calculational model is inadequate for small Bonner spheres. The details of the geometry of the central detector and possible differences in densities of the polyethylene used for moderators and internal fittings should be included. BS response generally depends strongly on the moderator mass density. In the cases studied, a linear interpolation seems justified to account for moderate changes of the partial ^{3}He pressure.

The ENDF data for the ^{3}He capture as given in version V are more consistent with our results at 24 and 144 keV than the data specified in version VI.

At present the model used in the calculations is too idealised. However, from the results obtained for large spheres it may be expected that more realistic calculations could be used in future to estimate the partial ^{3}He pressure with acceptable standard deviations.

ACKNOWLEDGEMENT

This work was partly supported by the Commission of the European Communities under contract number Bi 7-0031 C, and sponsored by EURADOS Working group IV ('Numerical Dosimetry').

Table 1. Determination of the partial ^{3}He pressures, p_e, and its standard deviation, σ, by intercomparing measured and calculated Bonner sphere responses for various diameters, D_{mod}, and two moderator densities for two incident neutron energies, E_n.

(a) *High partial ^{3}He pressure*
Moderator density: 0.95 g.cm^{-3} , 0.945 g.cm^{-3}

	D_{mod} (cm)	E_n (keV)	p_e (kPa)	σ (kPa)	p_e (kPa)	σ (kPa)
3"	7.62	24	138	± 15	143	± 14
4.5"	11.43	24	184	± 13	184	± 13
6"	15.24	24	205	± 12	201	± 12
8"	20.32	24	212	± 12	208	± 15
4.5"	7.62	144	186	± 14	187	± 14
8"	20.32	144	217	± 15	208	± 15

(b) *Low partial ^{3}He pressure*
Moderator density: 0.95 g.cm^{-3}

D_{mod} (cm)	E_n (keV)	p_e (kPa)	σ (kPa)
11.43	144	12.3	± 1.2
20.32	144	17.7	± 0.8

Table 2. Determination of the partial ^{3}He pressure, p_e, using the bare counter in two neutron fields with nominal energies, E_n, and two versions of ENDF/B data for the analysis.

ENDF/B data version:		V		VI	
E_n (keV)	R_e (10^{-3} cm^2)	P_e (kPa)	σ (kPa)	P_e (kPa)	σ (kPa)
24	4.11	213	± 10	220	± 9
144	1.48	209	± 8	234	± 10

REFERENCES

1. Bramblett, R. L., Ewing, R. I. and Bonner, T. M. *A New Type of Neutron Spectrometer*. Nucl. Instrum. Methods **9**, 1-12 (1960).

2. Mares, V., Schraube, G. and Schraube, H. *Calculated Response of a Bonner Sphere Spectrometer with ^{3}He Counter.* Nucl. Instrum. Methods **A 307**, 398-412 (1991).
3. Hunt, J. B., Harrison, K. G. and Wilson, R. *Calibration of a De Pangher Precision Long Counter and Two Neutron Survey Monitors at 21.5 keV.* Nucl. Instrum. Methods **169**, 477-482 (1980).
4. Hunt, J. B. *Calibration of a De Pangher Precision Long Counter.* Private communication (1983).
5. Siebert, B. R. L., Alberts, W. G. and Bauer, B. W. *Computational Study of Phantoms for Individual Neutron Dosimetry.* External PTB Report PTB-N-6 (PTB, Braunschweig) (1990).

DETERMINATION OF A PHOTON RESPONSE MATRIX FOR SIMULTANEOUS MEASUREMENT OF NEUTRON AND PHOTON SPECTRA IN MIXED FIELDS

T. Novotný
Institute of Radiation Dosimetry
Czechoslovak Academy of Sciences
Na Truhlářce 39/64, CS-18086 Prague 8, Czechoslovakia

Abstract — The results of the calculation of the NE-213 scintillator photon response matrix by means of EGS4 and GRESP codes are presented, and the comparison of the measured and calculated responses. The application of the response matrix in two unfolding procedures based on the Bayes' theorem and on the singular value decomposition method gives good results in the energy interval 0.2–1.6 MeV. A reliable extension of the photon response matrix to energies above ~ 2 MeV would require a precise verification in known high energy photon spectra.

INTRODUCTION

The determination of dose equivalent in mixed neutron–photon fields with sufficient accuracy requires spectrometric information because of the strong energy dependence of the fluence to dose equivalent conversion factor. A spectrometer based on a NE-213 liquid scintillation detector can simultaneously give information on fast neutrons (with energy above ~ 0.8 MeV) and photons (with energy above ~ 0.2 MeV) because of its pulse shape discrimination capabilities.

To obtain energy spectra of photons (or neutrons) the measured pulse-height spectrum from the NE-213 detector should be unfolded. For this purpose the response matrix of the detector is required which describes the relationship between the measured pulse-height spectrum and the energy spectrum of incident photons (or neutrons). The determination of the photon spectra in the energy region 0.2 – 10 MeV for the fusion reactor experiment shielding suffers from the lack of knowledge of the photon response matrices and therefore we focused our attention on their determination.

The photon response matrix cannot be obtained experimentally because it is not possible to realise a set of monoenergetic photon sources covering the required energy interval with an acceptable density. Therefore it must be calculated by means of a transport code. We chose the EGS4 code[1] which is a general code for the Monte Carlo simulation of the coupled transport of electrons and photons in an arbitrary geometry for particles with energies above several keV up to several TeV.

CALCULATION OF THE PHOTON RESPONSE MATRIX

The energy deposition of photons in the NE-213 is primarily caused by Compton electrons. Because of the linear response of the NE-213 to electrons[2] it is usual to express the light output L (or the pulse height) of the NE-213 in the so called light units (LU):

$$L = \xi(E - E_0) \quad \xi = 1 \text{ MeV}^{-1} \quad E_0 = 0.005 \text{ MeV} \tag{1}$$

It means that 1 LU is the difference of the light outputs when the increase in the electron energy is 1 MeV.

For the calculation of the energy deposited in the NE-213 by means of the EGS4 a user code has been prepared consisting of a main program, two user procedures and the rest from the EGS4 package. The main program sets up parameters of the calculation and handles input and output. The first user procedure (called the scoring) sums the energy deposited in the scintillator whereas the second one describes the geometry.

Figure 1 shows the simplified model of the Nuclear Enterprises BA1 cell encapsulating the liquid scintillator used in the calculation and Table 1 presents material data. The photon source was modelled as a parallel beam. Preliminary calculations proved that the influence of the face glass of the photomultiplier, the gap for the expansion tube and the Al cylinder covering the light pipe can be neglected.

To make a compromise between the desired accuracy of the results and the computing time needed it is necessary to choose energy cut-offs AE, AP, ECUT and PCUT[1]. Each electron interaction which produces a δ ray with total energy of at least AE, or a photon with energy of at least AP, is considered to be a discrete event. Remaining interactions are considered to be continuous. Tracing of the electrons or photons is finished when their total energy drops below ECUT or PCUT. The result calculated by the EGS4 code, i.e. distribution of the deposited energy, was transformed to the light output scale using Relation 1.

Preliminary verification of the results obtained by the EGS4 was made by comparison with the GRESP code[2]. This code is designed for the calculation of the response of the packed NE-213 scintillator and can be used only for photons with energy up to 3 MeV because of its simplified description of the electron transport. The responses for ^{137}Cs photons (661 keV) calculated by both codes are in good agreement (Figure 2). Discrepancies below 0.1 LU are probably caused by the different description of the transport of electrons.

COMPARISON OF MEASURED AND CALCULATED RESPONSE FUNCTIONS

The response function calculated using the photon transport code cannot be compared directly with the measured one because it does not include the processes of the light collection, the photoelectric conversion, and charge amplification. These processes can be taken into account by smearing of the calculated response function R(L) by means of a smearing function G(L,L'). The smeared response function R'(L) is then given by

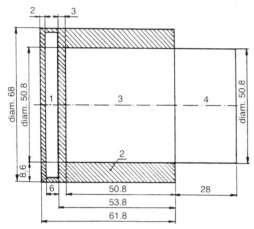

Figure 1. The simplified model of the BA1 cell used in the calculation. 1, air; 2, Al cover; 3, NE-213; 4, light pipe.

$$R'(L) = \int_0^\infty R(L') \, G(L,L') dL' \qquad (2)$$

The smearing function is assumed to be Gaussian with a full width at half maximum (FWHM) depending on L:

$$G(L,L') = \frac{2\sqrt{\ln 2}}{\sqrt{\pi} \, \text{FWHM}(L')} \exp\left(-4\ln 2 \, \frac{(L-L')^2}{\text{FWHM}^2(L')}\right) \qquad (3)$$

Let L'_i is the light output corresponding to the Compton energy of the monoenergetic photons. FWHM(L'), as a slowly varying function of L', can be approximated by FWHM(L'_i) for L' close to the L'_i. The parameter FWHM(L'_i) can then be found as the best fit of the calculated response R'(L) to the measured one in the interval bounded by the Compton edge. Values of FWHM(L_i) obtained by this procedure for a set of sources were fitted by Formula 2:

$$\text{FWHM}(L) = L \left(a^2 + \frac{b^2}{L} + \frac{c^2}{L^2} \right)^{1/2} \qquad (4)$$

where a = 0.0150, b = 0.1000, c = 0.0346 and (L) = LU. The same fits were used for the determination of the calibration constant relating the channels of the multichannel analyser to the light output L.

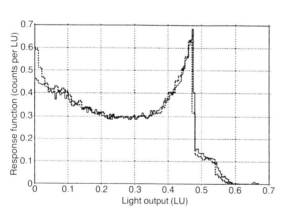

Figure 2. The response functions to ^{137}Cs photons (662 keV) calculated by EGS4 (——) and GRESP (---) codes.

Table 1. Material data used in calculation.

Material	NE-213	Light pipe	Cover	Air
Density (g.cm^{-3})	0.874	1.18	2.702	1.23.10^{-3}
Composition	H(54.8%) C(45.2%)	H(52.4%) C(47.6%)	Al(100%)	O(75.5%) N(23.2%) Ar(1.3%)

The smeared-calculated and measured response functions of ^{22}Na, ^{60}Co, ^{65}Zn and ^{137}Cs sources are in good agreement (e.g. see Figure 3); there are only small differences in the shape of the response functions below 0.2 LU. Calculated and measured efficiencies (i.e. the integrals of the response functions) for pulse heights higher than 0.2 LU are compared in the Table 2. They are in agreement within 7%.

Figure 4 shows measured and smeared-calculated response functions to the photons from a Pu-Be neutron source which emits mainly 4.44 MeV photons. The measured response function (curve A) is considerably higher than the calculated one to 4.44 MeV photons only (curve B). To remove this discrepancy we utilised the EGS4 code to model a more realistic photon output from the Pu-Be source. Due to the lack of information about the construction of the Pu-Be source we used a simplified model inside the source and its steel container assuming that the 4.44 MeV photons are uniformly emitted from the whole volume of the source. The calculated response function for such a photon source (curve C in Figure 4) is closer to the measurement. To achieve better agreement it would be necessary to include neutron-induced photons from nuclear reactions in the source materials into the model.

CALCULATED RESPONSE MATRICES AND THEIR TESTS

Two response matrices were calculated:

(i) The response matrix (dim 181×75) for photons with energy in the range 200 keV – 1.6 MeV (step width 20 keV). The energy cut-offs were set up as: ECUT = AE = 551 keV and PCUT = AP = 10 keV.
(ii) The response matrix 236×40 covering the interval of photon energies from 200 keV to 10 MeV with a step width of 250 keV. In this calculation higher energy cut-offs were set up: ECUT = AE = 1.511 MeV and PCUT = AP = 1 MeV.

To test the reliability of both matrices a series of unfolding procedures have been performed employing two unfolding codes DIFBAS[3] and UNSVD[4]. The DIFBAS is based on the Bayes' theorem and the UNSVD on the singular value decomposition method.

Using the measured pulse height spectra covering a range (0.2 – 1.6) MeV of the response matrix and code DIFBAS the photon spectra from ^{137}Cs, ^{65}Zn, ^{60}Co, ^{22}Na and a combination of ^{133}Ba, ^{137}Cs and ^{60}Co sources were unfolded. Shapes of spectra are reproduced acceptably (see Figure 5) except in the low energy region, where the disproportion is probably caused partly by the unfolding method and partly by an improper description of the real response by the calculated one.

Table 3 presents comparison of fluences in the centre of the detector derived from the spectra unfolded by the DIFBAS and the fluences calculated from the source strength. The UNSVD code gave in this case oscillating results because of a too subtle scale.

The measured pulse height spectrum and the response matrix for the range 0.2 – 10 MeV were

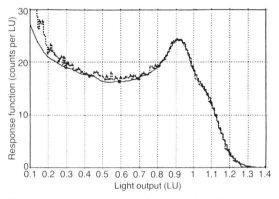

Figure 3. The smeared-calculated (——) and measured (---) response functions to photons from ^{60}Co source.

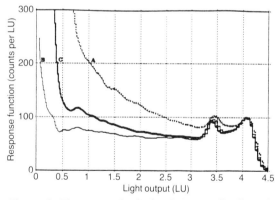

Figure 4. The smeared-calculated (curves B, C) and measured (curve A) response functions to photons from the Pu-Be source.

Table 2. Comparison of calculated and measured efficiencies for light output higher than 0.2 LU.

Source	^{137}Cs	^{22}Na	^{60}Co	^{65}Zn
ε_{cal}	0.0931	0.0870	0.1029	0.1024
ε_{mea}	0.0900	0.0817	0.1003	0.1071
$\varepsilon_{cal} / \varepsilon_{mea}$	1.034	1.065	1.026	0.956

used for the unfolding of the photon spectrum from the Pu-Be neutron source. Results of both DIFBAS and UNSVD codes (Figure 6) are in good agreement and thus we conclude that using this matrix gives numerically stable results. Unfortunately there are insufficient data about the spectrum of photons emitted from the Pu-Be source and therefore a comparison with other experimental data could not be done as in the case of low energy photon sources.

CONCLUSIONS

The EGS4 code was used for the determination of two photon response matrices of a liquid scintillator NE-213 (5 cm × 5 cm). The matrices cover the energy intervals 0.2 – 1.6 and 0.2 – 10 MeV. The comparison of both calculated and measured responses revealed that calculated responses are slightly lower for L<0.2 LU.

The application of these two calculated response matrices in unfolding procedures proved that they reproduced the real spectra satisfactorily in the energy interval 0.2 – 1.6 MeV. Tests for higher energies were limited to photons from a Pu-Be source. In this case it was not possible to draw definitive conclusions on the quality of the response matrix.

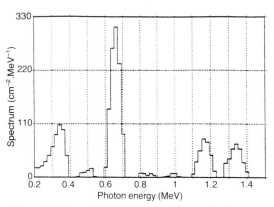

Figure 5. The photon spectrum of combined source ^{133}Ba, ^{137}Cs, ^{60}Co obtained by the unfolding of the measured response function using DIFBAS.

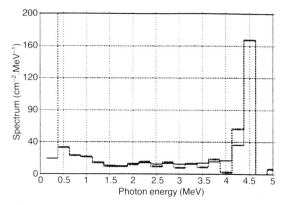

Figure 6. The photon spectrum of the Pu-Be source obtained by the unfolding of the measured response function using DIFBAS (——) and UNSVD (---).

Table 3. Comparison of measured and calculated fluences.

Source	^{137}Cs	^{65}Zn	^{60}Co	^{22}Na	^{133}Ba	^{137}Cs	^{60}Co	
Energy (MeV)	0.662	1.115	1.173 1.332	1.274	0.2–0.4	0.662	1.173	1.332
F_{th} (cm^{-2})	7.44	1.61	4.32	0.444	6.95	19.3	5.82	5.82
F_{mea} (cm^{-2})	7.76	1.51	4.01	0.412	10.49	21.1	5.97	5.84
$\frac{F_{th}}{F_{mea}}$	0.959	1.066	1.076	1.077	0.662	0.913	0.975	0.998

REFERENCES

1. Nelson, W. R., Hirayama, H. and Rogers, D. W. O. *The EGS4 Code System.* Slac-265, Stanford (December 1985).
2. Dietze, G. *Energy Calibration of NE-213 Scintillation Counters by Gamma-Rays.* IEEE Trans. Nucl. Sci. **NS-26**, 398-402 (1979).
3. Tichý, M. *Bayesian Unfolding of Pulse Height Spectra.* Jad. Energ. **36**, 265-271 (1990).
4. Půlpán, J. and Králík, M. *Application of the Singular Value Decomposition Method to the Unfolding of Neutron Spectra.* Research report 344/92 (Institute of Radiation Dosimetry) (1992).

ABSOLUTE NEUTRON FLUENCE DETERMINATION WITH A SPHERICAL PROTON RECOIL PROPORTIONAL COUNTER

M. Weyrauch and K. Knauf
Physikalisch-Technische Bundesanstalt
D-W-3300 Braunschweig, Germany

Abstract — A spherical proton recoil proportional counter is used to measure neutron fluences. This is now possible with improved precision since the response functions of that counter have recently been calculated with sufficient accuracy, taking into account gas amplification and wall effects. Neutron scattering from the counter walls is estimated using the MCNP code. Results are compared with data obtained independently with a cylindrical proton recoil proportional counter. At present the numbers obtained with the two methods differ by 5%. More experiments and detailed investigations are called for to study this discrepancy.

INTRODUCTION

Proton recoil proportional counters are widely used to determine neutron fluences and neutron spectra. Since the neutron–proton total and differential cross sections are better known than any other neutron cross section (about 1% uncertainty in the energy range below 1 MeV), proton recoil proportional counters are potentially best suited for absolute neutron fluence determinations. Hitherto, mostly cylindrical counters have been employed; they are best understood and response functions can be calculated relatively easily[1].

However, cylindrical counters do not respond isotropically. This fact prompted the construction of a spherical proton recoil proportional counter[2] with the advantage that its response is independent of the incident beam direction. It is therefore suitable for measuring neutron fields of unknown directional distribution. Unfortunately use of that counter was hampered by a lack of precise knowledge of the response functions of the counter. Indeed, semiphenomenenological response functions (which are measured in a certain proton recoil energy domain and then extrapolated to zero proton energy) are normally used[3].

Recently, significant progress has been made in quantitatively understanding the response of the spherical counter[4]: it could be shown that including the variation of gas amplification along the counting wire, as a result of the variation of the electric field strength, is sufficient to explain experimental data. Gas amplification is modelled in terms of Townsend's theory[4,5]. A comparison between calculated and measured response functions is shown in Figure 1 for an incident neutron energy of 144 keV (filtered reactor beam). In the proton energy range from about 25 keV to 100 keV the shape of the calculated and measured responses agree on the 0.1% level. The low energy part of the response (0–25 keV), which is sharply peaked at zero proton recoil energy, contains about 25% of the proton recoil events and cannot be measured, since photon induced pulses and electronic noise are contributing. However, the calculation can be employed to extrapolate the measured response into the low energy region. This fact makes absolute fluence determinations possible.

The procedure for the data analysis is presented in this paper and it is applied to neutron fluences produced at a proton accelerator. Results are compared with neutron fluences obtained independently[6] with a cylindrical counter using the procedures recommended by Parker et al[1].

ABSOLUTE FLUENCE DETERMINATION

The measured pulse height distribution $P(E_p)$ is

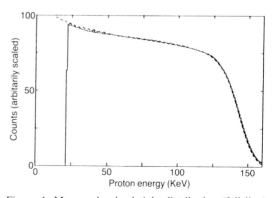

Figure 1. Measured pulse height distribution (full line) compared with the calculated response (dashed line). The SP2 counter was exposed to reactor neutrons filtered through 1372 mm Si and 60 mm Ti, which produces a nearly monoenergetic neutron line at 144 keV. Both curves normalised to unity.

related to the spectral neutron fluence $\Phi(E_n)$ by

$$P(E_p) = N \int_0^\infty R(E_p - E_n, E_n)\sigma(E_n)\Phi(E_n) \, dE_n \quad (1)$$

where E_p is the recoil proton energy and E_n the energy of the incident neutron; N denotes the total number of protons in the counter, and $\sigma(E_n)$ is the neutron–proton total cross section. The response function $R(E_p - E_n, E_n)$ is normalised, so that the total number of recoil protons Z is obtained from

$$Z = \int_0^\infty P(E_p) \, dE_p = N \int_0^\infty \sigma(E_n)\Phi(E_n) \, dE_n \quad (2)$$

One method to obtain the absolute fluence distribution is by unfolding Equation 2. This is the only way of choice if the fluence is completely unknown. If, however, the relative spectral distribution of neutrons $\lambda(E_n)$ is known (e.g. a δ function for monoenergetic neutrons), then unfolding is not necessary, and Equation 2 can be used to obtain the fluence $\Phi(E_n) = \Phi_0 \lambda(E_n)$. In this case only Φ_0 remains to be determined. Normalising $\lambda(E_n)$,

$$\int_0^\infty \lambda(E_n) \, dE_n = 1 \quad (3)$$

one obtains from Equation 2

$$\Phi_0 = \frac{Z}{N \int_0^\infty \sigma(E_n)\lambda(E_n) \, dE_n} \quad (4)$$

As an application of this procedure, the neutron fluence produced by a target at a proton accelerator is determined. Results are then compared with data obtained completely independently with a cylindrical counter. ($\lambda(E_n)$ is assumed to be known from calculations of the neutron production target system.)

At low energies the measured pulse height spectrum is spoiled by photon induced pulses and electronic noise, so that below a certain recoil proton energy the measured response is unusable. Therefore, for energies below this recoil proton energy, the calculated response function is used with the condition that measured and calculated response agree above that energy. By this means Z can be determined, and then Φ_0 can be calculated from Equation 4.

In principle, the calculated response function must take into account neutron scattering from the counter walls. We have estimated this effect using the MCNP code[7]. It is found that scattering from the walls of the particular counter used is important. However, the calculation indicates that on the average as many neutrons are scattered away from the sensitive volume as are scattered into the volume. Therefore neutron attenuation in the counter wall has not been included in the calculation.

In Figure 2 the measured and calculated responses of the spherical counter are compared. They disagree for energies below E_p^t unlike the case shown in Figure 1. This can be explained mainly by a tail of lower energy neutrons produced by the neutron production target, which is, however, not predicted by the code available to us to simulate this target. Very recent experiments[6] confirm this observation. In Table 1 absolute fluences determined with the spherical counter are compared with measurements with the cylindrical counter. These measurements have been made at three different accelerator beams. To separate off the tail of lower energy neutrons mentioned above, an adequately high E_p^t is used. The numbers obtained with the spherical counter are systematically about 5% smaller than those measured with the cyclindrical counter. The reason for this discrepancy will be studied in forthcoming experiments.

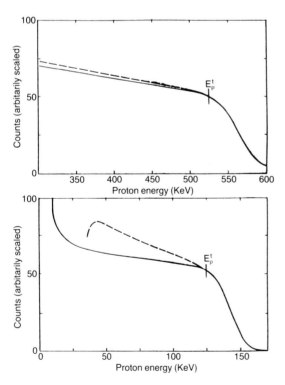

Figure 2. Measured (dashed line) and calculated (full line) recoil proton spectra for 145 keV (top) and 561 keV (bottom) neutron energy at the accelerator produced neutron fields. Statistical uncertainties, which are not shown, are smaller than 0.1%.

ESTIMATE OF UNCERTAINTIES

The determination of the relative uncertainty $s(\Phi_0)/\Phi_0$ of the absolute neutron fluence starts from Equation 2 yielding

$$\left(\frac{s(\Phi_0)}{\Phi_0}\right)^2 = \left(\frac{s(Z)}{Z}\right)^2 + \left(\frac{s(N)}{N}\right)^2 + \left(\frac{s(\hat{\sigma})}{\hat{\sigma}}\right)^2 \quad (3.1)$$

where $\sigma = \int_0^\infty \sigma(E_n)\lambda(E_n)\,dE_n$. The contributions to the uncertainty will now be studied separately. The uncertainty in the number of protons N arises from uncertainties in the determination of the counter volume and the measurements of the filling pressure p and the temperature T. If it is assumed that the radius of the counter can be determined to ±0.1 mm and that $s(p/T)/(p/T) = 10^{-3}$, it is estimated that $s(N)/N < 0.4\%$, i.e. negligible.

To determine the uncertainty of $\hat{\sigma}$ it is noted that the neutron–proton cross section is known to about 1% in the energy region of interest here; $\lambda(E_n)$ is relatively sharply peaked, but it contains lower energy events which are not sufficiently known at present. A rough estimate indicates a relative uncertainty for $\hat{\sigma}$ of up to 2%.

The uncertainty of Z is very difficult to assess at the present time, since it involves the fitting of the data to a calculation. At present it is simply not clear how well, for example, the gas amplification model (cf. first section) entering the calculation actually performs. In a previous publication[4] it was found that the shape of the response function measured with a filtered reactor beam is reproduced by the calculation on the 0.1% level (Figure 1). But more measurements and comparisons would be necessary to assess possible uncertainties arising from neutron background effects which are not fully understood at present. If these systematic influences are disgarded for the time being, and consider statistical contributions only, then the relative uncertainty for Φ_0 is about 3%.

ACKNOWLEDGEMENT

The authors would like to thank S. Guldebakke and D. Schlegel-Bickmann for providing us with the results of the cylindrical counter and S. Jetzke for showing us how to use the MCNP code. The assistance of W. Wittstock and W. Hobach in running the experiments is gratefully acknowledged. This work is supported by the Commission of the European Community (CEC) under contract BI7-0031-C.

Table 1. Absolute fluences (in units of $10^9 sr^{-1}$). The measurements were made with an SP2 counter filled with 98% H_2 and 2% CH_4 at a pressure of 0.588 MPa. See the main text concerning a discussion of the quoted uncertainties.

Measurement number	Neutron energy (keV)	Cylindrical counter (keV)	Spherical counter (keV)	Difference (%)
1	145.5 ± 3	145 ± 7	142 ± 5	−2
2	246.9 ± 5	123 ± 5	117 ± 5	−5
3	560.9 ± 11	552 ± 22	530 ± 16	−4

REFERENCES

1. Parker, J. B., White, P. H. and Webster, R. J. *The Interpretation of Recoil Proton Spectra*. Nucl. Instrum. Methods **23**, 61 (1963).
2. Benjamin, P. W., Kemshall, C. D. and Redfearn, J. *A High Resolution Spherical Proportional Counter*. AWRE Report NR1/64 (1964). Also in Nucl. Instrum. Methods **59**, 77 (1968).
3. Birch, R., Peaple, L. H. J. and Delafield, H. J. *Measurement of Neutron Spectra with Hydrogen Proportional Counters, Part I: Spectrometry System and Calibration*. AERE-R 11397 (AERE, Harwell) (1984).
4. Weise, K., Weyrauch, M. and Knauf, K. *Neutron Response of a Spherical Proton Recoil Proportional Counter*. Nucl. Instrum. Methods **A 309**, 287 (1991).
5. Nasser, E. *Fundamentals of Gaseous Ionization and Plasma Electronics* (New York: Wiley-Interscience) (1971).
6. Guldbakke, S. and Schlegel-Bickmann, D. Private communication.
7. *MCNP — A General Monte Carlo Code for Neutron and Photon Transport-Version 3A*. Ed. J. F. Briesmeister. LA-7396-M, Rev. 2 (Los Alamos National Laboratory, Los Alamos, NM, USA) (1986).

MEASUREMENT OF HIGH ENERGY NEUTRON FLUENCE WITH SCINTILLATION DETECTOR AND PROTON RECOIL TELESCOPE

R. Nolte, H. Schuhmacher, H. J. Brede and U. J. Schrewe
Physikalisch-Technische Bundesanstalt (PTB)
W-3300 Braunschweig, Federal Republic of Germany

Abstract — The consistency of spectral neutron fluence determination with a liquid scintillation detector using time-of-flight techniques and fluence measurements with a proton recoil telescope is investigated in the neutron energy region between 20 and 70 MeV. If the full detector response including contributions from neutron reactions with carbon nuclei is used for the fluence determination with the scintillation detector, the results deviate up to 15% from measurements with a proton recoil telescope which is based exclusively on n-p scattering. Good agreement has been found if only events due to n-p scattering are used for the analysis of scintillation detector measurements.

INTRODUCTION

Absolute measurements of neutron fluence are an essential prerequisite of dosimetric investigations. For neutron energies above 1 MeV it has become common practice to use proton recoil telescopes (PRT) as reference devices for fluence determination in monoenergetic neutron beams[1]. Above 20 MeV, 'monoenergetic' neutron sources produce a spectrum characterised by a prominent high energy peak and a background extending to very low energies. Additional information about the spectral fluence of the neutron beam is therefore required. If the neutron source has a suitably pulsed timing structure time-of-flight (TOF) spectroscopy with liquid scintillation detectors can be used.

The application of two different detection systems obviously raises the question of the consistency and uncertainty of their results. The accuracy of PRT fluence measurements is essentially limited by the uncertainty of the angular differential n-p scattering cross section at backward angles in the CM system. In the case of scintillation detectors the angular n-p cross section must be known over a larger energy range because of multiple scattering contributions in the scintillation volume. In addition to n-p scattering, neutron reactions with carbon nuclei also contribute to the response.

By a careful modelling of the detection processes in both detector systems[2,3] it has been demonstrated that fluences measured with PRTs and scintillation detectors in monoenergetic neutron fields below 16 MeV agree with 2%[4]. This situation changes drastically above 20 MeV. Here, the contribution of multiparticle break-up reactions to the scintillation detector response rapidly increases[5]. Precise nuclear data on such reactions are very scarce. Although empirical adjustments of cross sections have been made to better reproduce measured detection efficiencies for fixed detection thresholds, it has not been possible to describe accurately the scintillator response in general.

MEASUREMENTS

The neutron beams were produced by bombarding a two mm thick Be target with 50.0 and 71.3 MeV protons from the Philips cyclotron of the Paul-Scherrer-Institute (PSI) at Villigen, Switzerland. A PRT of the Los Alamos type[1] and an NE213 scintillation detector 5.1 cm in diameter and 10.2 cm in length were used to determine the neutron fluence.

The measurements with the PRT were carried out using the proton beam with the normal cyclotron repetition frequency of about 15 MHz. The average beam current during these measurements was 2 µA. The PRT was positioned at a distance of 7.789 m (reference position) downstream from the Be target.

For the TOF measurements with the scintillation detector, the proton beam repetition frequency was reduced to about 1 MHz by means of an external electrostatic deflector in the beam-line system. The average beam current in this case was about 10 nA. The NE213 detector was positioned at a distance of about 10 m from the Be target. Pulse-shape discrimination techniques were used to separate neutron from gamma ray events in the detector.

An important aspect for the comparison of the two measurements is the normalisation to the same neutron yield of the Be target. In principle, the beam charge can be used for this purpose. However, a strong, time-dependent leakage current contribution was found during the scintillation detector measurements in which the beam current was reduced by almost three orders of magnitude compared with the PRT measurements. The count rate of a gain stabilised 0.5 mm thick NE102 detector which was positioned behind the collimator exit and operated in transmission mode therefore served to monitor both measurements. For the normalisation to the beam charge on the Be target, the average ratio of the number of NE102 events to the beam charge determined during the PRT measurements was used. The uncertainty of the monitoring procedure is estimated to be less than 2%.

All scintillation detector data were related to the reference position assuming a $1/r^2$ dependence of the neutron fluence. Dead-time losses, beam current due to incomplete suppression of unwanted proton beam pulses by the external deflector system, neutron absorption in the detector housing and in the air between scintillation detector and PRT as well as non-perfect timing due to amplitude walk effects in the detector were corrected for.

DATA ANALYSIS

The analysis of the PRT data has been described in detail elsewhere[6]. In the present work angular differential n-p scattering cross sections derived from a recent phase-shift analysis (VL35)[7], which used most of the experimental data available to date has been used. In the PRT used for the present experiment only n-p scattering events at angles greater than 148° in the CM system contribute to the response. The uncertainty introduced in the determination of neutron fluences by the n-p cross section was estimated by comparing experimental n-p scattering data to the predictions of the phase shift analysis[6]. In the present case this uncertainty amounts to 4.7% and 5.7% for the measurements with the 50.0 MeV and the 71.3 MeV proton beam, respectively.

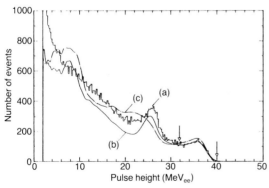

Figure 1. Response spectrum of the scintillation detector for incident neutrons with an energy of 60.59±0.78 MeV: (a) experimental spectrum, (b) calculated with the modified SCINFUL[10] code and normalised to the experimental spectrum in the region indicated by the arrows, (c) calculated with the KENT01[8,9] code for the same neutron fluence. The pulse height is indicated in units of electron equivalent energy.

The scintillator data were analysed with two Monte Carlo codes. The KENT01 code[8,9] is widely used for the calculation of detector efficiencies. The SCINFUL code[10] contains a more detailed description of the nuclear physics than older codes like KENT01 and reproduces experimental and theoretical charged particle emission spectra for carbon with an acceptable overall accuracy[11]. SCINFUL was tested extensively[10,11] for neutron energies below 20 MeV and found to be in agreement with experimental detector efficiencies and response spectra to monoenergetic neutrons. However only a few tests above 20 MeV were reported[10].

In order to obtain results comparable with the PRT measurements, the n-p scattering cross section data sets in the SCINFUL code were replaced by the cross sections which were also used in the PRT analysis. At 180° in the CM system the differences between the original SCINFUL data set and the results of the phase shift analysis amount to 20% at the most. In the KENT01 code the original nuclear data set was kept. In both codes the original light output functions for protons and deuterons were replaced by experimental ones determined from the measured pulse-height spectra. For the alpha particle extrapolations from lower energies were used[2].

RESULTS

Figure 1 shows an experimental response spectrum for neutrons selected from an energy interval between 59.81 and 61.37 MeV with the TOF technique. The region at high pulse heights is predominantly due to n-p scattering events. The adjacent prominent edge is caused by protons from the $^{12}C(n, p\ x)$ reaction and by deuterons from the $^{12}C(n, d\ x)$ reaction. At lower pulse heights the $^{12}C(n, \alpha\ x)$, $^{12}C(n, t\ x)$ and $^{12}C(n, {}^3He\ x)$ reactions also contribute. The experimental spectrum is compared with the predictions of KENT01 and the modified SCINFUL code. The spectrum calculated with SCINFUL was normalised to the experimental one in the region indicated by the arrows in Figure 1. From the normalisation factor, the incident neutron fluence in the selected energy interval can be determined. The second spectrum was calculated using the KENT01 code and the same neutron fluence. It is obvious that neither code reproduces the experimental spectrum at lower pulse heights, i.e. the events due to neutron reactions on carbon. The SCINFUL code accurately predicts the shape of the n-p part of the response whereas the KENT01 code overestimates this contribution by about 15% due to the inadequate n-p cross section used in this code.

For a 'conventional' determination of the

spectral neutron fluence from TOF spectra, detector efficiencies were calculated for various fixed detection thresholds with the modified SCINFUL code and the KENT01 code. In Figure 2 the results obtained for a detection threshold of 2 MeV$_{ee}$ (1 MeV$_{ee}$ denotes the light yield produced by an electron of 1 MeV in NE213) are depicted. The predictions from both codes deviate by about 15% in the peak region. Figure 3 shows the differences in spectral fluences evaluated for higher detection thresholds relative to the fluence determined for a threshold of 2 MeV$_{ee}$. The discrepancies of ±8% clearly demonstrate the threshold dependence of the predicted neutron fluence caused by an inadequate description of the carbon-dominated part of the detector response. Not only the absolute normalisation but also the relative shape of the measured spectral fluence is clearly affected.

Alternatively, spectral fluences were derived by normalising calculated pulse-height spectra for monoenergetic neutrons to experimental ones in the same way as in Figure 1. The neutron energies were selected by placing small adjacent windows in the TOF spectra. The normalisation was always restricted to that part of detector response which is affected by n-p scattering. Hence the uncertainty of the normalisation is mainly determined by the accuracy of the n-p scattering cross section. The spectral fluences obtained in this way are also depicted in Figure 2. In Table 1 a comparison of the fluences Φ_p integrated over the peak region of the neutron spectrum and measured with the scintillation detector and the PRT is shown. For the scintillator measurements, the uncertainty Δ_1 includes all contributions except those for the n-p cross section, i.e. those from statistics, the distance to the target, the hydrogen content of the detector, corrections for neutron absorption and dead-time losses, neutron energy determination and the uncertainty of the current integration. The uncertainties of the PRT measurements have been discussed elsewhere[6]. The total uncertainties are denoted by Δ_2.

As expected, the fluences measured with the two detectors are consistent with respect to the uncertainties Δ_1. This is essentially due to the analysis procedure, which avoids any problems with unknown neutron reactions with carbon nuclei.

CONCLUSION

A comparison of the PRT and NE213 scintillation detector measurements has shown that the uncertainty in determining the absolute and spectral neutron fluence of high energy neutron beams with an NE213 detector can rise to 15% if an

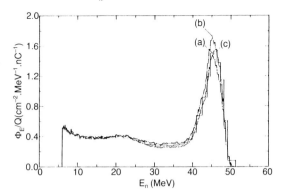

Figure 2. Spectral neutron fluence normalised to the beam charge and measured at a distance of 7.789 m from the Be target for a proton energy of 50.0 MeV: (a) analysis restricted to the n-p part of the scintillation detector response, (b) detector efficiency calculated with the modified SCINFUL[10] code for a detection threshold of 2 MeV$_{ee}$, (c) efficiency calculated with the KENT01[8,9] code for a threshold of 2 MeV$_{ee}$.

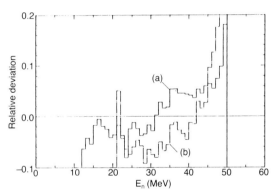

Figure 3. Relative deviation of the spectral neutron fluence evaluated for detection thresholds of (a) 6 MeV$_{ee}$, and (b) 12 MeV$_{ee}$, from the spectral fluence determined for a 2 MeV$_{ee}$ detection threshold. The efficiencies were calculated with the modified SCINFUL[10] code.

Table 1. Integrated neutron fluences (Φ_p/Q), measured with the PRT and the NE213 detector and normalised to the beam charge. The analysis of the scintillator data used the modified SCINFUL code and was restricted to the n-p part of detector response. The values refer to a distance of 7.789 m from the Be target. The uncertainty (1 SD) without the contribution from the normalisation to the n-p cross section is denoted by Δ_1, the total uncertainty (1 SD) by Δ_2.

Energy interval	PRT (cm^{-2}.nC^{-1})			NE213 (cm^{-2}.nC^{-1})		
(MeV)	Φ_p/Q	Δ_1	Δ_2	Φ_p/Q	Δ_1	Δ_2
38 - 50	10.5	0.5	0.7	10.1	0.6	0.8
58 - 75	10.3	0.5	0.8	10.3	0.7	0.9

analysis technique with fixed detection thresholds is used. Not only the absolute fluence measurement but also the measured relative shape of the spectral fluence is affected. A restriction of the analysis to a region of the scintillator response which is exclusively determined by n-p scattering can reduce the uncertainty to about 8% for neutron energies up to 70 MeV.

ACKNOWLEDGEMENT

This work has been partly supported by the Commission of the European Communities, contract No. B17-0030-C.

REFERENCES

1. Bame, S. J., Haddad, E., Perry, J. E. and Smith, R. K. *Absolute Determination of Monoenergetic Neutron Flux in the Energy Range 1 to 30 MeV.* Rev. Sci. Instrum. **28**, 997 (1957).
2. Dietze, G. and Klein, H. *NRESP4 and NEFF4.* PTB-Report PTB-ND-22 (1982).
3. Siebert, B. R. L., Brede, H. J. and Lesiecki, H. *Corrections and Uncertainties for Neutron Fluence Measurements with Proton Recoil Telescopes in Anisotropic Fields.* Nucl. Instrum. Methods **A235**, 542 (1985).
4. Börker, G., Böttger, R., Brede, H. J., Klein, H., Mannhart, W. and Siebert, B. R. L. *The Differential Neutron Scattering Cross Section of Oxygen between 6 and 15 MeV.* In: Nuclear Data for Science and Technology, Ed. S. Igarasi (Tokyo: Saikon Publ. Co.) pp. 193-196 (1988).
5. del Guerra, A. *A Compilation of n-p and n-C Cross Sections and their Use in a Monte Carlo Program to Calculate the Detection Efficiency in Plastic Scintillator in the Energy Range 1-300 MeV.* Nucl. Instrum. Methods **135**, 337 (1976).
6. Schuhmacher, H., Siebert, B. R. L. and Brede, H. J. *Measurement of Neutron Fluence for Energies between 20 MeV and 65 MeV using a Proton Recoil Telescope.* Contribution to the NEANDC Specialists' Meeting on Neutron Cross Section Standards for the Energy Region above 20 MeV, Uppsala 1991. To be published as a NEANDC report.
7. Arndt, R. A., Hyslop, J. S. and Roper, L. D. *Nucleon–Nucleon Partial Wave Analysis to 1100 MeV.* Phys. Rev. **D35**, 128 (1987) and Arndt, R. A., private communication.
8. Cecil, R. A., Anderson, B. D. and Madey, R. *Improved Predictions of Neutron Detection Efficiency for Hydrocarbon Scintillators from 1 MeV to about 300 MeV.* Nucl. Instrum. Methods **161**, 439 (1979).
9. Sailor, W. C., Byrd, R. C. and Yariv, Y. *Calculation of the Pulse-Height Response of Organic Scintillators for Neutron Energies $28 < E_n < 492$ MeV.* Nucl. Instrum. Methods **A277**, 599 (1989).
10. Dickens, J. K. *SCINFUL: A Monte Carlo Based Computer Program to Determine a Scintillator Full Energy Response to Neutron Detection for E_n between 0.1 and 80 MeV: Program Development and Comparison of Program Predictions with Experimental Data.* Oak Ridge National Laboratory Report ORNL 6463 (1988).
11. Dickens, J. K. *Computed Secondary-Particle Energy Spectra Following Nonelastic Neutron Interactions with ^{12}C for E_n between 15 and 60 MeV: Comparison of Results from two Calculational Methods.* Oak Ridge National Laboratory Report ORNL/TM-11812 (1991).

NEUTRON MEASUREMENT INTERCOMPARISONS SPONSORED BY CCEMRI*, SECTION III (NEUTRON MEASUREMENTS)

R. S. Caswell† and V. E. Lewis‡
†National Institute of Standards and Technology#
Gaithersburg, MD 20899, USA
‡National Physical Laboratory, Teddington, Middlesex TW11 0LW, UK

Abstract — A large number of neutron measurement comparisons has been carried out under the sponsorship of Section III (Neutron Measurements) of the Consultative Committee for Measurement Standards of Ionizing Radiations (CCEMRI) since Section III became active in 1961. These have included a series of comparisons of radionuclide neutron source emission rate, neutron fluence, and neutron dose measurements with the participation of laboratories worldwide. This paper discusses these comparisons and what has been learned from them. Future plans of Section III and neutron measurement services available from BIPM to national laboratories are also given.

INTRODUCTION

Measurement intercomparisons are carried out for several reasons: (1) to achieve a consistent international measurement system, (2) to determine systematic errors, (3) to study changes in a standard with time, (4) to compare measurement methods or instruments, (5) to compare corrections as evaluated in different laboratories, (6) to evaluate the state-of-the-art in measurement, (7) to provide confidence in standards, and finally (8) to encourage scientists to make better measurements. Since the formation of CCEMRI in 1961, Section III has sponsored a major series of comparisons of radionuclide neutron source emission rate, neutron fluence, and neutron dose measurements with the participation of laboratories worldwide. The purpose of this paper is to summarise a broad ranging series of neutron measurement intercomparisons carried out by many world laboratories under the auspices of CCEMRI Section III and its predecessor organisation, the Neutron Working Group. The Neutron Working Group was formed in 1961 under the chairmanship of Klaus W. Geiger, and became CCEMRI Section III in 1969.

RESULTS

Comparison of measurements of the neutron emission rate of a Ra-Be (α, n) source, organised by BIPM, 1962-1965[1]

The Ra-Be (α, n) neutron source, No. 200-1, of the National Research Council of Canada was circulated to 10 laboratories by BIPM to carry out a comparison of measurements of neutron emission rates. Most laboratories used a manganese bath as a method for calibration of the laboratory standard source, although other methods were used as well by some laboratories: activation of gold or indium foils in a water tank, or comparison with a ^{3}H(d, n)^{4}He source in a graphite pile using boron counters. The maximum spread among the 11 results (an earlier 1959 result being included) is about 3%, slightly improved from three earlier comparisons. The intercomparison pointed up the uncertainty in the correction for the absorption of fast neutrons by sulphur and oxygen in the manganese sulphate bath for which the laboratory corrections varied from 2% to 3%.

International comparison of determinations of the neutron emission rate of a ^{252}Cf neutron source, organised by NPL, 1979-1984

A ^{252}Cf neutron source was furnished by the NBS as the principal intercomparison source. The neutron emission rate of the source was about 4.5×10^7 s^{-1} in May 1978 and fell to about 10^7 s^{-1} by February 1984. To permit the participation of additional laboratories, a second ^{252}Cf neutron source was provided by the NPL, and a third by INEL. Fourteen laboratories participated, and four different methods of absolute emission rate determination were used. The manganese bath method predominated.

As a test the results were normalised to standard cross sections (the capture cross sections of H, Mn, S, and Au), and to a common basis for corrections (shown in Figure 1)[2]. This produced a real improvement even though it was incomplete.

---

* Comité Consultatif pour les Étalons de Mesure des Rayonnements Ionisants, Bureau International des Poids et Mesures, Sèvres, France. R.S.C., Chairman, CCEMRI Section III, 1969-1989. V.E.L., Chairman, CCEMRI Section III, 1989-present.
\# Ionizing Radiation Division, Physics Laboratory, Technology Administration, US Department of Commerce, Contribution of the National Institute of Standards and Technology. Not subject to copyright.

Excluding two apparently discrepant results, a standard deviation of the residuals of 0.57% was found, indicating a considerable advance in the measurement of neutron source emission rates since the previous intercomparison.

Intercomparison of thermal neutron flux density (fluence rate), organised by NPL, 1966-1968

At the time of the comparison, many laboratories possessed moderating geometries containing neutron sources, usually (α, n), in which the neutron flux density in a certain region served as a standard. The quantity compared was $n_{th}v_o$ where n_{th} is the neutron density below cadmium cut-off energy and $v_o = 2200$ m.s^{-1}. These 'standards' were intercompared by activation of gold foils which could then be airmailed to other laboratories for counting. Two laboratories, which did not possess

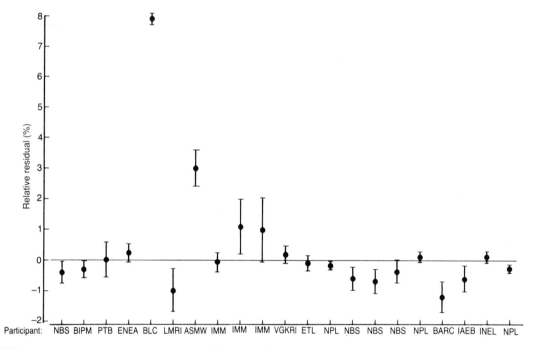

Figure 1. Results of the ^{252}Cf neutron source emission rate comparison, normalised to standard cross sections and to a common basis for corrections.

Key to participating laboratories

ASMW	Office for Standardization, Measurements, and Testing, Berlin, Germany
BARC	Bhabha Atomic Research Centre, Bombay, India
CBNM	Central Bureau for Nuclear Measurements (EURATOM), Geel, Belgium
BIPM	Bureau International des Poids et Mesures, Sèvres, France
BLC	Centre for Studies, Bruyères-le-Châtel, France
ENEA	National Committee for Nuclear Energy and Its Alternatives, Rome, Italy
ENDIP	European Neutron Dosimetry Intercomparison Project (H. Schraube and J. Zoetelief)
ETL	Electrotechnical Laboratory, Tokyo and Ibaraki, Japan
IAEB/AEI	Institute of Atomic Energy, Beijing, China
IMM	Institute of Metrology Mendeleev, Leningrad/St. Petersburg, Russia
INEL	Idaho National Engineering Laboratory, Idaho Falls, USA
IRK	Institute for Radium Research and Nuclear Physics, Vienna, Austria
LMRI	Laboratory for the Metrology of Ionizing Radiations, Saclay, France
MRC	Medical Research Council, Hammersmith Hospital, London, UK
NBS	National Bureau of Standards (now National Institute of Standards and Technology), Gaithersburg, MD, USA
NPL	National Physical Laboratory, Teddington, UK
NRC	National Research Council, Ottawa, Canada
PTB	Physikalisch-Technische Bundesanstalt, Braunschweig, Germany
VGKRI	V. G. Khlopin Radium Institute, Leningrad/St. Petersburg, Russia
WGH	Western General Hospital, Edinburgh, UK.

a standard of their own, made direct measurements of the NPL standard.

The results of the intercomparison were analysed by two different methods[3,4]. Both methods gave similar values. All individual values were within ±2% of the adjusted value, the absolute average deviation being about 1%. The thermal neutron flux density standards described here have now been largely replaced by standards in graphite thermal columns using reactor or accelerator neutron sources.

Comparison of measurements of 0.25, 0.565, 2.2, 2.5, and 14.8 MeV fluences, organised by BIPM, 1973-1978[5]

A Bonner sphere (203 mm diameter) and ^{3}He proportional counter were used at 250 and 565 keV, and the same Bonner sphere at 2.2 and 2.5 MeV. At 14.8 MeV a ^{238}U fission counter was used together with iron foil activation based on the ^{56}Fe (n,p)^{56}Mn reaction. It was found that the Bonner sphere transfer instrument response was badly affected by lower energy scattered neutrons. The fission chamber was much less affected by these, but its much lower sensitivity restricted its use to the highest energy and positions close to the target. The ^{3}He counter was sufficiently sensitive for only the lowest energies. Iron foil activation was successful.

Comparison of measurements of 144 and 565 keV neutron fluences using indium activation, organised by NPL, 1984/1985[6]

Indium foils were irradiated in measured fluences and the β rate of the induced activities due to the ^{115}In(n,γ)^{116}In^m reaction (half-life 0.9 h) were measured using a 4πβ proportional counter circulated to participants. This work showed the necessity for corrections for the effects of target-scattered neutrons. The transfer method is suitable for measurements close to the target, but at greater distances the statistical accuracy is much worse than that for Bonner spheres.

Comparison of measurements of 2.5 and 5.0 MeV fluences, using indium activation, organised by CBNM, 1981

Standard indium discs were irradiated in measured fluences and the gamma ray of the activity induced by the ^{115}In(n,n') ^{115}In^m reaction (half-life 4.5 h) was measured by the participating laboratories[7]. Calibrated ^{51}Cr sources were circulated by CBNM as gamma activity standards.

The uncertainties due to the transfer method did not increase the overall uncertainties significantly.

The agreement is generally good apart from one measurement made under adverse conditions. An attempt to use this transfer method at 14.8 MeV failed due to its much higher sensitivity to scattered neutrons.

Comparison of measurements of d + T neutron fluence and energy using niobium and zirconium activation, organised by NPL, 1981

Small ingots of niobium and zirconium were irradiated in d + T neutron fields and mailed to NPL for comparison of the induced gamma activities due to the ^{93}Nb(n,2n)^{92}Nb^m reaction (half-life 244 h) and the ^{90}Zr(n,2n)^{89}Zr^m reaction (half-life 78 h)[8]. Calculation of the quotient of the zirconium and niobium activities enabled the mean neutron energies to be measured. The results are shown in Figures 2 and 3.

The uncertainties due to the transfer method did not increase the overall uncertainties significantly. Agreement was generally good, and the few discrepancies are now understood. This method is excellent for comparing both d + T neutron fluence and energy. It forms the basis of an international postal secondary standard.

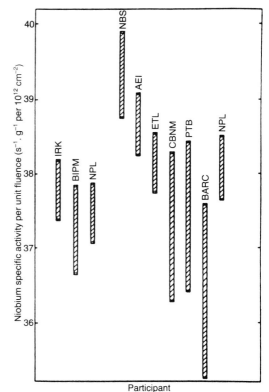

Figure 2. Summary of the results for the intercomparison of d + T neutron fluence by the activation of niobium.

Comparison of measurements of 0.25, 0.565, 2.2, 2.5 and 14.8 MeV fluences using fission chambers, organised by Harwell Laboratory, 1983-1988

The sensitivities of two large, specially-built, fission chambers, one containing ^{235}U and the other ^{238}U deposits, were measured at all five energies and the three highest energies, respectively[9]. At two laboratories pulsed 'white' neutron fields were used. The superior sensitivities enabled time-of-flight techniques to be used to correct for room-scattered neutrons. A summary of the results is given in Figure 4, in terms of differences between measured sensitivities of the ^{235}U chamber and the weighted mean values.

The fluence measurements were consistent within the estimated uncertainties. The uncertainty contribution from the transfer method was generally less than that from the fluence measurement. The fission chambers are suitable for use over a wide range of energy.

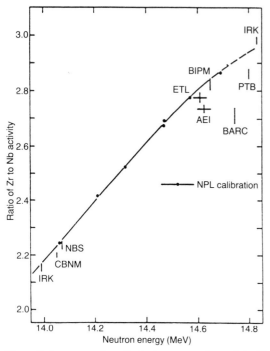

Figure 3. Summary of results for intercomparison of d + T neutron energy by the simultaneous activation of zirconium and niobium.

Comparison of measurements of 2.5 and 14.7 MeV fluences using two Bonner spheres, 1986-1990

The sensitivities of two Bonner spheres (89 and 241 mm diameters) were measured at 2.5 and 14.7 MeV[10]. This was a comparison of the type of measurement employed for the calibration of a neutron dosemeter. Corrections for target and air-plus-room scatter were made using both a shadow cone technique and a non-linear model (a negative exponential times a second-order polynomial in distance). The results of the comparison are shown in Figure 5 in terms of deviations from the weighted means.

Agreement was good where the size of sphere

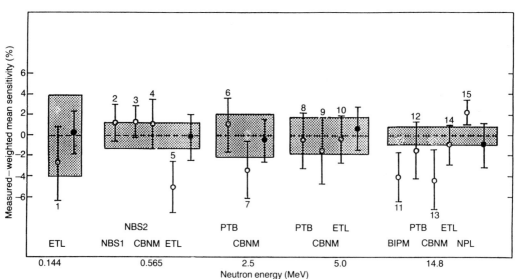

Figure 4. Differences between measured sensitivities of the ^{235}U chamber and the weighted mean values. The solid circles are values derived using the ENDF/B-VI evaluated fission cross section.

was appropriate for the energy concerned. The poor agreement for the small sphere at 14.7 MeV was due to its higher sensitivity to scattered neutrons. The size of the calibration hall affected the accuracy of the method.

Comparison of measurements of absorbed dose in collimated beam, organised by NPL, 1983

Visiting teams from participating laboratories made measurements in the NPL collimated 14.7 MeV neutron beam at positions free-in-air and at three depths inside a water phantom[11]. Tissue-equivalent plastic ionisation chambers were used along with Geiger–Müller counters or argon-filled magnesium ionisation chambers as the 'neutron-insensitive' devices. An example of the results is given in Figure 6, where the quotient of gamma to neutron absorbed dose, a difficult quantity to measure, is shown for two measurement methods: TE + GM counter, and TE + Mg/Ar chamber.

Agreement was excellent for measurements made with TE chambers plus GM counters, but variations in the results obtained using TE plus Mg/Ar chambers suggested anomalous variations in the latter's k_u value (i.e. neutron sensitivity).

Comparison of absorbed dose measurements using circulated set of transfer instrumentation, organised by BIPM, 1985-1988

Measurements were made in free-in-air conditions in the fields of participating laboratories using tissue-equivalent plastic and argon-filled magnesium ionisation chambers, and Geiger-Muller counters circulated by BIPM and corresponding 'local' dosemeters. The values obtained using both sets of instruments were compared[12]. Results are presented in another paper at this meeting[13].

Agreement was generally good for the combination of TE chamber and GM counter, but could vary markedly when the Mg/Ar chamber was used instead of the GM counter. However, no significant trends were observed.

FUTURE PLANS

Comparison of measurements of 25 keV neutron fluence using a set of three Bonner spheres is planned to start in 1992. Four types of neutron

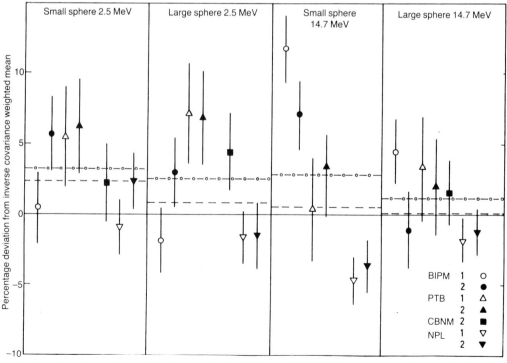

Figure 5. Deviations from the weighted means for the 2.5 and 14.7 MeV fluence comparisons using two Bonner spheres. The open symbols are values where the scatter correction was determined using shadow cones. The dash-circle horizontal lines show the positions of the unweighted means. The dashed lines show the positions of the inverse–variance weighted means without correlations.

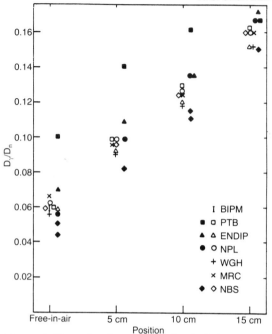

Figure 6. Example of results of absorbed dose comparison organised by NPL. Quotient of gamma to neutron absorbed dose measured using TE + GM counter (open symbols); TE + Mg/Ar chamber (solid symbols).

field will be used: filtered reactor beams, antimony-beryllium sources, the $^{45}Sc(p,n)^{45}Ti$, and the $^7Li(p,n)$ ^7Be reactions.

Comparison of thermal neutron fluence measurements proposed for after 1993.

For comparison of spectral fluence measurements it is planned to circulate Am-Be and Am-B sources.

For development of neutron spectrometry system at BIPM work is in progress on a liquid scintillator. It is hoped to add a proton recoil proportional counter and perhaps Bonner spheres in order to extend the energy range.

NEUTRON MEASUREMENT SERVICES OFFERED BY BIPM

BIPM offers the following services to national standards laboratories of countries adhering to the Convention of the Meter: (1) Calibration of neutron source emission rates. This service is available both to laboratories entering the field and to bilateral comparisons with national laboratories. (2) Loan of transfer instruments for measurement of neutron fluence rate and/or kerma rate. (3) Calibration of neutron detectors in the 2.50 and 14.65 MeV neutron beams.

REFERENCES

1. Naggiar, V. *Rapport sur la Comparaison Internationale de la Mesure du Taux d'Emission de la Source de Neutrons Ra-Be(a,n) du Conseil National de Recherches n° 200-1 par la Méthod de Ralentissement des Neutrons dans une Solution de Sulfate de Manganèse*. Recueil de Travaux du BIPM, Vol. 1 (1966-1967).
2. Axton, E. J. *Intercomparison of Neutron-Source Emission Rates (1979-1984)*. Metrologia **23** 129-144 (1987).
3. Axton, E. J. *Results of the Intercomparisons of the Thermal Neutron Flux Density Unit (1966-1968)*. Metrologia **6**, 25 (1970).
4. Murphey, W. M. and Caswell, R. S. *Analysis of Results of the Bureau International des Poids et Mesures Thermal Neutron Flux Density Intercomparison*. Metrologia **6**, 111 (1970).
5. Huynh, V. D. *International Comparison of Flux Density Measurements for Monoenergetic Fast Neutrons*. Metrologia **16**, 31-49 (1980).
6. Ryves, T. B. *International Fluence-Rate Intercomparison for 144 and 565 keV Neutrons*. Metrologia **24**, 27-37 (1987).
7. Liskien, H. *International Fluence-Rate Intercomparison for 2.5 and 5.0 MeV Neutrons*. Metrologia **20**, 55-59 (1984).
8. Lewis, V. E. *International Intercomparison of d + T Neutron Fluence and Energy Using Niobium and Zirconium Activation*. Metrologia **20**, 49-53 (1984).
9. Gayther, D. B. *International Intercomparison of Fast Neutron Fluence-Rate Measurements Using Fission Chamber Transfer Instruments*. Metrologia **27**, 221-231 (1990).
10. Axton, E. J. Report 91-4 to Section III, CCEMRI, 1991 (to be published).
11. Lewis, V. E. *Neutron Dosimetry Intercomparison at the National Physical Laboratory*. NPL Report RS(EXT) 79 (1983).
12. Huynh, V. D. *Report on BIPM Neutron Dosimetry Intercomparison*. Rapport BIPM-88/5 (1988).
13. Huynh, V. D. *BIPM Neutron Dosimetry Comparison Based on the Circulation of a Set of Transfer Instruments*. Radiat. Prot. Dosim. **44**(1-4), xxx (1992) (This issue).

BIPM NEUTRON DOSIMETRY COMPARISON BASED ON THE CIRCULATION OF A SET OF TRANSFER INSTRUMENTS

V. D. Huynh
Bureau International des Poids et Mesures
Pavillon de Breteuil, F-92312 Sèvres Cedex, France

Abstract — An international neutron dosimetry comparison has been organised by the Bureau International des Poids et Mesures (BIPM). The transfer instruments, supplied by the BIPM and consisting of two tissue-equivalent (TE) and one magnesium–argon ionisation chambers, and an energy compensated Geiger–Müller counter (GM), were sent to seven laboratories in turn. Stability checks were made at the BIPM, using a ^{60}Co field, before and after the use of the instruments at each participating laboratory. The purpose of the comparison was to compare, under specified conditions, the kerma values, in A-150 plastic, for the neutron and gamma components, K_N and K_G, per unit of reference monitor in the local (d+T) mixed field, when measured with the BIPM equipment and with the local dosimetry systems. The results of the comparison are given and discussed.

INTRODUCTION

This comparison started at the end of 1985 and was completed in 1987 with a total number of eight participating laboratories (LAB), listed in Table 1. One laboratory asked for additional measurements which were performed in 1988. All participants used the (d+T) reaction with average neutron energies ranging from 14.6 MeV to 15.0 MeV, except for one which had two groups of neutrons with average energies of 17.5 MeV and 4.7 MeV produced by the (d+T) and (d+D) reactions, respectively. The BIPM reference equipment continues to be available to laboratories entering the field of neutron dosimetry.

For determining the values of K_N and K_G the so-called 'twin dosemeter' method[1] was used. The quantities R'_T, R'_U, k_T and k_U have the same significance as in ICRU 26[1]; they were measured and corrected according to the ECNEU protocol[2].

RESULTS AND DISCUSSION

Photon calibration measurements

During the period of comparison the BIPM photon calibration factors for the BIPM TE chambers (Exradin, T2 type) varied over a range of 0.6%, the largest change being 0.4%. The corresponding figures for the Mg/Ar chamber (Exradin, MG2 type) were both 0.5%. The mean of the BIPM calibration for the BIPM TE chambers agreed well with that of the LAB calibrations; likewise, the means of the BIPM and LAB calibrations of the BIPM GM counter (ZP 1311 type) also agreed well, except those obtained with ^{137}Cs instead of ^{60}Co fields. However, the BIPM calibration factors of the BIPM Mg/Ar chamber appeared to be on average about 2% higher than those obtained by the LAB. The reason for this discrepancy has been clarified, after the comparison, by an investigation of the BIPM Mg/Ar chamber at the BIPM[3], by using different types of gas tubes, on the one hand, and by doubling the length of some tubes, on the other hand. As an example, Figure 1 shows that the responses with Voltalef type tubes, which are usually employed at the BIPM, and only by the BIPM, are lower than those obtained with all the other types of tubes used by the LAB. Moreover, for longer tubes one obtains larger differences.

Table 1. Participating laboratories and names of the scientists involved.

BIPM	Bureau International des Poids et Mesures, Sèvres, France (V.D. Huynh)
ETL	Electrotechnical Laboratory, Tsukuba, Japan (N. Takata)
IAEB	Institute of Atomic Energy, Beijing, People's Republic of China (J. Zheng and C. Rong)
NIM	National Institute of Metrology, Beijing, People's Republic of China (Z. Zhang)
NIST	National Institute of Standards and Technology, Gaithersburg, USA (L. J. Goodman)
NPL	National Physical Laboratory, Teddington, United Kingdom (V. E. Lewis)
PTB	Physikalisch-Technische Bundesanstalt, Braunschweig, Germany (D. Schlegel-Bickmann)
TNO*	Institute of Applied Radiobiology and Immunology, Rijswijk, The Netherlands (J. Zoetelief)

* Present name: ITRI-TNO.

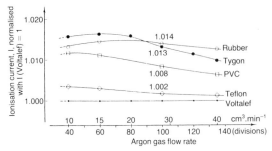

Figure 1. Response of the Mg chamber, in terms of Ar gas flow rate, for different types of tubes. ^{60}Co; Ch. 139 (Mg) + 2 mm cap; HV = +250 V. Usual length (2 × 2.3 m) employed at the BIPM.

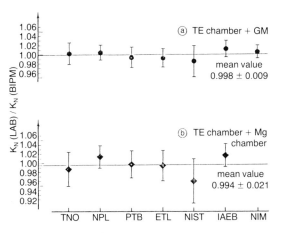

Figure 2. Ratios of K_N obtained with local dosemeters relative to those obtained with the BIPM dosemeters calibrated at BIPM.

Kerma measurements in local mixed fields

The results of the comparison showed that the neutron kerma components, K_N, of the local mixed fields, measured with the BIPM and LAB systems, generally agreed well for both the TE chamber plus GM and the TE chamber plus Mg/Ar chamber combinations (Figure 2). However, the spread of the values of K_N obtained from the (TE chamber + Mg/Ar chamber) combination was larger than that obtained from the (TE chamber + GM) combination. From Table 2 one can see that, in some cases, the values of K_G/K_{tot} differed significantly when they were measured by the BIPM equipment and with the local dosimetry systems. Fortunately, the contribution of the photon component to the total kerma is so small for all laboratories that its influence on the determination of K_N is not critical. The values of k_T and k_U for the local and BIPM dosemeters used in the comparison as well as the contributions of the photon component to the total kerma, K_G/K_{tot}, obtained by the (TE chamber + GM) and the (TE chamber + Mg/Ar chamber) combinations, respectively, are given in Table 2. An article which describes the comparison in more detail has been submitted to *Metrologia* for publication[4].

CONCLUSION

During the period of comparison a satisfactory long-term stability of the BIPM transfer instrument has been observed. The results of the comparison show that the neutron kerma values of the local mixed fields, measured with the BIPM and local dosimetry systems, generally agree well.

Table 2. Values of K_G/K_{tot}, in the local fields, obtained with local and BIPM dosemeters.

LAB	Combination of local dosemeters	k_T	Local dosemeters		BIPM dosemeters	
			k_U	K_G/K_{tot} (%)	k_U	K_G/K_{tot} (%)
TNO	T2 + GM	0.951	0.018	6.3	0.026	7.4
	T2 + MG2		0.150	5.8	0.160	5.5
NPL	T2 + GM	0.951	0.027	1.2	0.026	1.5
	T2 + MG2		0.171	1.4	0.157	2.3
PTB	T2 + GM	0.942	0.025	1.3	0.024	1.2
	T2 + MG2		0.165	1.2	0.162	1.6
ETL	T2 + GM	0.955	0.019	1.5	0.024	1.8
	T2 + MG2		0.159	1.4	0.162	1.9
NIST	T2 + GM	0.953	0.023	3.2	0.019	4.9
	T2 + MG2		0.140	−2.0**	0.140	−2.6**
IAEB	*TE + GM	0.949	0.017	1.5	0.024	1.6
	*TE + Mg		0.168	1.7	0.162	2.0
NIM	*TE + GM	0.955	0.017	4.2	0.024	4.7

* TE and Mg chambers constructed by the Institute of Radiation Medicine, Beijing, People's Republic of China.
** These negative values were probably due to incorrect values of k_U.

REFERENCES

1. International Commission on Radiation Units and Measurements. *Neutron Dosimetry for Biology and Medicine.* Report 26 (Bethesda, MD: ICRU Publications) (1977).
2. Broerse, J. J., Mijnheer, B. J. and William, J. R. *European Protocol for Neutron Dosimetry External Beam Therapy.* Br. J. Radiol. **54**, 882-898 (1981).
3. Huynh, V. D. and Lafaye, L. *Influence of Gas Tubes on the Current Measurements of the Mg/Ar Ionization Chambers.* Rapport BIPM-90/1 (BIPM, F-92312 Sèvres Cedex) (1990).
4. Huynh, V. D. *BIPM Neutron Dosimetry Intercomparison.* Metrologia (to be published).

CRITICAL ASSESSMENT OF CALIBRATION TECHNIQUES FOR LOW PRESSURE PROPORTIONAL COUNTERS USED IN RADIATION DOSIMETRY

P. Pihet*, S. Gerdung, R. E. Grillmaier, A. Kunz and H. G. Menzel**
Fachrichtung Biophysik und Physikalische Grundlagen der Medizin
Universität des Saarlandes, D-6650 Homburg (Saar), Germany

Abstract — To evaluate neutron kerma experiments, the calibration of tissue-equivalent and carbon low pressure proportional counters (TEPC, CPC) was critically investigated to reduce the overall uncertainty of the measurement of the absorbed dose in the wall material, in particular, systematic uncertainties due to the techniques and the basic physical data used. The proton and alpha edge techniques were combined with photon calibration, with an agreement found better than 2%. W ratios for the radiations used for calibration were investigated using published data in methane and propane TE gases. The calibration of the TEPC and the CPC could be related within 3% with regard to calculated edges. The analysis of the measured ion yield spectra and the neutron correction factors for the PCs were improved for neutrons above 20 MeV. The neutron absorbed dose measured with the PCs in a reference neutron field was found to be in excellent agreement with other methods.

INTRODUCTION

Low pressure proportional counters (PCs) simulating micrometer volumes have become a reference technique in radiation research. The combination of spectral information, in terms of distributions in lineal energy[1], and dose information, from the integral of the measured spectra, makes PCs suitable for measuring interaction quantities for radiation dosimetry[2]. For this purpose the calibration methods used must be critically investigated due to the demand for high accuracy, e.g. 5% for neutron kerma factors in dosimetry for therapy[3]. A high accuracy in pulse height calibration and absorbed dose measurement is required for comparing experimental and calculated ion yield spectra and kerma data[2,4,5].

Calibration techniques for PCs were described in earlier reports[6-10]. However different methods were seldom directly compared for analysing experimental data in order to check the validity of unavoidable assumptions, namely with regard to the basic physical data used[2,11]. The present analysis was performed to evaluate neutron kerma experiments[3,12,13] and shows the benefit obtained by combining different methods. It emphasises the role of W, the average energy required to produce an ion pair[14], and presents the evaluation of dose correction factors[15] for the PCs, which were improved taking into account the last experiments performed.

CALIBRATION OF PCs FOR DOSIMETRY

According to well established relations[9], the calibration of pulse height in units of energy imparted $\varepsilon^{(1)}$ can be achieved using a 'single event' technique (SC), to which the internal α source method refers[10] and the maximum energy loss or 'edge' technique[8,10,15], or a 'multiple event' technique (MC) by comparing the absorbed dose measured in a standard field with the reference dose. However, the W for a given event cannot be determined and only mean corrections can be applied to the integral. A true pulse height calibration in energy loss is therefore not possible, i.e. ε is approximated with the so-called 'quasi-energy'[4] assuming a constant W. Furthermore, MC and SC are only equivalent for the calibration of pulse height and dose if the cavity geometry and the mass of the gas (m) are accurately known, the problem being complicated by the uncertainty of different stopping power and W data.

To evaluate neutron kerma measurements the combination of independent methods of calibration may reduce the risk of systematic errors. To compare the methods used, including multiple correction factors, the procedure consisted first in accepting the approximation:

$$D = \frac{1}{m} \int \varepsilon\, n(\varepsilon)\, d\varepsilon \approx M_D = \frac{C}{m} \int h\, n(h)\, dh$$

with $C = \dfrac{\varepsilon_c}{h_c} = \dfrac{q_c}{g.e}\, W_c\, \dfrac{1}{h_c}$ (1)

where 'C' stands for the radiation used for calibration, h the pulse height, q the collected charge, g the gas amplification and M_D the dose reading of the PC. The calibration of neutron dose in the wall

---

* Present address: Institut de Protection et de Sureté Nucléaire, IPSN/DPHD/S.DOS, BP 6, F-92265 Fontenay aux Roses Cedex, France.
** Present address: Commission of the European Communities, DGXII/D/3, B-1200 Bruxelles, Belgium.

material is derived by applying to M_D the $\overline{(r_{m,g})_n}$ and $\overline{W}_n/W_c$ dose correction factors to take into account the spectrum of secondary particles[2].

BASIC PHYSICAL DATA

Stopping power (STP) and W determine the calibration of PCs and may introduce several systematic errors. Accurate data for the radiations used for calibration are required as well as comprehensive data for the broad range of secondary radiations. STP are used to calculate the maximum energy loss by charged particles in the cavity and to determine the gas-to-wall conversion factor $r_{m,g}$. The STP from Ziegler et al[16,17] are currently used[10,15].

W values present more difficulties due to the lack of data for the propane based TE gas used with the PCs. In the present work, the following approximations were used: (i) for neutron secondaries (e, p, α, ions) W values for methane TE gas were used, assuming a constant W ratio for given particles in the propane and methane mixtures[1]; (ii) for calibration purposes, ratios of W_c values were compared using available data for both gas mixtures (Table 1), they agree within quoted uncertainties and the weighted average was used; and (iii) for crossing particles differential w values for a given simulated diameter were calculated using W(E) functions[11,23] and range tables[16,17]. The assumption of constant W ratios in both gas mixtures is questionable at least for ions heavier than protons below 100 keV[21]. This was assumed to have negligible influence on $\overline{W}_n$ and the emphasis was put on comparing $\overline{W}_n$ from different W data sets for the methane mixture.

Table 1. W of radiations used for pulse height calibration with PCs (α,p, α of 5.7 MeV; p,m, p of 100 keV; α,m, α of 700 keV[10]). Values for ions for each gas mixture are averages from published data accounting for the quoted uncertainties. If not available at the required energy the data were extrapolated using W(E) functions[11] which were also used to derive the differential w values (d=1 μm). W_e equal to 29.3 and 26.1 eV for high energy electrons[22,23] were used as average values for the secondary electron spectra of high energy photons, for methane and propane based TE gas respectively.

	$w_{\alpha,p}$	$W_{p,m}$	$w_{\alpha,m}$
CH_4 TE	29.32±0.13	30.94±0.30	32.55±0.98
	(11,18)	(19,20)	(11)
C_3H_8 TE	26.66±1.75	27.90±1.05	27.76±0.98
	(11,14)	(21)	(11,21)
Ratio to $W_{p,m}$*	0.948±0.010	1	1.035±0.028
Ratio to W_e*	1.005±0.019	0.943±0.019	1.090±0.030

* Weighted average for both gases.

PULSE HEIGHT SPECTRA AND ABSORBED DOSE READING

The built-in α source was used during the measurements to control the stability of gas gain and the dynamic range. Due to significant deviations (10%) observed between different PCs and attributed to uncertainties of the energy and the path of the α particles[8], other calibration techniques were used. The PCs were exposed at the reference ^{60}Co source of the PTB (Braunschweig, FRG). To evaluate the neutron measurements ($14 < E_n < 70$ MeV) the position of the proton edge (p,m) and the α edge (α,m) (Figure 1) were determined by differentiation in the probability density f(ε) after subtracting the overlapping components. For CPCs the α edge can be used at neutron energies above 10 MeV (Figure 1). For TEPCs both p and α edges were used, offering an additional check for the consistency of the reference values. To relate each calibration to a single reference, the proton edge position was chosen due to the low uncertainty of the proton stopping power in the region of the Bragg peak[24] and $W_{p,m}$[19,20]. The photon and α edge calibrations were adjusted using the corresponding W_c ratios (Table 1, Figure 1). The uncertainty of the dose reading using these methods or their combination is shown in Table 2.

After independent calibrations at the photon source, the ratio of the reference dose and the dose reading after α source calibration, $M_D(w_{\alpha,p})$ times $W_e/w_{\alpha,p}$, was found for TEPCs 1 and 2 and CPC 1 to be equal to 1.085 ± 0.051, 0.955 ± 0.045 and 0.983 ± 0.046 respectively. For the same counters exposed to neutron fields[3,12,13] the ratio of $M_D(W_{p,m})$ and $M_D(w_{\alpha,p})$ times $W_{p,m}/w_{\alpha,p}$ was found to be 1.082 ± 0.024, 0.937 ± 0.035 and 1.066 ± 0.039 respectively, in agreement with photon measurements within quoted uncertainties. For TEPC 2 the larger variance was due to instabilities in gas gain due to ageing effect. For CPC 1 the apparent discrepancy may be explained by systematic errors of the lineal energy[10] $y_{\alpha,m}$ or/and $W_{p,m}/w_{\alpha,m}$. Indeed over all TEPC measurements the ratio of the observed p and α edges with regard to the reference values was found on average to be equal to 0.972 ± 0.022. The weighted average of edge and photon calibration factors was systematically used, leading to a reduced overall uncertainty of the dose reading (Table 2).

SPECTRAL ANALYSIS AT HIGH NEUTRON ENERGY

The measured spectra were analysed in terms of the relative dose contribution for the secondary

CALIBRATION TECHNIQUES FOR LOW PRESSURE PROPORTIONAL COUNTERS

Table 2. Overall uncertainties (1 SD) of absorbed dose reading by using different calibration methods as derived by quadratically adding the separate contributions for the technique used, the related W and stopping power data, and the mass of gas as listed in brackets.

Calibration technique	CPC (%)	TEPC (%)
Internal α source (w_α)	10.7 (9.4[a]; 2[b]; 4.0[c])	
Reference photon field (W_e)	5.1 (3.5; 2.4; –)	
p edge (W_p)		5.4 (2.1; 2.0; 4.0)
α edge (W_p)	6.4 (3.2; 3.4; 4.0)	
p/α edge combined (W_p)		4.9
p/α edge + photon combined (W_p)	3.5	3.4

[a]Experimental technique. [b]Basic data. [c]Mass of gas.

radiation components (d_i) and their mean energy ($\overline{E}_i$), this information being later used to evaluate the dose correction factors[15]. The analysis was first based on the decomposition of the measured spectra, accepting simple approximations[5] due to the overlap of the peaks corresponding to different secondaries. It was improved by comparing the spectra measured with the TEPC and the CPC exposed in the same conditions assuming the component above the proton edge in the TEPC spectrum to be due only to carbon interactions in the A-150 plastic[12,15]. For the present analysis the results were re-evaluated more systematically using the new set of TEPC/CPC neutron measurements up to 70 MeV. The major steps consisted in (i) relating more precisely the calibrations of the CPC and the TEPC accounting for the improved pulse height calibration, and (ii) improving the

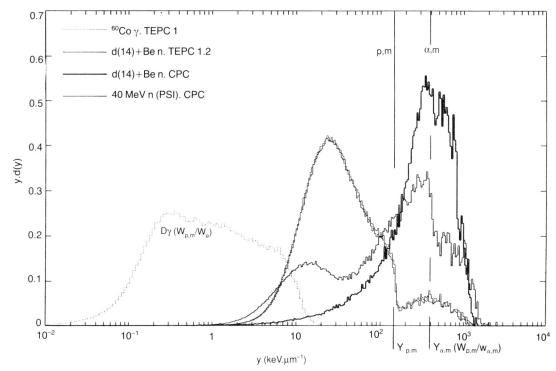

Figure 1. Ion yield spectra measured with TEPCs and CPCs (d=1 μm) in a gamma ray and a neutron reference field at the PTB Braunschweig (FRG)[27] presented in terms of lineal energy[1]. All spectra refer to the p edge value accounting for the corresponding W ratios. $y_{p,m}$ and $y_{\alpha,m}$ ($W_{p,m}/w_{\alpha,m}$) are equal to 144.9 and 386.4 keV.μm^{-1}[10,15]. The CPC spectrum in a 40 MeV neutron field at the PSI Villigen (Switzerland)[3] shows the capability of using the α edge for calibration value at high energies.

discrimination of the alpha peak by approximating the ion component from calculated spectra[4] and assuming its shape to be nearly constant with neutron energy. Calculated spectra were also used to assess secondary order corrections such as the contribution of interactions with oxygen and nitrogen and the influence of wall effects in the overlap interval of the α edge and the ion peak[5]. Finally, this careful study enabled assessment of the uncertainties in d_i and $\overline{E}_i$ due to the empirical approach used (Table 3(a)).

DOSE CORRECTION FACTORS

At a given neutron energy, average W values for neutrons $\overline{W}_n$ can be derived practically by calculating the secondary particle spectra, or energy deposition relative to ion yield spectra, and using the W values for the charged particles as a function of their energy[4,26]. The secondary particle spectra not being available for neutron energies above 20 MeV, the dose components d_i and the corresponding mean energies $\overline{E}_i$ were evaluated as described above allowing the relation:

$$\overline{W}_n^{-1} = \sum_i d_i \overline{w}_i^{-1} \text{ with } \overline{w}_i = w(\overline{E}_i) \quad (2)$$

to be used as approximation. The large uncertainties in secondary dose components and related mean energies inherent to the empirical method used, and the uncertainty in the W values were taken into account to estimate the overall uncertainty of $\overline{W}_n$ (Table 3(b)). $\overline{W}_n$ values were derived from different $\overline{w}_i$ data sets for the charged particle components[11,23,25] and, for a given data

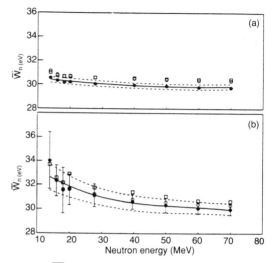

Figure 2. $\overline{W}_n$ as a function of neutron energy from the analysis of ion yield spectra and using different W data sets for the TEPC (a) and the CPC (b). The counters were filled with propane TE gas (d=1 μm). The figure indicates the overall uncertainty at each energy point (Table 3) and the confidence interval (dashed lines) for the fitted values (closed circles, solid line). (●) Bichsel 1982 (α, HI), Thomas 1985 (p); (○) Bichsel 1982 (p,α,HI); (□) Bichsel 1982 (p,α), Burger 1987 (HI)

set, they were fitted by a least squares polynomial (Figure 2). For the TEPC the main differences between different data sets are due to W for protons, the data from Thomas and Burke[25] being lower by about 3%. The single W(E) function proposed by Burger et al[23] for ions heavier than

Table 3. (a) Estimated dose components from secondary charged particles, mean energies and related uncertainties derived from the analysis of TEPC and CPC spectra, listed for two neutron energies and for the CPC to illustrate the empirical approach used. (b) Overall uncertainty of $\overline{W}_n$ from the fitted data (Figure 2) and separate contributions to account for W of charged particles, dose components and mean energy.

(a) Secondary radiation components [( ) = estimated uncertainty in %].					
E_n	p [C(n,p)]	α	Be	B	C
13.9 MeV d_i(%)	–	44.8 (30)	15.9 (50)	–	39.1 (50)
$\overline{E}_i$ (MeV)	–	1.0 (25)	2.2 (35)	–	1.5 (35)
70.0 MeV d_i(%)	48.2 (10)	31.1 (22)	3.8 (50)	2.1 (50)	14.8 (50)
$\overline{E}_i$ (MeV)	16.6 (40)	6.4 (30)	16.4 (35)	7.1 (35)	7.5 (35)

(b) Uncertainty of $\overline{W}_n$ (in %)				
Proportional counters	CPC		TEPC	
E_n (MeV)	13.9	70.0	13.9	70.0
W for charged particles, W_i	1.9	1.1	0.7	0.7
Dose component, d_i	7.2	1.3	0.8	0.3
Average energy, $\overline{E}_i$	2.3	0.8	<0.2	<0.2
Overall (after fitting)	3.4	1.8	0.8	0.7

α particles introduces no significant difference, as expected. For the CPC the influence of proton W only appears at neutron energies above 30 MeV due to the fast increase of the C(n,p) reaction[3,13] implying a decrease of $\overline{W}_n$ and the uncertainty of $\overline{W}_n$. The assumption of Burger et al for ions appears more critical above 20 MeV although within the confidence interval.

Average gas-to-wall conversion factors, $\overline{(r_{m,g})}_n$, for the PCs were derived using a similar approach[15]. A detailed discussion of the last evaluation is found elsewhere[12].

DISCUSSION

This investigation confirms the edge technique as the most appropriate method for pulse height calibration with the main advantages of being nearly independent of the cavity geometry and of reducing the influence of gas gain instabilities during the measurement. However, the main results from the experience gained with different PCs is the generally good agreement between edge and photon calibration, better than 1% with regard to the average for TEPCs, around 4% for CPCs. The larger deviations observed for the CPCs suggest remaining problems with regard to the basic data used for α particles rather than experimental errors due, for example, to the discrimination between the alpha and heavy ion overlapping components. This was confirmed by the observations of the peak/edge ratio in the pulse height spectrum of the α source for each PC[10] which provides an additional consistency check since the same deviation is expected for a given detector between (α,p) and (p,m), and between (α,p) and (α,m) calibration factors. A discrepancy in the order of 5 to 10% was generally observed which cannot be understood unless competing systematic errors of the basic physical data are invoked. However, the influences of the uncertainties of the stopping powers and W cannot be discriminated.

In addition to a lower uncertainty in the dose, the improvement of pulse height calibration enabled a more rigorous analysis of the measured spectra, in particular the combination of TEPC and CPC spectra, providing an alternative method to evaluate neutron correction factors at energies above 20 MeV for which no theoretical data exist. As a major step, this analysis provided the uncertainty in these mean correction factors. The remaining problems of the appropriate W data for charged particles and improved spectrum calculations are currently discussed within Eurados (European Radiation Dosimetry Group).

As an additional check for the methods used, the absorbed dose in A-150 plastic was measured with two TEPCs and compared with that measured with ion chambers (IC) in the d(13.4 MeV)+Be reference neutron field at the PTB[27]. Using a $\overline{W}_n$ value of 30.3 eV calculated for this field and the respective W_c values, PCs deviate from ICs from –0.7% and +1.5% for TEPCs 1 and 2 respectively. A similar agreement was observed by comparing IC with calorimetry measurements[27].

ACKNOWLEDGEMENTS

The authors are grateful to Dr H. Schuhmacher and U. J. Schrewe for their useful recommendations. The work was supported by the Commission of the European Communities (No Bi7 030) and the Bundes Minister Für Umwelt Naturschutz and Reaktorsicherheit (FRG). The investigation of basic physical data for gas detectors in particular is in keeping with the pattern of EURADOS activities.

REFERENCES

1. International Commission on Radiation Units and Measurements. *Microdosimetry.* Report 36 (Washington DC: ICRU Publications) (1983).
2. Menzel, H. G., Bühler, G., Schuhmacher, H., Muth, H., Dietze, G. and Guldbakke, S. *Ionisation Distributions and A-150 Plastic Kerma for Neutrons between 13.9 and 19 MeV Measured with a Low Pressure Proportional Counter.* Phys. Med. Biol. **29**, 1537-1554 (1984).
3. Schuhmacher, H., Brede, H. J., Henneck, R., Kunz, A., Meulders, J. P., Pihet, P. and Schrewe, U. J. *Measurement of Neutron Kerma Factors for Carbon and A-150 Plastic at Neutron Energies of 26.3 and 37.8 MeV.* Phys. Med. Biol. **37**, 1265-1281 (1992).
4. Caswell, R. S., Coyne, J. J. and Goodman, L. J. *Comparison of Experimental and Theoretical Ionisation Yield Spectra for Neutrons.* In: Proc. 4th Symp. on Neutron Dosimetry, München, 1981. Eds G. Burger and H. G. Eberts. CEC-EUR 7448, pp. 201-212 (1981).
5. Bühler, G., Menzel, H. G. and Schuhmacher, H. *Neutron Interaction Data in Carbon Derived from Measured and Calculated Ionisation Yield Spectra.* Radiat. Prot. Dosim. **13**(1–4), 13-17 (1985).
6. Menzel, H. G., Bühler, G. and Schuhmacher, H. *Investigation of Basic Uncertainties in the Experimental Determination of Microdosimetric Spectra.* In: Proc. 8th Symp. on Microdosimetry, Jülich, 1982. Eds J. Booz and H. G. Ebert. CEC-EUR 8395, pp. 1061-1072 (1983).

7. Varma, N. M. *Calibration of Proportional Counters in Microdosimetry.* In: Proc. 8th Symp. on Microdosimetry, Jülich, 1982. Eds. J. Booz and H. G. Ebert, CEC-EUR 8395. pp. 1051-1059 (1983).
8. Waker, A. J. *Experimental Uncertainties in Microdosimetric Measurements and an Examination of the Performance of Three Commercially Produced Proportional Counters.* Nucl. Instrum. Methods Phys. Res. **A234**, 354-360 (1985).
9. Dietze, G., Menzel, H. G. and Bühler, G. *Calibration of Tissue-equivalent Proportional Counters used as Radiation Protection Dosemeters.* Radiat. Prot. Dosim. **9**(3), 245-249 (1984).
10. Schrewe, U. J., Brede, H. J., Pihet, P. and Menzel, H. G. *Improvements on the Calibration of Tissue-equivalent Proportional Counters with Built-in Alpha Particle Sources.* Radiat. Prot. Dosim. **23**(1-4), 249-252 (1988).
11. Bichsel, H. and Rubach, A. *Neutron Dosimetry with Spherical Ionisation Chambers II. Basic Physical Data.* Phys. Med. Biol. **27**, 1003-1013 (1982).
12. Pihet, P., Guldbakke, S., Menzel, H. G. and Schuhmacher, H. *Measurement of Kerma Factors for Carbon and A-150 Plastic: Neutron Energies from 13.9 to 20.0 MeV.* Phys. Med. Biol. (in press).
13. Schrewe, U. J., Brede, H. J., Gerdung, S., Kunz, A., Meulders, J. P., Nolte, R., Pihet, P. and Schuhmacher, H. *Determination of Kerma Factors of A-150 Plastic and Carbon at Neutron Energies between 45 and 66 MeV.* Radiat. Prot. Dosim. **44**(1-4) 21-24 (1992). This issue.
14. International Commission on Radiation Units and Measurements. *Average Energy Required to Produce an Ion Pair.* Report 31 (Washington DC: ICRU Publications) (1979).
15. Pihet, P. and Menzel, H. G. *Atomic Data Required in Accurate Measurements of Kerma for Neutrons with Low Pressure Proportional Counters.* In: IAEA-TECDOC 506. Proc. of Symp. — Advisory Group Meeting on Atomic and Molecular Data for Radiotherapy, Vienna, June 1988. pp. 91-105 (1989).
16. Andersen, H. H. and Ziegler, J. F. *Hydrogen Stopping Powers and Ranges in all Elements.* In: The Stopping and Ranges of Ions in Matter. Ed. J. F. Ziegler (Oxford: Pergamon Press) Vol. 3 (1977).
17. Ziegler, J. F., Biersack, J. P. and Littmark, U. *The Stopping and Range of Ions in Solids.* In: The Stopping and Ranges of Ions in Matter. Ed. J. F. Ziegler (Oxford: Pergamon Press) Vol. 1 (1985).
18. Thomas, D. J. and Burke, M. *W Value Measurements for ^{241}Am Alpha Particles in Various Gases.* Phys. Med. Biol. **30**, 1215-1223 (1985).
19. Nguyen, V. D., Chemtob, M., Chary, J., Posny, F. and Parmentier, N. *Recent Experimental Results on W-values for Heavy Particles.* Phys. Med. Biol. **25**, 509-518 (1980).
20. Waibel, E. and Willems, G. *Ionisation Ranges and W Values for Low Energy Protons in Tissue-equivalent Gas.* Prot. Dosim. **13**(1-4), 79-81 (1985).
21. Posny, F., Chary, J. and Nguyen, V. D. *W Values for Heavy Particles in Propane and in TE Gas.* Phys. Med. Biol. **32**, 509-515 (1987).
22. Combecher, D. *Measurement of W Values of Low-energy Electrons in Several Gases.* Radiat. Res. **84**, 189-218 (1980).
23. Burger, G., Makarewicz, M. and Combecher, D. *Average Energy to Produce an Ion Pair.* GSF Internal Report (1987).
24. Waibel, E. and Willems, G. *Stopping Power and Ranges of Low-energy Protons in Tissue-equivalent Gas.* Phys. Med. Biol. **32**, 365-370 (1987).
25. Thomas, D. J. and Burke, M. *W Value Measurements for Protons in Tissue-equivalent Gas and its Constituent Gases.* Phys. Med. Biol. **30**, 1201-1213 (1985).
26. Goodman, L. J. and Coyne, J. J. W_n *and Neutron Kerma for Methane Based Tissue Equivalent Gas.* Radiat. Res. **82**, 13-26 (1980).
27. Brede, H. J., Schlegel-Bickmann, D., Dietze, G., Daures-Caumes, J. and Ostrowsky, A. *Determination of Absorbed Dose within a A-150 Plastic Phantom for a d(13.35 MeV)+Be Neutron Source.* Phys. Med. Biol. **33**(4), 413-426 (1988).

THE APPLICATION OF A NEW GEOMETRY CORRECTION FUNCTION FOR THE CALIBRATION OF NEUTRON SPHERICAL MEASURING DEVICES USING LARGE VOLUME NEUTRON SOURCES

M. Khoshnoodi and M. Sohrabi
National Radiation Protection Department
Atomic Energy Organization of Iran
PO Box 14155-4494, Tehran
Islamic Republic of Iran

Abstract — A new geometry correction function for the calibration of spherical dose equivalent survey meters, using large volume neutron sources, has been developed and is presented. This is a modification of the geometry factor formalism developed by Hunt for the correction of non-parallelism of the field which is usually produced by a small neutron source near a spherical detector. This function is used to obtain a functional distance parameter specifying the source virtual centre and the geometry correction factor. The consistency of this correction factor and the source virtual centre, derived in this study, have been investigated by analysing the data of close and far distances using two calibration models applicable in these ranges of distances.

INTRODUCTION

Neutron survey meters are usually calibrated in known neutron fields produced by physically small neutron sources. During the past decade, many calibration models or techniques have been proposed, originating from the inverse square law, and aimed at the determination of the calibration factors. The shadow cone technique[1,2] and the second-order polynomial fit model[3,4] have been shown to be applicable to the data for large separation distances between the source and the detector, where both source and detector are assumed to behave as a two-point measuring system. On the other hand, the scatter-free model developed by Hunt[3] fits the data for very small separation distances where the geometry effect arising because of the finite volume of the detector will influence the detector response remarkably. Very recently[4], the second-order polynomial fit and the scatter-free models have been combined into a generalised model applied to the calibration of neutron spherical dose rate meters using bare and D_2O moderated ^{252}Cf[5]. In this study, a new relationship based on the correction of the detector response for the effects of both source and detector geometry was developed. This correction function is followed by a functional length parameter which represents the position of the source virtual centre inside the source volume. The present work also attempts to test the validity of these relationships by analysis of the data obtained using the scatter-free and polynomial fit models, the results of which are presented and discussed.

THEORETICAL DEVELOPMENT

Inverse square law

For a spherical neutron device under irradiation by a spherical neutron source, it is possible to use the inverse square law to obtain $C(d_o)$ as the detector count rate:

$$C(d_o) = F_1(d_o) F_2(d_o) [1 + \Phi(d_e/d_o)^2]/d_o^2 \quad (1)$$

where

$$K = QF(\theta)\varepsilon/4\pi \quad (2)$$

$$d_e = d + r_s' + r_d' \quad (3)$$

K is the source–detector combination characteristic constant, Q is the source emission rate, $F(\theta)$ is the anisotropy factor[2]; ε is the detector efficiency; d_o is the spheres centre-to-centre separation distance; d_e is the source virtual centre position as measured from the detector effective centre which, in the case of spherical detectors, can be taken as the geometric centre, i.e. $r_d = r_d'$; d is the distance between the front faces of the source and the detector and r_d' and r_s' are respectively the positions of the detector effective centre and the source virtual centre, as measured from their front faces.

The function $F_1(d_o)$ in Equation 1 corrects the detector reading for the effects of air scattered and room reflected neutrons as defined below:

$$F_1(d_o) = 1 + Ad_o + Sd_o^2 \quad (4)$$

where A is a constant related to the net contribution of out scattered and inscattered neutrons[7,8]. The parameter S is the room reflected constant[9] and its value depends on the dimensions of the calibration room and the angular response of the detector under investigation. The function $F_2(d_o)$ is defined as the geometry correction factor and is given by:

$$F_2(d_o) = 1 + \delta F_3(d_o) \qquad (5)$$

This is the general form of the geometry correction factor[6,3] which, for a point source and a spherical detector, introduces the function:

$$F'_3(d_o) = 2(d_o/r_d)^2 \{1 - [1-(r_d/d_o)^2]^{1/2}\} - 1 \qquad (6)$$

The factor $F_3(d_o)$ accounts for the source–detector geometry effect which increases the detector reading over that expected according to the inverse square law. The function $F_3(d_o)$, similar to $F'_3(d_o)$, is also related to the additional fractional reading which is induced by non-parallel incidence of a neutron field. This function will be discussed and shown in the next section. The parameter δ accounts for the effectiveness of these additional number of neutrons in producing a response in the detector. The square bracketed term in Equation 1 corrects the detector reading for the scattered neutrons from the source surface and the function Φ is given by:

$$\Phi = F_3(d_o) - F'_3(d_o) \qquad (7)$$

where the function $F'_3(d_o)$ represents the additional fractional number of the direct neutrons which expose the detector surface from the central part of the source.

Derivation of the source virtual centre and geometry correction factor

When using a small neutron source to expose a spherical detector, Axton's theory[6] of geometry correction considers non-uniform illumination of the detector and predicts the relative extra number of neutrons entering the detector volume over those expected according to the inverse square law. Axton's theory may be expanded over the surface of a spherical neutron source provided that each element of the surface is taken as a point source. This consideration implies that part of the source surface irradiating the detector may have the same effect as if it were a point source at a certain point to be specified. We have defined this point as the 'virtual centre' of the source, being entirely a function of d_o, r_d and r_s. The function $F_3(d_o)$ is given by the relation:

$$F_3(d_o)/z = \{[1-(1-x)^{1/2}](2-x)-x\}/x^2 \\
- \{[1-(1-y)^{1/2}](2-y)-y\}/y^2 \\
+ \ln\{(y/x)^{1/2}[1+(1-x)^{1/2}]/[1+(1-y)^{1/2}]\} \qquad (8)$$

where

$$x = r_d^2/(d_o^2 - r_s^2) \qquad (8a)$$
$$y = r_d^2/(d_o - r_s)^2 \qquad (8b)$$
$$z = r_d^2/[4r_s(d_o - r_s)] \qquad (8c)$$

In the above relations, r_s and r_d are the source and detector spheres radii, respectively.

Assuming that the neutron emission by the source surface emerges from the virtual centre, then it can be written:

$$F_3(d_o) = 2(d_e/r_d)^2[1-(1-r_d^2/d_e^2)^{1/2}] - 1 \qquad (9)$$

and hence

$$d_e = (1/2)\,[1+F_3(d_o)]r_d/[F_3(d_o)]^{1/2} \qquad (10)$$

Figure 1 shows the variations of $F_3(d_o)$ as a function of d_o for a point source, i.e. Equation 6, and sources of different radii, i.e. Equation 8.

The virtual centre position from the detector centre, d_e, and the function, $F_3(d_o)$, which are derived in this study, can be applied to Hunt's scatter-free model and the second-order polynomial fit model[3]. These parameters modify the above models and provide necessary geometry corrections when using a large source.

EXISTING MODELS APPLIED

Although some different models have been proposed for calibration of neutron spherical detectors[7,8], the scatter-free and the second-order polynomial fit models seemed more appropriate for application of the new correction factors, introduced in this paper.

Scatter-free model[3]

By combining Equations 1, 5 and 8 and setting $F_1(d_o) = 1$, the relationship assigned to this model is given by:

$$C(d_o)d_o^2/[1+\Phi(d_e/d_o)^2] = K[1+\delta F_3(d_o)] \qquad (11)$$

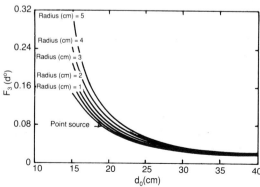

Figure 1. The variations of $F_3(d_o)$ for the assumed sources of different size. The curve referred as 'Point source' was plotted using earlier Hunt's derivation.

According to this model, the contribution of the reflected neutrons due to different sources of scattering may be neglected provided that the counting measurements are made at very small separation distances between the source and the detector.

Second-order polynomial fit model[3,4]

By combining Equations 1 and 4 and setting $F_2(d_o)=1$, it can be shown that

$$C(d_o)d_o^2/[1+\Phi(d_e/d_o)^2] = K(1+Ad_o+Sd_o^2) \quad (12)$$

This model is convenient for separation distances greater than one metre, and in this range of distance the factor $F_2(d_o)$ approaches unity, while the value of d_e should remain effective.

EXPERIMENTAL PROCEDURES

In order to reduce the scattering effects, the measurements were made in an open area outside the building with the source and the detector mounted at 180 cm height from the ground. Two Am–Be neutron sources, X3 and X14[10] having different emission rates and dimensions were used, the dimensions corresponding to equivalent spheres of radii 1.43 and 2.16 cm. The neutron dosemeter was a spherical dose equivalent rate meter type MK7 manufactured by Nuclear Enterprises. This detector consists of a 20.8 cm diameter sphere moderator surrounding a ^{3}He spherical proportional counter at its centre[11]. For close separation distances between the detector and source surfaces, the data were measured at distances 0.4, 1.4, 2.4, 3.4 and 4.4 cm for the X3 source and at 0.6, 1.6, 2.6, 3.6 and 4.6 cm for the X14 source. Another set of data was obtained at larger distances ranging from 1 to 2 metres.

RESULTS AND DATA ANALYSIS

The geometry correction relationship and the virtual centre of a spherical neutron source obtained in this study, corrects the detector reading for the deviation from the inverse square law. This correction expression, i.e. Equation 8, appears to have the same formalism as Equation 6. This is applicable to point sources and spherical detectors which may be shown by taking the limit of Equation 6 as r_s approaches zero, as shown in Figure 1. The virtual centre of a spherical neutron source accounts for the contribution of the source surface to the detector irradiation. This accordingly considers the induced deviation in the detector reading (from the inverse square law point of view). The position of the source virtual centre is always inside the source volume and approaches a fixed point as the distance is increased.

In order to check the validity of the correction factors, as discussed before, the data for close distances, corrected for the dead-time, were fitted to Equation 11 to obtain the source–detector characteristic constants and the neutron effectiveness parameters, δ. Table 1 shows the value of different constants derived using the least squares fitting technique. The uncertainties associated with the constants are statistical in nature and they were computed using error propagation formulae. The data obtained for greater distances were also fitted to the polynomial model[3] shown by Equation 12 for computing the constants K, A and S. Table 1 also shows the analysis of a series of data by Hunt[12] obtained at the NPL large calibration room using a D_2O moderated ^{252}Cf source, applying the same detector as used in this study. In Hunt's measurements, the minimum and maximum distances between the source and the detector front faces were 2 and 450 cm. For each Am–Be

Table 1. Intercomparison of the model parameters obtained for close and far distances using two Am–Be and a D_2O-moderated ^{252}Cf source.

Source	Scatter-free model			Polynomial fit model			
	$K\times10^{-5}$ $(cm^2.s^{-1})$	δ	ε (cm^2)	$K\times10^{-5}$ $(cm^2.s^{-1})$	$S\times10^6$ (cm^{-2})	$A\times10^3$ (cm^{-1})	ε (cm^2)
X3*	0.5886 ±0.0021	0.331 ±0.014	0.2670 ±0.0009	0.5713 ±0.0013	39.5 ±31.2	−5.23 ±5.11	0.2591 ±0.0013
X14*	4.6643 ±0.0101	0.177 ±0.007	0.2780 ±0.0006	4.7910 ±0.0168	64.5 ± 0.6	−8.04 ±0.09	0.2856 ±0.0010
^{252}Cf (Mod.)**	16.28 ±0.02	0.152 ±0.038		15.39 ±0.03	1.375 ±0.030	1.02 ±0.02	

* Am–Be neutron sources with the approximate emission rates of 2.69×10^6 and 2.04×10^7 s^{-1} for X3 and X14 sources, respectively.
** The calculations are based on the data taken at the NPL large calibration room using a D_2O-moderated ^{252}Cf neutron source.

neutron source, the values of K, obtained at close and far distances, are of the same order, as shown in Table 1. The discrepancy between the two values of K, at close and far distances, is not greater than 6%. The negative sign associated with the values of A can be interpreted as the major domination of the ground reflected component over the air scattering component. However, the corresponding values for S are in agreement within their uncertainties.

ACKNOWLEDGEMENT

The authors would like to thank Dr J. B. Hunt from NPL in England for his valuable contribution to this study.

REFERENCES

1. Hunt, J. B. and Roberston, J. C. *The Long Counter as a Secondary Standard for Neutron Flux Density.* In: Proc. 1st Symp. on Neutron Dosimetry in Medicine and Biology. EUR 4596 (Luxemburg: CEC) pp. 935-953 (1972).
2. Hunt, J. B. *The Calibration and Use of Long Counters for the Accurate Measurement of Flux Density.* NPL Report RS(EXT)5 (National Physical Laboratory, Teddington, England) (1976).
3. Hunt, J. B. *The Calibration of Neutron Sensitive Spherical Devices.* Radiat. Prot. Dosim. **8**, 239-251 (1984).
4. Kluge, H., Weise, K. and Hunt, J. B. *Calibration of Neutron Sensitive Spherical Devices with Bare and D_2O-moderated ^{252}Cf Sources in Rooms of Different Sizes.* Radiat. Prot. Dosim. **32**, 233-244 (1990).
5. Schwartz, R. B. and Eisenhauer, C. M. *The Design and Construction of a D_2O-moderated ^{252}Cf Source for Calibrating Neutron Personnel Dosimeters Used at Nuclear Power Reactors.* NUREG/CR-1204 January (1980).
6. Axton, E. J. *The Effective Center of a Moderating Sphere when Used as an Instrument for Fast Neutron Flux Measurements.* J. Nucl. Energy **26**, 581-583 (1972).
7. Eisenhauer, C. M. and Schwartz, R. B. *The Effect of Room-Scattered Neutrons on the Calibration of Radiation Protection Instruments.* In: Proc. Fourth Symp. on Neutron Dosimetry, Neuherberg, Munchen. EUR-7445 (Luxemburg: CEC) Vol. I, 421-430 (1981).
8. Schwartz, R. B. and Eisenhauer, C. M. *Procedures for Calibrating Neutron Personnel Dosimeters.* NBS Special Publication 633 (National Bureau of Standards, Washington, DC, USA) (1982).
9. Dietze, G., Jahr, R. and Schölermann, H. *Effect of Neutron Background on the Standardization of Neutron Fields.* In: Proc. 1st Symp. on Neutron Dosimetry in Medicine and Biology. EUR 4596 (Luxemburg: CEC) 915-933 (1972).
10. Eisenhauer, C. M., Hunt, J. B. and Schwartz, R. B. *Calibration Techniques for Neutron Personal Dosimetry.* Radiat. Prot. Dosim. **10**, 43-57 (1985).
11. Leake, J. W. *An Improved Spherical Dose-Equivalent Neutron Monitor.* Nucl. Instrum. Methods **63**, 329-332 (1968).
12. Hunt, J. B. Private communication (1991).

EXPERIMENTAL ASSEMBLY FOR THE SIMULATION OF REALISTIC NEUTRON SPECTRA

J. L. Chartier†, F. Posny† and M. Buxerolle‡
†CEA/IPSN/DPHD/S.DOS
Centre d'Etudes Nucléaires de Fontenay-aux-Roses
BP n°6 - 92265 Fontenay-aux-Roses Cedex, France
‡CEA/IPSN/DPHD/S.DOS/GDN
Centre d'Etudes Nucléaires de Cadarache
13108 St. Paul Lez Durance Cedex 147, France

Abstract — A set-up intended to replicate in the laboratory realistic neutron spectra encountered in practice at workplaces is presented. Such a facility will provide means of calibrating dosimetric systems in spectral conditions similar to those of their use for radiation protection purposes. The main results of a computational approach are compared with Bonner multisphere spectrometry measurements.

INTRODUCTION

The simulation of realistic neutron spectra at workplaces is currently an important field of research in radiation protection. With this technique, more accurate values should be obtained for neutron doses to exposed individuals working in nuclear facilities or laboratories.

The importance of the goal is justified by several reasons, for example, (i) the limited performances of dosemeters involved in individual and/or environmental monitoring, and (ii) the recent publication of ICRP recommendations[1], dealing with a decrease of primary dose limits and an increase of quality factors for neutrons. As a consequence, a few dosimetric systems have become still more inadequate, because of either their unsatisfactory neutron sensitivity, or their energy-dependent response, or even both.

Type testing and calibration are mandatory stages in the development of an instrument. In this field, much work has already been performed in the frame of the activities of ISO/TC 85/SC 2/WG 2 by the 'Neutron Sub-group', by finalising documents[2,3] dealing with reference radiations and calibration techniques for radiation protection instruments. However, for calibration purposes, the series of calibration neutron spectra consists of four samples which are not unanimously considered as appropriate. Therefore, these current standard calibration fields have to be complemented by other neutron spectra for which the similarity with situations encountered in practice has been established. A joint action by several laboratories has been initiated within the framework of CEC Contract Bi7-0031C in order to measure as extensively as possible the realistic spectral conditions and to replicate them in the laboratory.

An evaluation of data presently available in the literature has enabled the definition of the principles of an experimental facility, its characteristics being studied and optimised by Monte Carlo simulation using the MCNP code. In this contribution, it is intended to present a survey of the work in progress in the CEA-DPHD/SDOS and its most interesting results. Finally, the principles of the procedures for applying that set-up to the calibration of dosemeters and dose rate meters will be considered shortly.

PRINCIPLES OF THE EXPERIMENTAL ASSEMBLY

General considerations on realistic neutron spectra

From published experimental data[4-7], it has been verified that a simplified representation of practical situations can be proposed. It includes three neutron energy ranges : the thermal range, TH, ($E_n < 1$ eV); the intermediate range, INT, (1 eV $< E_n <$ 10 keV) and the high energy range, HE, (10 keV $< E_n <$ 15 MeV). In this paper, the lethargy representation, $E\phi(E)$ against $\ln(E)$, considered as convenient for neutron spectra, will be used.

The available spectral data generally derive from Bonner multisphere spectrometry (BMS), and such spectrometric information has to be handled cautiously[8]. From several sample spectra, it has been observed that the relative fluence contributions of TH, INT and HE components may vary strongly, and that the mean energy of the HE fluence distribution shifts according to the emission neutron spectrum of the radioactive materials, and the neutron absorbing materials of the biological shieldings.

As far as radiation protection is concerned, the quantity dose equivalent, has also to be considered, and the energy dependence of the fluence to dose equivalent conversion factor, $h_\phi(E)^{(9)}$, modifies the relative importances of the TH, INT and HE parts, and consequently their dosimetric 'weights'.

Main features of the assembly

In order to master as easily as possible the production of different neutron spectra, the concept of an assembly providing a collimated radiation beam, instead of an extended radiation field more or less modified by neutron scattering on the environment, has been preferred. By this means, specific calibration procedures having to take into account the influence of scattered neutron sources (walls, air) can be avoided because the available neutron spectra depend exclusively on the materials and geometries of the set-up elements.

Other objectives have also been targeted, in particular the production of realistic neutron spectra with a rather low mean energy, as usually encountered behind biological shieldings. Furthermore, for each sample spectrum, the simultaneous realisation of the three spectral components in adequate proportions and a low relative photon dose equivalent contribution have been considered.

In Figure 1, the main characteristics of the assembly are presented.

Basically, it is composed of a radiation source made of an almost spherical ^{238}U converter irradiated by the neutron emission of an accelerator target placed at its centre, a scattering chamber consisting of a cylindrical polyethylene duct and a series of additional shields. The calibration zone is shown schematically. To take into account the instruments or calibration phantom sizes to be irradiated, the shape of the calibration zone has been defined as that of an orthocylinder (height = diameter = 30 cm), with its centre at a 30 cm distance from the $(CH_2)_n$ duct exit.

The study of neutron and photon spectra in the calibration zone has been performed by Monte Carlo simulation with the MCNP-3A code$^{(10)}$, for two energies E_n^1 and E_n^2 of the primary neutron source.

Configuration (I) $E_n^1 = 14.6$ MeV - CFG (I)
Configuration (II) $E_n^2 = 2.8$ MeV - CFG (II)

A systematic check of the local fluence uniformity has been carried out for several cross sections of the calibration zone. The relative fluence variations are below 2%.

STUDY OF CONFIGURATION (I): $E_n^1 = 14.6$ MeV

Through the exoenergetic T(d,n) reaction, a 14.6 MeV neutron source can be realised by using a 'small' accelerator (HV < 150 kV). The emission rate, depending on the quality of the target, can reach 10^9 neutrons.s^{-1} in 4π. The characteristics of the neutron spectra have been investigated as a function of different parameters of the assembly. The main dimensions are derived from previous studies$^{(11)}$ with respect to converter thickness and from practical considerations.

Attenuation of the direct 14.6 MeV neutron contribution

In configuration (I), an important drawback is the direct 14.6 MeV neutron component leaking out of the converter. This contribution, which is not representative of practical situations, has to be drastically reduced. By adding a 15 cm thick iron shell over the converter and a 5 cm radius ^{238}U half sphere close to the target, the relative dose equivalent contribution has decreased to less than 1.7%.

A further decrease of that value can be performed by inserting an additional iron or copper shield on the assembly symmetry axis. According to the results, we are justified in neglecting the influence of the tiny residual 14.6 MeV neutron contribution. The symbolic representation of configuration (I) source will be denoted by $[E_n^1 + U + Fe]$.

Characteristics of CFG (I) neutron spectra

Examples of calculated neutron spectra are presented in Table 1 and plotted in Figure 2. The main characteristics, fluence and dose equivalent$^{(9)}$ contributions in the different energy ranges, and the mean energies $\bar{E}$ and $\bar{E}_H$ of the spectra are given for different experimental arrangements, demonstrating the flexibility of the CFG(I) set-up.

STUDY OF CONFIGURATION (II): $E_n^2 = 2.8$ MeV

The same type of experimental assembly can be equipped with a 2.8 MeV neutron source provided by the D(d,n) reaction. In this case, the 2.8 MeV

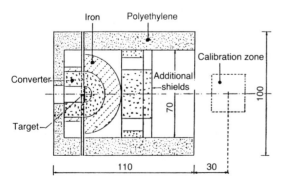

Figure 1. Diagram of the irradiation facility (dimensions in cm).

transmitted component belongs to the high energy range of the available neutron spectrum. Therefore, there is no need to absorb that monoenergetic radiation by an additional shield. Accordingly, the basic arrangement for CFG (II) will be denoted as $[E_n^2 + U]$.

Calculated neutron spectra obtained for several CFG (II) models are given in Table 2 and plotted in Figure 3.

COMMENTS ON THE CALCULATED NEUTRON SPECTRA

The purpose of this paper is not to present an exhaustive survey of neutron spectra generated by such an assembly. Instead, our main objective was to deal with practical situations in the nuclear industry, in which different amounts of biological shieldings and neutron scattering contribute to produce degraded spectra, namely, irradiated fuel containers, laboratories manufacturing fuel elements or sources, waste disposal sites, etc.

From our current results, it can be stated that two series of low energy neutron spectra can be reproduced, in the following mean energy intervals: 0.10 to 0.20 MeV for CFG (I), 0.25 to

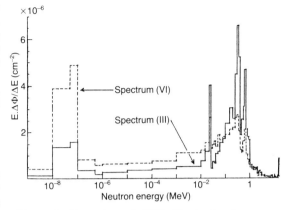

Figure 2. Examples of CFG (I) simulated neutron spectra. Normalised to 1 neutron /4π at the target.

0.35 MeV for CFG (II). However, the quantity $\bar{E}$ does not seem sufficient to characterise accurately a realistic neutron spectrum. A proposal would be to consider either the dose equivalent average neutron energy $\bar{E}_H$, and eventually, the high energy dose equivalent average neutron energy $[\bar{E}_H]_{HE}$ for $E_n > 10$ keV.

Table 1. Calculated data for several CFG (I) arrangements. l: length of the $(CH_2)_n$ duct; t = thickness of the additional shield; ϕ = relative fluence contribution; H = relative dose equivalent[9] contribution; *for a target emission rate of $5 \times 10^8 s^{-1}$ in 4π.

$$\bar{E} = \int E\phi(E)dE / \int \phi(E)dE$$

$$\bar{E}_H = \int h_\phi(E)E\phi(E)dE / \int h_\phi(E)\phi(E)dE$$

Spectrum number	Arrangement CFG (I) $[E_n^1 + U + Fe]$	ϕ_{TH} (%)	ϕ_{INT} (%)	ϕ_{HE} (%)	$\bar{E}_1$ (MeV)	H_{TH} (%)	H_{INT} (%)	H_{HE} (%)	$\bar{E}_H^1$ (MeV)	Typical dose equivalent rate (μSv.h^{-1})*
I	without duct	0	1	99	0.42	0	0	100	0.71	2850
II	+ duct $(CH_2)_n$ l = 90 cm	36	20	44	0.15	6	3	91	0.61	3700
III	+ duct $(CH_2)_n$ l = 50 cm	21	18	61	0.23	2	2	96	0.65	3200
IV	+ duct $(CH_2)_n$ l = 90 cm + Fe shield t = 10 cm	45	27	28	0.08	12	7	81	0.50	1700
V	+ duct $(CH_2)_n$ l = 90 cm + Fe shield t = 10 cm	30	28	42	0.14	5	5	90	0.62	1500
VI	+ duct $(CH_2)_n$ l = 90 cm + D_2O shield t = 5 cm	43	26	31	0.09	10	7	83	0.56	2100

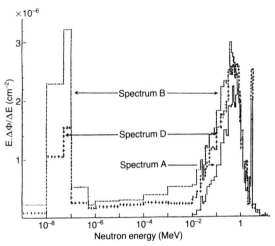

Figure 3. Examples of CFG (II) simulated neutron spectra. Normalised to 1 neutron/4π at the target.

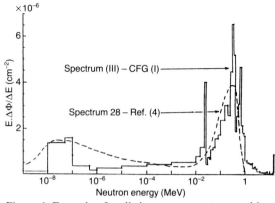

Figure 4. Example of realistic neutron spectrum and its CFG (I) simulation.

The quantity $[\overline{E}_H]_{HE}$ strongly depends on the HE energetic distribution. In that energy range, the conversion factor $h_\phi(E)$ increases at least by a factor of 10. Therefore, a good agreement between $\overline{E}$ values derived from Bonner multisphere (BMS) measurements and calculations is not necessarily maintained for the corresponding $[\overline{E}_H]_{HE}$ values. The relative dose equivalent contribution of HE neutrons is generally higher than 80%, and an accurate measurement of the neutron spectrum in this energy range is essential. When possible, recoil proton spectrometry should be preferred to BMS measurements.

We will now proceed with a comparison between calculations and *in situ* measurements published by other authors[4,5].

Configuration (I)

In Table 1, the modifications of the spectral distributions are mainly due to the length of the $(CH_2)_n$ duct altering the TH contribution and to the additional shields responsible for the attenuation of the HE component up to its entire absorption.

Although several working fields in power plants[4] (spectra 22-25) can be reproduced with our irradiation facility, more frequent handlings of fuel elements requiring neutron dosimetry around transport containers, give a strong interest in replicating the spectrum 28[4]. A satisfactory similarity with the calculated spectrum III is shown in Figure 4.

Configuration (II)

In Table 2, numerical data show that a higher

Table 2. Calculated data for several CFG (II) arrangements. l, t, φ, H, $\overline{E}$, $\overline{E}_H$ have the same meanings as in Table 1. * for a target emission rate of $5\times10^8 s^{-1}$ in 4π.

Spectrum number	Arrangement CFG (II) $[E_n^2 + U]$	ϕ_{TH} (%)	ϕ_{INT} (%)	ϕ_{HE} (%)	$\overline{E}_2$ (MeV)	H_{TH} (%)	H_{INT} (%)	H_{HE} (%)	$\overline{E}_H^2$ (MeV)	Typical dose equivalent rate ($\mu Sv.h^{-1}$)*
A	without duct	0	0	100	0.71	0	0	100	1.05	240
B	+ duct $(CH_2)_n$ l = 90 cm	37	17	46	0.24	4	3	93	0.85	600
C	+ duct $(CH_2)_n$ l = 90 cm + $(CH_2)_n$ shield t = 1 cm	36	16	48	0.29	4	2	94	0.98	600
D	+ duct $(CH_2)_n$ l = 60 cm + $(CH_2)_n$ shield t = 1 cm	25	15	60	0.36	2	1	97	1.00	540

mean energy $\bar{E}_2$ is obtained, providing the possibility of generating a series of other neutron spectra. Measurements of field spectra at TRU[5] have shown that the mean energies of several spectra belong to the interval 0.24 – 0.34 MeV. Among these spectra a sample is selected in order to be compared with calculation results. A general agreement is observed (Figure 5), but a complementary investigation of the influence of the unfolding procedure on the high energy part of the data should be valuable.

Parasitic photon spectrum

For CFG (I) and CFG (II), the calculated photon spectra are quite similar. The main feature is the 2.2 MeV photon emission due to neutron capture by hydrogen. The ratio photon/neutron dose equivalent remains low, ranging from 0.02 to 0.05 for CFG (I) and from 0.015 to 0.025 for CFG (II). Thanks to these experimental conditions, the opportunity is given to perform irradiation or calibration of dosemeters in very pure neutron fields.

THE IRRADIATION FACILITY IN THE CEA/S.DOS LABORATORY

A CFG (I) facility, realised in the S.DOS/Cadarache laboratory, is operating with a SAMES 150 kV accelerator. Three Si detectors, measuring the α particles emission rate, monitor the neutron flux independently from the environment. A retractable radioactive source allows an easy control of monitors. The scattering chamber is made of a series of $(CH_2)_n$ rings (thickness 10 cm). An experimental characterisation has been performed, the results of which are presented elsewhere[12].

REFERENCE QUANTITY — CALIBRATIONS

The reference quantity is the neutron fluence spectrum ϕ_E in the calibration zone, from which the dosimetric quantities can be derived, through appropriate conversion factors. When using such a device for calibration of instruments, the monitor indication I is the only information available. The traceability to primary references has to be ensured through adequate spectrometric techniques giving a relation between the flux density spectrum and I. By comparing calculations and spectral measurements, and after having calibrated the monitor as a function of the target emission

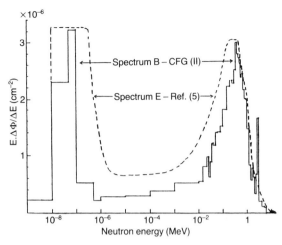

Figure 5. Example of realistic neutron spectrum and its CFG (II) simulation.

rate, it is intended to use in general, as a reference quantity, the corrected calculated data. Research work is presently in progress to achieve that objective.

CONCLUSIONS

After having presented the principle of an irradiation facility intended to replicate, in a calibration laboratory, realistic neutron spectra encountered in radiation protection situations, the characteristics of the calculated spectra are given. Comments on the field of application of that device are followed by a comparison with BMS measurements published in recent literature. With such a type of facility a simplification of the calibration procedures is to be obtained, the radiation field not being altered by the environment.

ACKNOWLEDGEMENTS

The authors appreciate the contributions of C. Itié and G. Audoin (S.DOS-FAR) for their participation in the spectrometric measurements, and the assistance of J. Kurkdjian and J. Pelcot (S.DOS-CAD) in operating the irradiation facility.

This work was partly supported by the Commission of the European Communities (Contract Bi7-0031C) and by the Bureau National de Métrologie (Convention de Recherche n° 89-2-46-0017).

REFERENCES

1. ICRP. *1990 Recommendations of the International Commission on Radiological Protection*. Publication 60 (Oxford: Pergamon Press) (1991).

2. ISO. *Neutron Reference Radiations for Calibrating Neutron Measuring Devices Used for Radiation Protection Purposes and for Determining their Response as a Function of Neutron Energy.* ISO 8529 International Standard (1986).
3. ISO. Draft Standard Proposal. *Procedures for Calibrating and Determining the Response of Neutron Measuring Devices Used for Radiation Protection.* ISO/CD 10647 (1991).
4. Buxerolle, M., Massoutié, M. and Kurkdjian, J. *Catalogue de Spectres de Neutrons.* Rapport CEA-R-5398 (1987).
5. Liu, J. C., Haynal, F., Sims, C. S. and Kuiper, J. *Neutron Spectra Measurements at ORNL.* Radiat. Prot. Dosim. **30**(3), 169-178 (1990).
6. Knauf, K., Alevra, A. V., Klein, H. and Wittstock, J. *Neutronen Spektrometrie in Strahlenschutz.* PTB Mitteilungen **99**, 02/89 pp. 101-106 (1989).
7. Griffith, R. V., Palfalvi, J. and Madhvanath, U. *Compendium of Neutron Spectra and Detector Responses for Radiation Protection Purposes.* IAEA Technical Report Series no 318 (Vienna: IAEA) (1990).
8. Alevra, A. V., Siebert, B. R. L., Aroua, A., Buxerolle, M., Grecescu, M., Matzke, M., Mourgues, M., Perks, C. A., Schraube, H., Thomas, D. J. and Zaborowski, H. L. *Unfolding Bonner-sphere Data: A European Intercomparison of Computer Codes.* PTB-7.22-90-1 (January 1990).
9. ICRP. *Data for Protection against Ionising Radiation from External Sources.* Publication 21 (Oxford: Pergamon Press) (1973).
10. Briesmeister, J. F. *MCNP – A General Monte Carlo Code for Neutron and Photon Transport (Version 3A).* LA-7396-M, Rev. 2 (Sept. 1986).
11. Benezech, G. *Rapport Interne SESR – CEN Cadarache* (1972).
12. Posny, F., Chartier, J. L. and Buxerolle, M. *Neutron Spectrometry for Radiation Protection: Measurements at Working Places and in Calibration Fields.* Radiat. Prot. Dosim. **44**(1-4) 239-242 (1992). (This issue).

EXTENDED USE OF A D$_2$O-MODERATED ^{252}Cf SOURCE FOR THE CALIBRATION OF NEUTRON DOSEMETERS

S. Jetzke, H. Kluge, R. Hollnagel and B. R. L. Siebert
Physikalisch-Technische Bundesanstalt
Bundesallee 100, D-3300 Braunschweig, FRG

Abstract — Two independent Monte Carlo programs were used to calculate the fluence spectrum of the D$_2$O-moderated ^{252}Cf source installed at the PTB. The results show a satisfactory agreement. Structures in the resulting spectra are caused by neutron scattering resonances of the moderator material. A method to produce a neutron field with a large thermal component is proposed.

INTRODUCTION

An irradiation facility[1] using radioactive reference neutron sources has been installed at the Physikalisch-Technische Bundesanstalt (PTB). Three bare ^{252}Cf sources of different source strength, a D$_2$O-moderated ^{252}Cf and a ^{214}Am-Be(α,n) source, can be placed at the centre of the irradiation room. These neutron sources are recommended by the International Organization for Standardization (ISO) for the routine calibration of neutron measuring devices used in radiation protection[2].

One aim of this paper is to determine that part of the neutron field which is produced by the D$_2$O-moderated ^{252}Cf source itself in the absence of the surroundings. Our interest in this question is twofold. First we would like to get a better knowledge of the field produced by the neutron source available at the PTB. This is of fundamental importance for calibration procedures. For this purpose the field emerging from a D$_2$O-filled sphere was studied with two independent computer codes: MCNP[3] and HLSI, a code developed in-house. Other effects which have to be taken into account additionally, such as the reflections from the wall and scattering by the air are discussed elsewhere[4]. Secondly the possibility of producing 'thermal' neutron fields with the special assembly used at the PTB was investigated.

After a short description of the set-up used for the experiment and in simulations, attention is focussed on the moderated spectra, including a comparison between the two codes used. The spectrum seen after the moderation by D$_2$O only, followed by a description of the 'real' spectra, i.e. the spectra obtained with the full source, including the iron and cadmium walls is discussed.

The last section is devoted to the possibility of producing a thermal neutron field with such a source and a critical discussion about problems connected with its use. It is shown that the source available at the PTB allows the production of three neutron fields of different characteristics.

SOURCE SPECTRA

The source used at the irradiation facility in the PTB is a ^{252}Cf source encapsulated by a hollow stainless steel cylinder which is brought into the centre of a sphere with a radius of 14.95 cm filled with D$_2$O. The D$_2$O is enclosed by two shells: one (1.05 mm thick) made of stainless steel, which for simulation purposes has been assumed to be pure iron, and a second, outer one of cadmium (1.08 mm thick). One half of the cadmium shell can be removed. This will be discussed in the chapter on the 'difference neutron field'. For the simulations the source was simplified in such a way that the cylinder has been replaced by a sphere having the same volume as the real cylinder and the ^{252}Cf source itself was assumed to be point-like in the centre of the sphere. Any additional equipment inside the D$_2$O was neglected. The D$_2$O was assumed to be pure.

The simulations were performed using two independent computer codes:

MCNP: This code was used without any modification and with the cross section data delivered with the code. The fluences given in this work were obtained using the surface estimator averaged over the full surface of the moderator assembly and point detectors for the spectral results at a distance of 1.70 m from the centre of the source. The full problem was treated as a one-step transport problem in all cases.

HLSI: This is an analogue Monte Carlo code to determine the fluence outside a D$_2$O sphere resolved with respect to energy and angle. The spectra given outside the sphere were calculated in a second step using the full information from the surface. The effect of the cadmium sphere was also added in an independent manner simply by including the transmission through the cadmium shell.

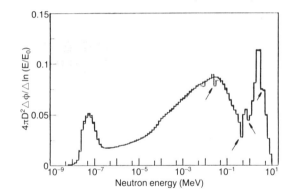

Figure 1. Spectra of the D_2O-moderated ^{252}Cf source at a distance of 170 cm from the centre of the source obtained with MCNP. The full line is the spectrum for the moderation only by D_2O, the results shown as a dotted line include the iron sphere. Arrows mark the positions where hole burning and the influence of the iron resonances can be seen.

Figure 1 shows the first neutron field of interest and compares two spectra calculated with MCNP for a distance of 1.70 m from the centre of the source. The full curve was obtained after moderation by pure D_2O and the dashed line was calculated including the iron sphere.

As the pure Maxwell distribution with its maximum at 0.7 MeV does not reproduce the ^{252}Cf spectrum in detail[5], simulations with two different Watt spectra[3,5] were performed. The results are summarised briefly. With the energy resolution used and the statistical uncertainties achievable, no differences were visible. Whether this is caused by the moderation or simply a result of the finite size of the energy intervals cannot be decided. In view of the limited energy resolution of our simulations the use of the Maxwell distribution is sufficient. It should be noted that for the energy bin structure chosen here, the Maxwell maximum for the group fluences lies in the interval 1.995 MeV $\leq E_n <$ 2.512 MeV.

The spectra shown in Figure 1 can be divided into four main parts:

(1) Neutrons that suffered none or one collision and having nearly the original energy distribution. In this spectrum an effect, similar to what is known in laser physics as *hole burning*, is indicated. Due to resonances in the neutron scattering cross section on oxygen at energies of about 1 MeV the high energy flank in the moderated spectrum is deformed, compared to the original Maxwell distribution. This resonance leads to a pronounced *hole* in the distribution of the unscattered neutrons and consequently to a distortion of the Maxwell distribution.

(2) The second peak in the high energy region below 1 MeV is dominated by neutrons which suffered a few collisions ($\leq$ 5). The pronounced dips are closely related to resonances in the neutron–oxygen scattering cross sections. A detailed analysis, however, cannot be given as the distribution seen is a complex superposition of elementary distributions.

(3) In the intermediate energy part neutrons that collided between 5 and about a few tens of times with the moderator material can be found. Their distribution depends on details of the moderator. This is nicely demonstrated by the deviation from the smooth behaviour near the maximum in this region. These structures are closely related to the resonances in neutron–iron scattering.

(4) At very low energies a well pronounced tail with a maximum approximately corresponding to the maximum of a Maxwell distribution of thermal neutrons at 300 K can be seen.

This general form remains unchanged when looking at the spectra at different distances D between the point of interest and the centre of the source, but the spectra become harder with increasing distance. Table 1 shows the total

Table 1. Total fluences as a function of the distance D from the (centre of) the source.

D (cm)	Φ_{D_2O} (cm^{-2})	Φ_{mod} (cm^{-2})	$\dfrac{\Phi_{mod}}{\Phi_{bar}}$
20	$(2.270 \pm 0.007) \times 10^{-4}$	$(1.986 \pm 0.004) \times 10^{-4}$	0.999
58	$(2.405 \pm 0.002) \times 10^{-5}$	$(2.097 \pm 0.002) \times 10^{-5}$	0.886
170	$(2.768 \pm 0.003) \times 10^{-6}$	$(2.414 \pm 0.002) \times 10^{-6}$	0.877
300	$(8.877 \pm 0.004) \times 10^{-7}$	$(7.744 \pm 0.003) \times 10^{-7}$	0.877

Φ_{D_2O}: fluence from the D_2O-moderated source.
Φ_{mod} : fluence from the complete source (iron and cadmium).
Φ_{bar} : fluence from the bare source: $\Phi_{bar} = 1/4\pi D^2$.
All fluences are normalised to one start neutron.

fluences for the moderated and the complete source. For distances larger than 1 m the fluence is proportional to D^{-2} and at a distance of 3 m the fluence of the purely D$_2$O moderated source, Φ_{D_2O}, is nearly equal to the fluence of the bare source (8.842×10^{-7} cm^{-2}).

The second neutron field of interest is shown in Figure 2. The outer cadmium sphere was also included. (The iron shell was neglected in HSLI). Absolute results from the two computer codes available are shown. The agreement in the medium energy region is very good, but with increasing energy small differences become visible. One reason for this might be the different modelling of the source itself.

From a dosimetric point of view, not only the fluence is important. In Table 2, some quantities of practical relevance are summarised: the mean energy $\bar{E}$, the effective energy $\tilde{E}^*$, weighted with the ambient dose equivalent spectrum and the mean fluence to ambient dose equivalent conversion factors $\bar{h}^*_\Phi(10)$, using the corresponding conversion factors for monoenergetic neutrons[6]. Numerical values were generally taken at the logarithmic mean value of the corresponding energy bin.

The agreement between the two calculations performed at the PTB is very satisfactory. Though the spectra with and without cadmium are qualitatively different, the corresponding values for $\bar{E}$ and $\tilde{E}^*$ differ only very little. However, the quantity $\bar{h}^*_\Phi(10)\Phi^{[\text{mod}]}$, relevant for calibrations, is higher than the value following from the ISO recommendations[2].

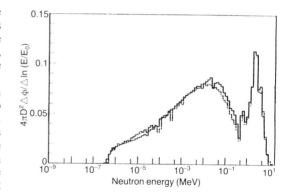

Figure 2. Spectra of the D$_2$O-moderated ^{252}Cf source at a distance of 170 cm from the centre of the source including iron and cadmium obtained with MCNP (full line) and including cadmium obtained with HLSI (dotted line), $E_0 = 1$ MeV.

DIFFERENCE NEUTRON FIELD

The third neutron field of interest, is a field with a large proportion of neutrons in the region of thermal energies. To develop such a field the difference between the fields obtained from a source that is shielded only by iron and a source shielded by iron and cadmium is taken. This can also be achieved experimentally by removing one half of the cadmium shell. The first experiment is then performed with the detector in front of the side covered with iron and a second experiment with the detector in front of the cadmium-covered side, and the difference of the two measurements

Table 2. Dosimetric quantities. The data used together with MCNP were taken at a distance of 1.70 m, those together with HLSI at a distance of 10 m. The difference from the values to be expected at the distance 'infinity' are negligible. MCNP (+ Fe, − Cd) stands for MCNP calculations including the iron conmtainment and not taking into account the cadmium sphere. $\Phi^{[\text{mod}]}$ stands for the corresponding quantity Φ_{D_2O} or Φ_{mod}, respectively.

	$\bar{E}^{(a)}$	$\tilde{E}^{*(a)}$	$\dfrac{\Phi^{[\text{mod}]}}{\Phi_{\text{bar}}}$	$\bar{h}^*_\Phi(10)$	$\bar{h}^*_\Phi(10) \times \dfrac{\Phi^{[\text{mod}]}}{\Phi_{\text{bar}}}$
	(MeV)	(MeV)		(pSv.cm^2)	(pSv.cm^2)
MCNP (− Fe, − Cd)	0.530	2.259	1.000	90.7 ± 0.1	90.7 ± 0.1
MCNP (+ Fe, − Cd)	0.527	2.253	0.997	90.4 ± 0.1	90.1 ± 0.1
MCNP (+ Fe, + Cd)	0.592	2.280	0.879	100.4 ± 0.1	88.3 ± 0.1
HLSI (− Fe, − Cd)	0.547	2.345	0.997	90.36	90.04
HLSI (− Fe, + Cd)	0.621	2.394	0.854	100.71	86.02
ISO 8529	0.539	2.219	0.885	93.26	82.54

All uncertainties are standard deviations. They only include the statistical uncertainties resulting from the Monte Carlo calculations.
(a) The uncertainties are of the order of 0.2%.

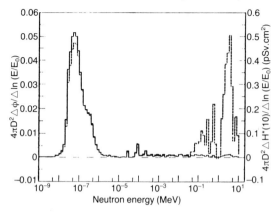

Figure 3. Spectrum obtained as the difference between the spectrum of the complete source and of the source without the cadmium shielding (full line). The dotted line shows the corresponding dose equivalent distribution.

is taken. A result for such a field obtained with MCNP is shown in Figure 3. One easily sees that the total fluence is predominantly determined by the thermal fluence, but with some contribution of high energy neutrons. However, these neutrons contribute considerably to the ambient dose equivalent. This is indicated by an effective energy of $\tilde{E}^* = (1.11 \pm 0.16)$ MeV and a mean conversion factor of $\bar{h}^*_\Phi(10) = (15.2 \pm 2.1)$ pSv.cm^2. The last value has to be compared with that for monoenergetic neutrons at the thermal maximum (E = 25 meV) $h^*_\Phi(10) = 8.4$ pSv.cm^2.

CONCLUSIONS

For the special environment available at the PTB we compared spectral fluences obtained by two independent Monte Carlo simulations leading to satisfactory agreement. Taking the results from these simulations, a sphere half covered by a cadmium shielding might be of practical interest. The fluence difference spectrum obtained in such a way shows a very pronounced thermal distribution with only small admixtures of higher energy neutrons. This moderator assembly could then be used in a twofold manner: The difference method to get a neutron field with a large amount of thermal neutrons could be used for calibration purposes with regard to the fluence response. The difference spectrum is predominantly suited for the calibration of dosemeters with a high dose equivalent response in the range of thermal and intermediate energies, i.e. albedo neutron dosemeters for the measurement of individual doses. Moreover by using the moderator without the cadmium shielding, dosemeters could be tested with a D$_2$O moderated ^{252}Cf spectrum which includes a component of thermalised neutrons. Difficulties which are closely related to the production of thermal neutrons and are caused by the scattering background from the walls will be discussed elsewhere[4].

REFERENCES

1. Kluge, H. and Seifert, H. *Kalibrierung und Prüfung von Ortsdosisleistungsmessern für Neutronen.* In: Strahlenschutz für Mensch und Umwelt, Fachverband für Strahlenschutz, Publikationsreihe 'Fortschritte im Strahlenschutz', Bd.2, FS-91-55-T, pp. 855-860 (1991).
2. International Organization for Standardization (ISO). *Neutron Reference Radiations for Calibrating Neutron Measuring Devices used for Radiation Protection Purposes and for Determining their Response as a Function of Neutron Energy.* International Standard ISO 8529 (1989).
3. Briesmeister, J. F. *MCNP — A General Monte Carlo Code for Neutron and Photon Transport – Version 3A.* LA-7396-M, Rev. 2 (1986).
4. Jetzke, S. and Kluge, H. PTB-Laborbericht, to be published.
5. Mannhart, W. *Status of the Cf-252 Fission Neutron Spectrum Evaluation with Regard to Recent Experiments.* In: Proc. IAEA Consultants' Meeting on the Physics of Neutron Emission in Fission, Mito, Japan, INDC(NDS)-220/L pp. 305-336 (1988).
6. Wagner, S. R., Grosswendt, B., Harvey, J. R., Mill, A. J., Selbach, H. J. and Siebert, B. R. L. *Unified Conversion Functions for the New ICRU Operational Radiation Protection Quantities.* Radiat. Prot. Dosim. **12**, 231-235 (1985).

A COMPUTER LIBRARY OF NEUTRON SPECTRA FOR RADIATION PROTECTION ENVIRONMENTS

B. R. L. Siebert†, H. Schraube‡ and D. J. Thomas§
†Physikalisch-Technische Bundesanstalt, Braunschweig, Germany
‡GSF – Forschungszentrum für Umwelt und Gesundheit GmbH, München, Germany
§National Physical Laboratory, Teddington, United Kingdom

Abstract — The spectral fluence response of practicable neutron dosemeters for routine use generally does not match the spectral fluence response of radiation protection quantities such as the ambient dose equivalent H*(10). A knowledge of the prevalent spectral fluence is required to provide appropriate calibration factors for operational dosemeters. A program package was developed which assists the collection, archiving and interpretation of neutron spectra relevant for working environments. The features of this program package are briefly outlined and examples for dosimetric applications are given.

INTRODUCTION

The response of routinely used neutron dosemeters in general and of personal neutron dosemeters in particular depends on the energy and angle of the incident neutrons. At least a rough knowledge of the spectrum and the angular distribution of the neutron field is required to obtain appropriate calibration factors for neutron dosemeters.

One possibility is to calibrate dosemeters in a neutron field similar to those encountered in working environments where measurements need to be made. Another approach is to calibrate the dosemeters in standard calibration fields and to derive appropriate calibration factors from 'practical' spectrometry performed within the radiation environment where the dosemeters will be used.

An important step in searching for optimal calibration procedures is to collect neutron spectra in relevant working environments. In this paper a computer program package will be presented which assists the collection, archiving and interpretation of neutron spectra.

In the following, the objectives of such a program package are discussed and a short description of already existing programs is given. Two examples are given to demonstrate the capabilities of such a program package. The programs can be used on simple personal computers, and their source versions are made available upon request.

DESIGN AIMS FOR A NEUTRON SPECTRA LIBRARY

(i) Neutron spectra collected are to be represented in a standardised manner. The experimental or calculational methods used are to be judged and reliable uncertainties to be determined.
(ii) A catalogue for the spectra is to be compiled which allows for quick access, display and intercomparison using appropriately selected qualifiers suitable for indexing.
(iii) Utilities to analyse the spectra, to derive dosimetric quantities and to test calibration procedures by simulation are to be provided.

THE PROGRAM PACKAGE SPKTBIB

The first and the third objective cannot be achieved by using simple computer programs as physical insight and judgement are required. Here, therefore, only 'tools' are provided to tackle these tasks. The second objective can be easily met by any data base system. However, in order to avoid copyright problems and to freely disseminate the program package a simple and portable FORTRAN program, SPKTBIB.FOR, was written to perform this task. The name of the package is an abbreviation for SPeKTren BIBliothek (library of spectra).

The capabilities of the catalogue program are simply described by the menus available:

1 = OVERVIEW (HEADING/AUTHOR/INSTITUTION/LOCATION)
2 = CREATE NEW ENTRY
3 = REMOVE OLD ENTRY to configure special LIBRARY
4 = SHOW ENTRY and ORIGINAL COMMENTS
5 = SEARCH IN ALL ENTRIES + optional USER-OPERATIONS
+6 = GENERATE DQ & RQ TABLE FOR ALL ENTRIES
7 = VIEW SPKTBIB.OLD VIA EDITOR
8 = ADD PLOT/CURVE to SPKTBIB.PLT
9 = QUIT

+ DQ and RQ are mean fluence to dose equivalent conversion factors and simulated detector readings.

EXAMPLES

One of the first steps in analysing spectra is to study the fractions of fluence and dose equivalent in energy bins selected, such that an analysis of various dosemeters is supported. To demonstrate this, MENU 5 is used to search for 'VVER440', i.e. for spectra at different positions at the power reactor Bohunice[1]. The program finds three spectra, numbered 63, 64 and 65, and provides the information summarised in Table 1. Using MENU 8 and the 'tool' SPKTPLT.FOR the spectra can be displayed as shown in Figure 1. The data obtained in this way allow one to judge the suitability of using specific dosemeters in a given environment.

For the spectra shown in Figure 1 it is seemingly impossible to derive calibration factors, which are applicable to all three spectra simultaneously, as long as one is restricted to any one dosemeter (c.f. bottom part of Table 1). However, an important problem in radiation protection dosimetry is the determination of a dosemeter's calibration factor for a specified 'class' of environments.

Again, MENU 5 is used to search for 'moderated fission spectra'. MENU 6 computes the DQ and RQ values (mean fluence to dose equivalent conversion factors and simulated detector readings) for the 39 spectra found.

The 'tool' SPKTCOR.FOR allows solution of an overdetermined system of linear equations by the method of least squares for any combination of Bonner spheres on the right hand side. In the present example one obtains 39 equations of the form

$$H^*_i/\Phi = (a_1 R_{i1} + a_2 R_{i2} + a_3 R_{i3} + a_4 R_{i4})/\Phi$$

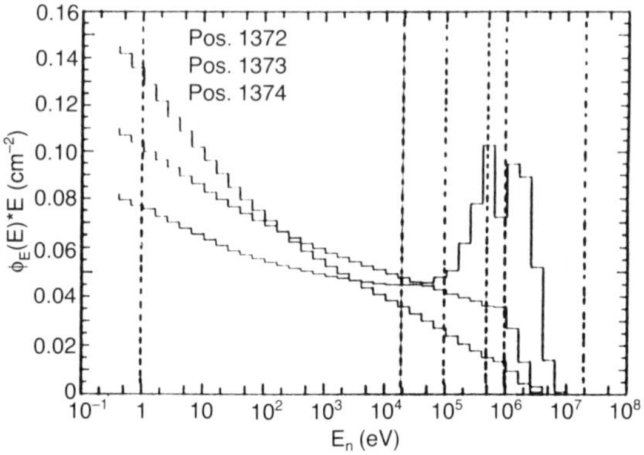

Figure 1. Sample of spectra extracted with the MENU 'ADD PLOT/CURVE' in SPKTBIB and plotted with the program SPKTPLT. Shown are spectra in different positions at the power reactor Bohunice[1]. The six energy regions used in Example 1 are indicated. The positions are ordered by the 'hardness' of their spectra.

Table 1. Example of analysing neutron spectra.

Analysis of spectra in energy regions:

From To		Thermal 1.0	1.0 10^4	10^4 10^5	10^5 5×10^5	$\times10^5$ 10^6	10^6 2×10^7
63 :	Φ:	0.088	0.520	0.102	0.097	0.062	0.131
	H^* :	1.118	5.312	5.356	29.017	27.771	63.788
	H^*/Φ:	12.7	10.2	52.4	299.6	448.1	487.0
64 :	Φ:	0.154	0.714	0.077	0.035	0.010	0.010
	H^* :	1.967	7.441	3.641	9.408	4.321	4.969
	H^*/Φ:	12.7	10.4	47.3	272.2	449.3	484.3
65 :	Φ:	0.118	0.660	0.106	0.064	0.024	0.029
	H^* :	1.501	6.769	5.267	17.957	10.860	14.001
	H^*/Φ:	12.7	10.3	49.9	281.1	450.0	484.6

Energy in eV, Φ in cm^{-2} and H^* in pSv.
Left hand column is number of spectrum.

Computer calibration factors, N = DQ/RQ

		NTA-FILM*	CR-39	Leake-C	Bare	3"	9"
63	N:	1.36×10^6	2.67×10^6	1.32×10^3	9.89×10^2	84.0	1.56×10^2
64	N:	3.64×10^6	4.93×10^6	0.74×10^3	1.44×10^2	15.4	0.71×10^2
65	N:	2.41×10^6	3.47×10^6	0.96×10^3	3.18×10^2	29.5	1.01×10^2

* Dosemeter responses as found in Reference 2 and Bonner sphere responses (bare, 3" and 9") as found in Reference 3.

for the weighting factors, a_j. The index j indicates the bare counter and the Bonner spheres with moderator diameters of 7.62 cm (3"), 22.86 cm (9") and 30.48 cm (12"). H^*_i/Φ is the fluence to ambient dose equivalent conversion factor for the ith spectrum (DQ). R_{ij}/Φ is the calculated mean fluence response for the jth Bonner sphere exposed to the ith spectrum (RQ).

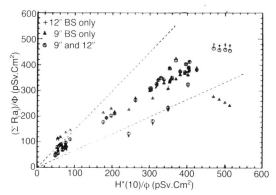

Figure 2. Example of using the MENU 'SEARCH IN ALL ENTRIES' in SPKTBIB and using the program SPKTCOR to study correlations. The abscissa represents the mean fluence to dose equivalent conversion factor. The ordinate shows the values indicated by appropriately calibrated Bonner spheres. The a_i are 'calibration coefficients" (least squares). The dashed lines indicate permissible over and underestimations.

In Figure 2, a group of three spectra (octagons and crosses) and of four spectra (triangles) indicate an underestimation of dose equivalent by more than 1/3. Upon looking up the source spectra it was found that an error had occurred in describing the spectra. These spectra do *not* pertain to moderated fission but to moderated 14 MeV neutrons.

These two examples may demonstrate the usefulness of the programming package presented here for dosimetry.

The program SPKTCOR.FOR could as well be used to expand spectra found in workplaces in terms of easily accessible calibration spectra.

The simplicity of the source code allows the user easily to add optional menus using his own written subroutines.

At present only spectra from existing compendia[1,2,4] are used.

The reader is kindly asked to assist the achievement of the objectives put forward here. If he has measured or calculated neutron spectra he is asked to communicate them to the authors. Questionnaires will be sent upon request. Any comments and suggestions for further applications are also welcome.

ACKNOWLEDGEMENT

This work is partly supported by the Commission of the European Communities under contract No. BI 7 - 0031 - C.

REFERENCES

1. Britvitch, G. I., Makagnov, A. V. and Flyamer, G. V. *Typical Neutron Spectra in Work and Auxiliary Environments of Reactors, Accelerators and Installations with Isotopes.* (In Russian) Report, Inst. of High Energy Physics, Dept. of Radiation Research, Serpouchow, UdSSR (1985).
2. Griffith, R. V., Palfalvi, J. and Madhvavath, U. *Compendium of Neutron Spectra and Detector Responses for Radiation Protection Purposes.* IAEA Technical Report Series No. 318 (Vienna: IAEA) (1990).
3. Marees, V., Schraube, G. and Schraube, S. *Calculated Response of a Bonner Sphere Spectrometer with ^{3}He Counter.* Nucl. Instrum. Methods **A 307** 398-412 (1991).
4. Ing, H. and Makra, S. *Compendium of Neutron Spectra in Criticality Accident Dosimetry.* IAEA Technical Report Series No. 180 (Vienna: IAEA) (1978).

DOSIMETRIC PARAMETERS OF SIMPLE NEUTRON + GAMMA FIELDS FOR CALIBRATION OF RADIATION PROTECTION INSTRUMENTS

K. Józefowicz, N. Golnik and M. Zielczyński
Institute of Atomic Energy, Radiation Protection Department
05-400 Otwock-Swierk, Poland

Abstract — Standard neutron fields for calibration and testing of detectors and instruments used in radiation protection were established based on calibrated sources of ^{252}Cf and ^{241}Am-Be. Applying also paraffin and iron filters and an additional ^{239}Pu-Be source, nine different fields of mixed radiation were obtained. The following parameters of these fields were determined: neutron emission rate; neutron emission anisotropy; neutron flux density at a particular point; dose equivalent rate vs distance from the source and its scattered component; total, neutron and gamma dose rates vs distance; gamma-to-total dose ratio; radiation quality factor; and total tissue kerma.

INTRODUCTION

The aim of this work was to establish (with simple, inexpensive means) the standard fields of neutron and gamma radiation for calibration and testing of detectors and instruments applied in radiation protection.

NEUTRON SOURCES AND FIELDS

^{252}Cf and ^{241}Am-Be have been chosen as the main standard sources, being internationally (ISO) recommended standards[1], mainly of neutron fluence. The sources have been manufactured by Amersham International Ltd and calibrated in the National Physical Laboratory, Teddington, Great Britain (Primary Standard Laboratory). Additionally a ^{239}Pu-Be neutron source, made in the USSR, served for routine calibration. According to the literature, the neutron spectrum of the ^{239}Pu-Be source is similar to that of the ^{241}Am-Be[2]; the neutron emission increases up to 1–2% per year.

The sources have been exposed either free-in-air, or surrounded with filters, to obtain modified fields of mixed neutron+gamma radiation, with different neutron spectra and gamma-to-total dose ratios. Nearly spherical paraffin and iron filters with 10 cm wall thickness have been chosen. The paraffin filter, absorbing and thermalising neutrons, increases the thermal neutron flux, seriously decreases the fast neutron flux (without serious change in its energy spectrum shape) and decreases the neutron doses. The iron filter shifts the fast neutron energy spectrum towards lower values and seriously decreases the gamma doses.

Nine different fields have been obtained, using the ^{241}Am-Be, the ^{252}Cf and the ^{239}Pu-Be neutron sources in free air or with filters.

GAMMA FIELDS

Several ^{137}Cs sources and a ^{60}Co source produce the gamma fields for calibration of gamma measuring instruments and mixed radiation dosimetry. The fields have been standardised by the Primary Standard Laboratory, Polish Committee for Normalization, Measures and Quality.

PARAMETERS TO BE DETERMINED

For the fields of neutron sources (i.e. fields of mixed radiation) the following parameters have been determined:
1. Neutron emission rate.
2. Neutron emission anisotropy.
3. Neutron flux density in a particular point.
4. Dose equivalent rate relative to distance from the source; scattered component.
5. Total, neutron and gamma dose rates relative to distance; gamma-to-total dose ratio.
6. Radiation quality factor.
7. Total tissue kerma.

EXPERIMENTAL EQUIPMENT

The laboratory has at its disposal a 4 m × 4 m × 16 m calibration room with simple devices for sources exposure and instrument alignment. The instruments used are: Studsvik-Alnor 2202D neutron remmeter[3], ionisation chambers and recombination chambers[4]. A computer aided device has been added for automatic measurements[5]. To measure scattered radiation a shadow cone has been constructed, similar to those used by Hunt[6], having an iron front part and paraffin + boric acid mixture (5% of B) at the rear. The cone dimensions (l = 500 mm, front diameter = 60 mm and rear diameter = 180 mm) enable measurements, with proper shielding, in the distance range 0.8–3 m. Because of the front iron part the shadow cone shields the detectors not only from direct neutrons, but from direct gammas as well, which was experimentally checked.

MEASUREMENTS

Neutron emission rates of standard sources have been measured in the National Physical Laboratory (NPL), using the manganese bath method. The results, given below (corrected for radioactive decay), are considered as secondary standards.

^{241}Am-Be $(1.124 \pm 0.010) \times 10^7$ s^{-1} on 31 December 1991

^{252}Cf $(0.753 \pm 0.006) \times 10^8$ s^{-1} on 31 December 1991

The total uncertainties are larger than those stated by NPL because of the uncertainties in the decay constants.

The anisotropy factor of the ^{241}Am-Be neutron emission, resulting mainly from the cylindrical shape of the source, has been measured to be 1.035 ± 0.007. For ^{252}Cf the value of 1.012 ± 0.006 has been taken, according to Hunt[6].

The flux densities of direct neutrons (φ), given in Tables 1 and 2 for standard sources, have been calculated for the distance of 1 m, taking into account anisotropy factors and air attenuation. The uncertainties combine the uncertainties of flux determination, the anisotropy factor and the air attenuation.

The measured neutron response of the 2202D instrument, i.e. the ratio of counting rate to the flux density of direct neutrons, is:

for ^{241}Am-Be $\quad 0.445 \pm 0.008$ cm^2

for ^{252}Cf $\quad 0.467 \pm 0.007$ cm^2

These values are typical for Studsvik's design instruments. The flux density of direct neutrons of the ^{239}Pu-Be neutron source at 1 m distance, shown in Table 3, has been determined in comparison with the standard ^{241}Am-Be source, assuming similar spectra and using the rem meter.

The neutron dose equivalent rate of direct (non-scattered) neutrons at a distance of 1 m from the source have been calculated for two standard sources, using the values given above for direct neutron flux density and fluence-dose conversion factors, taken from the literature. For MADE (maximum dose equivalent) the conversion factors, recommended by ISO[1], are:

for ^{241}Am-Be $\quad 3.8 \times 10^{-10}$ Sv.cm^2

for ^{252}Cf $\quad 3.4 \times 10^{-10}$ Sv.cm^2

For neutron ambient dose equivalent, $H^*_n(10)$, the conversion factors, calculated by various authors, differ slightly, but a reasonable approximation gives values equal to those for MADE.

The calculated dose equivalent rate values, $\dot{H}^*_n(10)_d$, caused by direct neutrons (identical for MADE and for $H^*_n(10)$), at a distance of 1 m, are listed for bare standard sources in Tables 1 and 2. The uncertainties result from uncertainties of flux density determination.

The dose equivalent response of 2202D, i.e. number of counts per dose equivalent, equals:

for ^{241}Am-Be $\quad 1.17 \pm 0.02 \times 10^9$ Sv^{-1}

for ^{252}Cf $\quad 1.37 \pm 0.02 \times 10^9$ Sv^{-1}

Again, these values are typical for Studsvik neutron rem meters — see Reference 3. The uncertainties

Table 1. Parameters of ^{241}Am-Be neutron source at 1 m distance; 31 December 1991.

Parameter		Bare	Paraffin filter	Iron filter
φ_{direct}	(cm^{-2}.s^{-1})	91.7 ± 1		
$\dot{D}_n$	(μGy.h^{-1})	13.7 ± 1	4.6 ± 1	8.5 ± 1
$\dot{D}_\gamma$	(μGy.h^{-1})	4.3 ± 0.3	4.3 ± 1	1.4 ± 0.1
$\dot{D}$	(μGy.h^{-1})	18 ± 1	9 ± 1	10 ± 1
$\dot{D}_\gamma/\dot{D}$	(%)	24 ± 2	48 ± 8	14 ± 2
Q		6.5 ± 0.3	5.3 ± 0.6	8.8 ± 0.8
Q_n		8.2 ± 0.4	9.3 ± 1.5	10 ± 1
$\dot{H}^*_n(10)_d$	(μSv.h^{-1})	126 ± 1	49 ± 4	111 ± 5
$\dot{H}^*_n(10)_t$	(μSv.h^{-1})	156 ± 2	65 ± 2	135 ± 3
$\dot{K}_{tissue}$	(μGy.h^{-1})	20 ± 1.5	9.6 ± 1.3	11 ± 1.3

Table 2. Parameters of ^{252}Cf neutron source at 1 m distance; 31 December 1991.

Parameter		Bare	Paraffin filter	Iron filter
φ_{direct}	(cm^{-2}.s^{-1})	600 ± 7		
$\dot{D}_n$	(μGy.h^{-1})	71 ± 6	15 ± 8	49 ± 3
$\dot{D}_\gamma$	(μGy.h^{-1})	38 ± 3	35 ± 8	5.6 ± 0.5
$\dot{D}$	(μGy.h^{-1})	109 ± 6	50 ± 3	55 ± 3
$\dot{D}_\gamma/\dot{D}$	(%)	35 ± 2	70 ± 15	10 ± 1
Q		6.8 ± 0.2	4 ± 0.2	10.1 ± 0.5
Q_n		9.9 ± 0.4	11 ± 5	11.1 ± 0.6
$\dot{H}^*_n(10)_d$	(μSv.h^{-1})	735 ± 10	170 ± 10	620 ± 30
$\dot{H}^*_n(10)_t$	(μSv.h^{-1})	930 ± 10	220 ± 6	790 ± 15
$\dot{K}_{tissue}$	(μGy.h^{-1})	115 ± 10	53 ± 6	62 ± 6

Table 3. Parameters of ^{239}Pu-Be neutron source at 1 m distance; 31 December 1991.

Parameter		Bare	Paraffin filter	Iron filter
φ_{direct}	(cm^{-2}.s^{-1})	215 ± 5		
$\dot{D}_n$	(μGy.h^{-1})	32 ± 2	12.6 ± 2	21.5 ± 1
$\dot{D}_\gamma$	(μGy.h^{-1})	10 ± 0.6	12.4 ± 2	3.6 ± 0.3
$\dot{D}$	(μGy.h^{-1})	42 ± 2	25 ± 1	25 ± 1
$\dot{D}_\gamma/\dot{D}$	(%)	24 ± 2	50 ± 8	14 ± 1.5
Q		6.5 ± 0.3	5.2 ± 0.3	9.2 ± 0.4
Q_n		8.3 ± 0.5	9.4 ± 1.5	10.5 ± 0.5
$\dot{H}^*_n(10)_t$	(μSv.h^{-1})	370 ± 10	150 ± 4	315 ± 8
$\dot{K}_{tissue}$	(μGy.h^{-1})	46 ± 3	27 ± 2.2	28 ± 2

combine the uncertainties of dose equivalent determination and of the counting rate.

The measurements of neutron dose equivalent rate relative to distance from the neutron source (ranging from 0.5 to 4 m) have been performed for all fields using the 2202D neutron rem meter. The shadow cone technique has been used for bare sources in the distance range 0.8 - 3 m to estimate the scattered component. An example of these measurements is shown in Figure 1 as $\dot{H}^*_n(10) \times r^2$ plotted against distance r. The measured values are extrapolated to the calculated value for direct neutrons at the distance r = 0. For the sources in filters the values of dose equivalent for unscattered neutrons have been determined by graphic extrapolation of $\dot{H}^*_n(10) \times r^2$ to the distance r = 0. The results for all neutron fields, at a distance of 1 m, are given in Tables 1–3; $\dot{H}^*_n(10)_d$ and $\dot{H}^*_n(10)_t$ being values for direct and total neutrons, respectively. The uncertainties combine the uncertainties of distance determination, of counting rate and of rem meter response. The systematic error, connected with the energy dependence of rem meter response — see, for example, Reference 7 — is not included; a rough estimation results in ±3% error for neutrons from bare sources (direct + scattered) and ±12% for sources in filters.

Total absorbed dose rate, $\dot{D}$, and its neutron and gamma components, $\dot{D}_n$ and $\dot{D}_\gamma$, have been measured using the pair of ionisation chambers. One of them, tissue-equivalent, filled with gas mixture containing 11% of hydrogen, has nearly the same sensitivity for gamma radiation and for neutrons. The other, a hydrogen-free, high pressure chamber, has a negligible neutron sensitivity, compared to that for gamma radiation[8]. The measurements with both chambers at various distances from the source have been performed for distances above 0.5 m. The results for all fields are presented in Tables 1–3 for 1 m distance. The value D means the total dose absorbed in the tissue-equivalent material of the chamber electrodes and can be considered as the total dose absorbed at the depth of about 2 cm in an object weighing 6 kg. This quantity gives the basis for the calculation of other dose-related parameters, such as kerma, ambient dose and dose absorbed in specific materials. Some measurements have been also performed with a small ionisation chamber, to confirm the calculations of kerma.

The main factors influencing the uncertainty of dose determination are :
(i) uncertainty of the calibration field of ^{137}Cs gamma radiation;
(ii) different absorption and scattering properties of tissue-equivalent and hydrogen-free ionisation chambers, due to different wall material;
(iii) uncertain energy correction of the chamber sensitivity, due to the not well known energy spectrum of scattered radiation.

The radiation quality factor, as defined in ICRP 21[9], has been approximated by the recombination index of radiation quality, $Q_4^{(10)}$, which was measured directly, using the tissue-equivalent recombination chamber. The results are shown in Tables 1–3 as quality factors for mixed radiation, Q, and for neutrons alone, Q_n. The accuracy of the latter depends mainly on uncertainties of the gamma to total dose rate ratio.

As one can see from the Tables, the product of the neutron absorbed dose, $\dot{D}_n$, and the neutron quality factor, Q_n, is lower than the neutron ambient dose equivalent $\dot{H}^*_n(10)$. The main reason is that the latter is related to the depth of 1 cm in the ICRU sphere, the former to the depth of 2 cm in the chamber (smaller than the sphere).

The total kerma in ICRU tissue, $\dot{K}_t$, was calculated from the measured values of the absorbed dose, taking into account the absorption of radiation in the chamber walls and scattering on chamber elements. The results are presented in Tables 1–3.

SUMMARY

Standard neutron and gamma fields have been established having sources traceable to Primary Standard Laboratories.

The dosimetric parameters of a number of mixed neutron+gamma radiation fields have been determined, with careful analysis of uncertainties. Quality factors range from 4 to 10; the ratio of gamma-to-total dose ranges from 0.10 to 0.70.

Simple and inexpensive means provided standard fields with good calibration possibilities for radiation protection instruments.

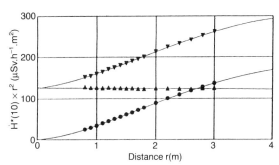

Figure 1. Neutron ambient dose equivalent rate plotted against distance from a ^{241}Am-Be neutron source. (▼) total, (●) scattered, (▲) direct.

ACKNOWLEDGMENTS

This work was partly supported by the International Atomic Energy Agency Technical Co-operation Project POL/1/005.

REFERENCES

1. ISO 8529. *Neutron Reference Radiations for Calibrating Neutron-measuring Devices Used for Radiation Protection Purposes and for Determining their Response as a Function of Neutron Energy* (1989).
2. Anderson, M. E. and Neff, R. A. *Neutron Energy Spectra of Different Size ^{239}Pu-Be (α,n) Sources*. Nucl. Instrum. Methods **99**, 231 (1972).
3. Widell, C. O. and Svansson, L. *Neutron Monitor for Radiation Protection.* In: Neutron Monitoring for Radiation Protection Purposes. STI/PUB/318 (Vienna: IAEA) (1973).
4. Zielczyński, M. and Zarnowiecki, K. *Differential Recombination Chamber.* Neutron Monitoring. STI/PUB/136 (Vienna: IAEA) p. 125-133 (in Russian) (1967).
5. Golnik, N. *Automation of Measurements with Ionization Chambers.* Report of the Institute of Atomic Energy, IAE-2123, Swierk, October 1991.
6. Hunt, J. B. *The Calibration and Use of Long Counters for the Accurate measurement of Neutron Flux Density.* NPL Report RS5 (1976).
7. Cosack, M. and Lesiecki, H. *Dependence of the Response of Eight Neutron Dose Equivalent Survey Meters with Regard to the Energy and Direction of Incident Neutrons.* In: Proc. 4th Symp. on Neutron Dosimetry, Neuherberg, 1981 (Luxembourg: CEC) EUR-7448 Vol. 1, pp. 407-420 (1981).
8. Golnik, N., Pliszczyński, T., Wysocka, A. and Zielczyński, M. *Determination of Dose Components in Mixed Gamma Neutron Fields by Use of High Pressure Ionization Chambers.* In: Proc. 5th Symp. on Neutron Dosimetry, Neuherberg,1984, EUR 9762 (Luxembourg: CEC) Vol.2, pp. 717-725 (1985).
9. ICRP. *Data for Protection against Ionizing Radiation from External Sources.* Publication 21, Suppl. to ICRP 15 (Oxford: Pergamon) (1973).
10. Golnik, N., Wilczyńska-Kitowska, T. and Zielczyński, M. *Determination of the Recombination Index of Quality for Neutrons and Charged Particles employing High Pressure Ionization Chambers.* Radiat. Prot. Dosim. **23**(1/4), 273-276 (1988).

ESTABLISHMENT OF A PROCEDURE FOR CALIBRATING NEUTRON MONITORS AT THE PHYSICS INSTITUTE OF THE UNIVERSITY OF SÃO PAULO, BRAZIL

M. T. Cruz and L. Fratin
Instituto de Física, Universidade de São Paulo
C.P. 20516, CEP 01498, São Paulo, Brazil

Abstract — A facility for neutron irradiation was designed and built in order to apply the ISO draft procedures for calibrating and determining the response of neutron measuring devices to Brazilian neutron monitors and dosemeters used for radiation protection. As a test for the system, the three methods proposed by the draft were applied. A calibrated Am–Be source, a Li(Eu) sensor, a Bonner multisphere detector, and three shadow cones were used. Based on the data from the Bonner spheres, the free field fluence response, the fractional room scattering contribution at unit calibration distance, and the source detector characteristic constant were determined. The unfolded neutron spectrum was also determined and the corresponding fluence, dose equivalent, and mean energy were calculated. All of them proved to be in good agreement with values expected from the spectrum of the neutron source.

INTRODUCTION

A procedure for calibrating a neutron measuring device consists of determining a calibration factor between the reading of the device and free field quantities in the absence of scattering and background effects. This means that the factor shall be independent of the facility where the irradiations are performed. The effects of gamma rays must be separately treated.

The ISO draft[1] describes three methods (semi-empirical method, polynomial fit method and shadow cone method) to correct the effects of scattered neutrons, that is, neutrons scattered by the air and by the walls, floor and ceiling of the calibration room. These effects depend on the type of the device, the source to device distance, and the dimensions of the calibration room.

Two basic assumptions of the methods are that the room scattered component is constant in the region close to the source and that the total air scattering component (in-scatter minus out-scatter) increases linearly with the source–detector distance[1–4].

Applying these methods to a Bonner multisphere spectrometer one gets the calibration factor for each sphere. Using the Bonner multisphere system it is possible to obtain an unfolded source spectrum and the corresponding neutron fluence rate, neutron dose equivalent rate, and mean energy, which shall all agree with those based on the true source spectrum. Such agreement implies that the free field quantities and the calibration factors are appropriate.

THE FACILITY

The measurements took place in a 14 m × 10 m × 4 m large room which has a stainless steel ceiling. This room was designed and built to achieve low neutron scattering.

The irradiation system consists of a 0.7 m underground rotary drum where five neutron sources can be placed. An 0.038 m diameter and 4.0 m height vertical aluminium tube allows the selected source to be hung to the irradiation height, which is 3.20 m above the floor and at least 5.0 m away from the walls of the room.

Aligned with the aluminium tube there is a thin iron structure, above which the monitors can be moved on holders made of low scattering materials. The distance from the centre of the source to the centre of the detector can vary between 0.300 m and 2.500 m with a 0.003 m uncertainty in position. The alignment of the system is done by an optical device placed on another iron structure in the end of the room.

A 185 GBq Am–Be source, calibrated by the manganese bath method at the Instituto de Radioproteção e Dosimetria, the Brazilian Secondary Standard Laboratory, was used in order that the measurements acquire traceability. This calibration indicates a neutron source strength of 1.13×10^7 s^{-1} with a 2% uncertainty. At 1 m from the source, corresponding to the unit calibration distance, the free neutron fluence rate is 90×10^4 m^{-2}.s^{-1} resulting in a neutron dose equivalent rate[5] of 3.42×10^{-8} Sv.s^{-1}.

A Bonner multisphere spectrometer, using a 4 mm diameter × 4 mm long right cylinder Li(Eu) sensor with 0.051 m, 0.076 m, 0.127 m, 0.203 m, 0.254 m and 0.305 m diameters polyethylene spheres, was used to apply the proposed procedure to the irradiation system.

Three 0.50 m long shadow cones were made

with two cone sections, one made of iron and 0.20 m long and the other made of borated paraffin and 0.29 m long, ended by a 1 mm cadmium disc[1,6]. The cones were designed to cover, at most, twice the solid angle determined by the Bonner spheres when they are at 2.002 m from the source. The following front and end diameters with their corresponding half opening angles resulted from this condition:

cone 1, 0.395 m and 0.628 m; $\theta_1/2 = 3.98°$
cone 2, 0.226 m and 0.286 m; $\theta_2/2 = 1.07°$
cone 3, 0.182 m and 0.292 m; $\theta_3/2 = 0.35°$

The operational parameters for the Bonner multisphere spectrometer were determined as indicated in the work of Bramblett et al[7]. Because of its great area to volume ratio, the Li(Eu) sensor has an excellent gamma ray discrimination, as required when a radionuclide neutron source is used. ^{60}Co and ^{137}Cs sources were employed to verify that the detector response to gamma rays is negligible.

CALIBRATION PROCEDURE

The calibration of the Bonner sphere spectrometer requires knowledge of the calibration factors for each sphere–bare detector combination and also for the bare detector.

In order to get these factors, the readings of the scaler counting pulses transformed into pulse rate, $M_T(l)$, at sixteen different source to detector distances (from 0.300 m to 2.500 m) were registered when the Am–Be and the different sphere–detector combinations were placed at the irradiation facility described above. Measurements with shadow cones were only made at 2.020 m.

Each of the three methods proposed by the ISO draft[1] (semi-empirical method, polynomial fit method and shadow cone method) allows the determination of the free field fluence response, $R\varphi$, and the source detector characteristic constant, K. The fractional room scatter contribution at unit calibration distance, S, can be determined by the semi-empirical method.

The free field fluence response, $R\varphi$, is given by:

$$R_\varphi = M_C/\phi \qquad (1)$$

where M_C is the pulse rate (s^{-1}) corrected for all extraneous effects, and ϕ is the free field fluence rate that is determined from the source strength, corrected for source anisotropy, and from the distance at which the detector has been exposed.

The source–detector characteristic constant, K, is defined as:

$$K = M_C \, l^2 \qquad (2)$$

where l is the distance from the centre of the source to the centre of the detector.

The semi-empirical method determines the fluence response, R_φ, and the room scattering component, S, from the plot of the total pulse rate, M_T, corrected for total air scattering, factor $(1+Al)^{(1)}$, and for the effect of the finite size of the detector or source, geometry factor, $F_1(l)^{(1)}$, as a function of l^2. One has:

$$\frac{M_T(l)}{\phi \, F_1(l)\,(1+Al)} = R_\varphi (1 + Sl^2) \qquad (3)$$

The fractional room scattering component, S, for each sphere, is identified by evaluating the slope of the straight line fitted to Equation 3.

In the shadow cone method, the following relationship holds:

$$(M_T(l) - M_S(l))\, F_A(l)\, l^2 = K \qquad (4)$$

where $M_T(l)$ and $M_S(l)$ are the detector counting rate obtained without and with the shadow cone respectively, and $F_A(l)^{(1)}$ is the air attenuation correction.

In the polynomial fit method the total detector counting rate, $M_T(l)$ is related to source–detector distance by the expression:

$$\frac{M_T(l)}{\phi \, F_1(l)} = R_\varphi (1 + al + bl^2) \qquad (5)$$

The measurements of total pulse rate, $M_T(l)$, relative to distance, l, can be fitted to Equation 5 to obtain R_φ.

FREE FIELD QUANTITIES DETERMINED

Table 1 shows the results of the analysis of the data. A least square method was employed to fit the curves defined by Equations 3 and 5. The source–detector characteristic constant, K, and the free field fluence response, R_φ, are listed for the three methods. The fractional room scatter contribution at unit calibration distance was found by the semi-empirical method and is given in the third column.

The determined quantities must actually be the same for any method employed. For the bare sensor, there is a difference in the source–detector characteristic constant determined by the shadow cone method, which was attributed to the fact that the shadow cone covered more than twice the solid angle determined by the detector. For the 0.051 m and 0.127 m spheres, there are small differences in the shadow cone method, which may be justified by the incorrect positioning of the shadow cone.

UNFOLDED SPECTRUM

In the previous sections, the characteristic

constants of a set of detectors belonging to a Bonner sphere spectrometer were determined.

A convenient unfolding procedure allows the determination of the neutron energy spectrum, if some assumptions about the energy dependence response of each sphere–detector combination are made.

As an example, it was decided to use the multisphere spectrometer response matrix for the polyethylene moderators with the 4 mm × 4 mm ^{6}Li detector described by Awschalom et al[8] and the BUNKI code[9] to calculate the neutron spectrum at 1 m from the Am–Be source.

To use the response matrix mentioned above the pulse rate corrected for all extraneous effects, M_C must be normalised[7]. To do this a convolution of this response matrix with the neutron spectrum of the Am–Be source[5] was made and a normalisation factor for each sphere–detector combination was determined.

The BUNKI code was utilised for getting the unfolding spectra that can be seen in Figure 1. The source detector characteristic constants obtained by the semi-empirical method were used in the unfolding procedure. The neutron spectrum considering the room and air scattering was also

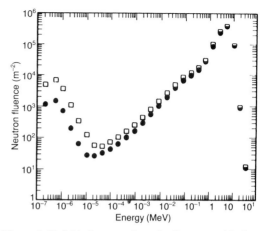

Figure 1. Unfolded spectra from the Bonner multisphere detector data with (□) and without (●) scattering obtained from the BUNKI unfolding code.

obtained using the data from the Bonner sphere spectrometer without any correction at unit distance from the source.

Table 2 shows the values of neutron fluence rate, neutron dose equivalent mean energy

Table 1. Summary of the results obtained. For each sphere–detector arrangement, the first line refers to the semi-empirical method, the second line to the polynomial fit method and the third line to the shadow cone method.

Sphere diameter (m)	Characteristic source–detector constant (count.m^2.s^{-1})	Free field fluence response (count.m^2)	Fractional room scattering at 1 m (%)
Bare	0.169 ± 0.004 0.160 ± 0.008 0.181 ± 0.007*	1.88 ± 0.02×10^{-7} 1.78 ± 0.08×10^{-7} 2.01 ± 0.09×10^{-7}	50 ± 1
0.051	0.56 ± 0.01 0.53 ± 0.02 0.52 ± 0.02	6.23 ± 0.08×10^{-7} 5.8 ± 0.2×10^{-7} 5.77 ± 0.02×10^{-7}	39.8 ± 0.9
0.076	2.58 ± 0.06 2.5 ± 0.1 2.49 ± 0.04	2.87 ± 0.03×10^{-6} 2.8 ± 0.1×10^{-6} 2.76 ± 0.07×10^{-6}	16.6 ± 0.6
0.127	10.2 ± 0.2 10.2 ± 0.4 9.71 ± 0.07	1.13 ± 0.01×10^{-5} 1.14 ± 0.04×10^{-5} 1.08 ± 0.02×10^{-5}	5.7 ± 0.5
0.203	15.5 ± 0.3 15.7 ± 0.6 15.3 ± 0.5	1.73 ± 0.02×10^{-5} 1.75 ± 0.05×10^{-5} 1.70 ± 0.07×10^{-5}	2.6 ± 0.5
0.254	14.4 ± 0.3 14.8 ± 0.6 14.3 ± 0.1	1.60 ± 0.02×10^{-5} 1.65 ± 0.05×10^{-5} 1.59 ± 0.03×10^{-5}	1.7 ± 0.5
0.305	11.9 ± 0.3 12.1 ± 0.4 11.80 ± 0.05	1.32 ± 0.01×10^{-5} 1.35 ± 0.04×10^{-5} 1.31 ± 0.03×10^{-5}	1.4 ± 0.4

* The solid angle in the shadow was more than twice that determined by the detector.

Table 2. Comparison between expected values from the calibrated neutron source and those obtained from the unfolded spectra.

	Expected values without scattering	From unfolded spectrum	
		Without scattering	With scattering
Neutron fluence rate at 1 m × 10^4 ($m^{-2}.s^{-1}$)	90	88	95
Neutron dose equivalent rate at 1 m × 10^{-8} ($Sv.s^{-1}$)	3.42*	3.3	3.5
Mean energy (MeV)	4.4*	4.3**	4.1**

* From Reference 2.
** The thermal neutrons were not considered.

CONCLUSION

The three methods applied to the Bonner multisphere detector enabled testing of the neutron irradiation facility built, and determination of the free field quantities and the room scattering component for the source–detector system.

It is therefore possible to obtain a calibration factor for neutron devices by using the proposed procedures and the neutron irradiation facility available.

The unfolded spectra for the Am–Be source, the neutron fluence and the dose equivalent rates calculated using BUNKI code at unit distance agree within 4% with those obtained from source strength and ISO[5] mean neutron fluence to dose equivalent conversion factors.

REFERENCES

1. Draft Standard Proposal. *Procedures for Calibrating and Determining the Response of Neutron Measuring Devices Used for Radiation Protection.* ISO/TC 85/SC2/WG2/SG 3(B) (1989).
2. Schwartz, R. B. and Eisenhauer, C. M. *Procedures for Calibrating Neutron Personnel Dosemeters.* NBS Special Publication 633 (National Bureau of Standards, Washington, DC) (1982).
3. Hunt, J. B. *The Calibration of Neutron Sensitive Spherical Devices.* Radiat. Prot. Dosim. **8**(4), 239-251 (1984).
4. Schwartz, R. B. *Neutron Personnel Dosimetry.* NBS Special Publication 250-12 (National Bureau of Standards, Washington, DC) (1987).
5. International Standard ISO 8529. *Neutron Reference Radiations for Calibrating Neutron–Measuring Devices used for Radiation Protection Purposes and for Determining their Response as a Function of Energy* (1989).
6. Hunt, J. B. *The Calibration and Use of Long Counters for the Accurate Measurement of Neutron Flux Density.* NPL report RS5 (1976).
7. Bramblett, R. L., Ewing, R. I. and Bonner, T. W. *A New Type of Neutron Spectrometer.* Nucl. Instrum. Methods **9**, 1-12 (1960).
8. Awschalom, M. and Sanna, R. S. *Applications of Bonner Sphere Detectors in Neutron Field Dosimetry.* Radiat. Prot. Dosim. **10**(1-4), 89-101 (1985).
9. Lowry, K. A. and Johnson, I. L. *Modifications to Iterative Recursion Unfolding Algorithms and Computer Codes to Find More Appropriate Neutron Spectra.* NRL Memorandum Report 5340 (Naval Research Laboratory, Washington, DC) (1984).

DEVELOPMENT OF A NEUTRON CALIBRATION FACILITY AT NRPB

T. M. Francis
National Radiological Protection Board
Chilton, Didcot, Oxfordshire OX11 0RQ, UK

Abstract — A neutron calibration facility that conforms to the specifications for calibration and testing laboratories laid down by the UK National Measurement Accreditation Service (NAMAS) and the International Organisation for Standardisation (ISO) is being developed at the National Radiological Protection Board. In the first phase of this programme a medium-sized room located amidst a suite of calibration laboratories has been adapted for the purpose. The methods used for achieving reduction in room-scattered neutrons and leakage of radiation to the adjoining laboratories/access areas are described. The results of measurements carried out to demonstrate the improvements achieved are presented.

INTRODUCTION

NRPB provides a calibration service for both passive and active neutron monitors using ^{252}Cf and ^{241}Am-Be neutron sources. The room assigned for this purpose is located amidst laboratories where calibrations with other types of radiations are also carried out (Figure 1). The suitability of this room as a facility for neutron calibration was investigated in order to reconcile dosimetric specifications with requirements of safety and low leakage radiation levels in adjacent laboratories. A detailed theoretical study previously carried out on neutron scattering levels in the room indicated the need for remedial measures to be taken if the room was to be made suitable for calibration work with neutron radiations to standards that would conform to the specifications for neutron calibration laboratories laid down by the National Measurement Accreditation Service (NAMAS)[1] and International Organisation for Standardisation (ISO)[2]. Furthermore, the situation of the laboratory amidst other work areas, and it not having appropriate shielding on the walls and door, caused unacceptable levels of background radiation in those areas. The computation of the scattering conditions within the room predicted that improvements could be made by lining the scattering surfaces with hydrogenous materials doped with lithium which captures low energy neutrons without emission of appreciable gamma radiation. The objectives of this paper are to describe the methods used for achieving sufficient reduction of neutron scattering in the room and radiation leakage to the adjoining laboratories/access areas, and to present the results of measurements carried out to demonstrate these improvements.

IMPROVEMENTS EFFECTED

The room (originally measured 8.1 m × 5.2 m × 2.7 m high) has solid concrete walls approximately 30 cm thick with no windows. The work of lining the laboratory surfaces was carried out in two stages.

Lining of the ceiling and floor was carried out in early 1984. The hydrogenous material used for the lining was a material originally developed for shielding neutron source containers[3] and has since found wider applications in shielding. It is marketed by Premise Engineering Ltd* under the trademark 'Premadex'. The elemental composition

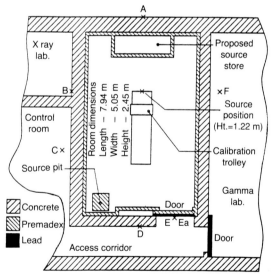

Figure 1. Layout of the laboratory. Measurement positions referred to in Table 3 are indicated by letters A - F. Position Ea denotes measurements with door open.

---

* In this paper certain commercially available products are referred to by their tradenames. Such references are made only for information purposes and should not be taken as endorsement by the Board as the only or the best product available for the intended application.

of the material[4] is given in Table 1. Premadex blocks, 7.5 cm thick, were sandwiched between existing concrete surfaces and newly created plywood false surfaces. The ceiling surface was painted and the floor surface was covered with a proprietary plastic floor covering.

The second stage of development took place in early 1991 and consisted of lining the walls with Premadex and installing a new door. Soft wood frames were fixed on to the concrete surface of walls and Premadex blocks were placed in the framework, forming a lining throughout the entire wall surface of the room. The surface was then finished off in plywood.

The new door measures 170 cm wide × 195 cm high × 13.5 cm thick. It consists of a mild steel frame with a 3 mm mild steel plate covering the outside surface. A 1 cm thick lead sheet was added on the inner surface followed by 7.5 cm thick Premadex lining. Both surfaces of the door were finished in plastic laminates. The door hangs from an overhead track and the bottom of the door moves in a channel cut into the floor.

MEASUREMENTS

Measurements were carried out to assess the reduction achieved in neutron scatter levels in the laboratory at different stages of its development. Scattering levels in terms of the reading of an Andersson-Braun rem counter (Studsvik Model 2202D) were measured before and after the installation of lining on the ceiling and floor. A 185 GBq ^{241}Am–Be α-n source (total neutron emission rate: $\approx 1.2 \times 10^7$ s^{-1}) in a 2 mm thick lead jacket was used for the measurements. In order to quantify the room-scattered component in terms of the reading of the rem counter, measurements were made with and without a 40 cm long polyethylene shadow cone intercepting the primary neutron fluence. The magnitude of the primary fluence at 150 cm (in terms of the instrument reading) is given by the difference between readings obtained with and without the shadow cone. The magnitude of room-scattered neutrons is the value obtained with the shadow cone in place, and was expressed as a percentage of the reading to primary neutrons for the distance of measurement.

After the completion of lining the walls and installing the new door a second series of measurements of room scatter was carried out, using similar methods as before, but with a different shadow cone. The shadow cone used[5] is a 50 cm long truncated cone with a front section (20 cm long) made entirely of iron. The rear section, 30 cm long, is a hollow iron truncated cone filled with a homogeneous mixture of approximately 50% paraffin wax, 10% polyethylene pellets and 40% boron oxide powder, made by adding the powder to a melt of the polyethylene and the wax. The plane surfaces at the front and rear have diameters of 9 cm and 17 cm respectively with a half angle of 4.57°. The optimum distance between the centre of the source used (^{241}Am–Be) and the front end of the cone just shadowing the detector from the primary radiation, was calculated to be 18.6 cm.

The leakage of both neutron and gamma radiations to the adjoining laboratories/access areas was measured at five monitoring points: four at various locations on the outside walls of the laboratory and one on the door. These points and the source position lie on a plane parallel to the floor and are shown in Figure 1. Measurements were carried out before and after the installation of the lining on the walls and door. Neutron leakage in terms of dose equivalent rate was deduced from counts measured over long periods of time with an Andersson-Braun rem counter (Studsvik Model 2202D). Gamma leakage measurements were carried out with a photon dose rate meter employing an energy compensated GM tube (PRM 300) in the integrating mode. Measurements were carried out over periods of several hours and were repeated several times with the ^{241}Am–Be source in normal calibration position.

RESULTS

From the shadow cone measurements, the scatter component as a percentage of the primary component was deduced. Table 2 gives the results

Table 1. Elemental composition of the hydrogenous material (Premadex) used for lining surfaces.

Element	Mass fraction (%)	Atom number density (cm^{-3})
Hydrogen	11.4	6.81×10^{22}
Lithium	1.3	0.11×10^{22}
Oxygen	39.9	1.51×10^{22}
Carbon	47.4	2.38×10^{22}

Table 2. Reading of an Andersson-Braun rem counter to room-scattered neutrons as a percentage of the reading to the primary neutrons (Am–Be) at a source–detector distance of 1.5 m, measured at different stages of room development. The uncertainties are standard deviations.

Stages of room development	Reading ratio 'scattered/primary' (%)
Original room	44 ± 1.5
With floor and ceiling lined	26 ± 1
With all surfaces lined	21 ± 1

for the three stages of room development. Table 3 gives the results of measurements carried out to assess leakage of both neutron and gamma radiations to the adjoining laboratories/access areas. The neutron dose equivalent rate only due to the primary from the source at the position Ea (door open) was calculated to be 3.65 $\mu Sv.h^{-1}$. If this value is subtracted from the values of neutron dose equivalent rates measured at this position before and after lining the walls (Table 3) a comparison of corresponding scatter levels can be made. Such a comparison suggests that there is approximately 30% reduction of scatter at that position after lining the walls.

DISCUSSION AND CONCLUSIONS

The scatter component as a fraction of the primary component (in terms of the reading of an Andersson-Braun rem counter) at the calibration position has been reduced by 52% as a result of the change from the original to the present state of the room. The NAMAS/ISO requirement for an acceptable room scattered component of the measured quantity is now satisfied.

Lining of walls and door has reduced the dose equivalent rate from leakage radiation in areas outside the laboratory, particularly in the access corridor where previously dose rate levels were deemed unacceptable from safety considerations. The detailed measurements performed indicate that the reduction in room scatter achieved as a result of lining the walls and door is significantly less in comparison with that achieved after lining the ceiling and floor. The main reason for this is believed to be the influence of the larger solid angle presented at the calibration position by the ceiling and floor as compared with that presented by the walls. The main benefit derived from lining the walls is the reduction of leakage radiation in areas outside the laboratory.

ACKNOWLEDGEMENT

The author would like to thank Dr David Thomas (National Physical Laboratory, Teddington) for making the shadow cone available for my use and Mr Jocelyn Martin for enthusiastically carrying out the measurements.

Table 3. Neutron and gamma dose equivalent rates measured before and after the last stage of room development. Measurement points are indicated by letters A–F. Position Ea denotes measurements with door open.

Measurement positions	Neutron dose equivalent rates				Gamma dose equivalent rates				Reduction in combined neutron and gamma dose equivalent rates (%)
	Before lining walls and door		After lining walls and door		Before lining walls and door		After lining walls and door		
(See Figure 1)	($\mu Sv.h^{-1}$)	Relative uncertainty 1 SD (%)	($\mu Sv.h^{-1}$)	Relative uncertainty 1 SD (%)	($\mu Sv.h^{-1}$)	Relative uncertainty 1 SD (%)	($\mu Sv.h^{-1}$)	Relative uncertainty 1 SD (%)	
A	0.06	± 16	0.03	± 8	–	–	–	–	–
B	0.32	± 5	0.16	± 10	0.64	± 3	0.18	± 3	65
C	0.09	± 9	0.05	± 11	0.53	± 3	0.15	± 6	68
D	0.63	± 3	0.25	± 5	0.43	± 3	0.19	± 2	58
E	6.20	± 2	2.40	± 2	0.40	± 2	0.29	± 2	59
Ea (Door open)	8.06	± 2	6.69	± 2	–	–	–	–	–
F	1.06	± 3	0.47	± 4	0.96	± 3	0.30	± 6	62

REFERENCES

1. National Measurement Accreditation Service (NAMAS). Information Sheet B0813. *Calibration of Radiological Protection Level Instruments: Neutrons*. (NAMAS Executive, National Physical Laboratory, Teddington, Middlesex, England TW11 0LW) (1989).
2. ISO. Draft Standard Proposal. *Procedures for Calibrating and Determining the Reading of Neutron Measuring Devices Used for Radiation Protection*. ISO/TC 85/SC 2/WG 2/SG 3 (B), International Organisation for Standardisation (1990).
3. Sherwin, A. G. *Neutron Shielding — Some Recent Developments*. Radiol. Prot. Bull. **57**, 19-24 (1984).
4. Premise Engineering Ltd. *Premadex Neutron Shielding* (Premise Engineering Ltd, Abbots Langley, Watford, Herts, WD5 0DR, England).
5. Hunt, J. B. *The Calibration and Use of Long Counters for the Accurate Measurement of Neutron Flux Density*. NPL Report RS(EXT)5 (National Physical Laboratory, Teddington) (1976).

NEW QUANTITIES FOR USE IN RADIATION PROTECTION

C. B. Meinhold
National Council on Radiation Protection and Measurements
7910 Woodmont Avenue, Bethesda, Maryland, USA

INVITED PAPER

Abstract — The ICRP has made significant changes in its recommendations relative to quantities and units reflecting concern over the precision implied by the Q-LET relationship as given in ICRU 40. The commission has adopted a radiation weighting factor, w_R, to modify the average absorbed dose, D_T, in an organ or a tissue. This results in the equivalent dose, $H_{T,R}$, in that tissue or organ. The equivalent dose is then modified by use of a tissue weighting factor which reflects the relative sensitivity of that tissue or organ to serious stochastic effects. Summing all the doubly modified tissue or organ absorbed doses results in the effective dose. The ICRU metrological quantities, such as the ambient dose equivalent, are expected to provide adequate estimates of the effective dose.

INTRODUCTION

In its Publication 60, the International Commission on Radiological Protection introduced the radiation weighting factor, w_R. This decision was based on a number of considerations which are described briefly here. Since ICRP Publication 60, namely, Chapters 2 and 3 and Annexes A and B, is the only appropriate source for 'The Commission Said' statements, the paragraphs given here are those of the author and should be ascribed to him and not to the ICRP *per se*.

First, with rare exception, the Commission based its recommendation on the fundamental assumption of a linear relationship between the dose and probability of fatal cancer or other stochastic detrimental effect. It therefore follows that once the sensitivity of an individual organ or occasionally that of a distinct tissue within an organ is established, it is only the absorbed dose in that tissue or organ that is of interest. The Commission then explicitly defined the absorbed dose as $D = d\bar{\varepsilon}/dm$ and the average tissue dose as $D_T = \varepsilon_T/m_T$.

Second, the Commission was concerned about the implications of the L or $\bar{y}$ relative to Q formulation given in ICRU 40. This elegant and extraordinarily important relationship certainly helps with the understanding of the mechanism of biological damage from a biophysical point of view. However, as can be seen in Table 1, due to the great variability in RBE values reported for each of the endpoints, the variations in these values between endpoints, the variation in the physical size of the various biological systems tested and because we have no human data, the Commission believes that essentially a multi-attribute approach is required in the selection of weighting factors needed to account for the influence of LET (or $\bar{y}$) is needed.

NEW QUANTITIES

Recognising the influence of the ICRU quantities in the measurement of radiation exposure (in its broadest sense) and to assist in its own selection process of these weighting factors the Commission defined a new set of Q against LET values drawing heavily on ICRU Report 40. This relationship is shown in Table 2.

The three aspects of this table are important: first, the lower level cut off of 10 keV.μm^{-1}. This was done in recognition that although there is radiobiological and biophysical reason to suggest there might be an RBE of two or three between 250 kV X rays and ^{60}Co gamma rays, distinction between these in operational radiation protection is unwarranted. The basic relationship was devel-

Table 1. RBE$_M$ for fission (or optimum energy) neutrons versus fractionated γ rays.

Type of effect	RBE$_M$
Tumour induction	≈15 – 60
Life shortening	15 – 45
Transformation	35 – 70
Cytogenetic studies	40 – 50
Genetic endpoints in mammalian systems	10 – 45
Other endpoints	
Lens opacification	25 – 200
Micronucleus assay	6 – 60
Testes weight loss	5 – 20

Table 2. Q-L relationship.

Unrestricted linear energy transfer, L(keV.μm^{-1})	Q(L)*
< 10	1
> 10 – 100	0.32L – 2.2
> 100	300/√L

oped by Kellerer who showed that one could interpret the ICRU 40 relationship in L as well as in $\bar{y}$. The selection of 10 keV.µm^{-1} ensured that the ambient dose equivalent calculated for the same absorbed dose of X or gamma rays be nearly equal; again, a practical matter in radiation protection. The upper level cut off of 100 keV.µm^{-1} was accepted by the Commission as the LET with a maximum potential for deleterious effects. As Kellerer points out, the basic data might suggest 1/L dependence, in the interest of caution, based primarily on the sparse data, 1/√L was adopted. Third, the Commission's selections of values of weighting factors are intended to reflect the lack of precise radiobiological knowledge that exists. This is also the reason the Commission recommends discrete values for the weighting factors for a range of radiation qualities as can be seen in Table 3.

Realising that many of those who perform calculations would prefer a smooth function for applying the neutron w_R recommendations, the Commission provides a simple smooth function with the expression:

$$w_R = 5 + 17 \exp - [\ln(2E)]^2/6$$

Primarily because of the suggestion that the present state of affairs arose partly as a result of the introduction of the smooth curve given in ICRP 21 for this purpose, the new recommendations contain a specific caution to the effect that there is no radiobiological significance to any aspect of this relationship. It is provided as a calculational tool which should enable a common basis to assist in comparing calculational results. To summarise to this point, the Commission fully intends that the effective dose, E, and the equivalent dose in organ or tissues, H_T, are the primary quantities in the application of its basic recommendations, i.e. using the w_R values as given in Table 2 and/or the smooth curve in Figure 1 will result in the correct values of the effective dose and the equivalent dose.

SECONDARY QUANTITIES

However, the Commission has traditionally recommended secondary quantities (ALIs for internal exposure and the ICRU quantities for external exposure). In Publication 26, the Commission suggested the use of the dose equivalent index for this purpose and Publication 60 contains the explicit statement in paragraph A-24 from Appendix A:

'The use of the ICRU quantities as given in ICRU Report 39 (ICRU 1985) are expected to give reasonable approximations of the effective dose and the equivalent dose to the skin when these quantities are calculated using the Q-L relationship given in Table A-1, the Commission will be re-examining these dosimetric quantities in detail as part of a general revision of ICRP Publication 51 (ICRP 1987) which will incorporate new radiation weighting factors.'

Although the Commission does not intend to constrain the development of other metrological quantities to establish compliance with the dose limiting objectives, it is just as clearly not discouraging the use of the ICRU quantities as given in its Publication 39. With regard to the revision of ICRP 51, it has been decided that it would be very helpful if this could be in conjunction with the Revision of ICRU 39. Consequently, a joint task group of the ICRP and the ICRU has been established to provide data for use in radiation protection against external radiation. They have been asked (1) to provide fluence to effective dose calculations for a variety of radiations and energies and for reference man, women, 15 year, 5 year, and 3 month old children, (2) to provide fluence to ambient dose equivalent, directional dose, individual dose equivalent penetrating, and individual dose superficial, and (3) to provide a detailed discussion of the relationship between these two sets of calculations.

When these comparisons have been completed, the Commission will provide guidance as to the application of these quantities. For example, if under some radiation geometries for some irradiation qualities the ambient dose equivalent underestimates or greatly overestimates the effective dose by some factor the Commission may suggest that this is acceptable or unacceptable. It is not necessarily true that the metrological quantities must overestimate the effective dose.

The question of the legal or regulatory aspects of these relationships have been raised, but in fact the Commission's previous recommendations on

Table 3. Radiation weighting factors.

Type and energy range	Radiation weighting factor, w_R
Photons, all energies	1
Electrons and muons, all energies	1
Neutron, energy < 10 keV	5
10 keV to 100 keV	10
> 100 keV to 2 MeV	20
> 2 MeV to 20 MeV	10
> 20 MeV (See also Figure A-1)	5
Protons, other than recoil protons, energy > 2 MeV	5
Alpha particles, fission fragments, heavy nuclei	20

dose equivalent limits have been met not by precise measurements of that quantity but on the reliance on metrological quantities such as the ambient dose equivalent, etc. It is hoped that the community of dosimetrists and metrologists might not limit themselves to the ICRU previously defined quantities but examine other measurable quantities for meeting these new recommendations; keeping in mind the lack of precision implied by the adoption of the discreet values of w_R.

WEIGHTING FACTORS

It should be recognised that similar consideration went into the derivation of the tissue weighting factors. In fact, the incidence of fatal cancers in the Japanese population is measured. Projecting that information to obtain the total stochastic related detriment to an average population of the world requires a number of assumptions. The results of all of these assumptions can be seen in Table 4.

What of course is remarkable here, is the number of significant figures in the column headed 'total detriment'. Table 5 lists the weighting factors which result from a review of this data.

Here you will note that there are tissues with rather high sensitivity, tissues with rather moderate sensitivity, and tissues with rather low sensitivity rather like the lack of precision given in Table 2. It is true that this argument would be a little stronger if a number of tissues such as bone marrow had a weighting factor of 0.1 rather than 0.12. In this table, you will notice that the remainder tissues are given a value of 0.05. As part of its overall approach to simplification the Commission would have liked to have simply distributed that remainder weighting factor against all of the remaining mass of the body; that is, that part of the 70 kg reference man which was not used up by listed organs. That seemed to be a good idea until it was recognised that part of the 70 kg is made up of bladder and stomach contents,

Table 5. Tissue weighting factors.

Tissue or organ	Tissue weighting factor, w_T
Skin	0.01
Bone surface	0.01
Bladder	0.05
Breast	0.05
Liver	0.05
Oesophagus	0.05
Thyroid	0.05
Remainder	0.05
Bone marrow (red)	0.12
Colon	0.12
Lung	0.12
Stomach	0.12
Gonads	0.20

Table 6. Remainder tissues.

Adrenals
Brain
Upper large intestine
Small intestine
Kidney
Muscle
Pancreas
Spleen
Thymus
Uterus

Table 4. Relative contribution of organs to the total detriment.

	Probability of fatal cancer [a]	Relative length of life lost	Relative non-fatal contribution	Total detriment[a]	Relative contribution
Bladder	30	0.62	1.50	27.9	0.040
Bone marrow	50	1.97	1.01	98.6	0.142
Bone surface	5	1.00	1.30	6.5	0.009
Breast	20	1.12	1.50	33.6	0.048
Colon	85	0.80	1.45	98.6	0.142
Liver	15	1.00	1.05	15.8	0.023
Lung	85	0.86	1.05	76.8	0.111
Oesophagus	30	0.73	1.05	23.0	0.033
*Ovary	10	1.07	1.30	13.9	0.020
Skin	2	1.00	2.00	4.0	0.006
Stomach	110	0.80	1.10	96.8	0.139
Thyroid	8	1.00	1.90	15.2	0.022
Remainder	50	0.87	1.29	56.1	0.081
*Gonads		0.27	–	127.0	0.183
Total	500			693.8	0.999

[a] Per 10,00 people per Sv.
* Gonads (including cancer in ovary).

mineral bone, fat, and connective tissues, none of which are sensitive to induction of cancer. As a result, the Commission has defined the following tissues as being the remainder tissues (Table 6).

The calculation for the remainder tissues is simplified, however, in that it is simply the total dose distributed to the tissues divided by the total mass of the tissues.

The effective dose then is simply the sum of the weighted equivalent doses in all the tissue organs of the body given as:

$$E = \sum_T w_T H_T$$

where H_T is the equivalent dose in tissue or organ t and w_T is the weighting factor for tissue T. Perhaps the most important summary statement is that the w_T values are entirely independent of the radiation quality and more importantly the w_R values are entirely independent of the tissue or organ irradiated, i.e.

$$E = \sum_R w_R \sum_T w_T D_{T,R} = \sum_T w_T \sum_R w_R D_{T,R}$$

SUMMARY

In summary, Q and w_R are considered to be complementary, but it will be the Joint Task Group of ICRP and ICRU and workers in the field who will provide the definitive answer to this question.

CALCULATED EFFECTIVE DOSES IN ANTHROPOID PHANTOMS FOR BROAD NEUTRON BEAMS WITH ENERGIES FROM THERMAL TO 19 MeV

R. A. Hollnagel
Physikalisch-Technische Bundesanstalt
Braunschweig, Germany

Abstract—Calculations are presented of the equivalent organ doses H_T and the corresponding effective doses E for neutrons according to the new proposals in ICRP Publication 60. Five broad neutron beam irradiations of a standing anthropoid male phantom were chosen: frontally, from the back, laterally, rotationally around the vertical phantom axis and isotropically. The neutron energies range from thermal to 19 MeV. Various numerical aspects of equivalent doses and the effective dose are charted.

INTRODUCTION

In their Publication 60[1] (henceforth abbreviated to ICRP 60) the International Commission on Radiological Protection has recently recommended new concepts for 'equivalent doses' of organs, H_T, introducing radiation weighting factors, w_R, instead of the formerly used quality factor. The weighted sum of the equivalent organ doses was given the name 'effective dose', E. ICRP 60 contains two major changes in the definition of the effective dose, E, compared with ICRP Publication 26[2]: first, the equivalent dose H_T in tissue or organ T was introduced,

$$H_T = \sum_R w_R D_{T,R} \quad (1)$$

where $D_{T,R}$ is the absorbed dose averaged over the organ T, due to radiation R. The radiation weighting factor w_R is selected for the type and energy of the radiation incident on the body, which is an important alteration as the radiation quality is no longer attributed to the radiation scattered in the body; this also includes photons induced within the body by neutron transport and contributing to $D_{T,R}$, a rather unusual idea for those active in the field of neutron dosimetry for a long time.

Second, E is then computed using the familiar formula

$$E = \sum_T w_T H_T \quad (2)$$

the sum of weighted equivalent doses in tissues or organs. The tissue weighting factors w_T are given for twelve organs instead of six, plus a 'remainder' for which a further ten organs are listed. The weighting factor of the remainder was reduced from 0.3 to a value of 0.05 and its dose is now easier to compute. The weighting factor for the remainder was interpreted to mean that each of the ten organ doses in the remainder had in effect a $w_T = 0.005$ when adding according to Equation 2. Since E refers to the exposed indivdual, no specification for an anthropoid phantom is given in ICRP 60.

In the present paper, results for a male phantom alone are presented and those for a female phantom will be added later together with tables of E and H_T in the male phantom. The phantom was irradiated with parallel neutron beams covering it homogeneously. Five beam geometries were selected as representative of a wide range of practical applications, some of them of composite directions. The monoenergetic neutron energies E_n ranged from thermal to 19 MeV, inclusive the bare ^{252}Cf neutron spectrum. Below E_n = 100 keV the induced photon contribution to E rose considerably in relation to that for the earlier effective dose equivalent, H_E. The mutual ratios of E for the various beam geometries are of interest because they show roughly how E depends on the beam geometry. The ratios of E for the five beam geometries to the operational quantity ambient dose equivalent H*[3] computed with the new Q(L) relationship[1], is of major significance in investigating whether H* is a conservative estimate of E, at least for the phantom used here. The share of the organs to which a w_T was assigned in ICRP 60, plus the remainder in the effective dose, indicates, for instance, the sensitivity of E to future changes of the tissue weighting factors w_T, the possibility of which is expressed in ICRP 60. However, these shares depend on energy E_n and beam geometry; any inferences from dosemeter indications (e.g. for epidemiological research) to organ doses are possibly more difficult than the estimate of E would be. In the present paper no interpretations or inferences based on curve shapes or given numerical values have been attempted.

PHANTOM AND COMPUTER PROGRAM

The mathematical model of a human male was the same as that in a report[4], which had already been used in an earlier paper on calculations of

$H_E^{(5)}$. Almost all the organs quoted in Table 2 of ICRP 60 are to be found in the report, with the exception of the oesophagus and muscles. The oesophagus was then taken from Reference 6 and fitted into the ADAM model[4]. A small section of

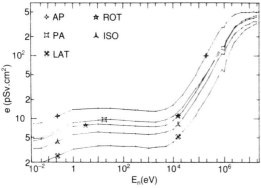

Figure 1. Effective dose per unit incident fluence ϕ, e = E/ϕ, over neutron energy E_n for five irradiation geometries.

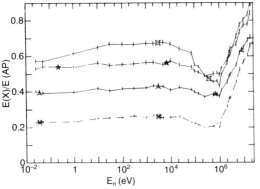

Figure 2. Ratios of the effective doses E at the incidences X = PA, LAT, ROT, ISO to the effective dose at AP. Symbols as Figure 1.

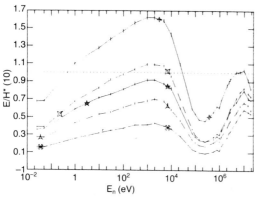

Figure 3. The ratios of the effective doses E to the ambient dose equivalent H*(10) at five incidences Symbols as Figure 1.

the heart had to be removed because the oesophagus intersected the heart, and likewise a small piece of the oesophagus was cut away at its outlet into the stomach. After establishing that these penetrations of organs by the oesophagus lead to program errors, the remedy was simply to redefine the combination of simple geometrical bodies[5]. The gullet's new volume was determined through scanning the MIRD phantom with a uniform mesh of pixels. The volume calculated was almost exactly the same in cm³ as the mass given in grams[6], namely 43.5 cm³. The heart's volume was only slightly reduced from 603 cm³ to 599.3 cm³. A section in the trunk of height h= 45 cm and volume V= 2794 cm³ between skin and rib cage not including arm bones, and the legs not including leg bones, were selected for the 'muscle' tissue. Here the skin tissue was the outer layer, about 0.1 cm thick, on the ellipsoidal surface of the MIRD trunk with a volume of 657.5 cm³. The other organs were as taken for the previous calculations of $H_E^{(5)}$, and the transport model and data base were also the same. Since the male phantom has no breasts ($w_{T=breast}$ = 0), the weighting factors w_T were renormalised to unity. The computational uncertainties are now much better than they were in the H_E paper, and range from 1% to 3% for E.

IRRADIATION CONDITIONS

For the upright MIRD phantom, this was: (1) monodirectional on the front of the body (anterior → posterior = AP), (2) monodirectional on the back (PA), (3) lateral with equal incidence from each side (LAT), (4) rotational around the vertical phantom axis (ROT) and finally (5) isotropic (ISO). The five beam geometries were computed at each of the 28 neutron energies E_n. The first one represents a thermal spectrum with the parameter kT= 0.0253 eV, whilst the second, E_n= 0.0506 eV, is for monoenergetic start neutrons with the average thermal energy; the remaining energies are all monoenergetic. They are marked in the figures by small vertical dashes.

RESULTS

All the results given below were normalised to unit fluence, i.e. are fluence to effective dose conversion functions. Nevertheless they will still be denoted as effective dose for the sake of coherence in the text. Of the five irradiation geometries investigated, the AP incidence had the maximum effective dose. Lateral incidence leads to the lowest E and the second lowest E is at ISO incidence. The effective doses for PA and ROT incidence run relatively close together, intersecting at about 200 keV (Figures 1, 2).

Figure 3 demonstrates that H*(10), the ambient dose equivalent, is not a conservative estimate of E calculated here. An analytical fit for H*(10) has been taken from Schuhmacher and Siebert[3]. At AP incidence E is partly underestimated, but, for instance, for broad spectra with isotropic incidence E could be overestimated considerably in practical workplaces. Figure 4 shows what would happen if only the heavy particle doses were to be multiplied by w_R to obtain E. Here the effective dose equivalent H_E as calculated by Wittmann et al[7] according to ICRP 26 has been put into Figure 4. A comparison of Figures 1 and 4 points out that the remarkable rise of E below 100 keV can be attributed to the ICRP proposal to endow a radiation weighting factor $w_R > 1$ not only to neutrons but also to induced photons. The photon portion in E, weighted with the w_R relating to the incident neutron energy, is dominant (95%) below $E_n = 10$ keV and starts to decrease steeply at about 20 keV. At $E_n = 100$ keV the photon share ranges from 60% (AP) to close on 80% (PA and LAT). The computation of E for a bare ^{252}Cf source revealed that even with this spectrum, the influence of the induced photons was not insignificant: not using the factors w_R on its photon doses would reduce E by 9.1%, 13.0%, 10.7%, 10.7%, 10.2% respectively, for our five beam geometries.

The relevance of annotation no. 3 to Table 2 in ICRP 60[1] for neutron exposure was also investigated: in Figure 5, the 'cases in which a single one of the remainder tissues or organs receives an equivalent dose in excess of the highest dose in any of the twelve organs for which a weighting factor is specified, a weighting factor of 0.025 should be applied to that tissue or organ and a weighting factor of 0.025 to the average dose in the rest of the remainder' are plotted as isolated symbols; the curves give the usual remainder dose. The effect on E is still small (mostly under 3%), so when the ED is calculated for irradiation with broad neutron beams, these possible cases can be neglected. Figure 6 gives the partial sum

$$\sum_{T=1}^{i} w_T H_T$$

i.e. the relative contributions of the organ doses H_T to the effective dose E, successively summed up in the sequence given in the figure captions, for AP incidence. Skin and bone surfaces were combined. At AP incidence the highest contributions are made by the testes and the stomach. In general

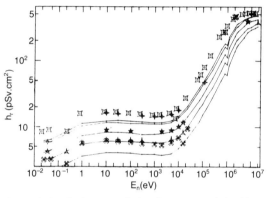

Figure 5. Equivalent remainder doses per unit incident fluence, h_r, at five incidences. The curves show dose averages over the ten remainder organs multiplied by the respective radiation weighting factors w_R, the isolated symbols refer to those cases where footnote 3 to Table 2 in ICRP 60 applies. Symbols as Figure 1.

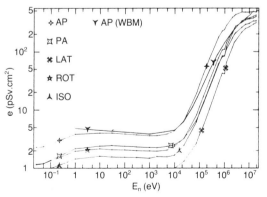

Figure 4. Effective doses per unit incident fluence, e, when the radiation weighting factor w_R is applied only to the neutron dose but not to the dose of the induced photons. At AP incidence the effective dose equivalent h_E from Wittmann et al[7] is shown for comparison.

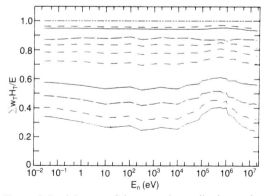

Figure 6. Partial sums of the organs' contribution to the effective dose E for AP incidence. The organs are successively: testes (lowest curve), red bone marrow, colon, lung, stomach, bladder, liver, oesophagus, thyroid, skin and bone surface together, remainder.

bone surfaces and skin together make the smallest contribution to E. The trend for isotropic incidence in relation to AP is a more even contribution of all organs. For both incidences, no sharp changes with neutron energy become apparent.

CONCLUSIONS

The results of the Monte Carlo estimates for the five different neutron incidences on an anthropoid male phantom suggest that attention should be paid not only to frontal incidence but also to neutron field geometries encountered in workplaces, in order to avoid a source of an unnecessary overestimation of the effective dose, E. Moreover, the operational quantity ambient dose equivalent H*(10) is not a conservative estimate of E at every neutron energy; E exceeds H*(10) at frontal incidence by up to 60% for neutron energies between 1 eV and 30 keV. The role of the induced photons in E is predominant below a neutron energy of about 10 keV.

REFERENCES

1. ICRP. *1990 Recommendations of the International Commission on Radiological Protection.* Publication 60 (Oxford: Pergamon Press) (1991).
2. ICRP. *Recommendations of the International Commission on Radiological Protection.* Publication 26 (Oxford: Pergamon Press) (1977).
3. Schuhmacher, H. and Siebert, B.R.L. *Quality Factors and Ambient Dose Equivalent for Neutrons Based on the New ICRP Recommendations.* Radiat. Prot. Dosim. **40**, 85 - 89 (1992).
4. Zankl, M., Williams, G. and Drexler, G. *The Calculation of Dose from External Photon Exposures using Reference Human Phantoms and Monte Carlo Methods, Part I: The Male (ADAM) and Female (EVA) Adult Mathematical Phantoms.* Report GSF S-885 (1982).
5. Hollnagel, R. *Effective Dose Equivalent and Organ Doses for Neutrons from Thermal to 14 MeV.* Radiat. Prot. Dosim **30**, 149-159 (1990).
6. Lewis, C.A. and Ellis, R.E. *Additions to the Snyder Mathematical Phantom,* Phys. Med. Biol. **24**, 1019-1024 (1979).
7. Wittmann, A., Morhart, A. and Burger, G. *Organ Doses and Effective Dose Equivalent.* Radiat. Prot. Dosim. **12**, 101-106 (1985).

EQUIVALENT DOSE VERSUS DOSE EQUIVALENT FOR NEUTRONS BASED ON NEW ICRP RECOMMENDATIONS

K. Morstin†, M. Kopec† and Th. Schmitz‡
†Institute of Physics and Nuclear Techniques AGH
al. Mickiewicza 30, PL-30059 Krakow, Poland
‡Institute of Medicine, Research Center KFA Jülich
POB 1913, W-5170 Jülich, Germany

Abstract — Recent ICRP recommendations include changes in the concept of radiation quality and weighting factors. This justifies new calculations of primary limiting quantities. The male (ADAM) and female (EVA) phantoms, based on MIRD specifications, seem to be well suited standards for this purpose. 3-D Monte Carlo calculations of radiation transport throughout the ADAM phantom and of energy release in particular organs have been performed for neutrons of energies from thermal up to 20 MeV, and for different irradiation geometries. Energy transfer by neutron-induced charged secondaries has been described using the continuous slowing down approximation (csda). Resulting LET spectra are the physical measure of radiation quality. They differ considerably from organ to organ and depend on irradiation geometry, especially for low energy neutrons. Newly introduced quantities, i.e. equivalent doses in organs and the effective dose in man, assessed without taking into account these variations, do significantly overestimate the dose equivalent values evaluated using LET based quality factors.

INTRODUCTION

As the result of recent re-evaluations of carcinogenic risk estimates, the International Commission on Radiological Protection issued new recommendations[1], including changes in the concept of radiation quality and weighting factors. Apart from reformulating the relationship of quality factor and LET and changing the number and magnitude of tissue weighting factors, which make obsolete previous calculations of phantom related dose equivalents[2-5], ICRP introduced new quantities, the equivalent dose in tissue or organ, H_T, and the effective dose, E. They are supposedly thought to replace presently used organ and effective dose equivalents (H_T and H_E), based on quality factor concept. The equivalent dose in tissue is defined as

$$H_T = \sum w_R \cdot D_{T,R} \qquad (1)$$

where $D_{T,R}$ are the fractions of the absorbed dose, averaged over the tissue or organ, T, due to the incident radiation fractions, R, the quality of which is characterised by the radiation weighting factors, w_R. (It is worth mentioning here that spectrometry of an undisturbed field is an intricate metrological task compared with direct monitoring of radiation quality with tissue-equivalent instruments[6,7]). The effective dose defined as

$$E = \sum w_T H_T \qquad (2)$$

where w_T are the tissue weighting factors, can thus be determined from knowledge of the undisturbed field, i.e.

$$E = \sum w_R \sum w_T D_{T,R} \qquad (3)$$

provided the organ-related dose fractions, $D_{T,R}$ are otherwise known. In fact, the latter are absolutely non-measurable quantities. On the other hand, information required to calculate $D_{T,R}$ is usually sufficient also to evaluate the organ-related mean quality factors, $\bar{Q}_{T,R}$. The intentional simplification, with respect to the concept of dose equivalent, lies in the fact that radiation weighting factors, w_R, being assigned to the undisturbed field do not depend on exposure geometry and do not vary from organ to organ. For neutrons, for example, ICRP actually proposed[1] three values of w_R factors: 5, 10 and 20, based on a rough interpretation of the results of the revised calculations[8] of the ambient dose equivalent[9], H*(10), thus defining a somewhat artificial step function over the incident neutron energy. This is considered an acceptable approximation and implies the assumption that the radiation quality at 1 cm depth in the ICRU sphere exposed to a broad parallel beam is reasonably representative for the whole human body, regardless of the actual irradiation geometry.

To quantify the impact of such simplifications we calculated, for a male anthropoid phantom, the newly defined quantities (equivalent and effective doses), as well as dose equivalents (organ and effective) which are still referred to in as yet unrevised national regulations. The latter were calculated with both recently[1] and previously[10] proposed Q-L and w_T specifications. The calculations were performed for varying incident neutron energies and for several irradiation geometries. Due to the lack of space only a fraction of the results can be presented here. The more comprehensive report is in preparation[11].

CALCULATIONS

The applied mathematical phantom is a slightly modified version of the male adult phantom, ADAM, developed at GSF[12] based on MIRD specifications[13]. The oesophagus has been added to satisfy the new ICRP requirements, and some other less important changes in geometry have been introduced to avoid organ intersections (probably unintentional). The tissues and their material composition follow exactly the MIRD-5 specifications, except for the trace amounts of Zn, Rb and Sr being neglected. The phantom is now composed of 32 organs, including all those to which ICRP assigned the new w_T factors. The remainder has been re-defined accordingly. Sex-averaged w_T factors were actually slightly modified in calculations (divided by 0.95) to account for lack of breast ($w_T = 0.05$) in the phantom. The layout of particular organs is illustrated in Figure 1, in which several intersections are overlapped in order to mimic the 3-D perspective.

The Monte Carlo method was applied to calculate neutron transport as well as induction of secondary gamma rays and their subsequent propagation in such a complex 3-D geometry. The MCNP transport code has been selected for this purpose[14]. It has been associated with the pointwise nuclear data sets derived from ENDF, ENDL and LASL evaluations. Thermalisation of neutrons was described with the $S(\alpha,\beta)$ scattering law[15], thus allowing for a complete representation of elastic and inelastic interactions of thermal neutrons with molecules (water-bound hydrogen at 300 K was assumed for all tissues).

Several irradiation geometries were investigated, including: anterior–posterior (AP), posterior–anterior (PA), left lateral (LL) and right lateral (RL), for which a broad parallel beam impinges on the front, back, and from both sides of the phantom, respectively. In each case the calculations were performed for 33 incident neutron energies, from thermal up to 20 MeV. In the case of the thermal beam, the Maxwellian spectrum at room ambient temperature (293 K) was assumed.

The resulting neutron spectra in particular organs were used in the subsequent calculations of charged particle production and their slowing down. This has been done with the NESLES program[16] which makes use of UKAEA cross sections and realises the csda approach with stopping powers allowing for molecular effects. The results were used to derive LET spectra and, subsequently, the organ-averaged quality factors (with both considered Q-L specifications). LET spectra are the physical measure of radiation quality. Figure 2 shows the LET spectra averaged over red bone marrow (RBM) and stomach, calculated for incident 2 keV neutrons, AP and PA exposures. The contributions due to recoil protons, protons from the exoenergetic (n,p) reaction with ^{14}N, and related ^{14}C recoils can be clearly distin-

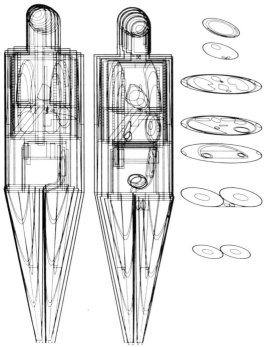

Figure 1. Schematic representation of the modified ADAM phantom by superpositions of vertical and planar sections.

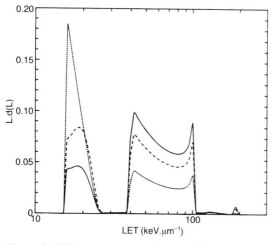

Figure 2. LET spectra averaged over red bone marrow and stomach due to 2 keV neutrons incident on the ADAM phantom, calculated for AP and PA exposures. Photon-induced contributions not included. (—) RBM, AP and PA: (– – –) stomach, AP: (- - - -) stomach, PA.

guished. Photon-induced electrons are not included in Figure 2; had they been, the differences between organs and/or exposure geometries would have been even more pronounced. Obviously, these local variations in radiation quality were not taken into account in the evaluation of newly introduced quantities, for which w_R factors in the form of the ICRP-defined step function were applied.

When evaluating the organ-averaged values of the absorbed dose, the ENDF-based kerma factors[17] (rather than those obtained from NESLES) were convoluted with the neutron fluence spectra in particular organs, which assured consistency with cross sections used in transport calculations. The kerma approach was also applied when evaluating photon fractions of the absorbed dose, this time, however, directly by MCNP (referring to the kinematics of photon interactions). Quality factors for photons were assigned to unity.

RESULTS

Results of calculations of primary limiting quantities in the male humanoid phantom ADAM exposed to neutrons are summarised in Table 1, for selected energies and organs, and for the AP incidence only. The complete set of the results will be given elsewhere[11]. All items in Table 1 are normalised per unit incident fluence. Computational uncertainties are given (in parentheses) in the form of relative standard deviations in per cent. The values of equivalent and effective doses are denoted by the symbol 60w, whereas dose equivalents are

Table 1. Equivalent doses (60w) and dose equivalents (60q and 26q) per incident neutron fluence (in pSv.cm^2), effective and for main organs, AP exposure. Relative computational uncertainties in per cent. See text for further explanations.

Energy		Effective	Testes	RBM	Lung	Stomach	Colon
20 MeV	60w	703.84 (0.9)	781.19 (3.6)	538.17 (0.4)	725.23 (0.8)	787.61 (1.5)	723.49 (1.1)
	60q	632.01 (1.0)	726.94 (3.6)	511.59 (0.4)	633.65 (0.8)	697.75 (1.5)	629.97 (1.1)
	26q	560.39 (1.2)	634.72 (3.6)	472.66 (0.4)	565.36 (0.8)	625.40 (1.5)	568.86 (1.1)
14 MeV	60w	629.16 (0.9)	707.73 (3.5)	463.50 (0.4)	647.65 (0.8)	714.60 (1.5)	643.58 (1.1)
	60q	462.59 (0.9)	534.33 (3.6)	375.73 (0.4)	461.90 (0.8)	514.94 (1.5)	454.42 (1.1)
	26q	456.23 (1.3)	522.89 (3.6)	379.93 (0.4)	459.91 (0.8)	515.08 (1.5)	458.23 (1.1)
10 MeV	60w	533.53 (0.9)	618.52 (3.4)	377.86 (0.4)	543.49 (0.8)	616.07 (1.5)	539.39 (1.1)
	60q	448.45 (1.0)	530.17 (3.5)	331.86 (0.4)	457.50 (0.8)	511.32 (1.5)	437.09 (1.1)
	26q	391.88 (1.3)	462.47 (3.5)	301.20 (0.4)	399.60 (0.8)	451.19 (1.5)	388.04 (1.1)
5 MeV	60w	403.24 (1.0)	498.34 (3.4)	262.30 (0.4)	411.14 (0.8)	464.06 (1.5)	391.43 (1.1)
	60q	336.33 (1.0)	422.82 (3.5)	227.23 (0.4)	342.46 (0.8)	383.23 (1.5)	315.56 (1.1)
	26q	301.82 (1.3)	378.16 (3.5)	211.55 (0.4)	306.06 (0.8)	345.17 (1.5)	285.30 (1.1)
2 MeV	60w	492.35 (1.1)	723.33 (3.3)	277.10 (0.4)	463.56 (0.9)	571.57 (1.5)	420.98 (1.1)
	60q	342.81 (1.2)	532.98 (3.4)	184.74 (0.4)	325.36 (0.9)	393.84 (1.6)	263.36 (1.3)
	26q	242.02 (1.5)	360.54 (3.4)	130.12 (0.4)	222.19 (0.9)	269.85 (1.6)	182.31 (1.3)
1 MeV	60w	280.69 (1.4)	506.70 (3.4)	141.72 (0.5)	216.54 (1.0)	314.57 (1.8)	184.65 (1.5)
	60q	253.50 (1.6)	506.85 (3.6)	114.11 (0.5)	193.42 (1.2)	271.01 (1.9)	128.61 (1.8)
	26q	169.78 (1.9)	303.21 (3.6)	70.64 (0.5)	116.57 (1.2)	163.91 (1.9)	78.78 (1.8)
500 keV	60w	189.94 (1.3)	329.40 (3.3)	102.93 (0.6)	148.69 (1.0)	211.31 (1.8)	136.34 (1.6)
	60q	154.06 (1.6)	311.48 (3.6)	70.24 (0.5)	119.82 (1.2)	164.67 (2.0)	72.02 (2.0)
	26q	100.91 (2.0)	181.50 (3.6)	41.96 (0.5)	69.55 (1.2)	96.08 (2.0)	42.78 (1.9)
200 keV	60w	130.24 (1.4)	227.30 (3.5)	74.52 (0.7)	91.87 (1.2)	144.34 (2.1)	94.03 (1.9)
	60q	86.30 (1.9)	199.15 (3.7)	40.12 (0.5)	54.50 (1.4)	80.67 (2.1)	27.07 (2.1)
	26q	60.17 (2.1)	115.43 (3.7)	23.97 (0.5)	31.78 (1.4)	47.37 (2.0)	16.84 (1.9)
100 kev	60w	45.80 (1.5)	70.84 (4.3)	28.51 (0.8)	33.87 (1.4)	54.46 (2.7)	39.86 (2.1)
	60q	34.46 (1.7)	75.28 (3.6)	18.03 (0.5)	21.71 (1.3)	32.79 (1.9)	13.40 (1.6)
	26q	25.22 (2.0)	46.94 (3.5)	11.63 (0.5)	13.77 (1.2)	20.99 (1.8)	9.28 (1.5)
50 keV	60w	37.74 (1.7)	53.73 (5.1)	25.00 (0.9)	29.13 (1.6)	42.62 (2.8)	35.96 (2.1)
	60q	15.35 (1.5)	30.13 (3.4)	9.21 (0.5)	10.24 (1.1)	15.05 (1.8)	8.39 (1.5)
	26q	12.47 (1.8)	21.91 (3.3)	6.92 (0.5)	7.63 (1.1)	11.23 (1.8)	6.64 (1.5)
20 keV	60w	34.24 (2.0)	44.13 (6.8)	23.29 (1.0)	26.49 (1.7)	42.52 (3.3)	36.55 (2.3)
	60q	7.92 (1.3)	12.89 (3.4)	5.38 (0.6)	5.92 (1.1)	8.93 (2.0)	6.14 (1.6)
	26q	6.88 (1.7)	11.01 (3.6)	4.68 (0.6)	5.14 (1.1)	7.82 (2.1)	5.53 (1.7)

Table 1. continued.

Energy		Effective	Testes	RBM	Lung	Stomach	Colon
10 keV	60w	32.88 (2.2)	42.96 (7.5)	22.35 (1.0)	25.36 (1.8)	40.42 (3.2)	36.30 (2.5)
	60q	5.38 (1.5)	7.62 (4.6)	3.87 (0.7)	4.26 (1.3)	6.47 (2.2)	5.03 (1.9)
	26q	4.91 (2.1)	7.33 (4.7)	3.68 (0.7)	4.08 (1.3)	6.21 (2.3)	4.87 (2.0)
5 keV	60w	16.09 (1.8)	19.40 (6.1)	11.29 (1.0)	12.73 (1.7)	20.52 (3.2)	17.03 (2.3)
	60q	6.47 (1.1)	8.69 (3.6)	5.42 (0.7)	5.47 (1.2)	7.93 (2.1)	5.57 (1.7)
	26q	5.23 (1.6)	7.38 (3.8)	4.34 (0.7)	4.60 (1.2)	6.84 (2.2)	4.94 (1.7)
2 keV	60w	16.09 (1.8)	18.89 (6.5)	11.30 (1.0)	13.14 (1.7)	21.01 (3.2)	17.11 (2.3)
	60q	7.25 (1.2)	9.20 (3.9)	6.15 (0.7)	6.55 (1.2)	9.22 (2.0)	6.52 (1.6)
	26q	5.35 (1.7)	7.27 (4.1)	4.62 (0.7)	5.09 (1.2)	7.39 (2.1)	5.38 (1.7)
1 keV	60w	15.94 (1.8)	17.71 (6.9)	11.37 (1.0)	13.49 (1.7)	20.70 (3.2)	18.01 (2.3)
	60q	7.49 (1.2)	9.20 (4.2)	6.25 (0.7)	6.87 (1.2)	9.59 (2.0)	6.89 (1.6)
	26q	5.35 (1.8)	7.06 (4.4)	4.66 (0.7)	5.25 (1.2)	7.51 (2.1)	5.63 (1.7)
100 eV	60w	16.27 (1.9)	19.42 (6.6)	11.48 (1.0)	13.54 (1.7)	20.33 (3.3)	17.11 (2.3)
	60q	7.84 (1.2)	9.80 (4.0)	6.51 (0.7)	7.28 (1.2)	9.83 (2.0)	6.69 (1.7)
	26q	5.64 (1.7)	7.50 (4.3)	4.79 (0.7)	5.46 (1.2)	7.56 (2.2)	5.39 (1.8)
10 eV	60w	17.34 (2.1)	22.38 (7.0)	11.67 (1.0)	13.69 (1.7)	22.29 (3.2)	17.29 (2.4)
	60q	8.27 (1.3)	11.40 (4.0)	6.56 (0.7)	7.11 (1.2)	10.53 (2.0)	6.38 (1.8)
	26q	6.06 (1.9)	8.66 (4.3)	4.83 (0.7)	5.36 (1.2)	8.12 (2.1)	5.22 (1.9)
1 eV	60w	15.78 (2.0)	21.57 (6.3)	10.59 (1.0)	12.09 (1.9)	19.98 (3.4)	14.66 (2.5)
	60q	8.14 (1.3)	12.89 (3.7)	6.23 (0.7)	6.56 (1.3)	9.56 (2.2)	5.40 (1.9)
	26q	6.21 (1.8)	9.59 (3.8)	4.62 (0.7)	4.96 (1.3)	7.42 (2.3)	4.45 (1.9)
Thermal	60w	8.76 (3.3)	13.97 (9.1)	5.69 (1.4)	6.20 (2.6)	10.14 (5.0)	6.94 (3.6)
	60q	4.36 (2.1)	8.28 (4.9)	3.30 (0.9)	2.87 (2.0)	4.27 (3.4)	2.25 (2.9)
	26q	3.64 (2.7)	6.17 (5.3)	2.46 (0.9)	2.24 (2.0)	3.4 (3.5)	1.92 (3.0)

symbolised as 60q and 26q, depending whether recently recommended (ICRP 60[1]) or still being in use (ICRP 26[10]) Q-L and w_T specifications have been applied. Consistently, the ICRP 26 specification of the remainder was used when deriving the 26q dose equivalents, which made them comparable with the results of previous calculations[2-5]. Interestingly, the best agreement at lower neutron energies has been found with respect to the older calculations[2], possibly due to compatible treatment of neutron thermalisation. The calculated values of effective doses (60w) are not quite comparable with the preliminary data of Alberts et al[18], since the smoothed $w_R(E_n)$ relationship and somewhat different phantom specifications were used when deriving the latter.

As one could expect, the newly defined equivalent and effective doses most frequently exceed the corresponding dose equivalent values. This is visualised in Figure 3 which compares, for AP geometry, the effective doses (60w) with the new (60q) and old (26q) values of the effective dose equivalent relative to incident neutron energy. The ratios of effective doses (60w) and effective dose equivalents (60q) for AP, PA and RL exposure geometries are given in Figure 4. Incompatibility of the two approaches is evident, although less pronounced than for deep-seated and/or asymmetrically located organs. Figure 5 compares the $w_R(E_n)$ step function, as recommended by ICRP[1], with the effective quality factor for an adult male, $\tilde{Q}_m(E_n)$, averaged over all organs and geometries. The latter has been derived as the dose-weighted mean of effective dose

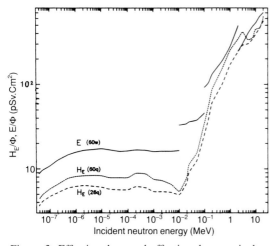

Figure 3. Effective dose and effective dose equivalent per incident neutron fluence plotted against incident neutron energy, calculated for the ADAM phantom, AP exposure, according to recent[1] and previous[10] ICRP recommendations. See text for further explanation.

equivalents (60q) for particular geometries divided by correspondingly tissue-weighted absorbed doses. The dashed lines delineate the variability range of quality factors for particular organs and geometries: for example, for 144 keV neutrons, organ-averaged quality factors vary from 1.2 (stomach, RL) to 18.3 (skin, RL and LL), compared to $\tilde{Q}_m$ of 8.5 and w_R of 20.

CONCLUSIONS

The results achieved indicate that a simplified concept of assigning radiation quality to an undisturbed field, and thus of neglecting quality variations within the human body, is highly incompatible with the LET-based quality factor concept. The proposed radiation weighting factors, w_R, and the related equivalent and effective doses, are by no means consistent with the simultaneously recommended reformulation of the Q-L relationship. The latter is supposed to reflect the advance in understanding low dose radiation effects. For neutrons, equivalent and effective doses can usually be considered only as rough but inconsistent (in terms of magnitude) overestimates of corresponding dose equivalent values. The latter can be evaluated based essentially on the same information that is required to determine organ-averaged values of the absorbed dose: thus the simplification itself appears questionable. It is not very practical, either, due to the problems with monitoring undisturbed fields. However, if the concept of radiation weighting factors were to be accepted, replacing the presently recommended w_R values for neutrons (5, 10 and 20) by sex-averaged effective $\tilde{Q}$ factors, similar to those calculated here for the male phantom, would bring the incompatible quantities of effective dose and effective dose equivalent numerically close together.

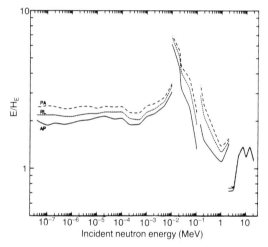

Figure 4. Ratio of effective dose to effective dose equivalent calculated relative to incident neutron energy for the ADAM phantom and different exposure geometries with recent Q-L and w_T specifications[1].

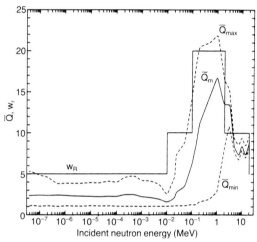

Figure 5. Comparison of ICRP defined radiation weighting factors for neutrons with the effective quality factor, calculated relative to incident neutron energy for the ADAM phantom using recent Q-L and w_T specifications[1]. Dashed lines: maximum and minimum quality factors for particular organs and exposures.

REFERENCES

1. ICRP. *1990 Recommendations of the International Commission on Radiological Protection*. Publication 60 (Oxford: Pergamon Press) (1990).
2. Burger, G., Morhart, A., Nagarajan, P. S. and Wittmann, A. *Conversion Functions for Primary and Operational Quantities in Neutron Radiation Protection*. In: Proc. 4th Symp. on Neutron Dosimetry. EUR 7448 (Luxembourg: CEC) pp. 33-48 (1981).
3. Wittmann, A., Morhart, A. and Burger, G. *Organ Doses and Effective Dose Equivalent*. Radiat. Prot. Dosim. **12**, 101-106 (1985).
4. Endres, G. W. R., Tanner, J. E., Scherpelz, R. I. and Hadlock, D. E. *Calculation Methods for Determining Dose Equivalent*. In: Proc. 7th Int. Congress of IRPA, Sydney, 10-17 April 1988, pp. 277-279 (1988).
5. Hollnagel, R. A. *Effective Dose Equivalent and Organ Doses for Neutrons from Thermal to 14 MeV*. Radiat. Prot. Dosim. **30**, 149-159 (1990).
6. Schmitz, Th., Smit, Th., Morstin, K., Müller, K. D. and Booz, J. *Construction and First Application of a TEPC Dose-Equivalent Meter for Area Monitoring*. Radiat. Prot. Dosim. **13**, 335-339 (1985).

7. Golnik, N., Wilczynska-Kitowska, T. and Zielczynski, M. *Determination of the Recombination Index of Quality for Neutrons and Charged Particles Employing High Pressure Ionisation Chambers.* Radiat. Prot. Dosim. **23**, 273-276 (1988).
8. Drexler, G., Veit, R. and Leuthold, G. Personal communication to ICRP, partly reflected in Ref. 1.
9. ICRU. *Determination of Dose Equivalents Resulting from External Radiation Sources.* Report 39 (Bethesda: ICRU Publications) (1985).
10. ICRP. *Recommendations of the International Commission on Radiological Protection.* Publication 26 (Oxford: Pergamon Press) (1977).
11. Kopec, M., Morstin, K. and Schmitz, Th. *Conversion Function to Primary Limiting Quantities for Neutrons and Photons.* Rep. KFA (in preparation).
12. Kramer, R., Zankl, M., Williams, G. and Drexler, G. *The Male (ADAM) and Female (EVA) Adult Mathematical Phantoms.* Rep. GSF S-885 (Munich: GSF) (1982).
13. Snyder, W. S., Ford, M. R. and Gordon, G. G. *Estimates of Specific Absorbed Fractions for Photon Sources Uniformly Distributed in Various Organs.* MIRD Pamphlet No. 5, Revised (New York: Soc. Nucl. Med.) (1978).
14. Briesmeister, J. F., Thompson, W. L. et al. *MCNP — A General Monte Carlo Code for Neutron and Photon Transport.* Rep. LA-7396-M, Rev. 2 (Los Alamos: LASL) (1986).
15. Koppel, J. U. and Houston, D. H. *Reference Manual for ENDF Thermal Neutron Scattering Data.* Rep. GA-8744, Rev. (ENDF-269) (1978).
16. Edwards, A. A. and Dennis, J. A. *NESLES — A Computer Program which Calculates Charged Particle Spectra in Materials Irradiated by Neutrons.* Rep. NRPB-M43 (Harwell: NRPB) (1979).
17. Caswell, R. S., Coyne, J. J. and Randolph, M. L. *Kerma Factors for Neutron Energies below 30 MeV.* Radiat. Res. **83**, 217-254 (1980).
18. Alberts, W. G., Hollnagel, R. A. and Siebert, B. R. L. *Some Comments on the New Quantity Effective Dose and Preliminary Data for Neutrons.* Radiat. Prot. Dosim. **37**, 201-202 (1991).

IMPLICATIONS OF NEW ICRP AND ICRU RECOMMENDATIONS FOR NEUTRON DOSIMETRY

G. Portal† and G. Dietze‡
†CEA -IPSN-DPHD-S.DOS, BP N6
F-92265 Fontenay aux Roses, France
‡Physikalisch-Technische Bundesanstalt Postfach 3345, D-3300 Braunschweig, Germany

INVITED PAPER

Abstract — Publication 60 of ICRP, which includes new recommendations on the quality factor concept and on annual equivalent dose limits, will introduce some changes in neutron monitoring procedures. The implications of the new 'radiation weighting factors', defined in that document, combined with the decrease in the annual dose limits are analysed with respect to the new ICRU operational quantities. It is shown that the energy response of radiation protection instruments and the angular response of individual dosemeters have to be improved, and that the lower limit of detection of portable monitors and individual dosemeters must be decreased by a large factor (up to 10), as a result of which difficulties will be encountered in individual dosimetry.

INTRODUCTION

Publication 60 of ICRP[1], which includes new recommendations on the quality factor concept and on annual equivalent dose limits, will introduce some changes in neutron monitoring procedures. The implications of the new 'radiation weighting factors' defined in that document combined with the decrease in the annual dose limits are analysed with respect to the new ICRU operational quantities. Their influence on the lower limit of detection, on the energy response and on the angular response of radiation protection instrumentation are considered.

MODIFICATIONS OF ICRP AND ICRU RECOMMENDATIONS

Equivalent dose, effective dose

In comparison with previous ICRP recommendations (ICRP 26 in 1977[2] and ICRP 35 in 1982[3]), the following items related to the subject of this report have been modified by the Commission in their publication 60:

(i) In 1985 in publication 51[4], the quality factor, Q(L), as a function of the unrestricted lineal energy transfer, L, is tabulated. This function has been modified resulting in new values of Q(L) accounting for the higher RBE values for intermediate energy neutrons.

(ii) H_T, which may be considered to represent the average value of dose equivalent over a tissue or organ, is now called the 'equivalent dose' and defined by:

$$H_T = \sum_R w_R D_{T,R} \qquad (1)$$

where $D_{T,R}$ is the mean absorbed dose in the tissue or organ T due to radiation R and w_R is a weighting factor.

(iii) w_R, the 'radiation weighting factor', which applies to the mean tissue or organ dose, is representative of the corresponding RBE values. It replaces $\bar{Q}$, the effective quality factor at a point of interest for a given spectrum of radiation calculated from the values of the quality factor Q. The values of w_R for neutrons taken from the ICRP document are shown in Figure 1. They are broadly compatible with the quality factor Q and are approximately equal to the effective quality factor calculated at a point at 10 mm depth in the ICRU sphere. The results of recent calculations from Schuhmacher and Siebert[5] and Leuthold et al[6] are also shown

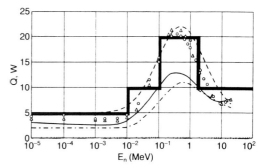

Figure 1. Quality factors Q and radiation weighting factors w for neutrons as a function of neutron energy. Shaded line, w_R table data from ICRP 60; (- - -) $w_{R,s}$ smooth function values from ICRP 60; (- . - .) effective quality factor $\bar{Q}_{21}$ from ICRP 21[1]; (—) effective quality factor $\bar{Q}_{51}$ from ICRP 51[4]. Quality factor $\bar{Q}_{91}$ in 10 mm depth of the ICRU sphere with new Q(L) values, (Δ) data from Schuhmacher[5], (o) data from Leuthold[6].

in Figure 1. When neutron spectra have to be considered in practice, $w_{R,S}$ can be deduced from the relationship:

$$w_{R,S} = 5 + 17 \exp[-(\ln(2E))^2/6] \quad (2)$$

with E in MeV, which gives a smooth fit of the w_R values for neutrons as a function of energy, also shown in Figure 1. Compared with the old $\bar{Q}$ values, the w_R values are increased by a factor of about 2 for neutrons.

(iv) The number of specified organs has been extended and the values of the weighting factors w_T have been modified for some tissues and organs.

(v) H_E, the 'effective dose equivalent', has been replaced by E, the 'effective dose', which is the weighted sum of the equivalent doses in all the tissues or organs of the body:

$$E = \sum_T w_T H_{T,R} \quad (3)$$

where $H_{T,R}$ is the mean equivalent dose in the tissue or organ T delivered by radiation R and w_T is the tissue weighting factor.

New dose limits

An effective dose limit of 100 mSv over a period of 5 years, with an upper limit of 50 mSv during any single year is now recommended by the ICRP for occupationally exposed persons. It is implicit in these recommended dose limits that the dose constraint for the optimisation of radiation protection should not exceed 20 mSv per year (instead of 50 mSv per year recommended previously in ICRP 26[2]). Furthermore, the annual limit of equivalent dose for the lens of the eye is fixed at 150 mSv, while the dose limit for the skin remains at 500 mSv.

All these limits apply to the sum of the doses from external exposures over the specified period and the 50 year committed dose from intakes over the same period.

Accuracy of measurements

According to the ICRP 60 recommendations, uncertainties of a factor of up to 1.5 will not be unusual in the monitoring of individual workers for external exposure when the energy and orientation of a radiation field are not known. In the estimation of intakes and the associated committed equivalent dose and effective dose, uncertainties of a factor of up to 3 may well have to be accepted.

In these new recommendations the factor 2 is no longer an acceptable uncertainty factor for low dose measurements (according to ICRP 35, paragraph 109, an uncertainty within a factor of 2 for dose values of 1/5 of the annual limit was acceptable). We will consider the impact on the detection threshold of instruments when the uncertainty factor is reduced from 2 to 1.5.

According to ICRP 35, paragraph 109, these uncertainties include errors due to variations in dosemeter sensitivity with incident energy and direction of incidence, as well as intrinsic errors in the dosemeter readings and its calibration.

New ICRU operational quantities

The experimental determination of the former defined index quantities $H_{I,d}$ and $H_{I,s}$ (see ICRP 26, paragraph 108[2]) is not easy. The ICRU has therefore defined a set of new operational quantities which allows an estimation of the effective dose to be obtained directly from radiation protection measurements. In ICRU report 39[7] the Commission defined these quantities and in ICRU report 43[8] set out the detailed considerations underlying its formulation.

In publication 51[4] the ICRP proposed a set of conversion coefficients for these operational quantities and introduced their definitions in publication 60[1]. The ICRP considers that the use of these quantities provides a good approximation of the effective dose and the equivalent dose to the skin.

The ICRU has recently prepared a new document on dose equivalent measurements for external gamma and electron radiations which will be published in the near future. Three types of information are given in this document:

(i) new definitions of the new operational quantities,
(ii) guidelines for the design and calibration of instruments,
(iii) physical data for the calibration of reference beams (conversion factors).

Some remarks on the new definitions of these operational quantities and further points relevant for their application are presented in Appendix 1. A new document should be prepared by ICRU in the near future for neutron radiations. From the available documents, it can be deduced that for neutron radiation:

(a) environmental monitors should be calibrated in terms of ambient dose equivalent, H*(10),
(b) individual dosemeters should be calibrated in terms of personal dose equivalent, $H_p(10)$, in a phantom having the composition of the ICRU tissue and of the same size and shape as the calibration phantom,
(c) environmental monitors should have an isotropic response,
(d) individual dosemeters should have an

isodirectional response (see definition in Appendix 1).

IMPLICATIONS OF THE NEW ICRP AND ICRU RECOMMENDATIONS FOR NEUTRON DOSIMETRY

Three main points will be considered in the following; the impact of these new recommendations on: the detection threshold (DT) of dosemeters, their energy response, and their angular response.

Detection threshold of dosemeters

The requirements on the detection threshold can be deduced from the ICRP recommendations on reference levels.

Recording level

The recording level has been defined in paragraph 150 of ICRP 26 as the dose level above which the result is of sufficient interest to be worth recording and keeping. The unrecorded results should be treated as null for assessing 'annual dose equivalents'.

The recording level has been set by the Commission (par. 181 of ICRP 26) to be one-tenth of the annual dose equivalent limit or intake limit. The usefulness of this concept has been confirmed in ICRP 60 but the value of the corresponding level has been neither confirmed nor rejected. We will assume in this paper that the value initially adopted for this level is still valid.

As a result of the reduction in the mean annual dose equivalent limit from 50 mSv to 20 mSv prescribed in ICRP publication 60, the annual recording level would be lowered from 5 mSv to 2 mSv. This may also favour the use of electronic devices, which will be useful to verify whether or not requirements of the ALARA principle are being fulfilled.

Detection threshold

It will be assumed here that the detection threshold of a dosemeter should be less than or equal to the recording level. When more than one detector is used for the evaluation of the effective dose, the sum of the weighted detection thresholds of the detectors used (weighted according to their respective contributions to the measurement of the dose) should be less than or equal to the recording level. This is particularly the case in neutron fields which are always mixed with photons. Both types are often measured with separate instruments. If we consider, for example, the case where the contribution of each radiation represents 50% of the total equivalent dose, the minimum detectable dose of a neutron dosimetry system should be divided by a factor of two (1 mSv per year, 0.08 mSv per month and 0.5 µSv per hour). In practice, due to the increase in the effective quality factor, i.e. radiation weighting factor, the contribution of neutrons to the total equivalent dose will often be of the order of magnitude given above,

For simultaneous internal and external exposures, the detection threshold of each instrument or method used to evaluate the doses from internal or external exposure must also be taken into account (the weighted sum according to the relative contribution of the different components to the total effective dose). Under these conditions it seems reasonable to assume that the detection threshold of a neutron dosemeter should not be higher than the above given values, i.e. 1 mSv per year, 0.08 mSv per month, 0.5 µSv per hour; when there is a large neutron contribution to the total effective dose or when there are risks of simultaneous internal contamination, detection thresholds should be lowered depending on the various relative contributions to the equivalent dose.

Factors influencing the detection threshold

When the detection threshold of a dosemeter is just sufficient for the evaluation of the corresponding recording level, it is called the minimum detection threshold.

As a first approximation, the detection threshold of existing dosemeters must be reduced by a factor of:

(i) 2.5 to take into account the reduction in the annual dose limit,
(ii) about 1.5 to 2 (see Figure 1) to take into account the increase in the effective quality factor (or radiation weighting factor),
(iii) approximately 2 (1.73 according to Chemtob[9]) to take into account the possible lowering of the acceptable uncertainty for low doses by the ICRP.

Consequently, the minimum detection threshold of the instrument must be reduced by a factor of 5 to 10, depending on whether or not the ICRP Commission intends to lower the factor of 2 for acceptable uncertainties in low dose measurements.

Detection threshold of existing dosemeters

Many area monitors have a sufficiently low detection threshold to satisfy these new requirements. However, individual dosemeters are a different matter. Many passive integrating detectors (nuclear emulsions, solid-state dosemeters) or electronic devices presently have a detection threshold around 0.2 mSv[10] which, after the introduction of the new values of quality factors or radiation weighting factors, corresponds to about 0.3 to 0.4 mSv; it will be really difficult to reduce their detection threshold to 0.08 mSv per month. Only bubble detectors achieve a sufficiently low detection threshold but due to their temperature sensitivity, their use is restricted to specific measurements at present; they still need further improvements before being used for routine monitoring (see Refs 11 and 12).

Energy response of dosemeters

When referred to the absorbed dose, the energy response curve of an instrument intended to measure dose equivalent (or equivalent dose) should be similar to the curves in Figure 1 representing the following quantities as functions of neutron energy:

(a) the effective quality factor, $\bar{Q}_{21}$, related to the maximum dose equivalent in a tissue-equivalent cylinder (MADE) (ICRP publication 21[13]),
(b) the effective quality factor, $\bar{Q}_{51}$, related to the ICRU operational quantities and extracted from ICRP publication 51,
(c) the radiation weighting factor, w_R,
(d) the results of Schuhmacher and Leuthold's

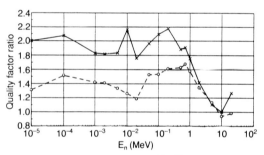

Figure 2. Variation of the ratios $\bar{Q}_{91}/\bar{Q}_{21}$ (full line) and $\bar{Q}_{91}/\bar{Q}_{51}$ (broken line) with neutron energy. ($\bar{Q}_{91}$ mean values of data from Schuhmacher and Leuthold).

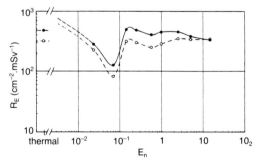

Figure 3. Energy response curve of the PTB type CR-39 neutron dosemeter[16]. Full line: response as given in Ref. 16. Broken line: response corrected for $\bar{Q}_{91}$.

recent calculations of the effective quality factor, $\bar{Q}_{91}$.

The ratios, $\bar{Q}_{91}/\bar{Q}_{21}$ and $\bar{Q}_{91}/\bar{Q}_{51}$ are shown in Figure 2, where $\bar{Q}_{91}$ represents mean values of data from Schuhmacher and Leuthold. These curves show that the energy response of radiation protection instruments must be considerably modified, particularly for the most frequently used dosemeters which are adjusted to $\bar{Q}_{21}$.

Instruments including a microprocessor and which take into account the energy or the LET of the incident radiation to calculate the equivalent dose will be easily modified by changing the parameters used in the calculation programme (e.g. Bonner sphere spectrometers, the Dineutron[14,15] and tissue-equivalent proportional counters). The response curves of other instruments with readings directly scaled in equivalent dose will, however, not be so easy to fit (e.g. the diameter of the scattering sphere of conventional rem counters will have to be reduced).

Passive individual dosemeters (nuclear emulsions, etched track detectors etc.) must be improved; for example the composition and size of hydrogenous converters, and the conditions of processing must be modified. An example is given in Figure 3 where an original energy response curve of a particular etched track dosemeter[16] is shown together with the curve modified to take into account the change of the quality factor (the correction factors in Figure 2 have been applied for this purpose). It can be seen that in this particular case, the modified energy response is more acceptable than the original, except for the energy range from 10 to 100 keV.

Angular response of dosemeters

To achieve an isotropic response, a portable monitor for measuring H*(10) should ideally consist of spherical detecting elements; an orthospherical detector surrounded by a spherical or an orthospherical converter, is convenient from a practical point of view. This is generally the case for rem counters where the detecting element is a proportional or GM counter and the converter is a polyethylene sphere.

To yield an isodirectional response with respect to H_p (definition see Appendix 1), an individual dosemeter for measuring $H_p(d)$ should have a flat converter and a flat active detector volume. This is usually the case, for example, for dosemeters based on the use of a nuclear emulsion, a solid-state detector (e.g. CR-39 foils) or a silicon diode covered with hydrogenous converters, the thickness of which, however, often varies considerably from d=10 mm, and none of the above dosemeters is really isodirectional. This can also be seen on

Figure 4, where the ideal angular response of an individual dosemeter with respect to fluence[17] for various neutron energies is compared with experimental results for CR-39 dosemeters obtained by Hankins et al[18].

Furthermore, dosemeters based on a bubble detector, a proportional or GM counter (cylindrical geometry) or albedo dosemeters are not isodirectional. In conclusion, portable monitors generally have a convenient isotropic angular response, but none of the individual dosemeters has a suitable isodirectional angular response. This problem is in fact not due to the introduction of new operational quantities but existed prior to this, when the dosemeters were forced to have an isotropic response which was not really well adapted to the needs of individual dosimetry.

GENERAL CONCLUSION

The implications of the new ICRP recommendations on radiation protection instrumentation are important; the energy response of all types of dosemeters will have to be improved, the detection threshold, i.e. the minimum detectable dose of integrating individual dosemeters (excluding bubble detectors) will be insufficiently low and their angular response will not be suitable.

The decrease in the annual dose limit combined with the increase in the quality factor for neutrons will favour the use of electronic devices in individual dosimetry which will be necessary to fulfill the requirements of the ALARA principle. Since this instrumentation has not yet been developed, much research work would be necessary to adapt future dosimetry instrumentation to the new requirements. Furthermore, the position of the ICRP on the acceptable uncertainty for low dose detection should be clearly defined in future recommendations.

ACKNOWLEDGEMENTS

We would like to thank Dr R. B. Schwartz from NIST (Washington) and Dr M. Chemtob from CEA (Fontenay aux Roses) for their assistance in the preparation of this document. This work has been carried out in the framework of the CEC contract N° B170020C.

APPENDIX 1

Remarks on new definitions of ICRU operational quantities and related points

Modifications have been introduced in the definitions of the operational quantities to take into account the experience gained in five years of use. These new definitions have been prepared by the ICRU Working Committee for 'Measurement of Dose Equivalents' and 'Quantities and Units' and will be published in the near future.

Individual monitoring

In order to simplify the concept of individual monitoring, H_p (individual dose equivalent, penetrating) and H_s (individual dose equivalent, superficial) have been combined in a unique quantity, the personal dose equivalent, $H_p(d)$. This quantity can be used for both weakly and strongly penetrating radiations with recommended depths of 0.07 mm and 10 mm, respectively; other depths may also be appropriate, such as 3 mm for the lens of the eye.

Calibration

For calibration of individual dosemeters, reference to the ICRU sphere is deleted; the quantity $H_p(d)$ may be used to specify the dose equivalent at a point in a phantom representing the body.

Angular response

According to the definition of H*, the response of portable instruments should not vary with the angle of incidence of the radiation; they should have an *isotropic* response.

According to the definition of H'(d) and $H_p(d)$, which are both quantities suitable for a non-aligned field, the instrument (or dosemeter) reading should vary with the direction of the incident radiation in accordance with the operational quantity under consideration. This means that the ratio R/H'(d) (where R is the reading of the dosemeter) should not vary with the angle of incidence of the radiation; the angular response of the detector is thus termed *isodirectional*.

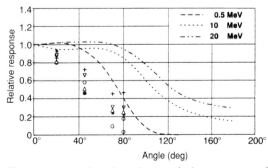

Figure 4. Angular dependence of the response of individual dosemeters with respect to fluence for various neutron energies. Ideal theoretical curves (approximated by H'(10)/φ from Ref. 17) and experimental results with CR-39 neutron dosemeters from Ref. 18. Key: (+) 14 to 16 MeV, (∇) 3 to 5 MeV, (Δ) 1.3 to 1.6 MeV, (*) 0.45 to 1.1 MeV, (o) 0.13 to 0.4 MeV.

REFERENCES

1. ICRP. *Recommendations of the ICRP.* Publication 60. Ann. ICRP **21**(1-3) (Oxford: Pergamon) (1990).
2. ICRP. *Recommendations of the ICRP.* Publication 26. Ann. ICRP **1**(3) (1977).
3. ICRP. *General Principles of Monitoring for Radiation Protection of Workers.* Publication 35. Ann. ICRP **9**(4) (1982).
4. ICRP. *Data for Use in Protection Against External Radiation.* Publication 51. Ann. ICRP **17**(2-3) (1987).
5. Schuhmacher, H. and Siebert, B. R. L. *Quality Factors and Ambient Dose Equivalent for Neutrons Based on the New ICRP Recommendations.* Radiat. Prot. Dosim. **40**, 85-89 (1992).
6. Leuthold, G., Mares, V. and Schraube, H. *Calculation of the Neutron Ambient Dose Equivalent on the Basis of the ICRP Revised Quality Factors.* Radiat. Prot. Dosim. **40**, 77-84 (1991).
7. ICRU. *Determination of Dose Equivalents Resulting from External Radiation Sources.* ICRU report 39 (Bethesda, Maryland: ICRU Publications) (1985).
8. ICRU. *Determination of Dose Equivalents from External Radiation Sources, Part 2.* ICRU report 43 (Bethesda, Maryland: ICRU Publications) (1988).
9. Chemtob, M. CEA-IPSN-STESN-Fontenay aux Roses, private communication (1991).
10. Gibson, J. A. B. *Individual Neutron Dosimetry.* Radiat. Prot. Dosim. **23**, 110-115 (1988).
11. Schwartz, R. B. and Hunt, J. B. *Measurement of the Energy Response of Superheated Drop Neutron Detectors.* Radiat. Prot. Dosim. **34**, (1/4), 377-380 (1990).
12. Schwartz, R. B. and Boswell, E. W. *Test of Temperature Compensated Bubble Dosimeters.* private communication (1991).
13. ICRP. *Data for Protection against Ionizing Radiation from External Sources: Supplement to ICRP Publication 15.* Publication 21 (Oxford: Pergamon Press) (1971).
14. Mourgues, M., Carossi, J. C. and Portal, G. *A Light Rem-Counter of Advanced Technology.* In: Proc. Fifth Symp. on Neutron Dosimetry, EUR 9762 EN (Luxembourg: CEC) Vol. 1 (1985).
15. Chemtob, M., Hunt, J. B., Mourgues, M. and Portal, G. *The Dineutron Area Survey Meter.* In: Proc. 22nd Midyear Topical Meeting on Instrumentation, San Antonio Health Physics Society pp. 172-185 (1988).
16. Luszik-Bhadra, M., Alberts, W. G., Dietz, E. and Guldbakke, S. *A Track-etch Dosemeter with Flat Response and Spectrometric Properties.* Nucl. Tracks Radiat. Meas. **19**, 485-488 (1991).
17. Jahr, R., Hollnagel, R. and Siebert, B. R. L. *Calculations of Specified Depth Dose Equivalent in the ICRU Sphere Resulting from External Neutron Radiations.* Radiat. Prot. Dosim. **10**, 75-87 (1985).
18. Hankins, D., Hommann, S. and Westermark, J. *The LLNL CR-39 Personnel Neutron Dosemeter.* Radiat. Prot. Dosim. **23**, 195-198 (1988).

VERIFICATON OF AN EFFECTIVE DOSE EQUIVALENT MODEL FOR NEUTRONS

J. E. Tanner, R. K. Piper, J. A. Leonowich and L. G. Faust
Pacific Northwest Laboratory
PO Box 999, Richland, WA 99352, USA

Abstract — Since the effective dose equivalent, based on the weighted sum of organ dose equivalents, is not a directly measurable quantity, it must be estimated with the assistance of computer modelling techniques and a knowledge of the incident radiation field. Although extreme accuracy is not necessary for radiation protection purposes, a few well chosen measurements are required to confirm the theoretical models. Neutron doses and dose equivalents were measured in a RANDO phantom at specific locations using thermoluminescence dosemeters, etched track dosemeters, and a 1.27 cm (1/2 in) tissue-equivalent proportional counter. The phantom was exposed to a bare and a D_2O-moderated ^{252}Cf neutron source at the Pacific Northwest Laboratory's Low Scatter Facility. The Monte Carlo code MCNP with the MIRD-V mathematical phantom was used to model the human body and to calculate the organ doses and dose equivalents. The experimental methods are described and the results of the measurements are compared with the calculations.

INTRODUCTION

The determination of effective dose equivalent, a concept first introduced by the International Commission on Radiological Protection (ICRP) in 1977, is required by most national and international authoritative bodies to show compliance with occupational exposure limits. The determination of effective dose equivalent requires a knowledge of the dose equivalent to certain organs which, in most cases, must be calculated and derived from other measurements. Any calculational method used must be shown through measurements to be as accurate as current technology allows; this has been fairly straightforward for photons but has proved to be much more difficult for neutrons. The object of this study was to identify and perform the measurements necessary to verify the calculational models used at the Pacific Northwest Laboratory (PNL) in support of implementing an effective dose equivalent system.

For the purpose of an experimental verification, simple exposure conditions were deemed necessary. A ^{252}Cf source was used for its traceability to the National Institute of Standards and Technology (NIST) and simple point source geometry. Measurements were performed with the bare source and with the source in the 15 cm radius, cadmium-covered D_2O sphere. The goal was to show that the calculated organ doses and dose equivalents agreed with the measured values to within 20%. This criterion was judged acceptable based on predicted measurement errors of 10–15% for current neutron dosemeters and a precision of 5–10% in the Monte Carlo calculations. The errors associated with the neutron cross sections used in the Monte Carlo calculations are small (1–2%) compared to the other sources of error.

NEUTRON CALCULATIONAL MODEL

The calculations were performed using the Monte Carlo general purpose radiation transport code, MCNP, developed by Los Alamos National Laboratory[1]. Continuous-energy neutron cross sections from the Evaluated Nuclear Data File B-V (ENDF/B-V) were used in the transport calculations, along with the MIRD-V mathematical anthropomorphic phantom[2]. The source terms were modelled as isotropic point sources with a Watt fission energy spectrum describing the bare ^{252}Cf source and the ISO energy histogram for the D_2O-moderated ^{252}Cf source[3].

Assuming neutrons incident uniformly on the surface of the phantom, the MCNP code calculates the volume-averaged, energy-dependent fluences of neutrons and neutron-induced photons to specified organs and tissues. A separate post-processing code is used to read the output from MCNP and apply the appropriate energy-dependent kerma factors, quality factors and tissue weighting factors. The result is a table of organ doses, dose equivalents, and a single effective dose equivalent. The contributions from neutrons and neutron-induced photons to the organ doses and dose equivalents were tracked separately to allow for comparison with the neutron measurements. The main goal of this study was to verify the Monte Carlo technique and input parameters that result in estimates of the organ doses and dose equivalents. By definition, the effective dose equivalent is accurate if the contributing organ dose equivalents are accurate.

NEUTRON MEASUREMENTS

All measurements were performed at PNL's

Low Scatter Facility with a RANDO Average Man phantom and a 1.47 mg ^{252}Cf neutron source. The RANDO phantom is composed of Alderson Muscle tissue substitute material moulded about an actual human skeleton with 5 mm diameter holes drilled in a 3 cm × 3 cm array for holding dosemeters. These holes are filled with tissue-equivalent (TE) plastic plugs when not being used. The phantom consists of 36 separate slices, 2.54 cm (1 in) thick, numbered 0 to 35 from the top of the head down. During the exposures, the distance from the source to the front surface of the phantom at mid-torso height (slice 21) was 1 m. Figure 1 shows the phantom set up at 1 m from the source. As can be seen, one disadvantage of using a RANDO phantom is the fact that although its torso and head dimensions are very realistic, it has no arms or legs.

The criteria used for selecting the measurement techniques for this study were: (1) sensitivity to neutrons, (2) accuracy over neutron energies from thermal to 10 MeV, (3) small enough to fit at a point of interest inside a phantom without perturbing the flux, and (4) exhibiting little or no directional dependence to simplify the interpretation of the results.

An initial measurement was performed using thermoluminescence dosemeters (TLDs) and etched track dosemeters because of their small size and availability. The TLDs consisted of ^{7}LiF chips used in conjunction with ^{6}LiF chips to correct for the gamma component of the response. Sturdy plastic capsules were used to hold a total of six chips, three ^{7}LiF chips and three ^{6}LiF chips. Sets of chips were exposed bare and covered with a cadmium filter to discriminate between the slow and fast neutron components. The etched track dosemeters (CR-39), 2.86 cm × 1.59 cm × 0.635 mm, were placed horizontally between phantom slices. In this initial set, only four organs were chosen for preliminary characterisation due to the large effort involved in the data processing and analysis. The locations of the specific holes and slices corresponding to a particular organ were based on the previous work by Huda and Sandison[4] and Golikov and Nikitin[5]. Also, since only one TLD could be accommodated in a single hole, the bare and cadmium-covered TLDs were staggered, with a piece of CR-39 placed between.

Another set of measurements was made with a tissue-equivalent proportional counter (TEPC) to obtain direct measurements of the absorbed neutron dose within the phantom. The TEPC used was a 1.27 cm (1/2 in) diameter tissue-equivalent (TE) sphere enclosed in a cylindrical aluminium housing at the end of a long aluminium stem manufactured by Far West Technology. A schematic diagram of the detector is shown in Figure 2. The sphere was filled with propane TE gas at a pressure of 4452 Pa (33.4 torr) to simulate a 1 μm site of tissue. To avoid drilling large holes in the RANDO phantom to accommodate the TEPC, a 2.54 cm (1 in) thick slab of polyethylene in the cross-sectional shape of the phantom torso was substituted for an actual slice of the phantom at several different measurement heights. The polyethylene slice had a 1.9 cm (3/4 in) hole drilled through from the back of the torso to the middle of the slice. The TEPC was inserted from the back into the phantom in this way and the centre of the TEPC's active volume was in the centre of the substitute slice. Pulse height distributions were collected on a multichannel analyser during the measurements and stored in computer files. The data were analysed with the computer code, tepc_ng, written at PNL[6]. Since the TEPC is unable to discriminate between incident and neutron-induced photons, only the neutron component of the measured spectra has been used in this study.

Measurements were attempted at three different elevations within the phantom torso. Although the outer shape of the torso changes with elevation, the front of the substitute slice was always flush with the front of the surrounding torso slices and the distances from the centre of the detector to the front of the torso and to the source were carefully measured and recorded. The measurements were performed at slices 11, 21, and 31 with the phantom exposed to both the bare and D$_2$O-moderated ^{252}Cf sources.

Figure 1. RANDO phantom exposed to the D$_2$O-moderated ^{252}Cf source in the Low Scatter Facility.

RESULTS AND DISCUSSION

The results of the etched track detector and TLD measurements and calculations for the thyroid, heart, liver, and kidneys are shown in Table 1 for the bare ^{252}Cf exposure. The phantom was exposed to the bare source for a reference dose equivalent, based on NIST's calibration methodology[7], of 5.0 mSv at 1 m. Table 2 shows the results for the D_2O-moderated ^{252}Cf exposure. The phantom was exposed to the moderated source for a reference dose equivalent of 7.5 mSv at 1 m. The etched track dosemeters and the fast component of the TLDs were calibrated using the bare and D_2O-moderated ^{252}Cf sources for the bare and D_2O-moderated ^{252}Cf exposures, respectively. The slow component of the TLDs was calibrated using the NIST thermal neutron beam. All calibrations were carried out with the dosemeters mounted on a phantom.

The etched track detector and TLD results varied significantly from the calculated organ dose equivalents. The etched track detector results varied from 27% to 60% of the calculated value. Most of the under-response can be attributed to the angular dependence of the etched track detector which causes the dosemeter response to decrease to roughly one-third when the incident neutrons are parallel to the front face of the dosemeter instead of perpendicular. This under-response is only about one-half if the etched track detector is within a highly scattering material such as the phantom. In Tables 1 and 2, the thyroid, which is the organ closest to the front of the phantom, had the largest under-response.

The TLDs, being very close to point detectors, do not suffer from the same angular dependence as the etched track detectors. However, the problem of calibrating the slow and fast neutron components is not trivial. The neutron source used to calibrate the slow neutron component alone can cause order of magnitude differences in the resulting dose equivalent. Calibrations were attempted using the D_2O-moderated ^{252}Cf source and a graphite-moderated Pu–Be source in addition to the NIST thermal beam. All three calibration sources produced results that were far above or below the calculated values.

To compare the calculated absorbed dose to the absorbed dose measured by the TEPC, the mathematical phantom was modified to include a model of the TEPC inside the phantom in place of tissue/organ material. The TEPC was modelled as a 1.27 cm inside diameter, A-150 TE plastic sphere, 0.127 cm thick with the TE gas inside and outside the sphere but inside the 1.9 cm diameter aluminium housing. The aluminium itself was not modelled because of its transparency to neutrons. Figure 3 shows the modified mathematical model as drawn by the MCNP code at slice 21 in the phantom. The phantom composition was also changed from ICRP tissue (10.47% H, 23.02% C, 2.34% N, 63.21% O) to Alderson Muscle (8.8% H, 64.4% C, 4.1% N, 20.4% O)[8] and the legs were deleted several inches below the torso to match the actual RANDO phantom. Instead of calculating organ-averaged fluences, a direct calculation of the energy deposited in the active volume of the detector was performed to compare to the absorbed dose measured by the TEPC.

The results of the TEPC measurements and the MCNP calculations are compared in Table 3.

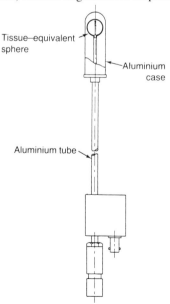

Figure 2. 1/2 inch tissue-equivalent proportional counter.

Table 1. Organ dose equivalents for bare ^{252}Cf (mSv).

Organ	Etched track detectors	TLDs	Calculated
Thyroid	0.960	4.240	3.555
Heart	0.693	5.694	2.375
Liver	0.880	3.506	2.264
Kidneys	0.360	2.366	0.599

Table 2. Organ dose equivalents for D_2O-moderated ^{252}Cf (mSv).

Organ	Etched track detectors	TLDs	Calculated
Thyroid	1.710	3.980	4.836
Heart	1.370	1.782	3.650
Liver	1.365	1.537	3.274
Kidneys	0.320	0.610	0.885

Although the calculated and measured values agree to within 18%, the calculated values are consistently higher than the measured values. This may partially be explained by the higher hydrogen density of the polyethylene slice, which was not modelled explicitly in the MCNP calculations, over the Alderson muscle (14.29% rather than 8.8%). The polyethylene is more effective at scattering and absorbing the incoming neutrons, thereby causing the measured values within the phantom to be lower.

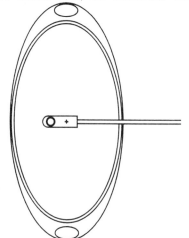

Figure 3. Mathematical model of 1/2 inch TEPC in MIRD-V phantom. The regions representing the ribs and the arm bones, as well as the TEPC, are shown.

Table 3. Neutron doses for bare and D$_2$O-moderated ^{252}Cf (mGy.h^{-1}).

Location	TEPC Measured ± 1σ	MCNP/MIRD Calculated ± 1σ	Calculated/Measured
Bare ^{252}Cf			
Slice 11	0.941 ± 0.019	1.107 ± 0.055	1.18
Slice 21	1.086 ± 0.035	1.269 ± 0.050	1.17
Slice 31	1.009 ± 0.030	1.080 ± 0.054	1.07
D$_2$O-moderated ^{252}Cf			
Slice 11	0.233 ± 0.013	0.260 ± 0.021	1.12
Slice 21	0.262 ± 0.023	0.294 ± 0.023	1.12
Slice 31	0.199 ± 0.007	0.218 ± 0.019	1.10

CONCLUSIONS

The usefulness of the etched track detectors and TLDs was limited due to the difficulty in calibrating the TLD chip response for the correct neutron energies and the angular dependence and energy threshold of the etched track detectors. The TEPC measurements and calculations of absorbed dose agreed within the 20% criterion for verification of this technique. It is assumed that a correct estimation of the absorbed dose at these locations within the phantom translates into correct estimations of absorbed doses at organ locations. After careful studies of the precision of the TEPC measurements and the Monte Carlo calculations, the observed differences can be attributed to the differences still remaining between the physical phantom and the mathematical phantom, which are also typical of the differences one might find between body types.

These results allow us to proceed to model more complex exposure situations in the future, such as different sizes of phantoms and non-uniform geometries. The same organ doses have also been used to estimate the new ICRP 60 quantity, effective dose, by applying the ICRP 60 radiation weighting factors and tissue weighting factors.

REFERENCES

1. *MCNP — A General Monte Carlo Code for Neutron and Photon Transport. Version 3A*. LA-6396-M, Revision 2 (September 1986). MCNP3B Newsletter (July 1988).
2. Medical Internal Radiation Dose (MIRD) Pamphlet No. 5, Revised. *Estimates of Specific Absorbed Fractions for Photon Sources Uniformly Distributed in Various Organs of a Heterogeneous Phantom* (New York: Society of Nuclear Medicine) (1978).
3. ISO, International Organization for Standardization. *Neutron Reference Radiations for Calibrating Neutron Measuring Devices Used for Radiation Protection Purposes and for Determining Their Response as a Function of Neutron Energy*. ISO 8529 (Geneva, Switzerland: ISO) (1986).
4. Huda, W. and Sandison, G. A. *Estimation of Mean Organ Doses in Diagnostic Radiology from Rando Phantom Measurements*. Health Phys. **47**(3), 463-467 (1984).
5. Golikov, V. Y. and Nikitin, V. V. *Estimation of the Mean Organ Doses and the Effective Dose Equivalent from Rando Phantom Measurements*. Health Phys. **56**(1), 111-115 (1989).
6. Brackenbush, L. W., Stroud, C. M., Faust, L. G. and Vallario, E. J. *Personnel Neutron Dose Assessment Upgrade — Volume 2: Field Neutron Spectrometer for Health Physics Applications*. PNL-6620 Vol. 2 (Richland, WA: Pacific Northwest Laboratory) (1988).
7. Schwartz, R. B. and Eisenhauer, C. M. *Procedures for Calibrating Neutron Personnel Dosimeters*. NBS Special Publication 633 (US Department of Commerce/National Bureau of Standards, Washington, DC) (1982).
8. ICRU, International Commission on Radiation Units and Measurements. *Tissue Substitutes in Radiation Dosimetry and Measurement*. Report 44 (Bethesda, MD: ICRU Publications) (1989).

A MODERATOR TYPE DOSE EQUIVALENT MONITOR FOR ENVIRONMENTAL NEUTRON DOSIMETRY

A. Esposito[†], C. Manfredotti[‡], M. Pelliccioni[†], C. Ongaro[§] and A. Zanini[§]
[†]INFN, Laboratori Nazionali di Frascati, Italy
[‡]Dipartimento di Fisica Sperimentale, Universita' di Torino, Italy
[§]INFN, Sezione di Torino, Italy

Abstract — The neutron monitors used daily in working areas around particle accelerators must be practical, economical, easy to read and to install and sensitive to low neutron dose equivalents. An environmental neutron detector, which responds satisfactorily to these requirements, was developed at LNF (Laboratori Nazionali di Frascati, Roma, Italy), for neutron monitoring in the area of the storage ring ADONE. It consists of a cylindrical moderator (height 15 cm, diameter 15 cm) of polyethylene with five TLD-600, sensitive to n+γ radiation and five TLD-700, sensitive to γ radiation, placed in the centre. A simulation of the instrument was performed using the 3-D Monte Carlo computer code MCNP (Monte Carlo Neutron and Photon transport), in order to obtain the theoretical fluence and dose equivalent response as a function of neutron energy. An accurate calibration of the instrument was carried out at the PTB (Physikalisch-Technische Bundesanstalt, Braunschweig, FRG) exposing it to monoenergetic neutron beams (with energies of 14.2, 5, 0.57, 0.144 MeV and thermal neutrons). A further calibration has been carried out at the CRE ENEA (Frascati, Roma, Italy), with a continuous spectrum from a ^{252}Cf fission neutron source ($\bar{E}$ = 2.3 MeV). From the results of the calibration, an average instrument response of 1.3–1.5 nC.mSv^{-1} seems to represent a sufficiently conservative choice in the evaluation of dose equivalent for high energy neutrons (over 5 MeV); however, for neutrons highly degraded in energy, such a factor could give an overestimation of the dose equivalent of more than a factor two.

INTRODUCTION

The systematisation of neutron protection dosimetry, despite being satisfactory from a security point of view, presents complex and still unsolved problems at a fundamental level, especially during the current transition period that radio protection is going through.

The introduction of radiation quality, by means of the Quality Factor to the definition of the dose equivalent, in fact presents practical difficulties in the experimental determination of this quantity. The changes which the quality factor's numerical values may undergo with time, on the basis of radiobiological and epidemiological data, also make the dose equivalent a rather unstable quantity.

As far as the neutrons are concerned, it should be remembered that the recommended values for such factors in ICRP Publication 21[1] were modified by a factor two in 1985[2], were then re-examined by a joint ICRP–ICRU task group in 1986[3] and were further re-evaluated in ICRP Publication 60[4], following the change which occurred in the Q–LET relationship.

Even the choice of the quantities to be measured has been subjected to continuous change over the past few years. The traditional dose equivalent was superseded by the index quantities[5] and then by the ICRU operative quantities[6], with further modifications in the ICRP recent recommendations[4] where the quality factors were replaced by 'radiation weighting factors', and the dose equivalent was replaced by the equivalent dose.

In the present work the quantity dose equivalent and the neutron quality factors recommended by ICRP 21 have been used, as this is the convention still followed by most of the operators in the radiation protection field.

NEUTRON ENVIRONMENTAL MONITORING SYSTEMS

For practical and economical reasons, in neutron environmental monitoring systems, the use of passive detection techniques is often preferred. This usually requires moderator type monitors, which consist of high efficiency thermal neutron detectors placed inside suitable spherical or cylindrical moderators.

As the doses which this type of application tries to determine are low, the size of the moderators tends to be reduced, thus increasing the sensitivity of the system, but lowering the quality of the energy response; consequently, these instruments rarely show the characteristic behaviour of a rem counter.

When the neutron spectrum does not vary significantly, this technique is acceptable and the choice of the calibration factor does not present great problems. If this condition is not fulfilled, a careful intercalibration with a rem counter is needed for each detector, carried out in the same position in which it will be used. Otherwise, a conservative calibration factor has to be chosen.

The most common moderator type monitor consists of thermoluminescence dosemeters ^{6}LiF

TLD–600 (sensitive to n + γ radiation) coupled with ^{7}LiF TLD–700 (sensitive to γ radiation only), placed inside polyethylene spheres or cylinders of different dimensions.

The simplicity and low cost of this technique have lead to its routine use at numerous research laboratories, such as CERN[7] (cylindrical moderator, diameter 12.5 cm, height 12.5 cm), DESY[8] (cylindrical moderator, diameter 15 cm, height 20 cm), etc.

At the LNF (Frascati, Italy) a detector consisting of a polyethylene cylinder (diameter 15 cm, height 15 cm), inside which are placed five TLD–600 and five TLD–700 (Figure 1), is used for neutron environmental monitoring in the area of the storage ring ADONE. The moderator size was determined by a simulation with the 3-D Monte Carlo computer code MCNP[9] in order to obtain an instrument response that differs as little as possible from the ICRP fluence to dose equivalent curve.

SYSTEM CALIBRATION

The system calibration was carried out at PTB (Physikalisch-Technische Bundesanstalt, Braunschweig; FRG) by exposing a detector prototype to monoenergetic neutron beams of various energies (0.025 eV, 144 keV, 570 keV, 1.2 MeV, 5 MeV, 14.8 MeV), perpendicular to the counter axis. An additional calibration was performed at the CRE ENEA (Frascati, Italy), by exposure to a ^{252}Cf ($\bar{E}$ = 2.3 MeV) neutron source, which provides a neutron energy spectrum roughly similar to the one produced by giant resonance reactions occurring in the high Z materials of the accelerator structures.

Figure 2 shows the instrument response in terms of fluence as a function of neutron energy.

As can be seen, the maximum of the response lies at about 1 MeV; at about the same energy the peak occurs of giant resonance photoproduction neutron spectra, which represents the main neutron component around low and medium energy electron accelerators.

Figure 3 shows the theoretical response calculated by the MCNP code both in terms of dose equivalent, evaluated using quality factors from ICRP 21 and of equivalent dose, evaluated using radiation weighting factors recently proposed in ICRP Publication 60[4].

Figure 4 compares the experimental results with the values obtained using MCNP simulation and the results of the Monte Carlo calculation for an 8" diameter spherical detector[10] (the cylindrical detector corresponding to an equivalent sphere of about 7" diameter). The consistency between the curves is good.

Figure 5 shows the experimental data per unit fluence, as a function of neutron energy, compared with the ICRP fluence to dose equivalent curve and with the values calculated with the MCNP code. For reasons of comparison, the curves have been normalised by equating the experimental values at 1.2 MeV, expressed in nC.cm^2, with the ICRP response at 1 MeV.

It can be seen that as the energy increases the detector response differs from the ICRP data,

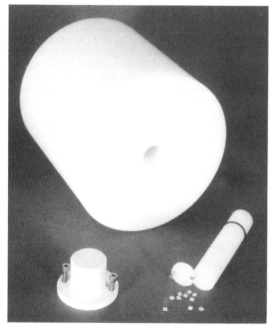

Figure 1. Neutron detector in use at LNF (Frascati, Italy).

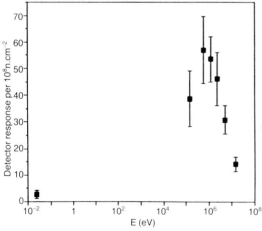

Figure 2. Experimentally determined detector response, in terms of unit fluence, as a function of neutron energy.

particularly over 5 MeV. However, above this energy the spectral component of giant resonance neutrons outside the shields is of minor importance.

In conclusion, Figures 3 and 4 would seem to confirm that an average conversion factor of 1.3–1.5 nC.mSv^{-1} represents a sufficiently conservative choice in the energy range of interest for neutrons produced in electron accelerators. For spectra highly degraded in energy, such as those expected at tunnel exits or, by skyshine effect, at long distances from the machine, such a conversion factor could overestimate the dose equivalent by more than a factor two.

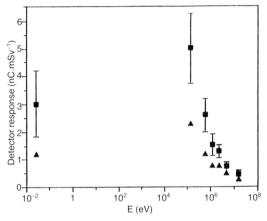

Figure 3. Calculated detector response, in terms of unit dose equivalent, as a function of neutron energy: (■) evaluated using Q values (ICRP 21)[1], (▲) evaluated using w_R values (ICRP 60)[4].

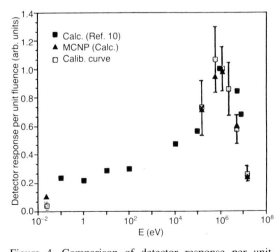

Figure 4. Comparison of detector response per unit fluence with values obtained by MCNP simulation. The results obtained with Monte Carlo code for 8" diameter spherical detector[10] are also shown.

PRELIMINARY MEASUREMENTS IN THE ADONE AREA

The environmental neutron monitoring system in the ADONE area consists of 12 measurement points; in each, one of the described detectors is positioned, as shown in the layout in Figure 6.

In two locations, marked A and B, two active Andersson-Braun rem counters were placed next to two detectors, in order to carry out an intercalibration. A and B are situated above the 50 cm thick concrete roof installed on top of the machine, corresponding to the electron and positron deflectors.

Table 1 shows the results of the measurements after an exposure of more than six months, compared with the results of preliminary comparative measurements at positions A and B. The latter, however, need further examinations. The response of the passive device obtained by the comparison were indeed unexpectedly different

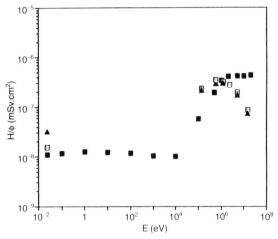

Figure 5. Experimental and calculated detector response compared with ICRP 21 data. (■) mSv.cm^2, ICRP. (▲) MCNP calculation. (□) nC.cm^2. (exp.)

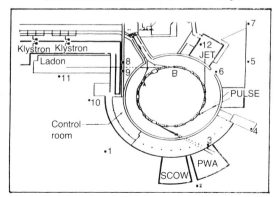

Figure 6. Neutron monitoring location in the ADONE area.

Table 1. Results of preliminary measurements in the ADONE area.

Detector location	TL signal (nC)	Error (nC)	Dose equivalent (mSv)	Error (mSv)
1	0.004	0.046	0.002	0.027
2	0.004	0.040	0.002	0.024
3	0.017	0.026	0.010	0.015
4	0.028	0.047	0.016	0.028
5	0.021	0.062	0.012	0.036
6	0.074	0.056	0.044	0.033
7	0.011	0.055	0.006	0.032
8	0.065	0.035	0.038	0.020
9	0.043	0.039	0.025	0.023
10	0.051	0.043	0.030	0.025
11	0.077	0.054	0.045	0.032
12	−0.007	0.026	−0.004	0.015
*A	2.714	0.452	1.596	0.266
*B	0.377	0.077	0.221	0.045

* A rem counter signal = 0.73 mSv.
  B rem counter signal = 0.22 mSv.

(1.7 $nC.mSv^{-1}$ and 3.7 $nC.mSv^{-1}$) and can be justified only in terms of neutron spectra highly degraded in energy.

The monitoring system data were interpreted using the most conservative of the two coefficients (1.7 $nC.mSv^{-1}$), which do not differ greatly from the one obtained from the PTB calibration curve.

In any case, the results of the readings cannot easily be separated from the background, thus confirming the excellent level of protection achieved by means of the new shielding disposition installed around ADONE.

REFERENCES

1. ICRP. *Data for Protection against Ionizing Radiation from External Sources*. Publication 21, (Oxford: Pergamon Press) (1971).
2. ICRP. *Statement from the 1985 meeting of the ICRP*. Radiat. Prot. Dosim. **11**, 134 (1985).
3. ICRU. The Quality Factor in Radiation Protection. Report 40 (Bethesda, MD: ICRU Publications) (1986).
4. ICRP. *1990 Recommendation of the International Commission of Radiological Protection*. Publication 60, Ann. ICRP **21** (Oxford: Pergamon Press) (1991).
5. ICRP. *Recommendation of the International Commission of Radiological Protection*. Publication 26. Ann. ICRP, **1** (Oxford: Pergamon Press) (1977).
6. ICRU. *Determination of Dose Equivalent Resulting from External Radiation Sources*. Report 39 (Bethesda, MD: ICRU Publications) (1985).
7. *The Use of Thermoluminescent Dosemeters for Environmental Monitoring around High Energy Proton Accelerators*. CERN,HS-RP/012/CF (1977).
8. Amsinck, K. *Neutronendosimetrie mit Festkorperdosimetern - Vergleich verschiedener Messverfahren*. DESY D3/17 (1974).
9. Briesmeister,J . *MCNP—A General Monte Carlo Code for Neutron and Photon Transport, Version 3A*. LA-7396-M, Rev. 2 (Los Alamos National Laboratory: New Mexico) (1986).
10. Dhairyawan, M. P., Nagarajan, P. S. and Venkataraman, G. *A Theoretical Study of the Response of Spherical Moderator Neutron Detectors*. Private communication (1979).

NEUTRON DOSE EQUIVALENT RATE METER ON THE BASIS OF THE SINGLE SPHERE ALBEDO TECHNIQUE

B. Burgkhardt, E. Piesch and M. I. Al-Jarallah*
Karlsruhe Nuclear Research Centre, Health Physics Division
P.O.B. 3640, W-7500 Karlsruhe 1, FRG

Abstract — In area monitoring there is a need for a more accurate neutron reference dose rate meter, especially for the purpose of albedo dosemeter calibrations in neutron stray radiation fields of interest. The so-called Single Sphere Albedo Counter makes use of three active ^{3}He proportional counters as thermal neutron detectors positioned in the centre and on the surface (albedo dosemeter configurations) of a polyethylene sphere. The linear combination of the detector readings allows for the indication of different quantities like $H^*(10), D, \phi$, and reduces the energy dependence significantly. The paper describes the dosimetric properties of a prototype instrument and its application in routine monitoring.

INTRODUCTION

In area monitoring there is a need for a more accurate reference dose rate meter which may replace the old type of rem counter with an energy dependence of a factor of about 10. Improvements in the energy response have been found by low pressure tissue-equivalent proportional counters (TEPC) or the application of Bonner spheres. These research instruments, having the additional properties of field diagnosis, are not alternatives in routine monitoring[1].

In order to provide a more accurate reference dosemeter, especially for the purpose of albedo dosemeter calibrations in neutron stray radiation fields of interest, the so-called Single Sphere Albedo Technique was originally adopted for TL detectors by combining a passive rem counter with the KfK albedo neutron dosemeter. Using three thermal neutron detectors positioned in the centre and on the surface of a polyethylene (PE) sphere, the linear combination of the detector readings reduced the energy dependence significantly[2].

The Single Sphere Albedo Counter, termed counter in the following, replaces the TL detectors by active ^{3}He proportional counters. The properties of this dose rate meter were optimised in the moderator design[3,4] and investigated with respect to their ability to indicate different quantities in the neutron radiation field. The paper describes the dosimetric properties of a prototype instrument and its application in routine monitoring.

DETECTOR SYSTEM

The counter makes use of three ^{3}He tubes, one in the centre of a PE sphere, designated 'c', and two on the surface, one just inside the moderator (albedo detector configuration, designated 'i', front-shielded, and the other just outside the moderator (detector configuration, designated 'a', for thermal field neutrons), back-shielded by a boron-plastic absorber (Figure 1). Detector positions, size and thickness of the boron absorber have been optimised to simulate the KfK albedo dosemeter[3]. Small ^{3}He proportional counter tubes of the type Thomson 0.5 NH 1/1K with an effective height of 10 mm

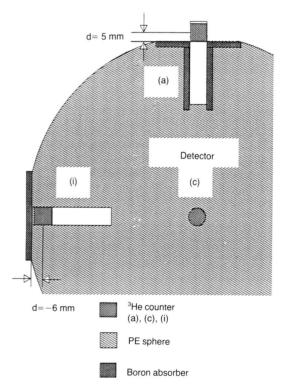

Figure 1. Cross section of the dose rate meter using ^{3}He detectors in the boron–plastic absorber configurations (i), (a), and (c) in a polyethylene sphere.

---

* On leave from the King Fahd University of Petroleum and Minerals, Physics Department, Dharan, Saudi Arabia.

and a diameter of 10 mm combined with a KfK improved Thomson ACH amplifier, offer a dose rate range from 1 µSv.h^{-1} to 100 mSv.h^{-1} with a background rate of ≤0.01 µSv.h^{-1} and a gamma discrimination factor of 3×10^{-5}. The count rates of the three detectors are indicated simultaneously.

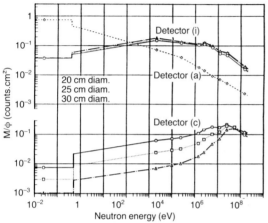

Figure 2. Energy dependence of the fluence response M/Φ for the ^{3}He detectors for frontal (AP) irradiation of the detector configuration (i) and (a) and for (c) in the PE spheres of different diameters.

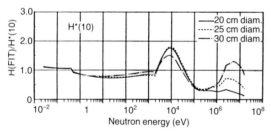

Figure 3. Energy dependence of the dose equivalent response H(FIT)/H*(10) of the counter using different sphere diameters and fits of the detector readings and frontal irradiation, based on the new Q–L relationship of ICRP 60[5].

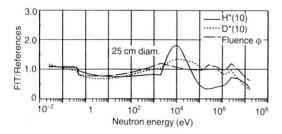

Figure 4. Energy dependence of the counter for the measurement of neutron fluence, absorbed dose and dose equivalent using different fits for frontal irradiation.

DOSIMETRIC PROPERTIES

The prototype counter has been calibrated at the PTB with monoenergetic neutrons between 144 keV and 14.8 MeV and with reactor neutrons at 2 keV, 24 keV and 143 keV as well as with thermal neutrons. After optimisation of the detector/absorber geometry, the PE sphere was varied in the diameter. The counter has been irradiated using the detector configurations (i) and (a) in a front position with the detector facing the source.

On the basis of the energy dependent fluence response for frontal irradiation of the detectors of the configurations (a), (i) and (c) (Figure 2) and a least squares fit, we estimated the weighting factors k_i for the linear combination of the detector readings in order to indicate alternatively the neutron fluence Φ, the absorbed dose D*(10) or, for instance, the neutron dose equivalent H*(10),

$$H*(10) = k_1 M(a) + k_2 M(i) + k_3 M(c)$$

For the estimation of the ambient dose equivalent, fluence to dose equivalent conversion factors have been used based on the new Q–L relationship recommended by ICRP 60[4]. The results in Figure 3 show that a sphere diameter of 25 cm would result in a sufficient energy response for H*(10), whereas a diameter of 27 cm would improve the underestimation of the 30 cm sphere in the energy range 0.1 to 1 MeV. The prototype with a diameter of 25 cm is a compromise of energy response and weight of the sphere. Different fits allow for the indication of the quantities Φ(E), D*(10), H*(10) and H_{made}, respectively (Figure 4).

For the simulation of the rotational and isotropic radiation incidence the mean readings of the detector configurations (i) and (a) have been used for the frontal, rear, right and left irradiation positions (rotational incidence), and once more right and left, respectively (isotropic incidence using six-step read-out at 90°). Due to the low scatter condition of the calibration field the readings of the remaining top and bottom positions of the sphere have been found to be equal to the right and left positions. Compared with the frontal detector readings it has been found that only the k factors change but the resulting energy dependence for the different quantities is practically equal to that of the frontal irradiation as presented in Figures 3 and 4.

For the application of the counter in isotropic radiation fields the mean readings of the detector configuration (a) and (i) for a six-step read-out were used to estimate the field quantity on the basis of the linear combination. Instead of stepwise rotation a continuous rotation and/or more than two detectors (up to six) may be used for routine application.

Figure 5 shows the energy dependence of the detector (c) (rem counter) and of the corresponding counter using a PE sphere of 25 cm and 30 cm. Here, the weighting factors k have been optimised to obtain a minimum deviation in the energy range above 100 keV resulting in a maximum deviation of about ±60% (factor 4.5) in the energy range E_{th} to 10 MeV, which is lower compared to the rem counter (factor 10) (Figure 5) or the TEPC type HANDI, for instance, with a typical underresponse of a factor of 5 at 24 keV[6]. The error bars represent the resulting variation in response for a 5% variation of the detector readings (i) and (c).

Using the prototype counter, experiments in PWR radiation fields have shown that a rem counter sphere with a diameter of 30 cm and a BF_3 counter may overestimate the neutron dose equivalent by about +20%[7]. Thus the prototype counter provides calibration factors for neutron albedo dosemeters which were more accurate and consistent for an application in individual monitoring.

CONCLUSION

For an application in stray radiation fields the single sphere albedo counter was found to be a dose rate meter with an improved energy and angular dependence.

The disadvantages of the single sphere albedo technique, namely the costs, the use of three proportional counters and the need for a correction of the angular response by rotating the counter, will be balanced by its advantages. In comparison with more sophisticated spectrometers, such as the TEPC or the Bonner spheres, the single sphere albedo counter has similar advantages to be used as:

(1) a dose rate meter for area monitoring with a reduced energy dependence in stray radiation fields;
(2) a reference instrument, for instance, for the calibration of albedo dosemeters in stray radiation fields of interest by indicating the calibration factor directly;
(3) an instrument for the interpretation of radiation fields in terms of neutron fluence, absorbed dose, mean quality factor or other quantities. New dose quantities can easily be adopted just by a change of the weighting factors of the linear combination.

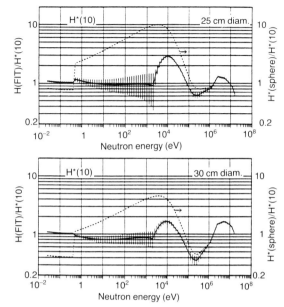

Figure 5. Energy dependence of the dose equivalent response H(FIT)/H*(10) of the counter for two sphere diameters of 25 cm and 30 cm, respectively, and the rem counter sphere, using detector (c) only, and an isotropic radiation incidence; special fit for a flat response in the energy range of fast and low energy neutrons.

ACKNOWLEDGEMENTS

This work was sponsored by the Bundesministerium des Inneren under contract No. St.Sch. 1.015.

REFERENCES

1. Piesch, E. *Requirements for the Performance of Dose Equivalent Meters in Area and Individual Monitoring: Conclusions of the Discussion.* Radiat. Prot. Dosim. **29**(1/2), 149-152 (1989).
2. Burgkhardt, B. and Piesch, E. *Field Calibration Technique for Albedo Neutron Dosemeters.* Radiat. Prot. Dosim. **23**, 121-126 (1988).
3. Al-Jarallah, M. I. *Optimisation of a Single Sphere Albedo System Using ^{3}He Counters for the Measurement of Neutron Dose Equivalent Rates and the Field Calibration of Personnel Albedo Neutron Dosimeters.* KfK 4632 (1989).
4. Burgkhardt, B., Al-Jarallah, M. I. and Piesch, E. *Neutronen-Dosisleistungsmesser nach dem Einkugelalbedomeßverfahren.* KfK 4631 (1990).
5. Schuhmacher, H. and Siebert, B. R. L. *Quality Factors and Ambient Dose Equivalent for Neutrons Based on the New ICRP Recommendation.* Radiat. Prot. Dosim. **40**(2), 85-89 (1992).
6. Alberts, W. G., Dietze, E. Guldbakke, S., Kluge, H. and Schumacher, H. *International Intercomparison of TEPC Systems Used for Radiation Protection.* Radiat. Prot. Dosim. **29**, 47-53 (1989).
7. Hofmann, I., Schwarz, W., Burgkhardt, B. and Piesch, E. *Neutronendosimetrie in Kernkraftwerken mit Leichtwasserreaktoren.* KfK 4499 (1989).

STUDY OF THE RESPONSE OF TWO NEUTRON MONITORS IN DIFFERENT NEUTRON FIELDS

A. Aroua†, M. Boschung‡, F. Cartier§, K. Gmür‡, M. Grecescu†, S. Prêtre§, J.-F. Valley† and Ch. Wernli‡
†Institut de Radiophysique Appliquée
Centre Universitaire, CH-1015 Lausanne, Switzerland
‡Paul Scherrer Institut, CH-5232 Villigen-PSI, Switzerland
§Swiss Nuclear Safety Inspectorate, CH-5232 Villigen-HSK, Switzerland

Abstract — The response of two commercial neutron monitors, a rem counter of the Andersson-Braun type and a Dineutron, has been investigated in a variety of fields around calibrated neutron sources and inside nuclear power plants. The reference instrument used to characterise these fields is a multisphere spectrometer. The response of the rem counter has been also calculated by convolution of the measured spectra and some 107 others taken from IAEA publications, with the response function of the monitor to monoenergetic neutrons. Measured and calculated values are in good agreement. The results obtained indicate that both instruments overestimate the neutron dose equivalent for the type of neutron spectra encountered inside nuclear power plants. In such spectra the overestimation is less than a factor 2 for the rem counter, but much more pronounced for the Dineutron.

INTRODUCTION

In operational radiation protection, the availability of monitoring instruments which are portable, user-friendly and have a constant energy response is highly desirable. The rem counter is a neutron monitor commonly used in the nuclear power plants. The Dineutron is another type of instrument, recently developed for neutron monitoring. The response of these devices is dependent on the spectral distribution of the measured neutrons.

The response of the Andersson-Braun rem counter to monoenergetic neutrons has been extensively investigated; a literature review is found in Reference 1. The results show that the rem counter tends to overestimate dosimetric quantities such as the ambient dose equivalent in the energy range from 100 eV to several keV if it is calibrated with ^{252}Cf spontaneous fission or Am-Be neutron sources; the overestimation reaches a factor 3 – 5, the maximum overestimation being located around 8–10 keV. Both at low energies (below 10 eV) and at high energies (beyond a few MeV) the rem counter underestimates the ambient dose equivalent. The underestimation may reach a factor of two at the extremities (thermal and 14 MeV neutrons).

Figure 1 presents the variation of the relative response of the Andersson-Braun rem counter with neutron energy; it shows:

(a) the original response curve given by Andersson and Braun in 1964[2] who used the dose equivalent produced in an infinite 30 cm thick slab of tissue, and
(b) the response curves established from data published by the IAEA in 1990[3], using the dose equivalent as defined in publication ICRP 21[4] and the ambient dose equivalent (old Q–L relation), the expression of the fluence to ambient dose equivalent conversion factors given by Wagner et al[5] is used.

Figure 1 shows that the rem counter gives an accurate indication of the ambient dose equivalent only between 100 keV and several MeV.

The rem counter's behaviour with respect to wide spectra is hardly described in the literature. Endres et al[6] have performed measurements, either by multisphere spectrometry or by using low pressure tissue-equivalent proportional counters, in six nuclear power plants (5 PWRs and 1 BWR) and used for comparison, an Andersson-

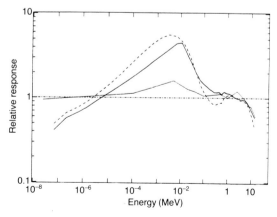

Figure 1. Variation of the Andersson-Braun rem counter's relative response for monoenergetic neutrons. (.....) Andersson and Braun (1964), H in an infinite 30 cm thick slab. (——) IAEA (1990), H according to ICRP 21. (– – –) IAEA (1990), H*(10).

Braun rem counter (SNOOPY model). The results given by this instrument show that it overestimates the neutron dose equivalent by a factor 1.5 – 2. Rantanen[7] determined by calculation the response of the rem counter in a variety of wide neutron spectra and found that the indication of the instrument is accurate within ± 30% for hard spectra, but presents an overestimation up to a factor 2 for soft ones.

The present work is intended to contribute to the understanding of the behaviour of the rem counter and that of the Dineutron in wide neutron spectra. It presents the results of the measurements performed with these monitors in a variety of neutron fields : around reference sources, inside the containments of nuclear power plants and close to nuclear fuel storage and transport facilities. It also presents calculated responses in several neutron fields with known spectra.

MATERIAL AND METHODS

The neutron monitors

The Andersson-Braun rem counter

The Andersson-Braun rem counter used is the 2202D model manufactured by Studsvik, AB Atomenergi (Sweden). It consists of a BF_3 proportional counter surrounded by polyethylene and boron plastic, and has one rounded end for a better angular response. The instrument was designed[2] in order to have a constant dose equivalent response over a wide energy range.

Due to its cylindrical form, the rem counter presents an anisotropy. The reference irradiation geometry corresponds to the case where the neutrons are incident from the side at a right angle to the instrument's axis of symmetry. The manufacturer indicates that if the instrument is irradiated from the front, then its response deviates from the reference irradiation geometry by ± 10% according to the neutron energy. If the instrument is irradiated from the back then it presents an underestimation of a factor 2 relative to the reference irradiation geometry. The anisotropy effect is unavoidable when multidirectional fields are measured. During the measurements the rem counter has been rotated in order to locate the position where it indicated a maximum and thus to avoid a significant underestimation.

The rem counter gives an analogue display of the dose equivalent but at low rates the reading is not very accurate due to fluctuations. Consequently the pulse output was used, coupled to a scaler. The neutron sensitivity of the rem counter in terms of the ambient dose equivalent, H*(10), has been measured for the Am–Be, Cf and $Cf(D_2O)$ sources[1].

For all the measurements inside the nuclear power plants, where the spectra are in general soft, it is convenient to calibrate the rem counter with intermediate neutrons. Among the calibration sources available the $Cf(D_2O)$, i.e. ^{252}Cf inside a φ 30 cm D_2O sphere, is the most suitable[8,9] one. The calibration factor corresponding to this source has been adopted in this work.

The Dineutron

The Dineutron is a portable neutron survey meter designed by the CEA (France)[10] and manufactured by Nardeux (France). It consists of two ^3He thermal neutron counters surrounded by two spherical moderators made of polyethylene of diameters 2.5" (6.35 cm) and 4.2" (10.67 cm). On the basis of the count rates given by the two counters, an integrated microprocessor computes different dosimetric quantities which are then displayed digitally.

The Dineutron presents an anisotropy within the plane containing the axis common to the two spheres. In the reference irradiation geometry the neutrons are incident from a direction perpendicular to the axis of the spheres. Mourgues has shown[11] that the anisotropy induces a response variation within the interval [–21%, +25%]. The anisotropy effect is most pronounced in multidirectional fields such as those encountered inside nuclear power plants.

According to the manufacturer the instrument indicates the dose equivalent with an accuracy of ± 30% for all neutron energies. However, Hankins[12] reported that at low energies the instrument's response deviates significantly from reference values. Hunt *et al*[13] reported an overestimation of a factor 1.4 for bare Cf and a factor 2 for $Cf(D_2O)$.

The neutron fields

The measured fields

Measurements with the monitors have been performed at :

(a) seven points near neutron sources of Am–Be, Cf and $Cf(D_2O)$, in different geometries;
(b) twenty-two points inside four nuclear power plants (two PWRs and two BWRs) in different locations : inside the containments, in the fuel storage facility and around transport flasks.

For all the measured points the neutron spectrum and the associated dosimetric quantities have been established by a multisphere spectrometer[1]. The spheres' indications are considered in this work as reference values.

The fields used for calculation of the rem counter's response

The response of the Andersson-Braun rem counter has been calculated by convolution of different wide spectra with the response function of the monitor to monoenergetic neutrons (Figure 1) as follows:

$$R = \int R_\Phi(E)\, \Phi_E(E)\, dE$$

where R is the reading in Sv of the Andersson-Braun rem counter in a neutron field of spectral distribution $\Phi_E(E)$ and $R_\Phi(E)$ is its energy response expressed in Sv per unit fluence.

The calculation includes the measured spectra and some 107 others taken from IAEA publications Nos 180[14] and 318[3]. The 107 spectra are likely to be encountered in the field of radiation protection and cover a wide variety of neutron fields associated with research reactors, power plants of different types, experimental and medical accelerators, calibration sources and other miscellaneous fields. For each of these spectra the 9" to 3" ratio (9"/3"), called spectral index and which is convenient for the characterisation of the neutron field, has been computed by convolution with the spheres' response functions.

RESULTS

Figure 2 presents the variation of the relative response of the rem counter with the 9" to 3" ratio for all measured spectra. The values established by calculation for the same spectra are also shown. The relative response represents here the ratio of the response for the given neutron field to that associated with Cf(D$_2$O), taken as reference field.

It can be seen that for most of the spectra measured inside the containments of nuclear power plants the rem counter tends to overestimate the neutron ambient dose equivalent and that the overestimation is less than a factor 2. One can also see a slight underestimation for hard spectra which is less than 25%.

The agreement between measured and calculated values is quite good. When comparing calculation to measurements, one should take into account the angular response of the instrument. The response of the rem counter is not isotropic, and the response function presented in Figure 1 and used in the convolution has been established for a side irradiation where the instrument is most sensitive. Since the measurements have been performed in multidirectional fields and the results referred to values established by the multisphere system which is quasi-isotropic, it is not surprising to find measured results lower than the calculated ones.

In Figure 3 the results of calculation for the 107 spectra taken from the literature are added to those derived from measurements. The agreement between them is good.

The great dispersion of the points between 9"/3" = 0.15 and 9"/3" = 0.50 is due to the fact that two completely different spectra can have the same 9" to 3" ratio: one consisting mainly of intermediate neutrons (overestimation area), and the other comprising only thermal and fast components (underestimation areas). This is a major limitation of this spectral index.

In Figure 3 the response curve for monoenergetic neutrons is also shown for comparison. It can be seen that the high overestimation observed in the case of monoenergetic neutrons for certain energies is attenuated in the case of wide spectra, in agreement with other reports[6, 7, 15, 16].

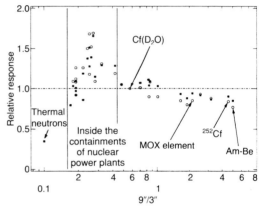

Figure 2. Measured (o) and calculated (■) relative responses of the Andersson-Braun rem counter for all the measured spectra; their variation with the 9" to 3" ratio is shown.

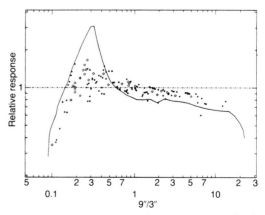

Figure 3. Calculated relative responses of the Andersson-Braun rem counter for the measured spectra (◊) and those taken from IAEA Reports 180 and 318 (■). (——) Response to monoenergetic neutrons.

All the neutron spectra encountered inside the containments of nuclear power plants, measured or taken from the literature, are located in Figure 3 in an area where the rem counter tends to overestimate the dose equivalent and this is due to the fact that this type of spectra contains a great proportion of intermediate neutrons.

Figure 4 presents the variation of the relative response of the Dineutron with the 9" to 3" ratio for all the measured spectra. The behaviour of this instrument is similar to that of the rem counter, but the overestimation of the dose equivalent for soft spectra is here more pronounced and can reach a factor of 4.

Figure 4 shows that the $Cf(D_2O)$ is situated in an overestimation area. The use of this source for calibration would lead to a significant underestimation of the dose equivalent for most spectra.

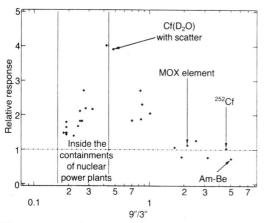

Figure 4. Variation of the relative response of the Dineutron with the 9" to 3" ratio.

Therefore it is recommended to calibrate this instrument with ^{252}Cf or Am–Be and to apply a correction factor to account for the over-response in certain types of radiation fields.

SUMMARY AND CONCLUSIONS

The responses of an Andersson-Braun rem counter and a Dineutron have been measured and calculated for a wide variety of neutron spectra. The results indicate that if the rem counter is calibrated with $Cf(D_2O)$, it tends to overestimate the dose equivalent in soft neutron spectra such as those encountered in the containments of nuclear power plants, by a factor less than 2. However, for hard spectra such as those encountered in the vicinity of nuclear fuel, the instrument presents an underestimation less than 25% which is acceptable. The reasonable accuracy and the fact that the rem counter is easy to handle justify its choice as an instrument for neutron field monitoring inside the nuclear power plants.

Concerning the Dineutron, the overestimation factor in soft spectra can reach a value of 4 if the instrument is calibrated with a fast neutron source such as ^{252}Cf. This is the price to pay for its merits which are mainly the practical dimensions, the light weight and the user-friendly handling. The use of $Cf(D_2O)$ for the calibration of the Dineutron is not recommended because it would lead to a significant underestimation of the dose equivalent for most spectra.

ACKNOWLEDGEMENT

This work has been supported by the Swiss Nuclear Safety Inspectorate.

REFERENCES

1. Aroua, A., *Etude des Champs Neutroniques dans les Centrales Nucléaires et de l'Influence de leur Diversité sur la Détermination des Grandeurs de la Protection Radiologique*. Thèse No 942, EPFL, Lausanne (1991).
2. Andersson, I. O. and Braun, J. *A Neutron rem counter*. Nukleonik, **6**(5) 237 (1964).
3. IAEA. *Compendium of Neutron Spectra and Detector Responses for Radiation Protection Purposes*, Technical reports series No 318 (Vienna: IAEA) (1990).
4. ICRP. *Data for Protection Against Ionizing Radiation from External Sources*. Publication 21 (Oxford: Pergamon Press) (1973).
5. Wagner, S. R., Grosswendt, B., Harvey, J. R., Mill, A. J., Sebach, H.-J. and Siebert, B. R. L. *Unified Conversion Functions for the New ICRU Operational Radiation Protection Quantities*, Radiat. Prot. Dosim. **12**(2), 231-235 (1985).
6. Endres G. W. R. *et al. Neutron Dosimetry at Commercial Nuclear Plants. Subtask A : Reactor Containment Measurements*. NUREG/CR-1789 (Pacific Northwest Laboratory) PNL-3585 (1981).
7. Rantanen, E. *Response of an Anderson-Braun Neutron Rem-counter to Various Neutron Spectra*. Health Phys. **41**(4), 671-674 (1981).
8. Schwartz, R. B. and Eisenhauer, C. M. *The Design and Construction of a D_2O-Moderated ^{252}Cf Source for Calibrating Neutron Personnel Dosimeters Used at Nuclear Power Reactors*, NUREG/CR-1204 (1980).
9. Kluge, H., Weise, K. and Hunt, J. B. *Calibration of Neutron Sensitive Spherical Devices with Bare and D_2O-Moderated ^{252}Cf Sources in Rooms of Different Sizes*. Radiat. Prot. Dosim. **32**(4), 233-244 (1990).

10. Mourgues, M., Carossi, J. C. and Portal, G. *A Light Rem-counter of Advanced Technology*. In: Proc. 5th Symp. on Neutron Dosimetry, Munich/Neuherberg, September 1984, EUR 9762 (Luxembourg:CEC), **1**, pp. 387-401 (1985).

11. Mourgues, M. *Essais physiques du "Dineutron", Version Prototype Industriel*. Rapport Technique SERE/165/MM/mtb (Commissariat à l'Energie Atomique, Cadarache) (1985).

12. Hankins, D.E. *Problems of Practical Neutron Health Physics Monitoring* Radiat. Prot. Dosim. **23**(1-4), 488 (1988).

13. Hunt, J. B., Champlong, P., Chemtob, M., Kluge, H. and Schwartz, R. B. *International Intercomparison of Neutron Survey Instrument Calibrations*, Radiat. Prot. Dosim. **27**(2), 103-110 (1989).

14. IAEA. *Compendium of Neutron Spectra in Criticality Accident Dosimetry*. Technical reports series No 180 (Vienna: IAEA) (1978).

15. Schraube, H. and Knöfel, T. M. J. *Derivation of Field Parameters and Aspects Concerning the Calibration and Use of Neutron Dosemeters in Slowing Down Neutron Fields*. In: Proc. 4th Symposium on Personnel Neutron Dosimetry, Munich-Neuherberg (Luxembourg Commission of the European Communities) Report EUR 7448 EN, pp. 561-573 (1981).

16. Birch, R., Delafield, H. J. and Perks, C. A. *Measurement of the Neutron Spectrum inside the Containment Building of a PWR*. Radiat. Prot. Dosim. **23**(1-4), 281-284 (1988).

DOSE EQUIVALENT RESPONSE OF NEUTRON SURVEY METERS FOR SEVERAL NEUTRON FIELDS

A. Rimpler
Bundesamt für Strahlenschutz (BfS)
Waldowallee 117 0-1157 Berlin, Germany

Abstract — Neutron survey meters should be able to read the dose equivalent rate with total measuring uncertainties of −50% to +100% that are commonly accepted in routine monitoring. On that account, the dose equivalent responses of a monitor of advanced technology (Dineutron) along with two Andersson-Braun type, a Leake type and a 10" spherical monitor were determined for numerous fields of shielded and bare ^{252}Cf and Pu–Be sources, having mean energies between 0.2 and 4 MeV, and around a transport cask for spent fuel elements. Reference data on the dose rate were revealed from neutron spectra measurements with Bonner spheres. The Dineutron response to neutrons from radionuclide sources was found lying between 0.4 and 1.7. At the spent fuel container, however, a distinct dose overestimation by more than a factor of 4 was revealed. The Andersson-Braun and Leake counters generally met the expectations, overresponding by a factor of 2 at maximum, both at the ^{252}Cf and Pu–Be sources and around the fuel cask. Surprisingly, the dose rates obtained when using the 10" sphere did not exceed the reference values by more than 50% in all fields investigated.

INTRODUCTION

Commercial survey meters generally use relatively large spherical or cylindrical moderators, partly with inner thermal neutron absorbers, and thermal neutron detectors (^{3}He, BF$_3$ proportional or ^{6}LiI scintillation counters). These rem counters have principally remained unchanged for many years.

The main disadvantages of all these single moderator monitors of the Hankins, Leake or Andersson-Braun type are their relatively high weight (6 to 13 kg) and a dose equivalent response between 0.3 and 6 in the energy range from 10^{-8} to 20 MeV.

To overcome these shortcomings, several suggestions have recently been made. One promising concept is the use of multidetector assemblies. A commercial version of a double-sphere survey meter, Dineutron 3.1, has been provided by Nardeux (France) for some years.

Since appropriate response data for this monitor have been missing, its dose equivalent response was investigated and compared with that of conventional area monitors such as Andersson-Braun or Leake type and moderating spheres.

The investigations were performed for a variety of fields of radionuclide neutron sources and recently supplemented by comparative measurements at a transport container for spent fuel elements.

MEASURING TECHNIQUE USED

The Dineutron consists of two ^{3}He proportional counters at the centre of moderating spheres with diameters of 2.5" and 4.2", respectively. The count rates of both detectors are converted to dosimetric quantities by means of a microprocessor using non linear transfer functions, described in detail elsewhere[1,2]. It enables measurements of the dose, dose equivalent and their rates, both in old and SI units. Additionally, the quality factor, exposure time and statistical uncertainties may be indicated.

The main advantages of this survey meter are its low weight of only 3.5 kg and easy handling that favours the Dineutron for field measurements. With regard to the most crucial point in neutron dosimetry, the energy dependence of the dose equivalent response, the manufacturer promises in his prospectus a maximum uncertainty of ±30% irrespective of neutron energy[3]. In the instruction manual a +50% to −40% response uncertainty is admitted, based on experimental results for numerous polyenergetic spectra[2]. However, an international intercomparison performed by four standard laboratories in the radiation fields of bare and D$_2$O-moderated ^{252}Cf sources has shown that Dineutron over-reads the dose equivalent by a factor of about 1.4 and 2.0, respectively[4]. According to the manufacturer's recommendations the device should preferably be used in stray neutron fields, e.g. at nuclear facilities.

The following survey meters have been used along with the Dineutron:

1. Andersson-Braun type monitors: 2202 D (Alnor) and NM2 (NE Technology).
2. Leake type monitor: 95/0949 (Harwell).
3. 10" spherical counter.

The No. 1 and 2 monitors are widespread in routine dosimetry and will not be described in detail; for their features see, for example, Reference 5. They are calibrated for Am–Be neutrons ($\bar{E} \approx 4.5$ MeV) by the manufacturer.

The 10" counter belongs to a set of six Bonner spheres (diameter 2" to 12") with a ^{6}LiI scintillation counter and a multichannel analyser, that has been used to measure reference values of the dose equivalent rate (MADE). These were derived from unfolding the neutron spectra by means of a SAND II type program with an estimated accuracy and precision of 10% (1 SD) at most. Further details such as spectrometer calibration and data evaluation have been described elsewhere[6]. The count rates of the ^{6}LiI detector in the 10" sphere were converted to dose equivalent rates using calibration factors determined by irradiations at a bare ^{252}Cf source.

The quantity to be estimated for a particular survey meter, the dose equivalent response, is the ratio of its dose equivalent reading in a special field and the reference value of the dose rate determined with the multisphere spectrometer.

NEUTRON FIELDS INVESTIGATED

The measurement of the exposure to neutrons originating from the utilisation of radionuclide sources (mainly ^{252}Cf and Be(α,n) sources) in research, technical sciences and medicine is an essential task in routine monitoring practice. A survey meter should be capable of reading dose rates with uncertainties of –50% to +100% that are commonly accepted in area dosimetry. On that account the suitability of the Dineutron along with the A–B meters and the 10" sphere was investigated by subsequent irradiations in various reference fields of ^{252}Cf and Pu–Be sources, both free-in-air and behind common shieldings, as well as under conditions representative for storing those sources. The characteristics of these fields are briefly described in Table 1.

The second series of comparative measurements, using the Dineutron, Leake counter and 10" spherical monitor was performed at several locations around a transport container for spent fuel elements. This cylindrical steel cask (gross weight 86 tons) was loaded with 30 highly burnt-up fuel elements from a pressurised water reactor and filled with water as coolant. The neutrons mainly arise from spontaneous fission of ^{244}Cm (95%), primarily generating a typical fission spectrum. The primary spectrum is modified by the fuel assembly, the water and mainly the outer container wall (32 cm iron), thus having mean energies of about 0.2 MeV. A more detailed description that includes measured spectra is given elsewhere[7].

To achieve small measuring uncertainties (≤10%) arising from counting statistics, the rem counters were either run in the dose reading mode for sufficient time (Dineutron, Leake counter) or with external pulse counting (A–B meters, 10" sphere). The random uncertainties of the count

Table 1. Field characteristics and survey meters responses for ^{252}Cf and Pu–Be sources.

Source	Shielding/Thickness (cm)	Distance (cm)	Dose equivalent response		
			Dineutron	A–B meters	10"sphere
^{252}Cf	Without	50 ... 200	1.19 ... 1.42	1.00 ... 1.22	0.95 ... 1.03
	PE spheres/radius				
	5	100	1.13	1.25 ... 1.37	1.09
	14		0.51	1.19 ... 1.35	1.08
	Borated PE				
	8		1.42	1.26 ... 1.34	1.02
	16	100	1.13	1.32 ... 1.33	1.03
	32		1.09	1.19 ... 1.38	1.12
	Concrete 30	100	0.84	1.12 ... 1.35	1.13
	Graphite				
	16	95	1.00	1.12	1.14
	35	92	0.39	1.14 ... 1.20	1.14
	80	120	0.71	2.06 ... 2.16	1.23
	Lead 15	100	1.74	1.18 ... 1.26	1.08
Pu–Be	Without	50 ... 200	0.79 ... 0.87	0.91 ... 1.02	0.91 ... 0.98
	Lead 15	100	1.74	1.18	1.08
^{252}Cf	*Storage cask*		0.72	1.13	1.08
	Wax 18	70			
Pu–Be			0.51	1.20 ... 1.29	0.94
Several	*Storage bunker*				
^{252}Cf	Wax ≤ 25	Inside	0.80	0.95	1.08
and Pu–Be	Wax/concrete ≈60	Outside	1.07	0.90 ... 1.05	1.11

rate measured with the 10" sphere, in almost all cases lay below 3% (1 SD).

RESULTS AND DISCUSSION

Selected results for the response of the Dineutron, A–B meters and 10" spherical monitor in several fields of ^{252}Cf and Pu–Be sources are summarised in Table 1. The total relative uncertainties may reach 20%, caused by the uncertainties (added linearly) of 10% at maximum both for the monitor readings and the reference dose rate. The latter has been determined by Bonner sphere spectrometry for every single configuration (source, shielding, distance). The responses of both A–B monitors are summarised in one column. The readings of the NM2 exceed those of the 2202 D by about 10% on average.

The response variation of all monitors at different distances to the bare Cf and Pu–Be sources is due to different contributions of direct and room-scattered neutrons to their reading (room size $9\times18\times3$ m^3, distance source/detector to floor/ceiling 1.5 cm).

Obviously, the Dineutron response changes considerably in the reference fields of the ^{252}Cf source, ranging from 60% underestimation behind graphite shielding up to 70% overestimation behind lead. Nearly the same variation (–50% to +70%) was found for Pu–Be sources. Actually these uncertainties are still acceptable; however, the strong dependence of the response on shielding thickness is unfavourable for monitoring practice.

For the A–B meters responses were derived lying between –10% and +40%, except the heavily moderated ^{252}Cf spectrum behind 80 cm graphite.

Surprisingly, the dose rates obtained from the 10" sphere of the Bonner spectrometer agree rather well with the reference values (–10% to +20%). To verify this unexpected result responses were calculated by folding the measured spectra with the detector response data used[8] showing agreement with the measured values within ±5%.

Response data derived from the operational measurements at the spent fuel transport cask are shown in Table 2. Analysing these results, a mean dose overestimation characterised by a factor of 4.5 must be stated when using the Dineutron around the fuel cask. A similar over-reading was found by other authors for typical fields inside a nuclear power plant[9]. At the measuring point on top of the cask where the moderation by the coolant predominates, only 80% over-response was revealed.

The Leake counter, calibrated for Am–Be neutrons ($\bar{E} \approx 4.5$ MeV), over-reads the dose equivalent by about 100% on average, as is to be expected from its well known energy-dependent response.

Even at the transport cask, the 10" sphere of the Bonner spectrometer provides acceptable over-estimation of 50% only, when using it with a calibration factor for unmoderated ^{252}Cf neutrons.

CONCLUSIONS

It has been found that the Dineutron survey meter may considerably under- and over-read the dose equivalent when used at radionuclide neutron sources. Its response just meets the commonly accepted uncertainty range (+100% to –50%) but varies strongly with type and thickness of shielding. Unacceptable over-responses were revealed at a spent fuel cask that do not correspond with the manufacturer's data.

Conventional rem counters of the Andersson-Braun and Leake type, as expected, provide conservative dose readings up to about +100%. If spectral information is available the well known response functions of these monitors allow them

Table 2. Measuring characteristics and survey meter responses at a spent fuel transport cask.

Measuring conditions	Location[1]	Distance to cask surface (cm)	Dose equivalent response		
			Dineutron	Leake	10" sphere
Isolated cask[2]	I – 11	45	4.6	–	1.4
	I – 12	100	4.1	2.4	1.3
	I – 13	200	4.0	1.7	1.4
	I – 2	45	4.0	1.7	1.4
	On top	25	1.8	2.0	1.5
Cask on transport wagon[3]	I – 11	45	5.7	2.0	1.3
	I – 13	200	4.3	1.3	1.5

[1] 2 m above floor, i.e. at highest burn-up level (except on top).
[2] Low-scatter environment (storage hall).
[3] Medium-scatter environment (rail hall).

to be used with corrected calibration factors if more accurate dose rates are needed.

Finally, the measurements have shown that even a simple 10" spherical counter may be successfully applied in numerous neutron fields if the counter is carefully operated and calibrated.

REFERENCES

1. Mourgues, M., Carossi, J. C. and Portal, G. *A Light Remcounter of Advanced Technology.* In: Proc. Fifth Symp. on Neutron Dosimetry, Munich/Neuherberg, September 1984. EUR 7962 (Luxemburg: CEC) Vol. 1, pp. 387-401 (1985).
2. *Dineutron — Technical Manual.* Ref. Nardeux 1883603, 1 (1987).
3. *Dineutron — Prospectus.* Ref. Nardeux 1883603, 7 (1987).
4. Hunt, J. B., Champlong, P., Chemtob, M., Kluge, H. and Schwartz, R. B. *International Intercomparison of Neutron Survey Instrument Calibrations.* Radiat. Prot. Dosim. **22**, 103-110 (1989).
5. Cosack, M. and Lesiecki, H. *Dose Equivalent Survey Meters.* Radiat. Prot. Dosim. **10**, 111-120 (1985).
6. Rimpler, A., Hermanska, J. and Prouza, Z. *Calibration and Interpretation of Neutron Moderation Spectrometers.* Report SAAS-367, Berlin (1989).
7. Börst, F.-M. and Rimpler, A. *Strahlungsmessungen an einem Transportcontainer für die Beförderung abgebrannter Brennelemente.* Report BfS (to be published).
8. Distenfeld, H. *Improvements and Tests of the Bonner Multisphere Spectrometer.* BNL 2193 (1978).
9. Aroua, A. et al. *Study of the Response of Two Neutron Monitors in Different Neutron Fields.* Radiat. Prot. Dosim. **44**(1-4), xxx-xxx (1992). This issue.

A NEUTRON SURVEY METER WITH SENSITIVITY EXTENDED UP TO 400 MeV

C. Birattari†, A. Esposito‡, A. Ferrari§, M. Pelliccioni‡ and M. Silari||
† Università di Milano, Dipartimento di Fisica, LASA
Via Fratelli Cervi 201, 20090 Segrate, Italy
‡ Istituto Nazionale di Fisica Nucleare
Laboratori Nazionali di Frascati, 00044 Frascati, Italy
§ Istituto Nazionale di Fisica Nucleare - Sezione di Milano, LASA
Via Fratelli Cervi 201, 20090 Segrate, Italy
|| Consiglio Nazionale delle Ricerche, Istituto Tecnologie Biomediche Avanzate
Via Ampère 56, 20131 Milano, Italy

Abstract — The well-known Andersson-Braun rem counter is widely employed for radiation protection purposes, but its efficiency shows a marked decrease for neutron energies above about 10 MeV. Since the availability of a survey meter with a good sensitivity to higher energies can be very useful, for instance, at many particle accelerator facilities, a neutron monitor with a response function extended up to 400 MeV has been achieved by modifying the structure of the moderator-attenuator of a commercial instrument. The first experimental tests carried out to verify the response of the new monitor both to low and high energy neutrons are reported. A comparison with the response function of three conventional commercial rem counters is presented.

INTRODUCTION

For the determination of the old concept of maximum dose equivalent (MADE) a so-called rem counter is usually employed. A rem counter consists of a detector with a high efficiency to thermal neutrons placed inside a moderator–attenuator whose structure is such that the response function of the instrument reproduces the curve of the conversion coefficient from neutron fluence to dose equivalent over a wide energy range. The best instrument of this type is the Andersson-Braun (A–B) rem counter[1], commercially available in a number of versions and of widespread diffusion. Its response is considered acceptable, for radiation protection purposes, for neutron energies from thermal to 14 MeV even if, in practice, it is affected by uncertainties which can be quite large for specific energy values. Although the performance of the A–B rem counter is judged satisfactory for most of the experimental conditions (i.e. range of energy) which are usually met, there are situations in which the availability of an instrument with a good sensitivity to higher energies becomes essential. This is the case, in particular, of particle accelerator facilities, where the energy spectrum of the neutrons produced shows a consistent high energy tail and therefore the abrupt decrease of the monitor efficiency with increasing energy makes it unsuitable for reliable measurements above about 10 MeV.

A numerical investigation had previously been carried out to assess the possibility of modifying the structure of the moderator–attenuator of a standard neutron survey meter of the A–B type in order to extend its sensitivity up to 400 MeV[2]. Among the different solutions which could be adopted for the realisation of such an instrument, a particularly simple one involving a limited modification to the moderator–attenuator structure was chosen for the experimental investigations. The present paper reports on the first tests carried out to verify the response of the new monitor both to low and high energy neutrons. A comparison with the response function of three conventional commercial rem counters is presented.

EXPERIMENTAL

The Tracerlab NP-1 Portable Neutron Monitor Snoopy was chosen as reference instrument for the design of the extended range version since it is a good representative of commercially manufactured neutron rem counters and it was readily available to us. Snoopy consists of a cylindrical BF_3 proportional counter of 2.54 cm diameter and 5.08 cm active length (95% ^{10}B enrichment), filled to a pressure of 8.0×10^4 Pa, surrounded by: (1) an inner polyethylene moderator 1.9 cm thick; (2) a boron doped synthetic rubber attenuator of 7.6 cm outer diameter, 14 cm length and 0.6 cm thickness, and (3) an outer polyethylene moderator of 21.7 cm outer diameter and 23.9 cm length. A number of holes are drilled both in the lateral surface and the front and end surfaces of the boron plastic attenuator to allow some of the thermal neutrons to penetrate.

Following the results reported by Birattari et al[2], a new moderator–attenuator structure was constructed to house the same BF_3 proportional counter. The new structure is obtained by adding a layer of 1 cm of Pb around the boron plastic attenuator (at a cost of about 8 kg of extra weight) and shifting the outer polyethylene moderator 1 cm outward (i.e. its thickness remains unchanged). Figure 1 shows a longitudinal cross section of the instrument. The response function of the modified rem counter shows a marked increase at high energies and no changes in the other regions. We shall therefore call this modified instrument Long Interval NeUtron Survey meter (LINUS). Figure 2 shows its calculated response function; the response of the standard A–B rem counter and the H*(10) curve[3] are also shown for comparison.

The response plotted in Figure 2 refers to lateral irradiation of the monitor by a uniform and parallel radiation field, i.e. the conditions under which H*(10) is calculated. A response of 275 s^{-1} per mSv.h^{-1} has been adopted[4]. The conversion coefficients from neutron fluence to ambient dose equivalent were taken from Stevenson[3]. These conversion coefficients are not significantly different from those adopted by ICRP 51[5]. The former were preferred because: (1) data were available up to high energies (while ICRP 51 only reports values up to 20 MeV) and (2) a larger number of data were given. The group structure of the cross section data sets employed is apparent from the histogram representation of the response function. The width of the dashed area corresponds to a statistical error of the Monte Carlo simulation of ± 1σ. The energy scale on the horizontal axis has been changed at 10^5 eV in order to expand and show the high energy region better.

Experimental investigations were first carried out with an Am–Be source to check that for this type of spectrum the response of LINUS coincides with that of conventional A–B rem counters. Next, tests with high energy neutrons were performed to verify the expected improved response of the new instrument. Comparisons were made with A–B monitors from different manufacturers. Besides the Tracerlab model NP-1 Snoopy, a number of other instruments were used. First of all, a spare BF_3 proportional counter identical to that from the NP-1 (model G-10-2A by N. Wood Counter Laboratory, Chicago, Illinois, fill pressure 8.0×10^4 Pa) was purchased to be used with the LINUS moderator–attenuator. A direct comparison between the standard A–B monitor and the new instrument is made by replacing the original BF_3 proportional counter of Snoopy with the new detector. A second monitor available for the intercomparison was the Snoopy Portable Neutron Monitor model NP-2 marketed by Nuclear Research Corporation, Warrington, Pennsylvania (BF_3 diameter of 2.54 cm, active length of 5.08 cm, fill pressure 8.0×10^4 Pa). A third instrument employed was the Alnor Neutron Radiation Meter 2002 B marketed by Studsvik AB Atomenergy, Sweden (maximum diameter of the BF_3 tube 3 cm, effective length 8 cm, fill pressure 11.7×10^4 Pa) routinely used at the Laboratori Nazionali di Frascati. Since the instruments have slightly different responses, cross comparisons have been carried out to normalise the data.

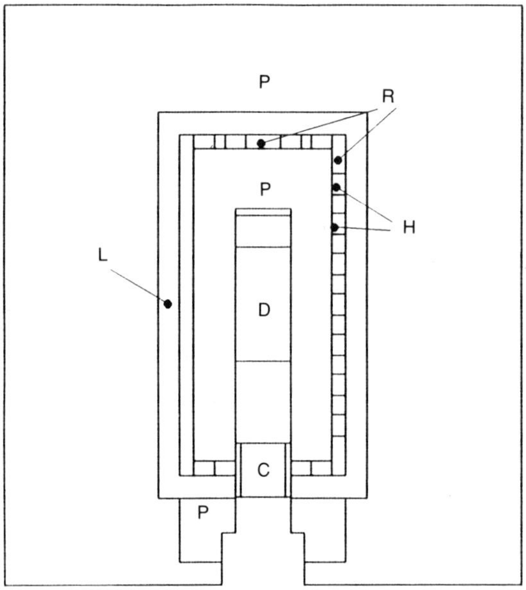

Figure 1. Longitudinal cross section of LINUS. P, polyethylene; R, boron doped synthetic rubber; L, lead; H, holes; D, detector; C, connectors.

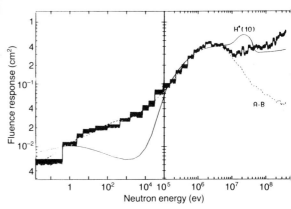

Figure 2. Response to lateral irradiation of LINUS. The response of a conventional A–B rem counter (the Tracerlab NP-1 Snoopy) and the H*(10) curve are shown for comparison.

RESULTS AND DISCUSSION

A comparison of the response of the different instruments to lateral irradiation by low energy neutrons was carried out with an Am–Be source with a certified neutron emission rate of 2.4×10^6 s^{-1}. Measurements were performed with the source positioned at the centre of a large room provided with a grid floor, to avoid any contribution from scattered radiation. The source was placed at 2 m from the geometrical centre of the monitor under test and the signal sent to a scaler after processing by conventional electronics.

The responses of the monitors are compared in Table 1. Different combinations of the moderator–attenuators and BF$_3$ tubes were tested, i.e. the detector from one monitor was also used with the moderating structure of another. This procedure allows one to distinguish between differences in the responses due to the BF$_3$ tube rather than to the moderating structure. The values reported in Table 1 are relative figures, normalised to the response of LINUS + the G-10-2A BF$_3$ proportional counter.

From data in Table 1 it can be deduced that the BF$_3$ detector of Snoopy NP-2 and the spare proportional counter have about the same sensitivity (their relative efficiency is 1.0 if tested with the LINUS moderator–attenuator and 1.04 when used with the Snoopy NP-2 structure), while the BF$_3$ detector of Snoopy NP-1 is about 10% less efficient. This reduced efficiency may simply be due to the fact that this detector is about 15 years old. No direct comparison can be made with the Alnor proportional counter since its specifications (active volume and fill pressure) are substantially different.

From the same table a direct comparison between the different moderating structures can be deduced. Their relative responses are: NP-1/LINUS = 1.03 ± 0.01, NP-2/LINUS = 1.08 ± 0.02 and NP-2/NP-1=1.05. It is therefore apparent that adding the lead layer does not produce any significant disturbance to the response of the moderator–attenuator for this Am–Be neutron spectrum. Actually, a variation as large as 5% is already observed between the moderator–attenuator structures of the two commercial models of the Snoopy rem counter. This variation may well be due to slight differences either in the composition of the plastic attenuator or in the density of the polyethylene moderator. In fact, the polyethylene cylinder from which the LINUS moderator has been machined has a density of 0.96 g.cm^{-3} against a density of 0.94 g.cm^{-3} of the polyethylene constituting our reference instrument, the Snoopy NP-1. This has of course been taken into account in the development of LINUS by appropriately scaling the radius of the outer moderator.

Finally, it should be mentioned that whilst LINUS and the Snoopy NP-1 and NP-2 models proved to have about the same response (893 μSv^{-1}, 920 μSv^{-1} and 905 μSv^{-1} respectively), Alnor showed a slightly higher sensitivity (1200 μSv^{-1}).

The response of the monitors to high energy neutrons was investigated in the course of an experimental run at CERN, where neutrons were produced by 205 GeV/c protons from the SPS

Table 1. Relative responses of the monitors to lateral irradiation by Am–Be neutrons, normalised to the response of LINUS + the G-10-2A proportional counter taken as 1. Where not otherwise indicated the moderator–attenuator is used in conjunction with its original BF$_3$ detector.

Moderator-attenuator	+ proportional counter	Relative response
LINUS	+ BF$_3$ detector from Snoopy NP-1	0.91
LINUS	+ BF$_3$ detector from Snoopy NP-2	1.00
Snoopy NP-1		0.93
Snoopy NP-1	+ G-10-2A BF$_3$ detector	1.04
Snoopy NP-2		1.06
Snoopy NP-2	+ G-10-2A BF$_3$ detector	1.10

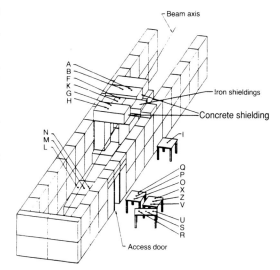

Figure 3. View of the target area and shielding at the beam line H6 from the SPS in the CERN north experimental area. The dimensions of the large concrete blocks are 240 cm x 80 cm x 160 cm. The smaller concrete blocks placed horizontally above the target area measure 160 cm x 80 cm x 40 cm. The letters indicate the locations where measurements were performed. At positions I,R,S,U,V,Z,X,O,P,Q the monitors were placed at about 1 m from the floor.

bombarding a target made of either iron or a sandwich of alternating iron and polyethylene layers. Although the neutron spectrum is unknown, a large number of high energy neutrons is expected[6]. This has been confirmed by contemporary measurements made by Bonner spheres[7]. Measurements were performed at a number of locations inside and outside the irradiation cell enclosing the target. An axonometric view of the target area and its surroundings is sketched in Figure 3. The positions where the monitors were placed are indicated by the letters. All the shielding blocks were made of concrete, except those ones corresponding to measurement positions F and K which were made of iron.

The results of the measurements are reported in Table 2. Here Alnor and Snoopy model NP-2 were operating with their original BF_3 detectors, while LINUS and Snoopy model NP-1 were used in conjunction with the G-10-2A BF_3 proportional counter. As for the Am–Be measurements the instruments were connected to a scaler, but a linear signal was also taken before the amplification stage within the monitor electronics and processed to carry out a spectrometry of the pulse produced by the recoiling alpha particles. This allows a control of the working conditions of the monitor.

Except at positions F and K, the response of LINUS is always larger than that of the other instruments, up to about a factor 3 at positions L,M,N (forward direction with respect to the proton beam and no shielding between the target and the neutron monitors). At positions G and H (on the top of the concrete block above the cell placed in the forward direction with respect to the incident beam) and I (outside the shield at about 90° to the beam direction) the difference is almost a factor 2. Corresponding to the access door to the irradiation cell (positions O,P,Q,R,S,U,V,Z,X) LINUS measures a neutron dose equivalent about 20-30% higher than the conventional A–B monitors and about 10% higher at positions A and B (on the top of the concrete block just above the target, with B corresponding within centimetres to the position of the front end of the beam dump). At positions F and K the measured neutron dose equivalent is considerably higher because the shielding thickness is lower as compared to the other positions. In addition, the monitors give the same reading as expected for a spectrum past an iron shielding which is well known to be dominated by neutrons with energy below about 1 MeV (compare Figures 5 and 6 of Ref. 3; see also Ref. 7).

It may also be noticed that positions I and A,B are both at about 90° with respect to the direction of the incident protons. At I the response of LINUS is significantly higher than that of the other monitors, whilst at positions A and B the rem counter readings differ only slightly, although the absolute dose equivalent is much higher. This is because at I the thickness of the concrete shield is larger than at points A,B and in addition measurements were carried out at about one metre from the shield, whilst at A,B the monitors were placed just in contact with the concrete surface. Therefore the neutron spectrum at I is more attenuated by the shielding structure but also richer in fast neutrons.

CONCLUSIONS

Although the experimental conditions under which measurements at CERN were carried out allow little knowledge of the neutron spectrum, the preliminary results obtained with the new monitor are encouraging and in good agreement with the Monte Carlo calculations reported earlier[2]. The measurements reported in the present paper have shown a much increased response of LINUS to high energy neutrons as compared to conventional A–B rem counters, while the response in the energy region from thermal to about 7-8 MeV is essentially not changed by the new structure of the moderator–attenuator. Because of the extra weight introduced by the lead layer, the instrument can

Table 2. Response of the monitors to neutrons produced by 205 GeV/c protons striking either an iron or iron + polyethylene target. Measurement positions are those indicated in Figure 3. LINUS and Snoopy model NP-1 were operating with the G-10-2A BF_3 proportional counter.

Position	Neutron dose equivalent (μSv per 10^{10} protons)			
	Alnor	LINUS	Snoopy NP-1	Snoopy NP-2
L	551	1444	—	494
M	509	1748	585	509
N	509	1550	—	532
I	7.1	12.5	—	7.7
O	66.9	77.5	65.4	63.1
P	72.2	85.1	—	70.7
Q	77.5	97.3	76.8	76.8
R,S,U	16.0	20.5	—	16.7
V,Z,X	16.0	20.5	—	15.2
A,B*	114	128	90.4	88.2
A,B**	156	179	—	114
G,H*	65.4	120	66.1	63.1
G,H**	91.2	173	103	91.2
F,K*	874	806	798	—
F,K**	1482	1467	1520	—

* Iron + polyethylene target, high proton dose rate.
** Iron target, low proton dose rate.

hardly still be regarded as 'portable'. More realistically, it can be either considered as a transportable survey meter to be moved around a facility by means of a little trolley or considered as a fixed instrument to be used for ambient monitoring at critical locations.

A complete calibration of the instrument with monoenergetic neutrons in the energy range from thermal to 19 MeV is already planned at PTB (Braunschweig, Germany). Further measurements at higher energies are also to be undertaken in the near future, in order to characterise completely the response function of the monitor. New numerical investigations will hopefully bring some indications about the possibility of obtaining a better response in the epithermal region. A careful redesign of the geometry of the moderator–attenuator may also significantly improve the angular response of the rem counter.

ACKNOWLEDGEMENTS

This work was financed by Istituto Nazionale di Fisica Nucleare, Italy.

REFERENCES

1. Andersson, I. O. and Braun, J. *Neutron Dosimetry.* STI/PUB/69 (Vienna: IAEA) Vol. II, p. 87 (1963).
2. Birattari, C., Ferrari, A., Nuccetelli, C., Pelliccioni, M. and Silari, M. *An Extended Range Neutron Rem Counter.* Nucl. Instrum. Methods **A297**, 250-257 (1990).
3. Stevenson, G. R. *Dose Equivalent per Star in Hadron Cascade Calculations.* CERN Divisional Report TIS-RP/173 (1986).
4. Thomas, R. H. and Stevenson G. R. *Radiological Safety Aspects of the Operation of Proton Accelerators.* IAEA Technical Report Series N. 283 (Vienna: IAEA) Ch. 3 (1988).
5. International Commission on Radiological Protection. *Data for Use in Protection against External Radiation.* ICRP Publication 51. Ann. ICRP **17**(2/3) (Oxford: Pergamon Press) (1987).
6. See, for example, Section 2.2.3 and Table 3.1 of Ref. 4.
7. Schmidt, P. *Application of a Bonner-sphere System in CERN Radiation Stray Fields.* CERN Technical Memorandum TIS-RP/TM/91-14 (1991).

TISSUE-EQUIVALENT PROPORTIONAL COUNTERS IN RADIATION PROTECTION DOSIMETRY: EXPECTATIONS AND PRESENT STATE

H. Schuhmacher
Physikalisch-Technische Bundesanstalt, Braunschweig, Germany

INVITED PAPER

Abstract — A survey is given of the application of low-pressure tissue-equivalent proportional counters (TEPC) in radiation protection dosimetry. The original motivation to use this technique and the related expectations are reviewed and the properties of instruments already available are evaluated. The numerous measurements carried out with such instruments in the past few years in various radiation fields show the large demand for 'intelligent' survey instruments. The implications of the new ICRP recommendation for the quality factor on the operational characteristics of TEPC-based instruments are investigated by re-evaluation of results from an intercomparison. It is shown that a simple modification of the evaluation procedure will make it possible to cope with the new situation.

INTRODUCTION

One of the very few attempts to employ new techniques for the improvement of radiation protection dosimetry of neutrons is based on low pressure tissue-equivalent proportional counters (TEPC). The principles of this technique have already been investigated nearly forty years ago when an 'LET spectrometer' was developed[1]. The difficult requirements for data acquisition with these detectors, in particular the need for pulse height analysis over more than four orders of magnitude, restricted the early applications of TEPC for a long time to experimental microdosimetry with specific and rather complex equipment for data analysis.

The decision of the ICRP[2] to define the quality factor, Q, of an ionising particle as a function of its unrestricted linear energy transfer in water, L, has led to the first attempt to construct a 'survey instrument suitable for measurements in mixed radiation fields around nuclear reactors and accelerators' based on a TEPC[3,4]. Many of the ideas now employed in TEPC radiation protection instruments have already been investigated in these early works, such as the use of non-linear amplifiers or few-channel pulse height analysis. However, the construction of an instrument suitable for routine applications in radiation protection failed.

Motivated by the availability of modern analogue and digital electronics, the interest of radiation protection in this technique was again aroused about ten years ago[5]. Prototype instruments were developed in the US[6,7] while in Europe the development of area monitors paralleled fundamental investigations into advantages and limitations of the TEPC method[8]. Many of the important features of TEPC-based instruments have been disclosed on the occasion of an intercomparison initiated by the European Radiation Dosimetry Group, EURADOS[9,10].

Applications include area monitoring, individual monitoring and investigations of the characteristics of radiation fields. Expected improvements as compared with existing instruments, such as the widely used moderator type detectors, concern the energy dependence of the dose equivalent response, practical aspects of the use of the instruments and the determination of 'diagnostic' information in unknown neutron–photon radiation fields, for example its mean quality factor or the discrimination between neutron and photon dose.

The paper summarises the motivations which have led to the development of TEPC radiation protection instruments, evaluates their properties and lists examples of area monitoring with TEPCs at nuclear reactors, nuclear fuel processing plants, particle accelerators, in spacecraft and aircraft. The feasibility of a transfer instrument for dose equivalent is also investigated. Finally, the implications of the new ICRP recommendation for the quality factor which makes new demands on neutron dosemeters are investigated.

The discussion of advantages and limitations of TEPC shows that most of the original expectations have been met.

MOTIVATION

The accuracy achievable in radiation protection dosimetry of mixed neutron–photon radiation fields is still unsatisfactory. For area monitoring, the heavy weight of the widely used moderator type instruments (rem counters) due to their massive moderator, and the strong energy dependence[11] of their dose equivalent response, R_H (the quotient of the instrument's reading, M, and the 'true' dose equivalent, H) are a disadvantage. The instruments become practically insensitive to neutrons with energies above 15 MeV. They are therefore not suited for radiation fields which contain a significant fraction of high energy neutrons, for example at particle accelerators, in aircraft and in spacecraft.

Individual monitoring is commonly performed with albedo detectors, etched track detectors, bubble detectors or a combination of these. Although significant progress has been achieved recently[12–14], in general the strong energy dependence of these measurement principles limits their ranges of application and requires a 'field calibration', i.e. their calibration for the neutron energy range in which the systems are to be used.

The original motivation to develop TEPC based devices came from the fact that a quantity can be derived from a TEPC which is similar to H. H can be written as[15]

$$H = \int Q\, D_L\, dL \qquad (1)$$

with the quality factor Q, defined as a function of $L^{(2)}$, and the differential distribution, D_L, of the absorbed dose with respect to L. The TEPC allows D_y to be measured, the differential distribution of the absorbed dose in tissue-equivalent material with respect to the lineal energy, y. Lineal energy, being defined as the ratio of the energy imparted to a volume (here, to the sensitive volume of the detector) and the average chord length in that volume[16], is closely related to L. Various procedures can be used to determine the dose equivalent reading of a TEPC, $M_H^{(9,10)}$. An obvious method is

$$M_H = \int q\, D_y\, dy \qquad (2)$$

with a weighting function, q, which approximates the quality factor Q, e.g. using q = Q and setting y = L. A corresponding reading for the mean quality factor can be obtained from

$$M_Q = M_H / \int D_y\, dy \qquad (3)$$

An instrument for the direct measurement of dose equivalent for any type of radiation seemed to be feasible if the relation between L and y, or more precisely between D_L and D_y, were known. This fact gave rise to investigations into the unfolding of the track-length distribution[17] and of the influence of energy loss straggling, delta rays and ranges of particles on the relationship between L and y[18]. It was shown that for many types of charged particles and energies relevant in irradiations with external sources, the difference between D_L and D_y is tolerable for the accuracy requirements in radiation protection dosimetry.

In neutron dosimetry, however, there is another aspect which was not taken into account in earlier investigations on the application of TEPCs in radiation protection. The neutron interaction with a person's body greatly influences the radiation field at a point of interest. In fact, for neutron energies below about 10 keV, H is dominated by the contribution of neutrons thermalised in the body while the contribution of the primary neutrons is negligible. The 'operational quantities', such as the ambient dose equivalent, penetrating, $H^*(10)^{(19)}$, are therefore defined at a reference point in a 30 cm diameter spherical phantom. They are supposed to give a conservative estimate of the primary limiting quantity effective dose. More recent work therefore concentrated on investigating the consequences of the differences in the phantom geometry and the irradiation conditions chosen for the definition of the quantity to be determined and the possible geometries of area monitors. These problems were treated in many contributions to two workshops[8,20] and by Dietze et al[21]. It was shown that the radiation absorption of a 10 mm tissue layer and the back scattering of a 30 cm diameter ICRU sphere cannot be realised simultaneously by an area monitor for which an isotropic response is required. Each instrument must therefore be adjusted to achieve an energy-independent dose equivalent response R_H.

PRESENT STATE

Properties of the Instruments

Many of the principal properties of TEPC area monitors were disclosed in the intercomparison measurements[9,10], in particular the energy dependence of R_H, the lower detection limit, the calibration of the instruments and their ability to determine average quality factors and to separate photon and neutron dose. Here, only the most important results are briefly summarised.

(1) The statistical uncertainty of the TEPC reading is a complex function of the neutron energy because it mainly reflects the stochastic nature of the energy deposition of ionising radiation. It depends on the size of the detector. For a sensitive volume of about 300 cm³ the lower detection limit of dose and dose rate is comparable to that of a moderator based dosemeter.

(2) The calibration of microdosimetric detectors in terms of kerma as used previously, leads to an underestimation of H. This problem can be easily solved by a calibration in terms of ambient dose equivalent in a neutron reference radiation field or by applying appropriate correction factors[22].

(3) The energy dependence of three typical detector systems is shown in Figure 1: a thick-walled TEPC (KFA), a thin-walled TEPC (BIO), and a moderator type dose equivalent rate meter (Leake). The systems were calibrated in the radiation field of a D_2O-moderated ^{252}Cf source in terms of the neutron ambient dose equivalent. While R_H varies for Leake by a factor of 20, for the other two systems the factor is 4 and 5, respectively. The reasons for the energy dependence of the TEPC systems are well understood: for the thick-walled TEPC the low R_H at about 100 keV is mainly explained by the neutron fluence attenuation in the detector wall, for the thin-walled TEPC the low R_H below 50 keV is mainly caused by the insufficient thermalisation of neutrons and the short range of the recoil protons produced.

Although the energy dependence is smaller than that of any other instrument for area monitoring, further improvements can be achieved by operating the detectors at lower gas pressure[23] and by using counting gases with additives which are sensitive to thermal neutrons[24]. A pragmatic solution can already be achieved with existing instruments by using the sensor under slightly modified irradiation conditions. If a thin-walled detector is used without a cap and with a cap, which results in the thickness being comparable to KFA, and the maximum of the two readings is taken, R_H values between about 0.5 and 1.5 are obtained (see Figure 1). In addition, the ratio of the two readings gives information on the average neutron energy. Another possibility is to design a wall which partly avoids the disadvantages of a thin wall (lack of thermalisation) and thick wall (high fluence attenuation). First calculations have shown that the use of a spherical TEPC with a wall containing conical holes, the axes of which point to the centre of the sphere, further improves the energy dependence.

While the response of TEPC, can be optimised, the requirement for a small energy dependence over the entire range of neutron energies conflicts to some extent with the demand for diagnostic properties. A high response for low-energy neutrons requires the thermalisation process in a rather massive detector wall be made use of. In this case the photons induced by neutrons in the detector cannot be distinguished from 'external' photons present in the radiation field.

Since calibration fields for H*(10) with neutron energies above 20 MeV are not available, experimental data of R_H do not exist in this energy range. Due to the measuring principle of a TEPC and our knowledge about the processes of neutron interaction, a small energy dependence can be expected for higher energy neutrons, even for neutron energies of 100 MeV and more.

Availability of the instruments

Most of the technical problems which were associated with the need for a large dynamic range in the pulse height analysis have been solved. Different solutions were found, such as the use of a logarithmic amplifier[25], miniaturised 'conventional' ADCs[26], ADCs with a logarithmic characteristic[27] or application of the variance method[28].

The measurement principle not only puts severe requirements on signal processing but also on the quality of the detector. This is due to the fact that extreme care must be taken to achieve high gas gains at very low noise. Although several detectors with the required properties exist, there are still problems in the production of detectors with predictable characteristics at comparatively little effort and hence low costs.

The main properties of systems which have been developed for and applied in radiation protection dosimetry are listed in Table 1. The prices listed are approximate values given by the manufacturers and comprise the complete systems. Depending on the intended field of application, very differing detector sizes were employed. Technical properties which are required for a type test, like linearity of the reading or dependence on ambient parameters such as temperature and electromagnetic fields, have not yet been investigated.

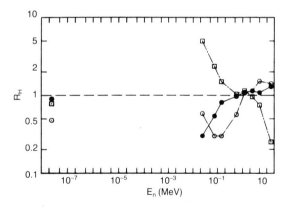

Figure 1. Ambient dose equivalent response, R_H, for monoenergetic neutrons as a function of neutron energy, E_n, for two TEPC systems (BIO: full circles, KFA: open circles) and a moderator type neutron dose equivalent meter (Leake: squares). The TEPC systems were calibrated in a $^{252}Cf(D_2O)$ field. The lines serve as eye guides. The results were obtained within the framework of an intercomparison[10].

APPLICATIONS IN VARIOUS RADIATION FIELDS

A large number of applications of TEPC systems, laboratory systems and area monitors, have been reported in the past few years. This fact clearly illustrates the need for investigations with dosemeters which allow additional information on the radiation field to be obtained. Many important details have been given in the various publications mentioned below.

Area monitoring with TEPC systems was performed at nuclear power plants[29–31], nuclear fuel processing plants[30], particle accelerators used in medicine[22,30,32–34] and in physics research[30,35]. In most of these surveys, in addition to the dose values, information on the radiation field obtained from the TEPC was used to judge the quality of shieldings and to interpret the readings of individual monitors. An extreme case was found in the survey at an accelerator where the radiation field changed drastically within a few metres from neutrons in the MeV range to neutrons in the 100 MeV range[30]. In many investigations the TEPC was compared with other methods. In general, differences observed could be explained by the differences in the energy dependence of the instruments.

The EG&G system was developed for individual monitoring[6]. No systematic investigations of its dosimetric features, such as its energy dependence, and into its practical use have been reported. Owing to the irradiation geometry, i.e. the fact that it is to be worn on a person's trunk, an energy dependence is to be expected which is similar to that of a transfer instrument (see below) and hence much smaller than that of an area monitor.

The complex radiation fields with photons, neutrons and charged particles encountered in high altitude aircraft and during space missions are typical examples of where 'intelligent' dosemeters are required. Two systems have been developed for space dosimetry[26,36]. The French system has been used on board the space station MIR since 1988[37] while the US system was used during a space shuttle flight in 1990[38]. Recent demands for radiation protection dosimetry in aircraft have led to first measurements during long-distance[39] and domestic flights[40].

TRANSFER INSTRUMENT

The use of a small TEPC introduced at a depth of 10 mm into a spherical phantom 30 cm in diameter, for the determination of ambient dose equivalent, $H^*(10)$, has been investigated theoretically[41]. In a unidirectional radiation field such a system would be very close to the conditions defined for the operational quantities for penetrating radiation and could therefore be used as a transfer instrument for H and for the 'field calibration' of individual dosemeters. In contrast to area monitors, its response would not be isotropic and it would have a smaller detection limit of dose and dose rate.

Systematic investigations using the neutron transport code MCNP[42] and energy deposition calculations[43] were performed in the energy region from thermal neutrons to 20 MeV. Figure 2 shows the energy dependence of R_H for a 12.7 mm diameter TEPC, filled with propane-based tissue-equivalent gas at a pressure of 4.5 kPa, and centred at 10 mm depth in a phantom made from A-150 tissue-equivalent plastic. The over-reading of about 30% for low-energy neutrons is due to the

Table 1. Technical properties of TEPC-based radiation protection dosemeters.

	CIRCE	CIRCEG	EG&G	HANDI	KFA	PNL	REM-402
m (kg)	5	10	0.6	6	7[1]	2	2
V (l)	6	30	0.8	18	10[1]	1.5	6.4
V_s (cm³)	98	27	109	110	270	1	97
T (h)	–	–	40	24	–	240	17
Price[2]	25000	16000[3]	2000	10000[4]	40000	–	5000
Application[5]	S	A	P	A	A	S	A
Reference	(36)	(48)	(6)	(27)	(22)	(26)	[6]

m = total mass of the system (detector and electronics)
V = total volume of the system (detector and electronics)
V_s = sensitive volume of detector
T = minimum time of operation with internal power supply (if available)
[1]Excluding a personal computer, required for data analysis.
[2]Approximate price in US$.
[3]Excluding the detector.
[4]A small number of prototypes were produced on a net-cost basis.
[5]Main area of application, A: area monitoring, P: personnel monitoring, S: space dosimetry.
[6]Data sheet from Far West Technologies, Goleta, CA.

fact that the detector wall contains 35% more nitrogen than the reference phantom material. A phantom material with a lower hydrogen density, such as polystyrene, would reduce the thermal neutron fluence at the detector position and hence this over-reading. The under-reading of about 20% in the 100 keV region is due to the fact that the recoil protons are stopped within the detector cavity. R_H can be increased by operating the TEPC at lower gas pressure. An energy dependence of less than 15% seems therefore to be achievable for the entire energy range.

FUTURE DEMANDS

In its recent report, the ICRP[44] recommends the continued use of the operational dose equivalent quantities as defined by the ICRU[19], but with a re-definition of the quality factor, Q. As in its earlier report[2], the ICRP defines Q as a function of L, but changes the Q–L relationship. This new situation puts new demands on the neutron dosemeters. In contrast to moderator type instruments, the new relation can be easily incorporated in the evaluation procedure of a TEPC dosemeter.

To illustrate the relevance of this decision, part of the results obtained in the intercomparison have been re-evaluated. The calculation of new neutron quality factors and fluence-to-ambient dose equivalent conversion factors[45] allowed the new reference quantities in the calibration fields to be determined. The spectra measured with two of the TEPC systems were re-evaluated with the new quality factor[46,47] to obtain the new readings of dose equivalent, M_H, and quality factor, M_Q. Figure 3 compares the old and new M_Q for BIO with the respective effective quality factors, $\bar{Q}$[45] (the quality factors at 10 mm depth in the ICRU sphere) as a function of the neutron energy. The changes introduced by ICRP 60 lead to an increase of $\bar{Q}$ of up to about 70%. A comparable increase is observed for M_Q while there is still an under-reading by about a factor of two in the energy region from 10 keV to 100 keV.

The situation for M_H is depicted in Figure 4 by plotting the new R_H for the BIO, KFA and Leake systems. They were calibrated in terms of H*(10) in the D_2O moderated ^{252}Cf field. While the energy dependence of the two TEPC systems is nearly unchanged (see Figure 1), the situation for Leake has slightly improved (maximum variation reduced from 20 to 17) because of the increased importance of neutrons for which moderator type detectors show an over-reading.

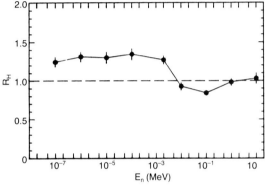

Figure 2. Calculated ambient dose equivalent response, R_H, for monoenergetic neutrons as a function of neutron energy, E_n, for a 12.7 mm TEPC positioned at 10 mm depth inside a spherical phantom. The lines serve as eye guides.

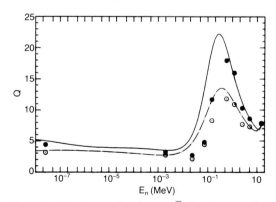

Figure 3. Effective quality factors, $\bar{Q}$, for the new (full line) and the old (dotted line) Q–L relation and the corresponding quality factor readings, M_Q, (new: full circles, old: open circles) for monoenergetic neutrons as a function of neutron energy, E_n, for the BIO system.

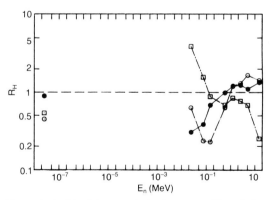

Figure 4. Ambient dose equivalent response, R_H, relative to the new H*(10) for monoenergetic neutrons as a function of neutron energy, E_n for two TEPC systems (BIO: full circles, KFA: open circles) and a moderator type neutron dose equivalent meter (Leake: squares). All systems were calibrated in a $^{252}Cf(D_2O)$ field. The lines serve as eye guides.

CONCLUSIONS

Several TEPC based radiation protection instruments are available which are suitable for routine use by inexperienced personnel. To date, these devices should be regarded as prototypes because many of their technical properties are yet unknown. Their dosimetric properties have been extensively investigated, with the result that they show the smallest energy dependence of all neutron dosemeters. Further improvements in the response are possible which may, however, influence the diagnostic capabilities. The large number of applications of TEPC systems reported in the past few years clearly illustrates the need for investigations with dosemeters which would allow additional information on the radiation field to be obtained. In unknown radiation fields TEPC systems have been used as reference instruments because of their superior properties. The new quality factors can be taken into account by a slight modification of the evaluation procedure to measure the new ambient dose equivalent and average quality factor. The responses of the modified instruments are comparable with those found for the old quantities.

REFERENCES

1. Rossi, H. H. and Rosenzweig, W. *A Device for the Measurement of the Dose as a Function of Specific Ionisation.* Radiology **64**, 404-410 (1955).
2. International Commission on Radiological Protection. *Protection Against Ionising Radiation from External Sources.* ICRP Publication 15 (Oxford: Pergamon) (1970).
3. Baum, J. W. *Nonlinear Amplifier for Use in Mixed Radiation Rem Responding Radiation Meters.* Health Phys. **13**, 775-781 (1967).
4. Baum, J. W., Kuehner, A. V. and Chase, R. L. *Dose Equivalent Meter Designs Based on Tissue Equivalent Proportional Counters.* Health Phys. **19**, 813-824 (1970).
5. Booz, J. *Advantages of Introducing Microdosimetric Instruments and Methods into Radiation Protection.* Radiat. Prot. Dosim. **9**, 175-183 (1984).
6. Quam, W., Del Duca, T., Plake, W., Graves, G., DeVore, T. and Warren, J. *Pocket Neutron Rem Meter.* In: Proc. Ninth DOE Workshop on Personnel Neutron Dosimetry, PNL-SA-10714 (Richland, WA: Battelle) pp. 138-149 (1982).
7. Braby, L. A., Ratcliffe, C. A. and Metting, N. F. *A Portable Dose Equivalent Monitor Based on Microdosimetry.* In: Proc. 8th Symp. on Microdosimetry, EUR8395 (Luxembourg: CEC) pp. 1075-1086 (1983).
8. *Microdosimetric Counters in Radiation Protection.* Proc. Workshop, Homburg (Saar), FRG, May 1984, Eds. J. Booz, A. A. Edwards and K. G. Harrison. Radiat. Prot. Dosim. **9** (1984).
9. Dietze, G., Edwards, A. A., Guldbakke, S., Kluge, H., Leroux, J. B., Lindborg, L., Menzel, H. G., Nguyen, V. D., Schmitz, T. and Schuhmacher, H. *Investigation of Radiation Protection Instruments Based on Tissue-Equivalent Proportional Counters, Results of a Eurados Intercomparison.* EUR 11876 (Luxembourg: CEC) (1988).
10. Menzel, H. G., Lindborg, L., Schmitz, Th., Schuhmacher, H. and Waker, A. J. *Intercomparison of Dose Equivalent Meters Based on Microdosimetric Techniques: Detailed Analysis and Conclusions.* Radiat. Prot. Dosim. **29**, 55-68 (1989).
11. Cosack, M. and Lesiecki, H. *Dependence of the Response of Eight Neutron Dose Equivalent Survey Meters with Regard to the Energy and Direction of Incident Neutrons.* In: Proc. 4th Symp. on Neutron Dosimetry, EUR 7448 (Luxembourg: CEC) pp. 407-417 (1981).
12. Griffith, R. V. *Review of the State of the Art in Personnel Neutron Monitoring with Solid State Detectors.* Radiat. Prot. Dosim. **23**, 155-160 (1988).
13. Gibson, J. A. B. *Individual Neutron Dosimetry.* Radiat. Prot. Dosim **23**, 109-115 (1988).
14. Luszik-Bhadra, M., Alberts, W. G., Dietz, E., Guldbakke, S. and Kluge, H. *A Simple Personal Dosemeter for Thermal, Intermediate and Fast Neutrons Based on CR-39 Etched Track Detectors.* Radiat. Prot. Dosim. **44**(1-4), 313-316 (1992). (This issue).
15. International Commission on Radiation Units and Measurements. *Radiation Quantities and Units.* Report 33 (Bethesda, MD: ICRU Publications) (1980).
16. International Commission on Radiation Units and Measurements. *Microdosimetry.* Report 36 (Bethesda, MD: ICRU Publications) (1983).
17. Rossi, H. H. *Microscopic Energy Distribution in Irradiated Matter.* In: Radiation Dosimetry. Eds F. H. Attix, W. C. Roesch, and E. Tochilin, (New York: Academic Press) pp. 43-92 (1968).
18. Kellerer, A. M. and Chmelevsky, D. *Criteria for the Applicability of LET.* Radiat. Res. **63**, 226-234 (1975).
19. International Commission on Radiation Units and Measurements. *Determination of Dose Equivalent for External Radiation Sources.* Report 39 (Bethesda, MD: ICRU Publications) (1985).

20. *Implementation of Dose-Equivalent Meters Based on Microdosimetric Techniques in Radiation Protection.* In: Proc. Workshop, Schloß Elmau, FRG, October 1988, Eds. H. G. Menzel, H. G. Paretzke and J. Booz. Radiat. Prot. Dosim. **29**, (1989).
21. Dietze, G., Menzel, H. G. and Schuhmacher, H. *Determination of Dose Equivalent with Tissue-Equivalent Proportional Counters.* Radiat. Prot. Dosim. **28**, 77-81 (1989).
22. Schmitz, Th., Morstin, K., Olko, P. and Booz, J. *The KFA Counter: a Dosimetry System for Use in Radiation Protection.* Radiat. Prot. Dosim **31**, 371-375 (1990).
23. Schuhmacher, H., Kunz, A., Menzel, H. G., Coyne, J. J. and Schwartz, R. B. *The Dose Equivalent Response of Tissue-Equivalent Proportional Counters to Low Energy Neutrons.* Radiat. Prot. Dosim **31**, 383-387 (1990).
24. Pihet, P., Menzel, H. G., Alberts, W .G. and Kluge, H. *Response of Tissue-Equivalent Proportional Counters to Low and Intermediate Energy Neutrons Using Modified TE-^{3}He Gas Mixtures.* Radiat. Prot. Dosim. **29**, 113-118 (1989).
25. Arbel, A., Booz, J., Müller, K. D. and Neuhaus, H. *Development of a Portable Microdosimetric Radiation Protection Monitor Covering a Dynamic Range of 120 dB Above Noise.* IEEE Trans. Nucl. Sci. **NS-31**, 691-696 (1984).
26. Brackenbush, L. W., Braby, L. A. and Anderson,G. A. *Characterising the Energy Deposition Events Produced by Trapped Protons in Low Earth Orbit.* Radiat. Prot. Dosim. **29**, 119-121 (1989).
27. Kunz, A., Pihet, P., Arend, E. and Menzel, H. G. *An Easy-to-Operate Portable Pulse-Height Analysis System for Area Monitoring with TEPC in Radiation Protection.* Nucl. Instrum. Methods in Phys. Res. **A299**, 696-701 (1990).
28. Tilikidis, A., Grindborg, J .E. and Lindblom, E. *A Variance Based Dose Equivalent Meter Device.* Radiat. Prot. Dosim **9**, 231-233 (1984).
29. Cummings, F. M., Endres, G. W. R. and Brackenbush, L. W. *Neutron Dosimetry at Commercial Nuclear Power Plants. Final Report of Subtask B: Dosimeter Response.* (Pacific Northwest Laboratories, Richland, WA.) Report PNL-4471 (1983).
30. Folkerts, K. H., Menzel, H. G., Schuhmacher, H. and Arend, E. *TEPC Radiation Protection Dosimetry in the Environment of Accelerators and at Nuclear Facilities.* Radiat. Prot. Dosim **23**, 261-264 (1988).
31. Barth, C. and Wernli, C. *Putting Microdosimetry to Work: the Measurement Campaigns of 1987 at KKG, KKL and PSI.* Radiat. Prot. Dosim. **23**, 257-260 (1988).
32. Schuhmacher, H. and Krauss, O. *Area Monitoring of Photons and Neutrons from Medical Electron Accelerators Using Tissue-Equivalent Proportional Counters.* Radiat. Prot. Dosim. **14**, 325-327 (1986).
33. Sabattier, R., Pihet, P., Bajard, J. C., Noale, M., Menzel, H. G. and Breteau, N. *Radioprotection à l'Unité de Neutronthérapie d'Orleans.* Radioprotection **24**, 123-132 (1989).
34. Bonnet, D. E., Waker, A. J., Sherwin, A. G., More, B. R. and Souliman, S. K . *Radiation Protection Metrology at a High-Energy Neutron Therapy Facility.* Radiat. Prot. Dosim. **37**, 95-101 (1991).
35. Schmitz, Th. *Microdosimetry Health Physics Instrumentation.* In: Advances in Radiation Protection. Ed. M. Oberhofer, (Bruxelles: ECSC, EEC, EAEC) pp. 171-197 (1991).
36. Nguyen, V. D. *A Dose Equivalent Meter Based on the Tissue-Equivalent Proportional Counter, and Problems Encountered in its Use.* Radiat. Prot. Dosim. **9**, 223-225 (1984) Nguyen, V. D. Private communication.
37. Nguyen, V. D., Bouisset, P., Parmentier, N., Akatov, I. A., Petrov, V. M., Kozlova, S. B., Kovalev, E. E., Katovskaia, A., Siegrist, M., Zwilling, J. F., Comet, B., Thoulouse, J., Chrétien, J. L. and Krikalev, S. K. *Real Time Quality Factor and Dose Equivalent Meter 'CIRCE' and its Use On-Board the Soviet Orbital Station 'MIR'.* Acta Astronaut. **23**, 217-226 (1991).
38. Badhwar, G. D., Braby, L. A. and Konradi, A. *Real-Time Measurements of Dose and Quality Factors on Board the Space Shuttle.* Nucl. Tracks Radiat. Meas. **17,** 591-594 (1990).
39. Schraube, H., Regulla, D., David, J. and Kunz, A. *Neutron Exposure at Civil Flight Altitudes.* Radiat. Prot. Dosim. **44**(1-4) xxx (1992). This issue.
40. Lindborg, L., Karlberg, J. and Elfhag, T. *Legislation and Dose Equivalents aboard Domestic Flights in Sweden.* (Swedish Radiation Protection Institute) SSI Report 91-12 (1991).
41. Schuhmacher, H. and Dietze, G. *Zur Realisierung der neuen Meßgrößen im Strahlenschutz in gemischten Photonen-Neutronen Feldern* In: Gemeinsame Jahrestagung 1990. Strahlenschutz im medizinischen Bereich und an Beschleunigern. Ed. D. Harder, pp. 106-107 (1990).
42. Briesmeister, J. F. (Ed.) *MCNP — A General Monte Carlo Code for Neutron and Photon Transport — Version 3A.* Los Alamos Report LA-7396-M Rev.2 (1986).
43. Coyne, J. J., and Caswell, R. S. *Microdosimetric Energy Deposition Spectra and their Averages for Bin-Averaged and Energy-Distributed Neutron Spectra.* In: Proc. Seventh Symp. on Microdosimetry 1981, EUR 7141, (Luxembourg: CEC) pp.689-696 (1981).
44. International Commission on Radiological Protection. *Recommendations of the International Commission on Radiological Protection.* ICRP Publication 60 (Oxford: Pergamon) (1990).

45. Schuhmacher, H. and Siebert, B. R. L. *Quality Factors and Ambient Dose Equivalent for Neutrons Based on the New ICRP Recommendations.* Radiat. Prot. Dosim. **40**, 85-89 (1992).
46. Kunz, A. *Entwicklung eines Meßgerätes zur Dosimetrie in gemischten Neutronen-Photonenstrahlenfeldern unter Anwendung eines gewebeäquivalenten Niederdruckproportionalzählers.* Thesis (Universität des Saarlandes) (1991).
47. Schmitz, Th. Private communication.
48. Marchetto, A., Leroux, J. B., Herbau, Y., Latu, M. and Tinelli, P. *CIRCEG, A Portable Device for Photon-Neutron Dosimetry.* Radiat. Prot. Dosim. **23**, 253-256 (1988).

INFLUENCE OF PHOTON RADIATION ON NEUTRON DOSE EQUIVALENT MEASUREMENT IN RADIATION PROTECTION WITH CIRCEG

A. Marchetto and Y. Herbaut
Service de Protection contre les Rayonnements
Centre d'Etudes Nucléaires de Grenoble BP 85 X, F- 38041 Grenoble cedex, France

Abstract — CIRCEG is a portable dose equivalent meter based on the microdosimetric technique. Associated with a tissue-equivalent proportional counter it gives absorbed dose and dose equivalent for photons and neutrons in a single measurement. It requires a validated threshold in terms of lineal energy to discriminate between photons and neutrons. Photon and neutron energy responses are studied and the results are compiled for different values of discrimination threshold. With the help of database software the mean energy neutron response and its uncertainty are computed, taking into account the influence of the photon field. In modifying some characteristics of this neutron–photon field such as photon energy, ratio of equivalent dose of photons and neutrons, and discrimination threshold, the mean energy neutron responses and its uncertainty are analysed. From these results the best choice of the threshold is determined.

INTRODUCTION

CIRCEG is a portable dose equivalent meter based on microdosimetric techniques[1]. This electronic device, associated with a tissue–equivalent proportional counter (TEPC), is used for workplace monitoring[2]. It detects photons and neutrons and gives the following information: absorbed dose, dose equivalent, and quality factor.

Moreover CIRCEG stores the last pulse height spectrum measured, from which it can produce the two microdosimetric spectra, one in dose and the other in dose equivalent. With the help of these spectra, it is possible to choose the best threshold, S.HL, for good neutron–photon discrimination and to calculate separately neutron and photon doses and dose equivalents. This diagnostic capacity is very useful for workplace monitoring in radiation protection. The usual separation technique attributes the lower part of the pulse height spectrum to the weakly ionising secondaries of photons, electrons and positrons, and the upper part to the secondaries of neutrons, recoil protons, α particles and heavier ions. Typically, photons correspond to $y < 20$ keV.μm^{-1}, and neutrons to $y > 1$ keV.μm^{-1}. With any photon–neutron mixed field these two parts overlap (Figure 1). For practical radiation protection it is, therefore, important to know the uncertainty introduced in the neutron dose equivalent reading due to the choice of a threshold, S.HL, to separate neutron and photon dose equivalent. This work explains a systematic method which takes into account neutron–photon field characteristics and gives the uncertainty in neutron dose equivalent due to the choice of S.HL.

The study takes into account the different parameters:

S.HL = photon–neutron discrimination threshold,
E_g = photon energy,
E_n = neutron energy,
(H^*_g/H^*_n) = photon and neutron ambient dose equivalent ratio.

A mean energy response is calculated in the neutron energy range studied. The uncertainty in this mean response is determined. From these results the best threshold for minimum uncertainty in neutron dose equivalent response will be chosen.

INSTRUMENTATION

Produced by NOVELEC (Grenoble, France) CIRCEG can utilise any microdosimetric tissue-equivalent proportional counter (TEPC) as the detector[2]. A 12.7 cm (5") diameter TEPC (Far West Technology) is used in the CIRCEG

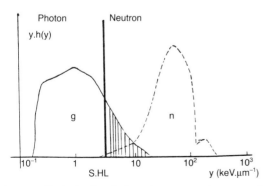

Figure 1. Typical microdosimetric spectrum in terms of lineal energy, y, for a mixed photon–neutron field. (——): dose equivalent spectrum due to photons, (------) dose equivalent spectrum due to neutrons. The photon–neutron discrimination threshold S.HL is easily adjustable.

microdosimetry system for the measurements of dose and dose equivalent. This counter has been chosen because of its large volume and consequently its good sensitivity to radiation. It is filled with a propane–based tissue–equivalent gas simulating a 2 μm site (P = 900 Pa). The wall of the counter is made of A–150 tissue-equivalent plastic 2.3 mm thick. The detector contains a ^{244}Cm alpha particle source which is used to calibrate in terms of lineal energy, y, the counter and the electronic system together.

EXPERIMENTAL RESULTS

The reference quantities that characterise the response of the dosemeter are ambient dose equivalent at a depth of 10 mm in the ICRU sphere[3,4].

H^*_n = ambient dose equivalent for neutron radiation,
H^*_g = ambient dose equivalent for photon radiation.

In the reference photon or neutron fields, one measurement is carried out from which the dose equivalents are evaluated for the four different values of S.HL of 3.5, 5, 10, 20 (keV.μm^{-1}).

Photon energy response

Photons are produced, according to ISO[5], in the energy range from 17.5 keV to 1.25 MeV, either from radioactive sources or X ray generators used in filtration or fluorescent mode. The response in terms of ambient dose equivalent is given in Figure 2.

Neutron energy response

A Van de Graaff ion accelerator was used to produce neutron beams in the energy range from

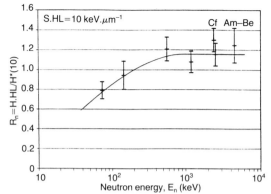

Figure 3. Neutron energy response: $R_n = H_n/H^*(10)$. Neutron energies for the values set are: 73, 144, 570, 1200 and 2500 keV.

73 keV to 2.5 MeV. They are produced according to the recommendations of ISO 8529[6], except for 73 keV. This is produced with a ^{7}Li target as 144 keV and 570 keV but at a different angle.

For neutron beams, the response in terms of ambient dose equivalent is presented in Figure 3. The variation of dose equivalent response with neutron energy over the defined energy range did not exceed ±21%. The response decreased below 150 keV, as explained by several authors[7]. For a given site size, the 'stoppers' become increasingly important as the neutron energy decreases. Table 1 displays ambient dose equivalent response for different neutron energies and different thresholds S.HL.

A correction term was introduced in dose and dose equivalent calculations for having a mean energy response equal to 1 for S.HL = 10 keV.μm^{-1}. This term, introduced in CIRCEG software, is taken into account for all dose or dose equivalent measurements in photon or neutron irradiation.

For S.HL = 3.5, 5 and 10 keV.μm^{-1}
$R_n = 1$, $I = 20\%$.

For S.HL = 20 keV.μm^{-1}
$R_n = 0.94$, $I = 27\%$.

where I is the relative uncertainty given in Equation 3.

Influence of photon radiation on neutron dose equivalent measurement

Method

Data for the analysis of the influence of photon radiation on neutron dose equivalent measurement are taken from the previous experimental results: $H_g(HL)$ = dose equivalent measured for y > S.HL in a photon field,

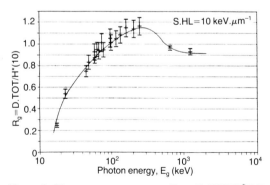

Figure 2. Photon energy response: $R_g = D.TOT/H^*(10)$. Photon energies for the values set are between 17 keV to 1.25 MeV. D.TOT is the measured dose for the whole lineal energies distribution.

$H_n(HL)$ = dose equivalent measured for y > S.HL in a neutron field.

Each of these readings is experimentally assessed for the four thresholds. They are stored in a microcomputer database from which it is possible to compute, for an arbitrary ratio (H^*_g/H^*_n), neutron energy response, R, influenced by photon radiation (Table 2),

$$R = (H_g(HL) + H_n(HL)) / H^*_n \quad (1)$$

This response depends on three parameters: E_g, E_n and S.HL. For each value of (H^*_g/H^*_n), the data table is modified consequently.

For a given case of photon irradiation (E_g and H^*_g/H^*_n) the mean neutron energy response $\bar{R}$ and its uncertainty are calculated:

$$\bar{R} = (R_{max} + R_{min}) / 2 \quad (2)$$

$$I = [(R_{max} - R_{min}) / (R_{max} + R_{min})] * 100 \quad (3)$$

R_{min} and R_{max} are respectively minimal and maximal values of the response R in the energy range studied.

With the possibility of making a change in (H^*_g/H^*_n) all the cases for which the dosemeter response is out of the chosen uncertainty can be seen.

Results

Independently of S.HL, the value of the neutron dose equivalent response, R, is maximum for the following conditions:

$E_n = 570$ keV and $E_g = 134$ keV

Table 3 gives the relative difference, S_e, between R_n, the mean neutron energy response without influence of photon radiation (Table 1) and R_{max}, the maximum value of R (Table 2) for a given value of (H^*_g/H^*_n),

$$S_e(H^*_g/H^*_n) = \frac{R_{max}(H^*_g/H^*_n) - R_n}{R_n} \quad (4)$$

Table 3 gives the maximum error Se which can occur in neutron dose equivalent measurements in an unknown mixed neutron–photon field. It shows how the choice of the discrimination neutron–photon threshold, S.HL is important.

CONCLUSION

For this type of instrumentation which integrates the possibility of drawing the microdosimetric spectra, the best choices for neutron–photon discrimination thresholds are:

S.HL = 10 keV.μm^{-1} when $H^*_g < H^*_n$
S.HL = 20 keV.μm when $H^*_g > H^*_n$.

Any instrumentation which has neither this possibility nor the capability to detect the discrimination threshold with fitted software, should use a discrimination threshold of 20 keV.μm^{-1}. This value warrants an uncertainty of less than 60% for $H^*_g/H^*_n < 20$.

This error leads to an over-estimation of H^*_n which is acceptable in radiation protection. All these results are only useful for this type of counter and in the energy range from 73 keV to 2.5 MeV.

ACKNOWLEDGEMENTS

The authors are grateful to PTN (Bruyères le Chatel) in making available their neutron irradiation accelerator and to Dr Chartier and his staff (IPSN/CEN-FAR) for their cooperation in the neutron experiments.

Table 1. Response in terms of ambient dose equivalent for different neutron energies, and different neutron–photon threshold values, S.HL.

E (keV)	S.HL (keV.μm^{-1})			
	3.5	5	10	20
73	0.83	0.81	0.79	0.68
144	0.96	0.96	0.94	0.89
570	1.21	1.21	1.21	1.19
1200	1.08	1.08	1.08	1.06
2500	1.16	1.16	1.15	1.10
R_n	1.02	1.01	1.00	0.94
I (%)	19	20	21	27

R_n = energy response in the range from 73 keV to 2.5 MeV.
I = relative uncertainty on R_n.

Table 3. Maximum error S_e on neutron dose equivalent measurement taking into account the neutron–photon discrimination threshold (S.HL), (H^*_g/H^*_n) and the characteristics of the photon field.

S.HL (keV.μm^{-1})	3.5	5	10	20
H^*_g/H^*_n				
0	19%	20%	21%	27%
0.1	27%	26%	21%	27%
1	90%	77%	27%	27%
2	>100%	>100%	33%	30%
5	>100%	>100%	51%	35%
10	>100%	>100%	81%	42%
20	>100%	>100%	>100%	57%
50	>100%	>100%	>100%	100%
100	>100%	>100%	>100%	>100%

Table 2. $R = [(H_n(HL) + H_g(HL))/H_n^*]$: neutron energy response of (CIRCEG + 5" counter) exposed to a mixed photon–neutron field.

Neutron energy		17.5			49.1			60			134			250			660			1250		
E_n (keV)	S.HL (keV·μm⁻¹)	H_n	H_g	R	H_n	H_g	R	H_n	H_g	R	H_n	H_g	R	H_n	H_g	R	H_n	H_g	R	H_n	H_g	R
73	3.5	0.83	0.25	1.08	0.83	0.46	1.29	0.83	0.70	1.53	0.83	0.81	1.64	0.83	0.63	1.46	0.83	0.43	1.26	0.83	0.37	1.20
	5	0.81	0.19	1.00	0.81	0.44	1.25	0.81	0.47	1.28	0.81	0.58	1.39	0.81	0.44	1.25	0.81	0.29	1.10	0.81	0.25	1.06
	10	0.79	0.02	0.81	0.79	0.04	0.83	0.79	0.04	0.83	0.79	0.06	0.85	0.79	0.05	0.84	0.79	0.03	0.82	0.79	0.03	0.82
	20	0.68	0.01	0.69	0.68	0.01	0.70	0.68	0.01	0.70	0.68	0.01	0.70	0.68	0.01	0.69	0.68	0.01	0.69	0.68	0.01	0.69
144	3.5	0.96	0.25	1.21	0.96	0.46	1.42	0.96	0.70	1.66	0.96	0.81	1.77	0.96	0.63	1.59	0.96	0.43	1.39	0.96	0.37	1.33
	5	0.96	0.19	1.14	0.96	0.44	1.39	0.96	0.47	1.43	0.96	0.58	1.53	0.96	0.44	1.40	0.96	0.29	1.25	0.96	0.25	1.20
	10	0.94	0.02	0.96	0.94	0.04	0.99	0.94	0.04	0.98	0.94	0.06	1.00	0.94	0.05	0.99	0.94	0.03	0.98	0.94	0.03	0.97
	20	0.89	0.01	0.90	0.89	0.01	0.91	0.89	0.01	0.91	0.89	0.01	0.91	0.89	0.01	0.90	0.89	0.01	0.90	0.89	0.01	0.90
570	3.5	1.21	0.25	1.46	1.21	0.46	1.67	1.21	0.70	1.91	1.21	0.81	2.02	1.21	0.63	1.84	1.21	0.43	1.64	1.21	0.37	1.58
	5	1.21	0.19	1.40	1.21	0.44	1.65	1.21	0.47	1.68	1.21	0.58	1.79	1.21	0.44	1.65	1.21	0.29	1.50	1.21	0.25	1.46
	10	1.21	0.02	1.23	1.21	0.04	1.25	1.21	0.04	1.25	1.21	0.06	1.27	1.21	0.05	1.26	1.21	0.03	1.24	1.21	0.03	1.23
	20	1.19	0.01	1.20	1.19	0.01	1.20	1.19	0.01	1.21	1.19	0.01	1.21	1.19	0.01	1.20	1.19	0.01	1.20	1.19	0.01	1.20
1200	3.5	1.08	0.25	1.33	1.08	0.46	1.54	1.08	0.70	1.78	1.08	0.81	1.89	1.08	0.63	1.71	1.08	0.43	1.51	1.08	0.37	1.45
	5	1.08	0.19	1.27	1.08	0.44	1.52	1.08	0.47	1.55	1.08	0.58	1.66	1.08	0.44	1.52	1.08	0.29	1.37	1.08	0.25	1.33
	10	1.08	0.02	1.10	1.08	0.04	1.12	1.08	0.04	1.12	1.08	0.06	1.14	1.08	0.05	1.13	1.08	0.03	1.11	1.08	0.03	1.11
	20	1.06	0.01	1.07	1.06	0.01	1.07	1.06	0.01	1.08	1.06	0.01	1.08	1.06	0.01	1.07	1.06	0.01	1.07	1.06	0.01	1.07
2500	3.5	1.16	0.25	1.41	1.16	0.46	1.62	1.16	0.70	1.86	1.16	0.81	1.97	1.16	0.63	1.79	1.16	0.43	1.59	1.16	0.37	1.53
	5	1.16	0.19	1.35	1.16	0.44	1.60	1.16	0.47	1.63	1.16	0.58	1.74	1.16	0.44	1.60	1.16	0.29	1.45	1.16	0.25	1.41
	10	1.15	0.02	1.18	1.15	0.04	1.20	1.15	0.04	1.19	1.15	0.06	1.21	1.15	0.05	1.20	1.15	0.03	1.19	1.15	0.03	1.18
	20	1.10	0.01	1.11	1.10	0.01	1.11	1.10	0.01	1.12	1.10	0.01	1.12	1.10	0.01	1.11	1.10	0.01	1.11	1.10	0.01	1.11
		$\bar{R}$		I(%)	$\bar{R}$		I(%)	$\bar{R}$		I(%)	$\bar{R}$		I(%)	$\bar{R}$		I(%)	$\bar{R}$		I(%)	$\bar{R}$		I(%)
	3.5	1.27		15	1.48		13	1.72		11	1.83		10	1.65		12	1.45		13	1.39		14
	5	1.20		17	1.45		14	1.48		14	1.59		13	1.45		14	1.30		15	1.26		16
	10	1.02		20	1.04		20	1.04		20	1.06		20	1.05		20	1.03		20	1.03		20
	20	0.94		27	0.95		27	0.95		27	0.95		27	0.95		27	0.95		27	0.95		27

$H_g(HL)$: dose equivalent measured for y > S.HL in a photon field.
$H_n(HL)$: dose equivalent measured for y > S.HL in a neutron field.
$\bar{R}$: mean energy response of CIRCEG.
I: relative uncertainty on $\bar{R}$.
$I = [(R_{max} - R_{min})/(R_{max} + R_{min})] \times 100$.

REFERENCES

1. International Commission on Radiation Units and Measurements. *Microdosimetry*. ICRU Report 36 (Bethesda, MD: ICRU Publications) (1983).
2. Marchetto, A., Leroux, J. B., Herbaut, Y., Latu, M. and Tinelli, P. *CIRCEG, a Portable Device for Photon Neutron Dosimetry*. Radiat. Prot. Dosim. **23**(1-4), 253-256 (1988).
3. International Commission on Radiation Units and Measurements. *Determination of Dose Equivalents Resulting from External Radiation Sources*. ICRU Report 39 (Bethesda, MD: ICRU Publications) (1985).
4. Wagner, S. R., Grosswendt, B., Harvey, J. R., Mill, A. J. and Siebert, B.R.L. *Unified Conversion Functions for the New ICRU Operational Radiation Protection Quantities*. Radiat. Prot. Dosim. **12**(2), 231-235 (1985).
5. International Organization for Standardization. *X and γ Reference Radiations for Calibrating Dosemeters and Dose Ratemeters and for Determining their Response as a Function of Photon Energy*. ISO 4037 (1979).
6. International Organization for Standardization. *Neutron Reference Radiation for Calibrating Neutron Measuring Devices used for Radiation Protection Purposes and for Determining their Response as a Function of Neutron Energy*. ISO 8529 (1989).
7. Caswell, R. S. *Deposition of Energy by Neutrons in Spherical Cavities*. Radiat. Res. **27**, 92 (1966).

THE HOMBURG AREA NEUTRON DOSEMETER HANDI: CHARACTERISTICS AND OPTIMISATION OF THE OPERATIONAL INSTRUMENT

A. Kunz†, E. Arend†, E. Dietz‡, S. Gerdung†, R. E. Grillmaier†, T. Lim† and P. Pihet†
†Fachrichtung Biophysik und Physikalische Grundlagen der Medizin
Universität des Saarlandes, D-6650 Homburg (Saar), Germany
‡Physikalisch-Technische Bundesanstalt, PTB
D-3300 Braunschweig, Germany

Abstract — Research aimed at the optimisation of the Homburg Area Neutron DosImeter (HANDI) was performed, in particular, to improve the technical performance of the radiation protection survey meter and its dose equivalent response for low and intermediate energy neutrons. The study is in progress and the paper reports on results which, although partial, confirm the high performance of TEPC systems and indicate potential improvements. Appropriate choice of simulated diameter and gas composition may improve the response to within 30% for about the whole neutron energy range. Additional work is in preparation based on calculations and investigations in practical radiation fields in combination with spectrometry studies. The consequence of the new ICRP recommendations are evaluated in terms of the change in the average quality factor determined with the TEPC in various practical radiation environments.

INTRODUCTION

The Homburg Area Neutron DosImeter (HANDI) was developed based on a low pressure tissue-equivalent proportional counter (TEPC) and newly designed electronics to be a portable instrument for health physics work[1–3]. A prototype of HANDI was used within an intercomparison of TEPC systems carried out by the EURADOS (European Radiation Dosimetry Group) in reference neutron fields from thermal energies to 14 MeV[4,5]. TEPC measurements were extended to neutron energies between 2 and 200 keV[6] and up to 70 MeV[7]. Experience was gained in photon–neutron fields of practical relevance for radiation protection[8–11]. The current work is concentrated on the optimisation of HANDI, namely its dose equivalent response for low and intermediate energy neutrons. This paper collects results from these investigations with the emphasis put on the practical performance and further improvement of HANDI.

THE HANDI INSTRUMENT

HANDI was specially designed to fulfil the requirements of operational survey instruments[12] with regard to low cost, weight, power consumption, simplicity in use while maintaining sufficient resolution. This goal was reached by developing the fast ADC circuitry using a pulse height analysis reduced to 16 channels[1] with quasi-logarithmic characteristics and the development of the adapted electronics for the analogue and digital pulse processing[1–3]. Using simple function keys, dose and dose equivalent rates are displayed in real time and the photon and neutron components are evaluated after the acquisition stops. For expertise work, the dose spectra are recorded. The system is powered with rechargeable batteries with a period of use of about 24 h.

A pre-series of five instruments is operational and currently being tested to determine technical specifications and the need of further adaptations. In parallel, research is continued to improve some important characteristics for their use in radiation protection work.

TEMPERATURE EFFECTS

Although the counters used operate at constant mass of gas, significant variations of gas gain with respect to temperature were observed at temperatures from 10 to 40°C, a similar temperature variation may be encountered in radiation protection environments. It was therefore essential to assess the amplitude of such instabilities and to search for more stable operating modes. The peak of the calibration alpha source was recorded for different spherical TEPCs maintained in a temperature-controlled chamber (Figure 1): (i) for TEPCs filled with propane-based TE gas, a decrease of gas gain with increasing temperature was found of about 0.7%.°C^{-1}; (ii) no significant variation was observed for counters filled with pure isobutane, indicating the influence of the CO_2 content; (iii) counters with metal walls and filled with propane-based TE gas also showed no significant variation, leading one to interpret the effect as due to exchange of CO_2 between the A-150 plastic wall and the cavity; and (iv) breakdown of signal is generally observed below 5°C and attributed to

DOSE EQUIVALENT RESPONSE

The ambient dose equivalent response, R_H (Table 1), of TEPCs is known to be nearly independent of neutron energy above 300 keV[5], but decreases steeply as neutron energy decreases[6]. This is due to the influence of different competing phenomena, including the range of low energy protons with respect to the simulated diameter, the neutron transport and the unavoidable differences between the definition of phantom conditions for determining the quantity dose equivalent and the real experimental conditions of the detector, all of which are responsible for this effect[13]. Different experiments[6,14,15] have indicated those parameters which can be chosen to compensate these effects and increase R_H substantially for the TEPC area monitor while preserving its diagnostic capabilities, in particular photon–neutron discrimination. The experiments were completed at the Research and Measurement Reactor FMRB of the PTB[16] to compare the response of 1/2" and 2" TEPCs exposed to 144 keV and thermal neutrons. The experimental data so far collected show that (Table 1): (i) the solution depends on a compromise between simulated diameter, wall thickness and gas composition; (ii) gas compounds with admixture of ^{3}He[15] in particular are appropriate below 50 keV with the advantage that rather thin walls can be used; (iii) this solution, as well as using thicker walls, is ineffective in the 100 keV region since thermalisation processes in the wall remain low compared with fluence attenuation; (iv) a lower diameter substantially improves the response and should be investigated first. Further analysis is required using calculations to interpret the results correctly, taking into account the different processes involved, namely to explain the differences observed in the response of small and large counters, i.e. larger mass of wall material, used to increase the sensitivity. The data available are nevertheless sufficiently consistent to project the result of the optimisation (Table 1) suggesting that a response of 1 ± 0.3 is feasible over about the whole energy range using an appropriate calibration method[5].

To evaluate the influence of the new ICRP recommendations[17] on the TEPC response, the whole spectra measured using the laboratory system and the HANDI in various radiation fields were re-evaluated by implementing the new Q(L) algorithm in the software[9]. The results for reference neutron fields are discussed elsewhere[13]. Table 2 summarises the variation of $\bar{Q}$ for practical fields due to the change in Q(L). The consequence of the new Q(L) for the optimisation of the TEPC response is under discussion. However, in a first analysis no major changes are expected as far as the shape of the response curve and the method of optimisation is concerned.

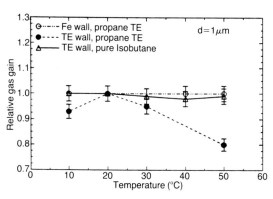

Figure 1. Temperature dependence of gas amplification observed for low pressure proportional counters (1/2" and 2" cavity size) with different wall materials and counting gases.

UNCERTAINTY OF DOSE EQUIVALENT MEASUREMENTS

The uncertainty of TEPC measurements raises specific problems due to the method used, i.e. electronic device, pulse height analysis, weighting

Table 1. Variation of the response of the TEPC at two neutron energies with respect to ambient dose equivalent as a function of different parameters and using different calibration procedures. The values were averaged from several experimental data sets available at these energies to show the improvement which can be expected by an appropriate adjustment of the TEPC characteristics.

E_n	$R_H = H / H^*(10)$ (absorbed dose calibration)*				$R_H = H / H^*(10)$ (dose equiv. cal.)**
	2.5 mm wall C_3H_8 TE d = 2 μm	2.5 mm wall C_3H_8 TE d = 1 μm	2.5 mm wall C_3H_8+1% ^{3}He d = 1 μm	8.5 m wall C_3H_8+1% ^{3}He d = 1 μm	8.5 mm wall C_3H_8+1% ^{3}He d = 1 μm
24 keV	0.21[6]	0.29[6]	0.42[15]	0.99[15]	1.30
144 keV	0.56[4]	0.69[p.w.]	0.77[p.w.][15]	0.64[p.w.][15]	0.84

* ^{60}Co source[4]. ** D_2O ^{252}Cf source[4,5].

broad LET distribution, separation of photon–neutron dose component. Pulse height analysis is limited by the detection threshold in lineal energy due to electronic noise. This threshold for HANDI is kept sufficiently low (< 0.09 keV.µm^{-1}) and leads to an underestimation of dose equivalent of less than 5% since the spectrum is not extrapolated to lower linear energy. Above this threshold any event is detected and the contribution of electronic noise to the resolution becomes negligible with increasing LET. TEPC area monitors are so called 'active' dosemeters within reasonable uncertainty levels[4]. Furthermore, the sensitivity of the instrument can be adapted by varying the size of the detector cavity. The reading of HANDI for natural background was typically 80 nSv.h^{-1} (<1 h measuring time), similar to that measured with a GM counter. In pure photon environments, however, a lower sensitivity was observed for several sensors due to the occurrence of spurious events corresponding to high lineal energy deposits. They can be attributed to poor gas quality and electrical properties. Neglecting this effect, the sensitivity in mixed fields is critically determined by the LET distribution, in particular the low frequency of high LET events as a function of the neutron dose rate[1]. To operate the detector in environments with unknown radiation components, a quantitative assessment of the uncertainty achieved is desirable and practical rules were sought to describe the uncertainty of dose equivalent as a function of measuring time cycle, count rate and radiation quality. The study aimed to provide a fast method for the user to determine measuring time required by knowing the count rate for a given detector.

Such rules can, in principle, be assessed analytically from the uncertainty of the absorbed

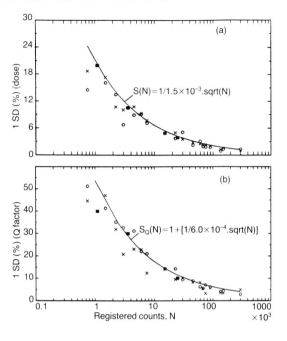

Figure 2. Uncertainty of (a) absorbed dose and (b) $\bar{Q}$, measured with HANDI in a D$_2$O-moderated ^{252}Cf reference field (10^3 counts = 0.3 µGy, $\bar{Q}$ = 3.2–3.6). The standard deviations were derived from computer simulations accounting for the event frequency spectra measured at (o) 50 cm and (x) 150 cm from the source[16]. The calculations are compared with the variance observed over 25 measurements at three different dose rate levels in both conditions (■). The data were smoothed (solid line) to provide practical s(N) functions to estimate the uncertainty of measurements in unknown practical fields.

Table 2. Comparison of mean quality factors derived from the measurements performed with the HANDI and the laboratory system in various radiation fields of practical relevance for radiation protection using the Q(LET) functions recommended in ICRP 21[18] and after re-evaluation using the new Q(LET) of ICRP 60[17].

Radiation protection environment	Short description	$\bar{Q}_{21}$ ICRP 21[18]	$\bar{Q}_{60}$ ICRP 60[17]	$\dfrac{\bar{Q}_{60} - \bar{Q}_{21}}{\bar{Q}_{21}}$ (%)
Cyclotrons used for neutron therapy[8]	p(45)+Be in the treatment room	4.89	6.57	34.4
	p(34)+Be in the maze	3.76	5.04	34.0
	p(45)+Be in the control room	1.61	1.82	13.0
	p(34)+Be in the control room	1.48	1.69	14.2
25 MeV X ray Linac[9]	In the treatment room	4.86	6.39	31.5
	In the treatment room, in front of the radiation protection door	2.35	2.91	23.8
High energy accelerator[10]	PSI 590 MeV p, above thick C target	3.92	4.20	7.1
	PSI 590 MeV p, above target for π production	1.78	1.65	−7.3
Nuclear fuel industry[12]	Workplace	2.95	3.29	11.5
	Pu-MOX fuel stock	1.73	1.75	1.2
Nuclear power plant[11]	Incore measurement channel	2.93	2.83	−3.4
	Incore measurement channel	1.79	1.99	11.2

dose as a function of the mean lineal energies and number of counts recorded, and that of the mean quality factor $\bar{Q}$, applying the error propagation law to the weighted distribution, assuming $Q(y) \sim Q(LET)$[4,5]. The problem is complicated by (i) the limited resolution and (ii) the, in general, unknown LET spectrum. The 16-channels resolution was shown to increase by about 10% the uncertainty of $\bar{Q}$ compared with the laboratory system[1]. In that study computer simulations and first experimental data were combined as the best method to account for the steep variation of event frequency with LET. This empirical approach was, for the present analysis, systematically applied to the whole range of data collected in reference and practical fields with the particular aim of deriving conservative functions for the uncertainty achievable as a function of the total number of counts recorded[9]. Comparing the results of computer simulations for the fields investigated the largest standard deviations of both the absorbed dose and $\bar{Q}$ were observed for measurements in a D_2O-moderated ^{252}Cf source (Figure 2). The uncertainty of $\bar{Q}$, however, dominates in this case. This figure was retained as a conservative estimate for the uncertainty in dose equivalent in practical fields. Using the information on count rate displayed by the instrument, the measuring time required can easily be determined. Typically, an uncertainty of dose equivalent of the order of 20% is estimated after 10 min response time at a dose rate of 60 $\mu Sv.h^{-1}$ using a 2" counter.

DISCUSSION

The progress achieved in the development of HANDI shows: (i) optimised electronics make the instrument suitable for operational dosimetry and offer a flexible system to meet practical radiation protection requirements; (ii) additional research is required to improve the properties of sensors, and (iii) the dose equivalent response in the low and intermediate energy range can be substantially improved, requiring additional research to operate the counter at lower pressure and to calculate the optimisation of the relevant parameters.

The performance and technical specifications of the present system are summarised in Table 3. The current work is focused on a reduction of the measuring device in size, weight (80%) and power consumption (50%) by improving the electronics using surface mounted devices and new microprocessor components[3]. The construction of cylindrical sensors with better properties and gas gain stability compared with commercial devices is investigated in collaboration with EURADOS (European Radiation Dosimetry Group).

The present work indicates that decreasing the simulated diameter is a determining factor to be combined with the optimum choice of wall thickness and gas composition. As far as the gas composition is concerned, isobutane with a small

Table 3. Technical data sheet for the present version of the HANDI area monitor.

Neutron energy range (investigated)	Thermal to 70 MeV
Angular response	Isotropic
Absorbed dose rate range	Up to 10 $mGy.h^{-1}$
Equivalent dose rate range	Up to 20 $mSv.h^{-1}$
Precision (1 SD) of H (H > 10 μSv)[a]	< 20%
(H > 100 μSv)[a]	< 5%
Sensitive volume[b]	Spherical, 59 mm diam.
External size	15.5 cm (L); 9.2 cm diam; (560 gr)
Gas gain stability (achieved at best)	2 years
Temperature range (Figure 1)	From +5°C to +40°C
Pulse height analysis system	Reprogrammable for changes in Q
Resolution	16 quasi-logarithmic channels
Real time display	$D, \dot{D}, H, \dot{H}, N, \dot{N}$
Stored data	Photon and neutron dose, quality factor, D and H spectra via serial 1
Response time	0.5 s, 5 s, or 60 s
Warm up time	30 s
Overload timing	70 times of the normal pulsewidth
Power consumption	5 W[c]
Instrument size	$32 \times 36 \times 15$ cm^3
Total weight	6 kg
Cost	8.000 ECU[d]

[a]Determined in a D_2O moderated ^{252}Cf field (Figure 2).
[b]Sensor currently used: Far West Technology (USA) model LET-2.
[c]24 h operation time with rechargeable batteries.
[d]Included detector and dedicated battery charger.

addition of ^{3}He seems promising since it provides better gas gain properties, a higher response to thermal neutrons and simultaneously the advantage of a negligible temperature dependence. The optimisation requires further effort, namely computational methods and investigating the response of the detectors in realistic fields which requires combined fluence spectrometry and TEPC measurements.

The revision performed according to the new quality factor demonstrates the advantage of using reprogrammable microprocessors, enabling a rapid and inexpensive adaptation of the system and the re-evaluation of existing dose equivalent measurements. It emphasises the importance of the experience accumulated with TEPCs in a broad range of fields directly relevant for radiation protection dosimetry[2,3,9].

ACKNOWLEDGEMENTS

This work was supported by the Commission of the European Communities (No Bi7 030) and the Bundesminister für Umwelt Naturschutz und Reaktorsicherheit (FRG) (No St. Sch. 1129). We express our thanks to the CERN (Geneva), GSF (Neuherberg), GSI (Darmstadt), KFK (Karlsruhe), PSI (Villigen), and PTB (Braunschweig) research institutes which currently test the HANDI for radiation protection work and contribute towards its improvement.

REFERENCES

1. Hartmann, G. H., Menzel, H. G., Schuhmacher, H. and Kraus, O. *Some Studies on Practical Aspects of the Applicability of Rossi-type Counters in Radiation Protection.* In: Proc. 8th Symp. on Microdosimetry, Jülich, 1982. Eds J. Booz and H. G. Ebert. EUR 8395 (London: Harwood for CEC) pp. 1117-1128 (1983).
2. Kunz, A., Menzel, H. G., Arend, E., Schuhmacher, H. and Grillmaier, R. E. *Practical Experience with a Prototype TEPC Area Monitor.* Radiat. Prot. Dosim. **29**(1-2), 99-104 (1989).
3. Kunz, A., Pihet, P., Arend, E. and Menzel, H. G. *An Easy-to-operate Portable Pulse-height Analysis System for Area Monitoring with TEPC in Radiation Protection.* Nucl. Instrum. Methods Phys. Res. **A229**, 696-701 (1990).
4. Dietze, G., Edwards, A. A., Guldbakke, S., Kluge, H., Leroux, J. B., Lindborg, L., Menzel, H. G., Nguyen, V. D., Schmitz, T. and Schuhmacher, H. *Investigation of Radiation Protection Instruments Based on Tissue-equivalent Proportional Counters. Results of a Eurados Intercomparison.* Report EUR 11867 (Luxembourg: CEC). (1988).
5. Menzel, H. G., Lindborg, L., Schmitz, T., Schuhmacher, H. and Waker, A. J. *Intercomparison of Dose Equivalent Meters Based on Microdosimetric Techniques: Detailed Analysis and Conclusions.* Radiat. Prot. Dosim. **29**(1-2), 55-68 (1989).
6. Schuhmacher, H., Kunz, A., Menzel, H. G., Coyne, J. J. and Schwartz, R. B. *The Dose Equivalent Response of Tissue Equivalent Proportional Counters to Low Energy Neutrons.* Radiat. Prot. Dosim. **31**(1-4), 383-387 (1990).
7. Schrewe, U. J., Brede, H. J., Gerdung, S., Kunz, A., Meulders, J. P., Nolte, R., Pihet, P. and Schuhmacher, H. *Determination of Kerma Factors of A-150 Plastic and Carbon at Neutron Energies Between 45 and 66 MeV.* Radiat. Prot. Dosim. **44**(1-4) 21-25 (1992) (This issue).
8. Vynckier, S., Sabattier, R., Kunz, A., Menzel, H. G. and Wambersie, A. *Determination of the Dose Equivalent and the Quality Factor in the Environment of Clinical Neutron Beams.* Radiat. Prot. Dosim. **23**(1-4), 269-272 (1988).
9. Kunz, A. *Entwicklung eines Meßgerätes zur Dosimetrie in gemischten Neutronen-Photonen-strahlenfeldern unter Anwendung eines gewebeäquivalenten Niederdruckproportionalzählers.* Universität des Saarlands (FRG), Ingenierwissenschaftliche Fakultät, Thesis (1992).
10. Boschung, M., Kunz, A. and Wernli, C. *Comparison of Different Neutron Area Monitors in the Field of High Energy Proton Accelerator.* Radiat. Prot. Dosim. **44**(1-4) 243-246 (1992) (This issue).
11. Asea Brown Boveri Reaktor GmbH. *Neutronendosimetrie in Kernkraftwerken mit Leichwasserreaktoren.* FE-Abschlußbericht. Dr. B. Hofmann, BMU RS II 2-510 322-463 St.Sch. 1000 (1988).
12. Gibson, J. A. B. and Delafield, H. J. *Requirements for Area Monitoring and Individual Dosemeters Based on Experiences in Operational Health Physics.* Radiat. Prot. Dosim. **29**(1-2), 143-148 (1989).
13. Schuhmacher, H. *Tissue-equivalent Proportional Counters in Radiation Protection Dosimetry: Expectations and Present State.* Radiat. Prot. Dosim. **44**(1-4) 199-206 (1992) (This issue).
14. Schuhmacher, H., Menzel, H. G., Bühler, G. and Alberts, W. G. *Experimental Basis for Optimisation of the Wall Thickness of Microdosimetric Counters in Radiation Protection.* Radiat. Prot. Dosim. **13**(1-4), 341-345 (1985).
15. Pihet, P., Menzel, H. G., Alberts, W. G. and Kluge, H. *Response of Tissue-equivalent Proportional Counters to Low and Intermediate Energy Neutrons Using a Modified TE-^{3}He-Gas Mixtures.* Radiat. Prot. Dosim. **29**(1-2), 113-118 (1989).

16. Alberts, W. G., Dietz, E., Guldbakke, S., Kluge, H. and Schuhmacher, H. *International Intercomparison of TEPC Systems used for Radiation Protection.* Radiat. Prot. Dosim. **29**(1-2), 47-53 (1989).
17. International Commission on Radiological Protection. *1990 Recommendations of the International Commission on Radiological Protection.* Publication 60, Ann. ICRP **21**(1-3). (Oxford: Pergamon Press) (1991).
18. International Commission on Radiological Protection. *Data for Protection Against Ionizing Radiation from External Sources.* Supplement to ICRP Publication 21 (Oxford: Pergamon Press) (1973).

AN INTERCOMPARISON OF NEUTRON FIELD DOSIMETRY SYSTEMS

D. J. Thomas, A. J. Waker†*, J. B. Hunt, A. G. Bardell and B. R. More**
Division of Radiation Science and Acoustics
National Physical Laboratory
Teddington, Middlesex TW11 0LW, UK
†Department of Medical Physics, The General Infirmary
Leeds, LS1 3EX, UK

Abstract — A Bonner sphere spectrometry system and a microdosimetric counter have been used to make measurements in the fields from three different neutron sources located in the working environment of a neutron source fabrication plant. Both the neutron spectra and the corresponding microdosimetric spectra are presented, quality factors are derived, and the measured dose equivalent values are compared with results from two different area survey instruments.

INTRODUCTION

To date it has proved impossible to design a neutron dosemeter, either for area monitoring or personal dosimetry, which responds correctly in all neutron fields. To make measurements to the required accuracy more elaborate neutron dosimetry systems have been developed which use neutron spectrometry or microdosimetric techniques. This paper describes an intercomparison of measurements with two such systems; a Bonner sphere (BS) spectrometer, and a tissue-equivalent proportional counter (TEPC). The results are compared with the readings of two area survey meters; the Harwell model 5/0949-5/6 area survey meter, and the Dineutron.

To test the systems in realistic situations Amersham International were approached to allow measurements to be made in their neutron source fabrication plant in environments typical of those where the dose equivalent needs to be known.

INSTRUMENTATION

The BS system consisted of nine polyethylene spheres, ranging in diameter from 7.62 cm (3") to 38.1 cm (15"), with a 3.2 cm diameter spherical proportional counter containing approximately 180 kPa of ^{3}He as the central thermal neutron detector. In addition the ^{3}He counter was used both bare and under a 1 mm thick cadmium cover to give a measure of the thermal component.

Response functions for the spheres were derived from a combination of measurements and calculations. Measurements were performed using monoenergetic neutrons, covering the energy range from 1 keV to 14.5 MeV[1], and also with thermal neutrons. The calculations were performed mainly with the discrete ordinates code ANISN[2]; although some additional calculations with the code MCNP were carried out in the low energy region where there were discrepancies between the ANISN calculations and thermal measurements. The neutron spectrum was unfolded using the program STAY'SL, which gives reliable results providing sufficient care is exercised in choosing a suitable 'start' spectrum.

The TEPC was a 12.7 cm (5") diameter, spherical, single wire proportional counter supplied by Far West Technology and operated with propane based tissue-equivalent gas at a pressure (6.5 torr) simulating a unit density tissue site of 2 μm in diameter. Signals from the TEPC were fed to a computer based pulse height analysis system enabling the microdosimetric data analysis to be carried out at the time of measurement. TEPCs record the absorbed dose as a spectrum of single events from which mean quality factors and the dose equivalent can be obtained[3]. It is also possible using the event size spectrum to separate the gamma component of a mixed field to give the dose equivalent of the neutron component alone.

Of the two area survey instruments used, the Harwell model 0949 is based on a 20.8 cm diameter moderating sphere of polyethylene with a central 3.2 cm diameter ^{3}He proportional counter. The instrument was adjusted according to the manufacturers instructions to give the correct reading for an Am–Be neutron source in the test jig. The second survey instrument, the Dineutron[4], has two polyethylene moderating spheres of diameters 6.4 cm and 10.7 cm, each with a small 1 cm right cylindrical ^{3}He counter. Information from both detectors is used to derive the dose equivalent using an algorithm which combines the count rates.

---

* Present address: AECL-Research, Chalk River Laboratories, Ontario, K0J 1J0, Canada.
** Present address: British Nuclear Fuels Plc, Capenhurst Works, Chester, Cheshire CH1 6ER, UK.

MEASUREMENTS

Measurements are reported here for three different locations. For the BS system each sphere was irradiated in turn, at a fixed point in the field, for sufficient time to ensure a statistical uncertainty of 1% or better for the count rate of each sphere. TEPC measurements were made with the centre of the TEPC at the same locations as the Bonner spheres. Calibrations of the TEPC were carried out using a built-in alpha calibration source. Measurements were carried out for sufficient time for the Poisson counting statistics to result in an uncertainty in the dose equivalent of between 5 and 6% for all three measurements.

The three different measurement sites, referred to as locations 1 to 3, are described below. Location 1 was an access corridor, approximately 2 m wide, behind a line of (empty) glove boxes. One side was the rear of the glove boxes, while the other was a thick concrete wall. The source consisted of about 3.2 TBq of americium oxide in a small cylindrical canister. This raw material for radionuclide source manufacture has a small but measurable neutron output. The source material was placed at the centre of the corridor about 1 m above the floor with the detectors placed 50 cm from the source on the centre line of the corridor. Location 2 was at the position of the trunk of an operator using a glove box containing about 260 GBq of ^{244}Cm. Detectors were mounted 125 cm above the ground, and about 50 cm from the source material. Location 3 was similar to 2, in being at the operator position for a glove box, however, this time the glove box contained a number of Am–Be sources, and was in a line of glove boxes containing further Am–Be sources.

RESULTS

The BS data were unfolded to derive neutron fluence spectra and these are shown at (a) in Figures 1 to 3 with the corresponding TEPC spectra at (b). Tables 1 and 2 give dose equivalent values and quality factors respectively derived from the neutron spectra, and these are compared with values obtained from the TEPC data and from the two survey instruments. Two sets of fluence to dose equivalent conversion factors were used for the quantity ambient dose equivalent, which is the most appropriate for comparison with the results of area monitors. These factors were obtained from ICRP 51[5] and the paper of Wagner et al[6]. For the quality factor the conversion factors of Wagner et al were used.

The quality factor and the dose equivalent for the neutron component of the radiation fields was derived from the TEPC measurements by subtracting the gamma component and then folding in the Q(L) relationship given in ICRP Report 21 to convert the dose distribution into a dose equivalent distribution. This method assumes an equivalence between lineal energy (y) and linear energy transfer (L). Subtraction of the gamma component was achieved by fitting the most appropriate pure photon event size spectrum to the lower end of the measured mixed field dose distributions. Best fits were obtained with 100 kV$_p$ photons for positions 1 and 2, and ^{60}Co photons for position 3.

It must be noted that the dose equivalent rates quoted here are not representative of rates encountered in normal operation. They had been significantly boosted, by maximising the amount of source material, to perform the measurements in the limited time available. The spectra are, however, typical of those in the working environments investigated.

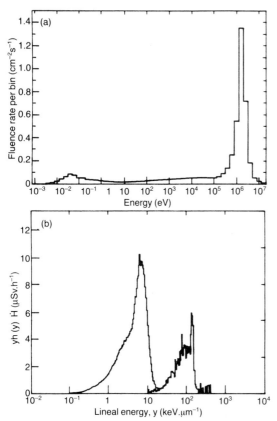

Figure 1. Location 1, 50 cm from 3.2 TBq of americium oxide, (a) neutron fluence spectrum, and (b) mixed field and pure neutron microdosimetric spectra giving the dose equivalent rate per logarithmic interval of lineal energy, (yh(y) is the probability density of the dose equivalent per logarithmic interval of lineal energy.

DISCUSSION AND CONCLUSIONS

The neutron spectra exhibit certain similarities in that all three have a high energy peak, a thermal peak, and a structureless component in the intermediate region. The energy of the high energy peak is greatest for location 3, where the source was Am–Be. The thermal component is smallest for location 1 which is perhaps surprising considering that these measurements were made close to a concrete wall in a somewhat confined space. Locations 2 and 3 were in reasonably open geometry, but exhibit larger thermal and intermediate energy components. This suggests that these components are due, not so much to moderation of neutrons from the primary source, as to neutrons from other more distant sources in the general vicinity.

The microdosimetric spectra differ most markedly in the low event size gamma region, with positions 1 and 2 being characterised by low energy fields, and position 3 by gammas of energy greater than 1 MeV. The increase in the proportion of events above the proton edge at around 140 keV.μm^{-1} for position 3 is an indication of the higher energy neutrons found in this field.

Table 1 shows that the TEPC and BS systems give very good agreement for neutron dose equivalent. This is despite the different gamma fractions which need to be subtracted from the microdosimetric spectra to derive the neutron component. The Harwell 0949 instrument also gives good results in these fields where the dose equivalent comes predominantly from the high energy neutrons. The Dineutron gives readings which are somewhat lower; this may be because it is designed primarily for lower energy neutron fields. Both area survey instruments were designed some years ago with the aim of matching their responses to the recommendations of ICRP 21. For the Dineutron the internal algorithm could be altered to give a response based on the quality factor of, for example, ICRP 51. This would increase the readings by a factor of around 1.10 to 1.15.

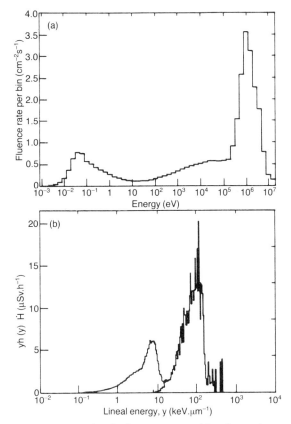

Figure 2. Location 2, the operator position for a glove box containing 260 GBq of ^{244}Cm, (a) neutron fluence spectrum, and (b) mixed field and pure neutron microdosimetric spectra giving the dose equivalent rate per logarithmic interval of lineal energy.

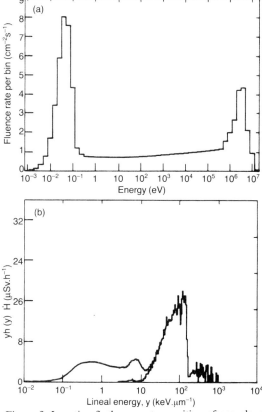

Figure 3. Location 3, the operator position of one glove box in a line of glove boxes containing Am–Be sources, (a) neutron fluence spectrum, and (b) mixed field and pure neutron microdosimetric spectra giving the dose equivalent rate per logarithmic interval of lineal energy.

The results shown in Table 2 for the neutron quality factors are somewhat at variance. Q values derived from the TEPC measurements are consistently some 10–40% higher than the BS values. The TEPC cannot discriminate between photon events generated by the external field and those that arise from thermal neutron capture in the wall and gas. If a significant component of thermal neutrons is present the method of gamma subtraction will lead to underestimation of the neutron dose and over-estimation of the neutron quality factor. However, the fact that for these spectra the total dose equivalent is dominated by the high energy component, means that the effect is very small for this quantity (< 3% even for location 3). An estimate of the magnitude of this effect on the Q value is obtained by comparing Q values derived from the BS measurements using the total spectrum with those obtained using only the region above 0.5 eV. The data are shown in Table 2, and although there is some slight increase in the Q value, the agreement with the TEPC results is still poor. This disagreement may be indicative of differences in the quality factor relationships used by the two techniques.

Table 2. Quality factors determined by the BS and TEPC systems for the three measurement locations.

Location	Quality factors		
	BS spectrum		TEPC
	Total	>0.5 eV	
1	9.8	9.9	14.1
2	9.6	9.7	10.8
3	8.1	8.5	9.7

ACKNOWLEDGEMENTS

We would like to thank Amersham International Plc, and in particular Mr David Page, for the opportunity to make these measurements. The work was partly supported by the CEC Radiation Protection Programme under contracts BI6-A-003-UK, and BI7-0031-C.

Table 1. Dose equivalent values determined by four different instruments at the three measurement locations.

Location	Neutron dose equivalent rate (μSv.h^{-1})					Gamma dose equivalent (%)
	BS system		TEPC	0949	Dineutron	
	ICRP 51	Wagner				
1	4.29	4.40	4.94	4.65	3.46	73.6
2	20.9	21.1	19.8	22.1	16.8	29.8
3	28.2	29.1	27.6	29.1	–	37.5

REFERENCES

1. Alevra, A. V., Cosack, M., Hunt, J. B., Thomas, D. J. and Schraube, H. *Experimental Determination of the Response of Four Bonner Sphere Sets to Monoenergetic Neutrons(II)*. Radiat. Prot. Dosim. **40**, 91-102 (1992).
2. Thomas, D. J. *Use of the Program ANISN to Calculate Response Functions for a Bonner Sphere Set with a ^{3}He Detector*. NPL Report RSA (EXT) 31, (1992).
3. Menzel, H. G., Lindborg, L., Schmitz, Th., Schuhmacher, H. and Waker, A. J. *Intercomparison of Dose Equivalent Meters Based on Microdosimetric Techniques: Detailed Analysis and Conclusions*. Radiat. Prot. Dosim. **29** (1-2), 55-68 (1989).
4. Mourges, M., Carossi, J. C. and Portal, G. *A Light Rem-counter of Advanced Technology*. In: Proc. Fifth Symp. on Neutron Dosimetry, EUR 9762 EN (Luxembourg: CEC) Vol. 1 (1985).
5. ICRP. *Data for Use in Protection Against External Radiation*. Publication 51. ICRP **17** (2/3) (1987).
6. Wagner, S. R., Grosswendt, B., Harvey, J. R., Mill, A. J., Selbach, H.-J. and Siebert, B. R. L. *Unified Conversion Functions for the New ICRU Operational Radiation Protection Quantities*. Radiat. Prot. Dosim. **12**, 231-235 (1985).

NEUTRON FIELD SPECTROMETRY FOR RADIATION PROTECTION DOSIMETRY PURPOSES

A. V. Alevra, H. Klein, K. Knauf and J. Wittstock
Physikalisch-Technische Bundesanstalt,
POB 3345, Bundesallee 100, W-3300 Braunschweig, Germany

Abstract — The PTB Bonner sphere spectrometer has been used to measure neutron fluence spectra under various conditions. The results are compared with other measurements or calculations.

INTRODUCTION

Reliable neutron dosimetry cannot be performed with the usual dosemeters without some knowledge of the spectral neutron fluence. For this reason the CEC supports projects aiming at the establishment of neutron fluence spectra relevant to routine surveillance. The areas of interest are workplaces at nuclear plants, fusion experiments, various accelerator facilities used for research, technical applications or medicine, installations using radioactive neutron sources, etc.

At PTB various experimental techniques using scintillators, proton recoil proportional counters or Bonner spheres have been developed to measure spectral fluences[1]. Among these techniques, Bonner spheres have the advantage of a nearly isotropic fluence response, and cover a wide energy range from thermal to tens of MeV. There are two sets of Bonner spheres at PTB, consisting of 12 and 14 spheres respectively, their efficiencies differing by about a factor of eight. This allows measurements to be performed in a great variety of fields as regards their spectra and intensities.

The fluence responses of the PTB Bonner spheres were recently determined experimentally[2] and completed from thermal energies to 25 MeV by fitting calculated responses to the experimental data.

Some interesting fields have recently been measured, mainly using Bonner spheres. A detailed analysis of the experiments will be given elsewhere. This paper is restricted to a presentation of the results, which illustrate the capability of the moderator technique to produce valuable spectrometric information under various conditions.

NATURAL BACKGROUND

The natural neutron background is chiefly due to neutrons produced by the interaction of primary and secondary cosmic rays with the earth's atmosphere, but also with the ground and any massive object present in the environment. The fluence rate and spectrum of neutrons produced by cosmic rays has been comprehensively studied in the past and the literature abounds in results from calculations and/or measurements[3-12].

The fluence rate of the atmospheric neutrons depends on altitude and geomagnetic latitude. Their spectral distribution changes mainly in altitudes of about 35 to 25 km, settling then into an equilibrium shape which is preserved up to a few hundred metres above the ground. The total intensity of the neutron background is proportional to the local neutron production whose maximum is attained at depths in the atmosphere between 60 and 100 g.cm^{-2} (altitudes of about 20–16.5 km), depending on the geomagnetic latitude. With increasing depth, the neutron background decreases nearly exponentially with a slope of about 140–170 g.cm^{-2} for an 1/e reduction in intensity.

The spectrum of the atmospheric neutrons itself extends from thermal energies to GeV, looking basically like a slowing-down spectrum, described as $\Phi_E(E) \sim 1/E^n$. According to various authors, n has a constant value of about 0.9 for neutron energies up to 300 MeV[11], or increases to values around 1.5 for energies over a few or ten MeV[3,7]. From spectrum calculations[8] or measurements[12] some authors obtain a slight broad bump around 50 MeV, but the largest contribution to the total fluence, and correspondingly to the total dose equivalent, is due to background neutrons with energies below 30 MeV[9]. In the atmosphere, in the low energy region of the spectrum the thermal peak is weak and is shifted from 0.023 eV to about 0.1 eV, because of the ^{14}N(n,p)^{14}C capture reaction. In contrast, at the air/ground interface an important thermal contribution is expected, due to earth-albedo neutrons. The energy region around 1 MeV is quite controversial in the literature. Although many authors have obtained from their measurements[3,7,12] or calculations[4-6] a clear Maxwellian distribution peaked at several hundred keV, interpreted as a softened evaporation spectrum, some results do not show this peak[8,10], or contradict its presence[9], or the authors express doubts about its existence[11]. The values given in the literature for the measured or calculated integral fluence of the background neutrons at air/ground interface, at sea level and geomagnetic latitudes between 44°N and 55°N, lie between 6 and 7×10^{-3} n.cm^{-2}.s^{-1}; the dose equivalent induced by these neutrons, calculated using the ICRP 21[13] conversion function, lies between approximately 6.5 and 7.8 nSv.h^{-1}.

The PTB in Braunschweig is located at a low altitude (practically sea level) and 53°N geomagnetic latitude. In a series of measurements, five Bonner spheres were used to determine the neutron spectrum of a free field at a place near the PTB research reactor (during a shut-down period). Another series of measurements was carried out in the basement of a two-storeyed concrete building at PTB, using nine Bonner spheres. In both cases the measuring times varied between three and eight days for each sphere. Special care was taken to distinguish the neutron induced events from the relatively high noise contribution by analysing the pulse height spectra accumulated during half-day periods. The spectral neutron fluences obtained from these measurements are shown in Figure 1. Besides a $1/E^{1.0}$ slowing-down part, both spectra contain a pronounced thermal peak (about 18% of the total fluence) and the evaporation contribution peaked at about 800 keV (25%). Any attempt to unfold the measured count rates by omitting the evaporation peak failed. All the neutrons with energies over 20 MeV were included in the histogram at the upper edge of the spectrum (about 17%). The quantity of neutrons in the histogram was determined using the BS fluence–response values at 25 MeV, the highest energy available in our response matrix. As the response values decrease in this region with increasing energy, the number of neutrons with energies over 20 MeV is somewhat underestimated. The integral values we obtained for the fluence and dose equivalent (using ICRP 21[13]) are 1.21×10^{-2} n.cm^{-2}.s^{-1} and 8.34 nSv.h^{-1} respectively, for the air/ground interface spectrum, and 4.70×10^{-3} n.cm^{-2}.s^{-1} and 2.79 nSv.h^{-1} respectively, for the spectrum measured inside the building. The estimated uncertainties, in terms of one standard deviation, are about 20% for the fluence and 35% for the dose equivalent. The thin dashed curve in Figure 1 shows the ratio $\Phi_E^{building}/\Phi_E^{air/ground\ intf.}$ in arbitrary units, indicating that there are no important changes in shape from one spectrum to another. The mean value of 0.39 for this ratio corresponds to a flux attenuation of 2.56 due to the concrete floors. Assuming that concrete, which contains light nuclei, behaves in the same way as air in neutron production, slowing down and capture, a total thickness of about 140 g.cm^{-2} is deduced for the floors and roof of the PTB building concerned. This result is in qualitative agreement with the building's constructional data and with the interpretation of Yamashita et al[7] of a similar experiment.

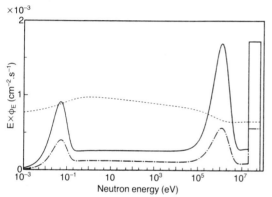

Figure 1. Spectral fluences obtained from Bonner sphere measurements of the free field neutron background at the air/ground interface (A, full line) and in the basement of a two-storeyed concrete building (B, dotted-dashed line) at PTB Braunschweig (nearly sea level, 53°N). The thin dashed line shows the ratio B/A in arbitrary units.

SIMULATION OF THE TEXTOR EXPERIMENT

An interesting radiation protection situation arises in plasma experiments. The intense but short 'shoots' make spectrometric measurements difficult. For this reason the neutron field of the TEXTOR plasma experiment in Jülich was partially simulated in the low-scatter area of the PTB accelerator facility, where 2.5 MeV monoenergetic neutrons were produced using the T(p,n) reaction. The geometry of this simulation is shown in Figure 2, where the target (T) simulates the plasma, the iron plate the vessel and the piece of wood the roof of the TEXTOR experimental hall, which had to be installed to reduce the neutron skyshine during discharges.

The fluence spectra obtained in three various

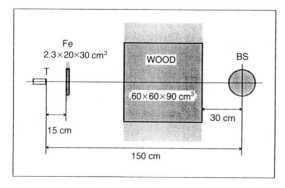

Figure 2. The geometry of the TEXTOR experiment simulation.

geometries are given in Figure 3. The target neutrons were measured in a free geometry, through 2.3 cm iron, and through both 2.3 cm iron and 60 cm wood.

The fluence of the 2.5 MeV neutrons as measured with a proton recoil telescope[14] is confirmed by the free-geometry BS measurement with less than 4% difference. A small background due to room-return neutrons (about 4.6% in fluence) is identified. The inelastic scattering in iron only degrades the neutron energy, while the wooden roof, through an efficient moderation and absorption, reduces the fluence by a factor of about eight and the dose equivalent by a factor of twenty.

FISSION NEUTRON SOURCES USED FOR DOSIMETRIC CALIBRATIONS

At PTB, bare or D_2O-moderated ^{252}Cf sources are employed for calibration purposes inside a bunker room (7m×7m×6.5 m). This room, with heavy concrete walls, is considerably smaller than the low-scatter experimental facility, and therefore the room-return neutrons contribute considerably to the total neutron fluence.

While complete information on the spectral distribution of the neutrons from both bare and D_2O-moderated ^{252}Cf sources in free geometry is available from the literature[15], information on the room-return in facilities similar to the PTB bunker is restricted to integral values, as a result of measurements[16–18] or calculations[19]. Using Bonner spheres up to 30.48 cm (12") in diameter, measurements were made in the bunker at a distance of 170 cm from the source centre. For each type of ^{252}Cf source (bare or moderated) the direct measurement was combined with one performed with a shadow object placed between the source and the Bonner spheres. The Figures 4 and 5 show some of the spectral results obtained for the bare and D_2O-moderated ^{252}Cf sources. For both sources the neutron fluence behind the shadow object, at the distance of 170 cm from source, was about the half the total fluence, producing about a quarter of the total dose equivalent. The value obtained for the strength of the bare ^{252}Cf source differs by only 1.8% from the value determined with the PTB water bath[20] (±1.2% uncertainty).

CONCLUSIONS

The PTB Bonner sphere spectrometer was employed to measure neutron spectra in various fields. Smooth spectra, but also spectra containing monoenergetic neutrons, were unfolded from the measured count rates. As lower limits, fluence and dose equivalent values as low as a few 10^{-3} cm^{-2}.s^{-1} and a few nSv.h^{-1}, respectively, were

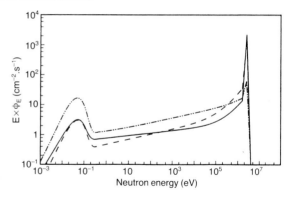

Figure 3. The fluence spectra as derived from Bonner sphere measurements in three geometries of the TEXTOR experiment simulation: free geometry (full line), through iron (dashed line) and through both iron and wood (dotted-dashed line).

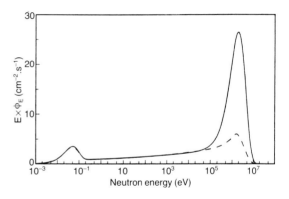

Figure 4. Fluence spectra as derived from Bonner sphere measurements in the bunker with a bare ^{252}Cf source at its centre: free geometry (full line) and behind a shadow cone (dashed line).

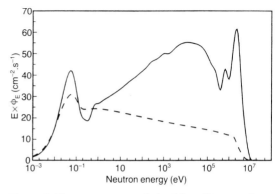

Figure 5. Fluence spectra as derived from Bonner sphere measurements in the bunker with a D_2O-moderated ^{252}Cf source in its centre: free geometry (full line) and behind a shadow object (dashed line).

measured. In all cases where reliable data from other methods were available, the results agreed within a few per cent, showing the Bonner sphere spectrometer to be a reliable instrument for use in radiation protection dosimetry and neutron metrology.

ACKNOWLEDGEMENTS

This work was supported by the Commission of the European Communities, Brussels, under contract No. BI7-0031-C.

REFERENCES

1. Alevra, A. V., Cosack, M., Klein, H., Knauf, K., Matzke, M., Plewnia, A. and Siebert, B. R. L. *Development of Neutron Spectrometers for Radiation Protection Practice.* EUR-Report 1326B, Vol. 1, pp. 132-142 (Luxembourg: CEC) (1991).
2. Alevra, A. V., Cosack, M., Hunt, J. B., Thomas, D. J. and Schraube, H. *Experimental Determination of the Response of Four Bonner Sphere Sets to Monoenergetic Neutrons (II).* Radiat. Prot. Dosim. **40**, 91-102 (1992).
3. Hess, W. N., Patterson, H. W., Wallace, R. and Chupp, E. L. *Cosmic-Ray Neutron Energy Spectrum.* Phys. Rev. **116**, 445-457 (1959).
4. Patterson, H. W., Hess, W. N., Moyer, B. J. and Wallace, R. W. *The Flux and Spectrum of Cosmic-Ray Produced Neutrons as a Function of Altitude.* Health Phys. **2**, 69-72 (1959).
5. Hess, W. N., Canfield, E. H. and Lingenfelter, R. E. *Cosmic-Ray Neutron Demography.* J. Geophys. Res. **66**, 665-677 (1961).
6. Lingenfelter, R. E. *The Cosmic Ray Neutron Leakage Flux.* J. Geophys. Res. **68**, 5633-5639 (1963).
7. Yamashita, M., Stephens, L. D. and Patterson, H. D. *Cosmic-Ray-Produced Neutrons at Ground Level: Neutron Production Rate and Flux Distribution.* J. Geophys. Res. **71**, 3817-3834 (1966).
8. Armstrong, T. W., Chandler, K. C. and Barish, J. *Calculations of Neutron Flux Spectra Induced in the Earth's Atmosphere by Galactic Cosmic Rays.* J. Geophys. Res. **78**, 2715-2726 (1973).
9. Merker, M. *The Contribution of Galactic Cosmic Rays to the Atmospheric Neutron Maximum Dose Equivalent as a Function of Neutron Energy and Altitude.* Health Phys. **25**, 524-527 (1973).
10. O'Brien, K., Sandmeier, H. A., Hansen, G. E. and Campbell, J. E. *Cosmic Ray Induced Neutron Background Sources and Fluxes for Geometries of Air Over Water, Ground, Iron, and Aluminium.* J. Geophys. Res. **83**, 114-120 (1978).
11. Hewitt, J. E., Hughes, L., Baum, J. W., Kuehner, A. V., McCasin, J. B., Rindi, A., Smith, A. R., Stephens, L. D., Thomas, R. H., Griffith, R. V. and Welles, C. G. *AMES Collaborative Study of Cosmic Ray Neutrons: Mid-Altitude Flights.* Health Phys. **34**, 375-384 (1978).
12. Nakamura, T., Uwamino, Y., Ohkubo, R. and Hara, A. *Altitude Variation of Cosmic-Ray Neutrons.* Health Phys. **53**, 509-517 (1987).
13. International Commission on Radiological Protection. *Data for Protection Against Ionizing Radiation from External Sources: Supplement to ICRP Publication 15.* ICRP Publication 21 (Oxford: Pergamon Press) (1973).
14. Guldbakke, S. PTB Braunschweig, Private communication (1991).
15. ISO. *Neutron Reference Radiations for Calibrating Neutron Measuring Devices used for Radiation Protection Purposes and for Determining their Response as a Function of Neutron Energy.* International Standard ISO 8529 (1989).
16. Eisenhauer, C. M., Schwartz, R. B. and Johnson, T. *Measurement of Neutrons Reflected from the Surfaces of a Calibration Room.* Health Phys. **42**, 489-495 (1982).
17. Dietze, G., Guldbakke, S., Kluge, H. and Schmitz, Th. *Intercomparison of Radiation Protection Instruments Based on Microdosimtric Principles.* Report PTB – ND – 29 (1986).
18. Bauer, B. W., Siebert, B. R. L., Alberts, W. G., Burgkhardt, B., Dietz, E., Guldbakke, S., Kluge, H., Medioni, R., Piesch, E. and Portal, G. *Experimental Investigation into the Influence of Neutron Energy, Angle of Incidence and Phantom Shape on the Response of Individual Neutron Dosemeters: Experimental Procedure and Summary of Results.* Report PTB – N – 5 (1990).
19. Liu, J. C., Sims, C. S., Casson, W. H., Murakami, H. and Francis, C. *Neutron Scattering in ORNL's Calibration Facility.* Radiat. Prot. Dosim. **35**, 13-21 (1991).
20. Kluge, H. PTB Braunschweig, Private communication (1991).

NEUTRON SPECTROMETRY AND DOSIMETRY MEASUREMENTS MADE AT NUCLEAR POWER STATIONS WITH DERIVED DOSEMETER RESPONSES

H. J. Delafield and C. A. Perks
AEA Environment and Energy
Harwell Laboratory, Oxon OX11 0RA, UK

Abstract — Neutron spectrometry measurements have been made in the working environments of three UK Nuclear Electric Magnox power stations, and of a PWR at Gosgen, Switzerland, using a high resolution spectrometry system based on high sensitivity cylindrical proton recoil counters and an alpha recoil counter. Multispheres were used to obtain dose equivalent information below the lower energy limit (~50 keV) of the recoil counter system. The derived H*(10) spectra show that, in general, the dose equivalent arises mainly from neutrons in two energy regions: thermal to 10 eV; and 10 keV to 1 MeV, although in certain situations a major contribution may arise from thermal neutrons. Quality factors for the H*(10) dose equivalent are derived and compared with the proposed ICRP 60 (1990) radiation weighting factors. The measured spectra are also interpreted in terms of the response of several individual (personnel) dosemeters and survey instruments.

INTRODUCTION

The dose equivalent responses of individual neutron dosemeters and survey meters are dependent upon neutron energy. Hence to overcome the well-known shortcomings in the energy responses of such devices, it is desirable to make field measurements of neutron spectra. These enable the best choice of dosemeter to be made, for a given working environment, and provide appropriate 'field calibration' factors to interpret subsequent dosemeter readings. Moreover, the spectrometry measurements provide a sound basis from which to implement future changes in defining the dose equivalent quantities or quality factor for neutrons[1].

To undertake this work, a high resolution neutron spectrometry system based on high sensitivity cylindrical proton recoil counters[2,3] and an alpha recoil counter[4] has been developed at Harwell Laboratory and calibrated at the National Physical Laboratory (NPL), UK. A multisphere spectrometry system[5] is used to provide dose equivalent information below the lower energy limit (~ 50 keV) of the recoil counter spectrometry system.

This transportable spectrometry system has been used to make measurements in the working environments at three UK Nuclear Electric Magnox power stations, and inside the containment building of a pressurised water reactor (PWR) at Gosgen, Switzerland[6]. The measured neutron fluence spectra have been converted to ambient dose equivalent[7], H*(10), spectra, which were used to derive the responses of several neutron dosemeters and survey meters. Finally, for each spectra, mean values of the quality factor for H*(10) were derived for comparison with the newly proposed ICRP 60[1] radiation weighting factors.

SPECTROMETRY SYSTEMS

A general description of both the proton recoil and multisphere spectrometers has been given previously by Birch et al[8]. The high resolution spectrometry system can be operated, either with a set of spherical hydrogen-filled proportional counters (type SP2, diameter 40 mm) or, with a set of higher sensitivity cylindrical counters (diameter 50 mm, length ~300 mm). Both sets of counters were calibrated using monoenergetic neutrons at the NPL and an unfolding procedure, generally SCOFH, was used to determine the neutron energy spectrum from the measured pulse height distributions. The hydrogen counters cover a neutron energy range from about 50 keV to 1.5 MeV.

The upper energy limit of measurement is extended to about 15 MeV using a specially developed alpha recoil counter[4]. The multisphere system, comprising a spherical neutron detector (^{3}He proportional counter) used with a set of eight polyethylene spheres having radii from 38.0 to 127.0 mm, is used to extend the measurements below the lower energy limit (~50 keV) of the combined recoil counter system. The multisphere measurements were unfolded with the code LOUHI78, using the neutron spectra measured by the recoil counters as input to constrain, and so improve, the spectral solution.

MEASUREMENTS

The principal measurements were made at locations where personnel perform occasional maintenance operations. The first measurements were made inside the containment building of a PWR (Gosgen, Switzerland) at positions either side of one of three primary pumps[6]. These were followed by measurements at UK Magnox power stations: in the Filter Gallery of the Burst Can Detector (BCD) Room at Trawsfynydd, inside a Boiler Cell at Dungeness 'A', and in an Upper Shield Filter Gallery and above a Pile Cap Charge Shute at Hinkley 'A'. In general, overnight measurements were made with the recoil counter spectrometer, employing electronics for simultaneous data acquisition for four counters, followed by daytime measurements with the multispheres.

In addition, measurements were made in the open air at Dungeness'A', at two low dose equivalent rate positions, on the roof of the boiler cell adjacent to the coolant ducts and on the roof walkway between the reactors, using only the more sensitive multisphere system.

RESULTS

Neutron spectra and quality factor

To make a direct comparison of these measurements, made over a period of a few years, the original measurements have been reformulated on a common basis. A definitive spectrum, in terms of neutron fluence per lethargy for discrete energy bins, is now taken as measured by the recoil counters for the upper energy region of the spectrum. Below the lower energy limit of the recoil counters (~ 50 keV), the spectrum, derived from a constrained solution of the multisphere measurements, is given in a similar format for energy bins at an arbitary resolution of 5 bins per decade of neutron energy.

These definitive neutron fluence spectra for > 0.5 eV, have been folded with the fluence-to-dose equivalent conversion factors[9] $H^*(10)/\Phi$ to give the corresponding dose equivalent spectra. The derived mean conversion factors $H^*(10)/\Phi$ (in ascending magnitude) and the relative additional dose equivalents attributable to thermal neutrons are given in Table 1. The fluence spectra were also folded with the fluence-to-absorbed dose conversion factors[9] $D^*(10)/\Phi$ to give the corresponding dose spectra. Then, following the procedure adopted by Jahr and Siebert[10], the mean quality factor $H^*(10)/D^*(10)$ was obtained for each spectrum (Table 2). For comparison, ICRP 60[1] proposed radiation weighting factors, W_R, by which the absorbed dose may be multiplied to obtain the 'equivalent dose', were also derived. Mean values of both $W_{R,step}$ and

Table 1. Fluence-to-ambient dose equivalent conversion factors $H^*(10)/\Phi$ and relative thermal neutron dose equivalent.

Neutron spectrum	Conversion factor for spectrum > 0.5 eV $H^*(10)/\Phi$ (pSv.cm^2)	$H^*(10)$ ratio Thermal neutron Spectrum (> 0.5 eV)
Dungeness 'A' Boiler Cell	15.3	0.33
Hinkley 'A' Shield Filter Gallery	19.0	1.11
Dungeness 'A' Walkway	21.5	0.19
Dungeness 'A' Roof	22.8	0.20
Gosgen PWR Position 2	26.1	0.27
Gosgen PWR Position 1	30.3	0.16
Trawsfynydd BCD Filter Gallery	31.6	0.25
Hinkley 'A' PC Charge Shute	145	0.004

Table 2. Comparison of quality factor $H^*(10)/D^*(10)$ with derived radiation weighting factors $W_{R,step}$ and $W_{R,smooth}$ for measured neutron spectra (> 0.5 eV).

Neutron spectrum	Quality factor, Q $H^*(10)/D^*(10)$	Ratio of ICRP 60 radiation weighting factor to quality factor	
		$W_{R,step}/Q$	$W_{R,smooth}/Q$
Dungeness 'A' Boiler Cell	4.8	1.61	1.64
Hinkley 'A' Shield Filter Gallery	5.3	1.58	1.61
Dungeness 'A' Walkway	5.8	1.59	1.63
Dungeness 'A' Roof	6.0	1.60	1.64
Gosgen PWR Position 2	6.6	1.59	1.63
Gosgen PWR Position 1	7.1	1.61	1.65
Trawsfynydd BCD Filter Gallery	7.2	1.60	1.64
Hinkley 'A' PC Charge Shute	12.0	1.59	1.64

$W_{R,smooth}$ were calculated by folding the absorbed dose $D^*(10)$ spectra with the step function and smooth function given for these factors by the ICRP. Ratios of the mean radiation weighting factors to the mean quality factors are given in Table 2. It should be noted that the radiation weighting factor and quality factor are not directly comparable quantities, and that this matter may be reviewed when ICRP 51[11] is revised.

Finally, to compare the dose equivalent spectra, they have all been converted into a 3 bin per decade format for presentation in this paper using a new computer program. The measured neutron dose equivalent in a given measured energy group was reallocated into the new energy groups by partitioning on the basis of the dose equivalent/lethargy times $\log_e$ (upper energy limit /lower energy limit). The reformatted data is presented in Figures 1-3 as histograms of the fluence per lethargy against the log of neutron energy. This presentation gives the dose equivalent as proportional to the area under the histogram. All spectra have been arbitrarily normalised to give an integrated dose equivalent of 1 mSv above 0.5 eV.

Dosemeter and survey instrument responses

Theoretical values, of the individual dose equivalent, penetrating[7], $H_p(10)$, response of typical neutron dosemeters and of the ambient dose equivalent $H^*(10)$ response of survey meters, were determined by folding the known fluence–energy response of the device with the measured fluence spectrum above 0.5 eV. For the neutron fluence-to-dose equivalent conversion, $H_p(10)$ was assumed to be equal to $H^*(10)$, that is the fluence was assumed to be unidirectional (AP incidence). The relative response values, given in Table 3, are based on the reference values of dose equivalent response normally used for that device ($1.97_7 \times 10^4$ tracks.mm^{-2}.Sv^{-1} for NTA film[12], $1.28_5 \times 10^5$ spots.cm^{-2}.Sv^{-1} for CR-39[13], 9.76×10^3 fissions.mg^{-1}.Sv^{-1} for ^{237}Np[14] free air, 8.54×10^8 counts.Sv^{-1} for Harwell 0075/0949[15] and 1.368×10^9 counts.Sv^{-1} for Studsvik 2202D). For the individual dosemeters and the Harwell 0075/0949, these reference responses are based on those used by Birch et al[8] for ICRP 21[16] dose equivalent, now revised to $H^*(10)$ dose equivalent by the ratio of the relevant fluence to dose equivalent conversion factors for ^{252}Cf fission neutrons. For the Studsvik 2202D, the reference response was that obtained by calibrating the instrument, at the Berkeley Nuclear Laboratories, to 2.7 MeV neutrons, for which the fluence rate was determined with a precision Long Counter with a calibration traceable to the National Physical Laboratory.

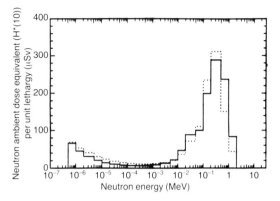

Figure 1. Neutron dose equivalent spectra for Gosgen (PWR) (normalised to give an integrated dose equivalent of 1 mSv above 0.5 eV). Position 1, solid line. Position 2, dotted line.

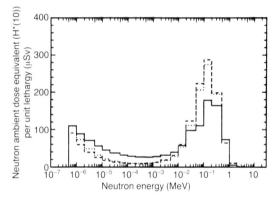

Figure 2. Neutron dose equivalent spectra for Dungeness 'A' (normalised as Figure 1). (——) boiler cell, (– – –) roof, (.....) walkway.

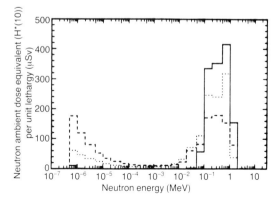

Figure 3. Neutron dose equivalent spectra for Hinkley Point (solid line, pile cap charge shute; dashed line, filter gallery) and Trawsfynydd (dotted line, burst can detector room). (Normalised as Figure 1.)

For Trawsfynydd and Dungeness, the relative responses were calculated by an earlier procedure based on a smoothed multisphere spectrum, but still constrained by the recoil spectrometer measurements. For the more recent measurements at Hinkley and for the PWR measurements (re-evaluated to H*(10)), the relative responses have been calculated directly from the definitive fluence spectra. A refolding of sample data by the new procedure was in good agreement with that formally given by the smoothed spectrum method.

DISCUSSION

Since the thermal neutron component of the total neutron dose equivalent is very dependent upon local scattering conditions, and may be measured directly, the spectra and the derived quantities are given for an energy above 0.5 eV. Nevertheless, in certain reactor environments, the thermal neutron component of the dose equivalent may be significant and even become the major component, as in the Upper Shield Filter Gallery at Hinkley 'A' (Table 1).

For the reactor sites measured, the ambient dose equivalent conversion factor (H*(10)/Φ) ranged from about 15 to 30 pSv.cm^2, the exception being for the measurement (145 pSv.cm^2) made above the Charge Shute on the Pile Cap at Hinkley. This was a special situation, giving rise to a small beam, for which it was only appropriate to use the four smallest multispheres in unfolding the spectrum.

For the more general situations, the hardest spectra were measured at Trawsfynydd (BCD Room, Filter Gallery) and at the Gosgen PWR, and the softest at Hinkley 'A' in the Upper Shield Filter Gallery (Figures 1-3). These spectra show that the dose equivalent arises mainly from two energy groups: from moderated neutrons from thermal to about 10 eV; and from intermediate and fast neutrons above 10 keV. Moreover for the Magnox stations, excepting the charge shute beam location, negligible dose equivalent arises from neutrons above 1 MeV. It is interesting to note that these moderated spectra show an increase in dose equivalent per unit lethargy with decreasing neutron energy below about 1 keV. This increase is greater than might be expected for a slowing down fluence spectrum (1/E), converted to H*(10), which increases the dose equivalent per lethargy below 1 keV by less than a factor of 2. This increase in low energy neutrons is thought to result from the albedo effect from scattering off surfaces. However, any uncertainties in the shape of the measured spectra in this energy range have a negligible effect on the dose equivalents derived from the spectra.

In terms of quality factor, Table 2 shows that, excluding the charge shute, the average quality factor for ambient dose equivalent is about 6. By comparison the average values of the proposed ICRP 60 radiation weighting factors are about 60% greater, irrespective of whether the step or smooth function is adopted. If a corresponding comparison is made with the older MADE dose equivalent (ICRP 21[16]); it is estimated that adoption of the new radiation weighting factors would represent an increase in the derived dose equivalent of from about 50% (Dungeness Boiler Cell), increasing with hardness of spectra, to about 70% (Trawsfynydd BCD Filter Gallery). For the special case of the charge shute spectrum, the increase would be by a factor of about two. Whilst these comparisons have been made for interest, it must be recognised that different concepts underlie the definitions of quality factor and radiation weighting factor and that moreover these matters are still under discussion.

Table 3. Calculated dose-equivalent responses of neutron dosemeters and survey instruments to measured spectra (> 0.5 eV).

	Response relative to ambient H*(10) of spectrum > 0.5 eV				
Neutron spectrum	Dosemeters*			Survey instruments	
	NTA film	CR-39	^{237}Np on body	Harwell 0075/0949	Studsvik 2202D
Dungeness 'A' Boiler Cell	0.03	0.23	1.37	3.3	1.18
Hinkley 'A' Shield Filter Gallery	0.08	0.35	1.03	2.5	0.93
Dungeness 'A' Walkway	0.03	0.26	0.77	2.6	0.98
Dungeness 'A' Roof	0.03	0.26	0.73	2.5	1.00
Gosgen PWR Position 2	0.08	0.46	0.85	2.2	0.93
Gosgen PWR Position 1	0.12	0.55	0.75	2.2	0.91
Trawsfynydd BCD Filter Gallery	0.13	0.55	0.80	2.1	0.94
Hinkley 'A' PC Charge Shute	0.21	0.83	0.50	1.1	0.69

* For individual dosemeters $H_p(10) = H*(10)$.

The calculated values for the relative responses of the individual dosemeters, given in Table 3, assume the fictitious situation of a normally incident and aligned field. In a practical situation, the dosemeters are expected to read less than these predicted values in most radiation fields, since the dosemeter may be partially shielded by the wearer and the response of some dosemeters, notably CR-39[17] and to a lesser extent NTA film, falls off for neutrons as the angle of incidence deviates from normal. Of the practical dosemeters (^{237}Np dosemeters are only suitable for special investigations) Table 3 shows CR-39, a dosemeter with a lower energy threshold (~100 keV) than that of NTA film (~500 keV), to be promising. This was confirmed by exposing CR-39 dosemeters on a phantom in the field at Trawsfynydd and at Dungeness (Boiler Cell). The measured responses, relative to the spectrum dose equivalent above 0.5 eV were 0.41 (calculated 0.55) at Trawsfynydd and 0.18 (calculated 0.23) at Dungeness. Albedo dosemeters (Berkeley Nuclear Laboratories type[18]), designed to measure thermal and intermediate energy neutrons, were exposed on a phantom at the same locations and gave measured responses of 0.38 (Trawsfynydd) and 0.58 (Dungeness) relative to the total dose equivalent, (spectrum + thermal neutrons).

Table 3 shows that the theoretical responses of the neutron survey instruments is type dependent, the Studsvik 2202D generally reading to within about ±20%, whilst the Harwell 0075/0949 significantly over-responds. This was confirmed by field measurements with the survey meters.

CONCLUSIONS

The neutron spectra measured at selected locations in reactor environments showed that generally the dose equivalent arises mainly from neutrons in two energy regions: less than 10 eV; and 10 keV to 1 MeV. However, special situations occur where major contributions to the dose equivalent may arise from fast neutrons (Hinkley, Pile Cap Charge Shute) or thermal neutrons (Hinkley, Upper Shield Filter Gallery). For such spectra, the combination of CR-39 and albedo type individual dosemeters provide the best overall coverage of the neutron energy range from thermal to fast neutrons. However, such dosemeters are sensitive to only part of the spectrum and so should be interpreted on the basis of 'field calibrations' as is done in practice. In contrast, neutron survey meters may read correctly or over-respond.

For these spectra, the average quality factor is about 6, for the ambient dose equivalent. By comparison, the average ICRP 60 weighting factor is about 60% greater, albeit that the underlying concepts are different, and that further interpretation is required.

ACKNOWLEDGEMENTS

The authors wish to thank the Managements of the UK and Swiss Power Stations cited for their assistance in making these measurements and for the support received from the Berkeley Nuclear Laboratories (BNL) of Nuclear Electric. Support was also received from the Commission of European Communities under contract No BI6 347-UK(H). Thanks are also due to Mr J. R. Harvey (BNL) and Mr J. A. B. Gibson (Harwell) for many helpful discussions, and Dr R. Birch (Harwell), Mr G. G. Gallacher (Harwell) and Mr C. D. Hart (BNL) for assistance with the measurements.

REFERENCES

1. ICRP. *1990 Recommendations of the International Commission on Radiological Protection*. Publication 60. (Oxford: Pergamon Press) Ann. ICRP **21**(1-3) (1990).
2. Delafield, H. J. and Birch, R. *Development and Calibration of Large Volume Proton-recoil Counters for Neutron Spectrometry in Radiological Protection*. (London: HMSO) Harwell Report AERE R-13010 (1988).
3. Delafield, H. J. and Birch, R. *Neutron Spectrometry Measurements with Large Volume Cylindrical Proton-recoil Counters Developed for Use in Radiological Protection*. (London: HMSO) Harwell Report AERE R-13103 (1989).
4. Birch, R. *An Alpha-recoil Proportional Counter to Measure Neutron Energy Spectra between 2 MeV and 15 MeV*. (London: HMSO) Harwell Report AERE R-13002 (1988).
5. Thomas, P. M., Harrison, K. G. and Scott, M. C. *A Multisphere Neutron Spectrometer Using a Central ^{3}He Detector*. Nucl. Instrum. Methods **224**, 225-232 (1984).
6. Birch, R., Delafield, H. J. and Perks, C. A. *Measurement of the Neutron Spectrum inside the Containment Building of a PWR*. Radiat. Prot. Dosim. **23**(1-4), 281-284 (1988).
7. International Commission on Radiation Units and Measurements. *Determination of Dose Equivalents Resulting from External Radiation Sources*. Report 39 (Bethesda, MD: ICRU Publications) (1985).
8. Birch, R., Delafield, H. J., Peaple, L. H. J. and Harrison, K. G. *The Neutron Leakage Spectra through the Steel Top Plates of Two Heavy-water-moderated Research Reactors*. Radiat. Prot. Dosim. **12**(3), 285-291 (1985).
9. Wagner, S. R., Grosswendt, B., Harvey, J. R., Mill, A. J. Selbach, H. J. and Siebert, B. R. L. *Unified Conversion Functions for the New ICRU Operational Radiation Protection Quantities*. Radiat. Prot. Dosim. **12**(2), 231-235 (1985).

10. Jahr, R. and Siebert, B. R. L. *Some Considerations on Radiation Weighting Factors for Neutrons.* Radiat. Prot. Dosim. **35**(4), 271-272 (1991).
11. ICRP. *Data for Use in Protection Against External Radiation.* Publication 51 (Oxford: Pergamon Press) Ann. ICRP **17**(2/3) (1987).
12. Bartlett, D. T., Bird, T. V. and Miles, J. C. H. *The NRPB Nuclear Emulsion Dosemeter.* NRPB Report R-99 (1980).
13. Bartlett, D. T. and Steele, J. D. *The Energy and Angular Dependence Response of the NRPB CR-39 Fast Neutron Dosemeter.* In: Proc. Fifth Symp. on Neutron Dosimetry, Neuherberg. EUR9762 EN (Luxembourg: CEC) pp. 511-518 (1985).
14. Harrison, K. G., Taylor, A. J. and Gibson, J. A. B. *The Neutron Response of a Neptunium Dosemeter worn on the Body.* Radiat. Prot. Dosim. **2**(2), 95-104 (1982).
15. Leake, J. W. *Spherical Dose Equivalent Neutron Detector Type 0075. A Recommendation for Change in Sensitivity.* Nucl. Instrum. Methods **178**, 287-288 (1980).
16. ICRP. *Data for Protection against Ionizing Radiation from External Sources: Supplement to ICRP Publication 15* Publication 21 (Oxford: Pergamon Press) (1973).
17. Harvey, J. R. and Weeks, A. R. *Progress Towards the Development of a Personal Neutron Dosimetry System based on the Chemical Etch of CR-39.* CEGB Report TPRD/B/0851/R86 (1986).
18. Harvey, J. R., Hudd, W. H. R. and Townsend, S. *A Personal Dosemeter which Measures Dose from Thermal and Intermediate Neutrons and from Gamma and Beta Radiation.* CEGB Report RD/B/N1547 (1969).

NEUTRON SPECTRA, RADIOLOGICAL QUANTITIES AND INSTRUMENT AND DOSEMETER RESPONSES AT A MAGNOX REACTOR AND A FUEL REPROCESSING INSTALLATION

D. T. Bartlett†, A. R. Britcher‡, A. G. Bardell§, D. J. Thomas§ and I. F. Hudson‡
†National Radiological Protection Board, Chilton, Didcot, Oxon OX11 0RQ, UK
‡British Nuclear Fuels plc, Sellafield, Seascale, Cumbria CA20 1PG, UK
§National Physical Laboratory, Teddington, Middlesex TW11 0LW, UK

Abstract — Measurements of the energy spectra and angle distributions have been carried out at five locations in a fuel reprocessing installation and at two locations at a Magnox reactor. The energy spectral distributions were determined using multispheres with either passive or active thermal neutron detection. Directional information was derived from the response of personal dosemeters mounted on phantoms. These evaluations enabled the estimation for the seven locations of the energy and angle weighted values of H_E, $H'(10)$ and personal dosemeter response, as well as of the isotropic quantity $H^*(10)$ and of the responses of NM2 and 0949 type rem meters. The results are analysed and discussed in terms of the relationships between primary and operational quantities, and between primary quantities and instrument and dosemeter responses. It is concluded that, in general, the operational quantity $H_p(10)$ will give a reasonable estimate of H_E, and that a PADC etched track personal dosemeter measuring thermal, epithermal and fast neutrons is suitable for use in the neutron radiation fields surveyed. Some estimates of the proposed new quantity, effective dose, are included.

INTRODUCTION

There has been growing interest in recent years in data, particularly measured data, for radiation fields, with especial interest in practical occupational fields. Such data allow the prediction of the response in these fields of radiation monitoring instrumentation, dose rate meters and personal dosemeters, in terms of the operational quantities which are required to be measured. This allows the assessment of the adequacy or otherwise of instrumentation, assists in the interpretation of instrument response and helps in the design of suitable instrumentation. In addition, such data enable relationships between quantities to be evaluated, for example, the relationships between limited quantities and operational quantities.

The British Nuclear Fuels Site at Sellafield in Cumbria consists of a wide range of facilities associated with the nuclear industry. The most important of these are Calder Hall Nuclear Power Generating Station which feeds electricity into the National Grid and the nuclear fuel reprocessing plant which processes irradiated fuel from both the UK and overseas reactors. There are also facilities for conditioning and storage of radioactive waste, as well as for research, dosimetry and other support services.

Occupational exposure to neutron radiation occurs near the operational Magnox reactors of Calder Hall, but apart from maintenance areas, neutron dose rates and neutron/gamma ratios are low, even taking into account the proposed change in the quality factor/linear energy transfer relationship and the introduction of radiation weighting factors[1]. The major potential for significant neutron exposures is in the plutonium processing plant where separated plutonium as nitrate is processed further. However, owing to increased burn-up of fuel and the move to reprocessing of oxide fuel, there is increased potential generally of exposure to neutrons in all fuel handling operations.

A number of surveys of the practical neutron radiation fields at Sellafield have been undertaken, one of which was reported at the 5th Neutron Symposium[2]. We report herein, the results of a recent set of measurements at seven locations, two at the Calder Hall reactors, and five in the plutonium processing facilities.

APPROACH, METHODOLOGY, TECHNIQUES

The general approach adopted was that used in a similar survey carried out in 1982/3; some results of this previous survey have been published[2]. At each of the seven measurement locations, two sets of measurements were made. One set to determine the neutron energy spectral distribution, the other to obtain information as to the angle spectral distribution. Knowledge of the energy spectral distribution enables the interrelationships of various radiometric, dosimetric and radiation protection quantities to be determined and the expression of instrument and dosemeter responses in terms of these quantities. Where the quantities or instrument responses are non-scalar, i.e. where there is dependence on direction, angle distribution data are needed in addition. A schematic of the inter-relationships of the various quantities and instrument and dosemeter responses is shown in Figure 1. The new ICRP basic dosimetric quantities[1], effective dose, E, and tissue-equivalent dose, H_T, are included, but no new data are yet available for the operational quantities calculated with the new quality factor relationship.

Both the relationships between the secondary (operational) calibration quantities ambient, directional dose equivalent, and individual dose equivalents, and the calculated effective and tissue dose equivalents (effective dose and tissue-equivalent doses) and between the response of an instrument or dosemeter and calibration quantity, are dependent on radiation energy and angle, particularly for neutron radiation fields. It is of value therefore to determine such relationships for practical working environments. These data enable better estimates to be made of the basic dosimetric quantities, if required, and to make corrections, as necessary, for the field dependence of instrument and dosemeter responses. It should be noted that the relationships with the basic dosimetric quantities are for the calculated quantities, E, (H_E) and H_T for which mean risk coefficients are available for populations[3]: the risk to a particular individual cannot be assessed.

For neutron energy spectral distribution measurements 'spectrometry', the multisphere system was used. This has the advantages that it can be operated as either a passive (using, for example, gold foils as thermal neutron detectors) or an active system (using, for example, ^{3}He proportional counters as thermal neutron detectors); the former method being suitable with little modification for use in contaminated areas. The system is adequately sensitive and can be used in radiological protection fields of a few μSv.h^{-1} (passive system), less than 1 μSv.h^{-1} (active system); covers the whole energy range of interest from thermal to 20 MeV; has an isotropic response; and gives good estimates of dose equivalent quantities as well as fluence. Its disadvantages are its ill-conditioned response functions which mean that the deconvolution procedure can lead to ambiguities in the deconvoluted spectrum and lack of resolution. This lack of resolution may be of particular importance in the 100 keV to 1 MeV energy range. Full details of the NPL systems may be found elsewhere[3]. The all important unfolding procedure used with the systems has recently been tested in a European intercomparison of multi-sphere systems organised by Eurados-Cendos (using constructed count rates) and produced extremely good results[4].

Data on the angle distributions were obtained with personal dosemeters mounted on a phantom which also acted as a shield. In this case, the dosemeter used was the NRPB track etch (PADC) dosemeter. The PADC dosemeter comprises a thin piece of the polymer poly-allyl-diglycol carbonate held in a laminated pouch with a piece of nylon in contact with the rear face, all within a polypropylene dosemeter holder. The piece of nylon provides a thermal neutron response via protons produced in the capture reaction on nitrogen. The fast neutron response is produced by recoil protons from elastic scattering of neutrons by the hydrogen in the PADC detector itself. The tracks of protons are made visible by an electrochemical etch procedure and counted by an automated image analysis system. The dosemeter responds to thermal neutrons, epithermal and neutrons up to a few keV and to fast neutron of energies greater than 100 keV. Full details of the dosemeter may be found elsewhere[5,6]. For some locations the dosemeters were mounted four per vertical face on a water filled polyethylene phantom of square cross-section (25 × 25 × 40 cm^3). At other locations

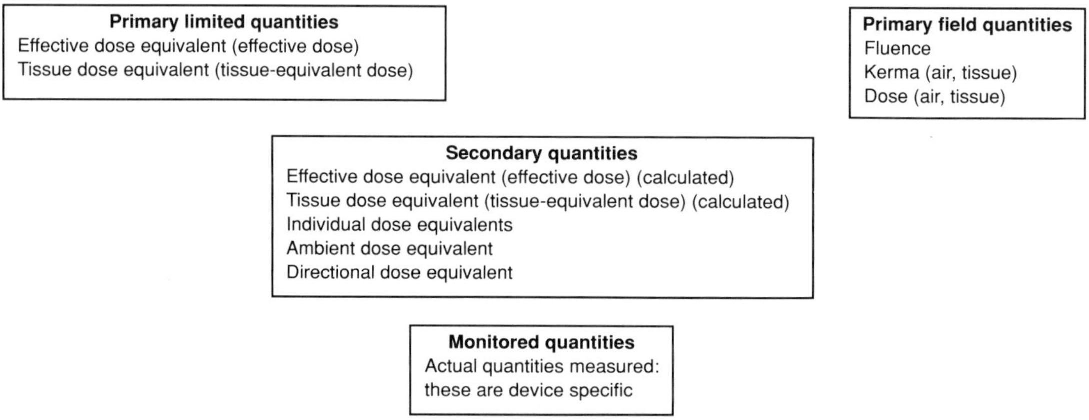

Primary limited quantities
Effective dose equivalent (effective dose)
Tissue dose equivalent (tissue-equivalent dose)

Primary field quantities
Fluence
Kerma (air, tissue)
Dose (air, tissue)

Secondary quantities
Effective dose equivalent (effective dose) (calculated)
Tissue dose equivalent (tissue-equivalent dose) (calculated)
Individual dose equivalents
Ambient dose equivalent
Directional dose equivalent

Monitored quantities
Actual quantities measured:
these are device specific

Figure 1. The inter-relationships of the various quantities and instrument and dosemeter responses.

the dosemeters were mounted vertically in four strips of four, orthogonally on the surface of an NM2 neutron monitor (Anderss and Braun type moderating instrument manufactured by NE Technology Ltd), long axis vertical. This enabled a simultaneous estimate to be made of ambient dose equivalent as derived from the NM2 instrument. The disadvantage of using devices which do not give energy spectral information and whose response needs to be corrected for energy dependence is that additional uncertainties are introduced owing to any change in energy spectrum with angle being neglected. Ideally one requires an energy and angle spectrometer.

The angular distributions of the radiation fields were derived by proposing the least number of unidirectional and planar isotropic fields which gave corrected dosemeter responses equal to those obtained (within the uncertainties), taking into account also the relative positioning of the sources of neutrons and the dosemeter/shield assembly. Corrections to dosemeter response to take account of its energy dependence of response in terms of ambient dose equivalent were made using the NPL measured energy spectral distributions. The field angle distributions obtained from the dosemeter response together with the energy spectral distributions enabled calculations to be made of the required quantities, directional dose equivalent, H'(10) (which can be assumed to be approximately equal to individual dose equivalent, penetrating, $H_p(10)$), effective dose equivalent and effective dose. Ambient dose equivalent is obtained directly for the multisphere responses or by correcting the response of the NM2 neutron monitor.

SUMMARY OF RESULTS AND DISCUSSION

The results of the energy spectral distributions of NPL are summarised in Table 1(a) and 1(b). The uncertainties in the total fluence rates and total dose equivalent rates are estimated to be 10 and 15% respectively. The uncertainties in the fractional contributions (the contents of the broad energy 'bins') are estimated to be up to 25%. Confidence in the results for the fractional contributions (the spectra in broad energy bins) is given by a comparison between measurements reported here and those made in 1982[2] using essentially the same approach and for approximately the same neutron field, at one location. Very close agreement is obtained: to within about 10% for any bin. The ambient dose equivalent rates shown are for the pre-ICRP 60 definition of the quality factor function. To a first approximation, the new values of ambient dose equivalent are anticipated to be 50% greater and the fractional contributions unchanged.

The assessments of the field neutron angle distributions are summarised in Table 2. These assessments are based on both the results obtained for the dosemeters mounted on the phantom/shield

Table 1. Summary of energy spectral distributions, all locations.

(a) Fluence rate.

Location	Fractional contribution to total fluence rate				
	Thermal	0.5 to 10^4 eV	10^4 to 10^5 eV	10^5 to 10^6 eV	10^6 to 1.26×10^7 eV
1	0.32	0.22	0.06	0.26	0.13
2	0.59	0.27	0.06	0.06	0.02
3	0.15	0.28	0.06	0.13	0.38
4	0.11	0.23	0.03	0.33	0.30
5	0.10	0.19	0.09	0.39	0.23
6	0.68	0.29	0.02	0.01	$<10^{-4}$
7	0.67	0.29	0.03	0.01	$<10^{-4}$

(b) Ambient dose equivalent rate.

Location	Fractional contribution to ambient dose equivalent rate				
	Thermal	0.5 to 10^4 eV	10^4 to 10^5 eV	10^5 to 10^6 eV	10^6 to 1.26×10^7 eV
1	0.03	0.02	0.02	0.51	0.43
2	0.17	0.08	0.06	0.41	0.28
3	0.01	0.01	0.01	0.17	0.80
4	0.005	0.01	0.005	0.46	0.52
5	0.005	0.01	0.02	0.46	0.51
6	0.57	0.26	0.05	0.12	0.001
7	0.53	0.23	0.07	0.17	0.001

and on the relative positions of sources of neutrons. The most intense sources of neutrons were in the same horizontal planes as the phantom locations and worker positions of interest. This enabled simple analyses into unidirectional and planar isotropic components. Planar isotropy (rotational isotropy) also describes the resultant field at a worker who is not static but moves in such a way that, in general, no specific orientation is preferred. This treatment neglects scattered radiation from above and below. It should also be noted that at present conversion coefficients to effective dose equivalent are available for neutrons incident in the horizontal plane (anterior-posterior, posterior-anterior, lateral and planar isotropy) but not from below or above or for an isotropic field. The uncertainties in the assessment given in Table 2 are estimated to be between 0.1 and 0.2.

Table 3(a) gives mean conversion coefficients between the primary field quantity, fluence, and other quantities of interest using ICRP 51 data[7]. Values for three standard fields are given for comparison. Recent data[8] for the new ICRP basic radiation quantity, effective dose, E, have also been folded with the spectra. Table 3(b) gives ratios of mean conversion coefficients for these quantities of interest. The relationships between the effective dose equivalent (calculated for a mathematical phantom) and the operational

Table 2. Summary of angle distributions, all locations.

Location	Fractional contribution to ambient dose equivalent rate	
	Unidirectional	Planar isotropic
1	–	1*
2	–	1
3	0.3**	0.7*
4	0.8	0.2
5	0.85	0.15
6	–	1
7	–	1

* From front only (semi-planar isotropic).
** Laterally.

Table 3. Relationships between fluence weighted mean conversion coefficients for quantities, all locations.

(a) Fluence weighted conversion coefficients.

Location	$H^*(10)/\Phi$ (pSv.cm^2)	$H'(10)(PLIS)/\Phi$ (pSv.cm^2)	$H_E(AP)/\Phi$ (pSv.cm^2)	$H_E(PLIS)/\Phi$ (pSv.cm^2)	$H_E(LAT)/\Phi$ (pSv.cm^2)	$H_E(PA)/\Phi$ (pSv.cm^2)	$E(AP)/\Phi$ (pSv.cm^2)	$E(ROT)/\Phi$ (pSv.cm^2)
1	111	50	51	27	14.9	24	105	58
2	28	11.7	12.8	6.7	3.6	6.3	29	15.4
3	178	94	111	67	42	68	196	120
4	198	88	89	44	23	38	181	95
5	170	80	86	47	28	45	166	116
6	9.7	3.3	4.6	2.5	1.38	2.7	11.1	6.0
7	10.4	3.6	4.8	2.6	1.41	2.8	11.8	6.3
Bare ^{252}Cf*	334	170	195	112	69	113	357	212
^{252}Cf + 15 cm D$_2$O*	90	45	51	29	18.1	30	100	59
Thermal*	8.0	2.6	4.0	2.3	1.3	2.6	8.2	4.3

* For comparison purposes.

(b) Relationship between quantities.

Location	Anterior-posterior			Planar isotropy			Field as assessed		
	$H_E/H^*(10)$	E/H_E	$H'(10)/H^*(10)$	$H_E/H^*(10)$	$H_EH'(10)$	E/H_E	$H'(10)/H^*(10)$	$H_E/H^*(10)$	$H_E/H'(10)$
1	0.46	2.1	0.45	0.24	0.54	2.1	0.81	0.30	0.37
2	0.45	2.2	0.41	0.24	0.57	2.3	0.41	0.24	0.57
3	0.63	1.8	0.53	0.37	0.71	1.8	0.78	0.37	0.47
4	0.45	2.0	0.44	0.22	0.50	2.2	0.89	0.40	0.45
5	0.51	1.9	0.47	0.28	0.59	2.5	0.93	0.48	0.51
6	0.47	2.4	0.34	0.26	0.77	2.4	0.34	0.26	0.77
7	0.46	2.5	0.34	0.25	0.73	2.4	0.34	0.25	0.73
Bare ^{252}Cf*	0.58	1.8	0.51	0.34	0.66	1.9			
^{252}Cf + 15 D$_2$O*	0.57	2.0	0.50	0.33	0.66	2.0			
Thermal*	0.50	2.1	0.33	0.29	0.88	1.9			

* For comparison purposes.

quantities, that is the degree of overestimation, are shown for each location for the field geometries, anterior-posterior, planar isotropy, and the assessed fields. Also shown are the ratios of directional dose equivalent, H'(10) (closely related to individual dose equivalent, penetrating, $H_p(10)$) and ambient dose equivalent. Broadly the fields measured could be grouped into two categories, 'soft' well moderated fields and 'hard' less well moderated fields. The first category includes the two reactor locations, 6 and 7 and a well shielded area within the old plutonium finishing plant, 2. For this category ambient dose equivalent overestimates effective dose equivalent, H_E, by a factor of 4 (uncertainty of about 20%) and directional dose equivalent H'(10) (and individual dose equivalent, penetrating, $H_p(10)$) overestimates effective dose equivalent by about a factor of 1.3 or 1.4 (reactor sites) and 1.7 (finishing plant) (uncertainty about 20%). The second category includes a second location in the old plutonium finishing plant, the fuel assembly plant and two locations in the new plutonium finishing plant. For these locations, ambient dose equivalent overestimates effective dose equivalent by about a factor of about 2.5 (uncertainty of about 20%) and directional dose equivalent, H'(10) (and individual dose equivalent, penetrating, $H_p(10)$) overestimates effective dose equivalent by about a factor of 2 (uncertainty of 20%).

Table 4 summarises the instrument and dosemeter response characteristics in the different neutron radiation fields in terms of field quantities and effective dose equivalent. The uncertainties in all the values are about 20%. (As is frequently the case, the estimation of uncertainties is difficult. This value of 20% represents a judgement based partly on accumulated experience, partly on repeated measurements and partly on intercomparison of different methods.) These values

Table 4. Summary of instrument and dosemeter response characteristics, all locations. (Energy and angle spectral distribution as in Tables 1 and 2).

(a) Field quantities.

Location	Normalised[a] response per unit ambient dose equivalent			Normalised[a] response per unit directional dose equivalent, planar isotropy			Normalised[a] response per unit directional dose equivalent, field as assessed		
	0949 type	NM2 type	PADC[b] dosemeter	0949 type	NM2 type	PADC[c] dosemeter	0949 type	NM2 type	PADC dosemeter
1	1.14	0.94	1.23	2.6	2.1	0.75	2.6	2.1	0.75
2	1.51	1.03	1.05	3.7	2.4	0.80	3.7	2.4	0.80
3	1.05	1.01	1.11	2.0	1.9	0.63	1.3	1.28	0.55
4	1.10	1.04	1.51	2.5	2.3	0.89	1.23	1.17	1.4
5	1.07	0.92	1.13	2.3	2.0	0.66	1.16	1.00	1.10
6	2.1	1.07	0.86	6.3	3.1	1.20	6.3	3.1	1.20
7	2.1	1.10	0.84	6.2	3.2	1.10	6.2	3.2	1.10

(b) Effective dose equivalent.

Location	Normalised[a] response per unit effective dose equivalent, anterior-posterior			Normalised[a] response per unit effective dose equivalent, planar isotropy			Normalised[a] response per unit effective dose equivalent, field as assessed		
	0949 type	NM2 type	PADC[b] dosemeter	0949 type	NM2 type	PADC[c] dosemeter	0949 type	NM2 type	PADC dosemeter
1	2.5	2.0	2.7	4.7	3.6	1.4	3.8	3.2	2.0
2	3.4	2.3	2.3	6.4	4.4	1.4	6.4	4.4	1.4
3	1.7	1.6	1.8	2.8	2.7	0.89	2.8	2.7	0.95
4	2.4	2.3	3.4	4.9	4.7	1.8	2.7	2.6	3.2
5	2.1	1.8	2.2	3.8	3.3	1.12	2.3	2.0	2.1
6	4.6	2.3	1.8	8.2	4.1	1.6	8.2	4.1	1.6
7	4.7	2.4	1.8	8.6	4.4	1.5	8.5	4.4	1.5

[a] Normalised to unity per ambient dose equivalent for the 0949 type and NM2 type instruments and to 1.25 for the PADC dosemeter for bare ^{252}Cf neutron field normally incident.
[b] Normal incidence response.
[c] Planar isotropic response.

enable location-specific correction factors to be applied to instrument and dosemeter response, if required, and show the relationships between uncorrected responses of instruments and dosemeters, and ambient and directional dose equivalents (Table 4(a)); or effective dose equivalent (Table 4(b)). The same grouping of fields into 'soft' well moderated and 'hard' less well moderated applies. In the 'soft' fields (locations 2, 6 and 7), the commonly used instrument, the 0949 type, would overestimate ambient dose equivalent by a factor of between 1.5 and 2, and effective dose equivalent by a factor between 6 and 9. In the 'hard' fields (locations 1,3,4,5) the overestimate of ambient dose equivalent would be between a factor of 1 and 1.2, and of effective dose equivalent by a factor of between 2.5 and 4. Uncertainties in factors are about 20–30%. For the NRPB PADC dosemeter, in the soft fields (locations 2, 6 and 7) directional dose equivalent (and also Hp(10)) would be estimated to within 0.8 to 1.2 of the 'true' value, and effective dose equivalent overestimated by a factor of between 1.4 and 1.6. In three of the 'hard' fields (locations 1, 4 and 5), directional dose equivalent (and $H_p(10)$) would be estimated to within 0.75 to 1.4 of the 'true', and effective dose equivalent overestimated by a factor of between 2 and 3. At location 3, a large component of the radiation field was incident laterally. For this field the dosemeter would underestimate directional dose equivalent by a factor of 0.55 and give an approximately true estimate of effective dose equivalent.

Of necessity, owing to the limited availability of data for the quantities as newly defined by ICRP[1], these analyses have been carried out in terms of previous definitions of both the basic dosimetric quantities and the operational quantities. Data[8] for effective dose, E, for AP and PLIS field components are shown in Table 3 where it can be seen that there is an increase by a factor of between 1.8 and 2.5. Ambient dose equivalent and directional dose equivalent PLIS are only anticipated to increase by a factor of about 1.5. A more detailed analysis will be carried out when data are available.

REFERENCES

1. International Commission on Radiological Protection. *1990 Recommendations of the International Commission on Radiological Protection*. Report 60 (Oxford: Pergamon Press) (1991).
2. Bartlett, D. T. and Bardell, A. G. *Field Measurements of Neutron Energy and Angle Spectral Distribution and their Interpretations in Terms of Relevant Radiological Protection Quantities*. In: Proc. Fifth Symp. on Neutron Dosimetry. EUR 9762 (CEC: Luxembourg) (1985).
3. Thomas, D. J. and Bardell, A. G. *Neutron Spectrometry in Working Environments*. In: Proc. BNES Conf. on Occupational Radiation Protection, Guernsey, April/May 1991.
4. Alvera, A. V., Siebert, B. R. L., Aroua, A., Buxerolle, M., Grecescu, M., Matzke, M., Mourges, M., Perks, C. A., Schraube, H., Thomas, D. J. and Zaborowski, H. L. *Unfolding Bonner Sphere Data: A European Intercomparison of Computer Codes*. PTB Report, PTB-7.22-90-01 (1990).
5. Gilvin, P. J., Bartlett, D. T. and Steele, J. D. *NRPB PADC Neutron Personal Dosimetry Service*. Radiat. Prot. Dosim. **20**(1/2), 99-102 (1987).
6. Bartlett, D. T., Steele, J. D. and Stubberfield, D. R. *Development of a Single Element Neutron Personal Dosemeter for Thermal, Epithermal and Fast Neutrons*. Nucl. Tracks **12**(1/6), 645-648 (1986).
7. International Commission on Radiological Protection. *Data for Use in Protection Against External Radiation*. Publication 51 (Oxford: Pergamon) (1987).
8. Hollnagel, R. A. *Calculated Effective Doses in Anthropoid Phantoms for Broad Neutron Beams with Energies from Thermal to 19 MeV*. Radiat. Prot. Dosim. **44**(1-4) 155-158 (1992) (This issue.)

NEUTRON SPECTROMETRY SYSTEM FOR RADIATION PROTECTION: MEASUREMENTS AT WORK PLACES AND IN CALIBRATION FIELDS

F. Posny†, J. L. Chartier† and M. Buxerolle‡
†CEA/IPSN/DPHD/S.DOS, Centre d'Etudes Nucléaires de Fontenay-aux-Roses
BP No 6 - 92265 Fontenay-aux-Roses Cedex, France
‡CEA/IPSN/DPHD/S.DOS/GDN, Centre d'Etudes Nucléaires de Cadarache
13108 St. Paul Lez Durance Cedex 147, France

Abstract — The determination of dose equivalent quantities in mixed photon–neutron fields is still a problem in radiation protection because the conventional environmental and individual dosemeters do not have the required energy dependence, especially if the new ICRP recommendations are taken into account. Furthermore, the available wide reference calibration spectra (ISO 8529) are not a satisfactory approximation to those met in practical situations. Therefore, a real need arises to improve the knowledge of the fields to which workers in the nuclear industry could be exposed, and to define new calibration fields. A sufficient accuracy in the characterisation of these two types of radiation field would be reached only if detector systems with spectrometric properties are employed. This paper presents an operational spectrometry system used for measurements of the spectral fluence both in working areas and in calibration fields. It describes two different techniques to cover the whole possible energy range: proton recoil detectors from 100 keV to 20 MeV and the multisphere technique for the lower energy range down to thermal neutron energies. The detectors and their main characteristics are briefly described, with the different unfolding codes. The dosimetric quantities are calculated from the measured spectrum. The improvement and comparison of the two techniques are presented and discussed for two ISO reference fields: Am–Be and ^{252}Cf. The whole spectrometry system is finally put in practice with measurements in: two radiation protection situations (source manufacturing work station, and irradiated fuel transport container, and a 'simulated realistic' neutron spectrum. This work participates in a larger European research project which has to gather measurements in working areas and finally provides, in the laboratory, neutron calibration fields close to those met in radiation protection practice.

INTRODUCTION

The determination of dose equivalent quantities in mixed photon–neutron fields is currently still a problem due to the fact that area monitors and personal dosemeters do not have the required energy response and sensitivity. Furthermore, the available wide calibration spectra recommended by ISO in the standard 8529[1] do not approximate well to all the spectra encountered in practical situations. Therefore a real need arises to improve the knowledge of the fields to which workers in the nuclear industry could be exposed, and to redefine more 'realistic' calibration fields.

A spectrometry system has been developed for measurements of the spectral fluence both in working areas and in calibration fields. The results presented in this paper concern two reference fields: ^{252}Cf, Am–Be, a so-called 'simulated realistic' neutron spectrum and two radiation protection situations.

SPECTROMETRY SYSTEM

Two different techniques to cover the whole energy range from 0.01 eV to 20 MeV are described: proton recoil detectors from 100 keV to 20 MeV and a multisphere system for the lower energy range down to the thermal neutron energies.

Detectors

The characterisation of neutron fields for calibration as well as those encountered in radiation protection requires detectors with the following qualities : an isotropic angular response, a good energy resolution and the ability to discriminate neutrons and photons. The equipment must be reliable, transportable, and the results should be available a short time after the measurements. These criteria have been approached as closely as possible by the use of three proton recoil detectors and unfolding codes running on a microcomputer. Two detectors are spherical gas filled proportional counters (type SP2 [2], 4 cm diameter) and the third one is an organic liquid scintillator probe (type NE 213, 1/2" × 1/2" cell).

The multisphere system is a classical set of nine polyethylene spheres of density 0.92 g.cm^{-3} with a central ^{3}He detector (LMT type 0.5 NH/KF1/1) which has been described in detail in previous papers[3,4].

Figure 1 shows the energy range covered by each detector and the overlapping regions used to link the results of the proton recoil detectors.

The characteristics of the proton recoil detectors between 120 keV and 3 MeV have first been determined with monoenergetic neutrons at the research centre of Bruyères-le-Châtel (Van-de-Graaf

4 MeV): response functions (shape and calibration), energy resolution and efficiency variation relative to the neutron energy. The NE 213 scintillator has also been checked at 14.7 MeV at the reference assembly of the BIPM in Sèvres (France).

The efficiency values determined with the measurement at CEA of Bruyères-le-Châtel led to a systematic overestimation of fluence in the field measurements. To reduce this discrepancy, we have measured the spectral distribution of a reference calibrated ^{252}Cf source at the CEN-Cadarache. The results are shown in Figure 2.

The Am–Be spectrum presented in Figure 3 is the result of the measurement of a calibrated reference source, also at the CEN-Cadarache. Only the part above 2 MeV has been used to check the characteristics of the NE 213 scintillator: it corresponds to the energy range where it is used. A good agreement with the ISO spectral distribution and with the total fluence has been found.

Unfolding codes

To evaluate the final spectral distribution three unfolding codes are used: one for each type of detector. For the liquid scintillator NE 213 the code MATXUFCORR[5] considers the shape of the experimental response functions before applying the mathematical process recommended by R. H. Johnson et al (folding product[6]). The efficiency calculation is based on the ANGEFF code of Dufold using two adjustable parameters: the 'proton' detection threshold of the detector and the angular distribution of the radiation incident on the detector. The analysis of the proton distributions measured by the proportional counters is carried out by a program based on the code SPEC4[7]. The response functions are calculated by the analytical expression of Snidow and Warren[8] but the slope is corrected by factors derived from the experimental response functions. This correction leads to a satisfactory agreement between the calculations and the measurements with monoenergetic neutrons. The experimental efficiency and energy resolution are also considered. The multisphere counts are handled by the MORELPA program[3,4] in which the sphere response functions are given at 170 energies by the three parameters of log–normal distributions. The unfolding of the neutron spectrum is achieved by using the 'model spectra' method which uses four basic mathematical functions as models[3,4].

MEASUREMENTS

With the described spectrometry system, two types of neutron fields have been studied. One (CANEL+) is a so-called 'realistic' spectrum produced in the laboratory at Cadarache, which is wide (0.01 eV to 15 MeV) and presents narrow

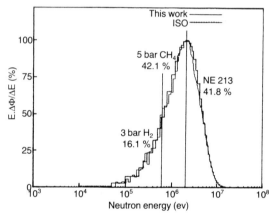

Figure 2. Neutron spectrum of the ^{252}Cf source. The two spectra have been normalised to the maximum value. The percentages indicate the contribution of each detector to the total fluence. Solid line, these measurements. Line with dots, ISO. Total fluence (in cm^{-2}), measurement = 11826, reference = 11873.

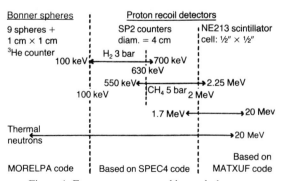

Figure 1. Energy range covered by each detector.

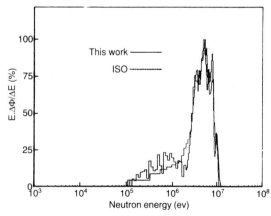

Figure 3. Neutron spectrum of the Am–Be source. The two spectra have been normalised to the maximum value.

structures. The second type (Am–Be source manufactoring work station and transport containers) concerns radiation protection fields which show generally a smoother shape but present the problem of low count rates leading to poor statistics. In both cases a parasitic photon contribution had to be considered.

CANEL+ spectrum: CEN-Cadarache

A fission spectrum resulting from the interaction of 14.7 MeV neutrons on a 12 cm thick ^{238}U shell is moderated by a 15 cm iron shell. This set-up is surrounded by polyethylene (detailed description in Reference 9). The parameters of the assembly are used as input data of the MCNP code[10]. The results of calculation and measurement are compared in Figure 4. A rather good agreement is observed between the two spectral distributions except in the ten keV energy region. As no discrimination between neutrons and photons is carried out for the proportional counters, the lower energy limit of the H_2 filled counter is 100 keV and the energy resolution of the multisphere technique is not sufficient to identify the 24 keV structure due to the iron shell.

Am-Be source manufacturing work station: CEN-Fontenay-aux-Roses

In order to reproduce the position of the chest of a working man, the measurement point was located at 35 cm in front of a glove-box in which the Am–Be sources are handled. Because of the low sensitivity of the proton recoil detectors, their measured proton distributions have been smoothed before the unfolding process. The spectrum shown in Figure 5 presents the general shape of an Am–Be source modified at the low energies by the scattered neutrons due to the walls of the glove-box and to the local environment.

Irradiated fuel transport container: La Hague

There are two main types of transport container: the most one often used is of LK100 type and the second type is TN12. The measured field was the one near a container type LK100 coming from a French nuclear plant to be reprocessed at La Hague. The measurement point was located at a distance of 30 cm and in the central area of the container. In Figure 6 the spectrum obtained shows structure in the MeV energy region, probably due to the shielding materials inside the container (steel, lead, wood, resin etc.). In that case, the measured proton distributions have also been smoothed because of the low count rates (0.3 to 0.8 per second).

Figure 7 shows the spectra measured by the multisphere technique near the LK100 container compared with the one of a TN 12 type container measured in Valognes in very different surroundings (small concrete room instead of a big hall with light walls). It appears that the spectra are different due to the type of the container but also due to the environment.

CONCLUSION

It could be debated that, in this spectrometry system, the low energy region is characterised with the multisphere technique which is known to have poor energy resolution, but if the part below 100 keV is not estimated in detail, its contribution to the total dose equivalent is generally small. Even if the spectrometry system as presented needs further investigations, the results obtained

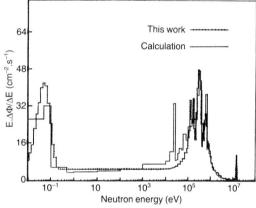

Figure 4. Neutron spectrum of the CANEL+ facility. The calculated spectrum has been normalised to the maximum value of the measured spectrum.

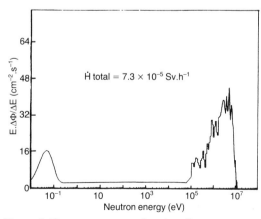

Figure 5. Neutron spectrum of an Am–Be source manufactoring work station. The dose equivalent quantity has been calculated using the convertion factors of ICRP 21.

are encouraging. This system is not easy to handle and the measurements in radiation protection fields are difficult because of the low count rates. This is not the case for the multisphere technique. But the real advantage of the proton recoil spectrometry is that there is much less uncertainty in the resulting spectrum. If a good accuracy is required for the shape of the spectral distribution, it is necessary to use this type of good energy resolution detector, especially for the fields with a high energy component. However there still remains the problem of the evaluation of the lowest limit of dose equivalent which can be measured with such a system.

ACKNOWLEDGEMENT

This work was partly supported by the European Communities (contract Bi7-0031-C) and the Bureau National de Métrologie (research agreement no. 89-2-46-0016).

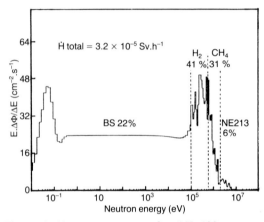

Figure 6. Neutron spectrum of a LK 100 transport container. The dose equivalent quantity has been calculated using the conversion factors of ICRP 21. The percentages indicate the contribution of each detector to the total dose equivalent rate.

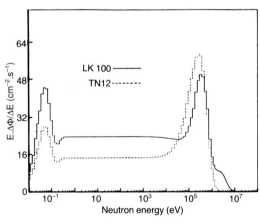

Figure 7. Neutron spectra of transport containers.

REFERENCES

1. International Standard. ISO 8529. *Neutron Reference Radiations for Calibrating Neutron-measuring Devices Used for Radiation Protection Purpose and for Determining their Response as a Function of Neutron Energy* (1989).
2. Kemshall, C. D. *The Use of Spherical Proportional Counters for Neutron Spectrum Measurements.* AWRE Report N° 031/73 (1973).
3. Buxerolle, M., Massoutié, M. and Kurkdjian, J. *Catalogue de Spectres de Neutrons.* Rapport CEA-R-5398 (1987).
4. Buxerolle, M. and Chartier, J. L. *Experimental Simulation and Characterisation of Neutron Spectra for Calibrating Radiation Protection Devices .* Radiat. Prot. Dosim, **23**, 285-288 (1988).
5. Coolbaugh, M. J., Faw, R. E. and Meyer W. *Fast Neutron Spectroscopy in Aqueous Media using an NE 213 Proton-recoil Neutron Spectrometer System.* USAEC, Doc. N° COO 2049-7 (1971).
6. Johnson, R. H., Wehring, B. W. and Dorning, J. J. *Smoothing in the Ferdor Method of Neutron Spectrum Unfolding.* Trans. Am. Nucl. Soc. **22**, 798 (1975).
7. Benjamin, P. W., Kemshall, C. D. and Brickstock, A. *The Analysis of Recoil Proton Spectra.* AWRE Report N° 09/68 (1968).
8. Snidow, N. L. and Warren, H. D. *Wall Effect Corrections in Proportional Counter Spectrometers.* Nucl. Instrum Methods **51**, 109-116 (1967).
9. Chartier, J. L., Posny, F. and Buxerolle, M. *Experimental Assembly for the Simulation of Realistic Neutron Spectra .* Radiat. Prot. Dosim. **44**(1-4) 125-130 (1992) (This issue).
10. Briesmeister, J. F. *et al.* *MCNP. A General Monte-Carlo Code for Neutron and Photon Transport (Version 3A).* LA-7396-M, Rev. 2 (1986).
11. ICRP. *Data for Protection Against Ionizing Radiation from External Sources.* Publication 21 (Oxford: Pergamon Press) (1973).

COMPARISON OF DIFFERENT NEUTRON AREA MONITORS AS ROUTINE RADIATION PROTECTION DEVICES AROUND A HIGH ENERGY ACCELERATOR

M. Boschung†, C. Wernli† and A. Kunz‡
†Radiation Hygiene Division, Paul Scherrer Institute
CH-5232 Villigen PSI, Switzerland
‡Fachrichtung Biophysik und Grundlagen der Medizin
Universität des Saarlandes, D-6650 Homburg/Saar, Germany

Abstract—The tissue-equivalent proportional counter (TEPC) is currently the only area monitor to assess reasonably the dose equivalent in unknown radiation fields. Two such systems were used in the environment of the high energy proton accelerator at the Paul Scherrer Institute in Villigen, Switzerland. The aim of the work was to compare results from the two TEPC counters to results obtained from the commercially available Dineutron and Andersson-Braun rem counter. Agreement between both TEPC systems is good. The readings of the rem counter and the Dineutron are lower than the TEPC results.

INTRODUCTION

The measurement of absorbed dose in tissue-like material and of energy deposition based on microdosimetric techniques has caused the tissue-equivalent proportional counter (TEPC) to be recognised as a useful dose equivalent meter. Although different designs match the ambient dose equivalent H*(10) fairly well, the use of such monitors for routine radiation protection purposes has not yet been implemented. In the past few years much effort has been expended on building and improving the TEPC system. These endeavours are motivated by the advantages TEPCs offer, such as the determination of absorbed dose and dose equivalent in tissue-like material, separation of photon and neutron fractions in mixed radiation fields, and assessment of radiation quality. Furthermore, the TEPC is currently the only area monitor to assess dose equivalent reasonably in unknown mixed radiation fields, such as are found around high energy accelerators. Radiation quality in such an environment is subject to large local variations, thus making the determination of dose equivalent difficult. This paper reports on the performance of two TEPC systems in the environment of the high energy proton accelerator at the Paul Scherrer Institute (PSI) in Villigen, Switzerland. The aim of the work was to compare their results with those obtained from commercially available neutron area monitors.

NEUTRON AREA MONITORING SYSTEMS

One of the neutron area monitors was a rem counter of the Andersson–Braun type[1] (STUDSVIK 2202 D). This rem counter is known to overestimate dose equivalent for neutron energies between 100 eV and 100 keV, and to underestimate it for neutron energies below 100 eV and above 2 MeV[2]. In the region between 100 keV and 2 MeV the rem counter matches the dose equivalent H*(10) fairly well. The second neutron area monitor was the Dineutron designed by CEA[3] (France) and distributed by Nardeux. The Dineutron will over-respond when irradiated from bare and heavy water moderated ^{252}Cf neutron sources by factors of 1.4 and 2.0, respectively[4].

The small and flexible TEPC system of Homburg, called HANDI[5] (Homburg Area Neutron Dosimeter), and the larger system of PSI complete the set of neutron area monitors used. Both of the latter two systems are based on microdosimetric techniques: absorbed dose measurement is based on the cavity chamber principles using tissue-equivalent counter wall material and gas. The detectors are operated at low gas pressure and the instruments analyse the pulse height of single events. The size of each event expressed in terms of lineal energy, y (keV.μm^{-1}), is related to the linear energy transfer, L_∞, of the secondary charged particles if their ranges are sufficiently large compared with the simulated diameter. The event size spectrum obtained initially is converted to a logarithmic frequency distribution, yf(y), weighted by the logarithmic channel and the calibration factor to obtain the dose distribution, yd(y). The approximation $y = L_\infty$ is used to calculate the quality factor, Q(y), as defined in ICRP 21[6]. By multiplying the dose with this approximated quality factor, Q(y), one gets the dose equivalent distribution, yh(y), of the measured field. Summation over the whole of yd(y) and yh(y) provides total dose and total dose equivalent.

The PSI counter was a copy from the KFA-Jülich design developed by Booz[7]. A full description of the design can be found elsewhere[8]. In view of the counter's use in the PSI accelerator environment and based on calculations by Morstin et al of tissue-equivalence of various materials at high energies[9], the outer wall consisted of 8 mm thick Plexiglas surrounded by an 0.8 mm thick aluminium cap. The counter was operated at a pressure of 1.35 kPa, simulating a tissue diameter of 1 μm. The electronics, which mainly compress the pulses from the counter into a pseudo-logarithmic event spectrum by a logarithmic preamplifier, were constructed in Jülich. To attribute absorbed dose and dose equivalent parts to photons and neutrons a fixed discrimination level at about 15 keV.μm^{-1} was used. The calibration was controlled in terms of dose equivalent with an external ^{137}Cs source. Prior to this, pulse heights are calibrated by the events of maximum linear energy transfer of protons at about 150 keV.μm^{-1} induced by the neutrons of a ^{252}Cf source, commonly called the proton edge.

The HANDI detector consisted of a spherical TEPC detector used at a gas pressure simulating a tissue–equivalent diameter of 2 μm. The A-150 counter wall was 2.5 mm thick and sat inside a cylindrical stainless steel housing with a wall thickness of 0.76 mm. The pulses were treated by a 16 channel ADC with logarithmic transfer characteristics. The quantities of absorbed dose (rate) and dose equivalent (rate) were evaluated on-line by a microprocessor and displayed in real time[10]. A pure microdosimetric spectrum for photons was fitted by a suitable procedure to the mixed radiation spectrum to separate the total absorbed dose into its photon and neutron components. The built-in α particle source (^{244}Cm) was used to check the gas gain stability. Calibration in terms of absorbed dose was carried out in a ^{60}Co reference field.

MEASUREMENTS

The PSI operates an accelerator of 590 MeV protons with normal current intensity of 200 μA. On its way, the high intensity beam twice strikes a carbon target from which many different particles for experimental purposes emerge. Part of the proton beam splits off and leads to a beryllium target where pions are produced for the biomedical department's radiation therapy. The beam is then stopped at the beam dump area. In collaboration with the radiation protection group of PSI four locations were chosen to perform the measurements. Both the HANDI and PSI systems were run at each position simultaneously and during identical acquisition periods, thus eliminating proton beam fluctuation effects. These conditions do not hold true for the rem counter and Dineutron, and therefore comparison of their results should be made with caution.

RESULTS

Figure 1 shows absorbed dose distributions, yd(y), as a function of lineal energy, y, measured with both TEPC systems at position 1, above the last target and beam dump area. The distributions measured at point 2, near the beryllium target of the biomedical installation, are not shown but have a similar shape. It can be seen that the PSI and HANDI systems agree very well. Pulses above the proton edge of 150 keV.μm^{-1} arise either from α particles, heavy recoil ions or multiple events, all produced by neutrons with energies above 20 MeV. The lack of a marked proton edge also indicates neutrons of high energy. The peak at 3 to 5 keV.μm^{-1} as well as the edge at about

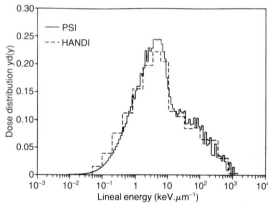

Figure 1. Dose distribution above the beam dump and target E area obtained with the PSI (straight line) and HANDI (dashed line) system.

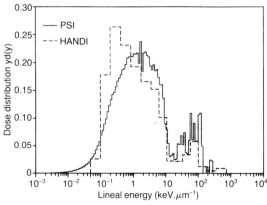

Figure 2. Dose distribution above the ventilation shaft of the biomedical target area obtained with the PSI (straight line) and HANDI (dashed line) system.

10 keV.µm^{-1} is due to fast recoil protons induced by high energy neutrons and therefore overlaps with the photon part in the microdosimetric distribution.

Figure 2 shows the same comparison for location 3, above the ventilation shaft in the biomedical target area. Distributions in location 4, near a radiation protection door of injector II, have the same shapes and are therefore not shown. From Figure 2 it can be seen that the PSI and HANDI system disagree in the low lineal energy region. There are not many events above the proton edge, therefore 5 MeV neutrons at most are present. The distinct proton edge, as well as the peak at 50 - 100 keV.µm^{-1}, suggest the existence of neutrons below 1 MeV. The edge at 10 keV.µm^{-1} and the peak between 0.5 and 10 keV.µm^{-1} is due to the photon part of the field. This distribution shows the typical shape of a multiple scattered neutron spectrum, producing a high photon component and neutron induced events up to some 100 keV.µm^{-1}.

Table 1 summarises the results obtained with the different neutron measuring devices, including total absorbed dose rate, total dose equivalent rate as well as photon and neutron components of dose rates of the investigated fields. The agreement of total absorbed dose rate is excellent, of total dose equivalent rate acceptable, and of photon and neutron components poor. The differences in the γ and neutron dose equivalent rate fractions between the HANDI and PSI systems in position 1 and 2 reflect the differences in the separation methods used to calculate neutron and photon components. For such distributions, the fitting method used by HANDI produces a 60 – 70% lower γ dose equivalent rate and a 15 – 20% higher neutron dose equivalent rate than the evaluation with a fixed discrimination level used by PSI. The fitting method used by the HANDI system in evaluating neutron and gamma fractions will meet, at those situations, reality closer than the discrimination method used in the PSI evaluation. At position 3 (Figure 2), where the distributions show a clear drop at about 10 keV.µm^{-1}, different methods of separation will produce only small differences in the evaluated doses. At position 4, because of the very low dose rates no statement can be made.

The differences in neutron dose equivalent reading of the TEPC systems and the rem counter and Dineutron can be attributed to the differences in the neutron energy response. As shown earlier[11], the rem counter gives a value three times that of the TEPC systems at positions with low mean neutron energy (ventilation shaft and near door of injector II), and 2.5 times lower at positions with high mean neutron energy (above target E and beam dump area and near beryllium target of biomedical installation). At point 1 the Dineutron value is 2.5 times lower than the values given by the TEPC system. At point 2, the underestimate by the Dineutron is only about 15% compared to the TEPC values. To understand these findings more detailed studies of the behaviour of the Dineutron in the environment of an accelerator is necessary.

A major advantage of the TEPC systems, apart from the possibility of assessing neutron and γ portions of unknown fields, is the flexibility of using different definitions of quality factor. Indeed, the effect of the new definition of quality factor by ICRP 60[12] can be checked very easily.

Table 1. Total, gamma, and neutron dose rates of the PSI measurements for different neutron area monitors with quality factor definition of ICRP 21 and ICRP 60.

No.	System	Absorbed dose rate (µGy.h^{-1})			Dose equivalent rate with quality factor definition of					
					ICRP 21[6] (µSv.h^{-1})			ICRP 60[12] (µSv.h^{-1})		
		Total	γ	Neutron	Total	γ	Neutron	Total	γ	Neutron
1	PSI	9.4	6.8	2.6	41	10	31	46	7	39
	HANDI	8.8	3.9	4.9	39	4	35	43	4	39
	Rem counter	-	-	-	-	-	11.8	-	-	-
	Dineutron	-	-	-	-	-	12.0	-	-	-
2	PSI	66	43	23	348	65	283	394	48	346
	HANDI	64	18	45	364	18	346	339	18	321
	Dineutron	-	-	-	-	-	270	-	-	-
3	PSI	21.6	18.7	2.9	54	22	32	70	19	51
	HANDI	17.2	15.1	2.1	35	15	20	39	15	24
4	PSI	1.0	0.97	0.03	1.5	1.1	0.4	1.5	1.0	0.5
	HANDI	0.95	0.65	0.30	2.4	0.7	1.8	2.3	0.7	1.6
	Rem counter	-	-	-	-	-	1.2	-	-	-

The results obtained with this new definition are summarised in Table 1. Comparison between both TEPC systems leads to the same remarks as given previously. Total dose equivalent is increased by 13 to 28% and the neutron part of dose equivalent is increased by 56% at position 3.

CONCLUSION

This work has shown that even though the HANDI system uses only 16 channels for data acquisition, its results were comparable to the more complex laboratory system of PSI. The measured dose equivalent rates agreed fairly well. At locations with low neutron dose rates, however, the precision of measurements and the agreement in neutron dose equivalent are worse. Differences in the photon and neutron dose fractions can be attributed to the different evaluation procedures. The separation between photon and neutron fraction with a fixed threshold event size is not adequate for such radiation environments. The somewhat more sophisticated fitting method seems to be more appropriate. The choice of the separation method should be governed by the shape of the measured microdosimetric distribution.

Contrary to the commercially available rem counter and Dineutron, the TEPC systems are able to measure and assess dose equivalent rates, both the neutron and photon parts, in unknown mixed fields. The lower readings of the rem counter and the Dineutron at position 1 and 2 illustrate the different responses of the investigated monitors in an environment of a high energy accelerator, where neutrons with energies far above 20 MeV are expected. In order to establish the TEPC detectors as routine monitoring devices around high energy accelerators further investigations have to be performed. Particularly, a comparison of the measured dose equivalent with a more sophisticated device would give a better appreciation of the TEPC detectors.

REFERENCES

1. Andersson, I. Ö. and Braun, J. *Neutron Dosimetry.* STI/PUB/69 (Vienna: IAEA) Vol. 2, p. 87 (1963).
2. Cosack, M. and Lesiecki, H. *Dependence of the Response of Eight Neutron Dose Equivalent Meters with Regard to the Energy and Direction of Incident Neutrons.* In: Proc. 4th. Symp. on Neutron Dosimetry, EUR 7448 (Luxembourg: CEC) pp. 407-417 (1983).
3. Mourgues, M., Carossi, J. C. and Portal, G. *A Light Rem-Counter of Advanced Technology.* In: Proc. Fifth Symp. on Neutron Dosimetry (Luxembourg: CEC) Vol. 1 pp. 387-401 EUR 9762 (1984).
4. Hunt, J. B., Champlong, P., Chemtob, M., Kluge, H. and Schwartz, R. B. *International Intercomparison of Neutron Survey Instrument Calibrations.* Radiat. Prot. Dosim. **27**, 103-110 (1989).
5. Kunz, A., Menzel, H. G., Arend, E., Schuhmacher, H. and Grillmaier, R. E. *Practical Experience with a Prototype TEPC Area Monitor.* Radiat. Prot. Dosim. **29**(1/2), 99-104 (1989).
6. ICRP. *Data for Protection against Ionizing Radiation from External Sources.* Publication 21, (Oxford: Pergamon Press) (1971).
7. Arbel, A., Booz, J., Müller, K. D. and Neuhaus, H. *Development of a Portable Microdosimetric Monitor, Covering a Dynamic Range of 106 above Noise Level.* IEEE Trans. Nucl. Sci. **NS-31,** 691-696 (1984).
8. Barth, C. and Wernli, C. *Putting Microdosimetry to Work: The Measurement Campaign of 1987 at KKG, KKL and PSI.* Radiat. Prot. Dosim. **23**, 257-260 (1988).
9. Morstin, K., Dydejczyk, A., Cartier, F. and Wernli C. *Tissue Equivalence of Materials for Possible Use in Area Monitoring of Neutron-enriched Radiation Fields around High-energy Accelerators.* EIR-Report 594 (1986).
10. Kunz, A. *An Easy-to-operate Portable Pulse-height Analysis System for Area Monitoring with TEPC in Radiation Protection.* Nuc. Instrum. Methods Phys. Res. (in press).
11. Menzel, H. G., Schuhmacher, H. and Cartier, F. *Radiation Protection Dosimetry of Neutrons and Photons at a High-energy Accelerator using a Low Pressure Proportional Counter.* In: Sixth Eur. IRPA Conf., Cologne, FRG 1986).
12. ICRP. *1990 Recommendations of the International Commission on Radiological Protection.* Publication 60. Ann. ICRP **21** (1-3) (Oxford: Pergamon Press) (1991).

MEASUREMENT OF NEUTRON DOSE EQUIVALENT AND PENETRATION IN CONCRETE FOR 230 MeV PROTON BOMBARDMENT OF Al, Fe, and Pb TARGETS

J. V. Siebers[†], P. M. DeLuca, Jr[†], D. W. Pearson[†] and G. Coutrakon[‡]
[†]Department of Medical Physics, University of Wisconsin – Madison
Madison, WI 53706, USA
[‡]Loma Linda University Medical Center
Loma Linda, CA 92354, USA

Abstract — Secondary neutron production from protons striking accelerator beam delivery components and the patient constitute the principal radiation hazard for 70–300 MeV accelerators used in proton radiation therapy. Because of the large mean free path of these high energy neutrons, neutron attenuation requires massive shields. To this end, we measured neutron dose as a function of emission angle and depth in concrete for the radiation environment produced by 230 MeV protons striking stopping targets of aluminium, iron, and lead. By using microdosimetric instrumentation, dose equivalent values were deduced. From these data, dose equivalent penetration as a function of depth in concrete and neutron emission angle were determined. Neutron production was found to vary rapidly with emission angle, while differences in dose equivalent values per incident proton as a function of depth and angle depended only slightly on target material.

INTRODUCTION

Fermi National Accelerator Laboratory recently engaged in the design, development, and construction of a variable energy, 70 to 250 MeV, proton synchrotron for proton radiation therapy. This was in support of the initiative by Loma Linda University Medical Center to perform heavy ion radiation therapy with protons. Beyond the complex problems associated with development of this accelerator and beam delivery system, the design of the facility radiation shield was a major uncertainty and ultimately the most expensive single component. In a large part, the scarcity of experimental data for thick target neutron production and subsequent dose attenuation in concrete resulted in a conservative and costly shield. As a consequence of this design difficulty, a unique opportunity arose to investigate the shielding needs for such facilities. While the accelerator was under performance testing at FermiLab, sufficient time was made available to conduct appropriate radiation attenuation measurements under controlled conditions.

A brief summary of the more important experimental results is presented. A complete discussion may be found in Siebers et al[1,2].

EXPERIMENTAL PROCEDURE

Neutrons were produced by 230 MeV proton bombardment of stopping aluminium, iron, and lead targets. The region surrounding the targets was encased in a concrete shield. Detector positioning tubes were located at various depths in the shield at angles of 0, 22, 45, and 90 degrees with respect to the proton direction. Unused detector tubes were filled with removable concrete plugs. Hence, a homogeneous concrete environment was established during data acquisition and is shown schematically in plan view in Figure 1.

Microdosimetric instruments were employed to investigate the gross features of the neutron spectrum and to determine absorbed dose, quality factor, and dose equivalent values. Low pressure spherical proportional counters, constructed of A-150 tissue-equivalent plastic, determined the absorbed dose by measuring the energy deposited in the counter gas by neutron generated charged particles produced in the detector walls and surrounding concrete. To encompass the range of absorbed dose encountered at different concrete penetrations, detectors of 10^3, 10^5, and 10^6 mm^3 nominal volume were used. A propane-based

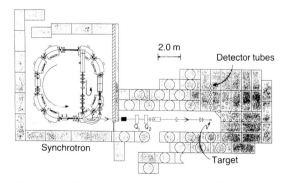

Figure 1. Schematic of concrete shield, neutron producing target, and detector locations. Q_1 and Q_2 are two quadrupole focusing elements to ensure all protons strike the target.

tissue-equivalent gas mixture filled the 1×10^3 mm^3 counter to 16 kPa and the 10^5 and 10^6 mm^3 counters to 1.6 kPa. In this manner unit density tissue thicknesses of 2 and 4 µm were simulated[3]. The signal from each counter was processed by two amplifiers with gains adjusted to span a dynamic range in energy deposition of 1:10000. Amplified linear signals were digitised, multiplexed, and stored on a microcomputer for display and real-time analysis.

Energy depositions in the counter gas by α particles from an internal ^{244}Cm source served to provide a dose calibration. A precision pulser related the different amplifier gains to the dose calibration. For those counters without internal sources, the location of the Bragg edge provided a dose calibration. This spectral feature is due to the sharp decrease in the frequency of energy depositions above the value corresponding to protons with an energy near the maximum in the proton stopping power. Protons of greater or lesser energy have lower stopping power and deposit less energy in the counter gas. From these calibrations, energy depositions in the gas were expressed as event size Y in units of keV.µm^{-1}. Since quality factors and dose equivalent values were also determined, dose spectra as a function of event size were converted to units of lineal energy by noting that $y = \frac{3}{2} Y$. As suggested by the ICRU[4], the quality factor Q is related to a lineal energy y by

$$Q(y) = \frac{5510}{y} [1 - \exp(-5 \times 10^{-5} y^2 - 2 \times 10^{-7} y^3)] \quad (1)$$

The dose equivalent is then given as

$$H = \int D(y) Q(y) \, dy \quad (2)$$

During data acquisition, three detectors located at different angles and depth in concrete were operated simultaneously. Using recommended techniques for the conversion of dose to dose equivalent[5], spectra were integrated to determine dose equivalent per incident proton values at all measured depths and angles. In this work, it is assumed that the conversion factor between dose to the tissue-equivalent counter gas and dose to the tissue-equivalent counter wall is unity, as is the conversion from dose to the A-150 tissue-equivalent wall to muscle.

Proton fluence on target was monitored by a transmission ion chamber positioned immediately before the target and one at the exit from the accelerator beam transport line. Multiwire proportional counters, also located proximal to the target and near the accelerator exit, determined the beam location and lateral dimensions. This ensured that the proton beam struck only the target. After traversing the various foils and beam monitoring devices, the beam energy at the target entrance face was 230 MeV.

During the irradiation phase, 10^{14} protons bombarded the Al, Fe, and Pb targets. A total of 171 two gain energy deposition spectra were acquired, 156 in the concrete shield and 15 measurements in air. At each location in the shield, several measurements were made with different detectors to minimise systematic errors. The maximum, minimum, and average dose rates encountered were 0.17, 3.25×10^{-6}, and 6.75×10^{-3} Sv.h^{-1} respectively.

RESULTS

From the event size spectra, total dose, mean quality factor, and dose equivalent values were determined. An example of the spectral data is given in Figure 2, which shows event-size weighted dose spectra measured at 0° and all depths for bombardment of the Al target. Dose values are normalised to the number of incident protons and the distance between target centre and the detector. These spectra are characterised by fast recoil protons producing energy depositions near 5 keV.µm^{-1}, corresponding to energetic neutron interactions in the media surrounding the detector. Neutron production intensity and energy distribution were observed to vary with angle.

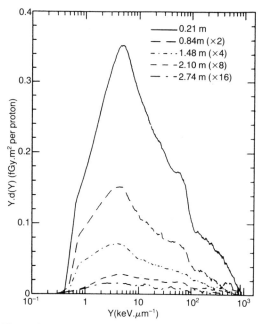

Figure 2. Event size weighted dose spectra observed at 0° and each depth into the shield for 230 MeV protons incident upon an Al target.

Figure 3 shows plots of event size weighted dose spectra acquired at 90° and all depths measured for the Al target. The neutron dose production is substantially reduced compared to 0° data. The spectra at shallow depths indicates a preponderance of lower energy neutrons as indicated by the increase frequency of events near the proton Bragg edge at 100 keV.µm^{-1}. At greater penetrations, the 90° neutron spectrum hardens and the dose distribution more closely resembles that observed at 0°. While the neutron dose spectra observed at significant depths in concrete were similar for all targets, this was not the case at small penetrations or in air. Figure 4 shows event size weighted dose spectra for each target measured at 90° in air. Relatively fewer events due to fast protons are observed, while enhanced production of lower energy neutrons occurs. This is particularly apparent for the lead target in-air measurements. These 0° and 90° spectra suggest that neutron production consists of two processes: (1) an energetic component that is forward peaked and seemingly target independent, and (2) an evaporative component that is target dependent and is largely isotropic.

From these spectra, event size weighted dose equivalent values (H) were calculated. After normalising to the target-to-detector distance (R) and the proton fluence, R^{2}H values were plotted against concrete thickness and fitted to an exponential. This served to determine the characteristic attenuation length for the dose equivalent for these neutrons. Figure 5 shows the results for the aluminium target. Beyond the attenuation

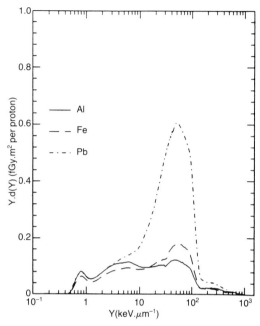

Figure 4. Event size weighted dose spectra observed in air at 90° for each target with 230 MeV protons incident.

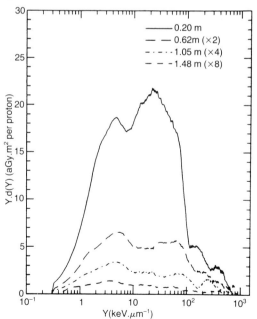

Figure 3. Event size weighted dose spectra observed at 90° and each depth into the shield for 230 MeV protons incident upon an Al target.

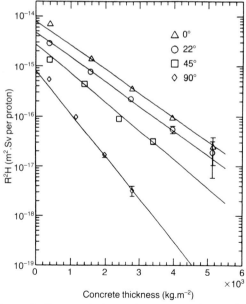

Figure 5. Dose equivalent values measured at various depths in concrete and at several angles for 230 MeV proton bombardment of the aluminium target. Values are normalised to the target-to-detector distance and to the number of protons.

Table 1. Summary of attenuation and dose equivalent production parameters measured using an aluminium target, deduced from the HETC-DO calculations, and using the estimation methods of the IAEA and Tesch. For the IAEA data, a concrete density of 2300 kg.m^{-3} was assumed.

	Production term (fSv.m^2.p^{-1})			
Angle	Measured	HETC-DO	IAEA	Tesch
0°	7.9± 0.5	37.2	8.7	–
22°	4.7± 0.5	12.9	–	–
45°	2.72.0.18	4.3	–	–
90°	0.89±0.05	0.49	1.6	0.45

	Attenuation length (kg.m^{-2})			
Angle	Measured	HETC-DO	IAEA	Tesch
0°	900±20	1150	930	–
22°	880±30	1020	–	–
45°	750±20	920	–	–
90°	506±8	790	730	760

Table 2. Concrete composition given in mass fractions.

Element	Ordinary	Present	HETC-DO
Hydrogen	0.0123	0.0170	0.0028
Carbon	0.0	0.0356	0.0
Oxygen	0.5351	0.5462	0.5236
Sodium	0.0049	0.0026	0.00
Magnesium	0.0017	0.0342	0.0
Aluminium	0.0324	0.0375	0.0094
Silicon	0.3118	0.1705	0.4211
Potassium	0.0083	0.0094	0.0
Calcium	0.0737	0.1276	0.0
Iron	0.0198	0.0193	0.0431

length, the intercept of such plots gives the characteristic neutron dose equivalent production. Similar results were observed with other targets.

CONCLUSIONS

Measured results may be compared to prior calculations and measurements. Considerable useful information for shielding purposes has been published by Alsmiller *et al* using the Monte Carlo technique High Energy Transport Code (HETC)[6–10]. These calculations are founded in the internuclear cascade model developed by Bertini[11]. More recently, Hagen *et al* applied HETC to estimate the shielding requirements for the Loma Linda Facility[12,13]. Braid also estimated dose equivalent production and attenuation in concrete and the results were summarised in a recommendation by the IAEA[14]. Finally, Tesch compiled lateral shielding data and calculations for 50 to 1000 MeV proton accelerators and estimated values of attenuation lengths and neutron dose equivalent production[15].

Table 1 compares our measured attenuation lengths and dose equivalent yield values with the results determined from the work of Hagen (labelled HETC-DO) and values reported by the IAEA and Tesch. Dose equivalent yields based on the HETC-DO result are substantially larger for forward emission angles. The HETC-DO determined attenuation lengths are also somewhat larger. Certainly some of this difference could be explained by the difference in the concrete compositions. The measured concrete composition used in this work as well as that used in the HETC-DO calculation are listed in Table 2 along with that for ordinary concrete. The IAEA and Tesch results for production and attenuation are in reasonable agreement with the present measurements.

The present work provides considerable shielding information for the shielding design of future accelerators employed in this energy region.

ACKNOWLEDGEMENT

This work was supported in part by the US Department of Energy through grant DE-FG02-86-ER60417 as well as by Fermilab and Loma Linda University Medical Center.

REFERENCES

1. Siebers, J. V. *Shielding Measurements for a 230-MeV Proton Beam*. Doctoral Thesis, Department of Medical Physics, University of Wisconsin-Madison (1990).
2. Siebers, J. V., DeLuca Jr, P. M. and Pearson, D. W. *Shielding Measurements for 230 MeV Protons*. Nucl. Sci. Eng. (submitted 1991).
3. Srdoc, D. *Experimental Technique of Measurement of Microscopic Energy Distribution in Irradiated Matter Using Rossi Counters*. Radiat. Res. **43**, 302 (1970).
4. International Commission on Radiation Units and Measurements. *The Quality Factor in Radiation Protection*. Report 40 (Bethesda, MD: ICRU Publications) (1986).
5. International Commission on Radiation Units and Measurements. *Microdosimetry*. Report 36 (Bethesda, MD: ICRU Publications) (1983).
6. Alsmiller, Jr R. G., Leimdorfer, M. and Barish, J. *Analytical Representation of Nonelastic Cross Sections and Particle-Emission Spectra from Nucleon–Nucleus Collisions in the Energy Range 25 to 400 MeV*. ORNL-4046 ORNL (April 1967).

7. Alsmiller, Jr, R. G., Barish, J., Boughner, R. T. and Engle, W. W. *Shielding Calculations for a 200 MeV Proton Accelerator.* ORNL-4336 (Dec. 1968).
8. Alsmiller, Jr, R. G. and Hermann, O. W. *Calculation of the Neutron Spectra from Proton-Nucleus Nonelastic Collisions in the Energy Range 15 to 18 MeV and Comparison with Experiment.* Nucl. Sci. Eng. **40**, 254 (1970).
9. Alsmiller, Jr, R. G., Santoro, R. T. and Barish, J. *Shielding Calculations for a 200 MeV Proton Accelerators and Comparisons with Experimental Data.* ORNL-TM-4754 February 1975 (also *Accelerators* **7**, 1 (1975)).
10. Alsmiller Jr, R. G., Barish, J., Barnes, J. M. and Santoro, R. T. *Calculated Neutron Production by 190 to 268 MeV Protons in a Water-Cooled Tantalum Target.* Nucl. Sci. Eng. **80**, 452 (1983).
11. Bertini, H. W. *Low-Energy Intranuclear Cascade Calculation.* Phys. Rev. **131**, 1801 (1963), with erratum Phys. Rev. **138**, AB2 (1965). ORNL-TM-1225 1965. ORNL-TM-3844 1966. Trans. Am. Nucl. Soc., Washington DC, Nov. 15-18 p. 634 (1965).
12. Hagen, W. K., Colborn, B. L. and Armstrong, T. W. *Radiation Shielding Calculations for the Loma Linda Proton Therapy Facility.* SAIC 87/1072 (1987).
13. Hagen, W. K., Colborn, B. L., Armstrong, T. W. and Allen, M. *Radiation Shielding Calculations for a 70-250 MeV Proton Therapy Facility.* Nucl. Sci. Eng. **98**, 172 (1988).
14. International Atomic Energy Agency. *Radiological Safety Aspects of the Operation of Proton Accelerators.* Technical Reports Series No. 283 (Vienna: International Atomic Energy Agency) (1988).
15. Tesch, K. *A Simple Estimation of the Lateral Shielding For Proton Accelerators in the Energy Range 50 to 1000 MeV.* Radiat. Prot. Dosim. **11**(3), 162-167 (1985).

HZE COSMIC RAYS IN SPACE. IS IT POSSIBLE THAT THEY ARE NOT THE MAJOR RADIATION HAZARD?

J. F. Dicello
Department of Physics, Clarkson University
8 Clarkson Ave, Potsdam, NY 13699, USA

INVITED PAPER

Abstract — The majority of hadrons in space are protons. Galactic cosmic rays have a small component of high energy heavy ions (HZE) which can be highly effective in initiating biological damage. That component is frequently assigned a large quality factor. Earlier risk analyses have indicated that individual heavy ions could be as much as 10,000 times more likely to induce cancer than protons. Although such large values are somewhat unusual and have been called into question, heavy ions do have the potential for being a major source of radiation risks to personnel in space. Nevertheless, estimates of such risk have large uncertainties associated with them. Previous procedures for establishing risk estimates are examined in the light of recently published microdosimetric spectra for heavy ions and protons. We conclude that higher quality factors appear to be unwarranted at this time. In fact, it is suggested that the present data do not exclude the possibility that biological risks from the direct ionisation density of primary galactic heavy ions may be less than that from protons when secondary delta rays, neutrons, mesons, heavy ions, and recoils from the protons and HZE particles are taken into account. Relatively straightforward ground based experiments are proposed to aid in addressing the issue.

INTRODUCTION

Assessments of radiation risks associated with personnel in space have proven not to be amenable to simple approaches. The problem is exacerbated by the fact that almost every known type of radiation is present with varying intensities and with varying environments. The primary particles, for example, produce a wealth of secondaries including neutrons and pions when they interact with a spacecraft or even with its occupants[1]. Measured intensity levels and known variations in dose rates could result in acute exposures to relatively high doses or significant chronic exposures over years. Research related to evaluating such effects is difficult, severely limited, and expensive. Most human data result from large single dose exposures or epidemiological studies with large uncertainties. Many of the particles thought to be relevant for risk assessments for space work have sufficiently high energies and atomic numbers that they are not easily available to ground-based researchers, if at all. The situation is sufficiently complex that risks must necessarily be represented on occasion by average parameters such as quality factors or dose equivalents, with relatively little biological data available even from cell and animal models. The variability of the radiation fields, their compositions, and their biological effects result in risk coefficients which can vary over two or three decades. These large uncertainties translate into unnecessary risks or high costs.

Evaluations of radiation hazards for lunar habitation and planetary exploration have been particularly difficult. Solar particle events, for example, are capable of delivering considerably more dose than HZE particles, although it should be kept in mind that the highest anticipated doses are to the skin. The consequences of such exposures could be tolerated better than those for whole-body doses of the same level and should not be evaluated by the same criteria as whole-body exposures of the same magnitude. Moreover, solar events are cyclic phenomena and solar cycles have periods of quiescence sufficiently long to allow interplanetary travel during average lulls. These minima of solar activity are sufficiently regular for the probability of a solar event during them to be relatively low with a corresponding reduction in risk. In contrast, HZE particles are always present, are subject to less variation than solar events, and are resistant to shielding. Such particles are known to have high relative biological effectiveness (RBE) relative to both photons and protons, although they can represent less than one per cent of the fluence because of the high linear energy transfer (LET). A recent calculation[1] suggested that potentially damaging neutrons and heavy charged secondaries from high energy protons may be present in levels comparable with those of HZE particles.

At least one conventional LET based risk assessment for an iron cosmic ray has produced a dose equivalent 13,000 times that for protons[2,3]. This resulted in the conclusion that a cosmic iron ion carries a risk for cancer which is correspondingly greater than that from a proton. Yet Fry and Lett[4] expressed concern that this value gives the impression that such dangers are 'not only frighteningly large but known with confidence'. In their rebuttal, they noted that measured RBEs for heavy ions are only in the range of 1 to 30. Letaw et al[2,3] indicated that the difference between the calculations and measurements was at least partially due to quality factors of 20 which were too conservative for certain acute exposures, i.e., too high. Since 1988 several relevant publications[5–10] have appeared which have stimulated renewed discussions among some researchers. This issue is relevant to space endeavours, and it is related to the applicability of LET, quality factors, and dose equivalents and to those procedures which employ them. This paper offers a potential explanation for apparent discrepancies while addressing the general issue of the significance of HZE particles.

BACKGROUND

An apparent dilemma can arise in the following way:
The dose equivalent, H, is

$$H = Q D \qquad (1)$$

where Q is the average quality factor and D is the macroscopic dose deposited. Recommended quality factors for energetic iron ions are in the neighbourhood of 20 to 25[11–13]. It is frequently assumed to be about 1 for high energy protons[13], although low energy protons are acknowledged to have higher values. The ratio of the dose equivalent for iron ions (particle 1) and protons (particle 2) would be

$$H_1/H_2 = (Q_1 D_1)/(Q_2 D_2) = (Q_1/Q_2)(D_1/D_2)$$
$$= (20/1)(D_1/D_2) \; H_1/H_2 = 20 \, (D_1/D_2). \qquad (2)$$

Letaw et al[2] went on to assume that the ratio of the doses could be approximated with the stopping powers or the LETs of the particles as:

$$D_1/D_2 \stackrel{?}{=} (K \, LET_1)/(K \, LET_2) = LET_1/LET_2 \qquad (3)$$

where K is the conversion factor from energy per unit path length to energy per unit mass (in tissue). The LET was assumed to be proportional to the square of the charge of the incident particles and this was assumed to be proportional to the square of the atomic number (this implies that the ions are fully stripped). It follows that:

$$D_1/D_2 \stackrel{?}{=} (Z_1/Z_2)^2 \qquad (4)$$

which, in the case of iron and protons, becomes

$$D_1/D_2 \stackrel{?}{=} (26/1)^2 = 676 \qquad (5)$$

and the ratio of the dose equivalents becomes

$$H_1/H_2 \stackrel{?}{=} 20(D_1/D_2) = 20(676) = 13,520 \qquad (6)$$

This last factor was then interpreted to mean that relativistic iron ions carry a risk of cancer 13,000 times greater than that of protons[3]. Herein lies an apparent dilemma, because measured RBEs tend to be below 30.

PRESENT ANALYSIS

It is informative to note that ICRU Report 40[12] recommended a quality factor for low energy protons of 25, not 1 as we have been assuming and which is generally assumed for space activities[13]. This higher value was intended to apply to low energy protons, because of their increased ionisation density as they approach the end of their range. Although the majority of events from protons are not as effective as heavier ions and their biological consequence is close to that for photons, there is nevertheless this small percentage of events from the secondaries which can be as effective as those from HZE particles. Furthermore, while the secondaries tend to increase the effectiveness of the protons, the same processes generally reduce the effectiveness of the HZE events[10].

Microdosimetry

With microdosimetric data for both protons[14] and heavy ions[10] it is possible to evaluate the relative biological equivalence of these secondaries without invoking an absolute value for a quality factor at all. Instead, it can be assumed, for example, that equivalent amounts of events of the same linear energy will produce equivalent effects (although this assumption is likewise subject to legitimate criticism, because the fractional contributions are dependent on target size and other geometrical factors[15]). In Figure 1 there is a decided separation in the microdosimetric spectra

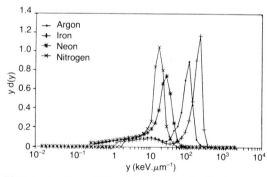

Figure 1 The differential dose distribution per logarithmic interval of lineal energy, y, as a function of the lineal energy for energetic heavy ion beams of nitrogen, neon, argon and iron (from Ref. 10). The distributions are plotted such that the area between two values of lineal energy is proportional to the fractional contribution to the dose from that region.

for the heavy ions at about 30 keV.μm^{-1} where the low LET delta rays and photons produce their maximum ionisation per unit volume. The distributions below 30 keV.μm^{-1} are similar in shape to that of a photon or electron beam. For simplicity, let us assume that events below about 30 keV.μm^{-1} produce equivalent effects at equivalent doses with an RBE (or a quality factor) of 1. Summing over all values above 30 keV.μm^{-1} (this implies a constant quality factor in that region and no saturation) allows the question, what percentage of high lineal energy dose would be produced by the protons and what percentage would be produced by the heavy ions, to be asked. By integrating the microdosimetric spectra, a value of about 0.2% for the percentage of the fluence that comes from events above 30 keV.μm^{-1} for protons of a couple hundred MeV is obtained. The corresponding percentage for the iron ions is about 10% in the wall-less case. The number of events per mGy calculated from the microdosimetric spectra are presented in Table 1 for different ions and site sizes.

For the sake of comparison with earlier studies, assume there is about one high LET ion for every 1000 protons and low LET ions as did Letaw et al[2,3]. Further assume that the ratio of protons and their equivalents to iron ions is about 4000 to one, and about 400 times as many protons per mGy of absorbed dose (the ratio of the stopping powers). Then the ratio of number of events above 30 keV.μm^{-1} is about:

[(0.016 events per mGy)(0.002)(4000 protons/ 0.043 protons per mGy)]

÷ [(0.00025 events per mGy)(0.1)(1 iron/0.00011 iron per mGy)]=13   (7)

Table 1. Some typical average number of events, N, in a specified volume per mGy of absorbed dose for ions as a function of energy (based upon data from Dicello et al[10,14]).

Particle Type	Energy (GeV)	Site diam (μm)	Type	Average N (×10^{-3})
Protons	0.2	2	Walled	16
Protons	Bragg	2	Walled	<6
Carbon	5	2	Walled	3.4
Neon	11	2	Walled	0.9
Neon	8	2	Walled	0.9
Neon	8	0.4	No wall	0.17
Neon	Bragg	2	Walled	2.2
Argon	23	2	Walled	0.7
Argon	23	0.4	No wall	0.12
Argon	Bragg	0.4	No wall	0.08
Iron	30	2	Walled	0.25
Iron	30	0.4	No wall	0.04
Iron	Bragg	2	Walled	0.24

In other words, about one tenth of the events above 30 keV.μm^{-1} would be produced by iron ions in comparison with protons. (The ratio is about 1/2 for the walled case.) On the basis of fluence, then, the correct ratio of protons to HZE particles may not be given by Equation 6.

The analysis implicitly assumes that multiple events are rare. Varma and Zaider[16] have noted that the single-event (or single-hit) approximations are valid only when the number of events per unit dose is much less than one. Table 1 shows the number of events per mGy in 2μm sites for various particles. The likelihood of multiple events for a few tens of mGy is small for most circumstances, even when the delta rays (measured by the wall-less detectors) are taken into consideration, a conclusion reached previously by Curtis and Letaw[7].

Equivalent doses

For equivalent fluences of events greater than 30 keV.μm^{-1}, the number of HZE particles would have to be greater than the fluence we assumed. To examine the same situation from a dose perspective, we used the Theory of Dual Radiation Action[17] to calculate relative survivals of a typical mammalian cell line with an RBE at a few Gy of about two, a reasonable value in comparison with the experimental data of Yang et al[18]. We found that it takes a physical dose of about 3 mGy of iron-type particles for an equivalent survival in comparison with a proton dose of 0.1 Gy, disregarding temporal effects and repopulation. At this survival or dose level we have a quality factor of about 30.

Quality factors

The ratio of the effective dose to mean lineal energies for iron to ^{60}Co would be approximately 40, and this ratio is considered in the linear–quadratic model to be an upper limit for the RBE at low doses. If an RBE for V79 cells is calculated, a value of about ten relative to ^{60}Co for cell survival at a few tens of mGy, in reasonable agreement with the data of Yang[18] is obtained. When the microdosimetric spectrum is used to calculate the quality factor directly, a value of 23 is obtained, in close agreement with the recommended value.

Fry and Lett[4] have already reviewed the literature with reference to the range of RBEs for HZE particles, as has NCRP 98[13]. In both cases an upper limit for RBEs for energetic heavy ions through iron is about 30. In many cases, for example certain acute effects in animal models, this may be too high by as much as a factor of ten or so. A value of twenty appears to be a

Linear energy transfer

Initial damage is usually assumed to be subcellular in origin, although the ultimate manifestation from such damage can be influenced by intercellular mechanisms and other temporal and environmental dose-modifying factors. Regardless of the size of the sensitive site, if the LETs of protons and iron ions are good approximations of the doses per particle, then the average number of events per unit dose should be inversely proportional to the LET. The ratio of the average number of events should be the ratio of the LETs, i.e., the average energy deposited in a site per event should equal the energy lost per incident particle traversing that site. This may seem to some readers a tautology. If so, an examination of the data tabulated in Table 1 will come as a surprise. Typical values of the average number of events per mGy calculated from experimental microdosimetric spectra are presented for a number of different ions at different site sizes. A comparison for example of values for 200 MeV protons for a 2 μm site with that for 535 MeV.amu^{-1} (30 GeV) iron ions show a ratio of not 676[2], but only 64. (For these initial energies, the ratio of the LETs is about 400.) There are, then, many more events in sites of 2 μm diameter per incident heavy ion and many less for the protons than expected on the basis of a calculation which uses the track-average LET. The actual number of events per particle is a strong function of the particle type, the site size, its shape, the delta ray distribution, and the energy distribution of the particles at the site.

Although damage in cells manifests itself with cross sections in the neighbourhood of square micrometres on occasion, there is little doubt that many of the relevant targets are molecules with much smaller crucial dimensions. As the site size decreases, the effectiveness of the heavier ions may very well decrease. There are few data for heavy particles as a function of site size, so it is difficult to illustrate this with experimental results. We have analysed the results of Kliauga and Dvorak[19] for photons to illustrate the point. We present the results in Table 2 to demonstrate the strong dependence of the number of events and their average energy deposited as a function of the dimensions of the sensitive site. One can draw an interesting conclusion by comparing the data in Tables 1 and 2. For small sites of dimensions of a fraction of a micrometre or less, such as the diameter of a DNA strand, low energy secondary electrons can be more effective biologically than HZE particles. This has been observed under specific conditions, so the conclusion is not new. The large fraction of events from delta rays from HZE particles, however, confounds attempts to establish a single effective cross section.

DISCUSSION AND CONCLUSIONS

The present analysis suggests that the nuclear byproducts, including the secondary neutrons, protons, and other charged secondaries and recoils, may be significant in the production of biological damage from galactic cosmic rays. This has been discussed in numerous previous publications[2,5–10,13]. Protons are roughly as effective per nucleon as heavy ions, for equal energies per nucleon, in producing nuclear secondaries. Because there are many more protons, they could be the major factor in macroscopic biological response for small initial sites. Any such response is strongly a function of the biological endpoint, the size of the initial targets, dose rates, and dose modifying facts. This discussion concentrated on sites with dimensions of a few micrometres, because that is where most of the data have been produced. Our argument, however, suggests that the differences between protons and heavier ions could become even less for smaller site sizes. In any case, the analysis is offered primarily at this point as a plausibility argument to stimulate

Table 2. The average number of events in the specified volume per mGy of absorbed dose for photons as a function of energy (based upon a analysis of data by Kliauga and Dvorak [14]).

Photon energy (keV)	Equivalent diameter of simulated sphere (μm)					
	0.24	0.48	0.96	1.9	3.9	7.7
12	7.3×10^{-5}	3.4×10^{-4}	1.5×10^{-3}	7.0×10^{-2}	4.05×10^{-2}	2.29×10^{-1}
25	1.3×10^{-4}	5.7×10^{-4}	2.4×10^{-3}	9.8×10^{-3}	4.68×10^{-2}	2.22×10^{1}
36	1.7×10^{-4}	7.6×10^{-4}	3.2×10^{-3}	1.27×10^{-2}	5.731×10^{-2}	2.64×10^{-1}
60	2.0×10^{-4}	9.3×10^{-4}	4.0×10^{-3}	1.7×10^{-2}	7.24×10^{-2}	3.23×10^{-1}
140	1.9×10^{-4}	9.0×10^{-4}	3.6×10^{-3}	1.49×10^{-2}	6.60×10^{-2}	2.88×10^{-1}
320	2.8×10^{-4}	1.65×10^{-3}	7.2×10^{-3}	2.9×10^{-3}	1.25×10^{-1}	4.90×10^{-1}
660	4.0×10^{-4}	2.5×10^{-3}	1.21×10^{-2}	4.98×10^{-2}	2.13×10^{-1}	8.23×10^{-1}
1250	5.1×10^{-4}	3.14×10^{-3}	1.16×10^{-2}	6.94×10^{-2}	2.91×10^{-1}	1.13

further investigations and discussions.

It is proposed that a definitive series of experiments would be the sequential irradiations of appropriate cell and animal models, first with protons and then with heavy ions. Total fluence ratios should be dictated by those expected in space. Two sets of irradiations should be performed, with and without the samples surrounded by shielding. Sufficiently sensitive biological models, such as the Harderian gland in rodents, exist, so that multiple events in sensitive sites can be kept small. If the responses are dominated by a per cent or so of HZE particles, the biological endpoints should be measurably altered even for acute exposures.

ACKNOWLEDGEMENTS

The author thanks Dr P. Deluca at the University of Wisconsin and Dr M. N. Varma of the Department of Energy for their interest in and support of this work. The support of Drs M. Jablin, W. Schimmerling, F. Sulzman, and R. White at NASA, Washington, and P. Russell at the American Institute of Biological Sciences in the preparation of this manuscript and the presentation upon which it was based is gratefully acknowledged. The impartial criticisms of Dr Stan Curtis, Lawrence Berkeley Laboratory are especially appreciated.

This work was partially supported by the University of Wisconsin, Madison, WI, and the Department of Energy, Washington, DC.

REFERENCES

1. Dicello, J. F., Schillaci, M. E. and Liu, L. *Cross Sections for Pion, Proton, and Heavy-Ion Production from 800 MeV Protons Incident upon Aluminum and Silicon*. Nucl. Instrum. Methods **B45**, 135-138 (1990).
2. Letaw, J. R., Silberberg, R., and Tsao, C. H. *Radiation Hazards on Space Missions*. Nature **330**, 709–710 (1987).
3. Letaw, J. R, Silberberg, R. and Tsao, C. H. *Reply to "Radiation Hazards on Space Missions in Space Put in Perspective"*. Nature **335**, 306 1988).
4. Fry, R. J. M.and Lett, J. T. *Radiation Hazards in Space Put in Perspective*. Nature **335**, 305-306 (1988).
5. Chuchkov, E. A., Pereslegina, N. V., Lyubimov, G. P., Tulupov, V. I., Ermakov, S. I., Kontor, N. N., Rozenthal, Yu. A., Kodobnov, V. B., Morozova, T. I., Pavlov, N. N., Rumkovskii, A. I., Rumkovskaya, G. A. and Ovsyannikova, M. A. *Investigation of Cosmic Rays on the Unmanned Interplanetary Spacecraft Venera-15 and Venera-16*. Kosmicheskie Issledovaniya **26**(5), 753-761 (1988).
6. Schimmerling, W. and Curtis, S. B. *Workshop Summary: Biomedical and Space-Related Research with Heavy Ions at the BEVALAC*. Radiat. Res. **119**, 193-204 (1989).
7. Curtis, S. B. and Letaw, J. R. *Galactic Cosmic Rays and Cell-Hit Frequencies Outside The Magnetosphere* Adv. Space Res. **9**, 293-298 (1989).
8. Badhwar, G. D, Braby, L. A. and Konradi, A. *Real -Time Measurements of Dose and Quality Factors on Board the Space Shuttle*. Nucl. Tracks Radiat. Meas. **17**(4), 591-594 (1990).
9. Wilson, J. W., Shinn, J. L, and Townsend, L. W. *Nuclear Reaction Effects in Conventional Risk Assessment for Energetic Ion Exposure*. Health Phys. **58**(6), 749-752 (1990).
10. Dicello, J. F., Wasiolek, M. and Zaider, M. *Measured Microdosimetric Spectra of Energetic Ion Beams of Fe, Ar, Ne, and C: Limitations of LET Distributions and Quality Factors in Space Research and Radiation Effects*. IEEE Trans. Nucl. Sci. **NS–38**, 1203-1209 (1991).
11. International Commission on Radiation Protection (ICRP). *Recommendations of the International Commission on Radiological Protection 1965*. Publication 9 (Oxford , Pergamon Press) (1966)
12. International Commission on Radiation Units and Measurements. *The Quality Factor in Radiation Protection*. ICRU Report **40** (Bethesda, MD: ICRU Publications) (1986).
13. National Council on Radiation Protection and Measurements. *Guidance on Radiation Received in Space Activities*. NCRP Report No. **98** (Bethesda, MD:NCRP Publications) (1989).
14. Dicello, J. F., Divadeenam, M., Wasiolek, M., Archambeau, J. O., Slater, J. M., Miller, D. M., Archambeau, M. H., Courtrakon, G. B., Moyers, M. F., Siebers, J. V., Young, P. E. and Robertson, J. B. *Quality Assurance for the Loma Linda Proton Therapy Facility: Microdosimetry*. Med. Phys **18**(3), 624 (abstract) (1990).
15. Zaider, M., Dicello, J. F. and Coyne, J. J. *The Effects of Geometrical Factors on Microdosimetric Probability Distributions of Energy Deposition*. Nucl. Instrum. Methods **B40/41**, 1261-1265 (1989).
16. Varma, M. N. and Zaider, M. *A Non-Parametric, Microdosimetric-Based Approach to the Evaluation of the Biological Effects of Low Doses of Ionizing Radiation*. In: Biophysical Modeling of Radiation Effects, (Bristol: Adam Hilger) (1992).
17. Kellerer, A. M. and Rossi, H. H. *The Theory of Dual Radiation Action*. Curr. Topics Radiat. Res. **8**, 85–158 (1972).
18. Yang, T. C., Craise, L. M., Mei, M. and Tobias, C. A. *Dose Protraction Studies with Low– and High-LET Radiationns on Neoplastic Cell Transformation in vitro*. Adv. Space Res. **6**, 137-147 (1986).
19. Kliauga, P. and Dvorak, R. *Microdosimetric Measurements of Ionization by Monoenergetic Photons*. Radiat. Res. **73**(1), 1–20 (1978).

INDIVIDUAL NEUTRON MONITORING — NEEDS FOR THE NINETIES

R. V. Griffith
International Atomic Energy Agency
PO Box 200, A-1400 Vienna, Austria

INVITED PAPER

Abstract — ICRP Report 60 has set a tone of change in radiation protection which will have an impact on neutron dosimetry for the next decade. The technical problems associated with existing neutron dosimetry techniques were not satisfactorily solved before its publication, and it only serves to add emphasis to the need for continued research. This comes at a time when a number of experienced neutron dosimetry researchers will be retiring, leaving an expertise gap. ICRP 60 is only one of the forces affecting this difficult area of radiation protection. Additional changes in recommendations from international standards organisations will have an impact on neutron dosimetry work. The impending publication of the third ICRU report on operational quantities, which will emphasise the practical aspects, does not deal with neutrons. Such guidance is vital for effective dosimetry, and priority must be given to its development. Moreover, weaknesses in ICRP Report 51 have been noted, and a revision can be expected.

INTRODUCTION

Earlier this year (1991), the International Commission on Radiological Protection (ICRP) published its long-awaited and controversial recommendations[1] which represent a revision of the guidance provided in publication 26[2]. Acceptance of Publication 60 principles will have widespread impact on radiation protection, including individual monitoring for neutrons. However, the recommendations that are likely to have the greatest impact are:

(1) Reduction of the annual dose limit averaged over 5 years from 50 to 20 mSv.
(2) Development of the radiation weighting factor, w_R, partly to replace the quality factor, Q, which is a point quantity.
(3) Assignment of increased values to w_R and Q, particularly for neutrons in the range 100 keV to 2 MeV (Figure 1).

From the metrological point of view, the International Commission on Radiation Units and Measurements (ICRU) has contributed, and will continue to contribute to the climate of change by recommendation of its 'new' set of operational quantities. These quantities were introduced and explained in more detail in ICRU Reports 39 and 43[3,4]. A third report, which provides more practically orientated guidance, has recently been published[5]. Unfortunately, however, for some very practical reasons the new report does not cover neutrons.

Changes in basic guidance for radiation protection always creates a 'domino effect', resulting in the need to revise existing standards and recommendations that were developed based on old principles. For example, largely as a result of the publication of ICRP 60, the International Atomic Energy Agency, together with the ILO, NEA and WHO, has begun a revision of the Interagency *Basic Safety Standards*[6]. However, partly because of the changes recommended by the ICRP, the revision process may take a few years.

For individual monitoring, it will be necessary to revise ICRP Publication 51[7]. The new radiation weighting factors must be incorporated into the fluence to equivalent dose conversion factors. In addition, it has been recognised that the current report lacks a summary of the sources of radiation exposures for mixed fields. Hopefully, any revision will also (1) address the confusion about the interrelationship of Q, w_R, LET and RBE, and (2) clarify the relationship between the ICRU

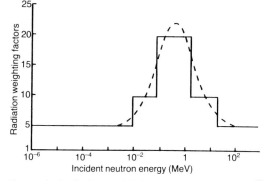

Figure 1. Radiation weighting factors for neutrons[1], with a smooth approximation (- - -).

operational quantities and effective dose, E. It is encouraging that the planned revision will be a cooperative effort between the ICRP and ICRU. This should provide important interagency consistency.

Although it is known that the number of workers monitored for neutrons is a small fraction of those occupationally exposed, it is unfortunate that there is no source of data on an international scale. This information would be valuable in assessment of the individual neutron monitoring problem, but such data is available only on a limited basis from a few national sources.

In this regard, the IAEA will be joining the Nuclear Energy Agency of the OECD and the CEC in a collaborative effort to develop an international Information System on Occupational Exposure (ISOE) for nuclear power reactors. Through ISOE, it may eventually be possible to obtain such data. It is also hoped that this effort can be expanded to include UNSCEAR and non-reactor information.

INDIVIDUAL MONITORING TECHNIQUES

Operational dosimetry systems

There has been little substantial change in the character and quality of operational dosimetry systems for individual monitoring since the previous symposium four years ago. The primary detection mechanisms in routine use are still TLD albedo detectors, NTA film and etched track detectors, primarily CR-39. The most significant changes are:

(1) Introduction of the operational use of bubble detectors, generally on a limited or experimental scale for special jobs.
(2) Refinement and introduction of CR-39 dosimetry by more users.
(3) General improvement in the quality of commercial CR-39.
(4) Continued, slow movement away from NTA film, particularly in power reactors.

Table 1 presents an admittedly subjective summary of the characteristics of the operational dosimetry systems, including bubble detectors. It should be noted that the lower limit of detection (LLD) reported by different users depends not only on the definition of LLD used, but the energy response and processing technique. For example, the 'minimum detectable' dose for ^{252}Cf reported from 25 participating systems in the recent multi-laboratory EURADOS–CENDOS experiment to study the response of etched track detectors to fast neutrons[8] ranged from approximately 0.02 mSv to about 1.5 mSv — a difference of a factor of 75!

The major problem presented by ICRP 60 will be a significant reduction in the LLD required of dosimetry systems. The need to detect 1/10 the annual dose limit will imply a minimum detectable dose for a service with a monthly dosemeter exchange of 0.17 mSv. Moreover, the increase in w_R for fast neutrons will make existing systems less sensitive by a factor of about 1.8. Even the albedo detector, prized for many years because of its sensitivity, will have difficulty meeting the challenge. Those that, by design, are wrapped in cadmium may not be sensitive enough for fission spectrum neutrons. It appears at this point that only bubble detectors will be sufficiently sensitive. However, in view of their cost, temperature and shock sensitivity, they do not at present represent a satisfactory general purpose personal neutron dosemeter.

Aside from the problems introduced by the ICRP recommendations, there are some basic questions about dosemeter performance. Two large international intercomparison programmes — the Oak Ridge National Laboratory Personnel Dosimetry Intercomparison Studies (PDIS)[9] and the joint multi-laboratory experiment organised by EURADOS–CENDOS to study the response of proton-sensitive etched track detectors to fast neutrons[8] — provide useful information on this question.

Sims, in a report to be published[10], finds that the accuracy of results reported in the PDIS is not

Table 1. Summary of advantages and disadvantages of operational neutron dosemeters. The LLD (lower limit of detection) and ULD (upper limit of detection) are approximate values that indicate the practical useful dose range of the dosimetry system.

	Albedo (TLD)	Bubble detectors	CR-39	NTA film
Energy response	Poor	Very good	Good	Fair
LLD (mSv)	0.005–0.2	0.005–0.02	0.02–0.3	0.3–0.8
ULD (mSv)	>100	1–10	>50	>100
Photon response	Yes	No	No	Some
Cost/detector	Moderate	High	Low	Low
Environmental sensitivity	Very little	Thermal, shock	Very little	Humidity, thermal
Ease of readout	Excellent	Good	Fair–Good	Tedious

increasing with time, and that for 'all dosimeter types in all PDIS, 67% of the measurements yielded results within ±50% of reference values'. He concludes from the PDIS results that 'It is obvious that the best dosemeter to use depends on the energy spectrum where the dose equivalent measurement is made.' Considering detectability, it is interesting to note that for the low dose exposures (0.5 mSv) for Health Physics Research Reactor (HPRR) spectra only 53% of track dosemeters had a non-zero response, while 96% on the participating albedo systems reported detectable exposures.

The EURADOS–CENDOS study is restricted to proton-sensitive track detectors. The first striking aspect of the study is that each of the 25 participating dosimetry systems essentially used a different recipe for processing and reading detectors. In some cases, the differences are relatively small and subtle, which leads one to question whether such differences are really significant. A quick look at the range of the minimum detectable doses shows that at least 10 participants had comparable results, using such different methods. In the interest of harmonisation, comparability of results and improvement of dosimetry, shouldn't we begin to standardise processing methods?

Developmental measurement techniques

In addition to the mechanisms already discussed, other possible solutions have been identified, and are currently under study:

Etched detectors with (n,α) radiators

Over the years, a number of researchers have suggested use of thin layers of material rich in ^{10}B or ^{7}Li as (n,α) radiators to balance the fast neutron response of CR-39, LR-115 or other etched track detectors[11,12]. However, perhaps the most thorough work to date has been recently reported[13] featuring a laminated detector (Figure 2) composed of five layers, including a 640 μm thick CR-39 track detector. The n,α radiator consists of a thin film (20 μm) of borated poly(vinyl alcohol). The alpha particles from the low energy neutron interactions not only increase the total track density, but it appears that the electrochemically etched alpha tracks are easily distinguished from the smaller recoil proton induced tracks. In concept, this also suggests that it may be useful in providing some very simple spectral information.

An optimum boron content has been shown to enhance the response to neutrons below 100 keV by a factor of ten and, for thermal neutrons, 100. Although the boron does not improve the fast neutron response, it does offer the possibility of a single, full energy range track detector.

Electronic dosemeters

There is currently a need for an active neutron dosemeter with an immediate indication or alarm to control short-term exposure[14]. Two basic detection mechanisms have been proposed for use in an electronic neutron dosemeter — tissue equivalent proportional counters (TEPC), and silicon surface barrier detectors.

TEPCs have been studied and developed relatively aggressively for a long time[15]. They have been valuable tools in characterising operational neutron fields. Moreover, they have been used for measurements in commercial aircraft and on satellites. The two major limitations for individual monitoring, however, are size and cost. Packaging in a size appropriate for use as an electronic dosemeter for individual monitoring is particularly difficult.

Various attempts have been made to develop an electronic dosemeter based on silicon surface barrier detectors for individual monitoring[16,17]. Limits of detection as low as 4 μSv for 8 h are possible[18]. However, limitations include cost and a relatively high energy threshold (about 1 MeV) which presents problems for use in applications where the neutron spectrum is degraded. There is also a significant angular dependence associated with this type of detector.

Optically stimulated luminescence (OSL)

The concept of adding TL material to hydrogenous matrices for fast neutron measurement is attractive and has been studied since the late 60s. However, because of the temperature required for TL readout and low melting temperatures of the hydrogen–containing plastics, this technique has

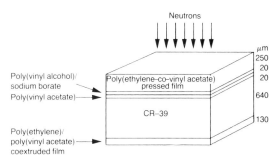

Figure 2. Schematic of CR-39 dosemeter with boron-doped layer as an n,α radiator for low energy neutron sensitivity enhancement[13].

not been particularly successful. Now, a low temperature technique has been developed[19,20] — cooled optically stimulated luminescence (COSL) — which may overcome this problem, resulting in a very sensitive dosemeter that is processed at low temperatures.

In concept, grains of OSL material are embedded in an hydrogenous matrix. As shown in Figure 3(a), the protons deposit nearly all their energy within the grain diameter, while the low LET electrons deposit energy more uniformly throughout the matrix. By microscanning individual grains (Figure 3(b)) with a laser, the differences in energy deposition would permit separation of the two types of radiation based only on the size of the pulse produced. Recently,

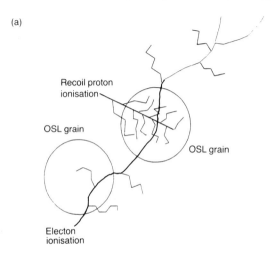

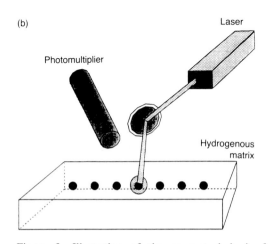

Figure 3. Illustration of the conceptual basis for neutron–photon discrimination using cooled optically stimulated luminescence (COSL)[21]. (a) Differential energy deposition for neutrons and photons and (b) schematic single grain readout technique.

promising work has been reported using CaF_2:Mn; however, neutron dosimetry using this technique will require considerable additional research.

Dynamic random access memories (DRAM)

The measurement of soft errors in DRAM[22,23] is attractive because of the analogy between an array of microscopic, neutron sensitive electronic elements and human cells. Neutron induced heavy charged particles can cause a random error to appear by generating a sufficient concentration of electrons near one of the electronic cells to change the stored charge, resulting in a bit flip. The range of energy necessary to cause the flip has been estimated to be 360 keV to 3.6 MeV. Because of the small size of the cell, it would be insensitive to photons. Recently, feasibility studies were reported using 64 K memory chips that demonstrated that a thermal neutron dose level as low as 25 μSv could be detected[24].

Hypersensitive TLD

As noted above, neutron dosimetry using direct proton recoil detection has not been successful. However, new high sensitivity TL materials such as LiF:Cu,Mg,P and Al_2O_3:C are up to 60 times more sensitive than TLD-100. The use of glow curve discrimination offers some interesting possibilities with these materials.

NEEDS FOR THE NINETIES

There are significant problems to be faced in individual neutron monitoring. Not all of them have been introduced by the publication of ICRP 60. The question is how to solve them in a systematic and organised way. The following, then, is a summary of my view of the Needs for the Nineties, if these problems are to be solved.

ICRP/ICRU guidance — completion and stability

It was noted in the introduction that the ICRP and ICRU are planning to provide some valuable new guidance related to individual neutron monitoring. ICRP Report 51 will undergo revision. The conversion coefficients and other information that it should contain will be necessary to the dosimetrist in meeting the challenge of establishing and maintaining an effective individual monitoring programme. A valuable element of the revision would be a clear explanation of the relationship between the various quantities, including Q, w_R, LET, RBE and the ICRU operational quantities.

In the next few years, the ICRU should be encouraged to finish its series on the operational quantities with a report on neutron measurement similar to the document on photons and electrons[5]. Perhaps this will be one outcome of the co-operative effort to prepare the Publication 51 revision.

Once this guidance has been provided, the most important need in standards development is stability. For many years there has been a succession of changing dosimetry standards. We need to stop these changes so that the real dosimetry system requirements can be established and solutions can be developed in a systematic manner.

People — experience, enthusiasm and creativity

Individual monitoring problems cannot be solved without qualified people. There has been a steady erosion of the number of available neutron dosimetry specialists with the skills and experience that are so important in solving the problems. This attrition is the result of a number of factors including promotion, retirement, and reassignment brought about by decreased funding or other programmatic requirements.

It is now very important that a core of qualified specialists be maintained and supported. Adequate, stable funding must be provided to meet the requirements of the new standards. This will be very necessary to maintain a cadre of capable neutron dosimetrists. In that regard, communications must be stimulated, in the form of workshops, conferences and intercomparisons. Networking — the informal communications system between specialists — is very important and should be encouraged. Many of us remember the neutron dosimetry community that developed in the 60s and 70s, stimulated by programmes such as the ORNL Accident Dosimetry Intercomparisons. This community was largely responsible for identifying practical solutions of the individual monitoring problems of that period.

New dosimetry mechanisms — sensitivity and accuracy

There is little doubt that no single detection technique can satisfactorily meet the challenge of the Nineties. In the long term, it is still necessary to find a detection mechanism that has sufficient sensitivity, appropriate energy response, and reasonable cost for deployment in large scale dosimetry systems. Other factors such as proper angular response are important, and will also have to be considered.

Use of existing resources — combination, optimisation and compensation

Since each of the current dosimetry techniques has limitations when used alone (Table 1), it is prudent to use a combination of techniques that compensates one weakness with another strength. A combination dosemeter has been proposed for a number of years[25,26]. TLD albedo offers the advantages of automation and sensitivity: therefore, the use of an albedo detector as a screening tool, together with etched track detectors having better fast neutron response characteristics is an attractive option. More recently the combination of an albedo detector with a bubble detector has also been described[27]. A number of dosimetry services avoid the use of combination detectors because of extra cost and, perhaps, effort. However, dosimetry system requirements have become more stringent and this alternative may become necessary until a fully satisfactory dosemeter can be developed.

Standardised dosimetry methods — harmonisation and cooperation

It was noted that the 25 participants in the multi-laboratory study of the response of proton-sensitive etched track detectors organised by EURADOS–CENDOS each used different processing techniques. A continuation of this practice simply prolongs the process of arriving at effective solutions to the individual monitoring problem. It is now time to analyse critically the results achieved using these different techniques and begin development of standardised methods so that dosimetry comparison and improvement can be approached more systematically. Perhaps this is a problem that ISO should be encouraged to address.

Field characterisation methods — simplicity and portability

The interpretation of dosemeter results can be improved significantly with knowledge of the characteristics of the neutron fields in operational areas. Spectral information can be used (1) in selecting a dosimetry system that is appropriate for the characteristics of the working environment, and (2) to assign appropriate energy dependent calibration factors. References are now available that present typical operational neutron spectra, and those that are characteristic of transport through various shielding materials[28–30]. Additional work in this area has been presented at this Symposium[31].

It is often important to make specific operational field measurements so that the spectral information can be obtained directly. When considering operational health physics neutron spectrum measurements, the vision of a set of large, heavy multispheres usually comes to mind. However, even two detectors can be used to obtain simple spectrometric information. This concept has been applied in the design of survey instruments which can be used directly to determine dose equivalent, or to improve interpretation of dosemeter readings. The design of these instruments has the additional advantage that the internal interpretation algorithms can be changed to match future changes in recommended flux-to-dose conversion coefficients.

The Dineutron consists of two polyethylene spheres, 107 mm (4.2 inch) and 63.5 m (2.5 inch) diameter, with a small cylindrical ^{3}He at the centre of each[32]. The ratio of count rates in the two detectors is a function of the spectrum, and is used to determine dose equivalent.

Another dual detector survey instrument has been designed that is intended specifically for low energy neutron measurements and field calibration of albedo dosemeters[33]. It features the use of two surface barrier detectors in contact with thin layers of ^{6}LiF. One detector is bare, while the other is located in the centre of a 76 mm (3 inch) diameter sphere of polyethylene. The readings give information on the thermal and intermediate components of the ambient dose equivalent.

The use of 'Bonner spheres' remains a common and valid field spectrometry technique. Personal computer based unfolding codes[34,35] make it possible to do desk-top processing that yields spectral results of a sufficient quality for radiation protection needs.

More compact and transportable systems are now being developed using liquid scintillators, ^{3}He proportional counters, or hydrogen proportional counters[36–38] to provide high resolution fast neutron spectra that supplement the low energy data that still must be obtained with moderated detector techniques. A complete spectrometry system including a rotating assembly of four ^{3}He detectors, microprocessor and unfolding software is now commercially available[39]. However, the price tag of nearly $120,000 is likely to prevent widespread sales.

Bubble detectors are available in different sensitivities. It is, therefore, possible to configure a simple spectrometer using an array of such detectors[40]. However, the statistical uncertainties and large variations in apparent responses between individual detectors may be a limitation.

Dosimetry intercomparisons — training, evaluation and improvement

Intercomparison programmes such as those at ORNL and supported by EURADOS–CENDOS provide a unique and vital opportunity to evaluate dosimetry systems, exchange information and train staff. It is encouraging to note that the first international Nuclear Accident Dosimetry Intercomparison in many years will be conducted at the Silene facility in France in 1993. Although they may be expensive, national and international programmes such as these are important and must be supported.

CONCLUSIONS

The number of radiation workers exposed to neutrons represents a small fraction of occupational exposures. However, in view of the differences in radiation weighting factors, and the technical difficulty in neutron measurement, the individual neutron monitoring problem is considerably more important than the simple numbers might suggest.

The challenge for individual neutron monitoring for the next decade is now becoming clear. Meeting the measurement requirements presented by current international recommendations will be a very demanding task. National competent authorities, regulatory bodies and management must recognise the magnitude of the technical issues and be prepared to provide appropriate support.

Organisations which set International Standards must be prepared to stabilise the guidance picture, and provide the clarification and details necessary properly to interpret, understand and implement their message.

Finally, it will be the responsibility of those charged with developing solutions to the dosimetry problems to work together in a spirit of co-operation, coordination and teamwork. We must be willing to discard our own, homegrown techniques and accept alternatives that help our community find a technical sound common ground. We must also work to stimulate commercial solutions to our problems that, in the long run, will provide efficient, cost effective technology for individual neutron monitoring.

REFERENCES

1. International Commission on Radiological Protection. *1990 Recommendations of the International Commission on Radiological Protection.* Publication 60 (Oxford: Pergamon) (1991).
2. International Commission on Radiological Protection. *Recommendations of the International Commission on Radiological Protection.* Publication 26 (Oxford: Pergamon) Reprinted (with additions) in 1987 (1977).

3. International Commission on Radiation Units and Measurements. *Determination of Dose Equivalents Resulting from External Radiation Sources*. Report 39 (Bethesda, MD: ICRU Publications) (1985).
4. International Commission on Radiation Units and Measurements. *Determination of Dose Equivalents from External Radiation Sources — Part 2*. Report 43 (Bethesda, MD: ICRU Publications) (1988).
5. International Commission on Radiation Units and Measurements. *Measurement of Dose Equivalents from External Photon and Electron Radiations*. Report 47 (Bethesda, MD: ICRU Publications) (1992).
6. International Atomic Energy Agency. *Basic Safety Standards for Radiation Protection*. Safety Series No. 9 (1982 Edition), Jointly Sponsored by IAEA, ILO, NEA(OECD), WHO (1982).
7. International Commission on Radiological Protection. *Data for Use in Protection Against External Radiation*. Publication 51 (Oxford: Pergamon) (1987).
8. EURADOS/CENDOS. *Response of Proton-Sensitive Etched Track Detectors to Fast Neutrons: Results of a Joint Multilaboratory Experiment*. EURADOS-CENDOS-Report 1990-01, Ed. H. Schraube (1990).
9. Swaja, R. E. and Sims, C. S. *Neutron Personnel Dosimetry Intercomparison Studies at the Oak Ridge National Laboratory: A Summary (1981-1986)*. Health Phys. **55**(3), 549-564 (1988).
10. Sims, C. S. *Neutron Personnel Dosimetry Intercomparison Studies*. In: Proc. Eleventh Department of Energy Workshop on Personnel Neutron Dosimetry, Las Vegas, Nevada, 4-7 June, 1991 (in press).
11. Oda, K., Michijima, M. and Miyake, H. *CR39-BN Detector for Thermal-Neutron Dosimetry*. J. Nucl. Sci. Technol. **24**, 129-134 (1987).
12. Matiullah, and Durrani, S. A. *A Mathematical Model for Thermal-Neutron Dosimetry Using Electrochemically Etched CR-39 Detectors with (n,p) and (n,α) Converters*. Nucl. Tracks Radiat. Meas. **15**, 511-514 (1988).
13. Koenig, M. F., Johnson, J. F., Huang, S. J. and Parkhurst, M. A. *CR-39 Personnel Neutron Dosimeters*. In: Proc. Eleventh Department of Energy Workshop on Personnel Neutron Dosimetry, Las Vegas, Nevada, 4-7 June, 1991 (in press).
14. Gibson, J. A. B. *Individual Neutron Dosimetry*. Radiat. Prot. Dosim. **23**(1-4), 109 (1988).
15. Booz, J., Dennis, J. A. and Menzel, H. (Eds) *Microdosimetry*. Proc. Tenth Symp. on Microdosimetry, Rome, Italy, 21-26 May 1989, Radiat. Prot. Dosim. **31**(1-4) (1990).
16. Falk, R. B. and Tyree, W. H. *Personnel Electronic Neutron Dosimeter*. In: Ninth DOE Workshop on Personnel Neutron Dosimetry. Las Vegas, Nevada, PNL-SA-10714, pp. 154-161 (1982).
17. Eisen, Y., Engler, G., Ovadia, E., Shamai, Y., Baum, Z. and Levi, Y. *A Small Size Neutron and Gamma Dosemeter with a Single Silicon Surface Barrier Detector*. Radiat. Prot. Dosim. **15**, 15-30 (1986).
18. Gibson, J. A. B. *Personal Alarm Neutron Dosemeters*. Radiat. Prot. Dosim. **10**, 197-205 (1985).
19. Miller, S. D., Endres, G. W. R., McDonald, J. C. and Swinth, K. L. *Cooled Optically Stimulated Luminescence in CaF_2:Mn*. Radiat. Prot. Dosim. **25**(3), 201-206 (1988).
20. Miller, S. D., Stahl, K. A., Endres, G. W. R. and McDonald, J. C. *Optical Annealing of CaF_2:Mn for Cooled Optically Stimulated Luminescence*. Radiat. Prot. Dosim. **29**(3), 195-198 (1989).
21. Miller, S. D. and Eschbach, P. A. *Neutron Dosimetry Using Stimulated Luminescence*. In: Proc. Eleventh Department of Energy Workshop on Personnel Neutron Dosimetry, Las Vegas, Nevada, 4-7 June, 1991 (in press).
22. Haque, A. K. M. M., Yates, J. and Stevens, D. *Soft Errors in Dynamic Random Access Memories — A Basis for Dosimetry*. Radiat. Prot. Dosim. **17**, 189-192 (1986).
23. Lund, J. C., Sinclair, F. and Entine, G. *Neutron Dosimeter Using a Dynamic Random Access Memory as a Sensor*. IEEE Trans. Nucl. Sci. **NS-33**(1), 620-623 (1986).
24. Haque, A. K. M. M. and Ali, M. H. *Neutron Dosimetry Employing Soft Errors in Dynamic Random Access Memories*. Phys. Med. Biol. **34**, 1195-1202 (1989).
25. Anderson, M. E. and Crain, S. L. *A Combination TLD–Film Personnel Neutron Dosimeter*. Health Phys. **36**(1), 76-79 (1979).
26. Griffith, R. V. and McMahon, T. *Development of an Operational Multicomponent Personnel Neutron Dosimeter/Spectrometer (DOSPEC)*. In: Proc. 6th Int. Cong. of the International Radiation Protection Association, Berlin, 7-12 May 1984. Vol. **III**. pp. 1239-1242 (1984).
27. Liu, J. C. and Sims, C. S. *Performance Evaluation of a New Combination Personnel Dosemeter*. Radiat. Prot. Dosim. **32**(1), 33-43 (1990).
28. International Atomic Energy Agency. *Compendium of Neutron Spectra and Detector Responses for Radiation Protection Purposes*. Technical Report Series No. 318 (Vienna: IAEA) (1990).
29. Buxerolle, M., Massoutie, M. and Kurdjian, J. *Catalogue de Spectres de Neutrons*. Commissariat A l'Enegie Atomique Report CEA-R-5398 (1987).
30. International Atomic Energy Agency. *Compendium of Neutron Spectra in Criticality Accident Dosimetry*. Technical Report Series No. 180 (Vienna: IAEA) (1990).

31. Siebert, B. R. L., Schraube, H. and Thomas, D. J. *A Computer Library of Neutron Spectra for Radiation Protection Environments.* In: Radiat. Prot. Dosim. **44**(1-4) 135-138 (1992). (This issue).
32. Mourges, M., Carossi, J. C. and Portal, G. *A Light Rem-counter of Advanced Technology.* In: Proc. 5th Symp. on Neutron Dosimetry. Eds H. Schraube, G. Burger and J. Booz (Luxembourg: Commission of the European Communities) EUR 9762, p. 387 (1984).
33. Harvey. J. R. and Hart, C. D. *A Neutron Survey Instrument which Gives Information in the Low Energy Neutron Spectrum and can be Used for Albedo Dosemeter Calibration.* In: Radiat. Prot. Dosim. **23**(1-4), 277 (1988).
34. Brackenbush, L. W. and Scherpelz, R. I. *SPUNIT, A Computer Code for Multisphere Unfolding.* PNL-SA-11645 (Pacific Northwest Laboratory, Richland, Washington) (1983).
35. Hertel, N. E. *BUNKI and Associated Plotting Routine, BUNKIPLT.* Health Phys. **50**(6), III [Software] (1986).
36. Brackenbush, L. W. *Field Neutron Spectrometer Using ^3He, TEPC, and Multisphere Detectors.* In: Proc. Eleventh Department of Energy Workshop on Personnel Neutron Dosimetry, Las Vegas, Nevada, 4-7 June 1991 (in press).
37. Clark, J. C. and Thorngate, J. H. *A Portable 0.5– to 16–MeV Neutron Spectrometer Using a Liquid Scintillator.* In: Proc. Eleventh Department of Energy Workshop on Personnel Neutron Dosimetry, Las Vegas, Nevada, 4-7 June 1991 (in press).
38. Posny, F., Chartier, J. L. and Buxerolle, M. *Neutron Spectrometry System for Radiation Protection: Measurements at Workplaces and in Calibration Fields.* Radiat. Prot. Dosim. **44**(1-4) 239-242 (1992) (This issue).
39. Bubble Technology Industries. *Rotating Neutron Spectrometer — ROSPEC.* Private Communication (1991).
40. White, B., Ebert, D. and Munno, F. *Use of the BD-100R as a Neutron Spectrometer Through Applied Pressure Variation.* Health Phys. **60**(5), 703-708 (1991).

PROPERTIES OF PERSONNEL NEUTRON DOSEMETERS ON THE BASIS OF INTERCOMPARISON RESULTS

E. Piesch, B. Burgkhardt and M. Vilgis
Karlsruhe Nuclear Research Centre, Health Physics Division
P.O.B. 3640 W-7500 Karlsruhe 1, FRG

Abstract — The results of various intercomparison experiments organised by Eurados–Cendos, ORNL and PTB have been used as a basis to compare the dosimetric properties of personnel neutron dosemeters such as various albedo dosemeter types, etched track detectors and a combination dosemeter with bubble detectors. The paper presents an actual review of different techniques currently under study and discussed as alternatives in routine monitoring. Data are given for the energy response and the lowest detectable dose as a function of neutron energy as well as the angular response and uncertainty of measurement in standard fields. For intercomparison experiments more realistic radiation fields are recommended; problems and consequences are pointed out.

INTRODUCTION

For more than 10 years, CR-39 etched track detectors with a flat energy response in the energy range of fast neutrons have, above all, been extensively investigated and refined by many laboratories in order to find the basis for a routine application[1–3]. Albedo neutron dosemeters, on the other hand, have been successfully used in personnel dosimetry for many years despite their extreme energy dependence. For routine monitoring, however, there is a need for an appropriate albedo dosemeter encapsulation and for calibration factors depending on application areas[4,5]. Highly sensitive bubble detectors, on the other hand, have found only limited application in intercomparison experiments and routine monitoring.

In order to investigate the dosimetric properties of routinely applied dosimetry systems, as well as experimental detectors still in the stage of development, intercomparison experiments are of special interest for the participating laboratories. On the other hand, the results from different dosimetry systems are available which have been irradiated under equal conditions. These data may therefore also be used to discuss the advantages and disadvantages of different dosimetric techniques. Although an interpretation of the results is restricted to those standardised calibration fields some observations may, nevertheless, be valid in general, such as statements about the energy dependence, the angular dependence, the lowest detectable dose and to some extent the uncertainty of measurement.

Intercomparison results of recent years, above all those of the KfK and the ORNL, are here analysed and interpreted in order to present an actual review of different techniques currently under study and discussed as alternatives in routine monitoring[1–10,12].

EXPERIMENTAL DETAILS AND DOSEMETER TYPES IN THE INTERCOMPARISONS

Most of the data to be discussed here are results of the following neutron intercomparison experiments:

(1) At the ORNL (1987-1991), personnel neutron dosemeters were irradiated yearly on slab phantoms using radionuclide sources[6]. The reference values were given in terms of H_{made}. For the irradiations the 9"/3" value and source information for every source were provided to participants.

(2) At the PTB (1988-1990), mainly universal albedo dosemeter types from government services in Germany were irradiated on slab phantoms using radionuclide sources and monoenergetic neutrons. The detector reading was corrected, however, for the albedo reading on the sphere phantom. The dose equivalents were expressed in terms of $H'(10)$[9,10].

(3) EURADOS–CENDOS (1987-1990) organised joint multilaboratory experiments for different CR-39 etched track detectors and etching procedures using mainly monoenergetic neutrons and the reference dose equivalent in terms of $H*(10)$[2].

Dosemeters

Results of three different types of albedo dosemeters with thermoluminescent detectors are available.

Type A

The KfK universal albedo dosemeter[4] used by four government services as the routine neutron dosemeter in Germany, consists of a boron-plastic encapsulation and two thermal neutron TLD-600 / TLD-700 detector pairs in one Alnor TLD card evaluated in the DOSACUS reader.

For the KfK albedo dosemeter the neutron radiation fields are, in general, divided into four application areas (N1 to N4), for which different calibration factors are used. The application areas of interest are heavily moderated neutron fields such as D_2O moderated ^{252}Cf sources (N1) and radionuclide sources such as ^{252}Cf sources (N3). The calibration factors for the different application areas are estimated by using calibration curves based on the reading ratio M(a)/M(i) of the albedo detector (i) and the thermal neutron detector for field neutrons (a). The actual calibration curves for the albedo dosemeter have been provided in the early eighties on the basis of calibrations in all available stray radiation fields. At this time, a polyethylene (PE) sphere of 30 cm in diameter was used as a phantom[5].

Type B

The Harshaw type under study at ORNL[8] makes use of two thermal neutron detectors, one of which is shielded in front by a cadmium filter. The calibration factor is given as a function of the actual reading ratio of the 9" diameter and 3" diameter Bonner spheres in the neutron field.

Type C

The Panasonic type used at ORNL[8] consists of three thermal neutron detectors shielded by a cadmium filter in front, at the back or on both sides. H_n is estimated by the spectrum-dependent actual set of calibration factors for the algorithm of the three neutron detector readings, provided by the data bank. The source information was known.

A combination personnel neutron dosemeter (CPND), recently investigated at ORNL[8], makes use of the Harshaw albedo dosemeter (type B) and two bubble detectors with energy thresholds at 0.1 MeV and 1 MeV, respectively. Using four neutron detector readings, the neutron dose equivalent H_n is estimated on the basis of a spectral stripping in four energy intervals and a computer supported evaluation algorithm.

Two types of etched track detectors using CR-39 (E_n>0.1 MeV) or PC (E_n>1.5 MeV) and a two-step electrochemical etching technique[1,2] are available here. In the case of CR-39 only one calibration factor has been used for all spectra.

RESULTS OF INTERCOMPARISON

Figure 1(a) shows the ORNL intercomparison results found for the KfK albedo dosemeter and the CR-39 detector. Except for the Pu–Be neutron irradiations of the albedo dosemeter (application area N3), the results are consistent within ±15%. The results of the PTB intercomparison experiments for the KfK albedo dosemeter presented in Figure 1(b) are also consistent within ±15% and indicate, in contrast to the ORNL results, that the actual calibration factors for the PE sphere phantom seem to be too low by about 10%. This nonconformity is small and may be explained by differences in the calculated backscattered neutron component of the reference fields and thus of the reference dose equivalents rather than by the different influences of the phantom geometry and the higher albedo reading expected on the slab phantom. Different neutron fluence-to-dose conversion factors, namely after ICRP 21 for ORNL and ICRU 39 for PTB, on the other hand, do not affect the results for the neutron spectra of interest here.

In principle, all types of albedo detectors have a similar energy dependence, except for low energy neutrons. They vary, however, in the spectral information used for the estimation of H_n. The ORNL intercomparison results of the various

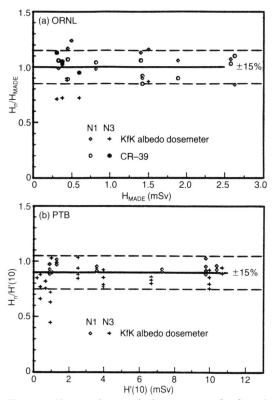

Figure 1. Neutron dose equivalent response for frontal irradiation incidence. (a) KfK albedo and CR-39 dosemeters, ORNL intercomparisons 1988-1991. (b) KfK albedo dosemeter, PTB intercomparisons 1988-90.

albedo dosemeter types are presented in Table 1. In comparison to type A, the types B and C show inconsistent results for runs 5 and 6, namely for type C a high 1 s value from 3 dosemeters of ±69% (this is equal to a scatter of a factor of 5.4) and for type B an over-response up to a factor of 1.9. Here, the capture gammas from the cadmium filters obviously affect the discrimination of the gamma contribution for low neutron doses already in the unmoderated radiation fields.

Figure 2 presents for the 1989 ORNL intercomparison the ratio H_n/H_{made} of the reported neutron dose H_n and the reference dose H_{made} which were found for the CPND, and the various albedo dosemeter types A to C and the track etched detectors. The results of the CPND and CR-39 detectors are directly comparable because at the time of evaluation, the neutron spectra were unknown for both systems. Excluding the irradiation at 60°, the CR-39 detector gives, for all neutron fields, more consistent results than the CPND using four neutron detectors. Including the irradiation under 60°, the mean scatter of the reported mean values was found to be comparable for the three systems, namely the CPND, the CR-39 and the KfK albedo dosemeter. The last one, however, makes use of spectral information.

The combination of four neutron detectors obviously suffers from the sensitivity of the evaluation algorithms with respect to small variations caused by the random and systematic uncertainty of the actual detector readings, which for bubble detectors exceeds 10% mainly because of the low bubble density ($\leq$ 100 bubbles).

On the basis of intercomparison results, we investigated for the neutron detectors of interest the energy range, the energy dependence and the lower limit of detection. Figure 3 compares the reading of the neutron detectors of interest expected for a neutron irradiation of 1 mSv as a multiple of the lowest detectable reading. The lowest detectable dose for photons indicated by the albedo detector is 0.03 mSv and of CR-39 detectors 0.05 mSv for ^{252}Cf neutrons. In the energy range of fast neutrons the bubble detector BR-100 shows the highest sensitivity[12]. In the energy range of 0.5 MeV, on the other hand, the neutron response of the albedo detector is equal to that of the CR-39 detector, but at 0.05 MeV already a factor of 10 higher.

As shown in Figure 4 the angular response of the neutron detectors is, in the case of ^{252}Cf + D$_2$O neutrons, significant for the CR-39 etched track detector, but uncritical for the albedo detector, whereas the bubble detector overestimates the new

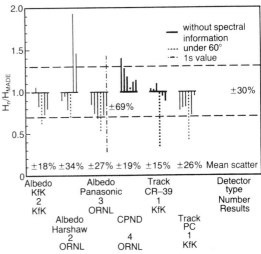

Figure 2. Intercomparison results for various personnel neutron dosemeters. 1989 ORNL Intercomparison data from irradiations No 1-6, see explanation in Table 1[6].

Table 1. Results of the ORNL intercomparison experiment (1989) for different albedo dosemeters[8].

No. Irradiation		H_n (ref) (mSv)	H_n/H_n (ref)[a]		
			A	B	C
1	^{252}Cf + 15 cm D$_2$O	0.82	0.99	0.90	0.85
2		2.58	1.06	0.95	0.74
3	Without Cd	2.63	0.83	0.79	0.68
4	Under 60°	2.58	0.62	0.70	0.53
5	^{252}Cf+15 cm PE+γ[b]	0.60[a]	0.73	1.93	0.72
6	^{238}Pu–Be	0.44	0.80	1.46	0.84[c]
	Results from		KfK	ORNL	ORNL

A, KfK Universal. B, Harshaw. C, Panasonic.
[a]Mean of 3 dosemeter readings (ORNL), one dosemeter reading (KfK).
[b]Additional photon dose of 1.47 mSv.
[c]1 s value ±69%.

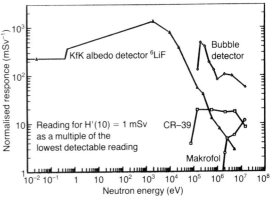

Figure 3. Energy dependence of the KfK albedo detector[5], the bubble detector BR-100[12], the CR-39 and PC track detectors[3,7].

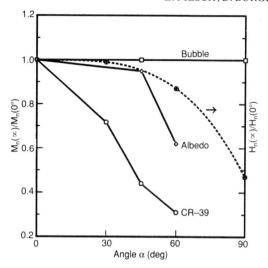

Figure 4. Angular dependence of neutron detectors for ^{252}Cf + D$_2$O neutrons.

dose quantity H'(10) at a high angle of incidence. The ratio H'(10,α)/H'(10,0°) is expected here to be 0.87 for 60° and 0.47 for 80°.

CONCLUSIONS AND CONSEQUENCES

In conclusion, the results of the intercomparison experiments in principle reflect only the situation in idealised calibration fields, but not the complexity of neutron fields and the orientation of the person in routine monitoring. Therefore the following observations in standard reference fields may not be valid in general:

(i) Two albedo dosemeter types disappointed because of the inappropriate capsule, and/or the evaluation/calibration technique.
(ii) CR-39 emerges as 'the best detector' because it happens to fit the intercomparison spectra.
(iii) Multi-detector systems may suffer from the sensitivity of the algorithms to random and systematic uncertainties of the actual detector readings.

Advantages and disadvantages of the dosimetry systems which, for most of the potential users, are alternatives in individual monitoring must be discussed in view of the fact that the ICRP recently increased the Q factors for neutrons to a factor 2 and reduced the annual dose limit to 20 mSv[11].

From the intercomparisons, it may be concluded that advantages of the CR-39 detector type are the gamma insensitivity, and for the two-step ECE technique[3] the energy independence (0.1–14 MeV): disadvantages are the significant angular dependence, the reduced dose and energy range and the relatively high random uncertainty. A truly dosimetry grade CR-39 plastic is still required. The KfK albedo dosemeter, on the other hand, was found to be a reliable neutron dosemeter for routine personal monitoring with an extended dose range from 0.1 mSv to 2 Sv and a tolerable uncertainty of measurement. In principle, the albedo detector in an appropriate encapsulation will in future serve as the basic neutron detector, combined, if necessary, with other detectors such as the etched track detectors with the advantages and disadvantages mentioned before.

The selected results of intercomparison experiments reflect the refinements of dosimetry systems found in recent years. As already mentioned at the last conference[13], there is no one detector that meets personnel dosimetry requirements for all operating situations or facilities. The limitations of detector types call, on the one hand, for an unsophisticated combination of detectors, and, on the other hand, for the use of more spectrometric data for a careful calibration of dosemeter systems in stray radiation fields.

As a future task to be recognised and solved in the 1990s, there is a need for representative intercomparisons in actual stray radiation fields at work places.

Reference radiation fields and phantom irradiations should be more realistic with respect to the irradiation conditions in routine monitoring, above all the angular distribution of the radiation field.

Better reference dosemeters should be provided, having an angular response comparable with that of the quantity 'personal dose equivalent' rather than that of the ambient dose equivalent, H*(10). These dosemeters should complement dose calculations which, up to now, have been performed using basic data of source strength and the scattered neutron components from the room.

REFERENCES

1. Lembo, L. (Ed.) *Result of the Survey of Neutron Dosimetry PADC Background Organised by EURADOS-CENDOS in 1988*. Report PAS-FIBI-DOSI (89) 1, Bologna (1989).
2. Schraube, H. (Ed.) *Response of Proton-Sensitive Etched Track Detectors to Fast Neutrons: Results of a Joint Multilaboratory Experiment*. GSF 22/90 (1990).
3. Alberts, W. G., Luszik-Bhadra, M., Piesch, E. and Vilgis, M. *Fast Netron Dosimetry with CR-39: Study of Various CR-39 Materials Using Electrochemical Etching*. Nucl. Tracks Radiat. Meas. **19**, 437-442 (1991).

4. Burgkhardt, B. and Piesch, E. *Albedo Dosimetry System for Routine Personnel Monitoring*. Radiat. Prot. Dosim. **23**, 117-120 (1988).
5. Burgkhardt, B. and Piesch, E. *Field Calibration Technique for Albedo Neutron Dosemeters*. Radiat. Prot. Dosim. **23**, 121-126 (1988).
6. Sims, C. S., Casson, W. H. and Patterson, G. R. *1989 Neutron and Gamma Personnel Dosimetry Intercomparison Study Using RADCAL Sources*. ORNL-6597 (1990).
7. Josefowicz, K. and Piesch, E. *Electrochemically Etched Makrofol DE as a Detector for Neutron Induced Recoils and Alpha Particles*. Radiat. Prot. Dosim. **34**, 25-28 (1990).
8. Liu, J. C. and Sims, C. S. *Performance Evaluation of a New Combination Personnel Neutron Dosemeter*. Radiat. Prot. Dosim. **32**, 33-43 (1990).
9. Alberts, W. G. and Kluge, H. *Ergebnisse einer probeweisen Vergleichsmessung von amtlichen Albedodosimetern im Jahre 1988*. PTB-7.32-89-1 (1989).
10. Bauer, B. W., Siebert, B. R. L., Alberts, W. G., Burgkhardt, B., Dietz, E., Guldbakke, S., Kluge, H., Medioni, R., Piesch, E. and Portal, G. *Experimental Investigation into the Influence of Neutron Energy, Angle of Incidence and Phantom Shape on the Response of Individual Neutron Dosemeters: Experimental Procedure and Summary of Results*. PTB Bericht, PTB-N-5 (1990).
11. ICRP. *1990–1991 Recommendations of the International Commission on Radiological Protection*. Publication 60. Ann. ICRP **21**(1-3) (Oxford: Pergamon) (1991).
12. Schwarz, R. P. and Hunt, J. B. *Measurement of the Energy Response of Superheated Drop Neutron Detectors*. Radiat. Prot. Dosim. **34**, 377-380 (1990).
13. Griffith, R. V. *Review of the State of the Art in Personnel Neutron Monitoring with Solid State Detectors*. Radiat. Prot. Dosim. **23**, 155-160 (1989).

THE ROLE OF PHANTOM PARAMETERS ON THE RESPONSE OF THE AEOI NEUTRIRAN ALBEDO NEUTRON PERSONNEL DOSEMETER

M. Sohrabi and M. Katouzi
National Radiation Protection Department
Atomic Energy Organization of Iran
PO Box 14155-4494, Tehran, Islamic Republic of Iran

Abstract — The response of the AEOI Neutriran Albedo Neutron Personnel Dosemeter (NANPD) which can also be used for other albedo dosemeter types was determined on 18 different phantom configurations. The effects of type, geometry, material, thickness, dosemeter-to-phantom angle in particular with the presence of legs were investigated using a Pu–Be neutron source. It was concluded that the slab phantoms (single or double) and circular and elliptical cylinder phantoms seemed to provide a better response, whereas the ICRU sphere geometry does not seem to be appropriate for the calibration of albedo dosemeters. It is interesting to note that the presence of legs maintains the constancy of the response in a situation when a radiation worker bends down during work.

INTRODUCTION

Recently, the AEOI Neutriran Albedo Neutron Personnel Dosemeter (NANPD) was developed and mass produced for a national service in Iran[1-4]. Since the albedo dosemeter response exists only in connection with a phantom, its proper calibration on a suitable phantom is essential. Many phantom studies have been carried out for TL-type albedo dosemeters[5-8]. However, a clear-cut and practical approach to using a proper phantom seemed to require further investigation, especially when albedo dosemeters of the NANPD type are used. Some important studies have focused on dosemeter calibration procedures[7-9], as well as on depth dose equivalent calculations[10], with the recommended ICRU sphere phantom[10]. However, using this phantom seems to have some limitations for practical dosemeter calibrations. Also, no literature seemed available to consider the effect of legs on the response when a radiation worker bends down during work. In this study, the response of the NANPD dosemeter was investigated on 18 phantoms of different geometry configurations as well as with legs included. The results of these studies are presented and discussed.

EXPERIMENTAL PROCEDURES

Dosemeter specifications

The optimum design and component dimensions of the three-component NANPD dosemeter have been reported elsewhere[2-3]. Figure 1 shows the components and cross-sectional view of the NANPD dosemeter. It is capable of measuring separately thermal and fast neutrons incident from radiation fields directly and albedo neutrons reflected from the body. Neutrons are detected by electrochemically etched (ECE) tracks of fast-neutron-induced recoils[12-13], and via the $^{10}B(n,\alpha)^7Li$ reaction in a radiator, thermal-neutron-induced α particles and albedo-neutron-induced α particles, in polycarbonate detectors (PC). For dosemeter calibration, the phantom parameters have a marked effect, in particular on the albedo neutron reading. Therefore, a single ^{10}B pellet and PC foil combination under a cadmium foil cover was used so that albedo neutrons pass the PC foil before interacting with ^{10}B.

For phantom irradiations, a ^{238}Pu–Be neutron source of 111 GBq activity was used. The exposed

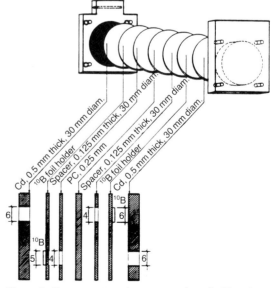

Figure 1. Components and a cross section of a Neutriran Albedo Neutron Personnel Dosemeter. Dimensions in mm.

PC detectors were etched electrochemically under the previously reported conditions[2–4]. The dosemeter on the phantom was at a height of 110 cm from the ground and 50 cm from the source in a 3.3 × 6.7 × 5.6 m^3 room. The conditions were the same for all phantom irradiations. The track counting was carried out by eye using a light microscope at 6.3 × 10 magnification. Each reading is the mean of seven independent exposures.

Phantom specifications

Different parameters, including type, geometry, material, thickness, dosemeter location on the phantom, dosemeter-to-phantom angle as well as the effect of legs, were studied. Phantoms of 18 different types, geometries and materials were used, including Michigan (front and side), double Michigan (front), cuboid (two sizes), circular cylinder (three sizes) and elliptical cylinder phantoms, all filled with water, as well as spheres (four sizes including the ICRU size), circular cylinder (end and side) and Studsvik rem counter (end and side) moderator as phantoms, all made of polyethylene. The type of phantoms, their specifications and irradiation directions are listed in Table 1.

RESULTS AND DISCUSSION

The effects of some phantom parameters on the NANPD albedo reading have been studied previously, leading to the following conclusions[2]:

(a) As the detector-to-phantom distance increases from zero to about 1 cm, the relative albedo neutron response increases by about 70% for a 160 cm dosemeter-to-source distance (due to the detection of more albedo neutrons from the phantom) reaching a broad maximum after which it decreases again.

(b) The albedo response is constant for dosemeter positions up to 8 cm from the centre of the front surface of the Michigan phantom. However, it was recommended that a phantom with larger dimensions (40 × 40 × 15 cm^3) could increase this distance for calibration.

(c) The dosemeter angular response R(α) on an elliptical phantom, measured by rotating the neutron source around the phantom, showed relative responses of R_{90}/R_o=60% and R_{180}/R_o=66%[2].

Dosemeter–phantom angle

The effect of dosemeter–phantom angle on the albedo dosemeter response was investigated by bending an elliptical phantom to resemble that of a radiation worker during work, as shown in Figure 2 for phantoms with and without the legs present. Without leg phantoms, by bending the phantom to increase the dosemeter–phantom angle, the response relative to 0° increases by about 5% up to 30° after which it decreases to about 30% at 90°. By adding two circular cylinder leg phantoms (15 cm diameter and 40 cm high, filled with water), the relative responses remain constant within ±5% with increasing angle, as shown in Figure 2. This indicates that production of more albedo neutrons generated by the legs compensates for losses.

Thickness of a hollow cylinder phantom

The role of phantom thickness on the NANPD dosemeter response was studied by inserting a hollow cylinder (15 cm diameter) into a circular cylinder phantom (30 cm diameter, 40 cm high), while changing the layer thickness between the cylinder and the phantom wall. Figure 3 shows the increase in response with increasing thickness compared with previous data of Alberts[6] using the same phantom geometry and material, and those of Hankins[5] using polyethylene slabs of varying thickness, both employing TL albedo dosemeters. The response increase with increasing thickness reaches a plateau at 7 cm. For this geometry, therefore, approximately 7 cm thickness is sufficient for the total moderation and scattering of neutrons necessary for a response, which is equivalent to that obtained by the full cylinder. This is in good agreement with the data of Hankins[5] and Alberts[6].

Phantom size and geometry

The dosemeter responses on 18 different phantom configurations are demonstrated in Table 1. By comparing the results, the following conclusions can be drawn:

(1) Slab phantoms of the Michigan (no. 1) and double Michigan (no. 2) type show similar responses, while the response on the side (no. 3) decreases by 22%. This indicates the importance of having phantoms of optimum area and thus enough surrounding material facing the neutron beam. As a result, phantoms with 15 cm depth are sufficient for the calibration of albedo dosemeters. Also, a larger slab phantom (no. 4) with the same response has the advantage of permitting more dosemeters to be exposed simultaneously. Dosemeter positions up to 12.5 cm from the phantom face centre provide equal dosemeter responses, which is only 8 cm for the Michigan phantom.

(2) The response on sphere phantoms increases by 33%, 12% and 4%, for increasing diameters from 20, 23 and 25, respectively, to 30 cm (nos. 6 to 9). The response on the 30 cm phantom is equal

THE ROLE OF PHANTOM PARAMETERS

Table 1. The relative albedo responses of the Neutriran albedo neutron personnel dosemeter to Pu–Be neutrons on 18 different phantom configurations.

No	Type of phantom	Geometry of phantom under irradiation	Dimensions h x w x d * (cm³)	Response (tracks.cm^{-2}. mSv^{-1})	Relative response** R/R$_1$	R/R$_2$
1	Michigan phantom		30 × 30 × 15	1891 ± 94	100	122
2	"		30 × 30 × 30	1954 ± 20	100	126
3	"		30 × 30 × 15	1503 ± 26	78	97
4	Cuboid		40 × 40 × 15	1901 ± 76	100	123
5	"		40 × 29 × 25	1828 ± 54	95	118
6	Sphere		20 cm diam.	1161 ± 20	61	75
7	"	" "	23 cm diam.	1376 ± 54	72	89
8	"	" "	25 cm diam.	1488 ± 11	78	96
9	"	" "	30 cm diam.	1541 ± 57	80	100
10	Circular cylinder phantom		40 (h) × 30 (diam.)	1540 ± 28	80	100
11	"	" " "	60 (h) × 30 (diam.)	1791 ± 61	93	116
12	"	" " "	40 (h) × 15 (diam.)	1266 ± 15	66	82
13	Elliptical cylinder phantom		20 × 36 cm axes 20 cm (h)	1548 ± 26	81	100
14	Thorax phantom		20 × 30 cm axes 40 cm (h)	1835 ± 58	96	119
15	Circular cylinder phantom (end)		21 (h) × 27.5 (diam.)	1919 ± 87	100	124
16	Circular cylinder phantom (side)		21 (h) × 27.5 (diam.)	1540 ± 74	80	100
17	Rem counter type Studvik (end)		22 (h) × (22, 9.5) (diam.)	1284 ± 63	67	83
18	Rem counter type Studvik (side)		22 (h) × (22, 9.5) (diam.)	1495 ± 29	78	95

Dosemeter   S - Neutron source

* Height (h) × width (w) × depth (d) in cm.
  Water phantoms: nos 1–5 and 10–14.
  Polyethylene phantoms: nos 6–9 and 15–18.
** Relative to the response R$_1$ (nos 1, 2, 4, 15 and R$_2$ (no 9).

to that on the ICRU sphere as well as on a circular cylinder of 30 cm diameter and 40 cm height (no. 10) and a side Michigan phantom (no. 3).

The ICRU sphere has some disadvantages for routine calibrations of albedo dosemeters in neutron radiation fields:

(a) The sphere geometry does not resemble the trunk of the body; thus the response is 20% lower than those of a slab phantom's (nos. 1, 2, 4).

(b) Simultaneous exposures of a number of dosemeters on the ICRU sphere are not possible.

(3) By increasing the height of cylindrical phantoms (nos. 10, 11) from 40 to 60 cm, the response increases by 16%. By decreasing, on the other hand, the 30 cm diameter (no. 10) to 15 cm (no. 12), the response decreases by 18%. Also for elliptical cylinder phantoms (no. 13 and 14), about a 19% increase is observed for increasing the height from 20 to 40 cm.

(4) On rectangular and circular phantoms of sufficient size (nos. 1, 2, 4, 15), the response increases by 25% in comparison with the ICRU sphere. On the other hand, the response on the phantom no. 1 having a 13.5 litre volume is about 22% greater than that of the phantom no. 10 with a 28 litre volume.

(5) Within the random uncertainty of the measurements, the albedo dosemeter response has been found to be similar for phantoms of the same geometry made from polyethylene or filled with water.

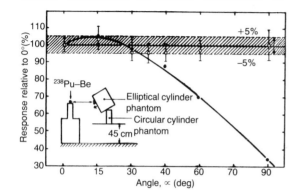

Figure 2. The relative albedo response of the dosemeter on an elliptical cylinder as a function of phantom-to-dosemeter angle with (○) and without (●) leg phantoms.

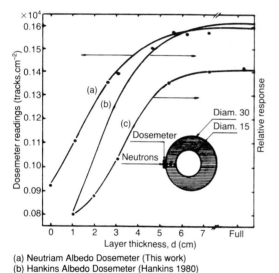

(a) Neutriam Albedo Dosemeter (This work)
(b) Hankins Albedo Dosemeter (Hankins 1980)
(c) Universal Albedo Dosemeter (Alberts 1988)

Figure 3. The dosemeter reading as a function of layer thickness, d, formed between a hollow cylinder inside a cylindrical water phantom (curves a and c) and a polyethylene slab phantom (curve b).

CONCLUSIONS

The phantom geometry is an essential parameter for the calibration of albedo dosemeters. A phantom should have enough moderator to resemble the trunk of an individual. While slab phantoms of at least 15 cm thickness and circular cylinder or elliptical phantom seem to provide a better response, the ICRU sphere geometry is not appropriate at all for albedo dosemeter calibrations. From a practical point of view, a $40 \times 40 \times 15$ cm^3 phantom is adequate for calibration of a number of dosemeters simultaneously as well as a suitable response. It was also shown that only 5 cm moderator obtained from a circular cylinder (with hollow cylinder inside) is enough to produce 90% of albedo neutrons in comparison to a 30 cm diameter cylinder. By adding legs to an elliptical phantom, a new important conclusion was drawn; there is constancy in angular response when a radiation worker bends during work.

REFERENCES

1. Sohrabi, M. *A New Dual Response Albedo Neutron Personnel Dosimeter.* Nucl. Instrum. Methods **165**, 135-138 (1979).
2. Sohrabi, M. and Katouzi, M. *Some Characteristics of the AEOI Neutriran Albedo Neutron Personnel Dosemeter.* Radiat. Prot. Dosim. **34**, 149-152 (1990).
3. Sohrabi, M. and Katouzi, M. *Design Characteristics of a Three-component AEOI Neutriran Albedo Neutron Personnel Dosemeter.* Nucl. Tracks Radiat. Meas. **19**, 537-540 (1991).

4. Katouzi, M. and Sohrabi, M. *Parametric and Dosimetric Investigation of Neutriran Albedo Neutron Personal Dosimeter.* AEOI Report No. AEOI-NRPD-DOS-2 (1989).
5. Hankins, D. *Phantoms for Calibration of Albedo Neutron Dosemeters.* Health Phys. **39**, 580-584 (1980).
6. Alberts, W. G. *Response of an Albedo Neutron Dosemeter to ^{252}Cf Neutrons on Various Phantoms.* Radiat. Prot. Dosim. **22**, 183-186 (1988).
7. Burgkhardt, B. and Piesch, E. *Field Calibration Technique for Albedo Neutron Dosemeters.* Radiat. Prot. Dosim. **23**, 121-126 (1988).
8. Bauer, B. W., Hollnagel, R. and Siebert, B. R. L. *Numerical Study of the Influence of Phantom Material and Shape on the Calibration of Individual Dosemeters for Neutrons.* Radiat. Prot. Dosim. **23**, 207-210 (1988).
9. Hollnagel, R. and Alberts, W. G. *Reading of Personal Neutron Dosemeters on the MIRD Phantom.* Radiat. Prot. Dosim. **34**, 153-156 (1990).
10. Jahr, R., Hollnagel, R. and Siebert, B. R. L. *Calculation of Specified Depth Dose Equivalent in the ICRU Sphere Resulting from External Neutron Irradiation with Energies Ranging from Thermal to 20 MeV.* Radiat. Prot. Dosim. **10**, 75-87 (1985).
11. ICRU. *Determination of Dose Equivalent for External Radiation Sources.* Report 39 (Bethesda, MD: ICRU Publications) (1985).
12. Sohrabi, M. *Electrochemical Etching Amplification of Recoil Particle Tracks in Polymers and its Application in Fast Neutron Personnel Dosimetry.* Health Phys. **27**, 598-600 (1974).
13. Sohrabi, M. and Morgan, K. Z. *A New Polycarbonate Fast Neutron Personnel Dosimeter.* Am. Ind. Hyg. Assoc. **39** (6), 438-447 (1978).

COMPARISON OF TWO TYPES OF ALBEDO DOSEMETERS IN SEVERAL MIXED NEUTRON-GAMMA FIELDS

A. Aroua†, M. Grecescu†, P. Lerch†, S. Prêtre‡ and J.-F. Valley†
† Institute of Applied Radiophysics, CH-1015 Lausanne, Switzerland
‡ Swiss Nuclear Safety Inspectorate, CH-5232 Villigen-HSK, Switzerland

Abstract — The response of two types of personnel albedo dosemeters, a prototype and a commercial one, has been investigated in mixed neutron-gamma fields by exposing them on a phantom simulating a human chest. The calibration of the dosemeters has been performed in several reference neutron fields. Additional values of the response were obtained in mixed neutron-gamma fields inside nuclear power plants. The variation of the response with the mean neutron energy characterised by two different spectral indices has been investigated. For the spectra encountered in the nuclear power plants, an average response may be employed. The usefulness of the albedo dosemeters for the simultaneous determination of the dose equivalent due to gamma rays has been also investigated.

INTRODUCTION

In the frame of a research and development programme in neutron dosimetry, two topics have first been investigated: (1) ambient dosimetry based on fluence and neutron spectra measurements with a Bonner spheres system[1,2]; and (2) personnel dosimetry based on a prototype albedo dosemeter. The behaviour of both systems in monoenergetic neutron beams has been studied.

Subsequently, a comparative evaluation of several types of personnel neutron dosemeters has been performed[3]. This evaluation included a comparison between the prototype albedo dosemeter and a commercial one. The comparison has been performed both with reference neutron spectra and in mixed neutron–gamma fields inside nuclear power plants.

The following problems were investigated:
(1) the response of both types of albedo dosemeter and its variation with neutron energy;
(2) the possibility of employing a unique calibration factor for personnel neutron dosimetry in nuclear power plants;
(3) the possibility of simultaneous evaluation of the dose equivalent due to both components in mixed neutron–gamma fields.

MATERIALS AND METHODS

Materials

The prototype albedo dosemeter has been in use at the Institute of Applied Radiophysics (IAR) since 1982. It consists of three pairs of Harshaw TLD–600 and TLD–700 chips with the dimensions $3.2 \times 3.2 \times 0.9$ mm^3 mounted in a polypropylene badge for conventional personnel dosemeters ($48 \times 36 \times 6$ mm^3). The neutron shielding is provided by two cadmium foils 0.7 mm thick. The dosemeter is divided into three compartments: A (open towards the wearer's body), E (open towards the radiation source); I (shielded on both sides). Each compartment contains a pair of TLD chips. The I compartment is similar to the Hankins albedo dosemeter[4]. A design with three compartments was adopted for the IAR albedo dosemeter for the purpose of obtaining some information on the neutron spectrum by combining the readings of the various compartments. It was hoped thus to avoid the need for independent information on the neutron spectra for a correct interpretation of albedo dosemeter measurements.

The commercial albedo dosemeter employed in this study is based on a prototype developed at the Kernforschungszentrum Karlsruhe (KfK)[5]. It contains a Harshaw 4NG6677 card with two TLD–600 and two TLD–700 chips mounted in a boron-loaded plastic badge for neutron shielding. The dosemeter is divided in two compartments (A and E according to the previous notation).

The dosemeters were irradiated at the surface of a phantom simulating a human chest. The phantom is an elliptic cylinder (base axes 31 cm and 23 cm, height 60 cm) with 3 mm thick polyethylene walls filled with water.

Various neutron dose equivalent quantities in mixed neutron–gamma fields were determined with a Bonner spheres neutron spectrometer[1,2].

The gamma dose equivalent in the same radiation fields was measured with a compensated GM counter (type ZP1320/PTFE, Alrad Instruments) having a low sensitivity to neutrons[1,2].

Methods

The response of the albedo dosemeters was measured by a procedure similar to the field calibration method developed by Burgkhardt and Piesch[6]. The measurements were performed in neutron fields including the scattered component. The dose equivalent was determined by the Bonner spheres spectrometer as reference instrument. Whenever necessary, fluctuations of the neutron beam were monitored for normalisation purposes. The influence of field anisotropy was estimated in a few typical situations by simultaneously exposing albedo dosemeters on both the front and the rear surfaces of the phantom. Eventually the albedo response was determined only from the reading of the front dosemeter which is largely predominant.

The response of the IAR albedo dosemeter was determined in monoenergetic neutron beams (thermal, 144 keV, 250 keV, 570 keV and 1.2 MeV) at the Physikalisch-Technische Bundesanstalt (Braunschweig) and in the following reactor spectra: unshielded HPRR spectrum and spectra through concrete, Lucite and steel shielding[7] at the Health Physics Research Reactor (HPRR) at Oak Ridge National Laboratory[8].

The responses of both types of albedo dosemeters (IAR and KfK) were compared by simultaneous irradiation in the field of reference neutron sources (^{241}Am-Be, ^{252}Cf and D_2O-moderated ^{252}Cf) at the Paul Scherrer Institute (Würenlingen).

The response of both albedo dosemeters (IAR and KfK) was also measured in mixed neutron-gamma fields at 17 locations inside nuclear power plants. The dose equivalent components at each point were determined with the Bonner spheres and the GM counter respectively.

RESULTS

The dose equivalent response (R_H) of the albedo dosemeters was determined according to the relation $R_H = M/H$ where M is the net reading of the TLD reader for the albedo neutron component of the radiation field expressed in arbitrary photon response units based on a TLD calibration at ^{60}Co. The net reading is evaluated according to the equation:

$$M = f_c(n_6 M_6 - n_7 M_7)$$

where M_6 and M_7 are the readings of the TLD-600 and TLD-700 detectors corrected for zero dose reading; n_6 and n_7 are the individual calibration factors of the two detectors for ^{60}Co photons and f_c is a correction factor for the daily reader calibration. H is the maximum neutron dose equivalent (MADE) at the location of the phantom on which the dosemeters have been exposed; its value has been determined from the fluence spectrum measured with Bonner spheres, using the $\hat{h}_\Phi$ conversion factor[9].

Since the determination is based on the experimental neutron spectrum, the albedo response for other dosimetric quantities may be readily obtained by using a suitable conversion factor.

The response is energy-dependent and consequently spectrum-dependent. A convenient characterisation of the neutron spectra by a spectral index has been proposed by Hankins[10]. This index is the 9"/3" ratio defined as the ratio of the counting rates of the Bonner spheres with diameters of 9 and 3 inches, respectively. As this ratio depends on the type of thermal neutron detector employed in the Bonner spheres, it is important to use a coherent set of 9"/3" ratios[8].

The variation of the response of both dosemeters relative to the 9"/3" ratio is shown in Figure 1. The response is almost constant for small 9"/3" ratios (corresponding to low energy neutrons) and then displays a linear decrease in a log-log plot.

DISCUSSION

The responses of both albedo dosemeters relative to the 9"/3" ratio are similar (Figure 1). The IAR dosemeter has a higher response by a factor of about three. This enhancement is presumably due to the fact that the holes of the A section (towards the body) are screened by a 2.8 mm plastic layer in the KfK dosemeter and only by a 0.1 mm paper sheet in the IAR dosemeter.

The normalisation of the response of both dosemeters to their respective values for the moderated ^{252}Cf spectrum leads to the superposition of the curves, which proves their similarity.

Since the dosemeter's response depends on neutron energy, spectral information is necessary in principle for more accurate personnel dosimetry with albedo detectors. Even limited information such as the 9"/3" spectral index requires additional

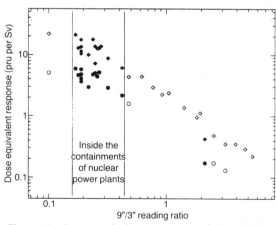

Figure 1. Dose equivalent response of the albedo neutron dosemeters relative to the Bonner spheres 9"/3" counting ratio (pru = photon response units, see text). Calibration spectra: (◊) IAR, (○) KfK. Spectra in power plants: (♦) IAR, (●) KfK.

TYPE TESTING AND ROUTINE CALIBRATION OF NEUTRON PERSONAL DOSEMETERS: PHANTOMS AND PHANTOM BACKSCATTER

R. J. Tanner, D. T. Bartlett, T. M. Francis and J. D. Steele
National Radiological Protection Board, Chilton
Didcot, Oxon OX11 0RQ, UK

Abstract — It is proposed that personal dosemeters be type tested and routinely calibrated in terms of the ICRU quantities $H_p(10)$ and $H_p(0.07)$: these quantities are defined as the dose equivalent at 10 mm and 0.07 mm respectively, in a $300 \times 300 \times 150$ mm^3 slab of ICRU 4-element tissue. In practice ICRU tissue cannot be fabricated so phantoms of various surrogate materials are used; for calibration and routine performance tests irradiation is sometimes performed free-in-air to facilitate the simultaneous exposure of large numbers of dosemeters to the same fluence. It may then be necessary to apply dosemeter-specific corrections to take into account the difference between the backscattered radiation field in the practical situation and that for which the calibration quantity is defined. It has been suggested that if a standard phantom of dimensions $300 \times 300 \times 150$ mm^3 composed of poly-methyl methacrylate (PMMA) is used, then no corrections for differences in backscatter are required. Results are presented for the differences in fluence response of routinely used neutron personal dosemeters irradiated on $300 \times 300 \times 150$ mm^3 phantoms composed of a tissue substitute material (MS20), PMMA and polyethylene. These are compared with free-air irradiations corrected for room scatter by the shadow cone technique. The contribution to the response due to backscatter from the phantom is thus derived and is discussed in terms of the appropriateness of phantom selection.

INTRODUCTION

Neutron personal dosemeters have, in general, been calibrated in terms of a quantity defined in a phantom. Until recently this quantity has been the maximum dose equivalent (MADE) in a cylinder of tissue-like composition (4-element: carbon, nitrogen, oxygen and hydrogen), with a height of 60 cm and a diameter of 30 cm[1]. Following the publication of the International Commission on Radiation Units and Measurements (ICRU) Report 39[2] and 43[3], however, there has been some use of the quantity directional dose equivalent at a depth of 10 mm, H'(10), which is defined in a 30 cm diameter sphere of ICRU 4-element tissue.

The use of a calibration quantity defined at a depth of 10 mm in a $300 \times 300 \times 150$ mm^3 slab of ICRU 4-element tissue is also under consideration[4]. In this case it has been suggested that for photons, dosemeters may be type tested using a calibration phantom made of poly-methyl methacrylate (PMMA) but using the calibration quantity defined in a tissue phantom.

The general principles of dosemeter calibration have been discussed elsewhere[5,6] and will not be repeated here. In this paper we report results of measurements of the effects in terms of dosemeter response of the use of different materials for the phantom composition. The contributions of the backscatter to the on-phantom response of the dosemeter are given.

Both the PMMA and tissue substitute (MS20) phantoms were used to test for interchangeability in an empirical sense. Additionally, a polyethylene phantom was used to test whether the experiment was sufficiently sensitive to detect small changes in phantom composition: polyethylene has a higher linear density of hydrogen and would be expected to produce a smaller rise in response when compared to free-air.

NRPB PADC NEUTRON PERSONAL DOSEMETER

The NRPB have been running a routine PADC neutron personal dosimetry service for approximately five years, its performance being well documented[7,8]. Additionally, in order to improve the performance of the dosimetry service, research continues into reducing the minimum detectable dose equivalent. Efforts to achieve this have resulted in a dual frequency etch process which has a response to ^{252}Cf neutrons which is greater than that of the existing service by a factor of ~ 4. The performance of this etch process is also well documented[9].

Both etch processes were used in this work since their very different low energy response functions may be expected to show differing responses to the neutrons backscattered from the phantom. The parameters of the etches are described in Table 1.

EXPERIMENTAL METHODS

The irradiations were performed in the NRPB's low scatter irradiation laboratory. This laboratory has been treated with 'Premadex' (lithium salt of a long chain acid) to reduce the room scattered component of the neutron field and thereby better simulate 'free-air'[10]: traceability is obtained with reference to irradiation in a low scatter environment at NPL which, in this context is assumed to approximate to 'free-air'.

The PMMA and polyethylene phantoms used each measure $300 \times 300 \times 150$ mm^3, the block of dosemeters being placed on one of the large faces as centrally as possible (see Figure 1). The MS20 phantom, however, is constructed from a hollow $300 \times 300 \times 300$ mm^3 square cross-sectioned tube. This has walls of thickness 20 mm and allows for the insertion of slabs of MS20 to produce the desired phantom dimensions ($300 \times 300 \times 150$ mm^3). 'Free-air' was simulated by suspending dosemeters on Sellotape drawn taut across an aluminium frame.

The three phantoms and the free-air frame were placed at a distance of 1.00 m from the ^{252}Cf neutron source at 90° intervals, thereby minimising the inter-phantom scattering. Twenty dosemeters were attached to each of the phantoms and the aluminium frame, ten of which were used for each of the etch processes. The final results were corrected for distance using the inverse square law to give a 'track production rate' (TPR) at 1.00 m. Account was not taken of the angle of incidence since the maximum angle was 5.75° and the response variation of the dosemeter is not well known over such small angles.

Since the 'free-air' irradiation will include a higher component from room scatter owing to the absence of shielding by a phantom, a shadow cone irradiation was also performed. For this 20 dosemeters from a different sheet of PADC were mounted on two identical aluminium frames and placed at 1.00 m from the source position to form an angle of 180° with the source. The shadow cone was then positioned so as to obscure completely one set of dosemeters from direct irradiation by the source, the backscatter from the shadow cone to the other frame having been estimated and considered to be negligible. These dosemeters were processed using the 50 Hz + 2 kHz etch since this should be capable of measuring a lower dose rate. The random uncertainties were hence also reduced by a factor of $\sqrt{2}$ by a doubling of the sample size.

RESULTS

The measurement of the 'unshadowed' track production rate yielded a value of 36.78 ± 0.24 h^{-1} whilst the 'shadowed' value was found to be 1.81 ± 0.24 h^{-1}. This gives a value of $4.92 \pm 0.66\%$ for the scattered component of the field as a percentage of direct. This result has been applied in full to both the routine and experimental etch data for free-air, but only half the final value (2.46%) has been applied to the on-phantom irradiations. This assumes complete shielding through 2π owing to the presence of the phantom.

The correction for room scatter should in reality be different for the routine and experimental etches. However, to measure its value for the routine etch would have required an irradiation with a duration of at least four times that used for the experimental etch. Such an irradiation was deemed unlikely to produce a statistically different result, but there remains the probability that a

Figure 1. The free-air frame, PMMA, polyethylene and MS20 phantoms (left to right) with dosemeters attached.

Table 1. Parameters of the two etch processes (ECE = electrochemical etch).

	Routine etch	Experimental etch
Etchant	5.00N NaOH	5.35N KOH
Pre-etch temperature (°C)	70	50
Pre-etch duration (h)	1	4
ECE temperature (°C)	30	50
RMS ECE field strength (kV.cm^{-1})	21	21
50 Hz duration (h)	–	12
2 kHz duration (h)	16	2

Table 2. Track production rates after correction for room scatter and ratios to free-air for the routine etch (the quoted uncertainties are standard errors of the mean).

Irradiation	Track production rate (h^{-1})	Ratio to free-air
PMMA	6.15 ± 0.468	1.120 ± 0.102
MS20	6.06 ± 0.16	1.102 ± 0.060
Polyethylene	5.75 ± 0.36	1.047 ± 0.082
Free-air	5.50 ± 0.26	

systematic uncertainty is present in the corrected data for the routine etch. The random uncertainty in the room scatter result would constitute a systematic uncertainty in the corrected values. Hence, the quoted uncertainties do not include this uncertainty (0.66%).

CONCLUSIONS

The presence of a phantom in most cases produced a statistically significant rise in track production rate, the exception being the routine etch with the polyethylene phantom which had a ratio of on-phantom to free-air of 1.047 ± 0.082. Otherwise, however, whilst the rise in response is statistically significant, the large uncertainties on the routine etch results obscure any differences between the etches.

MS20 is known to behave as a good imitation of ICRU 4-element tissue for γ rays and also β particles. In this work its influence on the response of the NRPB neutron personal dosemeter for neutrons is statistically identical to that of a PMMA phantom of the same dimensions. MS20, however, has a 8.1% elemental weight of hydrogen which is lower than that of ICRU 4-element tissue (10.1%). This is closer to that of PMMA (8.0%) which, assuming that it is the relative concentrations of hydrogen and higher atomic number nuclei which determine the degree of backscatter, explains why the results for the two materials are statistically identical.

Table 3. Track production rates after correction for room scatter and ratios to free-air for the experimental etch (the quoted uncertainties are standard errors of the mean).

Irradiation	Track production rate (h^{-1})	Ratio to free-air
PMMA	39.29 ± 0.68	1.104 ± 0.035
MS20	39.49 ± 0.78	1.10 ± 0.037
Polyethylene	38.70 ± 1.25	1.088 ± 0.046
Free-air	35.58 ± 0.96	

The polyethylene phantom has a much higher elemental weight of hydrogen (14.4%) which should lead to fewer neutrons being scattered through large enough angles with sufficient energy for detection: this is shown by the slightly lower rise in response with respect to 'free-air' on the polyethylene phantom.

The small change in on-phantom/'free-air' ratio between PMMA and MS20, and polyethylene, implies that the difference associated with ICRU 4-element tissue would not be measurable using PADC dosemeters. Thus this work supports the concept of using PMMA as a substitute for ICRU 4-element tissue for PADC dosemeters with a ^{252}Cf source, although the results may be very different for albedo neutron dosemeters: our dosemeters slightly overestimate the thermal dose equivalent when calibrated using a ^{252}Cf source, but albedo dosemeters would be more sensitive to the moderating ability of the phantom.

REFERENCES

1. International Commission on Radiological Protection. *Data for Protection against Ionising Radiation from External Sources: Supplement to ICRP Publication 15.* Publication 21 (Oxford: Pergamon Press) (1971).
2. International Commission on Radiation Units and Measurements. *Determination of Dose Equivalents Resulting from External Radiation Sources.* Report 39 (Bethesda, MD: ICRU Publications) (1985).
3. International Commission on Radiation Units and Measurements. *Determination of Dose Equivalents Resulting from External Radiation Sources. Part 2.* Report 43 (Bethesda, MD: ICRU Publications) (1988).
4. International Commission on Radiation Units and Measurements. Report 47 (in press).
5. Bartlett, D. T., Francis, T. M. and Dimbylow, P. D. *Methodology for the Calibration of Photon Personal Dosemeters: Calculations of Phantom Backscatter and Depth Dose Distributions.* Radiat. Prot. Dosim. **27**(4), 231-244 (1989).
6. Alberts, W. G. *Calibration of Individual Monitors Using the New ICRU Quantities.* Radiat. Prot. Dosim. **27**(4), 245-249 (1989).
7. Bartlett, D. T., Steele, J. D. and Stubberfield, D. R. *Development of a Single Element Neutron Personal Dosemeter for Thermal, Epithermal and Fast Neutrons.* Nucl. Tracks **12**, 645-648 (1986).
8. Gilvin, P. J., Bartlett, D. T. and Steele, J. D. *The NRPB Personal Neutron Dosimetry Service.* Radiat. Prot. Dosim. **20**, 99-102 (1987).
9. Tanner, R. J., Bartlett, D. T. and Steele, J. D. *A Two Frequency Electrochemical Etch for Routine Processing of Neutron Dosemeters at the NRPB?* Nucl. Tracks Radiat. Meas. **19** (1-4) 217-222 (1991).
10. Francis, T. M. *Development of a Neutron Calibration Facility at NRPB.* Radiat. Prot. Dosim. **44**(1-4) 147-149 (1992) (This issue).

CALIBRATION METHOD FOR PERSONNEL NEUTRON DOSEMETERS IN STRAY RADIATION FIELDS

A. V. Sannikov
Institute for High Energy Physics
142284 Protvino, Moscow Region, Russia

Abstract — A calibration method for personnel neutron dosemeters, based on measurement of neutron dose equivalent on the phantom surface by the passive neutron dosemeter–spectrometer (PNDS) with fission track detectors, is described. The main characteristics of PNDS are presented. Experimental results on the neutron personnel dosemeters response behind the shielding of the 70 GeV IHEP proton accelerator are given. For example, the data for MK-20 nuclear film, TLD albedo dosemeter, etched track detector CZ (similar to CR-39) and also for combined albedo–track personnel dosemeters with these detectors are presented. The results obtained show that combined dosemeters are the most promising in neutron personnel monitoring at charged particle accelerators where the neutron spectra are highly varied in space and time. The presented calibration method is recommended for practical use at high energy accelerators.

INTRODUCTION

The problem of neutron personnel dosemeters calibration in stray radiation fields has been widely discussed in recent years. For example, advantages and shortcomings of different calibration techniques have been described by Piesch[1]. Most of the neutron personnel dosemeters now being used have a great energy and/or angular dependence of dose equivalent response, hence there is a need for special calibration in the radiation fields of interest. The neutron dose equivalent is usually measured in this case by survey meters and other devices based on neutron moderation principles. These devices, with nearly isotropic response, do not allow adequate interpretation of the readings of personnel dosemeters irradiated on the phantom. Recently, a method has been proposed based on the measurement of the neutron dose equivalent on the phantom surface by the passive neutron dosemeter–spectrometer (PNDS)[2,3].

PASSIVE NEUTRON DOSEMETER–SPECTROMETER

The PNDS consists of ^{235}U in cadmium and boron (0.36 g.cm^{-2} of ^{10}B) filters, ^{237}Np, ^{238}U, ^{232}Th and ^{209}Bi fission converters. The fission cross sections of these nuclides are shown in Figure 1. Fission converters with masses of 1–2 mg and 1 cm^2 area, deposited on thin aluminium sheets, are placed in cadmium and boron filters, each layer being in contact with Mylar track detectors 6 μm thick. The dimensions of the cadmium and boron filters are 5 mm × 22 mm diam. and 15 mm × 30 mm diam., respectively. The counting of fission fragment tracks after chemical etching of Mylar detectors is performed by a spark counter.

Unfolding of neutron spectra and estimation of integral quantities from PNDS readings are performed by the computer code REDPAR based on parameterisation[3] and reduction[4] methods. The advantage of the reduction method is the strict mathematical formalism of error minimising developed for different models of measurement scheme and taking into account *a priori* additional information in optimal form. The neutron spectrum parameterisation method, realised in the program UNFOLD[3], provides a quick and effective algorithm of creation of the neutron spectrum library for derivation of additional information. The detailed description of the REDPAR code will be published later.

The advantages of PNDS are small dimensions, weak disturbance of the radiation field, low sensitivity to photons and charged particles and

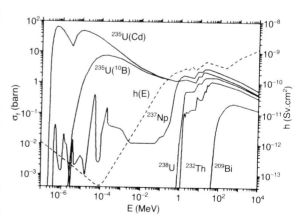

Figure 1. Neutron fission cross sections[3] of nuclides used in PNDS (solid curves) and heavy particle dose equivalent conversion factor[6,7] (dashed curve) as a function of neutron energy.

277

also a small anisotropy of the neutron response. The error of dose equivalent measurement behind accelerator shielding is less than 30% in most cases when H>1.5 mSv for a one month exposure. The use of passive detectors and spark track counting allows systematic calibration of personnel dosemeters to be carried out at a number of points at small expense.

Neutron spectra measurements behind the shielding of the 70 GeV IHEP proton accelerator have been performed for several years. The main results of investigations are presented elsewhere[2,3]. For calibration of personnel dosemeters, a number of points with high enough dose rates and different neutron spectra have been selected. The typical accelerator spectra measured by PNDS are shown in Figure 2 in comparison with the ^{252}Cf neutron spectrum. The highest energy spectrum is found behind the upper shielding at the point RM-43 (radiation monitor number). The spectrum at the point RM-7 is a typical one behind the side shielding in personnel workplaces. The softest spectrum is found near the entrance of the accelerator transport labyrinth (RM-13).

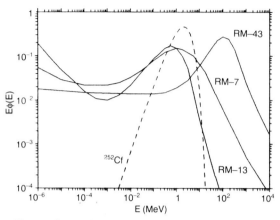

Figure 2. Typical neutron spectra behind IHEP accelerator shielding measured by PNDS (solid curves) in comparison with ^{252}Cf neutron spectrum (dashed curve).

CALIBRATION OF PERSONNEL DOSEMETERS BEHIND ACCELERATOR SHIELDING

According to the ICRU recommendations[5], the neutron personnel dosemeters should measure the individual dose equivalent, penetrating, $H_p(10)$ excluding secondary (n,γ) photon contribution measured by photon detectors. When performing a calibration, it is recommended that the ICRU 30 cm sphere be used as a phantom. As an approximation of the ICRU sphere, we use spherical polyethylene moderators of radiation monitors (RM) 25.4 cm in diameter. PNDS in a plastic container with a wall thickness of about 1 g.cm^{-2} is placed on the phantom surface near the personnel dosemeters. The readings of personnel dosemeters are interpreted in the term of heavy particle dose equivalent at a depth of PNDS location:

$$H = \int \phi(E) \, h(E) \, dE \quad (1)$$

where $\phi(E)$ is the neutron energy spectrum measured by PNDS. The conversion factor h(E) for the heavy particle component of neutron dose equivalent is equal to the product of the kerma factor and the quality factor:

$$h(E) = k(E) \, Q_k(E) \quad (2)$$

The values of h(E) below 20 MeV have been taken from Alekseev et al[6], at higher energies we have used the dose equivalent conversion factor at a depth of 5 mm in a 30 cm slab from Belogorlov et al[7]. The energy dependence of conversion factor h is shown in Figure 1.

The experimental data on dose equivalent

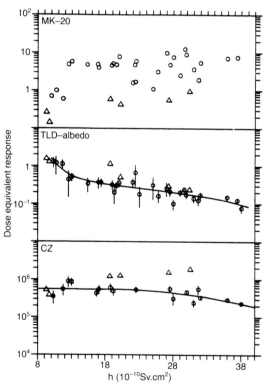

Figure 3. Dose equivalent response of MK-20 film (relative to ^{239}Pu–Be), TLD albedo dosemeter (Gy.Sv^{-1}) and etched track detector CZ (track.cm^{-2}.Sv^{-1}) as a function of neutron spectrum conversion factor $\bar{h}=H/\phi$. (Δ) reference fields[8]; (o) accelerator fields, (----) calibration curves.

response of different personnel dosemeters behind accelerator shielding are shown in Figure 3 as a function of the spectrum hardness parameter $\bar{h} = H/\phi$ (mean fluence-to-dose equivalent conversion factor). The choice of this parameter for specifying the neutron spectrum hardness is explained by much better accuracy of its estimation by PNDS as compared with an average energy $\bar{E}^{(3)}$. The data obtained in IHEP neutron reference fields[8] are also given in Figure 3. These reference fields have been established by using ^{239}Pu–Be and ^{252}Cf radionuclide sources (bare sources, ^{252}Cf in iron and polyethylene spheres 30 cm in diameter, ^{252}Cf in concrete labyrinth behind the iron slab 50 cm thick and round the corner of the first labyrinth section). The energy spectra of neutron fields have been measured by multisphere and organic scintillation spectrometers. The personnel dosemeters were exposed in this case on the surface of an IAEA water phantom. Their readings have been interpreted in terms of heavy particle dose equivalent at 1 cm depth in a 30 cm diameter cylindrical phantom calculated by Cross and Ing[9].

The data for the MK-20 nuclear film[10] (similar to NTA film), TLD albedo dosemeter and etched track detector CZ have been obtained in the experiments. The MK-20 film is a routine personnel neutron dosemeter used now in IHEP. Its disadvantage is a laborious visual track counting by microscope and the complication of this process automation. Albedo dosemeters on the basis of ^{6}LiF and ^{7}LiF TLD are very promising from this point of view. Figure 3 shows the experimental data for one of the albedo dosemeters studied by us recently. The etched track detector CR-39 is one of the most promising for neutron personnel monitoring. Recently, the etched track detector CZ, having characteristics close to CR-39, has been developed in the Institute of Chemical Physics (Moscow). For processing of CZ detectors, we used chemical etching at conditions 6N NaOH, 65°C, 9–11 h. Tracks were counted on both sides of the detector by an optical microscope. The results have been obtained by irradiation of CZ with a 2 mm thick polyethylene radiator.

The range of MK-20 dose equivalent response behind accelerator shielding is greater than a factor of 10. It should be noted that the correlation between spectrum hardness and MK-20 response is practically absent. The response of MK-20 film in accelerator neutron fields is much higher as compared with reference fields. This is explained by the high sensitivity of MK-20 film to charged particles in accelerator fields. The variation of albedo dosemeter response exceeds a factor of 10. The problem is also of high measurement errors in the radiation fields with evident charged particle contribution. In contrast to MK-20 film, albedo dosemeter response is well correlated with neutron spectrum hardness. The data for etched track detector CZ have a small scatter behind accelerator shielding in comparison with the nuclear film and albedo dosemeter. The CZ response in accelerator fields (stray radiation) slightly decreases when the neutron spectrum hardness increases, in contrast to the reference field data (normal irradiation). This is explained by the anisotropy of the CZ response and also by differences in the shapes of reference and accelerator spectra (Figure 2).

Disadvantages of the CZ detector are the limited range of dose equivalent measurement and rather high background ($\geq$400 tracks.cm^{-2} for two sided counting). The main shortcoming of the MK-20 film and albedo dosemeter is a great dependence of the dose response on the neutron spectrum and the component composition of radiation. Calibrations in personnel workplaces for taking into account such dependence are not acceptable at charged particle accelerators for which radiation fields are highly variable in space and time.

The most promising approach is the use of combined albedo–track dosemeters which provides an intrinsic correction of the calibration factor[2]. For example, the values of MK-20 response are given in Figure 4 as a function of MK-20 and albedo dosemeter reading ratio for the combined dosemeter MK+TLD. In spite of the absence of correlation between MK-20 response and spectrum

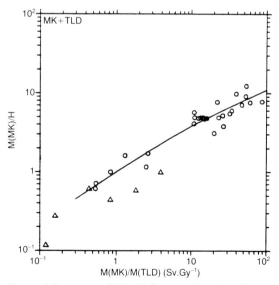

Figure 4. Response of MK-20 film as a function of MK-20 and albedo dosemeter reading ratio for combined dosemeter MK+TLD. (Δ) reference fields[8]; (O) accelerator fields; (----) calibration curve.

hardness (Figure 3), the data for accelerator fields are well described by a unique calibration curve with small scatter of points. Experimental data for CZ+TLD combined dosemeter are shown in Figure 5. A calibration curve is drawn through the points obtained behind the side accelerator shielding where the personnel workplaces are. The combined dosemeter technique provides, in this case, only a small decrease of measurement error because the CZ detector has a good accuracy itself behind the side shielding.

CONCLUSION

The calibration method described in this paper provided a number of new results on the response of personnel neutron dosemeters behind IHEP accelerator shielding. Its essential distinction from traditional calibration methods is the measurement of neutron dose equivalent on the phantom surface near personnel dosemeters. Such an approach allows us to take into account the contribution to the dose equivalent and personnel dosemeter reading of radiation scattered by the phantom in the same way independently of the angular distribution of incident neutrons. The calibration technique presented is mostly close to the real conditions of personnel dosemeters irradiation and can be recommended for practical use at high energy accelerators.

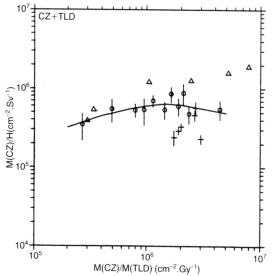

Figure 5. Response of etched track detector CZ as a function of CZ and albedo dosemeter reading ratio for combined dosemeter CZ+TLD. (Δ) reference fields[8]; accelerator fields: (ϕ) side shielding, (+) upper shielding; (----) calibration curve.

ACKNOWLEDGEMENTS

The author would like to thank Dr V. N. Lebedev for support of this work, E. P. Korshunova, V. A. Volkova and E. G. Spirov for help in performing measurements, A. V. Makagonov for the help in REDPAR code development.

REFERENCES

1. Piesch, E. *Calibration Techniques for Personnel Dosemeters in Stray Neutron Fields.* Radiat. Prot. Dosim. **10**(1-4), 159-173 (1985).
2. Sannikov, A. V., Vorob'ev, I. B., Korshunova, E. P. and Nikolaev, V. A. *Response of Combined Albedo-Track Neutron Personnel Dosemeters behind IHEP Proton Synchrotron Shielding.* IHEP Preprint 89-211, Protvino (1989).
3. Sannikov, A. V. *Passive Neutron Dosemeter–Spectrometer for High Energy Accelerators.* IHEP Preprint 90-133, Protvino (1990) (In Russian).
4. Pyt'ev, Yu, P. *Mathematical Methods of Experiment Interpretation.* (Highest School, Moscow) (1989) (In Russian).
5. ICRU. *Determination of Dose Equivalents from External Radiation Sources-Part 2.* Report 43 (Bethesda, MD: ICRU Publications) (1988).
6. Alekseev, A. G., Golovachik, V. T. and Savinsky, A. K. *Kerma Equivalent Factor for Photons and Neutrons up to 20 MeV.* Radiat. Prot. Dosim. **14**(4), 289-298 (1986).
7. Belogorlov, E. A., Golovachik, V. T., Lebedev, V. N. and Potjomkin, E. L. *Depth Dose and Depth Dose Equivalent Data for Neutrons with Energy from Thermal up to Several TeV.* Nucl. Instrum. Methods **199**, 563-572 (1982).
8. Britvich, G. I., Volkov, V. S., Kolevatov, Yu. I., Kremenetsky, A. K., Lebedev, V. N., Mayorov, V. D., Rastsvetalov, Ja. N., Trykov, L. A. and Chumakov, A. A. *Spectra and Integral Values of Reference Neutron Fields from Radionuclide Neutron Sources.* IHEP Preprint 90-48, Protvino (1990) (In Russian).
9. Cross, W. G. and Ing, H. *Conversion and Quality Factors Relating Neutron Fluence and Dosimetric Quantities.* Radiat. Prot. Dosim. **10**(1-4), 29-42 (1985).
10. Lebedev, V. N. and Sannikov, A. V. *Response Functions of IFK-2.3 and DVS-1 Personnel Dosemeters in the Neutron Energy Range from Thermal to 5 GeV.* IHEP Preprint 86-1. Serpukhov (1986) (In Russian).

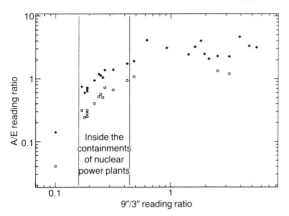

Figure 2. Dependence of the A/E reading ratio (see text) for the albedo dosemeters on the 9"/3" spectral index. (♦) IAR, (○) KfK.

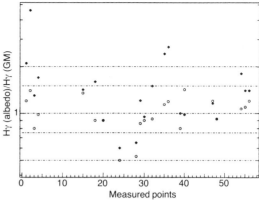

Figure 4. Comparison between the dose equivalent of the gamma component in mixed neutron–gamma fields measured with albedo neutron dosemeters and with a GM counter. (♦) IAR, (○) KfK.

measurements with Bonner spheres; thus the definition of an alternative spectral index based only on the albedo dosemeter results would be attractive. The ratio of the A/E compartment readings has been tested for this purpose, as it has been previously used by other authors[6]. The results are presented in Figure 2. A unique relationship between this quantity and the former spectral index is obtained up to a 9"/3" ratio of 0.5, which turns out to be just the useful range in the nuclear power plants. Each dosemeter has its own range of A/E ratios.

The response of both dosemeters has been plotted against the specific A/E ratio in Figure 3. For the range of ratios encountered in the nuclear power plants (Figure 2), an average value of the response may be used in personnel dosimetry with an accuracy of ± 50%. This result supports the published recommendations concerning the

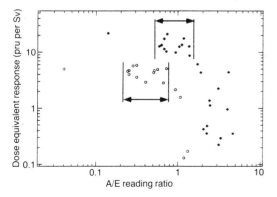

Figure 3. Dose equivalent response of the albedo neutron dosemeters plotted against the A/E reading ratio (pru = photon response units, see text); the arrows show the range of values in nuclear power plants. (♦) IAR, (○) KfK.

response of the KfK albedo dosemeter in heavily shielded areas, which are based on a field calibration[6]. The A/E ratio provides a simple check of the validity of the assumption concerning the constant sensitivity.

A practical albedo dosemeter should also be able to measure the dose equivalent of the gamma component in mixed fields using the indication of the TLD–700 chips in the E compartment. The response of both albedo dosemeters for the gamma component normalised to the directional dose equivalent measured by the GM counter is represented in Figure 4. The accuracy is within a factor of 2 for the KfK albedo dosemeter in the fields investigated, which is satisfactory for radiation protection purposes. The uncertainty is sometimes greater for the IAR albedo dosemeter because of the effect of neutron capture gamma rays in the cadmium shielding.

CONCLUSIONS

The comparison between the two types of albedo neutron dosemeters investigated in this study leads to the following conclusions:

(i) The IAR prototype albedo dosemeter has a higher response than the KfK dosemeter, which is an interesting feature when taking into account the recent ICRP 60 recommendations.
(ii) The dose equivalent due to the gamma component of a mixed radiation field can be determined with a better accuracy by the KfK albedo dosemeter.
(iii) Both types of albedo dosemeters may be employed for personnel neutron dosimetry in nuclear power plants with a reasonable accuracy by using an average calibration

factor, independent of the neutron spectrum.

(iv) The A/E readings ratio is a spectral index of limited usefulness, but it may be applied in nuclear power plants to verify the conditions allowing the use of a constant calibration factor.

ACKNOWLEDGEMENT

This work has been supported by the Swiss Nuclear Safety Inspectorate, Villigen.

REFERENCES

1. Vylet, V. *Détermination des Champs de Radiations Neutroniques par la Méthode des Sphères de Bonner.* Thèse No 671, Ecole Polytechnique Fédérale de Lausanne, Lausanne (1987).
2. Aroua, A. *Etude des Champs Neutroniques dans les Centrales Nucléaires et de l'Influence de leur Diversité sur la Détermination des Grandeurs de la Protection Radiologique.* Thèse No 942, Ecole Polytechnique Fédérale de Lausanne (1991).
3. Aroua, A., Azimi, D., Boschung, M., Cartier, F., Grecescu, M., Prêtre, S., Valley, J. F. and Wernli, Ch. *Rapport sur la Campagne de Mesure des Neutrons Effectuée dans les Centrales Nucléaires Suisses.* Joint IAR-PSI-HSK Report, Lausanne (1991).
4. Hankins, D. E. *Design of Albedo Neutron Dosimeters.* In: Neutron Monitoring for Radiation Protection Purposes, STI/PUB/318 (Vienna: IAEA) p.15 (1973).
5. Piesch, E. and Burgkhardt, B. *Albedo Dosimetry System for Routine Personnel Monitoring.* Radiat. Prot. Dosim. **23**, 117-120 (1988).
6. Burgkhardt, B. and Piesch, E. *Field Calibration Technique for Albedo Neutron Dosimetry.* Radiat. Prot. Dosim. **23**, 121-126 (1988).
7. Griffith, R.V., Palfavi, J. and Madhavanath, U. *Compendium of Neutron Spectra and Detector Responses for Radiation Protection Purposes.* Technical Reports Series No. 318 (Vienna: IAEA) (1990).
8. Vylet, V., Swaja, R. E., Prêtre, S., Valley, J.-F. and Lerch, P. *On the Use of the "9/3" Ratio with Albedo and Fission Track Neutron Personnel Dosemeters.* Radiat. Prot. Dosim. **27**, 29-33 (1989).
9. ICRP. *Data for Protection against Ionizing Radiation from External Sources.* Publication 21 (Oxford: Pergamon Press) (1973).
10. Hankins, D.E. *The Effect of Energy Dependence on the Evaluation of Albedo Neutron Dosimeters.* In:Proc. 9th Mid-Year Topical Symp. of the Health Physics Society, Denver, Colorado (1975).

PERSONAL ALBEDO NEUTRON DOSEMETER USING HIGHLY SENSITIVE LiF TL CHIPS

D. Nikodemová, A. Hrabovcová, M. Vičanová and S. Kaclík
Institute of Preventive and Clinical Medicine
Bratislava, Limbová 14, Czechoslovakia

Abstract — The reduction of the annual dose limits recommended in ICRP 60, together with the proposed higher radiation weighting factors of neutrons necessitates the introduction of a more sensitive personal neutron dosemeter, which would enable the separation of different kinds of radiation and would meet the requirements of accuracy. Further development of the albedo neutron dosemeter, used in this Institute in combination with Pershore CR-39 (500 μm) SSTD as an operative individual dosemeter, has therefore been carried out by using the new Chinese ^{6}LiF(Mg,Cu,P) and ^{7}LiF(Mg,Cu,P) TLD. The results of the dosimetric properties investigation, as well as the comparison of the new TLDs with other widely used pairs of ^{6}LiF and ^{7}LiF, are discussed. The possibilities for measuring neutron dose equivalents with sufficient sensitivity and accuracy are demonstrated in several examples.

INTRODUCTION

An albedo dosemeter using a pair of TLDs is used in this Institute in combination with CR-39 SSTD as an operative individual neutron dosemeter. Further development of this dosemeter has lead to the usage of new Chinese TL materials ^{6}LiF (Mg,Cu,P) GR-206 and ^{7}LiF (Mg,Cu,P) GR-207[1,2], instead of the Harshaw TL-600 and TL-700 LiF detectors.

The aim of this work is to describe the dosimetric characteristics and to compare the properties of the Chinese TL materials with the routinely used Harshaw TLDs. The advantages of the modified neutron dosemeter are presented.

MATERIALS AND METHODS

The responses of TLD-600 and TLD-700 Harshaw LiF chips ($3.2 \times 3.2 \times 0.13$ mm^3) as well as the responses of the GR-206 and GR-207 (Beijing Shiying Radiation Detector Works) chips (4.5 mm × 8 mm diam.) were measured with a Harshaw Model 2000A/B TLD reader, using an inert nitrogen flow of 4 to 5 l.min^{-1}. For annealing the TL detectors an oven with air circulation, whose temperature was controlled within ± 1°C was used. Annealing parameters for Harshaw TLD are 1 h at 400°C + 2 h at 100°C, for the Chinese TL materials 10 min at 240°C. The heating cycle consists of a pre-heat at 100°C and a linear heating rate of 6.43°C.s^{-1} up to 250°C. The total heating time was 45 s. Table 1 summarises the characteristics of the types of Tl materials investigated.

A calibration curve[3] obtained by measurements of neutron spectra in various neutron fields serves for the neutron dose estimation. The calibration factors are determined through the known ratio of the responses of thermal neutron detectors inside 76.2 mm and 254 mm polyethylene spheres.

DESCRIPTION OF EXPERIMENTS AND RESULTS

The albedo dosemeters using either the TL-600 and TL-700 Harshaw detectors or GR-206 and GR-207 Chinese chips were irradiated in mixed neutron gamma fields of a bare ^{252}Cf neutron source (Amersham type CVN 330, S=1.78×10^8s^{-1} to the date 20 Feb. 1991) and of the same ^{252}Cf source inside PE spheres of diameter 76.2 mm, 101.6 mm, 127 mm, 203.2 mm and 254 mm. In these experiments the TL-700 and GR-207 were used as gamma ray discriminators. The response of the tested materials to thermal neutrons was studied on a standard thermal neutron assembly, where standard thermal neutron flux density ϕ_{th} is 4760 s^{-1}.cm^{-2}. The responses are shown in Table 2 together with analogue responses for neutrons from a bare ^{252}Cf neutron source.

DISCUSSION AND FINAL REMARKS

(1) The experimental results show that

Table 1. Background, detection threshold, uncertainty and relative photon response (^{60}Co) for LiF (Mg,Ti) and LiF (Mg,Cu,P) chips.

	TLD-700	TLD-600	GR-207	GR-206
Background (μGy)	900	1000	40	40
Standard deviation of background (μGy)	60	40	1.6	2.5
Detection threshold (μGy)	180	120	5	7.5
Coefficient of variation (%) at tissue kerma (^{60}Co, k=7 mGy)	6	4	5	5
Relative photon response	1	0.91	24.4	22

^{6}LiF(Mg,Cu,P) and ^{7}LiF(Mg,Cu,P) are 22-24 times more sensitive for gamma rays than ^{6}LiF(Mg,Ti) and ^{7}LiF(Mg,Ti).

(2) The response of ^{6}LiF(Mg,Cu,P) for thermal neutrons was found to be only 4 times higher than that of ^{6}LiF(Mg,Ti). In various neutron fields (produced by ^{252}Cf source inside PE moderators of different diameters), the response (for ^{252}Cf bare) is between 3.4 and 7.2 times larger for ^{6}LiF(Mg,Cu,P) than for ^{6}LiF(Mg,Ti).

(3) Based on the analysis of the experimental results one can conclude that the use of the new pair of GR-206 and GR-207 detectors, in the combined TLD/CR-39 neutron dosemeter, has the advantage of low detection threshold and reduction of uncertainties at low dose measurements.

(4) In order to prove the acceptable photon to neutron dose ratio further experiments are needed. This should involve testing of the neutron dosemeters in more different mixed fields such as those to be found in routine practice.

Table 2. Responses of TL detectors to neutrons.

TL material	Thermal neutrons Neutron TL reading* per 10^8 cm^{-2}	Neutrons from ^{252}Cf bare source Neutron TL reading* per 10^8 cm^{-2}
TLD-600	7.85	0.84
GR-206	31.0	6.07

* Response R = M (nC per 10^8 cm^{-2}).

REFERENCES

1. Wang Shoushan, Chen Goulong, Wu Fang, L, Ynanfang, Zha Ziying and Zhu Sianhuan. *Newly Developed Highly Sensitive LiF(Mg,Cu,P) TL Chips with High Signal-to-Noise Ratio.* Radiat. Prot. Dosim. **14**(3), 223-227 (1986).
2. Horowitz, J. S. and Shachar, B. B. *Thermoluminescent LiF: Cu,Mg,P for Gamma Ray Dosimetry in Mixed Fast Neutron-Gamma Radiation Fields.* Radiat. Prot. Dosim. **23**(1/4), 401-404 (1988).
3. Nikodemová, D., Hrabovcouá, A., Salacká, A. and Komockov, M. M. *The Use of Albedo Dosemeters for Evaluating Personal Doses in Mixed (n,gamma) Radiation Fields).* Jad. Energ. **36**(2), 58-64 (1990).

CALIBRATION OF THE BRAZILIAN ALBEDO DOSEMETER AT A CV-28 CYCLOTRON

P. W. Fajardo† and C. L. P. Maurício‡
†Instituto de Engenharia Nuclear – CP 68550 CEP 21945, Brazil
‡Instituto de Radioproteção e Dosimetria – CP 33750, CEP 22602
Rio de Janeiro, Brazil

Abstract — This work deals with the analysis and application of neutron measurement techniques in order to gain information on the neutron spectrum and neutron dose equivalent at several representative work places of the cyclotron facility of the Instituto de Engenharia Nuclear at Rio de Janeiro, Brazil. The foil activation technique was used for the neutron spectra measurement and the neutron dose equivalent was measured with the single sphere albedo technique, using BF_3 and ^3He proportional detectors and a ^6LiI scintillation detector. The results from these different techniques were compared to determine the most appropriate method to calibrate the Brazilian albedo dosemeter for use in the particular radiation environment of particle accelerators.

INTRODUCTION

Particle accelerators produce neutron fields with broad energy spectra. As all personal neutron dosemeters available are strongly energy dependent, it is necessary to calibrate them directly at representative workplaces. Thus, it was very important to check the possibility of using the Brazilian albedo dosemeter in accelerators and, if possible, to establish a calibration procedure for it. Different neutron dose equivalent measurement techniques were compared in order to select the technique that would give the reference neutron dose equivalent value to calibrate the albedo dosemeter.

The first accelerator chosen for this work was the CV-28 cyclotron of the Instituto de Engenharia Nuclear (IEN), Rio de Janeiro, Brazil.

THE CYCLOTRON OF THE INSTITUTO DE ENGENHARIA NUCLEAR

The CV-28 cyclotron of IEN is installed inside a room with 1.8 m of ordinary concrete shielding (Figure 1). The main beam line is divided using a magnet to provide two beam exits inside the cyclotron room (numbers 2 and 6) and three beam exits in the experimental area (numbers 3, 4 and 5) arriving in three rooms shielded with 0.6 m of ordinary concrete. The most used are: number 6 for radioisotope production, mainly ^{123}I and ^{67}Ga for nuclear medicine, using the $^{124}Te(p,2n)$ ^{123}I and $^{67}Zn(p,n)^{67}Ga$ reactions respectively, and number 5 for neutron production studies using the $D(d,n)$ reaction.

The neutron spectra around the cyclotron come from a combination of these nuclear producing reactions with the nuclear reactions of the accelerated particles with the components of the cyclotron and the beam line, which are mainly made of copper, iron, aluminium and steel as well as neutrons scattered in the concrete shielding.

THE BRAZILIAN ALBEDO DOSEMETER SYSTEM

The Brazilian albedo dosemeter was developed by the Instituto de Radioproteção e Dosimetria (IRD), Rio de Janeiro, Brazil, in collaboration with the Gesellschaft für Strahlen- und Umweltforschung MbH (GSF), Munich, Germany. The response of the dosemeter had not so far been investigated in the mixed fields around a particle accelerator.

The Brazilian albedo dosemeter[1] consists of a nylon hemisphere (diam. 32 mm) and a cylindrical extension of 10 mm. A pair of commercial $^6LiF:Mg,Ti$ (TLD-600) and $^7LiF:Mg,Ti$ (TLD-700) $3\times3\times0.9$ mm^3 chips from the Harshaw Chemical Company is arranged at the symmetry axis. The nylon front curve surface is covered with 1 mm of a boron mixture which contains approximately 0.8 $g.cm^{-3}$ of boron. The encapsulation is made of aluminium. The dosemeter is used attached to a belt to maintain it in close contact with the body. The effect of angular dependence is minimised by using two symmetrically opposite dosemeters[2]. The TL dosemeters are evaluated in a Teledyne TLD reader 7300C connected to a clone IBM

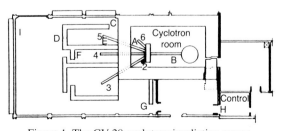

Figure 1. The CV-28 cyclotron irradiation rooms.

PC/XT microcomputer[3]. This home-made interface facilitates computerised TLD glow curve analysis.

In order to obtain information on the spectrum moderation which can be obtained from the ratio of incident to albedo neutron readings, a pair of TLD600/TLD700 was fixed to the front surface of each albedo capsule. This technique was successfully applied in the Karlsruhe albedo dosemeter to correct for energy dependence of the albedo neutron response[4].

NEUTRON SPECTRA MEASUREMENTS

The neutron spectra inside the cyclotron room were measured with the foil activation technique. Manganese, copper, indium, iron, nickel, and aluminium foils were irradiated at the surface of a spherical phantom positioned at points A and B (Figure 1). The activated foils were analysed by gamma spectrometry. The neutron spectra (Figure 2, a and b) were obtained using the computational system SAIPS which employs the SAND II code[5].

The neutron spectrum measured at position A shows a larger contribution of fast neutrons than that measured at position B, due to the proximity of the beam exit, but in both cases the presence of thermal neutrons is very marked due to the contribution of scattered neutrons. The result obtained indicates the possibility of using the Brazilian albedo dosemeter in this particular neutron field. It also shows that any equipment with metallic parts could be undesirably activated inside the cyclotron room during irradiation.

CALIBRATION OF THE ALBEDO DOSEMETER

Dose equivalent measurements

The decision was made to calibrate the albedo dosemeter directly at workplaces due to its high energy dependence. The performance of the neutron monitors, Snoopy of REM-RAD Monitoring Systems using a BF_3 detector, the Dineutron of Nardeux using a ^{3}He detector, the Nemo of Texas Nuclear using a ^{6}LiI detector and the single sphere albedo system[6] developed for field calibration of albedo dosemeters at the Kernforschungszentrum Karlsruhe, Germany, were compared in order to select the technique that would give the reference neutron dose equivalent value to calibrate the Brazilian albedo dosemeter. The single sphere albedo technique consists of a 30 cm spherical phantom, with the Karlsruhe albedo dosemeter fixed at the surface and a pair of TLD-600/TLD-700 positioned at the centre of the sphere. A linear combination of the response of the incident, albedo and centre components provides the dose equivalent[6]. All dose equivalent meters were exposed at position C outside the cyclotron room during the ^{124}Te(p,2n) ^{123}I irradiation at beam exit 6. In this moderated spectrum, Nemo and Snoopy gave double the value obtained by the single sphere albedo technique. This overestimation was expected as the neutron monitors are calibrated for Am–Be sources. In turn, the Dineutron gave lower results.

The single sphere albedo technique was selected as more appropriate for the calibration of the albedo dosemeter. This technique has some advantages: it integrates radiation for long periods of time at places with low neutron fluence rate and the Brazilian albedo dosemeter is exposed exactly in the same conditions as the sphere. This is especially important because of the high temporal and spatial variations of the neutron field. Also, the single sphere albedo system could be used inside the cyclotron room because it has no metallic parts to be activated by the high thermal neutron fluence present.

The selection of the calibration points to be taken as representative localisations for workplaces was made by taking into account the area occupation factor and the results of the routine

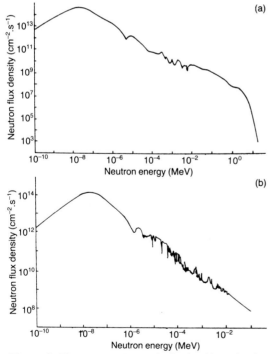

Figure 2. Neutron spectrum obtained with activation foils inside the cyclotron room during ^{124}Te(p,2n) ^{123}I, 24 MeV, 5 µA irradiation: (a) at position A, near the beam exit; (b) at position B, in the middle of the room.

radiological monitoring[7] which indicated the points with higher dose equivalent rate and provided the exposure time. Figure 1 shows the selected points A, B, C, D, E, F, G, H and I.

The neutron dose equivalent obtained with the single sphere albedo system at the cyclotron during specific irradiations and from a long-term integration that includes all types of irradiations are shown in Table 1.

Response of the albedo dosemeter

The relative dose equivalent responses (ratio of the ^{60}Co radiation response to the response to neutron radiation) for the Brazilian albedo dosemeter were found for the selected places and plotted against the ratio of incident to albedo neutron readings (Figure 3). Three distinct regions can be visualised: beam region (position D), scattered region inside the irradiation rooms (position A, B, and E) and shielded region (positions C, G, H and I). In routine applications, when the workers are protected by the concrete shielding, a single mean value of the calibration factors could be used at the different places, which much simplifies the individual dosimetry task. In the case of accidental exposure inside the irradiation rooms, as originally the dosemeter has no incident detector, there is need for a complementary device, such as an activation detector, to alert for the use of the correct calibration factor. The relative dose equivalent responses of the Brazilian albedo dosemeter at the cyclotron of IEN range from 0.34 to 4.40 (Figure 3).

CONCLUSION

The Brazilian albedo dosemeter was calibrated at the neutron field of the cyclotron of the Instituto de Engenharia Nuclear for routine and accidental applications. The reference dose equivalents were measured with the single sphere albedo technique that was shown to be the most appropriate for the variable and moderated neutron fields around the cyclotron. With this technique it was possible to calibrate the albedo dosemeter at the most representative positions around the cyclotron.

The calibration of the albedo dosemeter at the cyclotron of IEN enables an effective individual neutron monitoring of the workers in this area. The calibration factors found for the modified Brazilian albedo dosemeter can also be applied in all accelerators and neutron generators in Brazil if the reading ratio of incident and albedo neutrons is used. The results of this approach are in good agreement with those found with the Karlsruhe albedo dosemeter in similar neutron fields[4].

Table 1. Results of the single sphere albedo technique at the CV-28 cyclotron of IEN.

Nuclear reaction	Position	H_{MADE} (mSv)	Exposure time
Te (p,2n)I 24 MeV 1 μA Exit 6	A	26.0 ± 5.2	1 min
	B	18.0 ± 3.6	1 min
Zn (p,n)Ga 24 MeV 9 μA Exit 6	A	50.0 ± 10.0	1 min
	B	25.0 ± 5.0	1 min
D (d,n)He 11 MeV 0.8 μA Exit 5	D	3.3 ± 0.7	1 h 43 min
	E	45.0 ± 9.0	55 min
	I	0.08 ± 0.02	22 h 18 min
	C	5.0 ± 1.0	30 min
	F	0.39 ± 0.08	3 h 42 min
ALL	C	1.4 ± 0.3	400 h
	G	0.11 ± 0.02	430 h

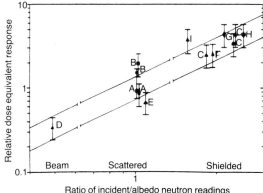

Figure 3. Relative dose equivalent response, i.e. the ratio of the photon to neutron response of the Brazilian albedo dosemeter at the CV-28 cyclotron of IEN in relation to the ratio of incident to albedo neutron readings. (■) Te(p,2n)I. (□) Zn(p,n) Ga. (▲) D(d,n). (♦) All.

REFERENCES

1. Carvalho, W. B. D. *Monitor Pessoal de Nêutrons baseado na Técnica de Albedo*. Master degree thesis. Rio de Janeiro, IME (1975).
2. Knöfel, T. M. J. and Schraube, H. *Status Report on Current Work in Neutron Health Physics Dosimetry in the Frame of the Brazilian-German Cooperation*. Private communication (1981).

3. Maurício, C. L. P., Becker, P. H. B. and Kasprzykowski, C. F. A. *TLD Data Acquisition and Analysis System for Neutron Individual Monitoring.* Radiat. Prot. Dosim. **34**(1/4), 161-163 (1990).
4. Burgkhardt, B. and Piesch, E. *Field Calibration Technique for Albedo Neutron Dosemeters.* Radiat. Prot. Dosim. **23**, 121-126 (1988).
5. International Nuclear Data Committee. *Neutron Dosimetry System SAIPS: Manual for Users and Programmers (vers. 87-02).* INDC(CCP)-285/GR (1987).
6. Piesch, E., Burgkhardt, B. and Comper, W. *The Single Sphere Albedo System — A Useful Technique in Neutron Dosimetry.* Radiat. Prot. Dosim. **10**(1/4), 147-157 (1985).
7. Fajardo, P. W., Teixeira, M. V., Santos, I. H. T. and Pujol Filho, S. V. *Radioproteção Ocupacional nos Laboratórios do Cíclotron e de Produção de Radioisótopos.* In: Congresso Geral de Energia Nuclear, 3. Rio de Janeiro (1990).

NEUTRON RESPONSE OF LiF TL DETECTORS

T. Hahn, J. Fellinger, J. Henniger, K. Hübner and P. Schmidt
Dresden University of Technology, Institute for Radiation Protection Physics
Mommsenstrasse 13, Dresden 0-8027, Germany

Abstract — For the application of thermoluminescent (TL) detectors in neutron dosimetry the detector response over the whole energy range must be known. Therefore the energy deposition by secondary ions and gamma radiation generated as a result of neutron interactions in the detector material or in the surrounding material (detector holder, radiator) is of fundamental importance. The different neutron response components can be calculated by means of computer programs. This paper deals with the method and results of response calculations. The effect of the absorbed dose caused by heavy charged particles to the detector reading (so-called light conversion factors) were experimentally determined for various ions and TL materials. The knowledge of these effects allows the calculation of the intrinsic response as well as of the response component caused by heavy charged particles from a hydrogen containing radiator. For different TL detectors the calculated responses are compared with experimental data in neutron fields. The experimental and theoretical results are in a good agreement.

INTRODUCTION

The application of thermoluminescent (TL) detectors for the dosimetry in mixed neutron–gamma fields requires knowledge of their response to neutrons over a broad energy range. The neutron fluence response R_Φ as function of the neutron energy E_n and the direction of incidence Ω is defined as the ratio

$$R_\Phi(E_n, \Omega) = \frac{M}{\Phi}(E_n, \Omega) \quad (1)$$

with the detector reading M for monoenergetic and monodirectional neutrons of the fluence Φ. R_Φ represents the sum of diverse components, such as the intrinsic component from secondary particles produced in the detector material via neutron reactions and the radiator component from neutron induced secondaries out of an adjacent radiator material. The amount of both depends on the cross section of the reactions involved and the energy of the secondary particles generated[1].

Table 1. Selected neutron interactions in LiF materials.

Reaction	Q value (MeV)
Elastic scattering ^{6}Li	–
Inelastic scattering ^{6}Li	–1.47128
^{6}Li(n,α)^{3}H	4.786
^{6}Li(n,p)^{6}He	–2.7336
Elastic scattering ^{7}Li	–
Inelastic scattering ^{7}Li	–0.477484
^{7}Li(n,d)^{6}He	–7.76382
Elastic scattering ^{19}F	–
Inelastic scattering ^{19}F	–0.11
^{19}F(n,p)^{19}O	–4.0363
^{19}F(n,d)^{18}O	–5.76892
^{19}F(n,t)^{17}O	–7.55613

According to Equation 1 the experimental determination of R_Φ necessitates irradiation experiments with monoenergetic neutrons. However, the feasibility of such experiments is limited, especially in the intermediate range of neutron energy[2,3]. We therefore decided to calculate the function $R_\Phi(E_n, \Omega)$. The calculation procedure developed involves two steps. The first describes the neutron induced production of the secondary charged particles and their energy transfer to the TL material and in the second step the conversion of the absorbed dose in the TL material to the TL reading is considered. The procedure can be applied for the calculation of both the intrinsic as well as the radiator response component[4,5].

CALCULATION OF THE NEUTRON RESPONSE

Neutron interactions

All neutron reactions in LiF used for the determination of the intrinsic response component[6] are summarised in Table 1. The function $R_\Phi(E_n)$ is considerably influenced by the 'giant' cross section σ and the Q value of the reaction ^{6}Li(n,α)^{3}H. Because of the 1/v dependence (v = neutron velocity) this cross section dominates for detectors containing ^{6}Li even for neutron fields with a small fraction of thermal neutrons. Combinations of ^{6}LiF and ^{7}LiF detectors are therefore used for a rough separation of the thermal neutron contribution in mixed neutron–gamma fields. TL materials investigated in this paper are given in Table 2.

Calculation procedure

For the calculation of the neutron response as a function of neutron energy the Monte Carlo program called NRES (neutron response) was developed[4]. This program enables the simulation of a large number of neutron life histories on their way in a LiF detector arrangement, including the radiator. For the description of the subsequent energy transfer of the produced secondary particles to the detector material the STOPOW code was used[7]. The energy deposited finally yields fluence-to-kerma conversion factors. Figure 1 shows the intrinsic fluence-to-kerma conversion factor for different materials. Whilst in the low neutron energy region the reaction ^{6}Li(n,α)^{3}H dominates, i.e. the different contents of ^{6}Li cause different factors, for high energy neutrons the functions approximate each other. The additional K/Φ component from the 1mm thick PE radiator to the sintered LiF detector is included in Figure 1.

TL response due to secondaries

For the calculation of the neutron fluence response the conversion of the energy deposited by the secondary heavy charged particles to the TL reading effect must be considered. For this purpose so-called light conversion factors for all secondaries as function of their energy must be known. The relative light conversion factor η is the ratio of the detector reading dM coming from a mass element where the dose dD was absorbed as a result of the energy deposition of the ion. This ratio is divided by the response R_γ for a gamma reference radiation, so that η values do not depend on TL equipment and evaluation parameters:

$$\eta = \frac{dM/dD}{R_\gamma} \quad (2)$$

Relative light conversion factors η were determined by experiment using proton, deuteron and alpha particle irradiation[4,8,9]. On the basis of these results η was derived as an unified function of $(Z_{eff}/\beta)^2$ as universal parameter (see Figure 2 for the sintered LiF detectors). From this function η values can also be estimated for those particles and particle energies which are experimentally not available. Results of this procedure are presented in Figure 3.

The TL reading M_{HCP} caused by a single heavy charged particle in a TL detector of the mass m_d is calculated according to

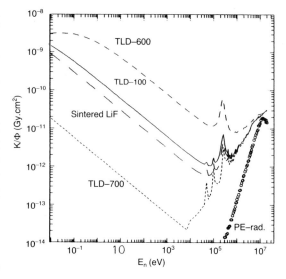

Figure 1. Kerma factors for LiF TLDs. (oooo) Additional kerma in sintered LiF TLD caused by a 1 mm thick polyethylene radiator.

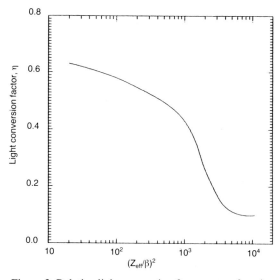

Figure 2. Relative light conversion factor η as a function of the parameter $(Z_{eff}/\beta)^2$.

Table 2. Properties of LiF TLDs.

Detector type	Material	Density (g.cm^{-3})	^{6}Li content (%)	Thickness (mm)	Manufacturer
TLD-100	LiF:Mg,Ti	2.64	7.5	0.89	Harshaw
TLD-600	LiF:Mg,Ti	2.64	95.62	0.89	Harshaw
TLD-700	LiF:Mg,Ti	2.64	0.07	0.89	Harshaw
LiF-sinter	LiF:Mg,Ti	2.21	3.7	0.6	Chemiew. Nünchritz

$$M_{HCP} = \frac{R_\gamma}{m_d} \int_{r_1}^{r_2} \eta(r) S(r) \, dr \qquad (3)$$

with S the actual stopping power along the path of the particle in the TL detector, r_1 and r_2 the starting and end points of neutron induced particle production or of particle incidence from a radiator. Finally, the sum over M_{HCP} values of all secondary particles produced by neutrons of the fluence Φ gives the detector reading M, and $R_\Phi(E_n,\Phi)$ can be calculated according to Equation 1.

RESULTS

In order to depict the neutron fluence response independent of readout parameters the presentation of the relative response R_{rel} has become standard. R_{rel} is given by

$$R_{rel} = \frac{R_\Phi}{R_\gamma} = \frac{1}{R_\gamma} \frac{M}{\Phi} \qquad (4)$$

and represents the gamma equivalent neutron dose reading M/R_γ related to the fluence Φ.

In Figure 4 the intrinsic component of R_{rel} for sintered LiF TL detectors is shown. Again, the function $R_{rel}(E_n)$ indicates the dominating role of the reaction $^6Li(n,\alpha)^3H$ in the thermal and intermediate neutron energy region. The radiator response component of R_{rel} (polyethylene radiator, thickness = 1 mm) for the same detector is also given in Figure 4. Due to the TL light self-attenuation during evaluation, two borderline cases can be distinguished. If the radiator covered plane of the detector faces the reader photocathode the upper curve is obtained. For the opposite case the curve below was calculated. With increasing neutron energy the range of recoil protons rises and the curves approximate one to another. The effect of TL light self-attenuation was treated by introducing an additional function in the integral kernel of Equation 3 on the base of the Bourger–Lambert law. Attenuation coefficients were determined experimentally[4].

In order to check the reliability of the calculation procedure the results were compared with experimentally determined neutron responses. Irradiations were carried out using thermal neutrons at the Physikalisch-Technische Bundesanstalt (PTB), Braunschweig, and with fast neutrons at the Rossendorf cyclotron. At the PTB reactor the thermal neutron beam is filtered by a bismuth block to reduce the accompanying gamma radiation[10]. The cyclotron neutrons produced via the reaction $^9Be(d,n)^{10}B$ are characterised by a mean neutron energy of 5.38 MeV (without shielding) and 2.7 MeV (behind 10cm iron), respectively.

In Table 3 experimental and theoretical data for the gamma-equivalent neutron dose readings are given. Obviously, there is an excellent agreement for the thermal neutron field. The results for the cyclotron neutrons differ by about 20%. It must thereby be taken into account, that uncertainties assumed for the neutron spectrum as well as for the gamma component could influence the results of the calculation. Also, in fields of fast neutrons

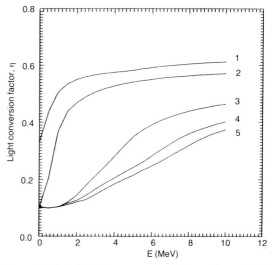

Figure 3. Relative light conversion factor η for selected ions in sintered LiF TLD (1: protons, 2: tritons, 3: alphas, 4: ^{6}Li-ions, 5: ^{7}Li-ions).

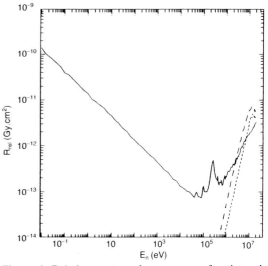

Figure 4. Relative neutron dose response for sintered LiF TLD. (——) intrinsic component, (— —) radiator component with the irradiated plane facing the photocathode, (-----) radiator component with the irradiated plane at the back.

more reaction channels are 'open', among them, reactions producing heavy particles like ^{19}F. For such particles the uncertainty of the relative light conversion factors η used in the calculations is larger. Nevertheless, the agreement between experiment and calculation is satisfactory.

CONCLUSION

The calculation method presented in this paper permits the determination of neutron responses in LiF TL detectors in a wide energy range. With the help of the Monte Carlo technique the generation of secondary ions is simulated and the energy transfer to the TL material is calculated. Using experimentally determined relative light conversion factors, the TL reading can be calculated as a function of neutron energies. The agreement between experimental and calculated results for the neutron response is reasonably good. If one were using relative light conversion factors for the evaluation of single glow peaks of a LiF detector, as measured by Schmidt et al[8,9], the neutron response for single peaks could be determined.

Table 3. Gamma equivalent neutron dose for sintered LiF.

E_n	D_{cal} (Gy)	D_{exp} ± 1 σ value (Gy)	$(D_{cal} - D_{exp})/D_{exp}$ (%)
n_{th}	0.117	0.117 ± 0.015	1 ± 15
2.7 MeV	0.0935	0.077 ± 0.0094	21 ± 21
7.8 MeV	0.142	0.111 ± 0.008	28 ± 10
with 1 mm thick polyethylene radiator:	0.174	0.154 ± 0.013	13.5 ± 95
contribution of radiator protons	0.0325	0.043 ± 0.021	25 ± 50

REFERENCES

1. Henniger, J., Hübner, K. and Pretzsch, G. *Calculation of the Neutron Sensitivity of TL-Detectors.* Nucl. Instrum. Methods **192**, 453-462 (1982).
2. Böttger, R., Guldbakke, S., Klein, H., Schölermann, H., Schuhmacher, H. and Strzelczyk, H. *Problems Associated with the Production of Monoenergetic Neutrons.* Nucl. Instrum. Methods **A282**, 358-367 (1989).
3. Gibson, J. A. B. *The Relative Tissue Kerma Sensitivity of Thermoluminescent Materials to Neutrons: A Review of Available Data.* CEC CENDOS Report EUR 10105 EN (Luxembourg: CEC) (1985).
4. Hahn, T. *Untersuchungen zum Einsatz von LiF-TL-Detektoren für die Havariedosimetrie in Kernkraftwerken,* Thesis, Technische Universität Dresden (1991).
5. Fellinger, J., Hahn, T., Henniger, J., Hübner, K. and Schmidt, P. *Fast Neutron Sensitivity of TL Detectors using the Proton Radiator Technique.* Isotopenpraxis **27**, in press (1991).
6. Pronyaev, V. G. *Japanese Evaluated Nuclear Data Library.* NDF-18 Rev. 2, Version 2, (1984).
7. Henniger, J. *Computer Program STOPOW 88.* Technische Universität Dresden, Sektion Physik, unpublished (1988).
8. Schmidt, P. *Untersuchungen zur Bestimmung der Nachweiseffektivität von TL-Detektoren bei der Bestrahlung mit schweren geladenen Teilchen.* Thesis, Technische Universität Dresden (1987).
9. Schmidt, P., Fellinger, J. and Hübner, K. *Experimental Determination of the TL Response for Protons and Deuterons in Various Detector Materials.* Radiat. Prot. Dosim. **33**, 171-173 (1990).
10. Alberts, W. G. and Dietz, E. *Filtered Neutron Beams at the FMRB — Review and Current Status.* PTB-Bericht PTB-FMRB-112 (1987).

TLD-300 DOSIMETRY AT CHIANG MAI 14 MeV NEUTRON BEAM

W. Hoffmann† and P. Songsiriritthigul‡
†Bergische Universität, Fachbereich Physik, Gaußstr. 20
5600 Wuppertal 1, Germany
‡Institute of Science and Technology Research and Development
Fast Neutron Research Facility
Chiang Mai University, Chiang Mai 50002, Thailand

Abstract — The different LET dependences of the low and high temperature glow peaks of CaF$_2$:Tm thermoluminescent material (TLD-300) allow the determination of neutron and gamma dose simultaneously. The method was calibrated for the 14 MeV neutron beam of the Fast Neutron Research Facility of Chiang Mai University. The relative neutron responses of both peaks in TLD-300 chips were found to be 0.10 and 0.32. Using this method various dose distributions of neutron and gamma dose in a water phantom were measured and compared with the results of GM counter measurements and Monte Carlo dose calculations.

INTRODUCTION

The 14 MeV continuous neutron beam of the Fast Neutron Research Facility (FNRF) of Chiang Mai University is generated from a 150 keV, 1.5 mA dueteron accelerator, where the deuterons hit a 300 GBq (8.5 Ci) tritium target. The beam is designed for neutron radiobiology experiments, determining survival curve parameters of cell cultures, chromosome abberations, micronuclei formations, etc.

Since neutrons in tissue or water phantoms are accompanied by photons, the resulting response or effect curves have to be unfolded to separate for the low LET effects of the secondary electrons produced by the photons and the high LET components like recoil protons and nuclear recoils produced by the neutrons. This requires the exact determination of neutron and gamma dose distributions.

The measurement of gamma dose in the presence of neutrons is normally done using GM techniques or proportional chambers used in micro-dosimetry. Recent studies at various neutron treatment facilities have shown[1-3], that the same measurement accuracy of about 10% can be achieved using TLD-300 dosemeters. Their advantage is fast analysis, high spatial resolution, high sensitivity and the possibility of *in vivo* dosimetry. Because of their small size, dose distribution is not affected and it is possible to measure a dose distribution in a single irradiation.

The possibility of separating γ and n doses with TLD-300 dosemeters results from the fact that the two main peaks of CaF$_2$:Tm (peaks 3 and 5) exhibit very different LET dependences[4] due to the different trap structures. Whereas the low temperature peak 1 at 140°C reduces its sensitivity ε_1 rapidly with increasing LET, the sensitivity ε_2 of the high temperature peak 2 at 230°C is constant up to 50 keV.μm^{-1} and then decreases.

The LET spectra measured in water for fast neutron fields[5] consist of a γ part up to y = 10 keV.μm^{-1} and a recoil proton part that extends from y = 10 keV.μm^{-1} to 125 keV.μm^{-1}. The normalised TLD readings corresponding to gammas and neutrons of the low and high temperature glow peaks should therefore be different according to

$$P_{1N} = \frac{P_1}{(P_1)_{60Co}} = D_\gamma + (\varepsilon_1 K)D_n \quad (1)$$

$$P_{2N} = \frac{P_2}{(P_2)_{60Co}} = D_\gamma + (\varepsilon_2 K)D_n \quad (2)$$

K = kerma − ratio CaF$_2$/H$_2$O = 0.27 at 14 MeV

From these equations the neutron, gamma and total dose can be derived:

$$D_n = \frac{P_{2N} - P_{1N}}{(\varepsilon_2 K) - (\varepsilon_1 K)}$$

$$D_\gamma = \frac{\frac{\varepsilon_2 K}{\varepsilon_1 K} P_{1N} - P_{2N}}{\frac{\varepsilon_2 K}{\varepsilon_1 K} - 1}$$

$$D_{total} = D_n + D_\gamma$$

The larger the difference $(\varepsilon_2 - \varepsilon_1)$ and the higher the kerma ratio K, the higher the accuracy of the dose determination. It should be realised, however, that the dose within the dosemeter can be different from the value KD$_1$(H$_2$O), because, depending on the size of the dosemeter and the energy of the neutrons, charged particle equilibrium may not be obtained and the dose contribution from secondary particles generated outside the dosemeter can be considerable. The effective neutron responses ($\varepsilon_1 K$) and ($\varepsilon_2 K$) therefore have to be determined for the actual dosemeter, phantom (or radiator) and neutron field.

EXPERIMENTAL METHODS AND PROCEDURES

The CaF$_2$:Tm (TLD-300) material was used as chips of size 3 mm × 3 mm × 0.9 mm. Before irradiation the chips were annealed at 400°C for 1 h. Each dosemeter was calibrated individually by ^{60}Co γ irradiation at an exposure level equivalent to 130 mGy (13 rad) in muscle tissue.

The radiation induced thermoluminescence was read out with a Harshaw 2080 pico processor, which was interfaced to a standard personal computer for fast analysis and storage of the glow curves. The heating rate was 10°C.s^{-1}. The TLDs were irradiated in a water phantom of size 30 × 30 × 30 cm^3, which was placed at a distance of 30 cm from the tritium target. The maximum dose rate at the surface of the phantom was 0.35 Gy.h^{-1}. The total dose was determined from standard Far West IC 18 and IC 17 chambers connected to Keithley 616 electrometers. A second chamber was placed close to the target to serve as a monitor chamber. The dose values from the air-filled chambers were determined using a stopping power ratio S^{TE}_{air} = 1.15 and a W value for the recoil protons of W_p = 36 eV.

The gamma dose was measured using a Far West GM 1 Geiger-Müller counter. The k_u value of the counter was taken to be 0.01 and the rad to Röntgen conversion factor for water 0.97. The dead time was measured to be 32.5 μs and the corresponding counting rate correction was applied.

Neutron and gamma dose were also calculated from the MORSE-CG computer code provided by KFA Jülich, Germany[6,7]. The code takes into account elastic neutron nucleus collisions including hydrogen collisions, non-elastic neutron nucleus collisions, inelastic collisions as well as capture and fission. The transport of the gammas produced is calculated including capture, fission, inelastic and elastic reactions. The multigroup cross section library was derived from the ENDF/B-IV data base[8]. The neutron dose is calculated using the kerma values from ICRU Report 26[9]. The number of energy bins was taken to be 118 between 14.9 MeV and 10^{-3} eV and the number of gamma energy bins 21 between 14 MeV and 10 keV. The code was installed on a VAX station 3200 and the number of events was so chosen that the neutron dose was calculated with a statistical error of better than 1% and the gamma dose better than 3%. The resulting computing time was 40 h for one dose distribution. It proved to be essential for the experimental setup simulation to include phantoms, holders, shielding walls and target flanges, because 10% of the γ dose in the phantom results from gammas produced outside the phantom.

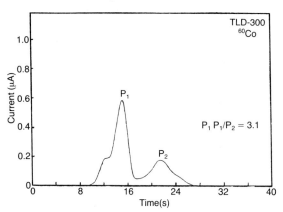

Figure 1. Glow curve of a TLD-300 dosemeter after ^{60}Co irradiation.

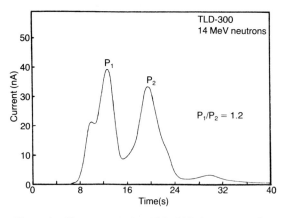

Figure 2. Glow curve of a TLD-300 dosemeter after irradiation in a 14 MeV neutron beam.

RESULTS

After neutron irradiation, the glow curves show a very different shape as compared to those induced by γ irradiation (see Figures 1 and 2). Due to the high LET component the response of the low temperature peak 1 is largely reduced, whereas the response of peak 2 remains relatively high. The peak height ratio therefore reduces from about 3 : 1 for gammas to 1.1 : 1 for free-air neutron irradiations.

However, free-air irradiations cannot be used for calibration of the TLD chips, since at 14 MeV the recoil protons from the tissue or water phantom contribute considerably to the TLD response. Because of this effect the free-air TLD dose response was found to be 0.68 ± 0.02 of that of the TLD dose response at 1 cm depth in water.

The normalised TLD responses P_{1N} and P_{2N} were therefore determined at 10 cm water depth, where the γ contribution D_γ/D_{total} was found to be 8.8% from the GM counter measurements. The resulting neutron responses could then be derived from Equations 1 and 2 to be

$$\varepsilon_2 K = 0.32 \quad \varepsilon_1 K = 0.10$$

The dose then was calculated according to

$$D_n = \frac{P_{1N} - P_{2N}}{0.22}$$

$$D_\gamma = \frac{3.2\, P_{1N} - P_{2N}}{2.2}$$

$$D_{total} = D_n + D_\gamma$$

where P_{1N} and P_{2N} are the normalised TL readings of both peaks measured as peak heights after irradiation in the neutron field.

Using these values, various dose distributions have been measured. Figure 3 shows the depth dose distribution. The error bars are plotted at ± 5%. The solid line is taken from the ion chamber measurements and the broken line gives the output of the Monte Carlo calculations. The agreement is good and an accuracy of ± 5% (2σ confidence level) can be achieved with the TLD as compared to ± 8% reported for 6 MeV neutron therapy beams[10]. It should be noted that this accuracy is achieved without special consideration of the annealing procedure or selection of TLD chips with extremely good reproducibility.

At greater depths the γ contribution increases. Figure 4 gives the γ contributions at various depths. The maximum contribution that could be measured in the phantom was D_γ/D_{total} = 15%. The accuracy of determining the γ contributions with the TLD is seen to be ± 0.5%. This corresponds at the 10% level to an accuracy ± 5% for the γ dose. This is comparable to the accuracies of GM counter or proportional chamber methods. For comparison, the GM counter measurements are shown (solid line) together with the results of the MORSE calculations (broken line). The calculated γ dose is considerably lower than the measured dose. Except for uncertainties in the k_u value of the GM counter this is probably due to the fact that not all possible sources of gammas in the large experimental hall could be included into the program. In particular, gammas produced by neutrons from d-d reactions in the target were not considered.

CONCLUSION

TLD-300 dosemeters can be used in 14 MeV neutron beams to measure γ and n dose simultaneously with an accuracy of ± 5%. The accuracy is higher than in 6 MeV neutron fields due to the higher neutron kerma. Distributions of γ and neutron dose can be determined in a single irradiation. The high response of the CaF_2:Tm material makes the method particularly competitive for the low dose rate beam produced at the FNRF neutron facility.

ACKNOWLEDGEMENTS

We are very grateful to Dr Filges, Dr Dagge and Ch. Reul of KFA, Jülich, for their help during the installation of the MORSE programs. We also wish to thank Prof. Rassow and Dr Meissner of

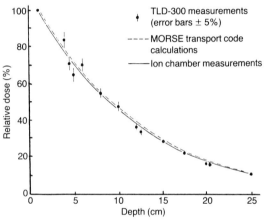

Figure 3. Comparison of determination of the depth dose $D_{total} = D_n + D_\gamma$ of the 14 MeV beam in water by the TLD-300 method, ion chamber measurement, and Monte Carlo transport code calculations.

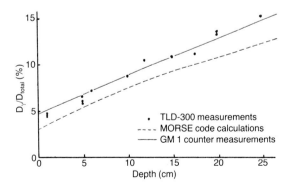

Figure 4. Comparison of determination of dose ratio D_γ/D_{total} along the depth dose of 14 MeV neutron beam in water by the TLD-300 method, Geiger-Müller counter, and Monte Carlo calculations.

the Fast Neutron Therapy Facility of the University of Essen for providing the GM counter. Part of this work was funded by the German Academic Exchange Service DAAD.

REFERENCES

1. Hoffmann, W. and Prediger, B. *Heavy Particle Dosimetry with High Temperature Peaks of CaF_2:Tm and Li^7 Phosphors*. Radiat. Prot. Dosim. **6**, 149-152 (1984).
2. Rassow, J., Temme, A., Baumhoer, U. and Meissner, P. *Dosimetrical Verification of Calculated Total and Gamma Dose Distribution*. Strahlentherapie **160**, 168-179 (1984).
3. Binder, J., Schmidt, R. and Scobel, W. *TLD-300 Dosimeter für 14 MeV Neutronen*. Med. Phys. **88**, 397-401 (1988).
4. Hoffmann, W. and Möller, G. *Heavy Particle Dosimetry with High Temperature Peaks of Thermoluminescent Materials*. Nucl. Instrum. Methods **175**, 205-207 (1980).
5. Menzel, H. G. and Schumacher, H. *Comparison of Microdosimetric Characteristics of Four Neutron Therapy Facilities*. In: Proc. 7th Symp. on Microdosimetry, Oxford, 1980. EUR 7147 (London: Harwood) pp. 1217-1231 (1981).
6. Emmett, M. B. *The MORSE Monte Carlo Radiation Transport Code System*. ORNL-4972 (1975).
7. Cloth, P., Filges, D., Neef, R. P., Sterzenbach, G., Reul, Ch., Armstrong, T. W., Colborn, B. L., Anders, B. and Brückmann, H. *HERMES, A Monte Carlo Program System for Beam Materials Interaction Studies*. KFA-Report Jül-2203 (May 1988).
8. Kinsey, R. *ENDF/B-IV Cross Section Measurement Standards*. Report BNL-NCS-50496 (1979).
9. International Commission on Radiation Units and Measurements. *Neutron Dosimetry for Biology and Medicine*. ICRU Report 26 (Washington, DC: ICRU Publications) (1977).
10. Pradhan, A. S., Rassow, J. and Hoffmann, W. *Fast Neutron Response of CaF_2:Tm Teflon Discs of Different Thicknesses*. Radiat. Prot. Dosim. **15**, 233-236 (1986).

THERMOLUMINESCENCE DOSIMETRY IN MIXED (n,γ) RADIATION FIELDS USING GLOW CURVE SUPERPOSITION

T. M. Piters[†], A. J. J. Bos[†] and J. Zoetelief[‡]
[†]Interfaculty Reactor Institute
Mekelweg 15, NL 2629 JB Delft, The Netherlands
[‡]Institute of Applied Radiobiology and Immunology TNO
PO Box 5815, NL 2280 HV Rijswijk, The Netherlands

Abstract — A new method (glow curve superposition) is introduced to measure absorbed doses in mixed (n,γ) radiation fields using thermoluminescence dosemeters. The new method is closely related to the so-called two peak method. The basic idea of the method is that the glow curve of a TL material irradiated in a mixed (n,γ) radiation field is a superposition of two glow curves obtained in pure photon and neutron radiation fields. The glow curve superposition method appears to be more accurate than the two peak method when one of the components of the mixed field strongly dominates.

INTRODUCTION

Thermoluminescent materials such as LiF(TLD-100), ^{7}LiF(TLD-700), ^{6}LiF(TLD-600) and CaF_2(TLD-300) show glow curves (TL intensity as a function of temperature, T) in which the different peaks exhibit a different dependence on the LET of the radiation. This allows the separate determination of photon and neutron doses in a mixed (n,γ) radiation field with the so-called two peak method[1–4]. In this paper a new method (glow curve superposition method) is introduced which is closely related to the two peak method. The new method is based on the different response of the whole shape of the glow curve to the different components of the mixed radiation field. The aim of the present contribution is to investigate the properties of this new method and to compare the results obtained with those from the two peak method.

METHODS

The basic idea of the glow curve superposition method is that the glow curve of a TL material irradiated in a mixed (n,γ) radiation field $I_m(T)$, can be expressed as a superposition of glow curves obtained in pure radiation fields:

$$I_m(T) = D_n^m J_n(T) + D_\gamma^m J_\gamma(T) \quad (1)$$

where D_n^m and D_γ^m are the neutron and photon absorbed dose in a TL material and $J_n(T)$ and $J_\gamma(T)$ are the glow curves per unit dose of a TL material irradiated in a pure neutron and in a pure photon radiation field respectively.

In practice, a neutron field is always contaminated by photons and $J_n(T)$ is therefore not directly measurable. This problem can be solved for a specific neutron energy (spectrum) by combining a calibration in a pure photon field with known photon absorbed dose D_γ^a, with that in a neutron field with known absorbed doses D_n^b and D_γ^b. The calibration curves $I_a(T)$ and $I_b(T)$ can be written as:

$$I_a(T) = D_\gamma^a J_\gamma(T) \quad (2)$$

$$I_b(T) = D_n^b J_n(T) + D_\gamma^b J_\gamma(T) \quad (3)$$

$I_m(T)$ can be expressed in terms of the calibration curves:

$$I_m(T) = A I_a(T) + B I_b(T) \quad (4)$$

where A and B are dimensionless constants. From Equations 2, 3 and 4 the unknown neutron and photon dose in a mixed (n,γ) radiation field can be derived from:

$$D_n^m = B\, D_n^b \text{ and } D_\gamma^m = A\, D_\gamma^a + B\, D_\gamma^b \quad (5)$$

In general, it is not possible to find single values for A and B so that Equation 4 holds true for every temperature T, because of experimental errors. However, the best possible values for A and B can be obtained by minimising Q^2 as a function of A and B, where Q^2 is given by:

$$Q^2 = \sum_i (I_m(T_i) - A I_a(T_i) - B I_b(T_i))^2 \quad (6)$$

where i refers to all measured points of the glow curves.

In practice, there are always small differences between the temperature profile of the measured glow curve $I_m(T)$: $T(t) = T_0 + \beta t$ (T_0 = temperature at time t = 0, β = heating rate) and the temperature profiles of the calibration curves $I_a(T)$: $T_a = (T_0 + dT_a) + (\beta + d\beta_a)t$ and $I_b(T)$: $T_b = (T_0 + dT_b) + (\beta + d\beta_b)t$. Such differences could be due to differences in thermal contact between TL detector and heating planchet. For small dT_a, dT_b, $d\beta_a$ and $d\beta_b$ the calibration curves, $I_a(T)$ and $I_b(T)$ can be corrected using:

$$I'_a(T(t)) = I_a(T(t) + dT_a + d\beta_a t) \frac{\beta + d\beta_a}{\beta} \quad (7)$$

$$I'_b(T(t)) = I_b(T(t) + dT_b + d\beta_b t) \frac{\beta + d\beta_b}{\beta} \quad (8)$$

The factors $(\beta+d\beta_a)/\beta$ and $(\beta+d\beta_b)/\beta$ have to be introduced to conserve the total TL response $\int I(T)dt$. With this correction Equation 6 becomes:

$$Q^2 = \sum_i (I(T_i) - A \frac{\beta+d\beta_a}{\beta} I_a(T_i+dT_a+d\beta_a t)$$
$$- B \frac{\beta+d\beta_b}{\beta} I_b(T_i+dT_b+d\beta_b t))^2 \quad (9)$$

The best values for A and B can now be found by minimising Q^2 as a function of A, B, dT_a, dT_b, $d\beta_a$ and $d\beta_b$.

For the two peak method the summation in Equation 6 is only carried out for two peak temperatures T_{m1} and T_{m2}. The parameters A and B can be derived from:

$$A = \frac{I_m(T_{m1}) I_b(T_{m2}) - I_b(T_{m1}) I_m(T_{m2})}{I_a(T_{m1}) I_b(T_{m2}) - I_a(T_{m2}) I_b(T_{m1})} \quad (10)$$

$$B = \frac{I_m(T_{m2}) I_a(T_{m1}) - I_m(T_{m1}) I_a(T_{m2})}{I_a(T_{m1}) I_b(T_{m2}) - I_a(T_{m2}) I_b(T_{m1})} \quad (11)$$

EXPERIMENTAL TECHNIQUES

The TL materials used were polycrystalline LiF(TLD-100), ^{6}LiF(TLD-600), ^{7}LiF(TLD-700) and CaF_2(TLD-300) chips with dimensions of 3.2×3.2×0.9 mm^3 (Harshaw Chemical Company, Ohio, USA). Before irradiation all chips were given a standard annealing procedure, i.e. 1 h at (400 ± 1)°C followed by fast cooling to room temperature. During neutron irradiation the chips were placed in a tray composed of A-150 tissue-equivalent plastic with a 1 mm thick layer in front of and behind the chips. The chips used for analysis were surrounded by chips of the same material. For the fast neutron irradiation a ^{6}LiF screen (1.8 mm thick) was used to reduce the contribution from thermal neutrons.

The irradiations with fast neutrons (with most likely energies of 1 MeV, 6 MeV and 15 MeV) were carried out at the TNO facility described elsewhere[5]. The contribution of photons to the total absorbed dose (after shielding with ^{6}LiF) were 1.5% for the 1 MeV beam, 5.6% for the 6 MeV beam and 6.8% for the 15 MeV beam. Irradiations with thermal neutrons (0.048 eV) were performed at beam line R3 of the Research Reactor at the IRI, Delft. The contribution of photons to the total dose was very low (0.006%). Irradiations with different contributions of neutrons and photons to the total absorbed dose were achieved by adding a photon irradiation to that in a neutron field. These irradiations are referred to as mixed fields. For the photon irradiations a ^{60}Co source at the IRI, Delft was used.

The TL glow curves were measured in a nitrogen atmosphere using a modified microprocessor-controlled Harshaw 2000 TL reader[6] using a heating rate of 6°C.s^{-1} from 40°C to 360°C. Differences in sensitivity of different chips were taken into account.

For the experiments with fast neutrons the calibration curves $I_a(T)$ and $I_b(T)$ were composed from the glow curves of three chips irradiated with photons (D^a_γ = 48 mGy) and neutrons ($D^b_n + D^b_\gamma \approx$ 300 mGy). For thermal neutrons the calibration curves $I_a(T)$ and $I_b(T)$ were obtained for each chip separately by an irradiation with photons (D^a_γ = 60 mGy) and neutrons ($D^b_n + D^b_\gamma \approx$ 2.3 Gy).

The absorbed photon and neutron doses of the chips used to obtain the test curves $I_m(T)$ were chosen in such a way that both components contribute significantly to the total glow curve. The TLD-100 chips were also used for investigating the performance of both methods in mixed (n,γ) fields (E_n=0.048 eV) with a small contribution to the total glow curve of one of the components (n or γ) in the mixed fields. For measurements with a small contribution of the photon component, the neutron dose was kept constant at approximately 2.5 Gy and the photon dose was varied, i.e. 0.2 mGy, 0.4 mGy and 1.2 mGy. For measurements with a small contribution of the neutron component, the photon dose was kept constant at 160 mGy and the neutron dose was varied, i.e. 0.34 Gy, 0.67 Gy and 1.24 Gy.

RESULTS AND DISCUSSION

The measured glow curves were analysed according to the glow curve superposition method and according to the two peak method. The peak heights were obtained by fitting the tops of the peaks to a Gaussian function. Figure 1 shows the results of a fit with the glow curve superposition method according to Equation 9. The peaks used for the two peak method are indicated by arrows.

The correction terms in Equation 8 have a significant effect on the fit results (see Figure 2). From the two terms given in Equation 8, $d\beta$ is the least significant one and may be omitted and still get good fit results. The temperature shifts dT_a and dT_b are of the order of 6°C which corresponds with a shift of 8 channels in Figures 1 and 2. The corrections in β are of the order of 2%.

For the situation where both calibration curves

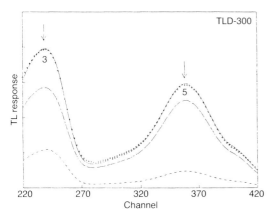

Figure 1. Glow curve of TLD-300 (+) after an irradiation in a mixed (n,γ) radiation field with actual dose D_γ = 10 mGy and D_n = 92 mGy (E_n = 6 MeV). The composed glow curve (solid line) is a superposition of the calibration curves I_a multiplied by A = 0.085 (dotted curve) and I_b multiplied by B = 0.32 (broken curve). The arrows indicate the peaks used in the two peak method. Results for superposition method: D_γ 9.4 mGy, D_n = 92 mGy. Results for the two peak method: D_γ = 9.6 mGy, D_n = 91 mGy.

contribute significantly to the whole glow curve that the results from both methods are similar. The deviations between the actual and derived doses are generally about 6% and 22% at maximum. The larger deviations appear simultaneously for both methods and are probably due to the inhomogeneity of the neutron beam. Table 1 shows the results of the dose determinations in mixed fields using TLD-600.

For TLD-100 irradiated with thermal neutrons the deviation between the actual and analysed photon and neutron dose as a function of the dose ratio D_γ/D_n are shown in Figure 3. For large dose

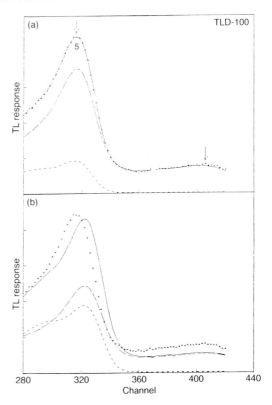

Figure 2. Glow curve of TLD-100 (+) irradiated in a mixed (γ,n) radiation field, D_γ = 3.5 mGy and D_n = 98.5 mGy (1 MeV) analysed with (a) and without (b) corrections in the heating profiles of the calibration curves. The dotted and broken curves are the contributions of the calibration curves I_a and I_b to the total glow curve (solid line). Results for (a): D_γ = 3.6 mGy, D_n = 92.7 mGy. Results for (b): D_γ = 4.8 mGy, D_n = 63.7 mGy.

Table 1. Results of dose determinations in mixed radiation fields using TLD-600.

Neutron energy	Photon dose			Neutron dose		
	Actual dose (mGy)	Analysing method		Actual Dose (mGy)	Analysing method	
		Superposition (mGy)	2-peak (mGy)		Superposition (mGy)	2-peak (mGy)
1 MeV	3.5	4.0 ± 0.4	4.4 ± 0.5	98.5	91.1 ± 1.1	91.0 ± 1.0
	5.5	5.9 ± 0.5	5.7 ± 0.5	98.5	94.3 ± 0.2	95.7 ± 0.6
6 MeV	10.2	10.2 ± 0.2	10.1 ± 0.3	97.2	90 ± 3	92 ± 3
	14.2	13.5 ± 0.2	12.9 ± 0.3	97.2	91 ± 2	92 ± 4
15 MeV	14.8	14.2 ± 0.7	13.2 ± 2.5	93.2	90.3 ± 0.3	94 ± 6
	22.8	22.1 ± 0.4	22.9 ± 0.6	93.2	89.8 ± 1.3	91.9 ± 0.8
				(Gy)	(Gy)	(Gy)
0.048 eV	16.1	17.1 ± 0.3	17.6 ± 0.1	1.22	1.10 ± 0.02	1.09 ± 0.02
	8.1	8.6 ± 0.2	9.0 ± 0.5	1.11	1.04 ± 0.04	1.04 ± 0.04
	4.1	4.9 ± 0.3	5.4 ± 0.3	1.06	0.95 ± 0.03	0.96 ± 0.04

ratios (small contribution of $I_b(T)$ to the total glow curve) the deviation between derived and actual neutron dose is significantly lower (factor 2) for the glow curve superposition method than for the two peak method. For the lowest dose ratio investigated (small contribution of $I_a(T)$ to the total glow curve) the glow curve superposition method still gives a reasonable result for the photon dose (deviation from actual dose: 24%) while the two peak method gives a photon dose which deviates by more than 100% from the actual dose.

CONCLUSIONS

In mixed fields where the contributions to the total glow curve of either the photon or of the neutron dose is small the glow curve superposition method provides better results than the two peak method.

It is necessary to apply corrections for differences in the heating profile between calibration curves and test curve. The corrections for temperature shifts appear to be more important than those for heating rate with the reader used.

ACKNOWLEDGEMENT

The authors are grateful to Messrs N. J. P. de Wit and F. S. Draaisma for their assistance in the neutron irradiations at TNO, Rijswijk.

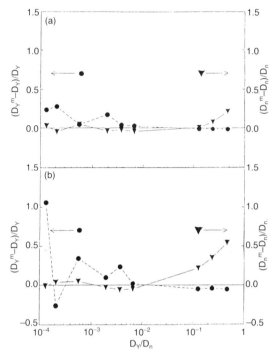

Figure 3. Difference between the derived dose and the actual dose, relative to the dose as a function of the dose ratio D_γ/D_n for neutrons (▼) and photons (•) for the glow curve superposition method (a) and the two peak method (b).

REFERENCES

1. Meissner, P. and Rassow, J. *Principles and Applicability of the Two-Peak-Method of Clinical Neutron Dosimetry.* In: Proc. Collection and Evaluation of Neutron Dosimetry Data, Ed. J. J. Broerse, CENDOS-Workshop on the Application of TLD-300 for Mixed Field Dosimetry, 21-22 May, Neuherberg (1984).
2. Pradhan, A. S., Rassow, J. and Meissner, P. *Dosimetry of d(14)+Be Neutrons with the Two-peak Method of LiF TLD-700.* Phys. Med. Biol. **30**(12), 1349-1354 (1985).
3. Rassow, J., Broerse, J. J., Duehr, R., Hensley, F. W., Marquebreucq, S., Olthoff-Muenter, K., Pradham, A. S., Temme, A., Vynckier, S. and Zoetelief, J. *Spectral Dependence of Response Coefficients and Applicability of the Two-Peak TLD Method in Mixed Neutron Photon Radiation Fields.* In: Proc. Fifth Symp. on Neutron Dosimetry, Neuherberg, EUR 9762 (London: Harwood) Vol. II, pp. 783-794 (1985).
4. Dielhof, J. B., Bos, A. J. J., Zoetelief, J. and Broerse, J. J. *Dosimetry in a Mixed (14.8 MeV Neutron, Gamma) Radiation Field with $CaF_2:Tm$ (TLD-300).* In: Proc. Symp. on Thermoluminescence and Dosimetry, Eds A. H. L. Aalbers, A. J. J. Bos and B. J. Mijnheer. NCS. Report 3, pp. 121-127 (1988).
5. Zoetelief, J., Engels, A. C., Bouts, C. J. and Broerse, J. J. *Experimental Arrangement and Monitoring Results at TNO.* In: European Dosimetry Intercomparison Project (ENDIP), Results and Evaluation. Eds. J. J. Broerse, G. Burger and M. Coppola (Luxembourg: Commission of the European Communities) EUR 6004 EN, pp. 32-43 (1978).
6. Wijngaarden, M. W. van, Plaisier, J. and Bos, A. J. J. *A Microprocessor Controlled Thermoluminescence Dosimeter Reader for Routine Use and Research.* Radiat. Prot. Dosim. **11**(3), 179-183 (1985).

CALIBRATION METHODS OF TLD-300 DOSEMETERS IN A CLINICAL 14 MeV NEUTRON BEAM

M. Kriens†, R. Schmidt‡, A. Hess‡ and W. Scobel†
†I. Institut für Experimentalphysik, Universität Hamburg, Germany
‡Abt. Strahlentherapie, Radiologische Universitätsklinik Hamburg-Eppendorf, Germany

Abstract — Calcium fluoride doped with thulium is a thermoluminescent material (TLD-300) having two dominant maxima of the glow curve (glow peaks) at 150°C and 240°C. Because of different sensitivities of the two glow peaks to low and high linear energy transfer (LET) of the incident radiation, calibrated TLD-300 dosemeters are suitable for separating the dose components D_N and D_γ due to neutrons and photons in a mixed n–γ radiation field. To calibrate TLD-300 dosemeters the individual sensitivities of the two glow peaks to neutrons and photons are experimentally determined. Different calibration methods are presented comparing TLD-300 measurements with ionisation chamber results. The measurements were performed at a single reference point in a phantom or by linear regression of multiple data sets measured for various photon dose components D_γ/D_T relative to the total absorbed dose D_T. The variation of the photon dose component was performed by different measuring conditions in different phantom depths, with different field sizes, by different additional exposures to ^{60}Co radiation or, because of the different attenuation of photons and neutrons in lead, by measurements behind lead layers of different thicknesses. For the calibration methods used the mean values of the relative neutron sensitivities of the two glow peaks were found to be 0.078 and 0.215.

INTRODUCTION

Neutron beams used in radiotherapy are contaminated with photons arising mainly from the capture of thermal neutrons by hydrogen in the irradiated medium. The photon component increases with the field size and depth in the phantom. As neutrons have a higher relative biological effectiveness (RBE) than photons the separate determination of the photon and neutron dose is necessary for clinical applications.

The separation is possible by the analysis of the responses of two dosemeters with different sensitivities to photons and neutrons exposed to a mixed n–γ beam. Commonly, neutron sensitive and neutron insensitive ionisation chambers (dual detector method) are used to evaluate the dose components separately[1], but any other dual response method is applicable. As for calcium fluoride doped with thulium (TLD-300) the two glow peaks at 150°C and 240°C respond selectively to low (photon) and high (neutron) linear energy transfer (LET) of the incident radiation[2], the responses of a single TLD-300 dosemeter provide information on both dose components separately.

The individual sensitivities of the two glow peaks of TLD-300 to photons and neutrons are experimentally determined by different calibration methods. The measurements were performed in a ^{60}Co beam for the photon calibration and in a collimated neutron beam (DT, 14 MeV).

CALIBRATION METHODS

Because of the different sensitivities of the two dominant glow peaks at 150°C and 240°C of TLD-300 material to photons and neutrons (low and high LET), the formalism applied for the dual detector method is applicable, using a single TLD-300 dosemeter[3]. The response R'_i of the glow peak i to the mixed n–γ field relative to its sensitivity to the γ rays used for calibration is given by:

$$R'_i = k_i D_N + h_i D_\gamma \qquad (1)$$

where k_i = sensitivity of the glow peak i to neutrons relative to its sensitivity to the γ rays used for calibration
h_i = sensitivity of the glow peak i to photons in the mixed n–γ field relative to its sensitivity to the γ rays used for calibration.

To calibrate these dosemeters the sensitivities of the two glow peaks are determined relative to their responses to γ rays used for calibration by the comparison of each TLD-300 glow peak response to the corresponding quantities of calibrated ionisation chambers. Different calibration methods were used.

Comparison at a single reference point

With the assumption that the sensitivities of the TLD-300 glow peaks to the photons in the mixed n–γ field are equal to their sensitivities to the γ rays used for calibration ($h_1 = h_2 = 1$)[4] the relative neutron sensitivities of the two peaks are given by:

$$k_{1,2} = \frac{(R'_{1,2} - D\gamma)}{D_N} \qquad (2)$$

The relative responses R'_1 or R'_2 of a single TLD-300 dosemeter irradiated in a mixed n–γ beam at a reference point with known dose components D_N and $D_γ$, therefore, leads to to the relative neutron sensitivities k_1 or k_2.

Linear regression of multiple data sets

Dividing Equation 1 by the total dose $D_T = D_N + D_γ$ yields:

$$\frac{R'_{1,2}}{D_T} = k_{1,2} + (h_{1,2} - k_{1,2})\frac{D_γ}{D_T} \quad (3)$$

The linear regression of multiple values R'_1/D_T or R'_2/D_T and $D_γ/D_T$ determines k_1 or k_2 and h_1 or h_2, if D_T and the ratios $D_γ/D_T$ are known for multiple measuring conditions. The variation of $D_γ/D_T$ is performed by three procedures:

(1) Different phantom depths
The measurements were performed at 10 different depths in the range from surface to 30 cm in a liquid phantom (water 60 cm × 60 cm × 50 cm) with a field size of 10 cm × 10 cm at the source-surface distance (SSD) of 80 cm. Values of $D_γ/D_T$ in the range of 0.06 to 0.18 were obtained.

(2) Different field sizes
The measurements were performed at a depth of 5.5 cm in a solid phantom (PMMA 30 cm × 30 cm × 30 cm) with 5 different field sizes in the range from 4 cm × 4 cm to 20 cm × 20 cm at 80 cm SSD. Values of $D_γ/D_T$ in the range of 0.06 to 0.12 were obtained.

(3) Additional exposure to ^{60}Co γ rays
The measurements were performed at first in the mixed n–γ field at a depth of 5.5 cm in a solid phantom (PMMA 30 cm × 30 cm × 30 cm) with a field size of 10 cm × 10 cm at 80 cm SSD. After this irradiation the TLD-300 were exposed to different gamma dose fractions of ^{60}Co. In this way the ratio $D_γ/D_T$ was increased up to 0.6.

Lead attenuation

The lead attenuation method is well known[5] for determining the relative neutron sensitivity of a neutron insensitive dosemeter. This method, based on the different attenuation of photons and neutrons in lead, is applicable also for TLD-300 if the relative response of each glow peak i is compared with the corresponding quantity of a neutron sensitive ionisation chamber with known relative neutron sensitivity k_T.

With the assumption that the photon dose $D_γ$ in the mixed n–γ field consists of a shielding component $D_γ^S$ originating mainly in the collimator and an aperture component $D_γ^A$ accompanying the collimated neutron beam

$$D_γ = D_γ^S + D_γ^A \quad (4)$$

and that $D_γ^S$ is independent of the attenuation of the neutron beam through different thicknesses x of lead, the relative responses of each single glow peak (i) as well as the neutron sensitive ionisation chamber (T) are, following Equation 1,

$$R'_{i,T}(x) = D_γ^S + D_γ^A e^{-μx} + k_{i,T} D_N(x) \quad (5)$$

with the photon mass attenuation coefficient $μ = 0.512$ cm^{-1} for the photons in the mixed n–γ beam with energy of 2.3 MeV[6].

From Equation 5 the quantities R_i^{Pb} and R_T^{Pb} are derived:

$$R_{i,T}^{Pb}(x) = \frac{R'_{i,T}(x) - R'_{i,T}(x=0) e^{-μx}}{(1 - e^{-μx})} \quad (6)$$

Assuming that the relative photon sensitivities $h_i = h_T = 1$ these quantities are bound by the linear relation:

$$R_i^{Pb}(x) = (1 - \frac{k_i}{k_T}) D_γ^S + \frac{k_i}{k_T} R_T^{Pb}(x) \quad (7)$$

The linear regression of multiple sets of R_1^{Pb} or R_2^{Pb} and R_T^{Pb} determines the relative neutron sensitivities k_1 or k_2 of each single TLD-300 glow peak.

RESULTS AND DISCUSSION

The measurements were performed in the collimated beam of ^{60}Co γ rays for the photon calibration and the collimated fast neutron beam at the Radiotherapy Department of the University Hospital at Hamburg-Eppendorf[7]. The monoenergetic 14 MeV neutrons are produced by bombarding a rotating TiT target[8] with accelerated deuterium ions (DT reaction).

As TLD-300 material, ribbons of size 3.2 mm × 3.2 mm × 0.9 mm were used. The dosemeters were analysed by a readout instrument (TL-Detector 2000 D, Harshaw), using a hot nitrogen gas stream at 350°C as the heating medium. The responses of the two glow peaks were reproducible within ±8%.

The results for all calibration methods used are given in Table 1.

With reference to the uncertainties, the method by lead attenuation is the most reliable, but the mass attenuation coefficient of the photons in the mixed n–γ beam must be known and the assumption $h_1 = h_2 = 1$ is necessary. No assumptions of that kind are demanded for the linear regression method of multiple data sets.

For all calibration methods used the mean value of the relative sensitivities of the two TLD-300 glow peaks were found to be $k_1 = (0.078 \pm 0.029)$ and $k_2 = (0.215 \pm 0.047)$ for fast neutrons as well as $h_1 = (1.02 \pm 0.36)$ and $h_2 = (1.02 \pm 0.26)$ for the photons of the mixed n–γ beam, comparable to recent measurements[9].

This result shows that for 14 MeV DT neutrons the relative neutron sensitivities k_1 and k_2 of TLD-300 material exceed remarkably the corresponding quantities reported for d(14)+Be neutron beams[3]. Within the tolerances the relative photon sensitivities obtained confirm the assumption that the sensitivities of the glow peaks to the photons in the mixed n–γ beam are nearly equal to their sensitivities to the γ rays used for calibration.

The uncertainties $\Delta k_i/k_i$ and $\Delta h_i/h_i$ have been calculated assuming that all uncertainties can be treated as if they were random errors. The uncertainties $\Delta k_i/k_i$ of the relative sensitivities are large, but for small k_i even large uncertainties in their values contribute only small uncertainties to the absorbed dose quantities D_T and D_T/D_γ. The given values in Table 1 yield an accuracy of about $\pm 10\%$ for the total absorbed dose D_T determination and $\pm 20\%$ for the photon dose fraction D_γ/D_T.

CONCLUSIONS

In neutron dosimetry, neutron sensitive ionisation chambers and neutron insensitive devices are commonly used to determine both the dose components in a mixed radiation field of neutrons and photons.

As an alternative to ionisation chambers calcium fluoride doped with thulium is a suitable thermoluminescent material (TLD-300) for determining dose distributions of photons and neutrons separately with an accuracy in the total dose D_T within $\pm 10\%$ and the photon dose fraction D_γ/D_T within $\pm 20\%$. TLD-300 dosemeters are in some cases advantageous, because of their small dimensions, the evaluation of the dose quantities after irradiation and the possibility of *in vivo* measurements. With the use of TLD-300 just one dosemeter provides information on both dose components.

Table 1. Relative neutron and photon sensitivities.

Method	k_1	$\Delta k_1/k_1$ (%)	k_2	$\Delta k_2/k_2$ (%)	h_1	$\Delta h_1/h_1$ (%)	h_2	$\Delta h_2/h_2$ (%)
Comparison at a single reference point	0.072	50	0.205	35	1.00*		1.00*	
Linear regression of multiple data sets								
1. Different phantom depths	0.079	35	0.188	20	1.07	35	1.20	25
2. Different field sizes	0.078	35	0.233	20	0.97	35	0.76	25
3. Additional irradiation	0.087	35	0.209	25	1.03	35	1.10	25
Lead attenuation	0.073	30	0.239	10	1.00*		1.00*	

* Assumed value.

REFERENCES

1. Broerse, J. J., Mijnheer, B. J. and Williams, J. R. *European Protocol for Neutron Dosimetry for External Beam Therapy*. Br. J. Radiol. **54**, 882-898 (1981).
2. Hoffmann, W. and Möller, G. *Heavy Particle Dosimetry with High Temperature Peaks of Thermoluminescent Materials*. Nucl. Intrum. Methods **175**, 205 (1980).
3. Rassow, J. and Temme, A. *Grundlagen und erste Ergebnisse eines neuen Thermolumineszenz-Messverfahrens mit TLD-300 für die klinische Dosimetrie in gemischten Neutronen-Photonen-Strahlenfeldern*. Medizinische Physik, **82**, 311-324 (1982).
4. McKinley, A. F. *Thermoluminescence Dosimetry*. Medical Physics Handbook 5 (Bristol: Adam Hilger) (1981).
5. Hough, J. H. *A Modified Lead Attenuation Method to Determine the fast Neutron Sensitivity k_u of a Photon Dosimeter*. Phys. Med. Biol. **24**, 734-747 (1979).
6. Schmidt, R., Magiera, E. and Scobel, W. *Neutron and Gamma Spectroscopy for Clinical Dosimetry*. Med. Phys. **7**(5) 507-513 (1980).
7. Franke, H. D., Hess, A. and Schmidt, R. *The DT Neutron Generator for Tumor Therapy in Hamburg-Eppendorf*. Atomkernenerg-Kerntech. **43**, 95-104 (1983).
8. Hess, A., Schmidt, R. and Franke, H. D. *Recent Developments of the Fast Neutron Therapy Facility at Hamburg-Eppendorf*. In: Proc. 4th Symp. on Neutron Dosimetry. EUR 7448 (Luxembourg: CEC) pp. 39-46 (1981).
9. Binder, J., Schmidt, R. and Scobel, W. *TLD-300 Dosimeter für 14 MeV Neutronen*. Medizinische Physik, **89**, 397-401 (1989).

A SIMPLE PERSONAL DOSEMETER FOR THERMAL, INTERMEDIATE AND FAST NEUTRONS BASED ON CR-39 ETCHED TRACK DETECTORS

M. Luszik-Bhadra, W. G. Alberts, E. Dietz, S. Guldbakke and H. Kluge
Physikalisch-Technische Bundesanstalt
Bundesallee 100, W-3300 Braunschweig, Germany

Abstract — A simple personal neutron dosemeter based on etched track detectors has been developed which shows a reasonably flat dose equivalent response for neutrons from thermal energies up to 15 MeV, and provides rough spectrometric information obtained by evaluating two additional track detectors. The dosemeter response was measured and tested in fields with monoenergetic neutrons and broad spectral distributions, some of them with additional photon irradiation, and at angles of incidence of up to 60 degrees.

INTRODUCTION

A simple personal dosemeter has been developed[1] with a reasonably flat dose equivalent response for neutrons in the energy range from thermal up to 15 MeV. It is based on CR-39 electrochemically etched track detectors using hydrogenous converters for fast neutrons and a 3 mm air gap as a converter for thermal and epithermal neutrons, making use of the $^{14}N(n,p)$ reaction. This paper describes further investigations concerning the dose equivalent response and dose linearity of this new dosemeter. The dosemeter response was tested by irradiations with neutrons from a ^{252}Cf source at low doses (0.2 mSv) and at high doses (10 mSv), in mixed neutron photon fields, by irradiations with a ^{252}Cf source (moderated by D_2O) and an ^{241}Am-Be (α,n) source. The evaluation of CR-39 sheets of two dosemeter stacks — one with and one without an air gap — and a Makrofol foil provides rough spectrometric information in the energy intervals from thermal to 70 keV, 70 keV to 1.5 MeV and energies higher than 1.5 MeV. In the final section measurements of the angle dependence of the dosemeter response in monoenergetic neutron fields are described and discussed with respect to calibration procedures for a simple universal personal neutron dosemeter.

THE DOSEMETER

Configuration and evaluation

The dosemeter configuration investigated is shown in Figure 1. It consists of two foil stacks which are separately sealed inside radon-tight aluminium/polyethylene pouches. The dosemeter is evaluated by etching the front sides of the CR-39 foils (material from American Acrylics) electrochemically in 6N KOH at 60°C and 20 ± 1 kV.cm^{-1} for 5 h at 100 Hz and 1 h at 2 kHz. The Makrofol foils are etched in a similar process in a 4:1 mixture of 6.85 N KOH and ethanol at 35°C and 26.7 kV.cm^{-1}. The tracks were counted by eye in an area of 1.12 cm^2 on a copy from a microfiche reader with a magnification of 15. The readings of the CR-39 detectors were corrected for field strength dependence and dose linearity, if necessary, as described in the next chapter.

Dose equivalent response

A measurement of the dose equivalent response as a function of neutron energy had been performed earlier[1] by using the cadmium shield only for the irradiation with thermal neutrons. These measurements were now repeated using cadmium shields at all energies and with higher doses to get more accurate results. An additional measurement at 2 keV was performed. In all cases three dosemeters were irradiated with normally incident neutrons on a 30 × 30 × 15 cm^3 PMMA slab phantom in monoenergetic neutron fields produced at the Van de Graaff accelerator of the PTB[2] from 24 keV to 14.8 MeV, and with thermal and 2 keV neutrons using the filtered beams of the PTB reactor[3]. The response was calculated from the track densities measured (background corrected) with respect to the directional dose equivalent H'(10)[4,5]; the results are shown in Figure 2 as a function of neutron energy. Within the uncertainties the same behaviour was found as before (see Reference 1).

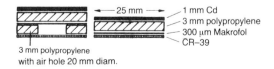

Figure 1. Sketch of the dosemeter configuration. The dosemeter is worn with the CR-39 foils next to the body.

The high energy response ($E_n > 144$ keV) of the CR-39 detector is one of the flattest measured for track detectors. This is achieved by the Makrofol foil in the chosen converter configuration which acts as neutron/proton converter for neutrons with energies around 2.5 MeV, and reduces the response maximum which is normally observed in this region when using a polypropylene converter only[6]. The small differences between the responses of the dosemeter with and without an air gap are similar to those found in the earlier measurement[1]. The smaller response measured for the dosemeter with air in the neutron energy region around 570 keV is due to the reduction of the effective hydrogenous converter thickness[6].

The predominant effect of the air converter in front of the CR-39 detector is the enhancement of response at intermediate energies and for thermal neutrons, including albedo neutrons, by detecting protons from the ^{14}N(n,p) reaction. An additional measuring point at 2 keV obtained from an irradiation at the scandium-filtered neutron beam of the PTB reactor[3] had a very large measurement uncertainty due to a large component of higher-energy neutrons in the beam. Measurements with a dosemeter containing a 6 mm air gap resulted in response values of 1091 ± 20 cm^{-2}.mSv^{-1} for thermal neutrons, 1080 ± 350 cm^{-2}.mSv^{-1} at 2 keV and 301 ± 39 cm^{-2}.mSv^{-1} at 24 keV. Since the values for thermal and 24 keV neutrons for the dosemeter with a 6 mm air gap were larger by about a factor of 2 than those for the dosemeter with a 3 mm air gap, a value of 540 ± 175 cm^{-2}.mSv^{-1} at 2 keV was deduced for the latter dosemeter. The shape of the response curve at lower energies is in general agreement with that of a typical albedo dosemeter.

DOSE LINEARITY

A deviation from linearity is usually observed for electrochemically etched tracks at high track densities. Spots near a big 'tree' are shielded from the full electric field and are usually not allowed to develop new 'trees', which causes a reduced registration at high track densities. The value of track density at which the decline sets in depends on the etching conditions, the type of detector, the nature of the incident particles and also on the type of counting.

We have measured the dose linearity up to doses of 15 mSv with a bare ^{252}Cf neutron source. Deviations from linearity start at track densities of about 1000 cm^{-2}, and have to be corrected by 15% at track densities of about 5000 cm^{-2}.

TEST OF THE DOSEMETER IN VARIOUS FIELDS

The reciprocal of the response value obtained for ^{252}Cf neutrons is a calibration factor which is suitable for the dosemeter with a 3 mm air converter (see Figure 2) for measurements at normal incidence. It was tested by determining doses from irradiations with ISO recommended[7] radioactive sources of the PTB: ^{252}Cf (bare), ^{252}Cf (D$_2$O moderated) and ^{241}Am–Be (α,n). In all cases four dosemeters were irradiated simultaneously in one irradiation run at normal incidence and evaluated. Small deviations of the CR-39 thickness were taken into account by an average field strength correction of 8% per kV.cm^{-1}. At high track densities, linearity corrections were applied. For all three neutron fields, all dose measurements in the dose range from 0.2 mSv to 10 mSv lie well within the limits in the preliminary recommendation of a German task group for individual neutron dosemeters (see Figure 3).

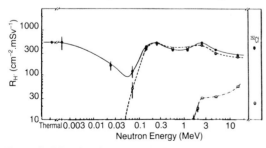

Figure 2. Directional dose equivalent response $R_{H'}$ for normally incident neutrons as a function of neutron energy for three track detectors. Lines are eye guides only. Only uncertainties larger than the full points are shown. CR-39 with 3 mm air converter (•) and without (○). (□) Makrofol E.

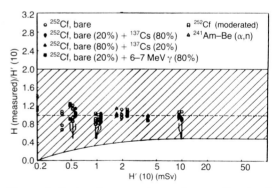

Figure 3. Ratio of measured to reference directional dose equivalent for various irradiations with neutrons from radioactive sources, some of them mixed with photon irradiation, at normal incidence as a function of the reference neutron dose equivalent. The hatched area between the solid lines denotes the acceptable region for routine dose determination.

One of the most outstanding properties of etched track detectors is their insensitivity to photon radiation. This is especially important for the measurement of low neutron doses. A mixed-field irradiation (80% photon dose, 20% neutron dose, a neutron contribution above which neutron monitoring might be required) was performed with ^{252}Cf neutrons at a neutron dose of 0.5 mSv. Besides photon irradiation with ^{137}Cs (energy 662 keV), the photon dose was generated in a 6–7 MeV photon field recently available at the PTB[8]. Here, photon-induced reactions are more likely to produce charged particles and thus additional tracks in the foils, but in all cases the neutron dose measured did not show any significant changes.

SPECTROMETRIC PROPERTIES

The evaluation of two CR-39 foils — with and without air converter — and a Makrofol foil, allows the total dose, the dose above a neutron energy of 70 keV and above 1.5 MeV to be determined, using average response values for all detectors. We have taken the CR-39 detector response of 425 cm^{-2}.mSv^{-1} for ^{252}Cf neutrons as reference response and 1.6 times the Makrofol response for ^{252}Cf neutrons, namely 38 cm^{-2}.mSv^{-1}, since for the latter only 63% of the dose contributes to the track density measured (energy cut-off at 1.5 MeV). The dose contributions measured in the neutron energy intervals from thermal to 70 keV, 70 keV to 1.5 MeV and 1.5 MeV to 15 MeV (and even higher) are shown in Figure 4 for ^{252}Cf (bare), ^{252}Cf (D$_2$O moderated) and ^{241}Am–Be (α,n), and compared with the corresponding values calculated from the group source strengths given by ISO[7]. Surprisingly good agreement was achieved. In cases where neutrons with even higher energies are mainly involved (for example 15 MeV), with the calibration used here the Makrofol foil should show a higher dose than the CR-39 foil, thus indicating especially high energy neutrons.

ANGLE DEPENDENCE

The angle dependence of the dosemeter response is shown in Figure 5 for monoenergetic neutrons of 144 keV, 570 keV and 5.0 MeV, and for ^{252}Cf neutrons with respect to the directional dose equivalent H'(10)[4,5]. The angle dependence does not vary much with the introduction of the 3 mm air layer.

According to ICRU 43[9] the angle dependence of personal dosemeters should follow that of the individual dose equivalent, penetrating, H$_p$(10), which is adequately represented by the directional dose equivalent. Thus a representation of the response with respect to H'(10) should show a flat angle dependence. This condition is not fulfilled for the track dosemeter, especially at 144 keV. In routine measurement, however, a single calibration factor for all energies and angles is desirable. If the reciprocal of 0.7 times the dose equivalent response to ^{252}Cf neutrons at normal incidence is used as a calibration factor, all measurements with neutron incidence between 0 and 60 degrees lie between the arbitrary limits of + 100% and – 50%, except for the point at 144 keV and 60°.

The angle dependence for thermal and intermediate neutrons has not been measured. An expected over-response, especially for thermal neutrons at lateral incidence, can be reduced by an appropriate shielding of the dosemeter sides against external thermal neutrons.

Since practical fields encountered in personal neutron dosimetry contain broad spectral distributions, under-responses in some limited energy regions and at some angles may be

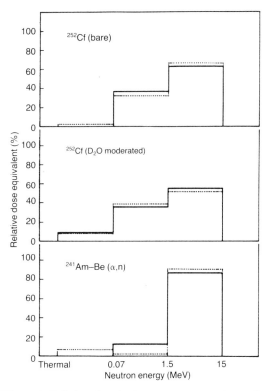

Figure 4. Relative dose in the neutron energy intervals from thermal to 70 keV, 70 keV to 1.5 MeV and 1.5 MeV to 15 MeV for various neutron fields with broad energy distributions as measured (dotted line) and calculated from spectral distributions (solid line).

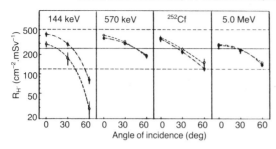

Figure 5. Directional dose equivalent response $R_{H'}$ as a function of angle of incidence for the CR-39 dosemeter with (•) and without (o) air converter, measured with 144 keV, 570 keV, 5.0 MeV and ^{252}Cf neutrons. The lines connecting the points are eye guides only. The proposed value for calibration (solid line) and arbitrary response limits (dashed lines at + 100% and − 50%) are indicated by horizontal lines.

compensated by over-responses in other regions. It is therefore expected that the dose measured with this simple dosemeter lies well within the recommended limits in neutron fields encountered in routine surveillance.

CONCLUDING REMARKS

A simple neutron dosemeter based on etched track detectors which can be used in neutron fields with broad energy distributions from thermal to 15 MeV and in mixed neutron–photon fields has been developed. Only one calibration factor is necessary to determine the dose of neutrons incident normally and at angles up to 60° to the normal. If desired, an evaluation of two additional foils gives rough spectrometric information on the dose distribution in the energy regions from thermal to 70 keV, 70 keV to 1.5 MeV and 1.5 MeV to 15 MeV.

ACKNOWLEDGEMENTS

Some of this work was performed within the framework of a research project sponsored by the Bundesminister für Umwelt, Naturschutz und Reaktorsicherheit, Contract No St. Sch. 1100. The assistance of Mrs E. Fiene during the preparation and evaluation of dosemeters and of H. Finger, W. Sosaat, H. Strzelczyk and G. Urbach during the irradiations is gratefully acknowledged.

REFERENCES

1. Luszik-Bhadra, M., Alberts, W. G., Dietz, E. and Guldbakke, S. *A Track-Etch Neutron Dosemeter with Flat Response and Spectrometric Properties.* Nucl. Tracks Radiat. Meas. **19**, 485-488 (1991).
2. Lesiecki, H., Cosack, M. and Schölermann, H. *Monoenergetic Neutron Fields for the Calibration of Neutron Dosemeters at the Accelerator Facility of the PTB.* PTB-Mitteilungen **97**, 373-376 (1987).
3. Alberts, W. G. and Dietz, E. *Filtered Neutron Beams at the FMRB — Review and Current Status.* Report FMRB-112 (PTB, Braunschweig) (1987).
4. Wagner, S. R., Großwendt, B., Harvey, J. R., Mill, A. J., Selbach, H.-J. and Siebert, B. R. L. *Unified Conversion Functions for the New ICRU Operational Radiation Protection Quantities.* Radiat. Prot. Dosim. **12**, 231-235 (1985).
5. Siebert, B. R. L. and Morhart, A. *A Proposed Procedure for Standardising the Relationship between the Directional Dose Equivalent and Neutron Fluence.* Radiat. Prot. Dosim. **28**, 47-51 (1989).
6. Luszik-Bhadra, M., Alberts, W. G., Guldbakke, S. and Kluge, H. *Influence of the Converter Configuration on the Angle Response of Track-Etched Neutron Dosemeters.* Radiat. Prot. Dosim. **38**, 271-277 (1991).
7. International Organisation for Standardization (ISO). *Neutron Reference Radiations for Calibrating Neutron-Measuring Devices Used for Radiation Protection Purposes and for Determining their Response as a Function of Neutron Energy.* International Standard ISO 8529 (1989).
8. Guldbakke, S., Rossiter, M. J., Schäffler, D. and Williams, T. T. *The Calibration of Secondary Standard Ionisation Chambers in High Energy Photon Fields.* Radiat. Prot. Dosim. **35**, 237-240 (1991).
9. ICRU. *Determination of Dose Equivalents from External Radiation Sources.* Report 43 (Bethesda, MD; ICRU Publications) (1988).

A COMPARISON OF THE NEUTRON RESPONSE OF CR-39 MADE BY DIFFERENT MANUFACTURERS

N. E. Ipe†, J. C. Liu†, B. R. Buddemeier‡, C. J. Miles§ and R. C. Yoder∥
†Stanford Linear Accelerator Center
Stanford University, Stanford, CA 94039, USA
‡Lawrence Livermore National Laboratory L-457
PO Box 5505, Livermore, CA 94550, USA
§Radiation Detection Company, 162 Wolfe Road, Sunnyvale, CA 94088, USA
∥Tech/OPS Landauer, Inc., 2 Science Road, Glenwood, IL 60426-1586, USA

Abstract — CR-39 obtained from American Acrylics and Plastics, Inc. (AA), N. E. Technology, Ltd. (NE), and Tech/Ops Landauer, Inc. (LT) were exposed to radioisotopic neutron sources at SLAC, and moderated ^{252}Cf at ORNL. The AA and NE detectors (0.06 cm thick) were electrochemically etched (a pre-etch for 1 h and 45 min in 6.5 N KOH at 60°C for a 5 h etch at 3000 V and 60 Hz, a 23 min blow-up step at 2 kHz and a post-etch for 15 min). The LT detectors were chemically etched in 5.5 N NaOH at 70°C for 15.5 h. and some AA, NE and LT detectors in 6.25 N NaOH at 70°C for 6 h. A pre-etch step in 60% methanol and 40% NaOH at 70°C for 1 h was added for some NE detectors. The results of the background track density and neutron dose equivalent response are reported.

INTRODUCTION

CR-39 is a polymeric solid-state nuclear track detector which is widely used for neutron dosimetry. CR-39 detects neutrons mainly by the damage trails from the nuclei of its constituent atoms, namely, hydrogen, carbon, and oxygen. The damage trails or tracks can subsequently be revealed by a suitable etching process: either chemical etching (CE) or electrochemical etching (ECE), or both combined. On the basis of calibration exposures, the track density can then be related to the neutron dose equivalent

The use of a hydrogenous radiator in contact with CR-39 can enhance its response, because of the additional proton recoils generated within the radiator. When combined with radiators CR-39 can detect neutrons over a fairly wide energy range from about 100 keV to 20 MeV. The energy range, dose equivalent response ($cm^{-2}.mSv^{-1}$), and lower limit of detection are dependent on the type of CR-39 material, the type and thickness of the radiator, and the particular etching process used.

This paper compares the neutron dose equivalent response of CR-39 plastics made by different manufacturers for both chemical and electrochemical etching.

EXPERIMENTAL METHODS

Plastics

CR-39 was obtained from three different manufacturers: (a) American Acrylics and Plastics, Inc., USA (AA); (b) NE Technology, Ltd, UK (NE); and (c) Tech/Ops Landauer, Inc., USA (LT).

The AA CR-39, dosimetry grade, was about 0.063 cm thick and covered on both sides with polyethylene of thickness 0.01 cm.

The NE CR-39 materials PN-3 and PN-4 were prototype materials. Two batches of PN-4 (NE B1, NE B2) were obtained. The PN-3 material was about 0.13 cm thick and recommended by the manufacturer for CE. The PN-4 material was about 0.06 cm thick and recommended for ECE. Both types were covered on both sides with Clingsol of thickness 0.002 cm.

The LT CR-39 material 'Lantrak', recommended by the manufacturer for CE, was 0.08 cm thick and covered on both sides with a protective layer of polyethylene of thickness 0.006 cm.

The AA and NE CR-39 were laser-cut into pieces of dimensions 1.59 × 2.85 cm^2. After removal of the protective layers, the detectors were placed inside a plastic badge whose front side, which serves as radiator, was about 0.05 cm thick. The LT CR-39 detectors had a polyethylene radiator of thickness 0.1 cm in front, in the 'Landauer badge'. It is important to note that the radiator thicknesses were not the same for the different plastics.

Neutron irradiations

At the Stanford Linear Accelerator Center (SLAC), the neutron irradiations were performed outdoors in a low scatter environment. Corrections were made for anisotropy of the neutron sources and for neutron scattering. The fluences were converted to dose equivalents using the methods outlined in NCRP 79[1].

All detectors were exposed on a cubic water phantom under identical irradiation conditions to radioisotopic neutron sources[2] (^{238}Pu-Be E_{av} = 4.2 MeV, ^{238}Pu-B E_{av} = 2.1 MeV, ^{238}PuF$_4$ E_{av}

= 1.5 MeV, ^{238}Pu-Li E_{av} = 0.5 MeV, and ^{252}Cf E_{av} = 2.2 MeV, where E_{av} is the fluence-averaged energy). The source strengths are all traceable indirectly back to standards at the National Institute of Standards and Technology (NIST). The uncertainties in source strengths are as follows: PuBe 10%, PuB 15%, PuF$_4$ 20%, PuLi 15%, and ^{252}Cf 3%.

As part of the Fifteenth Personnel Dosimetry Intercomparison Study (PDIS 15), the detectors were exposed on a Lucite slab phantom at the Oak Ridge National Laboratory (ORNL) to ^{238}PuBe and ^{252}Cf (D$_2$O). The D$_2$O moderator was 15 cm in radius.

ECE procedure

The AA and NE detectors were electrochemically etched using the following procedure: a pre-etch in 6.5 N KOH at 60°C for 1 h and 45 min, followed by a 5 h etch at 3000 V and 60 Hz, then a 23 min blow-up step at 2 kHz and, finally, a post-etch for 15 min. The detectors were etched in the standard Homann type etch chamber which can handle up to 24 detectors simultaneously[3].

This etch differs from Hankins' latest etch process in that it lasts for 5 h instead of 3 h, and the blow-up step lasts for only 23 min instead of 30 min[3]. The pre-etch reduces the background on the detectors. The blow-up step further amplifies the track size after the etch, thus improving the precision that can be attained with the image analyser which is used for track counting.

CE procedure

The NE detectors were chemically etched using the procedure recommended by the manufacturer: a pre-etch in 60% methanol and 40% 6.25 N NaOH at 70°C for 1 h, followed by an etch in 6.25 N NaOH at 70°C for 6 h. These detectors are designated as NE (pre-etch) in the text. About 25 µm were removed from one side during the pre-etch and 12 µm during the etch. The pre-etch was recommended for reduction in background caused by defects and radon.

Some NE, AA, and LT detectors were chemically etched, using the process mentioned above, but with the elimination of the pre-etch step. These detectors are designated as NE (6 h), AA (6 h), and LT (6 h) in the text. Some other LT detectors were chemically etched in 5.5 N NaOH at 70°C for 15.5 h as recommended by the manufacturer. These detectors are designated as LT (15.5 h).

Track counting

The Homann Track Size Image Analyzing System was used to count the tracks[3] for the electrochemically etched CR-39 detectors. The system consists of a computer and a camera connected to a standard microscope. A computer-controlled automatic stager moves the detector. The image analyser displays the track size distribution and the number of tracks. Three fields, each of an area of about 0.6 cm^2, were counted for each detector; The electrochemically etched AA detectors exposed at PDIS 15 were counted using a Zeiss optical microscope at a magnification of 210. A total of 10 fields of total area 0.11 cm^2 were counted.

Track densities for the chemically etched CR-39 detector were determined with a microscope at a magnification of 300. Nine fields of total area of 3.68 mm^2 were counted for each detector.

RESULTS AND DISCUSSION

Comparison of response of different materials

Table 1 compares the background and the ^{252}Cf neutron response of different materials for both electrochemical and chemical etching. The results are mean values and standard deviations (of the means) of three detectors except as otherwise indicated in brackets. The standard deviation includes the variation between the detectors and the uncertainty in the source strength.

Chemically etched LT (15.5 h) has the highest response and the lowest background. This response is about 2.5 times greater and the background is about 2.5 times lower than that of AA (ECE).

The response of AA (ECE) is about 1.5 to 2 times greater than that of NE B2 and NE B1. The background of AA (ECE) is about the same as that of NE B1. The response and background of NE B2 is slightly higher than that of NE B1. Track size distribution studies with the image analyser indicated that track sizes for AA (ECE) are larger than for NE (ECE) for both background and the ^{252}Cf neutron exposure. The lower response of NE B1 and NE B2 could be attributed to the presence

Table 1. Dose equivalent response, background and standard deviations of different CR-39 plastics.

Etch process	Material	Response to ^{252}Cf (cm^{-2}.mSv^{-1})	Background (cm^{-2})
ECE	AA	471 ± 26	69 ± 22 (2)
	NE B1	231 ± 9	65 ± 6 (6)
	NE B2	284 ± 11	82 ± 12 (6)
CE	AA (6 h)	333 ± 57	217 ± 70 (2)
	NE (6 h)	204 ± 49	1060 ± 250 (2)
	NE (pre-etch)	231 ± 42	571 ± 95
	LT (6 h)	181 ± 34	63 ± 4
	LT (15.5 h)	1148 ± 48 (2)	27 ± 27 (2)

Three detectors were used for each plastic except where otherwise indicated in brackets.

of additives such as DOS (dioctyl sebacate)[4]. The AA CR-39 does not contain any additives.

The other etching condition of Hankins for AA (ECE)[3] indicates a similar ^{252}Cf neutron response of 467 cm^{-2}.mSv^{-1}.

Within the uncertainty of measurement, chemically etched AA (6 h) CR-39 is about 1.5 times more sensitive than NE (pre-etch) and NE (6 h). NE (pre-etch) has a slightly higher response and a much lower background than NE (6 h). Data provided by the manufacturer for NE (pre-etch) indicates a response of 240 cm^{-2}.mSv^{-1} for ^{252}Cf neutrons, which is in good agreement with our measurement.

LT (6 h) has a similar but lower response than NE (6 h); however, its background is more than one order of magnitude lower.

Energy response of CR-39 (ECE)

Figure 1 shows the response of electrochemically etched CR-39 as a function of the average neutron energy which, for all the detectors, generally decreases with increasing neutron energy except for a lower response to the Pu-B neutrons. The response of AA to Pu-Be neutrons is about 15% lower than its response to ^{252}Cf neutrons, whereas a 30% reduction is reported by Hankins[3].

For all the neutron sources, AA has a higher response than NE and NE B2 has a slightly higher response than NE B1.

Energy response of CR-39 (CE)

Figure 2 shows the response of chemically etched CR-39 as a function of average neutron energy. For the information of the reader, it must be pointed out that Figure 2 has a broken scale. The scale in Figure 2 is also different from that in Figure 1. LT (15.5 h) shows the highest response for all the neutron sources. The response, in general, increases as the average neutron energy decreases.

LT (15.5 h) is about 3 to 8 times more sensitive than AA (6 h) and about 2 to 4 times more sensitive than AA (ECE), depending upon the neutron source.

The Am-Be response of LT (15.5 h) as provided by the manufacturer appears to be in reasonable agreement with the response for Pu-Be results.

As for LT (6 h), the response increases with decreasing average neutron energy by more than a factor of 2. Clearly, AA (6 h) does not show this trend, and appears to be fairly independent of the average neutron energy.

The chemically etched NE, with and without pre-etch, also appears to have responses fairly independent of average neutron energy within a factor of about 2. The response of NE (6 h) is about the same as that of AA (6 h). NE (pre-etch) is more sensitive than NE (6 h).

PDIS 15 results

The results of PDIS 15, held at ORNL, are shown in Table 2.

The first and second columns show the neutron source and the dose equivalent delivered and the last three columns show the mean response and its

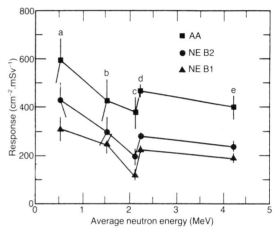

Figure 1. Response of CR-39 (ECE) as a function of average neutron energy for different radioisotopic neutron sources: (a) Pu-Li; (b) PuF$_4$; (c) Pu-B; (d) ^{252}Cf and (e) Pu-Be.

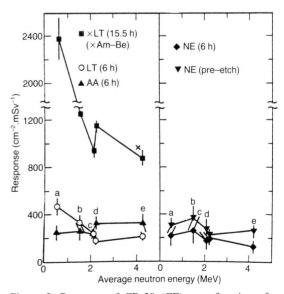

Figure 2. Response of CR-39 (CE) as a function of average neutron energy for different radioisotopic neutron sources: (a) Pu-Li; (b) PuF$_4$; (c) Pu-B; (d) ^{252}Cf and (e) Pu-Be.

standard deviation for the CR-39 plastics: AA (ECE), LT (15.5 h), and NE (pre-etch). The standard deviation includes only the variations in track density between the detectors since the uncertainty in source strength was not reported by ORNL. Three AA detectors were used for each exposure, except two for Exposure 1. Six detectors were used in each case for LT (15.5 h) and three for NE (6 h) except two for Exposure 6. Results for electrochemically etched NE are not reported because of the very high background track densities found on the controls. Since track counting was done over an area of only 0.11 cm^2, there is a high random uncertainty associated with the AA (ECE) results.

As expected, within the uncertainty of measurement, the responses are the same for each plastic for Exposures 1, 3 and 4.

In Exposure 2, the phantom was rotated through 45° about the vertical centreline. The LT (15.5 h) and AA (ECE) responses are about 60% and 70% of that for normal incidence, respectively, which is in good agreement with the 60% reported by Hankins for AA (ECE). Because of the large random uncertainty of track counting, the AA and NE results cannot be used for intercomparison.

For Exposure 6, about 56% of the contribution to dose equivalent is from moderated ^{252}Cf and 46% from bare ^{252}Cf neutrons. As is evident from Tables 1 and 2, the response to moderated and bare ^{252}Cf is about the same for AA (ECE). For LT (15.5 h), the response to moderated ^{252}Cf is about 20% higher than bare ^{252}Cf; for NE, however, the response is only 60% of that for bare ^{252}Cf. Thus, based on the sensitivities of the individual detectors to bare and moderated ^{252}Cf neutrons, these results are reasonable [within 11% for AA (ECE), 4% for NE (pre-etch) and 11% for LT (15.5 h)].

We are unable to explain the discrepancy that the responses of both AA (ECE) and LT (15.5 h) to Pu-Be are about 60% higher than that obtained with the exposures at SLAC.

CONCLUSIONS

The neutron response of CR-39 made by different manufacturers, when exposed under similar conditions, has been compared for both chemical and electrochemical etching.

The Lantrak 15.5 h chemically etched CR-39 has the highest dose equivalent response for all neutron sources and the lowest background. The response is energy (E_{av}) dependent and generally increases as the average energy decreases. The response to Pu-Li neutrons is about 3 times higher than the response to Pu-Be neutrons.

Electrochemically etched American Acrylics has a much lower response than Lantrak (LT (15.5 h)). It has a higher background and its energy response is similar to Lantrak, but not as pronounced. The less pronounced energy response is desirable for neutron dosimetry.

Electrochemically etched NE has a lower response than that of AA (ECE). The background and energy response are similar to AA (ECE).

Chemically etched NE has about the same response as electrochemically etched NE, but its background is higher. The response, however, is fairly independent of average energy.

ACKNOWLEDGEMENTS

We thank NE (formerly Vinten Analytical Systems, Ltd) for providing the CR-39 for this study. Acknowledgements are also due to the SLAC Publications Department, for help with the preparation of this manuscript. This work was supported by the Department of Energy under contract DE-AC03-76SF00515.

Table 2. Dose equivalent response and standard deviations for PDIS 15 intercomparison study.

Source	Dose equivalent (mSv)	Response (cm^{-2}.mSv^{-1})		
		AA (ECE)	LT (15.5 h)	NE (pre-etch)
(1) ^{252}Cf (D$_2$O)	1.40	467 ± 6	1305 ± 67	136 ± 38
(2) ^{252}Cf (D$_2$O) 45° irradiation	2.05	368 ± 33	829 ± 60	146 ± 78
(3) ^{252}Cf (D$_2$O) without cadmium	1.42	589 ± 74	1401 ± 49	115 ± 42
(4) ^{252}Cf(D$_2$O) + ^{137}Cs	0.45	547 ± 28	1444 ± 127	–
(5) ^{238}Pu-Be	0.38	631 ± 62	1313 ± 150	–
(6) ^{252}Cf (D$_2$O) + ^{252}Cf	3.49	567 ± 35	1444 ± 60	179 ± 21

REFERENCES

1. NCRP. *Neutron Contamination from Medical Electron Accelerators*. Report No. 79 (Bethesda, MD: NCRP Publications) pp. 42-46 (1984).
2. Liu, J. C., Jenkins, T. M., McCall, R. C. and Ipe, N. E. *Neutron Dosimetry at SLAC: Neutron Sources and Instrumentation*. SLAC-TN-91-3 (Stanford Linear Accelerator Center, Stanford, CA 94309) (October 1991).
3. Hankins, D. E., Homann, S. G. and Buddemeier, B. *Personnel Neutron Dosimetry Using Electrochemically Etched CR-39 Foils*. UCRL-53833, Rev. 1 (Lawrence Livermore National Laboratory, University of California, Livermore, CA 94550) (December 1989).
4. Portwood, T. and Stejny, J. *Analysis of CR-39 and the Effect of Additives*. Nucl. Tracks Radiat. Meas. **8**, 151-154 (1984).

INTERNATIONAL STUDY OF CR-39 ETCHED TRACK NEUTRON DOSEMETERS (EURADOS–CENDOS 1990)

W. G. Alberts
Physikalisch-Technische Bundesanstalt
Bundesallee 100, W-3300 Braunschweig, Germany

Abstract — The paper briefly reports on a multilaboratory experiment investigating the response of various etched track neutron dosemeters at neutron energies between 144 keV and 14.7 MeV. Some results of the angle dependence of the response are presented.

INTRODUCTION

In 1990, the EURADOS-CENDOS Working Committee 5, Subcommittee on the Application of Track Detectors to Neutron Dosimetry, carried through the fourth international irradiation experiment for dosemeters based on the proton-sensitive plastic PADC (CR-39) (see also Ref.1). The intention of this work was to provide reproducible irradiations for an international group of laboratories dealing with these dosemeters either routinely or within a research programme. The main aim of this investigation was to look at the energy and angle dependence of the dose equivalent response in view of use of the detectors for individual monitoring.

The irradiations were performed at the GSF - Forschungszentrum für Umwelt und Gesundheit, Neuherberg, and at the PTB, Braunschweig, with 2 to 3 mSv of monoenergetic neutrons of 144 keV, 570 keV, 5.3 MeV and 14.7 MeV and of neutrons from a ^{252}Cf neutron source. For the irradiations the dosemeters were fixed on PMMA phantoms 30 cm x 30 cm x 15 cm; angles of incidence of 0°, 30° and 60° were selected.

Fifteen laboratories from ten countries (Europe, Algeria, Canada and USA) participated in this study. They all used their established materials, dosemeter constructions and converter configurations, and etching and counting techniques; a total of 31 sets of results was obtained. Etching procedures ranged from a single-step chemical (CE) or electrochemical (ECE) etching to three-step etching consisting of two ECE steps preceded or followed by a CE step: CE (1 laboratory), ECE (1), CE-CE (1), CE-ECE (3), ECE-ECE (5), CE-ECE-ECE (3), ECE-ECE-CE (1). The participants were given the ambient dose equivalent as evaluated by the irradiating institutes and were asked to report track densities, background track densities, and an evaluated ambient dose equivalent response, together with their standard deviations.

In a forthcoming combined PTB and EURADOS-CENDOS report[2] graphical representations of the response and also linearity and background results will be presented together with the full results as reported by each participant. It is the aim of this note to review briefly some of the results.

EXPERIMENTAL RESULTS

At present the only phantom quantity suitable for use as a calibration quantity for individual dosemeters substituting the personal dose, $H_p(10)$, for irradiation experiments is the directional dose equivalent, $H'(10)$. Figure 1(a) shows the angle dependence of the fluence-to-directional dose equivalent conversion factor, $H'(10)/\Phi$, for the neutron energies encountered in this experiment and for angles up to 60°. Figures 1(b) and 1(c) show, for comparison, two examples of the angle dependence of the response with respect to the ambient dose equivalent, a quantity with no angle dependence itself.

The examples shown are typical for flat dosemeters, with the detector and phantom surfaces parallel. Two deviating dosemeter constructions (a flat detector, set at an angle of 30° with respect to the phantom surface, and a detector bent to a full cylinder) showed an expected difference in behaviour.

Background track densities, expressed in terms of ambient dose equivalent caused by ^{252}Cf neutrons, ranged from 0.04 mSv to over 1 mSv. A 'minimum detectable dose' was calculated as three standard deviations of the background track densities as reported by the participants. Values ranged from 0.01 mSv up to 0.5 mSv.

The linearity of the systems was tested with five irradiations of 5.3 MeV neutrons at about 0.5, 0.8, 1.2, 2 and 5 mSv. With a few exceptions the linearity observed was satisfactory; the large gap between 2 mSv and 5 mSv does not allow conclusions on the reasons for apparent saturation effects for some systems. No systematic dependences of the linearity on the sensitivity, the type of

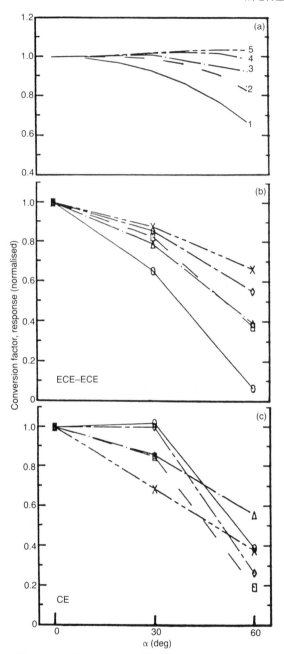

Figure 1. Fluence-to-directional dose equivalent conversion factor (a) and dose equivalent response of track dosemeters (b and c) plotted against angle of incidence α for the neutron energies: 1, (o) 144 keV. 2, (– – –, ☐) 565 keV. 3, (–·–·–, Δ) ^{252}Cf. 4, (◊) 5.3 MeV. 5, (x) 14.7 MeV.

etching procedure or counting technique could be determined from this study.

CONCLUDING REMARKS

Individual dosemeters should probably have a response of which the energy and angle dependence is that of the personal dose $H_p(10)$. The response of most of the track dosemeters investigated, however, falls off more steeply at large angles, especially for low energies.

For most systems the deviations from dose–track density linearity can be corrected for and should not cause any problems in routine radiation protection. More than one half of the results show minimum detectable dose equivalents below 0.2 mSv for ^{252}Cf neutrons, a value which will have to be discussed in the light of the new ICRP recommendations.

ACKNOWLEDGEMENTS

The following scientists participated in the joint experiment and contributed their results to the study: D. Azimi and C. Wernli, PSI Villigen, Switzerland; R. P. Bradley and F. N. Ryan, Health Prot. Ottawa, Canada; S. Djeffal and Z. Lounis, CRS Alger, Algeria; B. Dörschel, H. Hartmann, G. Streubel and A. Guhr, TU Dresden, Germany; F. Fernandez, C. Baixeras, C. Domingo and E. Luguera, Univ. Barcelona, Spain; P. J. Gilvin, P. V. Shaw and D. T. Bartlett, NRPB Chilton, UK; J. R. Harvey, A. R Weeks and C. W. Christie, Nucl. Electric BNL Berkeley, England; L. Lembo and M. Beozzo, ENEA Bologna, Italy; M. Luszik-Bhadra, PTB Braunschweig, Germany; B. Majborn, Risø Roskilde, Denmark; M. A. Parkhurst and E. P. Moen, PNL Richland, USA; E. Piesch, M. Vilgis, KfK Karlsruhe, Germany; R. J. Tanner, D. T. Bartlett and J. D. Steele, NRPB Chilton, UK; L. Tommasino, G. Torri, T. V. Giap, M. Riccardi and B. Flores, ENEA Roma, Italy; M. Zamani, D. Sampsonidis, U. Kimoundri and E. Savvidis, Univ. Thessaloniki, Greece.

I appreciate the opportunity of presenting this short summary of results on behalf of the participants; the contribution of S. Guldbakke (PTB) and H. Schraube (GSF) in performing the irradiations and providing the reference data is also gratefully acknowledged.

REFERENCES

1. Schraube, H. *Neutron Response of Etched Track Detectors in the Light of a Multilaboratory Experiment*. Int. J. Radiat. Appl. Instrum., Part D: Nucl. Tracks Radiat. Meas. **19**(1-4), 531-536 (1991).
2. Alberts, W. G. (Ed.) *Investigation of Individual Neutron Monitors on the Basis of Etched-track Detectors: The 1990 EURADOS-CENDOS Exercise*. PTB report N-10, EURADOS-CENDOS Report 1992-02 (1992).

A THREE ELEMENT ETCHED TRACK NEUTRON DOSEMETER WITH GOOD ANGULAR AND ENERGY RESPONSE CHARACTERISTICS

J. R. Harvey
Nuclear Electric plc, Technology Division
Berkeley Technology Centre
Berkeley, Gloucestershire GL13 9PB, England

Abstract — A weakness of all single element etched track neutron dosemeters is that the sensitivity falls off too rapidly with increasing angle of incidence. This can lead to significant errors in practical situations. A possible solution is to incorporate in a single dosemeter one or more planar etched track detectors set at an angle to the body surface so that the sensitivity to obliquely incident neutrons is enhanced. The response of a dosemeter in which three planar elements are set in a pyramid structure is investigated. Relationships are developed which allow the response to be estimated for any given direction of incidence and any given angle between face and base of the pyramid. The results indicate that the response is close to that required to measure $H_p(10)$ for any given direction of neutron incidence if the angle between face and base is between 30° and 40°.

INTRODUCTION

The International Commission on Radiation Units and Measurements advise[1] that a personal dosemeter for assessing dose equivalent to deep organs should measure the individual dose equivalent, penetrating, $H_p(10)$. This applies in the case of neutron radiation since in the normal irradiation situation dose to deep organs is limiting. This requires there to be some fall-off in dosemeter reading* as the angle of incidence to the body increases for a given fluence of unidirectional neutrons.

An etched track neutron dosemeter in which the plane of the etch plastic is parallel to the surface of the body has poor angular response characteristics because the $H_p(10)$ response* falls off too rapidly with increasing angle of incidence. This is particularly significant because the probability of a neutron entering the body at a given angle of incidence increases (because of the increase in the associated solid angle) as the sine of that angle.

It is possible to improve the $H_p(10)$ response by exposing two or more pieces of etch plastic at an angle to the body surface since this increases the sensitivity for oblique incidence. It has been suggested[2] that a composite dosemeter consisting of two planar dosemeters inclined at an angle of 35° to the surface of the body in a 'tent' structure would have a roughly correct $H_p(10)$ response if the readings of the two elements were simply added together. In the analysis which lead to this conclusion the angle of incidence of the neutrons was varied about an axis defined by the apex of the wedges (ridge of the tent). The predicted response was broadly confirmed in subsequent experimental irradiations[3].

In many practical situations it is not possible to assume that the direction of the incoming radiation is restricted in this way and in this paper the response of a dosemeter consisting of three elements set in a pyramid structure is assessed for radiation incident from any given direction.

Angular response data for a single CR-39 element

The analysis in this paper is based upon the measured response of single planar CR-39 etched track detectors. The detectors were chemically etched in sodium hydroxide and the number of etch tracks assessed with an optical system which magnifies and automatically counts the tracks[2,4]. This approach is the basis of an automatic reading system (produced by NE Technology Ltd, Bath Road, Beenham, Reading, England) which reads up to 60 elements in 30 min and is hence particularly suitable for processing the throughput of elements which might stem from the use of a multi-element dosemeter. The experimental information was taken from two different experiments[3,4] involving six neutron energies. The data at each neutron energy were fitted to the following mathematical relationship using non-linear regression analysis:

$$R^*/R^*_o = A(\cos^2 I)^B + C \qquad (1)$$

where R^* is the $H^*(10)$ response (see section

---
* The following terms have the given meanings:
  Reading: the output of the reading system in tracks per cm² (less background).
  Response: the quotient of the reading and the value of a quantity, either $H_p(10)$ or $H^*(10)$.

headed 'The required angular response of the dosemeter' for a discussion of quantities) for radiation incident at angle I to the plastic and A, B and C are the fitted constants. R^*_o is the response at normal incidence (zero I). For the neutron energies where the data points lie on a smooth curve (565 keV and 1242 keV) the relationship fits well. At the other energies the higher scatter leads to greater uncertainty in the values of the constants. However, the overall trend with increasing neutron energy is well defined and more accurate analysis will be possible as more data become available. The values of the constants with standard errors are given in Table 1.

RESPONSE OF A THREE ELEMENT 'PYRAMID' DOSEMETER

Relationship between the angle of incidence to the plastic and the angle of incidence to the body

It can be shown that the angle of incidence, I to the (normal to) the face of a pyramid is given by

$$\cos I = (\cos W \times \cos D) + (\sin W \times \sin D \times \cos Z) \quad (2)$$

where W, the 'wedge angle' is the angle between the face and the base plane. D is the declination and Z is the azimuth which together define the incoming direction. The meaning of the angles is illustrated in Figure 1.

The required angular response of the dosemeter

For the purposes of the analysis so far it has

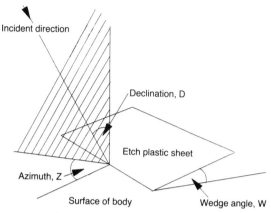

Figure 1. The meaning of the angles.

been convenient to discuss the $H^*(10)$ response since $H^*(10)$ has a value which is independent of the angle of incidence. The quantity which the dosemeter should measure, however, is $H_p(10)$. The ideal fall-off in $H^*(10)$ response with increasing angle of incidence is therefore given by $H_p(10)/H^*(10)$. The values of this ratio given in Table 2 stem from interpolation of calculated values[5] of $H'(10)$ in the ICRU sphere which is taken to be a reasonable approximation of $H_p(10)$ in the human trunk. The more internally consistent data from Jahr et al[6] were given precedence at 100 keV where there is a significant difference between authors. The data for ^{252}Cf were derived by folding the data for given energy bands weighted according to $H^*(10)$/fluence into a fission spectrum. Values of $H^*(10)$ were taken

Table 1. The constants from Equation 1.

Neutron energy	Value of A	Standard error, A	Value of B	Standard error, B	Value of C	Standard error, C
311 keV	0.95	0.17	4.3	2.3	0.051	0.096
565 keV	0.95	0.06	1.00	0.16	0.05	0.05
1242 keV	0.89	0.04	0.95	0.13	0.11	0.04
^{252}Cf	0.87	0.15	0.69	0.32	0.13	0.15
5.3 MeV	0.85	0.17	0.82	0.43	0.15	0.17
14.7 MeV	0.45	0.10	0.71	0.39	0.55	0.10

Table 2. Values of $H_p(10)/H^*(10)$ for the ICRU sphere as a function of declination (angle of incidence to the body).

Neutron energy	0°	15°	30°	45°	60°	75°	90°
311 keV	1.0	0.98	0.91	0.78	0.60	0.37	0.15
565 keV	1.0	0.98	0.92	0.84	0.65	0.44	0.20
1.242 MeV	1.0	1.0	0.98	0.88	0.73	0.53	0.25
^{252}Cf	1.0	1.0	0.98	0.92	0.81	0.65	0.40
5.3 MeV	1.0	1.0	1.0	1.0	1.0	0.90	0.68
14.7 MeV	1.0	1.0	1.0	1.0	1.0	0.95	0.90

from Wagner et al[7].

These factors, f, allow the $H_p(10)$ response of any given dosemeter to be estimated. Specifically, if T is the track density then the $H^*(10)$ response, $R^* = T/H^*(10)$, and the $H_p(10)$ response, $R_p = T/H_p(10)$, and so $R_p/R^* = H^*(10)/H_p(10) = 1/f$, i.e.

$$R_p = R^*/f \qquad (3)$$

The $H_p(10)$ response should ideally be independent of the angle of incidence. It is equal to the mean of the readings of the three elements of the pyramid dosemeter divided by $H_p(10)$.

Variation of response with angle of declination (angle of incidence to the body)

The $H^*(10)$ response for any given declination, azimuth and wedge angle is given by combining Equations 1 and 2:

$$R^*/R^*_o = A\{[(\cos W \times \cos D) + (\sin W \times \sin D \times \cos Z)]^2\}^B + C \qquad (4)$$

The $H_p(10)$ response was estimated from Equations 3 and 4 for given values of wedge angle and declination (angle of incidence to the body) and averaged over all azimuth. The averaging over azimuth means that these results apply to any dosemeter where the element or elements are inclined at the same angle, regardless of the number of elements.

The data suggest that any wedge angle between 30° and 40° will give an acceptable angular response. The spread of angles of incidence in any practical working environment would minimise the significance of the remaining imperfections in the angular response. A thirty-five degree wedge was assessed to have an acceptable overall response (see Figure 2) and is the basis of further calculations of azimuthal response.

Variation of response with azimuthal angle of incidence

The $H_p(10)$ response of the pyramid dosemeter with faces set at 35° to the base was calculated for a range of values of azimuth and declination. If the azimuthal angle of incidence to one element is Z then the azimuthal angles of incidence to the other two elements are (120–Z)° and (120+Z)°. The overall response is therefore the average response for azimuthal incidence angles of Z°, (120–Z)° and (120+Z)° as given by Equations 3 and 4. The $H_p(10)$ response for two values of declination and a range of azimuth is shown in Figure 3. Calculations are only required for variation of azimuth over 60° since the device has six-fold symmetry for the full variation of azimuth over 360°.

The small variation of response with azimuthal angle of incidence shown in Figure 3 stems from the remarkably adventitious fact that

$$(1+K \cos Z)^2 + (1+K \cos(120-Z))^2 + (1+K \cos(120+Z))^2 = 3(1+K^2/2)$$

i.e. has a value which is independent of the value of Z.

This means that the $H_p(10)$ response of the pyramid dosemeter is independent of azimuthal angle if B in Equation 1 is unity, i.e. if the reading of a single element for a given fluence of neutrons incident at angle I falls off as ($\cos^2 I$ + a constant). This applies for any given values of declination and wedge angle. As can be seen from Table 1, B is close to unity for neutrons of energy 565 keV and 1242 keV, fortuitously representative of the neutron energies common in reactor environments. Even at other energies the variation of $H_p(10)$ response with azimuthal angle of incidence is unlikely to be a problem in practice.

Response data for a range of values of declination and azimuth, which could not be

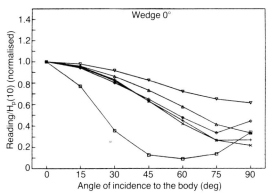
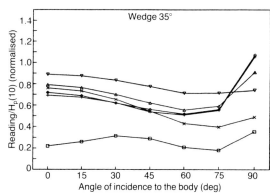

Figure 2. Response averaged over all azimuth for various wedge angles — normalised to normal incidence and wedge 0°. (□) 311 keV, (+) 565 keV, (◊) 1242 keV, (Δ) ^{252}Cf, (x) 5.3 MeV, (∇) 14.7 MeV.

included here because of space restrictions, are available on request.

CONCLUSION

The analysis reported in this paper suggests that a neutron dosemeter consisting of three etched track elements located in a pyramid structure, where the angle between each face and the base is 35°, has acceptable response characteristics for the measurement of $H_p(10)$ for neutron radiation incident from any direction over a wide range of neutron energies. Experiments are currently underway to test this prediction. Although the analysis leading to this conclusion is based on a particular etched track system it will apply to any etched track or electrochemical-etch system with similar angular response characteristics.

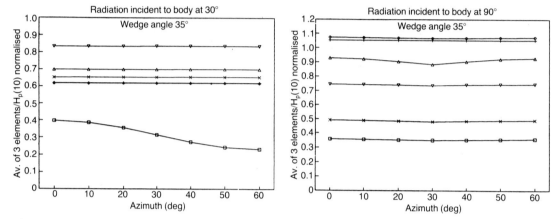

Figure 3. Response as a function of azimuthal angle — normalised to normal incidence and wedge 0°. Key as Figure 2.

REFERENCES

1. ICRU. *Determination of Dose Equivalents Resulting from External Radiation Sources*. Report 39 (Bethesda, MD: ICRU Publications) (1985).
2. Harvey, J. R. and Weeks, A. R. *Recent Developments in a Neutron Dosimetry System Based on the Chemical Etch of CR-39*. Radiat. Prot. Dosim. **20**(1/2), 89-93 (1987).
3. Alberts, W. G. *Results of the 1990 EURADOS-CENDOS Joint Neutron Irradiation*. PTB report (to be published).
4. Harvey, J. R. and Weeks, A. R. *Progress Towards the Development of a Personal Neutron Dosimetry System Based on the Chemical Etch of CR-39*. UK CEGB Report TPRD/B/0851/R86 (1986).
5. Harvey, J. R. *The Individual Monitoring Quantity for Neutrons and its Relationship with Fluence*. Radiat. Prot. Dosim. **20**(1/2), 19-24 (1987).
6. Jahr, R., Hollnagel, R. and Siebert, B. R. L. *Calculations of Specified Depth Dose Equivalent in the ICRU Sphere Resulting from External Neutron Irradiation with Energies Ranging from Thermal to 20 MeV*. Radiat. Prot. Dosim. **10**, 75-87 (1985).
7. Wagner, S. R., Grosswendt, B., Harvey, J. R., Mill, A. J., Selbach, H.-J. and Siebert, B. R. L. *Unified Conversion Functions for the New ICRU Operational Radiation Protection Quantities*. Radiat. Prot. Dosim. **12**(2), 231-235 (1985).

RECENT DEVELOPMENTS ON THE CRS PADC FAST NEUTRON PERSONAL DOSEMETER

S. Djeffal[†], Z. Lounis[†] and M. Allab[‡]
[†]Laboratoire de Dosimétrie, Centre de Radioprotection et de Sureté
2, Bd. Frantz Fanon, BP 1017, Alger, Algeria
[‡]Institut de Physique, Université des sciences et de la Technologie
El-Alia BP 32, Bab Ezzouar, Alger, Algeria

Abstract — The Dosimetry Laboratory of the Radiation Protection and Safety Centre in Algiers has developed a fast neutron personal dosemeter using a PADC detector. This dosemeter has been type tested in terms of the ICRU calibration quantity, directional dose equivalent at 10 mm, H'(10). This paper gives the results of type test measurements. In addition, some properties of five different PADC materials were investigated in this study.

INTRODUCTION

The Dosimetry Laboratory of the Radiation Protection and Safety Centre in Algiers has developed a fast neutron personal dosemeter using a PADC Detector. Some shortcomings in using PADC as a neutron detector still exist, the most important being the strong angular dependence, the background reproducibility and the ageing of commercially available PADC materials[1–3].

In order to limit the angular dependence, a cylindrical multidirectional dosemeter was used[4,5]. In this context, PADC detectors supplied by Pershore Mouldings Ltd were given a half cylindrical geometry using the technique of softening and bending[4]. These detectors were placed in good contact with the inner face of a plastic foil made of pure polyethylene (1000 μm thickness) acting as a radiator material. The developed dosemeter has been described elsewhere[6,7].

In our investigation, a study of the energy and angular dependence of response of the proposed dosemeter, which consists of two cylindrical detectors placed perpendicularly to each other, has been undertaken. These investigations have been carried out using chemical etching and electrochemical etching without pre-etching. The advantages and limitations of these two etching processes have been described in previous works[8,9].

Since the quality of PADC detectors so far produced has not yet reached the requirements for routine use[10,11], the changes of dosemeter properties with time, between manufacturing and irradiation (ageing) and between irradiation and processing (fading), have been investigated.

RESULTS OF THE TYPE TEST MEASUREMENTS

The energy and angular dependence of response of the developed dosemeter has been determined using two different etching procedures.

Irradiations were performed with the dosemeters mounted on a 30 cm × 30 cm × 15 cm polymethyl methacrylate (PMMA) slab phantom. Radiation fields were provided at different angles (i.e. 0°, 30°, and 60°) by PTB Braunschweig (144 keV, 14.7 MeV) and GSF München (0.565 MeV, 5.3 MeV and bare ^{252}Cf) under the aegis of EURADOS–CENDOS. The neutron dose equivalents applied ranged between 2 mSv and 2.9 mSv; in addition at 5.3 MeV a linearity test was performed with five dose equivalent values ranging from 0.5 mSv to 5 mSv.

Irradiated detectors were processed using the chemical etching procedure (CE) for 6 h in 6.25N KOH at 60°C. Another batch of irradiated detectors was processed using the single sided electrochemical etching procedure without pre-etching (ECE only), at the same temperature applying 30 kV.cm^{-1} at a low frequency (50 Hz) for 6 h. A circular area of about 1 cm^2 was electrochemically etched, and about 20 fields per area were evaluated.

Track density measurements have been performed using an image analyser (type AMS. 40–10) coupled to an optical microscope. The track density deduced from the average of the evaluated fields was corrected for the background. Neutron fluences were converted to directional dose equivalent at 10 mm, H'(10), using the conversion coefficients given by ICRP 51[12] taking into account the angle data from Harvey[13].

Energy and angle dependences of the response are shown in Figures 1 and 2, respectively for the CE and ECE only. The variation in response is nearly independent of angle of incidence whatever the etching procedure is, except for the low neutron energies (144 keV and 565 keV) where the only notable discrepancy occurs at 60°. In the case of the electrochemical etching, the dosemeter

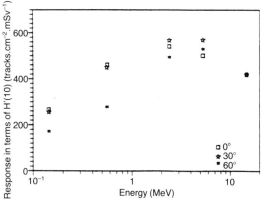

Figure 1. Energy and angle dependence of response in terms of H'(10) for the chemically etched detectors.

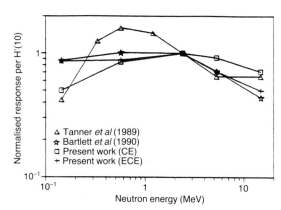

Figure 3. Response as a function of neutron energy for normally incident neutrons. Response normalised to ^{252}Cf.

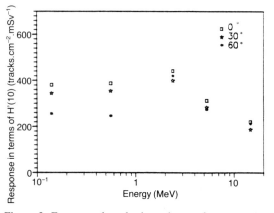

Figure 2. Energy and angle dependence of response in terms of H'(10) for the electrochemically etched detectors.

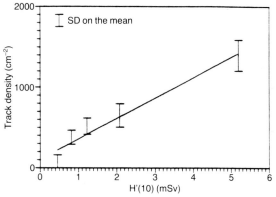

Figure 4. Dose response to 5.3 MeV monoenergetic neutrons (angle of incidence =0°).

response is nearly flat.

Response as a function of neutron energy, normalised to the spectrum of ^{252}Cf is shown in Figure 3 where the responses of the NRPB dosemeter are shown for comparison[14,15]. Despite the different etching procedures and the different PADC batches used, the results are of the same order of magnitude. Our results are very close to unity (within 20% when excluding the energy of 144 keV). Due to the presence of 1 mm polyethylene radiator in front of the detector, this response was improved, especially in the high energy region[16].

In Figure 4, the results of the linearity experiment are reported. One can see that the data obtained, in the dose range studied, exhibit a good linear dependence of the track density within the experimental errors. The error bars indicate the standard deviation on the mean.

CHANGES IN PADC PROPERTIES WITH TIME

An intercomparison study of different PADC materials was performed, investigating the background, ageing and fading in normal air at ambient temperature.

Storage periods up to 20 months in ambient laboratory conditions have been used to study changes in background and fading of the PADC materials. The PADC materials investigated and their corresponding specifications are listed in Table 1.

PADC samples from each manufacturer were investigated as follows: half of the samples from each batch were irradiated with 3 MeV alpha particles whereas the other half served for the background study. The samples were sealed in polyethylene pouches (1 mm thickness) to minimise effects of environmental alpha radiation.

Every month, 10 detectors (5 background and 5 fading) from each batch were chemically etched and counted.

The results obtained indicate that for all the tested materials used during the first six months period after manufacturing, the change in ageing of intrinsic background does not vary much. After one year storage in conditions of relatively elevated temperature and humidity, such as in Algiers, the background values become three times higher. The high values of background track density observed in American Acrylics and Pershore Mouldings materials are due to the fact that these PADC materials have been stored, in ambient laboratory conditions, for a long period after manufacture and delivered without radon tight packing.

The variation in response with time delay between alpha particle irradiation and processing is reported in Figure 5. The track density obtained was normalised to day zero before storage in ambient laboratory conditions. It indicates that there is some fading of latent signal. This observation was also reported by Tanner et al[14]. Over seven months, at ambient conditions, fading from 15 to 25% had been found in all the tested samples (American Acrylics (15%), Pershore Mouldings (25%), Lantrak (20%) and Patras (25%)).

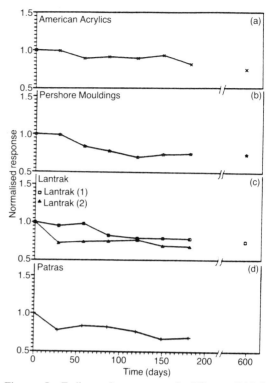

Figure 5. Fading of response of different PADC materials. Response normalised to day zero.

CONCLUSION

The energy and angle dependence of response of the developed dosemeter have been determined using two etching processes. The variation in response is found to be relatively independent of angle of incidence whatever the etching procedure

Table 1. Specifications of the studied PADC materials.

Manufacturer Type/additives	Protective layers (μm)	Thickness (μm)	Year of issue	Background[b] track density (cm^{-2})	Minimum[d] detect. dose (mSv)
American Acrylics & plastics Inc. (USA): Dosimetry grade	80–100 on both sides	600–612	1984	245 ± 48[c]	0.13
Pershore Mouldings Ltd (UK): Std. grade: 2.6% IPP, 0.2% DOP 32 h curing cycle	50–60 on both sides	580–620	1987	384 ± 59	0.20
Tech/Ops Landauer, Inc. (USA) Lantrak[a] 1	50 on both sides/radon-tight packing	800–900	1989	31 ± 9	0.05
Tech/Ops Landauer, Inc. (USA) Lantrak[a] 2	50 on both sides/radon-tight packing	350–400	1989	26 ± 8	0.05
Rathenower Optishe Werke/GDR Patras[a] : 3% IPP, 0.3% DOP Surfaces pre-etched	No prot. layer radon-tight packing	850–900	1989	35 ± 9	0.30

[a] Sample material supplied for tests.
[b] Five (05) samples were used to estimate the background.
[c] Standard deviation on the mean value.
[d] Minimum detectable dose : 2 σb/sensitivity for ^{252}Cf.

is.

The dose equivalent response as a function of neutron energy normalised to the spectrum of ^{252}Cf, obtained using 1 mm polyethylene radiator for the proposed dosemeter was improved in the high energy region.

The storage experiment in ambient laboratory conditions over periods of 20 months resulted in a decrease of the response of pre-irradiated PADC detectors and an increase of the background. Fading ranging from about 15% to 25% had been found in all the detectors tested: however, it is less pronounced for the PADC materials from American Acrylics and Lantrak.

ACKNOWLEDGEMENTS

The authors thank EURADOS–CENDOS and the staff at PTB (Braunschweig) and GSF (Neuherberg) for providing the radiation fields.

REFERENCES

1. *Proceedings of CEC/EURADOS Workshop on Etched Track Neutron Dosimetry.* Radiat. Prot. Dosim. **20**(1) (1987).
2. Lembo, L. (Ed.) *Results of a Survey of Backgrounds of Etched Track Neutron Dosimeters, organised by EURADOS–CENDOS in 1988.* Report PAS-FIBI-DOSI (89) 1, Bologna (1989).
3. Schraube, H. (Ed.) *Response of Proton-Sensitive Etched Track Detectors to Fast Neutrons: Results of a Joint Multilaboratory Experiment.* GSF-Bericht 22/90, EURADOS-CENDOS Report 1990-01.
4. Al-Najjar, S. A. R. and Piesch, E. *Angular Response Characteristic of CR-39 Detector of Cylindrical Geometry.* Radiat. Prot. Dosim. **20**(1), 67-70 (1987).
5. Lounis, Z. *Etude Experimentale d'un Dosimètre Individuel de Neutrons basé sur l'utilisation du P.A.D.C.* Thèse de Magister en Genie Nucléaire, Haut Commissariat à la Recherche, Alger (1991).
6. Djeffal, S. and Lounis, Z. *Progress Towards the Development of CR-39 Fast Neutron Personal Dosemeter.* Nucl. Tracks Radiat. Meas. **19**(1–4), 457-460 (1991).
7. Djeffal, S. and Lounis, Z. *EURADOS-CENDOS Neutron Dosimetry : 1990 Joint Irradiations* (Results from CRS-Dosimetry laboratory). EURADOS-CENDOS Report 1991 (will be edited by W. G. Alberts PTB).
8. Tommasino, L., Zapparoli, G., Djeffal, S. and Cross, W. G. *Protons and Neutrons Registration by Chemically and Electrochemically Etched CR-39 Detectors.* In: Proc. 5th Symp. on Neutron Dosimetry, Munich-Neuherberg, CEC Report EUR 9762 EN (Luxembourg: CEC) pp. 469-477 (1984).
9. Tommasino, L., Zapparoli, G., Griffith, R. V., Djeffal, S. and Spieza, P. *Personal Neutron Dosimetry by CR-39 Plastics with Chemical Etching, Electrochemical Etching and their Combination.* In: Proc. 6th Int. Congress of IRPA, Berlin West, pp. 1201-1202 (1984).
10. Alberts, W. G. and Luszik-Bhadra, M. *Fast Neutron Dosimetry with CR-39 : Study of Various Materials using Electrochemical Etching.* Nucl. Tracks Radiat. Meas. **19**(1–4), 437-442 (1991).
11. Birkholz, W., Winkler, C. and Baumbach, H. *Neutron Sensitivity of CR-39 SSNTD.* Nucl. Tracks Radiat. Meas. **19**(1-4), 453-456 (1991).
12. ICRP. *Data for Use in Protection Against External Radiation.* Publication 51. (Oxford: Pergamon Press) (1987).
13. Harvey, J. R. *The Individual Monitoring Quantity for Neutrons and its Relationship with Fluence.* Radiat. Prot. Dosim. **20**(1), 19-24 (1987).
14. Tanner, R. J., Gilvin, P. J., Steele, J. D., Bartlett, D. T. and Williams, S. M. *The NRPB PADC Neutron Personal Dosemeter: Recent Developments.* Radiat. Prot. Dosim. **34**, 17-20 (1990).
15. Bartlett, D. T., Tanner, R. J., Steele, J. D., Williams, S. M. and Gilvin, P. J. *The Energy and Angle Dependence of Response in Terms of the ICRU Quantities of the New Design of NRPB Neutron Personal Dosemeter.* Nucl. Tracks Radiat. Meas. **19**(1-4), 449-450 (1991).
16. Al-Najjar, S. A. R., Ninomiya, K. and Piesch, E. *Properties of Electrochemically Etched CR-39 Plastic for Fast Neutron Dosimetry.* Radiat. Prot. Dosim. **27**(4), 215-230 (1989).

ANGLE AND ENERGY RESPONSE TO FAST NEUTRONS OF CR-39 COVERED WITH A RADIATOR

E. Pitt, A. Scharmann and R. Simmer
I. Physikalisches Institut, Heinrich-Buff-Ring 16
W-6300 Giessen, Germany

Abstract — CR-39 detectors covered with polypropylene radiators were exposed to fast neutrons in the energy range from 0.5 to 5 MeV at varying angles of incidence. The response has been estimated for electrochemical etching (ECE) by calculations simulating the latent track formation. Their visualisation has been checked based on the criteria for an electrical breakdown at a conducting tip prolonged into an insulator. By comparison with the experimental data consequences are discussed for the critical angle of incidence in the ECE technique, etched track velocity and track geometry.

INTRODUCTION

Polymeric solid state nuclear track detectors (SSNTD) are widely used to detect charged particles by means of etched latent tracks. Improvements of the quality of the detector material and the availability of low cost image analysis systems entailed more and more applications of SSNTDs in routine dosimetry, especially in personnel neutron dosimetry and in radon measurements. Usually, electrochemical etching is used in order to increase the track diameter and to achieve easy readout.

Nevertheless, the electrochemical track etching mechanism is not entirely understood. Considerations of the electric field at a track tip based on Mason's equation gave rise to different models[1,2]. A careful consideration of the track geometry has shown that this equation is not applicable to the conditions of neutron dosimetry and radon measurement because of its presupposition of long tracks in thin detector sheets[3]. Even a novel model with a key-parameter 'specific ionisation' neglects the influence of etched track geometry on the amplification of the applied electric field at the track tip[4].

The real parameter determining the response of a SSNTD is the actual electric field at the tip of an etched track during ECE. At given etch conditions electrical treeing occurs and an incident particle is registered if it falls short of a critical angle of incidence. A model based on Smythe's formula enables an analytic description of this critical angle as a function of particle energy with respect to the parameters of the ECE process[5]. In a first step the response of CR-39, covered with a polypropylene radiator, has been calculated. Good agreement with experiment has been achieved for monoenergetic neutrons up to 5 MeV at normal incidence but modifications are necessary for non-normal incidence.

EXPERIMENTAL

Flat CR-39 detectors (American Acrylics, dosimetry grade, 635 μm thick) partly covered with step radiators (10 steps with a radiator thickness between 4.7 and 40 μm) have been irradiated at PTB in directed, monoenergetic neutron fields at various angles of incidence[5]. The step thickness was adapted to the range of recoil protons of interest. It was varied as a function of neutron energy. The step radiator effects an increasing track density with increasing radiator thickness up to saturation.

The detectors were electrochemically etched in 5N NaOH (70°C, $E = 2.25 \times 10^6$ V.m^{-1}, $\nu = 1$ kHz, etching time 5 h) and read out with a semi-automatic counting system.

MODEL CONSIDERATIONS

Calculations of the neutron response of a SSNTD covered with a radiator require knowledge of the recoil spectrum and angular distribution of the charged particles behind the radiator and of the track formation criteria.

Electrochemical etching of a track yields treeing when the actual electric field strength at the track tip exceeds the dielectric strength of the detector material E_D. It occurs if the charged particle has fallen short of the energy dependent critical angle of incidence α_{crit} to the detector normal. At given etch conditions α_{crit} is correlated to a critical value p_{crit} of the track parameter $p = l/r$ (l = actual track length projected on the detector normal, r = track tip radius). α_{crit} may be calculated by[5]

$$\cos(\alpha_{crit}) = \frac{v_B}{v_T} + \frac{p_{crit}\, r}{v_T\, t} \qquad (1)$$

when v_B = bulk etch rate, v_T = track etch rate and t = track etch time.

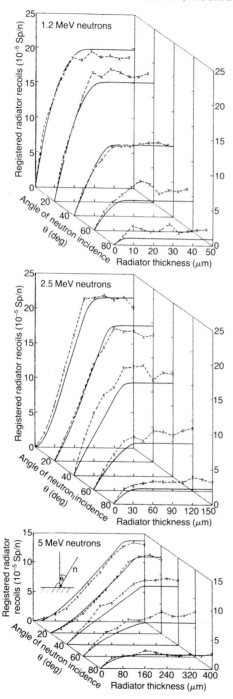

Figure 1. Measured (dashed line) and calculated (solid line) fluence response contribution due to radiator recoils as a function of angle of neutron incidence for the radiator thickness up to recoil saturation. (The fluence response is the ratio of track density and neutron fluence).

p_{crit} can be calculated by Smythe's formula

$$\frac{E_D}{E_O} = \frac{2 p_{crit}}{\ln(4 p_{crit}) - 2} \quad (2)$$

where E_O = applied electric field strength.

Several authors determined the ratio v_T/v_B in CR-39 of density ρ for chemical etching by measuring the track diameter of chemically etched tracks. Its increase as a function of the restricted linear energy transfer L_Δ is usually approximated by

$$v_T/v_B = A + B\, L_\Delta/\rho \quad (3)$$

where Δ was found to be 350 eV. Published data for the parameters A and B vary in the range A = 0.89–1, B = 0.00185–0.0038 MeV^{-1}.cm^{-2}.g[6,7], but there are no data available for ECE.

The track etch time t is the complete etch time t_e for long tracks whereas only the time $t_s = (R-S)/v_T$ is considered for short tracks with a range R < $v_T t_e + S$. The Bragg width S is the distance between the end of a latent track and the point of maximum amplification of the electric field strength during the etching of the latent track. Exceeding this point during the etch process causes an increasing track tip radius because of the enhanced material damage at the end of a latent track due to the Bragg peak. As a consequence the amplification of the electric field strength decreases.

The treeing criterion depends evidently on the parameters A, B, S, p_{crit} and the tip radius. Knowledge of this set of parameters enables the calculation of the response.

The recoil spectrum and angular distribution behind the radiator has been calculated as usual. The amount of recoils scattered from a radiator layer of 1/1000 of the radiator thickness into an angular element has been calculated with respect to the scattering neutron incidence. The recoils have been followed up to the rear radiator side and the residual energy has been determined from the energy–range curve.

Finally the latent proton tracks have been counted by numerical integration over the angular range of (n,p) scattering and over the radiator thickness with respect to the criterion for visualisation by ECE.

MODEL CALCULATIONS

In order to simplify the calculations only the contribution of radiator recoils to the track density has been simulated. The corresponding experimental data were calculated as the difference between the track density of a radiator covered detector area and an uncovered area. By this, only recoil protons and C recoils leaving the $(CH_2)_n$ radiator have been taken into account. There are no (n,α) reactions with carbon for neutron energies up to 5 MeV.

By methods of trial and error varying the parameters r, A, B, and S the radiator contribution to the fluence response has been calculated for neutrons of 1.2, 2.5 and 5 MeV at angular incidence. Good agreement with the experimental results (dotted line) has been achieved with the radiator thickness up to saturation (Figure 1) for the following parameters:

$p_{crit} = 108$, $v_B = 0.9$ μm.h^{-1}, $A = 0.12$,
$B = 0.007$ MeV^{-1}.cm^{-2}.g, $r = 5.4$ nm, $S = 3.4$ μm.

Only at 0.565 MeV an underestimation by a factor of 2 occurs. It may be caused by the simplifying presuppositions, especially the assumption of a constant track tip radius.

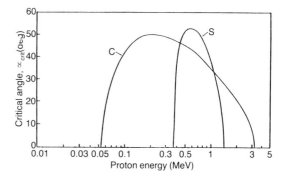

Figure 2. Critical angle, α_{crit}, for protons calculated for CR-39 etched in 5 N NaOH, 70°C, 5 h and 2.25×10^6 V.m^{-1} (S) in comparison to α_{crit} calculated for CR-39 etched in 5.4 N KOH, 60°C, 5 h and 2.5×10^6 V.m^{-1} [8] (C).

CONCLUSIONS

The improved model enables the simulation calculation of neutron response of ECE-treated CR-39 covered with a polypropylene radiator by means of a model for recoil release and track visualisation. The calculated number of registered radiator recoils at angular neutron incidence for a radiator thickness up to recoil saturation corresponds well with experimental results in the range of 1.2–5 MeV. This enables the making of some deductions about the etch process:

(a) The critical angle depends strongly on the etching conditions as shown by a comparison with published data obtained with CR-39 and different etching conditions[8] (Figure 2).

(b) The maximum amplification of the electric field strength during the track etching process was found at the increasing damage at $S = 3.4$ μm ahead from the end of the latent track, corresponding to the range of a 0.31 MeV proton.

(c) Transformation of Equation 1 with respect to Equation 3 enables the calculation of a critical L_{350}^{crit} required for the development of an electrical tree. For normally incident protons this lower limit is

$$L_{350}^{crit}/\rho = \frac{1}{B} + \frac{p_{crit}\, r}{B\, v_B\, t} - \frac{A}{B} = 144 \text{ MeV.cm}^2.g^{-1} \quad (4)$$

(d) The difference of A and B from published data obtained for chemical etching points to an influence of the electric field on the etch rates.

Further improvements of the response calculations of CR-39 require the measurement of the etch rates during the ECE process and the track geometry of ECE tracks.

ACKNOWLEDGEMENTS

The authors are grateful to Dr S. Guldbakke (PTB) for neutron irradiations, to Hoechst AG for the preparation of the radiator foils and to the Bundesminster für Umwelt, Naturschutz und Reaktorsicherheit for support.

REFERENCES

1. Al-Najjar, S. A. R. and Durrani, S. A. *Interpretation of Mason's Equation in Terms of Measurable Electrochemical-Etching Parameters Governing the Dielectric Breakdown Phenomenon*, Nucl. Tracks **12**, 37-42 (1986).
2. Li Xiangbao, *Study of the Relation between Relative Sensitivity of CR-39 and Electric Field Strength in Electrochemical Etching*. Nucl. Tracks **12**, 197-200 (1986).
3. Pitt, E., Scharmann, A. and Werner, B. *Model Calculations for Electrochemically Etched Neutron Detectors*. Radiat. Prot. Dosim. **23**, 179-182 (1988).
4. Doi, M., Fujimoto, K. and Kobayashi, S. *Etch Pit Formation Model during Chemical and Electrochemical Etching in Polycarbonate Foil*. Radiat. Prot. Dosim. **37**(1), 5-12 (1991).
5. Pitt, E., Scharmann, A. and Simmer, R. *Model Calculations for the Fast Neutron Response of a CR-39 Detector Covered with a Radiator*. Nucl. Tracks Radiat. Meas. **19**, 517-520 (1991).
6. Harvey, J. R., Ramli, A. G. and Weeks, A. R. *Measurement of the Etch Characteristics of Proton Tracks in CR-39 and the Implications for Neutron Dosimetry*. Nucl. Tracks **8**, 307-311 (1984).
7. Matiullah and Durrani, S. A. *Chemical and Electrochemical Registration in CR-39 — Implications for Neutron Dosimetry*. Nucl. Instrum. Methods **B29**, 508-514 (1987).
8. Cross, W. G., Arneja, A. and Ing, H. *The Response of Electrochemically-Etched CR-39 to Protons of 10 keV to 3 MeV*. Nucl. Tracks **12**, 649-652 (1986).

EXPERIMENTAL AND THEORETICAL DETERMINATION OF THE FAST NEUTRON RESPONSE USING CR-39 PLASTIC DETECTORS AND POLYETHYLENE RADIATORS

F. Fernández, C. Domingo, E. Luguera and C. Baixeras
Física de les Radiacions, Departament de Física
Universitat Autònoma de Barcelona. E-08193 Bellaterra, Barcelona, Spain

Abstract — A Monte Carlo method is used in order to calculate the dose equivalent response ($cm^{-2}.mSv^{-1}$) of a fast neutron dosemeter, composed of a polyethylene radiator and CR-39 plastic track detector, to monoenergetic neutron beams at different incidence angles as well as to neutron source beams, taking into account the electrochemical etching conditions. The method is applied to neutrons having energies from ~ 80 keV to ~ 5 MeV. The total number of protons emerging from the radiator is calculated, as well as their energetic and angular distribution. Detection efficiency as a function of proton energy and angle is taken into account. The theoretical calculation is compared with experimental results obtained from the EURADOS-CENDOS irradiation for protonic equilibrium radiator thicknesses. The agreement between measured and calculated values is good, within experimental errors, for neutrons in the above mentioned energy range.

INTRODUCTION

Since PADC (poly-allyl diglycol carbonate), conventionally known as CR-39, was used for the first time as a track detector more than ten years ago[1], it has been the most promising compound to be used as a fast neutron dosemeter. Protons originated by the neutron interactions in an hydrogenated radiator or into the same detector can be recorded over a wide energy range, specially when the electrochemical etching technique is used[2]. The PADC photon insensitivity allows selective neutron dosimetry to be carried out in radiation fields where neutrons come mixed with gamma rays.

Although CR-39 is widely used as a track detector in many different configurations (detector plus radiator of a given material) and etching techniques, fast neutron dosimetry is still not carried out under routine conditions in a satisfactory way. Some problems, like background levels that vary from batch to batch or a strong dependence of the response on the neutron incidence energy and angle[3] have been the subject of several studies[4,5] and intercomparisons between different laboratories[6,7]. Commercial CR-39 presents variable background levels which give rise to lower detection limits of the order of ~ 50 μSv[8] if great care is taken during dosemeter manufacture and during the etching process. Several solutions have been proposed for the problem of neutron response dependence on energy and angle, like mounting the CR-39 in non-planar geometries[9–11]. These non-planar dosemeters may be useful for research purposes, but they are not feasible for routine dosimetry because of the practical problems that arise for data acquisition and analysis. In addition, some new difficulties arise if thermal neutron dosimetry is desired: the use of a radiator for thermal neutrons besides the one for fast neutrons, and the elimination of the thermal neutron component which comes from the interaction of fast neutrons with air. All these facts suggest that finding and testing an optimal dosemeter configuration for a particular application may be difficult from the experimental point of view.

A Monte Carlo simulation method is presented as a very useful tool for optimising the response of a given dosemeter configuration if the detector is etched under known conditions. This simulation avoids the necessity of dealing with expensive and slow sets of irradiations and measurements, otherwise needed to set up the optimum dosemeter for the desired application. An analysis of the response of a fast neutron dosemeter with planar geometry is carried out, taking into account its geometrical configuration, its composition, and the parameters related to the etching conditions used. Values calculated from this analysis are given and compared with those obtained experimentally by this group and by other authors.

NEUTRON RESPONSE CALCULATION

A Monte Carlo method has been used to simulate the passage of a fast neutron beam (monoenergetic or emerging from a neutron source) through an hydrogenated material at any incidence angle. As a first approximation, the following considerations have been taken into account:

(i) Only elastic (n,p) scattering is relevant in the energy range studied.
(ii) A given neutron can interact only once in the radiator, as mean free paths for the energies involved are far larger than the normal radiator thicknesses.

In the case of neutron beams originating in non-monoenergetic sources, the energy spectrum of the source is divided into thin energy bands in order to obtain the emission probability of a neutron of a given energy. The number of neutrons of a given energy is obtained by multiplying this probability by the total neutron fluence. In this way, a non-monoenergetic neutron source is henceforth treated as a set of monoenergetic neutron beams of different fluences, according to the calculated probabilities.

The interaction probability of neutrons with radiator hydrogen nuclei is calculated from the mean free path, inferred from the elastic (n,p) cross sections. The cross section computation is carried out by means of a semiempirical formula[12]. The total number of protons originating when a neutron beam of a known fluence runs through a radiator is thus obtained. The distance between the neutron entry point in the radiator and its interaction point is randomly obtained, and the kinetic energy and emergence angle of every proton produced are calculated assuming that proton emergence is isotropic in the centre of mass reference system. In practice, this isotropy is imposed by dividing a spherical surface of unit radius, centred in the interaction point, into a great number (of the order of 3×10^5) of equal parts. Each part, which characterises a given direction in space, is assigned a positive integer number lying between 1 and the total number of divisions. A random routine picks up one of these numbers and, therefore, a given emergence direction is assigned to each proton. Conservation laws and knowledge of the geometrical conditions of the interaction in the centre of mass determine the recoil proton kinetic energy and momentum. All magnitudes are subsequently converted into the laboratory reference system, taking into account the neutron beam incidence angle and energy. All recoil protons originating in the radiator are individually followed.

The proton range–energy tables obtained from Steward's program[13] are used in order to find out whether a given proton emerges from or stops inside the radiator. The number of protons that emerge from the radiator and their emergence energy and angle can, therefore, be calculated. If the dosemeter is constructed with more than one radiator all calculations are repeated for each of them, and the proton energy loss in the succeeding radiators is considered. In such a way, the number of protons that reach the CR-39 detector can be obtained, as well as their incidence energy and angle. Our computer program also takes into account protons originating in the CR-39 layer removed during the etching process, as well as the energy loss in this layer of the protons proceeding from the radiators. If a neutron fluence of 10^7 cm^{-2} is used, the fluctuations in the number of protons that reach the detector are smaller than 1.5%.

A semiempirical function of the form

$$\varepsilon = \frac{1}{1 + e^{(\theta-\mu)/\tau}}$$

has been used to calculate detection efficiencies for proton registration as a function of their incidence angle (θ) and energy (E). This function has been fitted so that the critical angle (μ) for proton registration as a function of the proton energy behaves in a similar way to that presented by Cross et al[2,14], and that the registration efficiency for a given proton energy displays the same angular behaviour as data from Fernández et al[15]. Cut-offs for the critical angle at low and high proton energies have been determined so that our Monte Carlo method reproduces the normally incident neutron response of our dosemeter. These cut-offs account for our electro-chemical etching conditions. The following values for the parameters have been obtained if θ is given in degrees and E in MeV

$\tau = 5.1466$

$$\mu = \begin{cases} 159923E^6 - 193450E^5 + 61491.8E^4 + 9713.55E^3 \\ -9045.39E^2 + 1681.61E - 46.86 \\ \quad \text{if } E < 0.2859 \text{ MeV} \\ -10.44E^3 + 38.24E^2 - 58.07E + 63.92 \\ \quad \text{if } E \geq 0.2859 \text{ MeV} \end{cases}$$

The effective number of recorded protons (number of etched tracks) is calculated from this function, and conversion from number of tracks to dose equivalent response (cm^{-2}.mSv^{-1}) is performed by means of the H_{MADE}/Φ factors[16].

The number of protons that reach the detector when neutron beams at different incidence angles are used, calculated from our Monte Carlo method, as well as the angle and energy spectra of these protons, agree with the results obtained by other authors[17] using Monte Carlo and analytical methods.

EXPERIMENTAL PROCEDURE

Experimental studies have been carried out using Tastrak CR-39 (Track Analysis Systems Limited, Bristol, UK) standard grade, 250 μm and 600 μm nominal thickness and 32 h curing time, manufactured in July 1989. CR-39 plates have been stored in air, in darkness and inside a refrigerator prior to irradiation. Sample CR-39 strips of 10×2 cm^2 have been assembled in 21 irradiation 10×10 cm^2 cards, each containing two 600 µm and one 250 µm thick strips. One of the 600 µm strips and the 250 µm one were covered with a polyethylene radiator to ensure proton equilibrium. Nineteen cards have been exposed to monoenergetic neutron beams of 0.144, 0.565, 5.3 and 14.8 MeV and to a ^{252}Cf source with different incidence angles (between 0° and 60°). The two remaining ones have been sent respectively to PTB and GSF, together with the irradiated ones, for background studies. Each irradiated strip was cut in five approximately equal plates of 2×2 cm^2 before etching.

The 250 µm thick samples have been electrochemically etched in cylindrical methacrylate cells, equipped with electrodes whose distance to the sample can be varied[15]. Etching has been carried out with a 6 N KOH aqueous solution at 60°C, setting the distance between the electrodes at 2.2 cm and using the following steps:

(1) 20 kV.cm^{-1} (rms) at 50 Hz for 2.5 h
(2) 20 kV.cm^{-1} (rms) at 2 kHz for 1 h
(3) 15 min post-etching.

Tracks have been counted using a microfiche reader giving a magnification of 20×. An area of about 0.60 cm^2 has been analysed in every etched plate.

RESULTS AND DISCUSSION

The response of our detector to neutrons at several incidence angles has been calculated from the Monte Carlo method described above, taking into account that recoil protons with energy below 90 keV do not produce an etchable track[18] and that the thickness of the layer removed during etching is 3.5 µm. Figure 1 shows this response together with the experimental results obtained from the irradiation of 250 µm thick Tastrak samples by normally incident neutrons. Background values are similar for samples exposed in both the PTB Laboratory (114±36 tracks.cm^{-2}) and the GSF Laboratory (78±13 tracks.cm^{-2}). There is a good agreement between the measured and the calculated values, with the exception of the point at 14.8 MeV. The discrepancy in this point is explained because our Monte Carlo calculation has only considered elastic (n,p) scattering, but the contribution of (n,C) and (n,O) reactions is not negligible for neutron energies over 5 MeV.

Although we do not know the cause of the high response fluctuation in the results obtained for the ^{252}Cf source, the calculated and measured response values for 0° incident neutrons are consistent within the error range (experimental value 760 ± 99 cm^{-2}.mSv^{-1}, calculated value 650 ± 20 cm^{-2}.mSv^{-1}).

The angular dose equivalent response, normalised to the normal incidence value, estimated from our Monte Carlo method differ by less than 1% from those calculated by Cross et al[14] for neutrons in the energy range 0.5–5 MeV. At 144 keV there is a discrepancy greater than 50% that could be explained because our lower energy cut-off for proton registration (90 keV) is significantly greater than the one used by Cross et al (40 keV).

The poor statistics of our experimental data for 30° and 60° incidence angles does not allow verification of the relative response values estimated from the Monte Carlo method for these angles. Our dose equivalent response values estimated for 30° and 60° neutron incidence differ respectively by 11% and 30% from the experimental results given by Luszik et al[19]. The discrepancy that appears at 60° suggests that our proton registration efficiencies should be revised for high neutron incidence angles. New experiments, involving non-perpendicular neutron incidence are being studied by this group in order to obtain statistically significant data in this field. On the other hand, the introduction of the (n,C) and the (n,O) reaction cross sections in our program, without any need to vary its structure, will allow extension of its range of validity to incident neutrons of at least 14.8 MeV.

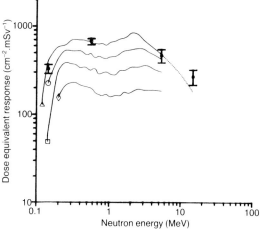

Figure 1. Dosemeter dose equivalent response estimated from the Monte Carlo method as a function of the incident neutron energy (monoenergetic beams) for different incidence angles. Also shown are our experimental results for normally incident neutrons. The dotted line shows the response calculated for normally incident neutrons having energies over 5 MeV assuming that only elastic (n,p) scattering is relevant. Angles of incidence: (△) 0°, (○) 30°, (□) 45°, (◊) 60°.

ACKNOWLEDGEMENTS

This work is partially supported by a contract with the Commision of the European Community (Bi7–020). The cooperation of the staff at the accelerator and neutron source facilities at PTB and GSF is gratefully recognised.

REFERENCES

1. Cartwright, B. G., Shirk, E. K. and Price, P. B. *A Nuclear Track Polymer of Unique Sensitivity and Resolution.* Nucl. Instrum. Methods **153**, 457-460 (1978).
2. Cross, W. G., Arneja, A. and Ing, H. *The Response of Electrochemically Etched CR-39 to Protons of 10 keV to 3 MeV.* Nucl. Tracks **12**, 649-652 (1986).
3. Griffith, R. V., Thorngate, J. H., Ruppel, D. W., Fisher, J. C., Tommasino, L. and Zapparoli, G. *Monoenergetic Neutron Response of Selected Etch Plastics for Personnel Neutron Dosimetry.* Radiat. Prot. Dosim. **1**, 61-71 (1981).
4. Ramli, A. G. and Durrani, S. A. *A Flat Dose-Equivalent Response Dosimeter for Fast Neutrons using CR-39.* Nucl. Tracks **8**, 327-334 (1984).
5. Matiullah and Durrani, S. A. *A Flat Dose-Equivalent Dosemeter for Fast Neutrons using a CR-39 Polymeric Track Detector with Front Radiators.* Radiat. Prot. Dosim. **17**, 149-152 (1986).
6. EURADOS-CENDOS Report 1987-01. *Neutron Irradiations of Proton-Sensitive Track Etch Detectors: Results of the Joint European/USA/Canadian Irradiations.* KfK 4305 (1987).
7. EURADOS-CENDOS Report 1989. *Results of a Survey of Neutron Dosimetry PADC Background Organized by EURADOS-CENDOS in 1988.* ENEA. PAS-FIBI-DOSI. 1 (1989).
8. Alberts, W. G., Luszik-Bhadra, M., Piesch, E. and Vilgis, M. *Fast Neutron Dosimetry with CR-39: Study of various Materials Using Electrochemical Etching.* Nucl. Tracks Radiat. Meas. **19**, 437-442 (1991).
9. Al-Najjar, S. A. R. and Piesch, E. *Angular Response Characteristics of a CR-39 Detector of Cylindrical Geometry.* Radiat. Prot. Dosim. **20**, 67-70 (1987).
10. Harvey, J. R. and Weeks, A. R. *Recent Developments in a Neutron Dosimetry System Based on the Chemical Etch of CR-39.* Radiat. Prot. Dosim. **20**, 89-93 (1987).
11. Matiullah and Durrani, S. A. *Further Progress Towards the Construction of a Cubical Fast-Neutron Dose Equivalent Dosimeter Based on the CR-39 Detector with Front Radiators.* Nucl. Tracks Radiat. Meas. **15**, 515-518 (1988).
12. Jung, M. and Ott, Ch. *Fast Neutron Analysis Code SAD1.* CRN-CPR 85-16 (Centre de Recherches Nucléaires, Strasburg) (1985).
13. Steward, P. G. *PhD Thesis.* Lawrence Radiation Laboratory, University of California at Berkeley (1968).
14. Cross, W. G., Arneja, A. and Kim, J. L. *The Neutron Energy and Angular Response of Electrochemically Etched CR-39 Dosemeters.* Radiat. Prot. Dosim. **20**, 49-55 (1987).
15. Fernández, F., Baixeras, C., Zamani, M., López, D., Jokic, S., Debeauvais, M. and Ralarosy, J. *CR-39 Registration Efficiency of Protons Using Electrochemical Etching.* Radiat. Prot. Dosim. **23**, 175-178 (1988).
16. ICRP. *Data for Protection against Ionizing Radiation from External Sources.* Report 21 (Oxford: Pergamon) (1971).
17. Makovicka, L., Sadaka, S., Vareille, J. C., Decossas, J. L. and Teyssier, J. L. *Study of Polyethylene and CR-39 Fast Neutron Dosemeters. I. Characteristics of Proton Fluxes Emitted by a Polyethylene Radiator.* Radiat. Prot. Dosim. **16**, 273-279 (1986).
18. Fernández, F., Domingo, C., Baixeras, C., Luguera, E., Zamani, M. and Debeuavais, M. *Fast Neutron Dosimetry with CR-39 Using Electrochemical Etching.* Nucl. Tracks Radiat. Meas. **19**, 467-470 (1991).
19. Luszik-Bhadra, M., Alberts, W. G., Guldbakke, S. and Kluge, H. *Influence of the Converter Configuration on the Angle Response of Etched Track Neutron Dosemeters.* Radiat. Prot. Dosim. **38**(4), 271-277 (1991).

A CR-39 FAST NEUTRON DOSEMETER BASED ON AN (n,α) CONVERTER

E. Savvidis, D. Sampsonidis and M. Zamani
University of Thessaloniki
Nuclear Physics Department
Thessaloniki 540 06, Greece

Abstract — CR-39 was used as a solid state nuclear track detector. Alpha particles from the ^{6}Li(n,α) reaction were counted at the same time with proton recoils for fast neutron detection. A neutron moderator in front of the detector increases the response of the system by making use of the ^{6}Li(n,α) reaction cross section for moderated intermediate neutrons. The linearity of the system at 2.5 MeV and 5.3 MeV neutrons was studied.

INTRODUCTION

The poly-allyl-diglycol carbonate, CR-39, etched track detector is a promising personal neutron dosemeter because of its high proton registration sensitivity[1]. The use of hydrogenous front radiators increases the detection response to fast neutrons[2]. Thermal neutrons, in particular, are usually detected by means of (n,α) converters (^{6}Li or ^{10}B) in contact with the detector. The range of (n,α) converters can be extended to fast neutron detection, based on the fraction of the neutron spectrum which is shifted towards intermediate energies by using a moderator in front of the detector[3]. The ^{6}Li(n,α) reaction cross section at these energies is comparable to that of neutron elastic scattering.

In this paper the linearity of a personal dosemeter based on a CR-39 track detector with a ^{6}LiF converter and a polyethylene moderator is investigated. The experiment was performed with 2.5 MeV and 5.3 MeV neutrons.

Comparison with the method based on proton recoils is also presented.

RESULTS AND DISCUSSION

The fast neutron dosemeter consists of Pershore CR-39 sheets, 500 μm thick and polyethylene (PE) plates of 0.5 cm and 1.0 cm as neutron moderators and proton recoil radiators. The thickness of the neutron moderator was chosen based on the analysis presented in recent work[3]. Two thicknesses of ^{6}Li converter were used: 300 μm (crystal) and 1.0 μm by evaporation of ^{6}LiF on the one surface of the detector. CR-39 was irradiated at the same time with 0.5 cm and 0.3 cm PE radiator in order to detect only proton recoils and to compare the characteristics of the two systems.

Irradiations have been carried out at the D-D neutron generator with 2.5 MeV neutrons in Thessaloniki's Nuclear Physics Laboratory. The distance between the centre of the dosemeter and the D target was 40 cm. The neutron fluence was measured using a long counter (BF$_3$) (Frieseke and Hoepfner GmbH, Type LB 6400). Irradiations with 5.3 MeV neutrons were made in the frame of the EURADOS-CENDOS joint irradiation programme (1989-90).

CR-39 sheets were chemically developed in 20% NaOH at 70°C for a time sufficient to obtain saturation of the number of tracks counted. The saturation starts from about 6 h of development, or for $V_g t = 10$ μm[3], V_g being the bulk etching velocity. For these conditions proton recoils of higher energies also start to be developed.

The linearity of the two detecting systems, with a ^{6}LiF converter and without, was tested for doses from 0.3 mSv up to about 5.0 mSv. The tracks counted are alphas and tritons from the ^{6}LiF converter as well as proton recoils from the moderator and the detector itself. Without the ^{6}LiF converter only proton recoils are counted.

In Figure 1 the track density is presented as a function of the dose equivalent for 5.3 MeV neutrons. It is shown that the two detecting systems are linear against dose equivalent. The dose equivalent response, defined as the detector reading per unit dose, is higher when the ^{6}LiF converter is used than in the case of proton recoils. This result is expected because with the ^{6}LiF converter some of the protons generated in the PE moderator are also counted. The resulting dose equivalent response of each detecting system is given in Table 1.

In the same table the response of the system with the LiF converter is presented, for 2.5 MeV neutrons. Two thicknesses were examined, a 1 μm evaporated layer and a 300 μm (crystal). The response with a ^{6}LiF crystal is higher than with a thin evaporated layer. Having in mind that 2 MeV alpha particles (coming from thermal plus epithermal neutron reactions) have a range of 3.3 μm in LiF it is evident why the LiF crystal gives a higher number of alpha particles than the thin LiF layer. 2.7 MeV tritons have a range of 10 μm in LiF (also from thermal and epithermal neutron reactions), and, therefore, contribute to the total track number counted. Finally, reactions with fast neutrons result in alphas and tritons of higher energies[4] which contribute according to the cross section of the reaction at each given energy. The number of ^{6}Li reactions from fast neutrons was tested by a separate experiment where 2 mm Cd foils were placed after the moderator. It was found that the number of particles counted was twice the observed level of proton recoils, under the same conditions of irradiation.

At 2.5 MeV the response of a system based on proton recoils only, for PE radiators of 0.2 and 0.5 cm has been tested. From Table 1 it is concluded that the dose equivalent response is the same for both PE radiator thicknesses because protonic equilibrium is already established. Thus, the combination of the two detecting systems, which is the case of the proposed dosemeter, is based on two equally linear systems, in the range of the dose equivalent examined.

CONCLUSIONS

(a) The neutron dosemeter based on (n,p) (n,α) converters has linear dependence on dose from 0.3 mSv up to about 5 mSv, for 2.5 MeV and 5.3 MeV neutrons. For this range of dose equivalent the dosemeter based on proton recoils is also linear.

(b) The dose equivalent response is higher when a ^{6}LiF converter is used than when using only PE radiators.

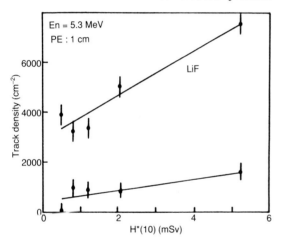

Figure 1. Track density as a function of the dose equivalent for neutron energy 5.3 MeV. The upper curve corresponds to a neutron dosemeter with ^{6}LiF converter and 1 cm PE moderator in front of the detector. The lower curve was taken with 1 cm PE radiator only, so is based on proton recoils.

Table 1. Response of various neutron detecting systems.

Detecting system	Neutron energy (MeV)	System response (tracks.cm^{-2}.mSv^{-1})
^{6}LiF converter 300 μm (crystal) 0.5 cm PE moderator	2.5	958±98
^{6}LiF converter 1 μm layer 0.5 cm PE moderator	2.5	422±34
^{6}LiF converter 300 μm (crystal) 1.0 cm PE moderator	5.3	903±139
1.0 cm PE radiator (proton recoils)	5.3	237±105
0.5 cm PE radiator (proton recoils)	2.5	283±21
0.2 cm PE radiator (proton recoils)	2.5	229±42

REFERENCES

1. Makovicka, L., Decossas, J. L. and Vareille, J. C. *Experimental Study of the Dosimetric Efficiency of a Radiator — CR-39 Fast Neutron Dosemeters.* Radiat. Prot. Dosim. **20**(1/2), 63-66 (1987).
2. Piesch, E., Al-Najjar, S. A. and Ninomiya, K. *Neutron Dosimetry with CR-39 Track Detector Using Electrochemical Etching. Recent Improvements, Dosimetric Characteristics, Aspects of Routine Application.* Radiat. Prot. Dosim. **27**(4), 215-230 (1989).
3. Savvidis, E., Zamani, M., Sampsonidis, D. and Charalambous, S. *Fast Neutron Detection Using (n-p), (n-α) Converter.* Nucl. Tracks Radiat. Meas. **19**(1/4), 527-530 (1991).
4. Savvidis, E., Zamani, M. and Charalambous, S. *An Approach to Fast and Thermal Neutron Spectroscopy Based on ^{6}Li(n,α)^{3}H Reaction.* Nucl. Tracks Radiat. Meas **15**(1/4), 495-498 (1988).

CHARACTERISATION OF NEW PASSIVE SUPERHEATED DROP (BUBBLE) DOSEMETERS

R. E. Apfel
Yale University, PO Box 2159, New Haven, CT 06520, and
Apfel Enterprises, Inc., 25 Science Park, New Haven, CT 06511, USA

Abstract — Superheated drop compositions have been employed in two new passive (non-electronic) direct reading detectors for neutron monitoring of personnel and in accelerator applications. The Neutrometer™ is a pen-sized tube partially filled with a superheated drop composition. Neutrons impinge on the superheated drops causing them to vaporise suddenly. The resulting bubbles push a piston up a calibrated tube, giving a direct indication of the neutron dose equivalent. The Neutrometer™ can be reused for several successive exposure periods, with each exposure adding an increment to the total accumulated dose. The Neutrometer™-HD is designed for higher dose ranges. Vials containing the superheated drop composition are placed at various locations in the irradiation area, such as near a high energy medical accelerator. Bubbles that are nucleated by neutrons force an indicator liquid up a calibrated pipette. This displacement gives the neutron dose equivalent. The spatial dependence of the dose equivalent can therefore be measured immediately, without recourse to activation foils, scintillation powders, or other similar systems that require extensive post-processing. Tests to characterise the performance of these detectors suggest that they are nearly dose equivalent, that batch-to-batch uniformity is good, and that they meet many of the demands required in their respective applications.

INTRODUCTION

In this article two passive (non-electronic) direct reading neutron measuring devices based on superheated drop technology are described. The Neutrometer™ is intended for dosimetry uses and the Neutrometer™-HD is intended for High Dose and dose rate application, such as found near high energy medical accelerators.

The photograph of Figure 1 shows the personnel neutron dosemeter, Neutrometer-100, tested for this study. It is 12.5 cm tall and weighs 10 g. The round glass tube which holds the superheated drop composition is held in a triangular plastic holder. The bottom section holds approximately 0.5 ml of aqueous gel composition with approximately 8000 drops of 100 μm diameter. Atop this composition is a gel piston, a white disc (which is what is measured against the calibrated scale), and a small amount of gel which acts as a lens so that the scale is magnified. The SDD-100 drop material has been characterised earlier with monoenergetic neutron sources from the RARAF accelerator of Nevis Laboratory (Irvington, NY) in the energy range 0.2 to 14 MeV, and from NIST research reactor (Gaithersberg, MD) for thermal energies and at 2, 24 and 144 keV[1]. The energy response is shown in Figure 2(a), compared with the fluence to ambient dose equivalent conversion factor[2]. In Figure 2(b), two modifications have been made. The albedo response of the detector has been estimated, comparing it with the fluence to individual dose equivalent, penetrating, conversion factor. From this factor the contribution from n-capture gamma rays has been subtracted, showing better agreement with the estimated albedo response based on

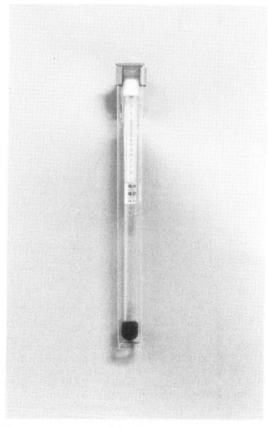

Figure 1. Photograph of the Neutrometer. The tube inside the outer holder contains the superheated drop composition and a piston which is pushed up the tube by bubbles nucleated by neutrons. The scale is calibrated in mSv.

measurements. This favourable comparison suggests that the superheated drop composition, which is insensitive to gammas below the photonuclear threshold, might measure something close to the heavy particle component of dose equivalent, whereas the gamma component could be read along with the external gamma exposure by a photon badge[2].

The Neutrometer was tested recently with an Am-Be neutron source (Amersham) rated at 37 GBq (1 Ci). Dose equivalent rates ranged from 0.05 to 2 mSv.h^{-1}. Total dose equivalents ranged from 5 μSv to 8 mSv in these tests. Tests performed include: (a) repeatability of response to the same fluence; (b) ability of Neutrometers to hold an accumulated dose reading over time and in various thermal environments; (c) use of the Neutrometer horizontally to allow for bubble count per μSv; (d) batch-to-batch repeatability of calibration factor. Control tests were performed to see what background doses were measured when Neutrometers sat at room temperature or at elevated temperatures (to 35°C).

Preliminary tests were also performed on a temperature-compensated version of the Neutrometer. This unit uses a volatile liquid to compensate partially for the temperature dependence of the composition with a pressure change. Readings at 22 and 32°C for 1 mSv exposures for both compensated and uncompensated Neutrometers were compared.

RESULTS—NEUTROMETERS

Several Neutrometers from different batches were tested to determine the calibration of the instrument. It was found that the displacement of the piston, d in mm, fits the following depletion equation (±10%).

$$d = 65(1 - e^{-0.27D}) \quad (1)$$

where D is the neutron dose equivalent in mSv. The practical limit of the Neutrometer was 57 mm corresponding to 8 mSv. The smallest calibration mark on the scale corresponded to 0.9 mm for a dose equivalent of 50 μSv (5 mrem).

By placing the Neutrometer horizontally, it was possible to count the bubbles that rose to the surface for a given exposure. For dose equivalents less than 0.5 mSv, the number of bubbles per 10 μSv averaged between 6 and 7, suggesting that dose equivalents as low as 3–5 μSv could be measured in this manner.

Successive measurements were performed with time intervals varying from one hour to one week. It was observed that over short time intervals, equal exposures led to equal readings (±10%) on the scale calibrated using Equation 1. For a time interval of a week, it was observed that the piston position in some Neutrometers had moved up by approximately 10–15% in dose equivalent. (Presumably, some gas from the gel comes out of solution, although the exact cause has been insufficiently researched.) Subsequent exposures, however, correctly recorded the incremental dose equivalent exposure. Thus, the Neutrometer is best used by recording the scale reading before and after each exposure and at least on a daily basis.

In initial tests of the temperature dependence of the Neutrometer, the readings were recorded for two temperatures, 22 and 32°C. In the uncompensated Neutrometer the second Neutrometer recorded a dose equivalent of 70% greater than the one at 22°C, consistent with earlier results showing an approximate 5.5%.°C^{-1} temperature coefficient for the SDD-100 composition exposed to an Am-Be source. Two temperature-compensated Neutrometers showed greatly improved performance, with only a 13% increase over the 10°C change, or a 1.2%.°C^{-1} temperature coefficient.

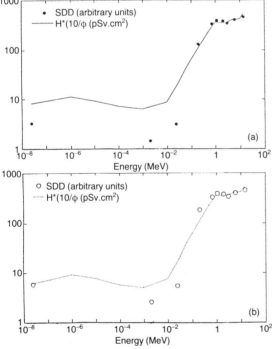

Figure 2. (a) The experimental results of exposure of SDD-100 material to monoenergetic neutrons from RARAF (Nevis Labs) Van de Graaff and the US National Institutes of Standards and Technology research reactor. The solid line is adapted from information provided in International Commission on Radiological Protection report 60, as discussed in Reference 2. (b) Data points have been modified to include estimated albedo response, and then compared to fluence to individual dose equivalent, penetrating, conversion factor, from which n-capture gamma ray contribution has been subtracted. See Reference 2.

When uncompensated Neutrometers were kept at room temperature for three weeks, without exposure to a neutron source, a small amount of bubble activity could be observed. (It is estimated that the natural background will produce one bubble every 3–6 days.) These bubbles, which rise to a position below the piston, will tend to coalesce and grow in size due to the tendency for gas in the gel solution to diffuse into the bubbles. The dose equivalent read on the calibration scale at the end of the three week period amounted to 0.05–0.1 mSv (5–10 mrem). Subsequent exposure intervals were properly recorded on the Neutrometer.

MATERIALS AND METHODS– NEUTROMETER™-HD

The same composition materials are used in the Neutrometer™-HD composition, as shown in Figure 3. However, in this case each vial has about 20,000 drops with the drop diameter being reasonably uniform at 60–70 μm. Because of the slightly smaller total volume of the sensitive superheated material (i.e. not including the gel), there are fewer bubbles per mSv, and the amount of vapour for each bubble event is much smaller than in the case of the Neutrometer. A 1 ml pipette, calibrated in units of 0.01 ml, is screwed on to the vial before each use. Several of these vials are placed at various locations in the radiation field.

In tests vials were irradiated by a Sagittaire 25 MV electron accelerator. Photon irradiation of the vials occurred when a tungsten target was used in the electron beam. Electron irradiations occurred when the tungsten target was removed. The parameter space for the tests is displayed in the data that follows.

Tests were carried out by Dr Ravinder Nath and his colleagues of the Department of Therapeutic Radiology of the Yale School of Medicine. Normal test procedures were as follows: (a) Vials were fixed in an inverted orientation in a Styrofoam holder and placed in the irradiation area. (b) The accelerator was set at a monitor count rate (in the photon irradiations) of 200–400 monitor units per minute (corresponding, in some field sizes, to 2–4 Gy.min^{-1}). Within one minute of the finished irradiation, the liquid position in the pipette was measured. (Each vial appears in a partially foamed state as the bubbles force liquid up the calibrated tube.) (c) Tests were sometimes repeated with the same vial if the total volume displacement was less than 0.2 ml. (d) The rate of drift of the liquid position in the pipette was monitored over time after the irradiation was finished.

RESULTS–NEUTROMETER™-HD

Calibration tests of Med-Pac vials were performed with an Am-Be neutron source at irradiation rates of up to 2 mSv.h^{-1}. Repeatability over several vials was within ±15%. The calibration factor used in the results given below was:

$$\text{Dose equivalent (mSv)} = -21.5 \ln(1 - 1.37 V) \quad (2)$$

where V is the volume displacement of indicator liquid in the pipette, in ml.

Results of tests by Dr Nath's group will be detailed in a subsequent publication. A preliminary evaluation of experiments performed in September 1991 at the Sagittaire accelerator is given in Figures 4 and 5. Vials were evaluated between 30 and 60 s after the irradiation. It was noted that after measurements were completed, the indicator liquid continued to drift up slowly, but at a rate of no more than 2% per minute. This drift continued, but at a much slower rate, for at least a day. Clearly, measurements should be taken within a minute or two of the exposure.

Figure 4 shows the neutron dose equivalent

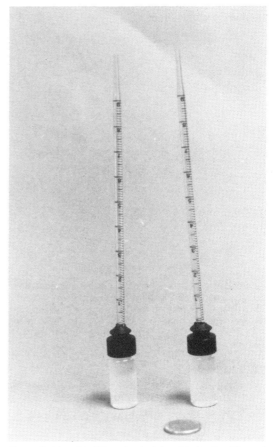

Figure 3. Two Med-Pac 4 ml vials with 1 ml calibrated pipettes for monitoring moderate to high dose and dose rate neutrons.

relative to distance from the beam axis for a 30×30 cm² field size, 25 MV X ray source. The open squares correspond to earlier results by Price et al with a phosphorus activation technique[3].

Figure 5 summarises the results with electron irradiation for energies ranging from 7 to 25 MV. Below 19 MV, the neutron contamination of the electron beam appears negligible.

CONCLUSIONS

Passive neutron detectors based on volume changes in superheated drop compositions and in two different dose ranges (5 µSv to 8 mSv and 0.5 mSv to 50 mSv) were tested over a wide range of irradiation and environmental conditions. Both showed good batch-to-batch repeatability and gave immediate readings that were nearly dose equivalent. The Neutrometer's sensitivity exceeds the requirements set forth in the new ICRP 60 report, and its nearly dose equivalent response for personnel dosimetry and its immediate readout makes it a strong candidate for several neutron monitoring applications.

The Neutrometer™-HD has applications in moderate to high dose and dose rate environments. Temperature compensated Neutrometers may greatly increase the practicality of this kind of instant readout detector.

ACKNOWLEDGEMENTS

The authors wish to acknowledge the assistance with monoenergetic neutron sources provided by Mr Steven Marino of RARAF (Nevis Labs) and Dr Robert Schwartz of NIST. We also thank Dr Ravinder Nath of the Yale School of Medicine for sharing preliminary data on tests with the Neutrometer™-HD system.

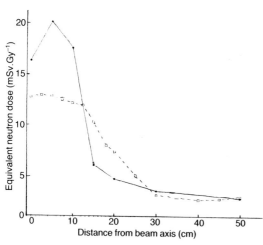

Figure 4. Spatial dependence of neutron dose equivalent per photon rad from Sagittaire accelerator (25 MV, X rays, 30×30 cm² field); preliminary data from Dr Ravinder Nath of the Yale School of Medicine. (•) this work, (□) Price et al[3].

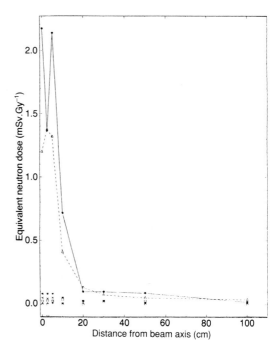

Figure 5. Spatial dependence of neutron dose equivalent per rad from electron irradiation from Sagittaire accelerator at energies ranging from 7 to 22 MV; preliminary data from Dr Ravinder Nath of the Yale School of Medicine. Energies in MV: □ 7, o 10, + 13, × 16, * 19, --Δ-- 22, --•-- 25.

REFERENCES

1. d'Errico, F. and Apfel, R. *A New Method for Neutron Depth Dosimetry with the Superheated Drop Detector*. Radiat. Prot. Dosim. **30**(2), 101-106 (1990).
2. Curzio, G. and D'Errico, F. *Rivelatori a bolle per la dosimetra neutronica*. In: Proc. XXVII Congresso Nazionale AIRP, Ferrara, Italy, 1991.
3. Price, K. W., Holeman, G. R. and Nath, R. *A Technique for Determining Fast and Thermal Neutron Flux Densities in Intense High-energy (8-30 MeV) Photon Fields*. Health Phys. **35**, 341 (1978).

BUBBLE DETECTORS IN FUSION DOSIMETRY

N. Smirnova, N. Semaschko and Y. Martinuk
KIAE, Plasma Physics Department
Kurchatov, sq. 46, Moscow, Russia

Abstract — Experience in the application of bubble detectors in fusion dosimetry is described. Results show how successful the measurements were on Tokamak-10, Tokamak-15 and pulsed 'plasma focus' devices. Requirements for individual dosimetry at fusion devices using bubble detectors are discussed.

INTRODUCTION

Strong electromagnetic fields, possible high intensity gamma radiation around fusion devices, and short duration of emitted radiation cause some troubles in routine dosimetry[1]. These specific conditions influence measurement results and make them incorrect.

In order to detect neutrons in the presence of affecting fields and radiations a non-electronic detector based on a Superheated Dispersed System (SDS) may be used, i.e. drops of superheated liquid dispersed in stable liquid immiscible with the first one. Such detectors have been named bubble detectors[2-4].

It is known that the characteristics of bubble detectors can be optimised to applications under various conditions[5]. For fusion dosimetry following advantages of bubble detector may be utilised:

(i) complete (if desirable) insensitivity to gamma radiation below photoneutron threshold in the detector material (Figure 1);
(ii) adjustable energy threshold of detectable neutrons,
(iii) high response to neutrons of various energies, and
(iv) an almost flat response curve depending on neutron energy above threshold (Figure 2).

BUBBLE DETECTORS WITH ISOBUTANE

For our purposes a bubble detector was developed here[4] which differs slightly from the ones developed in CRNL (Canada)[2] and AEI (USA)[3]. The stable phase in our detector was identical to the Canadian one: it was polyacrylamidic gel. The superheated liquid was isobutane. The drops in our detector were very small — about 1–5 μm in size. Because of the small size of the drops the detector response to fast neutrons could be increased to about 10^3 n per bubble or 4–7 μSv in dose. Such a response was, in the past, only obtained in bubble detectors with freon-12 and therefore with much higher superheating than with isobutane. Thus freon-12 bubble detectors were sensitive to thermal neutrons due to the protons generated in the reaction ^{35}Cl (n,p). It is important to note that the characteristics of our detector (except response) remained unchanged with decreasing drop size.

RESULTS OF MEASUREMENTS

The bubble detectors were used to measure

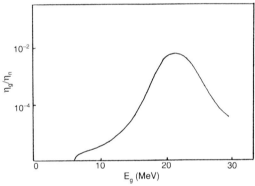

Figure 1. Calculated equivalent dose response of bubble detectors to gamma radiation with respect to experimental response to 1.5 MeV neutrons (η_g, gamma radiation response, η_n, neutron response).

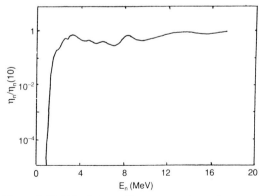

Figure 2. Bubble detector flux response to neutrons as a function of neutron energy ($\eta_n 10$, flux response to 10 MeV neutrons). The data were obtained in Van de Graaff generator experiments.

neutron fluxes and doses at the fusion devices Tokamak-10 and Tokamak-15 and at the pulsed 'plasma focus' devices. Neutron measurements at modern fusion devices are used for the estimation of the radiation risk for the fusion device personnel and also as initial data for plasma diagnostics. Thus measurements of radiation fields are usually carried out close to fusion devices. Plasma disruptions, which are specific to Tokamaks, lead to the appearance of sharply directed radiation sources located on the chamber surface where runaway electrons strike. The flux density of photoneutrons in the case of a disruption depends on the device configuration, plasma parameters and the target material, i.e. the material of the chamber surface of the diaphragm in the position of collisions with the electron beam[6].

For Tokamak-10 (work began in 1975) with major torus radius R=150 cm, plasma radius a= 35 cm, magnetic field induction B=3.5 T, ion temperature $T_i < 3$ keV, plasma density $\rho = 10^{13}$ cm^{-3}, the maximum total yield ratio of thermonuclear neutrons (10^9 n per pulse) and photoneutrons (10^{12} n per pulse) is K = 10^{-3} and neutron emission duration is 10^{-2}–10^{-1} s. The neutron fluence through the bubble detector placed at a distance of about a metre from the diaphragm under conditions described above was less than 10^5 n per pulse for the carbon diaphragm and F_g/F_n (gamma to neutron fluence) is equal to 10 in plasma disruption.

For Tokamak-15 (work began in 1988) R = 245 cm, a = 80 cm, B = 5 T, $T_i < 10$ keV, $\rho < 10^{14}$ cm^{-3}, the maximum neutron yield ratio of thermonuclear neutrons to photoneutrons is 10, and maximum neutron fluence through the detector is 10^{10} n.cm^{-2} at a distance of 1 m from the carbon diaphragm ($F_g/F_n > 1$). In the working period 1988 the fluence of photoneutrons has been measured[7] to be 10^4 n.cm^{-2}. This corresponds to a photoneutron source strength of ~ 10^9 n per pulse.

The advantages of bubble detectors were shown to be even more obvious in experiments on the pulsed 'plasma focus' devices. A 'plasma focus' is actually a point (in local extension) pulse neutron – X ray source with a pulse duration of 10^{-9} – 10^{-8} s. Both neutrons and X rays were emitted in a short discharge and dose rates were too high to be measured.

The measurements on the 'plasma focus' led to the conclusion that bubble detectors have been proved to be 'un-inertial' and can successfully detect very short neutron pulses. The neutron fluences measured on the 'plasma focus' were about 3×10^3 – 1.5×10^6 n.cm^{-2} per pulse at various distances from the pinch discharge. They were in good agreement with calculations and other indirect measurements. Values of the neutron fluxes shortly after the pulse had passed could be obtained. A few minutes delay is necessary to allow bubbles to grow to visible sizes (Figure 3). A high energy threshold cut off the low energy scattered neutrons and thus it was ensured that only thermonuclear neutrons directly from the source had been taken into account.

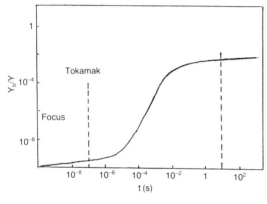

Figure 3. Bubble growth in bubble detector (V_b, bubble volume; V, detector volume) after the time of neutron emission. The emission durations for 'focus' and Tokamak are also indicated. Bubble growth for times below 1 s was calculated, above 1 s it was measured.

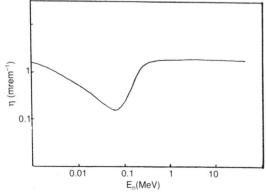

Figure 4. Dose response of an albedo bubble detector to neutrons as a function of neutron energy.

BUBBLE DETECTOR IN PERSONAL DOSIMETRY

Bubble detectors may also be used to estimate the radiation risk to the personnel caused by both photo and thermonuclear neutrons, either scattered or not. For this purpose a highly sensitive thresholdless detector with an appropriate energy–dose characteristic curve such as an albedo bubble detector is required.

The maximum permissible neutron dose for workers in the Tokamak-10 facility is about 5 μSv.d^{-1}. It is important to note that the average neutron energy at the workplaces is about 0.5 MeV. Therefore albedo corrections in case of bubble detectors (Figure 4) would not be necessary.

ACKNOWLEDGEMENTS

We appreciate the assistance of Dr M. Belukov in performing numerous irradiations at Tokamaks, and calibration of irradiations. Particular thanks are due to Mr S. Malinovski for his help in calculation. From 1986 the bubble dosimetry investigation in KIAE is supported by the Soviet ITER group.

REFERENCES

1. Martinuk, Y. N., Smirnova, N. S. and Fomin, G. V. *Biological Effects of Nonionized Irradiation of Fusion Devices.* Review, KIAE (Moscow, SU) (1988).
2. Ing, H. and Birnborn, H. C. *A Bubble-damage Polymer Detector for Neutrons.* Nucl. Tracks Radiat. Meas. **8**(1), 285-288 (1984).
3. Apfel, R. E. *The Superheated Drop Detector.* Nucl. Instrum. Methods **162**, 603-608 (1979).
4. Ivanov, V. I., Semaschko, N. N. and Smirnova, N. S. *Neutron Flux Dosimetry by Superheated Liquid.* At. Energ. **63**(1), 58-62 (1986).
5. Ing, H. *Bubble-damage Polymer Detector for Neutron Dosimetry.* Chalk River Nuclear Laboratory, Ontario, KOJ, 190 (1988).
6. Belukov, M. M. and Smirnova, N. S. *Radiation Aspects of Plasma Disruption in Tokamak-15.* KIAE, 4915/7 (Moscow, SU) (1989).
7. Alchimovich, V. A., Achtirsky, S. V., Babaev, I. V. *The Results of Setting into Operation of Tokamak-15.* VANT, 'Fusion' **3**, 3-16 (Moscow, SU) (1989).

MEASUREMENTS OF FAST NEUTRONS BY BUBBLE DETECTORS

J. Schulze†, W. Rosenstock† and H. L. Kronholz‡
†Fraunhofer-Institut für naturwissenschaftlich-technische
Trendanalysen (INT),
PO Box 1491, D 5350 Euskirchen, Germany
‡Radiologische Universitätsklinik
Albert Schweizer Str. 33, D 4400 Münster, Germany

Abstract — The properties of neutron bubble detectors (type BD-100 R*) have been investigated with fast neutrons from D-T and Am-Be sources. The experimentally determined responses of the bubble detectors have shown considerable fluctuations (up to 50%). The response after repeated recompressions has been examined. The detectors proved to be insensitive to gammas. Measurements at various dose rates have been performed in comparison with TE chambers. Since the bubble detectors are easy to use without any power supply they may be suited for detection and location of fissionable material.

INTRODUCTION AND AIMS

According to Griffith[1] the most attractive new neutron dosimetry technique is the bubble damage or superheated drop detector. We have studied the properties of bubble detectors with respect to (i) response after recompression, (ii) dose equivalent dependence with respect to dose rate and energy (especially behind shielding), (iii) gamma response, and (iv) behaviour in environmental situations (dose equivalent dependence on temperature, impact sensitivity).

A special advantage of bubble detectors is the high response to neutrons with energies above 100 keV. These detectors may thus be well suited for detecting neutrons from spontaneous fission of actinides at low intensities. Possible applications of the bubble detectors are radiation protection measurements in the vicinity of a nuclear material depot or locating of fissionable material.

^{3}He detectors are competitive detection devices in this low level flux application when the neutrons are moderated by shielding material. A comparison between bubble and ^{3}He detector response inside a shielded room has been done by both measurement and calculation. Behind concrete shielding the comparison has been done experimentally.

PROPERTIES OF THE BUBBLE DETECTORS

Neutron bubble detectors are easy to use and cheap neutron dosemeters having the advantage of direct reading. The bubble detectors (type BD-100 R) consist of a small test tube (length overall 80 mm, active length 45 mm, diameter 15 mm) filled with an elastic clear polymer. Interspersed in this polymer are superheated freon droplets. Recoil protons may be produced by neutron interaction with the polymer. If these protons strike such a droplet it may vaporise. This vaporised droplet remains trapped as a visible bubble in the polymer[2,3]. Recharging is accomplished by pressurising the bubble detectors considerably above the vapour pressure of the freon gas mixture, thereby reforming the bubbles to liquid droplets.

The bubble detectors are very sensitive with a detection limit of a few μSv. The neutron energy threshold of the standard option is 100 keV with a nearly constant efficiency of approximately 3.5×10^{-5} bubbles per (n.cm^{-2}) for higher energies (Figure 1). Furthermore they need no electronics

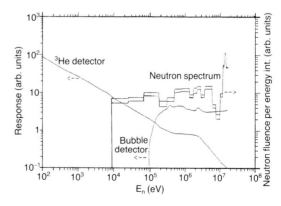

Figure 1. Energy spectrum of the D-T neutron generator (2 m from target and approximately 4 m from the walls, calculated with the Monte Carlo code MORSE) and energy dependence of the response of bubble detectors[4] and ^{3}He proportional counters[5].

---

* BD-100 R is a registered trademark of Bubble Technology Industries (BTI), Chalk River, Ontario, Canada.

or power to operate and therefore cannot be disturbed by electromagnetic interference. The main disadvantages are the strong temperature dependence (change in sensitivity approx. 4% per degree Celsius) and the severe shock sensitivity. Upon a sharp impact (e.g. when pushed hard on the table) the superheated drops vaporise, forming bubbles like those induced by neutrons.

The responses of about 40 bubble detectors (type BD-100 R) were investigated in many irradiation cycles with neutrons from Am-Be sources and 14 MeV D-T generators. We used bubble detectors with four different responses: 0.15 bubbles per µSv, 0.27 bubbles per µSv, 0.82 bubbles per µSv and 1.2 bubbles per µSv as quoted by the manufacturer. The bubbles were counted visually. Methods for computer-aided bubble counting and their limits are reported elsewhere[6]. The optimum number of bubbles proved to be 50 to 60 per bubble detector. In this case the statistics are sufficient and the counting error is still small. After counting, the bubble detectors were recompressed with a pressure of 5500 to 6000 kPa for at least three hours. Like Ipe et al[7], we found that the size of the bubbles grows with time: within 20 hours after irradiation the diameter of the bubbles increased by approximately 100%. This increase in size was independent of the temperature between 6°C and 22°C.

In order to test the response of the bubble detectors after repeated recompressions we irradiated a set of detectors with neutrons from an Am-Be source up to eleven times. The neutron dose equivalent evaluated with the bubble detectors was compared with the actual dose equivalent. The latter was consistently determined by calculation using data taken from Burger[8] and with a commercial rem counter, type Studsvik 2202D. Figure 2 shows the deviation of the dose equivalent measured with the bubble detectors as a function of the recompression cycles. The values from bubble detectors of the two low sensitivity classes as well as of the two high sensitivity classes were combined for greater clarity. In particular, the bubble detectors of the low sensitivity classes showed large deviations when used for the first time and the reading was significantly too high (mean +150%). After the third recompression (i.e. the fourth irradiation) all bubble detectors showed a 40% to 50% higher dose equivalent than that applied but the variations were still large.

In addition to the neutron irradiation we examined the gamma response to ^{60}Co and ^{137}Cs sources. The neutron bubble detectors were irradiated with gamma doses up to a maximum of 4 kGy; so that the gamma dose was many orders of magnitude larger than the maximum neutron dose (limited by the number of bubbles). New bubble detectors of all sensitivity classes were completely insensitive to gamma radiation. The same effect was found for bubble detectors of low and medium response already irradiated by neutrons. However, a few of the detectors with high response sometimes formed large bubbles when irradiated with gamma doses greater than 3 mGy. These gamma-induced bubbles looked completely different from those produced by neutrons: they looked like flat discs with a diameter 6 to 9 times larger than neutron induced bubbles. After recompression we found no change in neutron response.

From these measurements we conclude, with the restrictions mentioned above, that by a combination of bubble detectors with different responses it is possible to measure pure neutron dose equivalents in the range from 10 µSv to 800 µSv in mixed neutron-gamma fields from Am-Be sources.

COMPARISON WITH TE CHAMBERS IN DOSE EQUIVALENT MEASUREMENTS

The comparison between TE chambers and bubble detectors located side by side was carried out at the 14 MeV neutron therapy facility of the radiological clinic in Münster in a 6.0 × 4.5 m² concrete shielded room. The tissue-equivalent plastic ionisation chamber (TEPIC) used was type 3302 (PTW, Freiburg, Germany) with tissue-equivalent material A-150. The results of the measurements are shown in Table 1.

The bubble detector dose was calculated from the measured dose equivalent by applying a factor of 0.33 as used at the neutron therapy facility. The

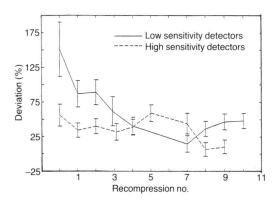

Figure 2. Deviation of the dose equivalents measured with bubble detectors from the dose equivalent as a function of the recompression cycles. Irradiation was done with an Am-Be source.

essential result should be taken from the ratio D_b/D_t because the spectral neutron fluence is not known precisely and therefore the absolute value of the bubble detector dose may differ by a constant factor.

Table 1 suggests that with decreasing dose rate the dose is overestimated by the bubble detectors, but further experiments are needed for confirmation.

COMPARISON WITH ^{3}He DETECTORS IN FISSIONABLE MATERIAL DETECTION

Bubble detectors are easy to handle without a power supply. We have therefore investigated the properties of these detectors for radiation protection measurements in the vicinity of a nuclear material depot or for locating fissionable material.

Devices frequently used for detecting neutrons with a high efficiency are, for example, the well known ^{3}He detectors. For a comparison with the bubble detectors we used ^{3}He proportional counters type 31 He-3/380/25 (Centronic) with an active length of 31 cm and a diameter of 2.5 cm filled with ^{3}He with a pressure of 0.5 MPa (5 atm).

As can be seen from Figure 1 the applicability of bubble detectors compared with ^{3}He detectors depends on the neutron energy spectrum. At 1 MeV a typical bubble detector needs about 25,000 neutrons.cm^{-2} to form one bubble and our ^{3}He detector needs about 140 neutrons.cm^{-2} for one count. The factor of about 200 may be compensated by extended measurements.

However, if the neutrons are moderated to lower energies by shielding or reflection the response of the ^{3}He detectors rises and the bubble detector response drops. In this case bubble detectors become increasingly less suitable compared with ^{3}He detectors.

For locating hidden fissionable material in a shielded room there has to be a distance dependence of the detector readings. We made calculations with the discrete ordinates code ANISN$^{(9)}$ (30 neutron energy groups, S_8, P_5) modelling a 14 MeV source in a spherical concrete shield at a radius from 8 to 10 m. The dimensions of our neutron generator room are 5.2 × 11.6 m^2. The spectral neutron fluence was weighted by the energy-dependent responses of the ^{3}He detector and the bubble detector and the fluence to dose equivalent conversion function in order to see the radial dependence in these three cases. As can be seen from Figure 3 the ^{3}He detector reading does not show a radial dependence whereas the two other curves decrease by a factor of a thousand, and between 3 and 5 m by a factor of two.

Behind a concrete shield the bubble detector reading decreases as expected. In contrast, a 15–20 cm concrete shield produces a higher count rate in a ^{3}He detector than does a bare neutron source at the same distance (Figure 4) because the shift of the neutrons to lower energies (i.e. to higher

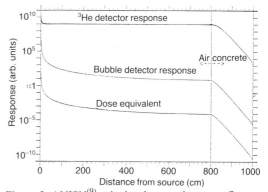

Figure 3. ANISN$^{(9)}$ calculated spectral neutron fluences weighted by energy response functions of ^{3}He and bubble detectors and by fluence-to-dose-equivalent conversion functions relative to radial distance from a 14 MeV source.

Table 1. Comparison of TEPIC and bubble detector results from an irradiation with 14 MeV D-T neutrons.

Irradiation no.	Dose rate (μGy.s^{-1})	Dose TEPIC D_t (mGy)	Dose bubble detector, D_b (mGy)	Ratio D_b/D_t
1	0.56	0.21	0.22	1.05
1	0.56	0.21	0.34	1.62
2	0.30	0.044	0.048	1.09
2	0.30	0.044	0.058	1.32
2	0.30	0.044	0.058	1.32
2	0.30	0.044	0.052	1.18
3	0.13	0.044	0.098	2.23
3	0.13	0.044	0.084	1.91
3	0.13	0.044	0.104	2.36

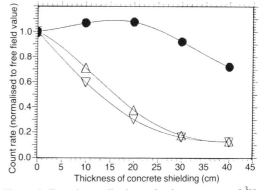

Figure 4. Experimentally determined count rates of ^{3}He and bubble detectors as a function of the shield thickness (circles, ^{3}He detector; triangles, bubble detectors, peak up high response, peak down low response). The lines are spline fits.

detector response) overcompensates the absorption.

CONCLUSION

By combination of bubble detectors with different sensitivities it was possible to measure neutron equivalent doses in the range from 10 μSv to 800 μSv (from an Am-Be neutron source) if an accuracy of 50% is sufficient. Temperature variations should be taken into account and even small shocks must be avoided.

For detecting neutrons from nuclear fissionable material inside or behind shielding, ^{3}He counters are superior to bubble detectors because of the moderated energy spectrum of the neutrons and the high (n, ^{3}He) cross section in the lower energy region. With a 15–20 cm concrete shield between source and detector there is a higher count rate in the ^{3}He detector than without any shielding of the source.

REFERENCES

1. Griffith, R. V. *Review of the State of the Art in Personnel Neutron Monitoring with Solid State Detectors.* Radiat. Prot. Dosim. **23**, 150-160 (1988).
2. Apfel, R. E. *The Superheated Drop Detector.* Nucl. Instrum. Methods **162**, 603-608 (1979).
3. Ing, H. and Birnboim, H. C. *A Bubble-Damage Polymer Detector for Neutrons.* Nucl. Tracks Radiat. Meas. **8**, 285-288 (1984).
4. Ing, H. (B.T.I., Chalk River, Ontario) private communication (1990).
5. *Evaluated Nuclear Data File.* ENDF/B IV, National Nuclear Data Center, Brookhaven National Laboratory (BNL), Upton, New York (1975).
6. Rosenstock,W., Schulze, J. and Kronholz, H. L. *Neutron Bubble Detectors — Measurements with fast Neutrons in Comparison with Customary Devices.* Strahlenschutz für Mensch und Umwelt Bd.II 849-854 (Publikationsreihe: Fortschritte im Strahlenschutz Verlag: TÜV Rheinland) (1991).
7. Ipe, N. E., Busick, D. D. and Pollock, R. W. *Factors Affecting the Response of the Bubble Detector BD-100 and a Comparison of its Response to CR-39.* Radiat.Prot. Dosim. **23**, 135-138 (1988).
8. Burger, G. and Schwartz, R. B. *Guidelines on Calibration of Neutron Measuring Devices.* STI/DOC/10/285 (Vienna: IAEA) (1988).
9. Engle, W. W. Jr *A Users Manual for ANISN.* ORNL-Report Nr. K-1693 (1973).

SOME NEW TECHNIQUES FOR NEUTRON RADIATION PROTECTION MEASUREMENTS

B. Dörschel, H. Seifert and G. Streubel
Dresden University of Technology
Institute of Radiation Protection Physics
D-O-8027 Dresden, Germany

INVITED PAPER

Abstract — The measurement techniques using neutron radiation protection dosemeters on the basis of new operation principles have been studied for some years. Among others, electret ionisation chambers and MOS breakdown sensors are of special interest. Whereas the latter are applicable in neutron area monitoring electret dosemeters were developed for neutron individual dosimetry. For each dosemeter type the characteristics, such as the energy dependence of the response, the dynamic range etc., must be determined under standard irradiation conditions. Furthermore, several dosemeter designs were studied which were adapted to different tasks of measurement in practice.

INTRODUCTION

Substantial tasks in neutron radiation protection are neutron area monitoring and neutron individual dosimetry. Area monitoring is aimed at the limitation of controlled areas and the estimation of personnel doses. The area dosemeters used for this purpose should yield the dose equivalent in soft tissue measured at a specified position. The relevant measuring quantity is the ambient dose equivalent H*(10). Individual dosimetry means the determination of the personnel dose, i.e. the dose equivalent in soft tissue measured at a specified position on the body surface. One of the measuring quantities considered in this case is the directional dose equivalent H'(10).

Thus, the energy and angular dependence of the fluence-related response of area and individual dosemeters, should be identical with that of the fluence-to-dose equivalent conversion factors H*(10)/Φ and H'(10)/Φ. The main problem in neutron dosimetry concerns the wide neutron energy range of more than 10 orders of magnitude and the distinct energy dependence of the neutron cross sections. The energy dependence of the detector responses usually differs quite substantially from the required energy dependence given by the conversion factors. Therefore some additional information on the neutron energy spectrum is necessary for the interpretation of dosemeter readings. For routine control, the determination of neutron fluences in selected energy groups is often sufficient. Furthermore, the additional sensitivity of several dosemeters to photons requires the use of a detector system consisting of detectors with different responses to neutrons and photons. Some applications also require the determination of the spatial distribution of the neutron fluence or flux density at neutron sources producing inhomogeneous radiation fields. In these cases detectors with a good spatial resolution are needed.

At present, numerous neutron detectors and dosemeters exist, but most of them do not fulfil all requirements and represent a compromise for a specified task of measurement. In order to improve the measuring technique in neutron radiation protection some alternative or supplementary dosemeters based on new operation principles have been suggested. For example, bubble damage dosemeters have been widely studied[1], and the use of dynamic RAMs in neutron dosimetry is under discussion[2].

At the Dresden University of Technology comprehensive investigations concerning electret ionisation chambers and MOS breakdown sensors have been carried out for several years. In the following, a survey of the most important characteristics of these detectors for application to neutron radiation protection measurements is given. Furthermore, special dosemeter designs are discussed which were adapted to different tasks of measurement in practice.

ELECTRET IONISATION CHAMBERS FOR INDIVIDUAL NEUTRON DOSIMETRY

Electrets are solids which show a quasi-permanent electrical charge and a surface potential after an appropriate formation procedure in which molecular dipoles are aligned or charge carriers are trapped. The surface potential is termed electret voltage U_E, which can be measured without contact using the capacitive field probe method[3]. Appropriate electret materials are polytetrafluoroethylene (PTFE) or its copolymer (FEP).

The principle of an electret ionisation chamber is shown in Figure 1. One electrode in a small air-filled ionisation chamber is replaced by an electret, while the opposite electrode is made of a thin graphite or aluminium layer. Since the electret generates an electric field within the chamber no external voltage supply is needed. Charge carriers produced by ionising radiation within the chamber gas are attracted by the electret, causing a lowering of the electret voltage. The dosemeter reading is the difference ΔU_E of the electret voltages before and after irradiation. In the saturation region this difference is proportional to the number of generated charge carriers of one sign which is in correlation with the absorbed radiant energy in the chamber gas. Applying the basic principles of dosimetry, other quantities, e.g. the absorbed dose in tissue or the dose equivalent, can be determined. The initial electret voltage amounts to about 800–1000 V. Electret stability is injured by a non-radiation electret voltage decay. In our case this decay was less than 1 V per day. Dosemeter readings caused by small doses must be corrected by this value.

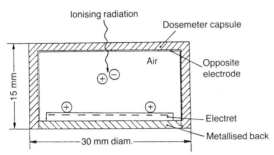

Figure 1. Scheme and operation principle of an electret ionisation chamber.

Electret ionisation chambers have been studied in particular with the aim of applying them in photon individual dosimetry[4–6]. The use of electret ionisation chambers in neutron dosimetry is also possible. In this case the dosemeter reading arises from the ionisation by neutron-induced secondary charged particles. However, the simultaneously existing sensitivity to gamma radiation does not allow a selective neutron detection in mixed neutron and gamma fields. Therefore, the gamma contribution to the dosemeter reading must be either compensated or separately determined.

For this purpose, a dual ionisation chamber[7,8] was developed (see Figure 2). It consists of two cylindrical chambers coaxially arranged in a dosemeter capsule. The electret of the inner chamber is positively charged whereas the annular outer electret carries a negative charge. The walls of the outer chamber are made from 7 mm thick polyethylene while the inner chamber is covered by a proton absorber made from 3 mm thick PTFE. Thus, neutron-induced recoil protons are not able to pass through the inner ionisation chamber. The measuring effect of the outer chamber is therefore caused by neutrons and photons whereas in the inner chamber the measuring effect is generated approximately by photons only. The geometrical dimensions have to be chosen in such a way that the responses of both chambers to gamma radiation are equal. Thus, the use of a charge reading probe covering the total surface of both electrets allows a selective determination of the neutron dose because the gamma component is compensated due to the different polarity of the electrets.

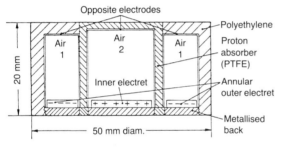

Figure 2. Cross section of the dual electret ionisation chamber for selective measurement of neutron dose without gamma background. 1, neutron and photon sensitive volume; 2, photon sensitive volume.

The absorbed gamma dose can be additionally determined by measuring the electret voltage of the inner chamber. In Figure 3, the calculated neutron energy dependence of the dosemeter response is presented along with experimental results obtained using three different neutron sources (^{252}Cf,

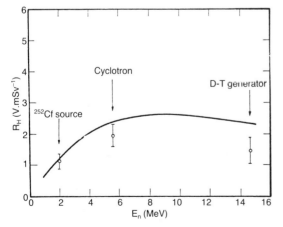

Figure 3. Response R_H of the dual electret ionisation chamber as a function of neutron energy E_n. (——) calculated, (o) measured. $R_H = \Delta U_E/H'(10)$.

cyclotron U-120 reaction $^{9}_{4}Be(d,n)^{10}_{5}B/E_d = 13$ MeV, d-t generator). The discrepancy between measured and calculated results at 14 MeV can be explained by the non-negligible neutron sensitivity of the inner chamber caused by neutron reactions with fluorene nuclei of the PTFE absorber. This effect was not taken into account in the computer program which only yielded the energy transfer by recoil protons. The results in Figure 3 show a nearly constant response in the neutron energy range from 2 to 15 MeV whereas the dosemeter is practically insensitive to thermal and intermediate neutrons.

In order to overcome this disadvantage a certain amount of natural boron was deposited on the electrode opposite to the outer electret. Thus, thermal neutrons can be measured by utilising the $^{10}B(n,\alpha)^{7}Li$ reaction. When the dosemeter is worn on the human body it acts as an albedo dosemeter, i.e. thermal neutrons backscattered from the human body also contribute to the measuring effect. The energy dependence of the dosemeter response is then similar to that of other albedo dosemeters. The response to thermal neutrons depends on the area of the boron layer deposited. Therefore, by varying this area the response of the dosemeter to thermal and intermediate neutrons could be adapted to that of fast neutrons. The result of such an optimisation is a dosemeter response in the whole energy range as shown in Figure 4. In this case, the deposited boron layer covers about 2.5% of the total opposite electrode. The response of the whole dosemeter is limited by the response to fast neutrons which is determined by the relatively small neutron cross sections. The minimum measurable electret voltage difference is about $(\Delta U)_{min} = 2$ V. The maximum usable voltage interval is determined by the linear part of the dose characteristics of the outer chamber, where the electret voltage decrease is caused by neutrons and photons. The maximum usable voltage range of the coaxial chamber with gamma compensation is about $(\Delta U)_{max} = 200$ V below the initial voltage. In this case no dose rate effect has been observed up to 10 Gy.h^{-1}. Considering the random and systematic uncertainties of the measured electret voltage and the chamber sensitivity, a dynamic range of neutron dose equivalent is deduced to extend from 1 mSv to about 100 mSv. Since the response to fast neutrons cannot be increased, the lower limit of the dynamic range cannot be reduced, substantially.

In several typical neutron fields met in practice, e.g. stray neutron fields at nuclear installations, fast neutrons with energies above 1 MeV contribute only little to the total dose equivalent. Therefore, it is more suitable in this case to use an electret ionisation chamber in the form of a simple albedo dosemeter instead of the dual chamber. The design is similar to the scheme in Figure 1. The walls of the chamber are made from 2 mm thick polystyrene covered on one side by boron-loaded plastic material whereas the opposite electrode is partly covered with boron. The neutron energy dependence corresponds to that given in Figure 4 in the low energy range up to about 0.5 MeV. In this case, however, the contribution to the response by recoil protons induced by fast neutrons is very small. The area of the deposited boron layer has been optimised so that the dose equivalent response to thermal and intermediate neutrons in stray neutron fields equals that of gamma radiation. The boron covered part of the opposite electrode amounts to about 20% in this case. Thus, the neutron response of the dosemeter is higher than in the case of the dual chamber. Furthermore, the reading of the electret ionisation chamber is a measure of the total dose equivalent in mixed neutron and photon fields. With this type of dosemeter a dynamic range from 0.1 mSv to 30 mSv could be achieved.

MOS BREAKDOWN SENSORS FOR NEUTRON FIELD ANALYSIS AND AREA DOSIMETRY

Several years ago the registration of heavy charged particles by means of MOS breakdown counters was suggested by Tommasino et al[9]. Starting from this idea we developed a neutron sensor which consists of a MOS thin film capacitor in combination with a fissile radiator (for scheme see Figure 5)[10,11]. The operation voltage applied to the MOS element must be chosen such that spontaneous breakdowns through the silicon dioxide layer can be neglected. Incident neutrons generate fission fragments within the

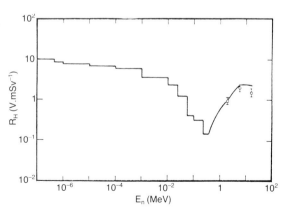

Figure 4. Response R_H as a function of neutron energy E_n for the dual electret ionisation chamber designed as albedo dosemeter. (——) calculated, (o) measured. $R_H = \Delta U_E/H'(10)$.

radiator which enter the silicon dioxide. Along their paths secondary charge carriers are produced in high concentration which gives rise to a local

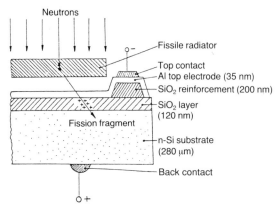

Figure 5. Scheme and operation principle of a MOS breakdown sensor.

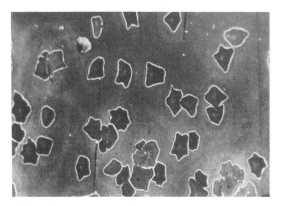

Figure 6. Vaporisation spots on the sensor surface (amplification: 750 x).

Figure 7. Measuring probe with and without aluminium protecting cap and device for pulse counting and data processing.

degradation of the breakdown field strength. The result is electric breakdowns at the sites of the fission fragment tracks which can be counted as voltage pulses by using a simple electronic circuit. The pulse rate z is proportional to the momentary neutron flux density in time-dependent neutron fields or to the corresponding derived dosimetric quantities. In the course of each breakdown a hole is created in the silicon dioxide and in the aluminium top electrode by vaporising material (see Figure 6). Therefore, the total number of vaporisation spots on the sensor surface is an additional measure of the fluence of incident neutrons — similar to the number of etched tracks in solid state nuclear track detectors.

MOS elements commonly used in microelectronics are not suitable for application as breakdown sensors. Therefore, a special preparation procedure must be developed which consists of cleaning the silicon wafers, oxidation, structuring, separation of the sensor chips and encapsulating them in cases of integrated circuits[12]. After bonding the MOS sensors and covering them with fissile radiators they can be used for neutron measurements. Figure 7 shows measuring probes with and without an aluminium protecting cap. For counting the voltage pulses and data processing a special one-chip microcomputer-controlled measuring device was developed in combination with a pulse shaper, which is also shown in Figure 7.

Since photons with energies below the photofission threshold do not induce breakdowns, MOS sensors measure the neutron component in mixed neutron–gamma fields practically without any background. The fission fragment registration efficiency also remains unchanged for high gamma doses up to 1000 Gy. Due to the evaporation of the aluminium top electrode, however, the effective sensor area decreases with increasing number of breakdowns. Furthermore, changes in the silicon structure occur after a large number of breakdowns[13]. Therefore, the registration efficiency gradually decreases. Since this degradation is reproducible the registration efficiency can be corrected without any problems. The total registration capacity of a MOS sensor with a sensitive area of 10^2 mm amounts to about 10^5 breakdowns.

The neutron response of MOS sensors is mainly determined by the radiator properties. Thus, the energy dependence of the flux density related response corresponds to that of the fission cross section of the radiator material. Using a combination of several MOS sensors with properly chosen radiators, the flux density or the dose equivalent rate of neutrons from different energy intervals can be measured. For example, the use of radiators made from ^{232}Th and ^{235}U, and the application of the cadmium difference method, allows neutron fields to be analysed roughly for the contributions of thermal, intermediate and fast neutrons. Figure 8 shows the flux densities of neutrons from these energy intervals measured by means of MOS sensors at a cyclotron U-120 (reaction $^9_4\text{Be}(d,n)^{10}_5\text{B}$, $E_d = 13$ MeV) in dependence on the target current. Previously determined flux densities measured by means of activation detectors are in good agreement with them. In this case a set of 22 activation detectors was used for measuring the neutron spectrum. From that the flux densities in certain energy intervals could be determined.

If the energy spectrum is approximately known a single MOS sensor can be calibrated so that its reading immediately yields the dose equivalent rate. In this way a simple gamma-insensitive area dosemeter is available. The dose equivalent rate related response depends on the radiator properties as well as on the properties of the neutron field. In Table 1, some results for the response are summarised, obtained by irradiations at the reactor WWR-S (Rossendorf Central Institute for Nuclear Research) and at a (d-t) neutron generator (Dresden University of Technology). MOS sensors with radiators of ^{235}U can also be used as gamma-insensitive detectors of thermalised neutrons within a moderator sphere which results in a special kind of a rem meter for measuring the neutron dose equivalent rate.

The dynamic range of MOS sensors covers approximately five orders of magnitude. A lower limit of 0.03 s^{-1} was derived from the tenfold background pulse rate by spontaneous breakdowns.

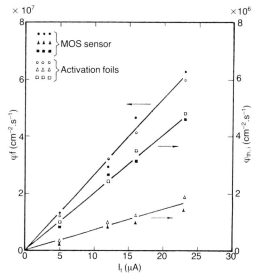

Figure 8. Flux density φ of thermal (▲△), intermediate (■□) and fast (●○) neutrons measured by means of MOS sensors and activation detectors at cyclotron U-120 as a function of the target current I_t (distance from the Be target: 1 m).

Table 1. Dose equivalent rate related response R_H of MOS sensors with a sensitive area of 10 mm^2 combined with different radiators (z = counting rate).

Radiator (material, thickness)	Neutron field	Energy threshold of registration	$R_H = \dfrac{z}{\dot{H}^*(10)}$ (Sv^{-1})
^{232}Th, d>R[1]	Reactor WWR-S[2] (d,t) generator	$E \approx 1$ MeV	97.9 242
^{238}U+Cd (99.6% depleted) d = 26 nm	Reactor WWR-S (d,t) generator	$E \approx 0.4$ eV	3.2 6.5
U$_{nat}$+Cd (surface polished) d>R	Reactor WWR-S (d,t) generator	$E \approx 0.4$ eV	264 500
U$_{nat}$+Cd (surface oxidised) d>R	Reactor WWR-S (d,t) generator	$E \approx 0.4$ eV	176 325
^{235}U (90% enriched) d>R	Reactor WWR-S (d,t) generator	$E \approx 0$	4.45×10^5 880
^{235}U (90% enriched) d = 28 nm	Reactor WWR-S (d,t) generator	$E \approx 0$	6531 12.9

[1] The radiator thickness d is larger than the range R of fission fragments in the radiator material.
[2] The neutron field is characterised by fluence fractions of 18.3% fast (E>1MeV), 39.7% intermediate (0.4eV ≤ E ≤ 1MeV) and 42.0% thermal (E < 0.4 eV) neutrons.

The upper limit results from the dead time quenching at high pulse rates. By assuming a quenching less than 10%, the maximum measurable pulse rate amounts to 3000 s^{-1}. The lower and upper limits of the neutron flux density or neutron dose equivalent rate depend just as the corresponding response on the fission yield of the radiator used. Therefore, variations are possible by using radiators made from different fissile materials and with different thicknesses. For example, the lower limit of the dynamic range of a MOS sensor combined with a thick ^{235}U radiator in cadmium shielding is about 750 µSv.h^{-1} in neutron fields at nuclear installations. The uncertainty of sensor calibration was about 10%. The response variation between different sensor batches also amounted to about 10%. When the relative statistical uncertainty of the pulse counting is assumed to be 5% a total measurement uncertainty of the dose equivalent rate of about 15% results.

For calibration of detectors and dosemeters not only the energy spectrum of neutrons is of interest, but also the spatial distribution of the neutron flux density must be known, especially for irradiations of detectors in narrow neutron beams. Because of their small geometrical dimensions MOS sensors are suitable for measuring neutron beam profiles with high spatial resolution. Figure 9 shows the profile of the thermal neutron beam of the Research and Measurement Reactor, Braunschweig, measured by means of a MOS sensor with a sensitive area of 18 mm^2. Here, the measurement uncertainty was about 5%.

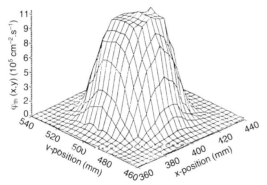

Figure 9. Thermal neutron beam profile measured by means of a MOS sensor at the Research and Measurement Reactor, Braunschweig.

Apart from these examples further applications of MOS neutron sensors for radiation protection measurements, e.g. at the teaching and research reactor AKR of the Dresden University of Technology, have been studied.

CONCLUSIONS

The problems in neutron dosimetry are of a principal nature. Therefore, an ideal dosemeter, which fulfils all requirements regarding energy dependence, dynamic range, measurement accuracy etc. cannot be expected. The development of dosemeters on the basis of new operation principles may, however, improve this situation. Under the aspect of increasing the reliability in neutron dosimetry every new method of neutron detection should be tested as to whether a utilisation in neutron radiation protection measurements seems to be practicable — at least as a supplementary method. Sometimes, special detector properties allow an adaptation of dosemeter designs to special measurement tasks. In this respect, the use of electret ionisation chambers and MOS breakdown sensors must also be considered. Electret ionisation chambers are integrating dosemeters with relatively high sensitivity and very simple measurement procedure. MOS breakdown sensors yield an on-line reading of dosimetric or field quantities. They register neutrons selectively in mixed neutron–gamma fields and have a very good spatial resolution. However, disadvantageous properties must also be considered such as the limited charge stability of electrets and the limited registration capacity of MOS sensors. Nevertheless, with special designs of these detectors it should be possible to achieve improvements in neutron dosimetry, just as with other new dosemeters, such as bubble damage detectors or dynamic RAMs. Therefore, scientific studies of all these detectors should be continued in more detail.

ACKNOWLEDGEMENT

The authors wish to thank R. Zimmermann (ZfK Rossendorf), Dr E. Dietz (PTB Braunschweig) and Ing. J. Jánsky (ÚJV Řež) for their assistance during the irradiation experiments as well as J. Reinhard (TU Dresden) for his help in MOS sensor preparation and measurements.

REFERENCES

1. Ing, H. and Birnboim, H. C. *Bubble-Damage Polymer Detectors for Neutron Dosimetry.* In: Proc. 5th Symp. on Neutron Dosimetry. Vol. II (Luxembourg: CEC) pp. 883-894 (1985).
2. Davis, J. L. *Use of Computer Memory Chips as the Basis for a Digital Albedo Neutron Dosimeter.* Health Phys. **49**(2), 259-265 (1985).

3. Sessler, G. M. (ed.) *Electrets*. Topics in Applied Physics. Vol. 33. (Berlin, Heidelberg, New York: Springer) (1980).
4. Gupta, P. C., Kotrappa, P. and Dua, S. K. *Electret Personnel Dosemeter*. Radiat. Prot. Dosim. **11**(2), 107-112 (1985).
5. Pela, C. A., Ghilardi, A. J. P. and Ghilardi Netto, T. *Long-Term Stability of Electret Dosimeters*. Health Phys. **54**(6), 669-672 (1988).
6. Dörschel, B., Prokert, K., Seifert, H. and Stoldt, C. *Individual Dosimetry Using Electret Detectors*. Radiat. Prot. Dosim. **34**(1-4), 141-144 (1990).
7. Seifert, H. and Dörschel, B. *Concepts for the Development of Neutron Dosemeters Using Electret Ionization Chambers*. Kernenergie **31**(4), 165-170 (1988).
8. Seifert, H. *Entwicklung von Elektretionisationskammern für die Photonen- und Neutronendosimetrie*. Thesis. Dresden University of Technology (1990).
9. Tommasino, L., Klein, N. and Solomon, P. *Thin-Film Breakdown Counter of Fission Fragments*. J. Appl. Phys. **46**(4), 1484 (1975).
10. Dörschel, B., Pfützner, F., Streubel, G., Knorr, J. and Paul, W. *Neutron Detection Using MOS Breakdown Counters Combined with Fissile Radiators*. Nucl. Instrum. Methods **216**, 227-229 (1983).
11. Streubel, G., Dörschel, B., Hanisch, T., Reinhard, J. and Schoop, K. *Properties of MOS Sensors for Application in Neutron Field Monitoring*. Nucl. Instrum. Methods **A274**(1-2), 194-202 (1989).
12. Reinhard, J. *Untersuchungen zum Nachweis hochenergetischer Photonenstrahlung mit MOS-Sensoren*. Thesis, TU Dresden, (in preparation).
13. Bernhadt, A. and Danziger, M. *Untersuchungen zum Mechanismus der Meßeffektentstehung in MOS-Neutronen-Sensoren unter Berücksichtigung des glasartigen Verhaltens der SiO_2 Schichten*. Thesis, TU Dresden (1987).

PRINCIPLES OF AN ELECTRONIC NEUTRON DOSEMETER USING A PIPS DETECTOR

B. Barelaud, D. Paul, B. Dubarry, L. Makovicka, J. L. Decossas and J. C. Vareille
Laboratoire d'Electronique des Polymères sous Faisceaux Ioniques
Faculté des Sciences, 123, avenue Albert Thomas
87060 Limoges Cedex, France

Abstract — This laboratory has already studied fast and thermal neutron dosimetry with the help of a passive dosemeter composed of a boron implanted polyethylene $(CH_2)_n$ converter placed on a solid state nuclear track detector (SSNTD), CR-39. With the development of semiconductor detector technology, real-time dosimetry is now being considered. The principle of the dosemeter is to detect the secondary particles from the converter using a PIPS (passivated implanted silicon) detector for which the registered pulses are processed to be proportional to the dose equivalent. Theory shows the neutron dose equivalent response (dosimetric efficiency) with the optimum thickness of converter and calculated for an isotropic incident neutrons beam, to be fairly independent of neutron energy (range 0.2 MeV – 5 MeV). Many experiments (accelerator, Am–Be source) show the response of the electronic device to be greater than that of a passive detector. The response of the device to gamma rays and background is overcome by adoption of a differential method, using a similar PIPS detector without converter.

INTRODUCTION

Personnel dosimetry in neutron fields using solid state nuclear track detectors (SSNTD) is a research topic which has been studied at LEPOFI using a boron implanted $(CH_2)_n$ converter and CR-39 detector[1-4] for the detection of thermal and fast neutrons. Owing to the evolution of semiconductor detector technology, real-time dosimetry is now considered[5-6]. This has many advantages in comparison with SSNTD. In addition, computed characterisation of the proton flux from the $(CH_2)_n$ converter is still applicable for an electronic detector.

ELECTRONIC DEVICE

The aim of this work was to obtain, as far as is practicable, a dose equivalent response for the device independent of neutron energy and angle of incidence. The response of the electronic device is obtained from transfer characteristics of its different parts. For such a formalism[7] — used by people who work on sensors and measuring devices — three input quantities can be considered:

(i) Quantities to be measured, G_d, are quantities for the measurement of which the device is specifically designed. They correspond to the protons emitted by the converter due to the (n,p) elastic scattering, α particles and ^7Li ions generated by thermal neutron reactions on ^{10}B as explained below[8].

(ii) Interfering quantities, G_i, are quantities to which the device is not intentionally sensitive. They disturb the measures and their effects add to those of the previous quantities to give the output signal. γ photons and direct neutron interactions with the silicon atoms of the diode are the most important of these quantities. Electronic background (detector, amplifier, etc) and natural radiations are also interfering quantities.

(iii) Modifying quantities or parameters, G_m, are quantities changing the transfer characteristics of the device. They have an influence on both the previous quantities. These are the detector bias which modifies the depth of the depleted zone, the gain of linear amplifier, the electronic threshold, and the converter characteristics.

The response of the device is due to the first two quantities. In order to eliminate as much as possible the interfering quantities, we apply a differential method which has been suggested by the laboratory and used with SSNTD by Makovicka[1]. The schematic diagram is presented in Figure 1. It is composed of two symmetrical channels; the first is a reference channel and is only sensitive to interfering quantities (detector without converter); the second, which is composed of a detector and a converter, gives a response which depends on all input quantities. The required neutron response is given by subtracting the measured pulse height spectrum, obtained for the detector without converter (interfering quantities), from that with the converter. Furthermore, interfering quantities concern mainly low energy channels of the analyser. With decreasing channel number, the resultant uncertainty on the differential counts in a channel increases. This determines the low energy threshold for neutron detection.

THE SENSOR

The sensor is a PIPS detector covered with a

^{10}B implanted polyethylene converter. Theoretical calculations for the proton spectra generated by the $(CH_2)_n$ converter were used to optimise the fast neutron response. These calculations are based on the Monte Carlo method. The number of protons from the converter, according to the energy and angle of incidence of the neutron flux, was calculated. Using the fluence to dose equivalent conversion factor, h_ϕ,[9] the intrinsic dosimetric efficiency of the converter was deduced, which is defined as the efficiency if all the emitted particles from the converter are detected[10]. Results are plotted in Figure 2. That value which is about 1 proton.cm^{-2}.μSv^{-1} is almost independent of energy in the range 0.2–5 MeV; these results were obtained for an isotropic incident neutron beam. After optimisation, we obtain the effective dosimetric response[10] which takes into account the detecting properties of the PIPS detector. The value, slightly lower than the intrinsic response, is expressed in pulses.cm^{-2}.μSv^{-1}. In this case, the sensor has a 35 μm thick polyethylene converter and an effective response of about 1 pulse.cm^{-2}.μSv^{-1}.

EXPERIMENTAL RESULTS

Response in monoenergetic fast neutron flux of normal incidence

A monoenergetic fast neutron flux was obtained with an accelerator, by the reaction D(d,n)^{3}He, the neutrons have an energy of 3.3 MeV.

Experiments have been performed for 5 mSv, 10 mSv and 15 mSv dose equivalent irradiations. The converter was 35μm thick. The depleted zone of the diode was adjusted in order to be deeper than the path of the recoil protons of maximum energy (3.3 MeV). Results are directly comparable with the results previously obtained at LEPOFI with SSNTD which had been made simultaneously[11]. Neutron dosimetry is based on the effective dosimetric response (pulses.cm^{-2}.μSv^{-1}) of the sensor. We note that the response is higher with diodes than with SSNTD (1 pulse.cm^{-2}.μSv^{-1} instead of 0.5 track.cm^{-2}.μSv^{-1}), and that this value is close to the intrinsic dosimetric response

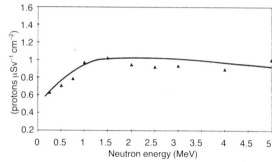

Figure 2. Intrinsic efficiency for $(CH_2)_n$ converter with 35 μm thickness plotted against energy range 0.2–5 MeV for an isotropic incident neutron beam.

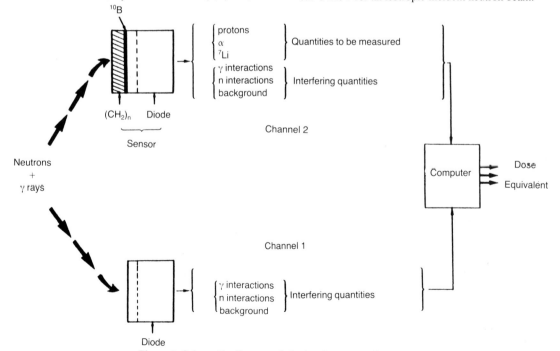

Figure 1. Schematic diagram of electronic neutron dosemeter.

of the converter.

Reducing the signal from interfering quantities

We have studied the response of our device as a function of the depth of the diode depleted zone in order to evaluate the influence of this parameter on the output signal. Irradiations were made with an Am-Be source (H*(10) = 1.19 mSv). A 1 mm thick $(CH_2)_n$ converter was used to obtain protonic equilibrium up to 10 MeV at normal incidence. Experimental results were obtained with a proton energy threshold of 390 keV and are summarised in Table 1.

From this table we can see that the differential response to neutrons (column 4) does not vary too much for a depleted zone range 35–80 μm and the response due to the interfering quantities (detector response without converter) changes more rapidly with the depth of depleted zone (column 3).

This first approach to the role of the depleted zone depth shows that an improvement of the discrimination between the quantities to be measured and the interfering quantities is possible. At the present time, the best compromise is achieved for this kind of detector with a converter thickness around 35μm. Further studies on this problem are now in progress and results concerning direct neutron interactions with silicon atoms of the diode and the role of γ rays are available[12,13].

CONCLUSION

This study has shown the feasibility of an electronic neutron dosemeter composed of two main parts: two sensors (silicon detector with or without a boron 10 implanted $(CH_2)_n$ converter), and two channels of signal analysis to achieve the differential method.

Our electronic system brings some improvements in neutron dosimetry in comparison with CR-39 based dosemeters: (i) The response is about twice that of the SSNTD, due to the lack of energy and angle limitations in recoil detection. (ii) Statistical studies have shown that the standard error of the mean for a series of experiments is low. For instance, at 3.3 MeV with an 80 μm depleted zone and a 35 μm thick converter, using a 400 keV threshold for the spectra integrations, the measured response is 0.865 ± 0.020 pulses.cm^{-2}μSv^{-1} in a 5 mSv beam. Even if the threshold is high and if accelerator neutron beams do not have important γ components, this result seems very encouraging.

Further studies are necessary to optimise the structure of the dosemeter. They are now in progress within the framework of a CEC contract.

Table 1. Registered responses on the two channels of the device, differential response to neutrons and (quantities to be measured/interfering quantities) ratio, against depleted zone. (σ_{n-1} = standard error of mean).

Depleted layer (μm) Col. 1	Detector response with converter ± σ_{n-1} Col. 2	Detector response without converter ± σ_{n-1} Col. 3	Differential response to neutrons (col. 2–col. 3) ± σ_{n-1} Col. 4	Ratio of neutron response to that from γ rays, etc. (Col. 4/Col. 3) Col. 5
80	4964 ± 74	1754 ± 61	3210 ± 96	1.83
76	4562 ± 30	1521 ± 49	3041 ± 57	2.00
71	4292 ± 24	1361 ± 91	2931 ± 94	2.15
65	3991 ± 49	849 ± 40	3142 ± 63	3.70
60	3723 ± 44	621 ± 24	3102 ± 50	5.00
53	3406 ± 87	474 ± 19	2932 ± 89	6.18
45	3401 ± 66	312 ± 7	3089 ± 66	9.40
35	3136 ± 59	254 ± 19	2882 ± 62	11.35

REFERENCES

1. Makovicka, L. *Contribution à la Dosimétrie Neutron-Gamma. Etude d'un Ensemble Radiateur-Détecteur Type CR 39*. Thèse n° 17-87, Université de Limoges (1987).
2. Makovicka, L., Barelaud, B., Decossas, J. L. and Vareille, J. C. *Detection of Thermal Neutrons by CR-39 Using a Boron Implanted Converter*. Radiat. Prot. Dosim. **23** (1/4), 191-194 (1988).
3. Makovicka, L., Decossas, J. L. and Vareille, J. C. *Experimental Study of the Dosimetry Efficiency of a Radiator — CR-39 Fast Neutron Dosemeter*. Radiat. Prot. Dosim. **20** (1/2), 63-66 (1987).
4. Sadaka, S. *Etudes Théoriques et Expérimentales d'un Dosimètre pour les Neutrons Rapides*. Thèse n° 30-83, Université de Toulouse (1984).
5. Barelaud, B. *Conception et Réalisation d'un Capteur pour les Neutrons Thermiques et Rapides*. Thèse n° 7-1989, Université de Limoges (1989).

6. Barelaud, B., Decossas, J. L., Makovicka, L. and Vareille, J. C. *Capteur Électronique pour la Dosimétrie des Neutrons.* Radioprotection **26**(2), 307-328 (1991).
7. Paratte, P. A. and Robert, P. *Système de Mesure* (Paris: Dunod) (1986).
8. Barelaud, B., Makovicka, L., Dubarry, B., Paul, D., Decossas, J. L. and Vareille, J. C. *Thermal Neutron Detection Using Si Detector and a Boron Implanted Polyethylene Converter.* In: Proc. 1st European Conf. RADECS 91, Montpellier (1991).
9. AFNOR. *Rayonnements Neutroniques de Référence Destinés à l'Etalonnage des Instruments de Mesure des Neutrons Utilisés en Radioprotection et á la Détermination de leur Résponse en Fonction de l'Energie des Neutrons.* Norme NF H-60-516, ISO-8529 (1989).
10. Sadaka, S., Makovicka, L., Vareille, J. C., Decossas, J. L. and Teyssier, J. L. *Study of a Polyethylene CR-39 Fast Neutron Dosemeter, Part II: Dosimetry Efficiency of the Device.* Radiat. Prot. Dosim. **16**(4), 281-287 (1987).
11. Paul, D. *Dosimétrie Neutronique par Ensemble "Convertisseur-CR 39". Comptage Automatique des Traces. Etude du Bruit de Fond.* D.E.A. de Physique Radiologique, Université de Toulouse (1988).
12. Dubarry, B., Barelaud, B., Paul, D., Makovicka, L., Decossas, J. L. and Vareille, J. C. *Elecronic Sensor Response in Neutron Beams.* Radiat. Prot. Dosim. **44**(1-4) 367-370 (1992) (This issue).
13. Paul, D., Barelaud, B., Dubarry, B., Decossas, J. L., Makovicka, L. and Vareille, J. C. *Gamma Interference on an Electronic Neutron Dosemeter Response in a Neutron Field.* Radiat. Prot. Dosim. **44**(1-4) 371-374 (1992) (This issue).

ELECTRONIC SENSOR RESPONSE IN NEUTRON BEAMS

B. Dubarry, B. Barelaud, J. L. Decossas, L. Makovicka, D. Paul and J. C. Vareille
Laboratoire d'Electronique des Polymères sous Faisceaux Ioniques
Faculté des Sciences, 123 Avenue Albert Thomas
87060 Limoges, France

Abstract — An electronic neutron sensor using a polyethylene converter associated with a PIPS detector has been tested in various neutron fields from 144 keV to 3.3 MeV. Measurements have confirmed the computed dosimetric response which is about one pulse per microsievert per cm² (twice solid state nuclear track detector sensitivity). It is almost independent of neutron energy. Computation of the interaction of neutron beams with the sensor should be verified experimentally. Since no computer transport code dealing with this specific problem is presently operational, we are developing a Monte Carlo model based on the methods already used for our calculation of the converter response. Initial results with this code show quite a good agreement between computed and experimental pulse height spectra, for the shape as well as for its integral. This model allows the study of parameters modifying the response to be studied and optimised.

INTRODUCTION

The need to improve neutron dosimetry, especially at low dose equivalent rates, has led our staff to study a real time electronic system. After a feasibility study[1], the sensor has been developed experimentally in order to determine its response in neutron beams. These experiments have mostly been carried out in close collaboration with the company Merlin Gerin Provence. They have brought to the fore the fact that the sensor response is modified by the 'interfering quantities'[2] which can be partly eliminated by the differential method. Improvement of the electronic sensor is continuing through a CEC contract (including comparison with SSNTD), theoretically as well as experimentally. The development of a code simulating the neutron interactions with the dosemeter allows us to hope for a better discrimination of the various contributions to the sensor response.

SYSTEM EXPERIMENTATION CONDITIONS

The sensor is based on a neutron charged particle converter (^{10}B implanted polyethylene) associated with an electronic detector (Si implanted and passivated diode)[1]. Recoil particles (protons from fast neutrons, α and Li particles from the reaction of thermal neutrons with ^{10}B) are detected in the diode sensitive layer; results with thermal neutrons will not be presented here. Acquisition and data processing are achieved in real-time[3]. The system uses two similar diodes and electronics, one of the diodes being covered with the converter, in order to apply the differential method for the elimination of interfering quantities (electronic background, γ pulses, direct neutron interactions with Si, etc).

This symmetry has been optimised firstly by selecting detectors with similar characteristics (on which the resolution depends) and secondly by an accurate electronic adjustment so that the energy calibration of the whole system (sensor and multichannel analyser) is the same for both paths. This calibration has been made using a pulser adjusted with a ^{233}U source, in order to verify the linearity of the two paths. The calibration factor obtained is 9.5 keV per channel.

The converter used is 35 μm thick. Theoretically[4], this thickness value leads to the intrinsic converter response (i.e. the number of protons coming out of the polyethylene per unit of dose equivalent and converter surface) which is almost independent of neutron energy, from 200 keV to about 5 MeV, and angle of incidence (isotropic incidence neutrons beam). The angular dependence of the whole sensor will be tested in the next step but previous calculations and experiments show that it will not be so pronounced as for CR-39 based dosemeters.

The irradiations were performed with a 4 MeV Van de Graaff accelerator (CEN Bruyères-le-Châtel, France) with 2.5 MeV, 1.2 MeV, 570 keV neutron beams (dose equivalent rate about 10 mSv.h^{-1}) and 144 keV neutrons (1 mSv.h^{-1}).

For all these experiments the target to detector surface distance is 75 cm and the neutron incidence angle is 0° (in these conditions the scattered neutron component is negligible). For these beams, the ratio of the neutron dose equivalent to the γ dose equivalent is about 10^3. Under these conditions the differential method is statistically valid, i.e. the γ pulses can be discriminated from the neutron signal.

RESULTS AND ANALYSIS

All the spectra are given in pulses per channel (9.5 keV). The dose equivalent is 1 mSv and the sensor surface 1.5 cm². Figure 1 illustrates the differential method for 1.2 MeV neutrons (normal incidence). Figure 2 gives the differential spectra for 2.5 MeV, 570 keV and 144 keV energies under the same experimental conditions.

The spectral shapes are in accordance with expectations, in particular for high neutron energies. For low energies, the signal is more perturbed. For 144 keV neutron energy, the low amplitude signal and electronic adjustment (threshold symmetry problem) mean that the differential method is being used at the limit of its validity. Computations will allow the various contributions to the response to be explained.

Background and γ pulses are registered in the low energy region. As the threshold adjustment cannot be exactly the same on the two sensor channels (this partly explains that the signal without converter is larger than that with, Figure 1), the statistical fluctuations mean that the differential method is unable to separate completely the useful pulses from the interfering quantities. An adjustable energy threshold is used in integrating the spectra; under these conditions the neutron dosemeter response is valid.

Spectra obtained at the different neutron energies have been integrated with a threshold varying from channel 10 to 25. The integral of the differential spectrum provides the dosemeter neutron response in pulses.mSv^{-1}.cm^{-2} (i.e. per cm²

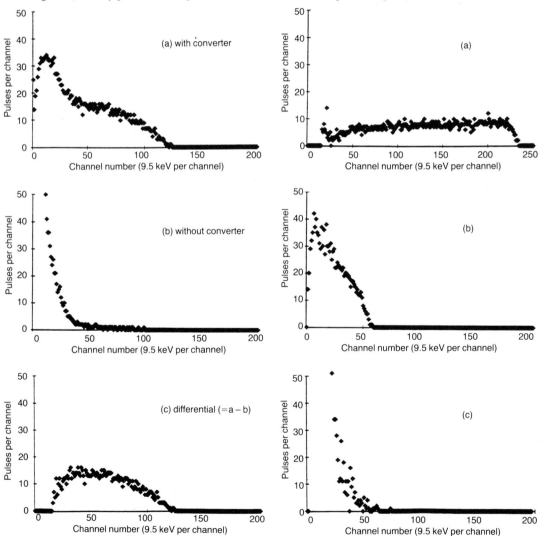

Figure 1. Sensor response in a 1.2 MeV neutron beam (1 mSv).

Figure 2. Differential method results for (a) 2.5 MeV neutrons, (b) 570 keV neutrons, (c) 144 keV neutrons.

of detector surface). Figure 3 gives the response as a function of energy for various thresholds.

We notice that a threshold between the 10th and the 15th channel, reduces the obvious hypersensitivity at low energies; then the response is about 1000 pulses.mSv^{-1}.cm^{-2}. This response is twice that obtained with CR-39 solid state detectors coupled with the same converter[4].

The preliminary experiments made with an 80 μm depleted layer lead to a response equal to 870 pulses.mSv^{-1}.cm^{-2} for 3.3 MeV neutrons (normal incidence) with a 400 keV low energy threshold. The use of a 30 μm depleted layer does not lead to a loss in proton registration, since they are likewise detected if they cross the depleted layer of silicon. It reduces the γ contribution[5] allowing the energy threshold to be lower (about 150 keV) and hence enabling the differential method to be applied down to lower neutron energies.

THEORETICAL ASPECT

From the above results, it is obvious that the differential method cannot totally eliminate the interfering quantities specially for low energies. One way to investigate the various contributions to the final spectra is to calculate them separately.

Computations have already been made of the SSNTD response to neutrons, including the converter contribution[4,6], so the number of protons emitted from the converter, as well as their energy and angle spectra for all kinds of incident neutron beams can be calculated. The optimum converter intrinsic response calculated as 1 proton.μSv^{-1}.cm^{-2} is still valid for our electronic system. This previous work has been used as a starting point for a theoretical study on the neutron interactions with our electronic sensor using a 'home made' Monte Carlo simulation. Considering neutron interactions with the whole sensor, it will enable the various contributions to the pulse spectra (converter, silicon diode, case, etc.) to be determined and with this information, the response can be optimised. γ studies are being made separately[5].

The bare sensor has been modelled (converter + detector). In a first step, only monoenergetic fast neutron fields of normal incidence were considered. The side wall effects were neglected. All possible reactions[7] have been taken into account but only particles from elastic scattering and nuclear reactions have been followed. The program provides information on the type of event producing each pulse. For the results presented here, 25 keV energy bins have been used.

The energy deposited by each charged particle in the depleted zone is calculated to determine the energy spectrum (channel spectrum). Computed spectra are given on Figure 4 for 1.2 and 0.57 MeV neutrons. The figure shows the pulse spectrum of protons from the converter and the total spectrum, including the elastic scattering of silicon atoms, compared with the experimental spectrum.

For these energies, notice that interactions with silicon give pulses only in very low channels (< 100 keV for 1.2 MeV neutrons; < 50 keV for 570 keV neutrons). The threshold which is set in the experiments eliminates most of these pulses. It can also be concluded that the computed spectral

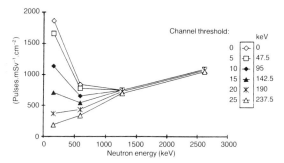

Figure 3. Detector response as a function of energy for various energy thresholds.

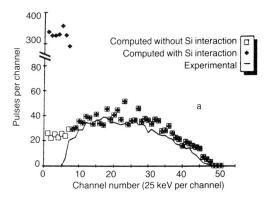

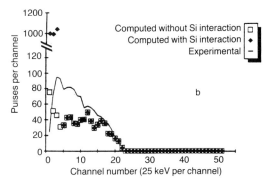

Figure 4. Computed spectra (1 mSv) for incident neutrons of (a) 1.2 MeV, (b) 570 keV.

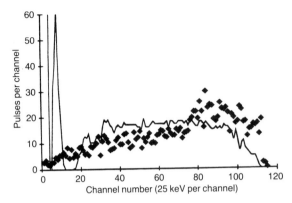

Figure 5. Comparison of computed (♦) and experimental (——) spectra for 3.3 MeV neutrons (1 mSv, normal incidence).

shapes are in accordance with experimental spectra for the same energies.

For neutron energies higher than 1.5 MeV, an accumulation of pulses in the high energy part of the spectrum were observed, due to the protons crossing the depleted zone. The peak observed corresponds to the maximum energy that a proton at normal incidence can deposit in 30 μm silicon.

The influence of the depleted layer is confirmed by the calculations. An 80 μm depleted layer spectrum, for 3.3 MeV neutrons (which does not consider direct interactions on silicon) is compared in Figure 5 with an experimental result. The spectral shapes are in good agreement. The peak observed around 2.25 MeV is the maximum energy deposited by a proton at normal incidence in an 80 μm layer of silicon.

Table 1 gives a summary of the responses (pulses.mSv^{-1}.cm^{-2}) obtained by integration of the computed and experimental spectra for various energies. Experimental and computed responses are in reasonable agreement especially for high energies where the differential method can be more strictly applied.

CONCLUSION

This work provides experimental verification of the theoretical predictions of the dosemeter response. The energy dependence shows little variation with the neutron energy, with an average value which is about one pulse per microsievert per cm^2. Computed results are in good agreement with experimental ones.

Further developments of the code will lead to an improvement of the sensor and probably to modifications of its structure in order to optimise the response. Results concerning the γ response will also be taken into account. The energy domain is now being extended to thermal neutrons thanks to a boron implanted converter and the albedo contribution will be studied.

ACKNOWLEDGEMENTS

Calculations presented in this paper are included in a broader experimental and theoretical study supported by CEC through BI 70020 C contract (Nov. 90).

Table 1. Response for various energies: 0.57, 1.2, 2.5 MeV (30 μm depleted zone, threshold at 100 keV) and 3.3 MeV (80 μm depleted zone, threshold at 400 keV).

Neutron energy (MeV)	0.57	1.2	2.5	3.3
Computations (number of pulses)	410	936	1063	975
Experiments (number of pulses) (differential spectrum)	648	748	1086	920

REFERENCES

1. Barelaud, B. *Conception et Réalisation d'un Capteur pour les Neutrons thermiques et rapides*. Thèse de 3ème cycle, Université de Limoges (1989).
2. Barelaud, B., Decossas, J. L., Makovicka, L. and Vareille, J. C. *Capteur électronique pour la Dosimétrie des Neutrons*. Radioprotection 26(2) 307-328 (1991).
3. Barelaud, B., Paul, D., Dubarry, B., Decossas, J. L., Makovicka, L. and Vareille, J. C. *Principles of an Electronic Neutron Dosemeter using a PIPS Detector*. Radiat. Prot. Dosim. 44(1-4) 363-366 (1992) (This issue).
4. Makovicka, L. *Contribution à la Dosimétrie Neutron Gamma. Etude d'un Ensemble Radiateur – Détecteur Type CR39*. Thèse n° 17-87. Université de Limoges (1987).
5. Paul, D., Barelaud, B., Dubarry, B., Decossas, J. L., Makovicka, L. and Vareille, J. C. *Gamma Interference on an Electronic Neutron Dosemeter Response in a Neutron Field*. Radiat. Prot. Dosim. 44(1-4) 371-374 (1992) (This issue).
6. Sadaka, S., Makovicka, L., Vareille, J. C., Decossas, J. L. and Teyssier, J. L. *Study of a Polyethylène CR-39 Fast Neutron Dosemeter. Part II: Dosimetry Efficiency of the Device*. Radiat. Prot. Dosim. 16(4), 281-287 (1987).
7. *Evaluated Nuclear Data File ENDF-6*. (National Nuclear Data Center – Brookhaven National Laboratory, Upton, New York).

GAMMA INTERFERENCE ON AN ELECTRONIC DOSEMETER RESPONSE IN A NEUTRON FIELD

D. Paul, B. Barelaud, B. Dubarry, L. Makovicka, J. C. Vareille and J. L. Decossas
Laboratoire d'Electronique des Polymères sous Faisceaux Ioniques
Faculté des Sciences, 123, avenue Albert Thomas
87060 Limoges Cedex, France

Abstract — It is a well-known phenomenon that neutron fields are generally contaminated by γ rays whatever the neutron energy. The neutron electronic dosemeter has a significant response to gamma rays, which is overcome by taking the differential response of two detectors, only one of which is sensitive to recoil protons, α particles, and ^{7}Li ions from neutron interactions in a polyethylene converter. The response ratio S_n/S_γ of the neutron detector is about 3×10^{-3}. Results in mixed fields, γ beams and from Monte Carlo simulation lead to a better knowledge of the sensor. The influences of the depleted layer, the γ energy, and the metallic components in dosemeter response are studied.

INTRODUCTION

Semiconductor devices are now commonly used in some fields of radiation dosimetry but few real-time systems for mixed fields (n,γ) have been studied. Eisen et al[1] and Matsumoto[2] have proposed analysing pulses for simultaneous dosimetry of neutrons and γ rays. The electronic dosemeter, described in previous papers[3–5], is designed for the measurement of neutrons, but has a significant inherent response to gamma rays. Hence, experiments have been done in France at Bruyères-le-Châtel for mixed fields, at Fontenay-aux-Roses (LCIE) and Lamanon (Merlin Gerin Provence) for γ beams. They were studied within the framework of a collaboration with Merlin Gerin Provence.

CHARACTERISATION OF MIXED FIELDS

Little information is available in the literature about the characteristics of realistic mixed fields, especially on the energy of the γ component. Table 1 gives a classification of the main γ rays associated with neutrons. Concerning the response of the dosemeter, two components must be distinguished: one is the external component (parts I, II, III of the table) from the target or source, the other is the internal one which is generated inside the dosemeter mainly by neutron interactions (part IV of the table) with the various dosemeter parts. It can be seen that the γ energy range is very wide.

EXPERIMENTS IN MIXED FIELDS

The description of these experiments at Bruyères-le-Châtel accelerator ($^3_1$T (p,n) $^3_2$He, $^7_3$Li (p,n)$^7_4$Be) is given in Dubarry's paper[11]. The γ component has been measured using a GM counter. The silicon detectors had been previously calibrated with ^{60}Co and ^{137}Cs sources so that the number of pulses due to γ rays could be evaluated (average response for ^{60}Co and ^{137}Cs: 321 pulses.cm^{-2}.μSv^{-1}). Results are given in Table 2 with a diode depleted layer of 30 μm and a converter thickness of 35 μm.

The general equation describing the dosemeter response in a mixed field (n,γ) can be simply written:

$$R_{n,\gamma} = R_\gamma + R_n = S_\gamma H_\gamma + S_n H_n = S_\gamma \frac{H_\gamma}{H_n} H_n + S_n H_n$$

$$= (S_\gamma' + S_n) H_n$$

where $R_{n,\gamma}$ is the total reading, S_γ and S_n the responses (pulses.cm^{-2}.μSv^{-1}) for γ and neutrons, H_γ the γ dose equivalent (=D_γ) and H_n the neutron dose equivalent.

Table 2 clearly shows a decrease of the ratio S_γ'/S_n as the neutron energy increases. From 2.17 (E_n = 159 keV) the ratio decreases to 0.34 (E_n = 2624 keV), showing that the γ response is weak for high energy neutrons but that it disturbs the neutron dosemeter response at low energies.

RESPONSE IN γ FIELDS

The pair of diodes is the same as before (30 μm of depleted layer, 15 V bias). Two sources have

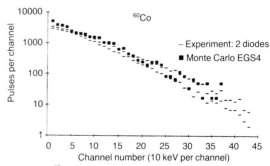

Figure 1. ^{60}Co experimental and computed spectra with a depleted layer of 30 μm (dose equivalent: 100 μSv).

been used; ^{60}Co, activity 6.956 GBq and ^{137}Cs, activity 3.108 GBq.

The cobalt spectrum is presented in Figure 1 with a dose equivalent of 100 µSv. A software threshold has been introduced at channel 15 for the integration of spectra. This leads to about the same response for ^{137}Cs and ^{60}Co: S_γ = 321 pulses.cm^{-2}.µSv^{-1}. The neutron response previously measured being about 1 pulse.cm^{-2}.µSv^{-1}, the ratio S_n/S_γ is about 3×10^{-3}.

CALCULATIONS AND DISCUSSION

Differential measurements for neutron dosimetry have been proposed and studied[3]. The validity and accuracy of the method depends on the ratio H_n/H_γ of the mixed field to be measured as seen in the previous sections of this paper. As the γ contribution to the pulse spectrum lies in the low energy region, a threshold is necessary in the integration. This leads to a limitation of the dosimetry towards the low neutron energies. Hence, the low energy threshold for the measurement of neutrons is dependent upon the H_n/H_γ ratio.

Monte Carlo calculations have been made to simulate the response of this dosemeter in γ fields in order to minimise the γ contribution. EGS4 computer code was used[12]. A computed spectrum for a diode of 30 µm depleted layer, irradiated by

Table 1. Gamma rays associated with neutrons (mixed fields).

	Mixed field	Reaction threshold		γ . Energy	Ref.
	Accelerators	T (p,n)^{3}He	0	E_γ = ?, 0.02 ≤ D_γ/D_n (tissue) ≤ 0.03 depend upon	6
		D (d,n)^{3}He	0	E_γ = ?, 0.16 ≤ D_γ/D_n (tissue) ≤ 1.5 E_d	6
		^{7}Li(p,n)^{7}Be	1.88 MeV	478 keV	6
		^{7}Li(p,γ)^{8}Be (associated)	0.44 MeV	16 & 19 MeV	6
I		Secondary reactions			7
		Fe (n,γ)		850 keV, 1.25 MeV, 1.8 MeV, 2.1 MeV	7
		C(p,γ) (target)		4.4 MeV	7
		Neutronic activation of tungsten		6–7 MeV	8
		Cyclotron		E_γ = ?, 5% ≤ D_γ/D_n ≤ 12% depending upon surroundings	6
	^{252}Cf source			D_γ/D_n = 1/18	9
	(γ,n) source	^{9}Be+γ→^{8}Be+n − 1.666 MeV		< 3 MeV	6
II		D + γ → ^{1}H+n − 2.226 MeV		< 3 MeV	6
	(Am-Be) source	Be (α,n) and excited state of the final nucleus ^{12}C		60 keV, 4.45 MeV, H_γ/H_n ≈ 1	6
	Reactors	Fuel element γ rays		0.7 MeV	8
		^{16}O activation (H$_2$O or CO$_2$)		6.13 MeV (68%), 7.12 MeV (5%)	10
	Nuclear weapons	^{24}Na activation		2.76 MeV	6
		^{56}Mn activation		0.845 MeV	6
III		Cobalt bomb		1.25 MeV	6
	Fission product γ rays	^{137}Cs		0.66 MeV	8
	n interactions inside the detector or the converter	neutron capture (thermal) ^{1}H (n,γ) D		2.2 MeV	6
IV	(CH$_2$)$_n$	Neutronic activation of aluminium, ^{27}Al (n,α)^{24}Na		1.369 MeV	8

Table 2. Relative gamma ray to neutron dose equivalent ratio for monoenergetic neutron beam and resultant response of the semiconductor detector (depleted layer: 30 µm).

Neutron energy (keV)	Ratio of gamma ray to neutron dose equivalent H_γ/H_n (%)	Detector	
		S_n (pulses.cm^{-2}.µSv^{-1})	S'_γ/S_n
73	0.88	1.43	1.95
159	0.48	0.71	2.17
593	0.1	0.54	0.58
1270	0.06	0.75	0.26
2624	0.11	1.09	0.34

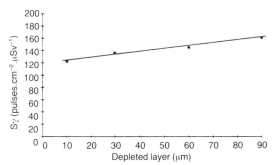

Figure 2. Computed response plotted against depth of the depleted layer for ^{60}Co.

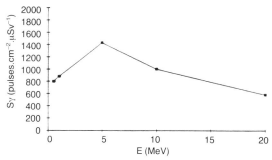

Figure 3. Computed response plotted against γ energy with a depleted layer of 30 μm.

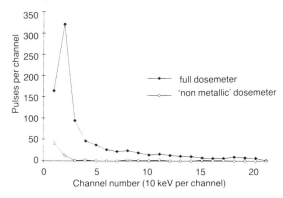

Figure 4. Computed modification of ^{60}Co spectrum by metallic components in the sensor (depleted layer: 30 μm, linear scale).

1.25 MeV γ rays is given in Figure 1 together with the experimental spectrum.

The general shapes of these spectra are in good agreement, as are their integrals.

The computed γ response is plotted in Figures 2 and 3. As one can see, it increases with the depleted layer (× 1.2 from 30 to 90 μm), the experimental value being × 1.4. We can observe in Figure 3 that S_γ does not vary strongly with the γ energy. The pulse spectrum is strongly modified by the metal components of the sensor (Figure 4), the aluminium case and front metallic layer. If they are suppressed from the model used in EGS4 the γ sensitivity is divided by 13. In the future this will lead us to modify the structure of the sensor, i.e. to reduce the thickness of the front metallic part, or to use an organic case.

CONCLUSION

The first studies on (n,γ) discrimination, which is the aim of this work, have clearly shown that several parameters of the dosemeter have to be optimised: electronic parameters (depth of the depleted layer, threshold voltage), constitution of the sensor (structure, materials used for its components), in order to eliminate the greatest number of γ pulses using the differential method. Further experiments and calibrations are now in progress.

ACKNOWLEDGEMENTS

We are grateful to Dr E. Grimaud (CEB/DPN, Arcueil, France) for his kind assistance in the installation and application of the EGS4 code. The results presented here are part of a more complete and theoretical work supported by CEC through the contract BI700200C.

REFERENCES

1. Eisen, Y., Engler, G., Ovadia, E., Shamai, Y., Baum, Z. and Levi, Y. *A Small Size Neutron and Gamma Dosimeter with a Simple Silicon Surface Barrier Detector*. In: DOE Workshop on Personal Neutron Dosimetry, Acapulco, Mexico. PNL-SA-12352, pp. 157-175 (1983).
2. Matsumoto, T. *PIN Diode for Real Time Dosimetry in a Mixed Field of Neutrons and Gamma Rays*. Radiat. Prot. Dosim. **35**(3), 193-197 (1991).
3. Barelaud, B. *Conception et Réalisation d'un Capteur pour les Neutrons Thermiques et Rapides*. Thèse no 7-1989. Université de Limoges (1989).
4. Barelaud, B., Decossas, J. L., Makovicka, L. and Vareille, J. C. *Capteur Électronique pour la Dosimétrie des Neutrons*. Radioprotection **26**(2) 307-328 (1991).

5. Barelaud, B., Paul, D., Dubarry, B., Makovicka, L., Decossas, J. L. and Vareille, J. C. *Principles of an Electronic Neutron Dosemeter using a PIPS Detector.* Radiat. Prot. Dosim. **44**(1-4) 363-366 (1992) (This issue).
6. Attix, F. H. and Tochilin, E. *Radiation Dosimetry* 2nd edn, Vol. 3 (1969).
7. Moyers, M. F. and Horton, J. L. *Determination of the Neutron and Photon Spectra of a Clinical Fast Neutron Beam.* Med. Phys. **17**(4), 607-614 (1990).
8. Messenger, G. C. and Ash, M. S. *The Effects of Radiation on Electronic Systems.* (New York: Van Nostrand Reinhold) (1986).
9. Colvett, R. D., Rossi, H. H. and Krishnaswamy, V. *Dose Distributions around a Californium-252 Needle.* Phys. Med. Biol. **17**, 356 (1972).
10. Bermann, F. and Portal, G. *Etalonnage des Détecteurs de Radioprotection avec des Gammas d'Énergie Supérieure à 1 MeV: Utilisation de Faisceaux de Gammas de Capture.* Radioprotection **26**(3), 493-513 (1991).
11. Dubarry, B., Barelaud, B., Paul, D., Makovicka, L., Decossas, J. L. and Vareille, J. C. *Electronic Sensor Response in Neutron Beams.* Radiat. Prot. Dosim. **44**(1-4) 367-370 (1992) (This issue).
12. Nelson, W. R., Hirayama, H. and Rogers, D. W. O. *The EGS4 Code System.* Stanford Linear Accelerator. Report SLAC-265 (Stanford, CA 94305) (1985).

DETECTION OF NEUTRON-INDUCED HEAVY CHARGED PARTICLE TRACKS IN RPL GLASSES

B. Lommler, E. Pitt and A. Scharmann
I. Physikalisches Institut, Heinrich-Buff-Ring 16
WD-6300 Giessen, Germany

Abstract — Tracks of heavy charged particles (HCP, with Z,A≥1) in solids are characterised by a strongly localised dose distribution around the path of the HCP. A method for the detection of these HCP-induced high-dose 'spots' in radiophotoluminescent (RPL) phosphate glasses is proposed. The relative HCP/γ luminescence efficiency of a single HCP track is calculated within the scope of a simple model, based on an analytical dose distribution formula and an adjusted γ response function. The relative luminescence output from a single HCP track depends on the ratio of the 'track volume' to the glass volume illuminated by the excitation light. Thus experimental detection of HCP tracks requires a small 'excitation volume' which may be achieved by using a focused laser beam and flat RPL glasses. In neutron dosimetry heavy charged particles mainly arise as recoils from elastic scattering by fast neutrons. The application of the proposed method to fast neutron detection is discussed.

INTRODUCTION

The passage of a heavy charged particle (HCP) through matter is accompanied by an intense production of secondary electrons. As a result a dose distribution limited to a microscopic area around the path of the HCP is to be expected. These high dose 'spots' are detectable in different ways depending on the type of the applied solid state detector. The etched track technique utilises the microscopic damage caused by doses exceeding a detector-dependent threshold to reveal tracks by means of an etching process in caustic solutions. Detectors which are based on thermal or optical stimulation of traps or centres occupied by radiation-released electrons may be used for track detection if stimulation is spatially limited to small volumes. In this way discrimination between the HCP signal and background from low ionising radiation (e.g. γ) and from the detector itself is achieved. Recently this has been shown for proton tracks in thin TL layers which were read out by scanning with a focused CO_2 laser beam[1].

In this paper a similar method utilising radio-photoluminescent glasses as track detectors is proposed. Commercially available flat RPL glasses assigned for use in automatic readout systems[2] seem to be suitable for a scanning readout technique using a focused UV (e.g. argon ion) laser beam. In order to estimate the signal yield from a single HCP track an attempt was made to develop a theoretical approach based on an analytical dose distribution formula and an adjusted detector response function. Computation method and results are presented in subsequent sections. A possible application of the proposed method as a tool for heavy recoil detection in fast neutron dosimetry is discussed.

RPL GLASS MATERIAL

Calculations were performed for flat RPL glasses FD-P10-7 from Toshiba, specified in Table 1.

METHOD OF CALCULATION

Suppose that a small volume V_L of a flat RPL glass of thickness d is illuminated by a focused laser beam perpendicular to the broad sides of the glass. The detector response function s(D) is the yield of luminescence light per unit intensity of excitation light and per unit glass volume as a function of dose. Without a HCP track in V_L the integral light emission from V_L is given by:

$$S_\gamma = \int_{V_L} s(D_\gamma) \, I \, dV \quad (1)$$

where

D_γ = sum of applied γ dose and detector background,

I = intensity of excitation light.

With a HCP track in V_L it is:

$$S_{HCP} = \int_{V_L} s[D(r)+D_\gamma] \, I \, dV \quad (2)$$

Table 1. Elemental composition, by weight fraction, and material constants of Toshiba FD-P10-7 glass[3,4].

Element	O	Na	Al	P	Ag
Weight fraction	0.512	0.110	0.061	0.3150	0.0017
Dimensions	$10 \pm 0.05 \times 10 \pm 0.05 \times 1.5 \pm 0.05$ mm^3				
Density	2580.8 ± 6.5 kg.m^{-3}				
Electron number density:	0.7638×10^{30} m^{-3}				

where $D(r)$ = radial dose distribution around the HCP's path. The relative HCP/γ luminescence efficiency is defined as

$$\varepsilon = (S_{HCP} - S_\gamma)/S_\gamma \quad (3)$$

Dose distribution function

The formula given by Zhang et al[5] for the radial dose distribution around the path of a heavy ion of charge Z in water is used. With adjusted values for mass density and electron number density and accommodated to SI units it is:

$$D(r) = \frac{N\,e^4\,(Z^*)^2\,[1-(r+\theta)/(T+\theta)]^{1/\alpha}}{(4\pi\,\varepsilon_0\,c\,\beta)^2\,\alpha\,\rho\,m_e\,r\,(r+\theta)} \quad (4)$$

where: ρ = density of glass (Table 1),
  N = electron number density in glass (Table 1)
  $e = 1.6022 \times 10^{-19}$ C, $m_e = 9.1095 \times 10^{-31}$ kg,
  $\varepsilon_0 = 8.8542 \times 10^{-12}$ A.s.V.m^{-1},
  $c = 2.9979 \times 10^8$ m.s^{-1},
  β = v/c with v = velocity of particle,
  Z^* = effective charge:
    $Z^* = Z[1-\exp(-125\,\beta\,Z^{-2/3})]$, $Z<18$,
  $\theta = R(I)$, $T = R(W)$, $I = 0.01$ keV,
  $W = 2\,m\,c^2\,\beta^2/(1-\beta^2)$

with electron range-energy relation $R = k\,w^\alpha/\rho$

where $k = 6 \times 10^{-5}$ kg.m^{-2}.keV$^{-\alpha}$,

and $\alpha = 1.079$ for w<1 keV, $\alpha = 1.667$ for w>1 keV.

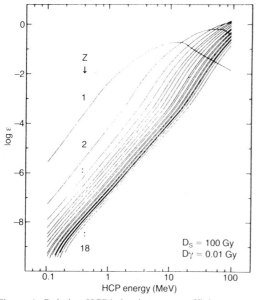

Figure 1. Relative HCP/γ luminescence efficiency as a function of HCP energy for HCP charge numbers from 1 to 18.

The formula matches experimental data within approximately 15% in the energy range 0.5–377 MeV per amu for $1 \leq Z \leq 53$ [5,6].

Detector response function

According to δ ray theory of track structure the response of a detector to heavy ions is determined by the radial distribution of dose deposited by secondary electrons and the γ response of this detector[7]. Measurements of the γ response of RPL glasses show a linear response up to about 10^2 Gy, a subsequent increasing non-linearity and a decrease of response beyond 10^3 Gy[8,9] which is attributed to increasing absorption of the excitation light by radiation-induced centres[8]. The degree of absorption is determined by the size of the RPL glasses[9]. By extrapolation it can easily be shown that absorption is negligible on the microscopic scale of HCP track volumes. For this reason a response function of the form of a saturating exponential is chosen[7,10]:

$$s(D) = s_\infty(1 - \exp(-D/D_s)) \quad (5)$$

where D_s is a characteristic saturation dose and s_∞ is the saturation response. In the present calculation D_s is treated as an adjustable parameter.

Laser beam intensity profile

A TEM$_{00}$ mode Gaussian beam profile is assumed[11]:

$$I(r,z) = 2/\pi\,P\,(r_z)^{-2} \exp(-2\,(r/r_z)^2) \quad (6)$$

where

$$r_z = r_0\,[1+(z/z_0)^2]^{1/2}, \quad z_0 = \pi(r_0)^2/\lambda$$

with P = laser beam power, λ = wavelength, $2r_0$ = focal ($1/e^2$) diameter.

With Equations 1–6 a straightforward calculation yields (assuming HCP track of length R from z=0 to z=R and focal plane at z=R/2):

$$\varepsilon = 2 \cdot \left(\frac{D_s}{D_\gamma}\right)\left(\frac{V_T^*}{V_L^*}\right) \quad (7)$$

where

$$V_L^* = \pi(r_0)^2\,d \quad (7a)$$

V_T^* = 'effective track volume':

$$V_T^* = \int_0^R (r_0/r_{z-R/2})^2\,\Sigma(z)\,dz \quad (7b)$$

with the 'cross section' Σ:

$$\Sigma(z) = 2\pi \int_0^{T(z)} [1-\exp(-D(r)/D_s)]$$
$$\exp[-2(r/r_{z-R/2})^2] \, r \, dr \tag{7c}$$

V_L^* in Equation 7a is smaller than the actual illuminated volume because of the quasi-infinite Gaussian beam profile and the beam broadening outside the focal plane. Range and stopping power data required for z–integration in Equation 7b were calculated by means of a published computer program[12]. The relative HCP/γ efficiency (Equation 7) has been calculated numerically. All calculations were performed for D_g = 10 mGy which includes the glass pre-dose for conventional (non–pulsed) UV excitation[13]. For the curves in Figures 1 and 3 a saturation dose D_S of 100 Gy is valid, which is a minimum value given by the measured γ response. In principle D_S may be derived from measured HCP responses. There are some old data for high-Z RPL glass[7], but to the authors' knowledge HCP responses for Toshiba FD-P10-7 glass do not exist. Calculations for determination of D_S from measured fast neutron sensitivities[4] are currently under way. The following laser beam parameters were chosen: wavelength λ= 350 nm and focal diameter $2r_0$ = 10 μm.

RESULTS

Figure 1 shows the relative HCP/γ luminescence efficiency as a function of HCP energy for various HCP charge numbers. Decrease of a curve indicates penetration of the RPL glass by the particle as in the case of Z=1 (p) and Z=2 (α).

Figure 2 shows the effect of varying the saturation dose D_S. The dose deposited in cylindrical shells around the path of the HCP decreases with decreasing particle LET, i.e. increasing energy of the particle. So the saturation dose is not reached in a growing number of shells resulting in convergence of the curves towards higher energies.

In neutron dosimetry heavy charged particles mainly arise as recoils from fast neutron scattering. In a central elastic collision a target nucleus of atomic weight A receives the kinetic energy E_R from a neutron of energy E_n, where E_R is given by:

$$E_R = E_n \frac{4A}{(A+1)^2} \tag{8}$$

With E_R as HCP energy the relative HCP/γ luminescence efficiency has been calculated for recoil nuclei present in Toshiba FD-P10-7 glass, O, Na, Al and P. In view of the widely used radiator–detector combinations for fast neutron sensitivity enhancement protons are also included. For a first estimation the relative HCP/γ luminescence efficiency of only the most energetic recoil protons (i.e. protons from central elastic collisions with monoenergetic neutrons (Equation 8) produced near the radiator–glass boundary) is calculated.

Results are shown in Figure 3. The rather flat

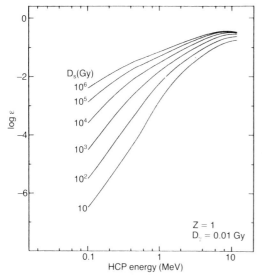

Figure 2. Relative HCP/γ luminescence efficiency as a function of HCP energy for various magnitudes of saturation dose.

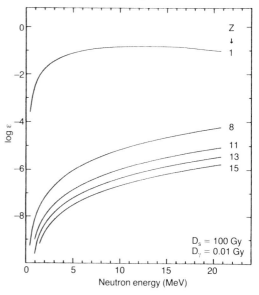

Figure 3. Relative HCP/γ luminescence efficiency as a function of HCP energy for central elastic collision recoil nuclei present in Toshiba FD-P10-7 glass and (most energetic) protons from a thin radiator foil covering the glass.

proton curve exceeds $\varepsilon=0.1$ for neutron energies > 5 MeV. A significant variation of the luminescence signal during a scanning readout of the RPL glass indicates the presence of a track. A value of ε of 0.1 or greater should be easily detectable. For the other recoils considered ε is to be expected several magnitudes below the proton value. Their detection requires sophisticated measuring equipment. An experimental set-up with an xy scanning stage and photon counting unit is currently under construction in the authors' institute.

CONCLUSION

The relative HCP/γ luminescence efficiency of heavy charged particle tracks can be estimated within the scope of simple theoretical considerations. As a result of calculation proton tracks turned out to be most easily detectable. Available flat RPL glasses are suitable for scanning readout technique using a focused UV laser beam. Thus a combination of a hydrogenous radiator foil and a flat RPL glass may find application in personnel dosimetry as a combined γ/fast neutron dosemeter.

REFERENCES

1. Bräunlich, P. and Tetzlaff, W. *Fast Neutron Dosimetry by Thermoluminescence Detection of Knock-on Protons.* Radiat. Prot. Dosim. **33**(1/4), 327-330 (1990).
2. Burgkhardt, B., Piesch, E., Vilgis, M., Ishidoya, T. and Ikegami, T. *Modern Automatic Readout Systems for Phosphate Glass Dosemeters using UV Laser Excitation.* Radiat. Prot. Dosim. **34**(1/4), 369-372 (1990).
3. Toshiba Glass Co. Ltd, Mori Building, Shinbashi Annex 2F., 35-10 Shinbashi 5-Chrome Minato-Ku, Tokyo 10 Japan.
4. Croft, S. and Weaver, D. R. *The Fast Neutron Sensitivities of SEI and Toshiba FD-P10-7 Radiophotoluminescent Glass Dosemeters.* Radiat. Prot. Dosim. **27**(4), 251-260 (1989).
5. Zhang, C., Dunn, D. E. and Katz, R. *Radial Distribution of Dose and Cross Sections for the Inactivation of D Enzymes and Viruses.* Radiat. Prot. Dosim. **13**(1/4), 215-218 (1985).
6. Waligorski, M. P. R., Hamm, R. N. and Katz, R. *The Radial Distribution of Dose Around the Path of a Heavy Ion in Liquid Water.* Nucl. Tracks Radiat. Meas. **11**(6), 309-319 (1986).
7. Katz, R., Sharma, S. C. and Homayoonfar, M. *Detection of Energetic Heavy Ions.* Nucl. Instrum. Methods 10 13-32 (1972).
8. Becker, K. *High γ-Dose Response of Recent Silver-Activated Phosphate Glasses.* Health Phys. **11**, 523-5 (1965).
9. Freytag, E. *Measurement of High Doses with Glass Dosimeters.* Health Phys. **20**, 94-97 (1971).
10. Zimmerman, D. W. *Relative Thermoluminescence Effects of Alpha- and Beta-Radiation.* Radiat. Effects **1** 81-92 (1972).
11. Das, P. *Lasers and Optical Engineering.* (New York, Berlin, Heidelberg: Springer-Verlag) (1991).
12. Henke, R. P. and Benton, E. V. *Charged Particle Tracks in Polymers: No 5 A Computer Code for the Computation of Heavy Ion Range-Energy Relationships in any Stopping Material.* USNRDL-TR-67-1 Report, San Francisco (1968).
13. Croft, S., Weaver, D. R. and Heffer, P. J. H. *The Application of Radiophotoluminescent Glass to γ Dosimetry mixed n-γ Fields.* Radiat. Prot. Dosim. **17**, 67-70 (1986).

NEUTRON THERAPY: FROM RADIOBIOLOGICAL EXPECTATION TO CLINICAL REALITY

A. Wambersie
Radiation Therapy, Neutron- and Curietherapy Department
Université Catholique de Louvain, Cliniques Universitaires St-Luc
1200 Bruxelles, Belgium

INVITED PAPER

Abstract — The radiobiological data at present available indicate that high LET radiations could bring a benefit in the treatment of some types of tumours (typically slowly growing, well differentiated). Radiobiology also suggests some mechanisms through which this benefit could be achieved: hypoxic gain factor, kinetics gain factor, etc. However, the need for patient selection is clearly apparent. Among the high LET radiations, fast neutrons are the least expensive and the most widely used in therapy. Seventeen centres are today actively involved in neutron therapy, and more than 15,000 patients have been treated so far. Two main difficulties are encountered when reviewing the clinical results of fast neutron therapy. Firstly, as indicated above, patient selection, which depends on the characteristics of the tumour (and of the normal tissues at risk) is important. A lack of proper selection can obscure or worsen the results. Secondly, there is a need for a high physical selectivity with neutrons. Throughout the history of radiation therapy, physical selectivity has been proved to be essential for the outcome of treatment. It becomes even more important with high LET than with low LET radiations, because of a reduction of the differential effect between the different cell populations when increasing LET. Neutron therapy started in the seventies in rather sub-optimal (or even poor) technical conditions. The situation has been significantly improved with the introduction of high energy, hospital-based cyclotrons, with isocentric mounting and variable collimators. However, in many neutron therapy centres today the level of physical selectivity does not yet fully reach the same level as modern photon therapy. The interpretation of the clinical results is then biased and it is difficult to separate what is really due to the high LET characteristics of the beams from that related to the technical treatment conditions. Among the clinical indications for fast neutrons, the following are most commonly recognised: (1) inoperable or recurrent salivary gland tumours, (2) locally extended tumours of the paranasal sinuses, (3) some other tumours of the head and neck area especially with fixed adenopathies, (4) slowly growing, well differentiated soft tissue sarcomas, osteosarcomas and chondrosarcomas, (5) locally extended prostatic adenocarcinomas, (6) palliative treatment of melanomas. The indications for neutron therapy represent about 10–15% of all patients currently referred to the radiation therapy departments.

INTRODUCTION — SHORT HISTORICAL REVIEW

When neutron therapy was started in the Hammersmith Hospital in the seventies and a few years later in several other centres, it generated a great deal of interest and hope in the radiotherapeutic community[1–4].

The rationale for introducing fast neutrons was radiobiological and could be summarised as follows:

(1) The existence of hypoxic cells in all (or in most of) malignant tumours.

(2) A specific radioresistance of these hypoxic cells to X rays, expressed by the oxygen enhancement ratio (OER) which is about 3. The OER is the dose ratio necessary to obtain a given biological effect when irradiation is performed under hypoxic or oxygenated conditions.

(3) From these two statements it is apparent that the presence of a small percentage of hypoxic cells (3 times more radioresistant to X rays) makes the tumour definitively radioresistant to X rays.

(4) Neutrons reduce the OER to about 1.6. To the extent that the hypoxic cells are really the cause of tumour radioresistance, it is then possible to evaluate the therapeutic gain which is the ratio of the OERs for X rays and neutrons respectively. The therapeutic gain factor is thus $3 : 1.6 = 1.9$.

Two points should be stressed when discussing the rationale for introducing fast neutrons: the gain factor of about 1.9 is high and it applies to all (or most of the) tumours since all (or most of them) contain hypoxic cells. In the seventies, when neutron therapy was started, it was expected that the radiobiological therapeutic gain would be high enough to overcome the difficult technical conditions which were nevertheless recognised.

Clinical applications were started at the Hammersmith Hospital after extensive and careful radiobiological experiments. These were justified by the negative conclusions derived from the clinical experience of Stone[5]. Some of the reasons for the late complications reported by Stone could be explained: e.g. increase of RBE with decreasing dose, higher RBE for late complications compared to early reactions, etc. The radiobiological information progressively accumulated was of great help in designing the therapeutic protocols.

In this review, starting from the initial radiobiological expectation, we shall consider the evolution of the situation during the past twenty years from the point of view of radiobiology, clinical data and technological developments.

RADIOBIOLOGICAL DATA

The hypoxic cells

The existence of hypoxic cells was confirmed in all tumours investigated[6]. In contrast, the relevance of the hypoxic cells in relation to tumour radioresistance was questioned after the discovery by van Putten and Kallman[7] of the phenomenon of tumour reoxygenation. During the course of a fractionated treatment, some of the hypoxic cells move from the hypoxic to the oxygenated compartment. However, the kinetics of the tumour reoxygenation varies to a large extent from tumour to tumour[8]. As indicated in Figure 1, for rapidly growing, rapidly shrinking tumours, the proportion of hypoxic cells decreases quickly after irradiation and, in a fractionated treatment, they never reach a level where they could become a factor of radioresistance.

Reoxygenation does not take place or takes place too slowly in slowly growing and slowly shrinking sarcomas, making the tumour progressively radioresistant. This was the case of the osteosarcoma studied by van Putten, where the percentage of hypoxic cells remained high for several days.

Without entering into detail, it can be assumed today that hypoxic cells do play a major role in the radioresistance of some tumours. In contrast, they do not play any role in other tumours because of efficient reoxygenation, but they probably play 'some' role in other tumour types. This raises the problem of the identification of these different groups of tumours or patients. The importance of patient selection will be discussed again later in this review.

A reduction in the differences in radiosensitivity

When comparing the effects produced by neutrons and X rays, differences other than a reduction in OER are also observed.

The situation could be summarised as follows: with neutrons there is a general reduction in the differences in radiosensitivity between cell populations. For example, Figure 2 illustrates the

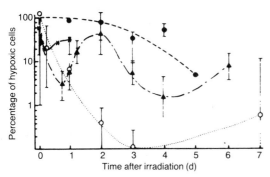

Figure 1. Tumour reoxygenation after irradiation: the percentage of hypoxic cells is plotted as a function of time (days) after irradiation. (o) For a mouse mammary carcinoma, reoxygenation is rapid and complete: 3 days after irradiation the percentage of hypoxic cells is lower than that seen before irradiation. (*) For a transplantable mouse sarcoma, reoxygenation is not complete by 24 h. (▲) For rat fibrosarcoma RIB5 there is at first a rapid reoxygenation: however, the proportion of hypoxic cells increases again at 2 days and then falls to about 5% at 3 and 4 days. (•) For a mouse osteogenic sarcoma reoxygenation is very slow. Comparison of the curves indicates large variations in the time course and extent of tumour reoxygenation from one tumour to another.
(After Thomlinson[8].)

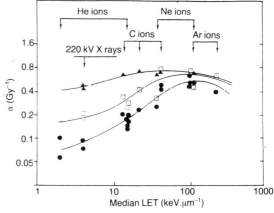

Figure 2. The differences in cell radiosensitivity related to the position in the mitotic cycle decreases with increasing LET. On the ordinate, the cell radiosensitivity is expressed by the parameter α (single-hit inaction coefficient). Synchronised populations of Chinese hamster cells are irradiated in (▲) mitosis, (□) G_1 phase and (•) stationary phase, with 220 kV X rays and various beams of charged particles. The α coefficients are plotted as a function of the median LET (in keV.μm^{-1}). (After Chapman[9].)

reduction in the difference of the radiosensitivity of the cells related to their position in the mitotic cycle[9]. Cell populations, synchronised *in vitro*, are irradiated in different phases of the mitotic cycle. The large differences which are observed with X rays (low LET radiation) are progressively but markedly reduced with increasing LET.

A reduction in the difference in intrinsic radiosensitivity between cell lines has also been observed[10], although other data suggest that the ranking of radiosensitivity of some cell lines could be altered when X rays are replaced by fast neutrons[11].

Finally, with increasing LET there is a reduction in the role of sub-lethal lesions. Differences in the capacity of accumulating and repairing sub-lethal lesions are then of less importance. In practice, this implies that the dose per fraction also becomes less critical.

It could thus be concluded that all cell populations in all conditions tend to respond in a more similar way to neutrons than to X rays. This can be logically related to the increase, by a factor of about 100, in the sizes of the individual energy deposits as can be derived from microdosimetric measurements (Figure 3)[12,13].

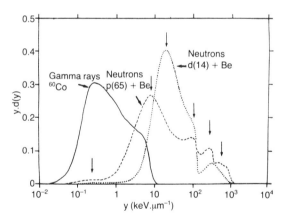

Figure 3. Comparison of energy depositions after irradiation with fast neutrons and γ rays. The curves indicate the distributions of individual energy deposition events in a simulated volume of tissue 2 μm in diameter. The parameter y (lineal energy) represents the energy deposited by a single charged particle traversing the sphere, divided by the mean cord length. The maximum with γ rays is at 0.3 keV.μm^{-1} and with d(14)+Be neutrons at 20 keV.μm^{-1}. The spectrum for p(65)+Be neutrons shows four peaks: the first is at 8 keV.μm^{-1} and corresponds to high energy protons, the second at 100 keV.μm^{-1} corresponds to low energy protons, the third at 300 keV.μm^{-1} is due to α particles and the last at 700 keV.μm^{-1} is due to recoil nuclei. (After Menzel *et al*[12] and Pihet *et al*[13].)

Practical conclusions for radiation therapy

Need for proper patient selection

A reduction in the differences in radiosensitivity related to the position of the cells in the mitotic cycle, cell line or repair capacity can be an advantage or a disadvantage depending on the characteristics of the tumours and of the normal tissues at risk. For example, neutrons should not be used for patients in whom, with X rays, a differential effect selectively protects the normal tissues. However, neutrons may be of advantage in inverse situations when tumour cells are more resistant to X rays than the normal cell populations.

This stresses the importance of patient selection: an incorrect choice of the radiation quality can worsen the clinical results. More generally, if a sub-group suitable for high LET radiation cannot be identified, and if the whole group is treated with neutrons, the advantage gained in the sub-group will be diluted or counterbalanced by the worse results obtained in the other subgroups which it would have been better to treat with photons. In practice this could lead to erroneous clinical conclusions and is discussed in more detail elsewhere in Tubiana *et al*[14].

The importance of physical selectivity

As a result of the reduction in the (radiobiological) differential effect, the therapeutic efficiency of the treatment will depend to a larger extent on physical selectivity (dose distribution). Thus, physical selectivity is at least as important with high LET as with low LET radiations. This is the second important conclusion which is derived from the radiobiological data.

REVIEW AND DISCUSSION OF THE CLINICAL DATA

The clinical data should be reviewed and interpreted bearing in mind the two main conclusions of the radiobiological analysis: the need for patient selection and the importance of physical selectivity for high LET radiations. The initial groups of patients were treated in far from optimal technical conditions. This has resulted in complications which in turn have influenced patient recruitment and have impaired the development of neutron therapy.

Salivary gland tumours

Locally extended inoperable salivary gland tumours are the first type of tumours for which the superiority of fast neutrons, compared to conventional low LET radiations, has been recognised.

Already, in 1979, Griffin *et al*[15] reported promising results from the University of

Table 1. Neutron therapy of salivary gland tumours. Loco-regional control rate relative to tumour histology (minimum one year follow-up).

Tumour histology	Number of patients	Loco-regional control (%)
Adenoid cystic	17	15
Mucoepidermoid	9	6
Malignant mixed	2	2
Undifferentiated	4	3
Overall	32	26 (81%)

Modified from Griffin et al (1988)[18].

Table 2. Review of the loco-regional control rates for malignant salivary gland tumours treated definitively with radiation therapy.

Fast neutrons

Authors	Number of patients*	Loco-regional control
Saroja et al (1987)	113	71 (63%)
Catterall and Errington (1987)	65	50 (77%)
Battermann and Mijnheer (1986)	32	21 (66%)
Griffin et al (1988)	32	26 (81%)
Duncan et al (1987)	22	12 (55%)
Tsunemoto et al (1989)	21	13 (62%)
Maor et al (1981)	9	6
Ornitz et al (1979)	8	3
Eichhorn (1981)	5	3
Skolyszewski (1982)	3	2
Overall	310	207 (67%)

Low LET radiotherapy photon and/or electron beams, and/or radioactive implants

Authors	Number of patients*	Loco-regional control
Fitzpatrick and Theriault (1986)	50	6 (12%)
Vikramet et al (1984)	49	2 (4%)
Borthne et al (1986)	35	8 (23%)
Rafla (1977)	25	9 (36%)
Fu et al (1977)	19	6 (32%)
Stewart et al (1968)	19	9 (47%)
Dobrowsky et al (1986)	17	7 (41%)
Shidnia et al (1980)	16	6 (38%)
Elkon et al (1978)	13	2 (15%)
Rossman (1975)	11	6 (54%)
Overall	254	61 (24%)

* Patients treated *de novo* and for gross disease after a post-surgical recurrence are included, but not patients who were treated postoperatively for miroscopic residual disease.
Updated from B. R. Griffin et al[18], T. W. Griffin et al[20] and Tsunemoto et al[19].

Washington in Seattle. In Europe, the two most important treatment series were obtained in Hammersmith[16] and in Amsterdam[17], with persistent local controls of 77% and 66% respectively. These results were confirmed by the recent publication of the Seattle data (Table 1)[18].

At the NIRS, 21 patients with inoperable or recurrent parotid gland tumours were treated with fast neutrons: local control was achieved in 13 cases (62%). In addition, 14 patients were treated after radical surgery: no local recurrence was observed. Of the total number of 35 patients treated with neutrons, 4 complications were recorded[19].

A recent review of all the patient series treated worldwide (Table 2) indicates an overall local control rate of 67%, while the overall local control rate for 'similar' patient series treated with low LET radiation (photons, electrons, interstitial therapy) is only 24%[18–20]. However, comparison of historical series is always questionable and only a randomised trial can provide a definite conclusion.

The normal tissue complication rates were marginally higher with low energy, fixed beam neutron generators, but were similar to megavoltage photons when high energy and/or isocentric hospital-based neutron generators were employed[16,21,22].

Subsequently, the National Cancer Institute (NCI) of the USA and the Medical Research Council (MRC) of Great Britain jointly sponsored a prospective randomised phase III clinical trial directly comparing fast neutrons with photons (and electrons) for patients with locally advanced, unresectable salivary gland tumours using both laboratory-based and hospital-based neutron generators (Table 3). The loco-regional control rates at 2 years were 67% for neutrons and 17% for photons in this group of patients with advanced tumours (up to 16 cm in maximum dimension). Although the number of patients was small (13 and 12 respectively), the study was closed in

Table 3. Neutron therapy of inoperable salivary gland tumours. Results of a RTOG/MRC prospective randomised trial.

	Photons	Neutrons
Number of evaluable patients	12	13
Loco-regional control		
at 1 year	17 ± 11%	67 ± 14%
at 2 year	17 ± 11%	67 ± 14%
Survival		
at 1 year	67 ± 12%	77 ± 12%
at 2 year	25 ± 14%	62 ± 14%

Modifed from Griffin et al 1988[20].

1986 for ethical reasons, when the statistical significance of the difference between treatments became apparent (p < 0.005)[20]. Attention is drawn to the fact that the local control rates observed in the randomised study are very similar to the average local control rates reported in the historical series. The normal tissue toxicities were not statistically significantly different between the two groups.

An analysis of the 52 patients treated at the University of Washington in Seattle has been published recently[22]. These patients had gross inoperable, residual unresectable, or recurrent disease. With a median follow-up of 42 months and a minimum follow-up of one year, local/regional tumour control within the treatment field was achieved in 92% (48/52) of the cases. An additional 8 patients had regional failures outside the treatment field (most in an untreated lower neck), resulting in an overall local/regional tumour control rate of 77% (40/52). Grouping patients according to treatment status, actuarial five-year local/regional tumour control rates were 92% for patients treated definitively with fast neutrons (without prior surgical procedure), 63% for patients treated postoperatively for gross (measurable by CT) residual disease, and 51% for patients treated for recurrent disease following a surgical procedure (Figure 4)[23]. There were no cases of radiation-induced facial nerve damage.

Taken as a whole, the results of the non-random clinical studies and of the prospective randomised trial overwhelmingly support the contention that fast neutrons offer a significant advance in the treatment of inoperable and unresectable primary or recurrent malignant salivary gland tumours. Fast neutron therapy alone should be the treatment of choice for advanced stage salivary gland tumours, and surgery should be limited to cases where there is a high likelihood of achieving a negative surgical margin and where the risk of facial nerve damage is small[23].

Paranasal sinuses

Remarkably good results have also been observed with neutron therapy for locally extended tumours of the paranasal sinuses. In the series treated at the Hammersmith Hospital, 86% (37/43) of the patients showed complete remission and relief of symptoms was noticed in all cases. Thirty percent of the patients survived at 3 years with a 50% local control rate[24]. Similar results were reported more recently by Errington at the Clatterbridge cyclotron (personal communication).

Several factors could explain these interesting results, indicating that paranasal sinuses could be a good indication for neutron therapy:

(i) the superficial location of these tumours (when poorly penetrating beams only are available);
(ii) the diversity of differentiated histology: in the Errington's series, there were 14 squamous cell carcinomas, but also 11 adenoid-cystic carcinomas and 8 adenocarcinomas (Table 4);
(iii) the presence of bone structures, in or near the target volume, which reduces the absorbed dose to the cells located in the osseous cavities[25,26].

Other head and neck tumours

In Europe conflicting results have been reported for neutron therapy of advanced squamous cell carcinomas of the head and neck. The first study conducted by Catterall[27] showed a highly significant advantage of neutrons over photons with respect to local control and survival. However, these results were not substantiated in a European multi-centre randomly controlled trial. The disease-free survival rate at 12 months was 34% (34/100) for the neutron and 38.9% (37/95) for the photon group. The recurrence rates were 37% and 39.7% respectively[28].

In Japan, 13 patients with tumour of the supraglottis were treated with fast neutrons at the NIRS. A local control was reported in 11 cases (84%), while, with photons for similar patients, local control was achieved in only 25% of the cases. At the same centre, no difference in local control after neutron or photon irradiation was reported for carcinoma of the glottis and subglottis[19].

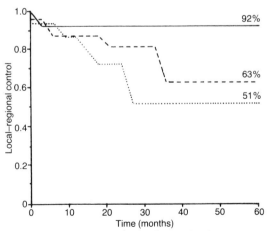

Figure 4. Neutron therapy of salivary gland tumours at the University of Washington, Seattle (1991). Loco-regional control as a function of presentation: (——) patients treated definitely with neutrons without prior surgery; (-----) patients treated postoperatively for gross residual disease; (.....) recurrence after surgery. (After Buchholz et al[22].)

In the United States, a RTOG trial, with a small number of patients with advanced disease, revealed a local control of 52% (12/23) for the neutron group, compared with 17% (2/12) for the photon group (p = 0.035). The actuarial survival rate at 2 years was 25% in the neutron group and 0% in the photon group[29].

Another RTOG randomised trial compared mixed schedule irradiation (2 neutron + 3 photon fractions per week) with conventional photon irradiation in unresectable squamous cell carcinomas of the head and neck[30]. A total number of 327 patients entered in the study: 163 patients of the mixed schedule group and 134 patients of the photon group were eligible for analysis. The minimum at-risk follow-up period was 6 years. Study results reveal no significant differences in overall loco-regional tumour control rates of survival. However, subgroup analysis reveals significant differences based on whether or not patients presented with positive lymph nodes. Loco-regional tumour control rates for patients presenting with positive lymph nodes were 30% for mixed schedule treated patients as against 18% for photon-treated patients (p=0.05). In contrast, loco-regional tumour control rates for patients presenting without positive lymph nodes were 64% for photon-treated patients and 33% for mixed-beam-treated patients (p=0.004). It is important to stress that control of the metastatic lymph nodes favoured mixed schedule over photons by a margin of 45% (49/109) to 26% (23/87) with a significance of p=0.004 (Figure 5)[30,31].

A possible explanation for the observed discrepancy of the results between patients presenting with or without positive lymph nodes could be related to the physical distribution of the dose. The suboptimal technical conditions (fixed horizontal beam, lack of adequate port verification, poor beam penetration) could be responsible for a geographic miss of part of the primary tumour in patients presented with negative lymph nodes (small planning target volumes and field sizes). In contrast, the large field sizes required for patients presenting with positive nodes could have increased the chance of adequate coverage of the primary site. On the other hand, the rather poor depth-dose characteristic of the neutron beams available for this study would have resulted in an increased irradiation of the neck (lymph nodes) compared to the deeper primary site. These physical characteristics of the neutron beams could explain an increased rate of nodal tumour control over primary tumour control. A similar difference of 12.5% in favour of the neutron treated patients was found in the Edinburgh series for the local control of lymph nodes metastases of more than 3 cm in diameter[32].

A last prospective, randomised, phase III study using the hospital-based cyclotrons was designed to answer definitively questions concerning the role of fast neutrons in treatment of squamous cell carcinomas of the head and neck. One hundred and seventy-eight patients were entered on this study directly comparing state-of-the-art fast neutron radiation therapy with state-of-the-art photon (and electron) radiation therapy; 89 patients were randomised to each treatment arm. Patients

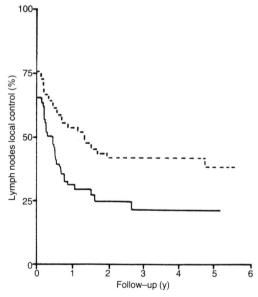

Figure 5. Neutron therapy of head and neck tumours. Results of a RTOG randomised trial. Comparison of tumour local control rates at the level of the lymph nodes after (——) photon only, 23/87, and (---) mixed schedule irradiation, 49/109. p = 0.004. (After Griffin et al[30].)

Table 4. Results of treatment with 7.5 MeV neutrons for advanced tumours of paranasal sinuses: histological types, responses and complications.

Histological type	Number of patients	Complete regression	Recurring	Patients with complications
Squamous	17	14	3	3
Adenoid cystic	11	10	4	4
Adenocarcinoma	8	6	–	1
Transitional cell	5	5	1	2
Undifferentiated	1	1	–	–
Malignant melanoma	1	1	–	–
Total	43	37 (86%)	8 (18%)	*10 (23%)

* 2 of them had received previous photon radiotherapy. From Errington[24].

Table 5. Review of the local control rates for soft-tissue sarcomas treated definitively with radiation therapy.

Neutrons

Institutions	Number of patients*	Local control	
Essen + Heidelberg, 1983	60	31	(52%)
Hammersmith, 1987	50	26	(52%)
Hamburg, 1987	45	27	(60%)
TAMVEC, 1980	29	18	(62%)
Fermilaboratory, 1984	26	13	(50%)
Seattle, 1986	21	15	(71%)**
Louvain-la-Neuve, 1982	19	4	(21%)
Amsterdam, 1981	13	8	(61%)
NIRS, 1979	12	7	(58%)
Edinburgh, 1986	12	5	(42%)
MANTA, 1980	10	4	(40%)
Overall	297	158	(53%)

Photons/Electrons

Authors	Number of patients*	Local control	
Tepper and Suit (1985)	51	17	(33%)
Duncan and Dewar (1985)	25	5	(20%)
McNeer et al (1968)	25	14	(56%)
Windeyer et al (1966)	22	13	(59%)
Leibel et al (1983)	5	0	
Overall	128	49	(38%)

* Patients treated *de novo* or for gross disease after surgery are included but not patients treated postoperatively for microscopic residual disease or for limited macroscopic residual disease.
** Two-year actuarial data.
Modifed from Laramore et al[36,37].

with T3 or T4 tumours, or T2N+ tumours originating in the oral cavity, oropharynx, hypopharynx, supraglottic larynx, and glottic larynx received either 70 Gy photons in 35 fractions over 7 weeks or 20.4 Gy neutrons in 12 fractions over 4 weeks. The complete response rate in the neutron treated group of patients is 70%, against 50% in the low LET treated group of patients ($p=0.003$). The loco/regional tumour control rates are 44% for neutrons against 35% for low LET radiations, and survival rates are 36% for neutrons against 42% for low LET radiations; and major complication rates are 20% for neutrons against 8% for low LET radiations. None of these differences are statistically significant.

It seems reasonable to conclude that fast neutrons can bring a significant benefit in well defined patient series with tumours in the head and neck area, especially locally advanced tumours and fixed metastatic lymph nodes. However, there is no argument at present for recommending neutron therapy as a general treatment policy for all tumours of the head and neck area. In particular, it seems reasonable to maintain the classical treatments for tumour types which are efficiently controlled with photons (such as T_{1-2} tumours of the larynx)[33].

Sarcomas of soft tissues, bone and cartilages

Slowly growing, well differentiated soft tissue sarcomas are treated in most of the neutron therapy centres, mainly because they are often resistant to X rays and also because of the excellent results reported from Hammersmith[2]. When evaluating the results of neutron therapy, comparison with historical series should be made very carefully, since the series may differ histologically, by degree of differentiation, local

Table 6. Indications for neutron (and/or photons) radiotherapy for low grade soft tissue sarcoma.

Type of surgery	Plane of dissection	Microscopic appearance	Local control after surgery	Indication for radiotherapy	Local control after combined modality
Intracapsular	Within lesion	Tumour at margin	0%	Neutrons (photons)	30–50%
Marginal	Within reactive zone, extracapsular	Reactive tissue, microsatellite tumour	10–20%	Neutrons (photons)	> 50%
Wide	Beyond reactive zone, through normal tissue, with compartment	Normal tissue	50–60%	Photons	90%
Radical	Normal tissue, extracompartmental	Normal tissue	80–90%	Photons (rare)	> 90%

(After Pötter et al[38].)

extent, localisation, etc. Furthermore, patient recruitment is influenced by the general treatment policy in a given centre (i.e. the relative place of surgery and/or chemotherapy). Therefore, ideally randomised trials are required, but so far have been difficult to achieve for practical reasons.

The largest patient series was treated in Essen. Neutrons only were used first and a 76.5% local control rate was achieved. However, a high percentage of complications was observed (22%), which may be related to poor beam penetration and high skin doses. Therefore, later on, neutrons were no longer used alone, but were applied as a boost: the local control rate was then 61.9% and the complication rate was reduced to 15%. The results of this study are reported in detail by Schmitt et al[34,35].

A review of the results reported from the different centres (Table 5) indicates an overall local control rate after neutron therapy of 53% for inoperable soft tissue sarcomas. This value is higher than the value of 38% currently observed after low LET radiation for similar patients series[36,37].

Taking account of the difficulties in initiating a randomised trial for soft tissue sarcomas, the German Neutron Therapy Group (M. Wannemacher) together with the EORTC-Heavy-Particle Therapy Group, initiated a collaborative study in order to collect all the data from the different participating neutron therapy centres. Strict rules govern data reporting with respect to tumour description, follow-up, treatment technique, dose specification, etc.

The proposed indications for neutron therapy (and/or photon therapy), for low grade soft tissue sarcomas, are presented in Table 6, after Pötter et al[38].

Conventional radiotherapy generally fails to control bulky primary bone tumours, as appropriate doses inevitably induce osteoradionecrosis. The low neutron kerma in bone reduces the absorbed dose by 25% or more to cells in osseous cavities[25] and allows the application of adequate doses with a reduced probability of late normal bone injury. Hence, differentiated primary bone tumours in adults were part of many clinical neutron programmes.

The review of published data indicates that for 88 patients with osteosarcoma treated at different institutions, a persisting local control of 54% (52/97) was achieved[36,39]. Most of these patients had inoperable tumours or refused amputation. An overall local control rate of 21% after photon irradiation is currently reported for similar patient series. However, due to the large treatment volumes and often preceding chemotherapy, a complication rate up to 36% was registered[40].

The review of the results reported from the same institutions indicates a persisting local control of differentiated chondrosarcomas after neutron therapy in 49% (25/51) of the patient[36,39]. This value compares well with the 33% (10/30) local control rate achieved after photon irradiation. Debulking surgery followed by appropriate neutron or neutron-boost irradiation then may become an alternative to ablative or mutilating surgery.

In conclusion, fast neutrons may be considered the best radiation quality for differentiated, slowly growing soft tissue sarcomas, especially locally extended, inoperable or recurrent tumours. A similar conclusion may apply to osteosarcomas and chondrosarcomas.

Prostatic adenocarcinomas

Prostatic adenocarcinomas, having in general a long doubling time, should be a good indication for neutron therapy, taking into account available radiobiological data[41]. In fact, the benefit of neutron therapy was rapidly recognised in several centres, initially in Hamburg by Franke[42]. At NIRS, in Chiba, for prostatic adenocarcinomas Stage A2, B and C, local controls at 3 years of 3/3, 3/5 and 8/14 respectively were reported[19]. Excellent results were also obtained at Louvain-la-Neuve with p(65)+Be neutrons used in mixed schedule following the RTOG protocol (but with 3n + 2ph fractions per week)[43].

The most convincing data are the result of a randomised trial, inititated by the RTOG, on locally advanced (C,D1) adenocarcinomas of the prostatic gland (Figure 6)[44]. The patients were randomised between photons and mixed beam therapy using physics-laboratory-based machines. The ten-year results are now available and demonstrate a statistically significant advantage for neutrons in terms of local/regional tumour control, survival and disease specific survival[45]. Ten-year local/regional control rates on this study were 70% for mixed beams against 58% for photons (p=0.03); ten-year survival rates were 46% for mixed beams against 29% for photons (p=0.04); and ten-year disease-specific survival rates were 58% for mixed beams against 43% for photons (p=0.05).

While these results are impressive, the study has been criticised on the basis that the photon local/regional control and survival rates were inferior to other photon retrospective series published in the medical literature. A confirmatory study was then designed using state-of-the art, hospital-based cyclotrons.

One hundred and seventy-eight patients were randomised on this follow-up study comparing

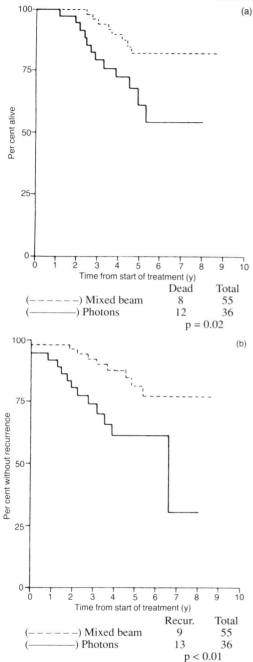

Figure 6. Neutron therapy of locally extended prostatic adenocarcinoma. RTOG randomised trial comparing a combination of fast neutrons and photons (mixed beam) and conventional photon irradiation alone. (a) The actuarial survival rates at 8 years are indicated, adjusted by exclusion of intercurrent non-cancer death (determinental survival rates). (b) The local control rates are indicated, combining clinical and biopsy criteria. (From Russell et al[44].)

20.4 Gy of neutrons delivered in 12 fractions over 4 weeks with 70 Gy of photons delivered in 35 fractions over 7 weeks. The depth-dose properties and isocentric delivery capabilities of the high energy neutron beams allowed treatment with neutrons alone to this deep-seated tumour. After stratification for stage, Gleason grade, and the presence or absence of surgical nodal staging, 89 patients were randomised to each treatment arm. Patients with high grade, stage B2, stage C, or stage D1 tumours were eligible for the study. The two treatment arms were balanced for all known prognostic factors.

Local/regional tumour control rates favour neutrons with a 94% control rate for neutron-treated patients against an 86% control rate for photon treated patients (p=0.04). It is too early (with this slow growing tumour system) to expect any significant differences to show up in survival rates, and at this time, no significant differences are seen. Survival rates are 80% for neutron treated patients against 85% for photon-treated patients (two deaths thus far can be attributed to prostate cancer both in the photon treatment arm) the rest have been due to intercurrent disease or homicide. There were no treatment complication related deaths. Local/regional failures are currently occurring in a ratio of 2.4:1 favouring neutrons.

The differences in major complication rates are statistically significant (p=0.03), and are primarily due to differences in large bowel toxicities resulting in surgical intervention and colostomies. The major complication rate in the neutron-treated group of patients was 8%, against 1% in the photon-treated group. The colostomy rates for neutron-treated patients are facility dependent (Table 7) and the differences in rates observed between the Cyclotron Corporation machines at M.D. Anderson and UCLA, and the Scanditronix machine at the University of Washington are statistically significant (p=0.01). Differences in these major complication rates are probably due to differences in beam collimation capabilities among the neutron facilities, but may also be due to differences in beam energy spectra[23,46].

Complications after neutron therapy have been studied in more detail by Russel et al[47]. Among 132 patients treated for prostatic adenocarcinoma at the Seattle facility (94 with neutrons, 16 with mixed schedule, 22 with photons) and with a median follow-up of 14 months (range 1-101 months), 31 have experienced either sciatica beginning during or shortly after treatment, or diminished bladder or bowel continence developing at a median time of 6.5 months after treatment (26/94 after neutron, 3/16 after mixed schedule and 2/22 after photon irradiation). Sciatica responded to oral steroids and was usually self-limited, whereas sphincter dysfunction appears to be permanent. Seven patients have moderate (5 pts) or severe (2 pts) residual problems, all in the groups receiving neutrons (6/7) or mixed schedule (1/7) irradiation. The total number of severely affected patients (2/110) represents a small percentage of the patients treated with fast neutrons only or with mixed schedule. It was concluded that these complications should not place a constraint on the use of neutrons, although it is conceivable that the incidence of complications will increase, as patient follow-up is still short. Because survival and local control of locally advanced prostate cancer achieved with neutrons or mixed schedule is superior to results with photons alone, this low incidence of severe neurological problems appears to be an acceptable risk to take for more effective treatment[47].

The data from Louvain-la-Neuve support this conclusion[43]. The early tolerance is excellent, and among more than 150 patients treated with mixed schedule (see above), only one late complication, scored grade 3, was observed (urethral stricture in a patient who underwent several surgical procedures).

Of course, account must be taken of the slow natural history of prostatic adenocarcinoma and caution is necessary before deriving definitive conclusions. However, the clinical data at present available indicate a significant benefit for fast neutrons (used alone or in mixed schedule?) compared to the current photon irradiation modalities for locally advanced cases[43,48,49]. They confirm the selective efficiency of neutrons against slowly growing, well differentiated tumours, as well as the importance of the physical selectivity when high LET radiations are used, as could be expected from the radiobiological data.

Melanomas

Although surgery, when feasible, is the treatment of choice for melanomas, radiation therapy may be required for some patients at a given stage of the disease (discussion of the value of adjuvant chemotherapy is outside the scope of this paper). Melanomas are often resistant to photon irradiation; this can be related, from a radiobiological point of view, to the broad shoulder of the survival curves for several cell lines[14,50]. Therefore neutron therapy could be an alternative.

Encouraging results were obtained at the Hammersmith Hospital (Catterall, personal communication) where 87 tumour sites, in 48 patients, were treated with fast neutrons. They consist of metastatic tumours, recurrences after surgery or sites unsuitable for surgery. Permanent local control was achieved in 62% of the sites (the minimum follow-up was 3 months). In addition, in 20 of the 25 patients good palliation was obtained. These results, as well as others (e.g. from Edinburgh), indicate that fast neutron therapy can be an alternative in the treatment of some melanomas, especially where surgery cannot be performed, and for metastatic tumours.

Conclusion of the clinical survey

The clinical indications for fast neutrons are summarised in Table 8. They represent about 10–15% of the patients currently referred to the radiation therapy departments.

In contrast, rather poor results were generally obtained for tumours of the central nervous system (CNS). This could have been predicted, to some extent, from the radiobiological data and especially from the high RBE value currently

Table 7. Neutron therapy of prostatic adenocarcinomas. Bowel morbidity by institution.

Institution	Colostomies
University of Washington, Seattle p(50)+Be neutrons multi-leaf collimator	0/49 (0%)
UCLA, Los Angeles p(45)+Be neutrons movable jaw collimator	2/25 (8%)
M.D. Anderson, Houston p(42)+Be neutrons fixed cone collimator	4/10 (40%)

After NCI Report 1991[23].

Table 8. Clinical indications for neutron therapy (summary).

1. *Salivary gland tumours*
   Locally extended, inoperable or recurrent/well differentiated
2. *Paranasal sinuses*
   Adenocarcinomas, adenoid cystic carcinomas, other histology (?)
3. *Some tumours of the head and neck area*
   Locally extended, metastatic adenopathies
4. *Soft tissue sarcomas, osteosarcomas, chondrosarcomas*
   Especially slowly growing/well differentiated
5. *Prostatic adenocarcinomas*
   Locally extended/well differentiated
6. *Melanomas*
   Inoperable/recurrent.

observed for late CNS tolerance[51–54]. However, a slight improvement in survival for grade IV astrocytoma was observed after neutron boost for patients with incomplete tumour resection or inoperable tumours[55]. Little benefit was observed after neutron therapy of inoperable pancreatic carcinoma[56,57].

Finally, other types of tumours have been investigated[58], but additional information is needed especially for uterine cervix[19,59], rectal adenocarcinoma[32,60–62], lung tumour and Pancoast[19,63], oesophagus[19], and bladder[32,60,61,64–66].

TECHNICAL DEVELOPMENTS

The technical conditions, in which fast neutron therapy is applied, have progressively been improved during the two past decades.

Briefly, as far as neutron energy is concerned, many of the first patient series were treated using low energy cyclotrons (16 MeV deuterons). Today, neutron beams produced by protons of 45 MeV (or more) are available at several facilities and in four centres neutrons are produced by 60 MeV protons, i.e. Clatterbridge, Faure, Fermilab and Louvain-la-Neuve. The depth doses and skin sparing are similar to that of a 10 MV photon linear accelerator.

A fixed horizontal beam was often a practical limitation for patient set-up when physics-laboratory-based cyclotrons were used. Today, a rotational gantry is available in several centres such as Clatterbridge, Detroit, Houston, Faure, Seoul, Seattle, UCLA, etc.

Variable collimators are used in Clatterbridge, Faure, UCLA, and multileaf collimators are used in Chiba, Detroit, Louvain-la-Neuve and Seattle (Figure 7).

The neutron therapy facilities operational today are listed in Table 9 with their main characteristics. Of course, the technical problems raised by the beam collimation and the isocentric gantry are far more complex and thus more expensive for neutrons than for photons.

Dosimetric data about neutron beams and protocols for neutron therapy, accepted at the international level, have been published[2,25]. In addition, several intercomparisons were performed between the different neutron therapy centres. These comparisons implied dosimetric, microdosimetric and radiobiological determinations[13,46,58,67].

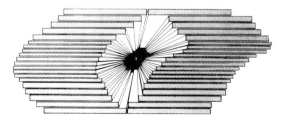

Figure 7. Variable multi-leaf collimator in neutron therapy. Diagram of a variable multi-leaf collimator showing the lower end of the leaves and the collimation surfaces which are all aligned with the proton target (symbolised by the +). Each leaf has its own motor drive and position readout (after Brahme, cited in Ref. 26).

SUMMARY AND CONCLUSIONS

Radiobiology

As indicated above, high LET radiations were introduced in therapy on the basis of radiobiological arguments, and more specifically the existence of hypoxic cells. The bulk of radiobiological data at present available confirm that high LET radiations indeed can bring a benefit for some tumour types or patient groups.

Radiobiology also suggests mechanisms through which this benefit could be achieved:

(1) The hypoxic gain factor (HGF) related to the reduction in OER.
(2) The kinetics gain factor (KGF) related to the position of the cells in the mitotic cycle.
(3) Less repair, and consequently less importance of fraction size. This factor could facilitate the application of concentrated irradiation (i.e. shorter overall time), which has been shown to bring a benefit especially when treating some fast proliferating tumours (short T_{pot})[68,69].

Two important practical conclusions can be derived from the radiobiological data. First is the need for proper patient selection and here radiobiology suggests some selection criteria:

(i) tumours for which hyperbaric oxygen, and hypoxic cell sensitisers were shown to bring a benefit[6,14];
(ii) in general, slowly growing, well differentiated tumours (see above KGF)[41,70];
(iii) rapidly proliferating tumours (possibility of reducing the overall time).

The two last arguments are to some extent in contradiction. This illustrates that the fact that radiobiology can only suggest selection criteria, but that, of course, clinical experience is needed to bring the definitive conclusions.

The second important practical conclusion which can be derived from radiobiology is the need for a high physical selectivity, which is at least as important with high as with low LET radiations. This is due to the general reduction of the differences in radiosensitivity between cell populations (i.e. less radiobiological differential effect).

Clinical results of fast neutron therapy

The review of the clinical data in general tends to confirm the predictions of radiobiology: indeed neutrons have brought a benefit, compared to photons, for several tumour types or sites: salivary gland tumours, prostatic adenocarcinomas, etc. (as summarised in Table 8).

Moreover, the tumour types for which neutrons were shown to bring a benefit correspond to a large extent to those predicted by the radiobiological data. Indeed, the most striking results were obtained for well differentiated, slowly

Table 9. The neutron therapy facilities in the world.

Centre		Neutron producing reaction	Comments
EUROPE			
UK	MRC, Clatterbridge	p(62)+Be	Rotational gantry Variable collimator
France	Orléans	p(34)+Be	Vertical beam
Belgium	UCL, Louvain-la-Neuve	p(65)+Be	Vertical beam Multi-leaf collimator (Horizontal beam in preparation)
Germany	Hamburg	(d + T)	Rotational gantry
	Heidelberg	(d + T)	Rotational gantry
	Munster	(d + T)	Rotational gantry
	Essen	d(14)+Be	Rotational gantry
	Garching – T.U. Münich	Reactor neutrons (av. energy 2 MeV)	Mixed beam
UNITED STATES			
Texas	M. D. Anderson, Houston	p(42)+Be	Rotational gantry
California	UCLA, Los Angeles	p(46)+Be	Rotational gantry Variable collimator
Washington	Seattle	p(50)+Be	Rotational gantry Multi-leaf collimator
Illinois	Fermilab	p(66)+Be	Horizontal beam
Michigan	Detroit	d(48)+Be	Isocentric mounting Multi-rod collimator
ASIA			
Japan	National Institute of Radiological Sciences (NIRS), Chiba	d(30)+Be	Vertical beam Multi-leaf collimator
	Institute for Medical Sciences (IMS), Tokyo	d(14)+Be	Horizontal beam
Korea	Korea Cancer Center Hospital (KCCH), Seoul	d(50.5)+Be	Rotational gantry
Saudi Arabia	King Faisal Hospital, Riyadh	p(26)+Be	Rotational gantry
AFRICA			
South Africa	National Accelerator Centre (NAC), Faure	p(66)+Be	Rotational gantry Variable collimator

growing tumours, often resistant to X rays as well as to chemotherapy.

On the other hand, there are still many tumour types for which additional information is needed, e.g. bladder, rectum, etc. A great difficulty in the interpretation of the results is due to the 'suboptimal' treatment conditions in which the neutron treatments were applied, especially for the early patient series (see next section). From these results conclusions cannot be derived concerning the value of high LET radiations in general or even concerning the value of neutrons applied under good technical conditions.

However, an important conclusion which can be derived from the rather poor results (especially the high rate of late complications) reported from some centres is that the benefit related to the improved radiobiological differential effect cannot compensate for a poor physical selectivity.

Lastly, there are tumour types for which neutrons do not bring a benefit, typically tumours of the central nervous system (CNS). This could also have been predicted from the radiobiological data, since a high neutrons RBE has been observed for late CNS tolerance.

Technical improvements

The improved technical conditions in which neutron therapy is applied today, at least in some centres, can no longer be compared with those existing in the seventies. Roughly, a similar improvement has been achieved as in the sixties, when 200 kV X rays were progressively replaced by high energy linear accelerators.

Progress has been made concerning:

(a) beam energy: with p(60)+Be neutrons, the same penetration and skin sparing is achieved as with a 10 MV linear accelerator;
(b) isocentric mounting; and
(c) variable collimation system and, in some centres, a multileaf collimator.

This does not mean that in each neutron therapy centre treatments are applied today in optimal technical conditions comparable to modern photon therapy. However, this situation is recognised by the local medical teams and, to take it into account, the clinical indications for neutron therapy are selected accordingly, sophisticated treatment plans are introduced and mixed schedule irradiation (or neutron boost) is applied when necessary.

However, with the modern (and already existing!) technology, it is possible today to reach with neutrons the same physical selectivity as with photons. It is only under these conditions that the real role of fast neutron therapy can be correctly evaluated.

The investment for a modern neutron therapy facility is approximately 3 times more expensive than for a modern linear accelerator. However, taking into account the reduction of the number of fractions in neutron therapy, the cost of full neutron treatment would then be about 1.5 times the cost of a photon treatment.

Difficulties and problems to be solved

The improvements and the achievements described in the previous sections do not imply that all problems are solved; two important questions remain open.

From a clinical point of view, one of the most important problems to solve is patient selection. If we are not able to identify the subgroups of patients suitable for neutron therapy, the benefit obtained will be diluted (or even not detectable) and, in addition, counterbalanced by the worse results obtained in other subgroups which would have been better treated with photons. This probably explains, at least in part, some of the discrepancies between the reported results.

The choice between low and high LET radiations is a radiobiological problem, related to tumour characteristics (and, of course, to the characteristics of the tissues at risk); it is not a machine or accelerator problem.

Our capacity for selecting the right patient for the right radiation quality will be improved partly with the progressive build-up of clinical experience, and especially with the conclusions of the randomised clinical trials.

However, today much hope is placed in the individual predictive tests, which could allow the therapist to select the optimal treatment modality on the basis of the characteristics of the individual tumours measured *in vitro* (biopsy). Several types of predictive tests are now being evaluated: intrinsic cell radiosensitivity measured after a test dose (e.g. survival at 2 Gy) *in vitro*, as well as the number of micronuclei, chromosome aberrations and Premature Chromosome Condensation (PCC). In addition, the potential doubling time, T_{pot}, now appears to be one of the best parameters to estimate the tumour cell proliferation capacity. Predictive tests will probably be applied more extensively in radiotherapy, not only to orientate adequately the patients between low and high LET radiations, but also to optimise other treatment parameters: dose level, fractionation, combination with drugs, etc.

The second important problem today is *patient recruitment*. On the assumption (see above and Table 8) that 10–15% of the patients referred today to the radiotherapy departments would require high LET therapy, one cyclotron would be needed (roughly) to every 10 electron linear accelerators. Under these conditions, only a few large centres will be able to use a cyclotron fully with their own patient recruitment. For centres of moderate or small size, the only solution would be to refer some of their patients to a high LET therapy facility, but this would imply difficult economical and psychological problems. An alternative is to set up a multicentre collaboration, but past experience has shown neither of these two solutions really work in practice.

In our opinion, the best hope for the future rests in the generalised application of predictive tests. Indeed if, for a given patient, predictive tests (routinely applied) indicate that high LET radiations would be better, the information would be difficult to ignore and there will be pressure to set up practical conditions for a collaboration between cancer centres.

The recruitment problem is probably the major reason for the slowing down of the neutron therapy programmes in the United States and in some European countries. There is, however, a need for high LET facilities. Taking, for example, the situation in Europe, it is expected that, in the year 2000, about 1 million new cancer cases will be diagnosed per year in the EC population. Among them, 60–70% will require radiation therapy (alone or in combination) at one or the other stage of the disease. Assuming that among them, 10–15% will require high LET therapy (fast neutrons or heavy ions), 60,000 – 100,000 patients from the EC population per year will need high LET therapy, and a similar figure can be assumed for North America.

REFERENCES

1. Breit, A., Burger, G., Scherer, E. and Wambersie, A. *Advances in Radiation Therapy with Heavy Particles.* Strahlentherapie **161**(12), 729-806 (1985).
2. Catterall, M. and Bewley, D. K. *Fast Neutrons in the Treatment of Cancer* (London: Academic Press) (1979).
3. Hall, E. J., Graves, R. G., Phillips, T. L. and Suit, H. D. *Proceedings of the CROS/RTOG Part III International Workshop 'Particle Accelerators in Radiation Therapy' Houston, 10-11 February 1982.;* Int. J. Radiat. Oncol. Biol. Phys. **8** (1982).
4. Wambersie, A., Bewley, D. K. and Lalanne, C. M. *Prospects for the Application of Fast Neutrons in Cancer Therapy. Radiobiological Bases and Survey of the Clinical Data.* Bull. Cancer (Paris) **73**, 546-561 (1986).
5. Stone, R. S. *Neutron Therapy and Specific Ionization.* Am. J. Roentgenol. **59**, 771-785 (1948).
6. Adams, G. E. *The Clinical Relevance of Tumour Hypoxia.* Eur. J. Cancer **26**, 420-421 (1990).
7. van Putten, L. M. and Kallman, R. F. *Oxygenation Status of a Transplantable Tumor during Fractionated Radiation Therapy.* J. Natl. Cancer Inst. **40**, 441-451 (1968).
8. Thomlinson, R. H. *Time and Dose Relationships in Radiation Biology as Applied to Radiotherapy.* In: Proceedings of N.C.I.-A.E.C. Conference, Carmel, California. USAEC Catalogue No BNL-50203 (C57) (Springfield, VA: US Dept of Commerce) (1969).
9. Chapman, J. D. *Biophysical Models of Mammalian Cell Inactivation by Radiation.* In: Radiation Biology in Cancer Research. Eds R. E. Meyn and H. R. Withers, pp. 21-32 (New York: Raven Press) (1988).
10. Barendsen, G. W. and Broerse, J. J. *Differences in Radiosensitivity of Cells from Various Types of Experimental Tumours in Relation to the RBE of 15 MeV Neutrons.* Int. J. Radiat. Oncol. Biol. Phys. **3**, 211-214 (1977).
11. Fertil, B., Deschavanne, P. J., Gueulette, J., Possoz, A., Wambersie, A. and Malaise, E. P. *In vitro Radiosensitivity of Six Human Cell Lines. Relation to the RBE of 50-MeV Neutrons.* Radiat. Res. **90**, 526-537 (1982).
12. Menzel, H. G., Pihet, P. and Wambersie, A. *Microdosimetric Specification of Radiation Quality in Neutron Radiation Therapy.* Int. J. Radiat. Biol. **57**, 865-883 (1990).
13. Pihet, P., Menzel, H. G., Schmidt, R., Beauduin, M. and Wambersie, A. *Evaluation of a Microdosimetric Intercomparison of European Neutron Therapy Centres,* Radiat. Prot. Dosim. **31**, 437-443 (1990).
14. Tubiana, M., Dutreix, J. and Wambersie, A. *Introduction to Radiobiology* (London: Taylor and Francis) (1990).
15. Griffin, T., Blasko, J. and Laramore, G. *Results of Fast Neutron Beam Radiotherapy, Pilot Studies at the University of Washington.* In: High-LET Radiations in Clinical Radiotherapy. Eds G. W. Barendsen, J. J. Broerse and K. Breur. Eur. J. Cancer Suppl. 23-29 (1979).
16. Catterall M. and Errington, R. D. *The Implications of Improved Treatment of Malignant Salivary Gland Tumors by Fast Neutron Radiotherapy.* Int. J. Radiat. Oncol. Biol. Phys. **13**, 1313-1318 (1987).
17. Battermann, J. J. and Mijnheer, B. J. *The Amsterdam Fast Neutron Radiotherapy Project: a Final Report.* Int. J. Radiat. Oncol. Biol. Phys. **12**, 2093-2099 (1986).
18. Griffin, B. R., Laramore, G. E., Russell, K. J., Griffin, T. W. and Eenmaa, J. *Fast Neutron Radiotherapy for Advanced Malignant Salivary Glands Tumors.* Radiother. Oncol. **12**, 105–111 (1988).

19. Tsunemoto, H., Morita, S., Satoh, S., Iino, Y. and Yul, Yoo. *Present Status of Fast Neutron Therapy in Asian Countries.* Strahlenther Onkol. **165**, 330-336 (1989).
20. Griffin, T. W., Pajak, T. F., Laramore, G. E., Duncan, W., Richter, M. P., Hendrickson, F. R. and Maor, M. H. *Neutron vs Photon Irradiation of Inoperable Salivary Gland Tumors: Results of an RTOG-MRC Cooperative Randomized Study.* Int. J. Rad. Oncol. Biol. Phys. **15**, 1085-1090 (1988).
21. Duncan, W., Orr, J. A., Arnott, S. J. and Jack, W. J. L. *Neutron Therapy for Malignant Tumors of the Salivary Glands: A Report of the Edinburgh Experience.* Radiother. Oncol. **8**, 97-104 (1987).
22. Buchholz, T. A., Laramore, G. E., Griffin, B. R., Koh, W. J. and Griffin, T. W. *The Role of Fast Neutron Radiotherapy in the Management of Advanced Salivary Gland Malignancies.* Cancer (in press).
23. National Cancer Institute (NCI). *Fast Neutron Radiation Therapy in the United States: A Twenty-year NCI Sponsored Research Program.* NTCWG Annual Report, 1991 (National Cancer Institute, Bethesda, MD 20892, USA) (1991).
24. Errington, R. D. *Advanced Carcinoma of the Paranasal Sinuses Treated with 7.5 MeV Fast Neutrons.* Bull. Cancer (Paris) **73**, 569-576 (1986).
25. Bewley, D. K. *The Physics and Radiobiology of Fast Neutron Beams* (Bristol: Adam Hilger) (1989).
26. International Commission on Radiation Units and Measurements. *Clinical Neutron Dosimetry. Part 1: Determination of Absorbed Dose in a Patient Treated by External Beams of Fast Neutrons.* ICRU Report 45 (Bethesda, MD: ICRU Publications) (1989).
27. Catterall, M., Bewley, D. K. and Sutherland, I. *Second Report on a Randomized Clinical Trial of Fast Neutrons Compared with X- or Gamma-rays in Treatment of Advanced Head and Neck Cancers.* Br. Med. J. **1**, 1942 (1977).
28. Duncan, W., Arnott, S. J., Battermann, J. J., Orr, J. A., Schmitt, G. and Kerr, G. R. *Fast Neutrons in the Treatment of Head and Neck Cancers: the Results of a Multi-centre Randomly Controlled Trial.* Radiother. Oncol. **2**, 293-301 (1984).
29. Griffin, T. W., Davis, R., Hendrickson, F. R., Maor, M. H., Laramore, G. E. and Davis, L. *Fast Neutron Radiation Therapy for Unresectable Squamous Cell Carcinomas of the Head and Neck: the Results for a randomized RTOG Study.* Int. J. Radiat. Oncol. Biol. Phys. **10**, 2217-2223 (1984).
30. Griffin, T. W., Pajak, T. F., Maor, M. H., Laramore, G. E. Hendrickson, F. R., Parker, R. G., Thomas, F. J. and Davis, L. W. *Mixed Neutron/Photon Irradiation of Unresectable Squamous Cell Carcinomas of the Head and Neck: the Final Report of a Randomized Cooperative Trial.* Int. J. Radiat. Oncol. Biol. Phys. **17**, 959-965 (1989).
31. Griffin, T. W., Davis, R., Laramore, G. E., Hussey, D. H., Hendrickson, F. R. and Rodriguez-Antunez, A. *Fast Neutron Irradiation of Metastatic Cervical Adenopathy. The Results of a Randomized RTOG Study.* Int. J. Radiat. Oncol. Biol. Phys. **9**, 1267-1270 (1983).
32. Duncan, W., Arnott, S. J., Orr, J. A. and Kerr, G. R. *The Edinburgh Experience of Fast Neutron Therapy.* Int. J. Radiat. Oncol. Biol. Phys. **8**, 2155-2157 (1982).
33. Wells, G., Koh, W., Pelton, J., Russell, K., Griffin, B., Laramore, G., Griffin, T., Parker, R., Peters, L. J., Davis, L. and Pajak, T. F. *Fast Neutron Teletherapy in Advanced Epidermoid Head and Neck Cancer.* Am. J. Clin. Oncol. **12**(4), 295-300 (1989).
34. Schmitt, G., Mills, E. E. D., Levin, V., Pape, H., Smit, B. J. and Zamboglou, N. *The Role of Neutrons in the Treatment of Soft Tissue Sarcomas.* Cancer **64**, 2064-2068 (1989).
35. Schmitt, G., Pape, H. and Zamboglou, N. *Long Term Results of Neutron- and Neutron-boost Irradiation of Soft Tissue Sarcomas.* Strahlenther. Onkol. **166**, 61-62 (1990).
36. Laramore, G. E., Griffeth, J. T., Boespflug, M., Pelton, J. G., Griffin, T. W., Griffin, B. R., Russell, K. J. and Koh, W. *Fast Neutron Radiotherapy for Sarcomas of Soft Tissue, Bone, and Cartilage* (Private communication).
37. Laramore, G. E. and Griffin, T. W. *High-LET Radiotherapy.* Int. J. Radiat. Oncol. Biol. Phys. **12**, Suppl. 1. Abstract no 505, p. 85 (1986).
38. Pötter, R., Knocke, T. H., Haverkamp, U. and Al-Dandashi, Chr. *Treatment Planning and Delivery in Neutron Radiotherapy of Soft Tissue Sarcomas.* Strahlenther. Onkol. **166**, 102-106 (1990).
39. Richter, M. P., Laramore, G. E., Griffin, Th. W. and Goodman, R. L. *Current Status of High Linear Energy Transfer Irradiation.* Cancer **54**, 2814-2822 (1984).
40. Schmitt, G., Rehwald, U. and Bamberg, M. *Neutron Irradiation of Primary Bone Tumours.* J. Eur. Radiother. **3**, 145-146 (1982).
41. Battermann, J. J. *Clinical Application of Fast Neutrons, the Amsterdam Experience.* Thesis, University of Amsterdam, The Netherlands (1981).
42. Franke, H. D., Langendorff, G. and Hess, A. *Die Strahlenbehandlung des Prostata-Carcinoms in Stadium C mit schnellen Neutronen.* Verh. Dtsch. Ges. Urol. **32**, 175-180 (1981).
43. Richard, F., Renard, L. and Wambersie, A. *Current Results of Neutron Therapy at the UCL, for Soft Tissue Sarcomas and Prostatic Adenocarcinomas.* Bull. Cancer (Paris) **73**, 562-568 (1986).

44. Russell, K. J., Laramore, G. E., Krall, J. M., Thomas, F. J., Maor, M. H., Hendrickson, F. R., Krieger, J. N. and Griffin, T. W. *Eight years Experience with Neutron Radiotherapy in the Treatment of Stages C and D Prostate Cancer: Updated Results of the RTOG 7704 Randomized Clinical Trial.* Prostate **11**, 183-193 (1987).

45. Laramore, G. E., Krall, J. M., Thomas, F. J., Russell, K. J., Maor, M. H., Hendrickson, F. R., Martz, K. L. and Griffin, T. W. *Fast Neutron Radiotherapy for Locally Advanced Prostate Cancer: Final Report of an RTOG Randomized Clinical Trial.* Radiother. Oncol. (in press).

46. Beauduin, M., Gueulette, J., Grégoire, V., De Coster, B., Vynckier, S. and Wambersie, A. *Radiobiological Comparison of Fast Neutron Beams used in Therapy. Survey of the Published Data.* Strahlenther. Onkol. **166**, 18-21 (1990).

47. Russell, K. J., Laramore, G. E., Krieger, J. N., Wiens, L. W., Griffeth, J. T., Koh, W. J., Griffin, B. R., Austin-Seymour, M. M., Griffin, T. W. and Davis, L. W. *Transient and Chronic Neurological Complications of Fast Neutron Radiation for Adenocarcinoma of the Prostate.* Radiother. Oncol. **18**, 257-265 (1990).

48. Russell, K. J., Laramore, G. E., Griffin, T. W., Parker, R. G., Maor, M. H., Davis, L. W. and Krall, J. M. *Fast Neutron Radiotherapy in the Treatment of Locally Advanced Adenocarcinoma of the Prostate. Clinical Experience and Future Directions.* Am. J. Clin. Oncol. **12**(4), 307-310 (1989).

49. Saroja, K. R., Cohen, L., Hendrickson, F. R. and Mansell, J. *Prostate Antigen in Locally Advanced Prostate Cancer Patients Treated with High Energy Neutrons at Fermilab.* (Private communication).

50. Malaise, E. P., Weininger, J., Joly, A. M. and Guichard, M. *Measurements in vitro with Three Cell Lines Derived from Melanomas.* In: Cell Survival after Low Doses of Radiation: Theoretical and Clinical Implications. T. Alper (John Wiley) pp. 223-225 (1975).

51. Griffin, B. R., Berger, M. S., Laramore, G. E., Griffin, T. W., Shuman, W. P., Parker, R. G., Davis, L. W. and Diener-West, M. *Neutron Radiotherapy for Malignant Gliomas.* Am. J. Clin. Oncol. **12**(4), 311-315 (1989).

52. Saroja, K. R., Mansell, J., Hendrickson, F. R., Cohen, L. and Lennox, A. *Failure of Accelerated Neutron Therapy to Control High Grade Astroctyomas.* Int. J. Radiat. Oncol. Biol. Phys. **17**, 1295-1297 (1989).

53. Schmitt, G., Bamberg, M. and Budach, V. *Preliminary Results of Neutron Irradiation of Patients with Spinal Gliomas.* Br. J. Radiol. **60**, 320-321 (1987).

54. Schmitt, G. and Wambersie, A. *Review of the Clinical Results of Fast Neutron Therapy.* Radiother. Oncol. **17**, 47-56 (1990).

55. Breteau, N., Destembert, B., Favre, A., Phéline, C. and Schlienger, M. *Fast Neutron Boost for the Treatment of Grade IV Astrocytomas.* Strahlenther. Onkol. **165**, 320-323 (1989).

56. Saroja, K. R., Cohen, L., Hendrickson, F. R. and Mansell, J. A. *Localized Unresectable Pancreatic Cancer Treated with High-energy Neutrons and Chemotherapy at Fermilab: Preliminary Results* (Private communication).

57. Thomas, F. J., Krall, J., Hendrickson, F., Griffin, T. W., Saxton, J. P., Parker, R. G. and Davis, L. W. *Evaluation of Neutron Irradiation of Pancreatic Cancer.* Am. J. Clin. Oncol. **12**(4), 283-289 (1989).

58. Wambersie, A. *Fast Neutron Therapy at the End of 1988 — a Survey of the Clinical Data.* Strahlenther. Onkol. **166**, 52-60 (1990).

59. Maor, M. H., Gillespie, B. W., Peters, L. J., Wambersie, A., Griffin, T. W., Thomas, F. J., Cohen, L., Conner, N. and Gardner, P. *Neutron Therapy in Cervical Cancer: Results of a Phase III RTOG Study.* Int. J. Radiat. Oncol. Biol. Phys. **14**, 885-891 (1988).

60. Battermann, J. J. *Results of d + T Fast Neutron Irradiation on Advanced Tumours of Bladder and Rectum.* Int. J. Radiat. Oncol. Biol. Phys. **8**, 2159-2164 (1982).

61. Battermann, J. J., Hart, G. A. M. and Breur, K. *Dose-Effect Relations for Tumour Control and Complication Rate after Fast Neutron Therapy for Pelvic Tumours.* Br. J. Radiol. **54**, 899-904 (1981).

62. Breteau, N., Destembert, B., Favre, A., Sabattier, R. and Schlienger, M. *An Interim Assessment of the Experience of Fast Neutron Boost in Inoperable Rectal Carcinomas in Orléans.* Bull. Cancer (Paris) **73**, 591-595 (1986).

63. Stewart, G., Griffin, T. W., Griffin, B. R., Laramore, G., Russell, K. J., Parker, R. G., Maor, M. N. and Davis, L. W. *Neutron Radiation Therapy for Unresectable Non-small Cell Carcinoma of the Lung.* Am. J. Clin. Oncol. **12**(4), 290-294 (1989).

64. Duncan, W., Arnott, S. J., Jack, W. J. L., MacDougall, R. H., Quilty, P. M., Rodger, A. Kerr, G. R. and Williams, J. R. *A Report of a Randomised Trial of d(15)+Be Neutrons Compared with Megavoltage X-ray Therapy of Bladder Cancer.* Int. J. Radiat. Oncol. Biol. Phys. **11**, 2043-2049 (1985).

65. Kirkove, C., Richard, F., Octave-Prignot, M. and Wambersie, A. *Neutron Therapy of Bladder Carcinoma at the UCL Cyclotron of Louvain-la-Neuve* Strahlenther. Onkol. (In press).

66. Russell, K. J., Laramore, G. E., Griffin, T. W., Parker, R. G., Davis, L. W. and Krall, J. M. *Fast Neutron Radiotherapy in the Treatment of Carcinoma of the Urinary Bladder. A Review of Clinical Trials.* Am. J. Clin. Oncol. **12**(4), 301-306 (1989).

67. Mijnheer, B. J. Battermann, J. J. and Wambersie, A. *What Degree of Accuracy is Required and can be Delivered in Photon and Neutron Therapy?* Radiother. Oncol. **8**, 237-252 (1987).

68. Horiot, J. C. *Etat des Essais Thérapeutiques du Groupe Coopérateur de Radiothérapie de l'EORTC.* Bull. Cancer/Radiother. **78**, 421-422 (1991).
69. Wambersie, A. *Le Facteur Temps en Radiothérapie Externe.* Bull. Cancer/Radiothér. **79**, 33-51 (1992).
70. Withers, H. R. and Peters, L. J. *The Application of RBE Value to Clinical Trials of High-LET Radiations.* In: High-LET Radiations in Clinical Radiotherapy. Eds G. W. Barendsen, J. J. Broerse and K. Breur, pp. 257-261 (Oxford: Pergamon Press) (1979).

DOSIMETRIC CHARACTERISTICS OF PROTON, NEUTRON AND NEGATIVE PION BEAMS AT THE PHASOTRON IN DUBNA

F. Spurný
Institute of Radiation Dosimetry (IRD)
Czechoslovak Academy of Sciences, Prague, Czechoslovakia

INVITED PAPER

Abstract — A clinical complex has been developed at the phasotron of the Joint Institute of Nuclear Research (JINR) at Dubna, in the USSR. High energy proton, neutron and pion beams were formed for use in medical-biological studies and for radiotherapy. Many dosimetric studies had already been performed in these beams also by co-workers from the Institute of Radiation Dosimetry of Czech. Acad. JG. Proton beams with energies from 250 MeV up to the region of Bragg peak were studied, with different depth–dose profiles. Other studies employed high energy neutron ($E_n \sim$ 340 MeV) and negative pion (90 to 130 MeV/c) beams. Absorbed dose values were established using thimble type ionisation chambers. Other measurements were carried out with thermoluminescent detectors and solid state nuclear track detectors. In some cases activation detectors were used to characterise the beams. A review is given of results obtained in studies since 1986.

INTRODUCTION

Photons and electrons have long been used for radiotherapy. However, it is well known that in some cases other types of radiation can have some advantages[1]:

(a) better depth–dose distribution (charged particles, negative pions), and/or
(b) higher linear energy transfer (LET) and lower oxygen enhancement ratio (heavy charged particles, negative pions).

Such types of radiation are not generally available, since high energy charged particle accelerators are necessary to produce them.

One such complex was built at the phasotron of the Laboratory of Nuclear Problems (LNP) at the Joint Institute of Nuclear Research (JINR) at Dubna, USSR. This paper briefly describes some possibilities available there for medical-biological irradiations, concentrating on the dosimetry of all available beams. Results presented were obtained in the period since 1986.

CLINICAL–PHYSICAL COMPLEX AT THE PHASOTRON OF LNP JINR

The phasotron of the LNP JINR accelerates protons to an energy of about 660 MeV. About 10^{13} protons can be produced per second[2]. The clinical–physical complex consists of three treatment rooms designated for proton therapy, one for neutron and another for negative pion irradiation[3].

Beams available

Protons

Proton beams with primary energies 100, 130,

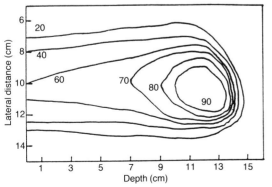

Figure 1. Isodose distribution (in horizontal plane) for a 200 MeV proton beam with modified Bragg curve in water.

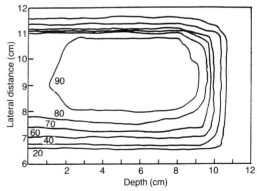

Figure 2. Isodose distribution (in horizontal plane) in water for a 130 MeV proton beam modified using a ridge type filter.

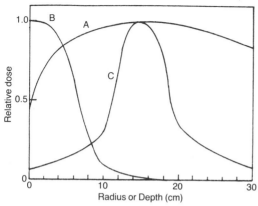

Figure 3. Absorbed dose distribution in water for a high energy neutron beam ($\bar{E}_n \sim 350$ MeV). Curves: A, depth dose distribution along the axis of the beam with diameter 100 mm; B, radial distribution of the dose at a depth of 120 mm; C, depth dose distributions in the case of full rotation of the beam with a diameter of 50 mm.

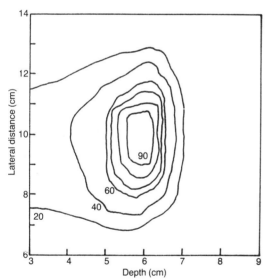

Figure 4. Isodose distribution in water for 40 MeV negative pions.

Table 1. Characteristics of negative pion beams formed at the phasotron of the LNP JINR.

Energy (MeV)	Range in H_2O (g.cm^{-2})	Range half-width (g.cm^{-2})	D_{max} (mGy)	D_{max}/D_{entr}
20	1.8	0.4	46	2.8
30	3.7	0.7	51	3.6
40	6.1	1.3	50	3.9
48	8.1	1.6	47	3.7
66	13.7	2.7	34	3.1
78	17.9	3.0	23	2.6

160, 180 and 250 MeV can be produced in the treatment rooms. A system of collimators, lenses, energy degraders and filters permits formation of beams of variable depth-dose and profile. Typical examples of depth–dose profiles for proton beams are presented in Figures 1 and 2.

Neutrons

Neutrons can be produced at the phasotron by the primary protons impinging on targets of different materials (Be, C, Al, Cu, Pb) and of different thicknesses[4]. It was found that the maximum neutron yield is obtained using a Be target with a thickness of 64 g.cm^{-2}. Other characteristics of the beam ($E_n \sim 340$ MeV) are also quite appropriate for radiotherapy applications. Typical dosimetric characteristics of the neutron beam formed are shown in Figure 3.

Negative pions

Negative pion beams are formed with the help of a large angle meson lens[5,6]. Negative pion beams with momenta between 76 and 170 Mev/c are generated. Two targets (W, Cu) were tested. The basic characteristics of the negative pion beams formed are given in Table 1, while a typical depth–dose distribution is shown in Figure 4.

Monitoring

Protons

Monitoring of proton beams is performed with plane parallel transmission ionisation chambers situated where the beam enters the treatment rooms[3]. These monitors report the total number of protons in a beam. The beam profiles are determined with multiwire ionisation chambers. Equipment involving a freely moving Si–Li detector ('isodosegraph') is used to measure beam profiles. Commercially available (GDR production) air-filled ionisation chambers type K-253 (sensitive spherical volume 1.5 cm^3) with a 'clinical dosemeter' VA-J-18 or KD-27012 are used to measure the absorbed dose in the beam, both free-in-air as well as in the phantom. One of the goals of our measurements was to verify the performance of this measuring equipment.

Neutrons

Neutron beam monitoring is performed by measuring the proton current on the Be target[4]; another monitor was also used, the K-253 ionisation chamber connected to the clinical dosemeters mentioned above.

Negative pions

The monitoring of the negative pion beam is performed primarily through measuring the proton current[5,6]. A plane parallel ionisation chamber placed where the beam enters the treatment room is used as a second monitor.

DETECTION METHODS USED AND TESTED
Ionisation chambers

Thimble-type ionisation chambers were constructed at the Nuclear Research Centre in Fontenay-aux-Roses (CEN FAR)[7,8]. One chamber is made of A-150 tissue-equivalent (TE) plastic, the second of aluminium. Both have a wall thickness of 0.5 mm; available build-up caps have thicknesses up to 6 mm for the TE chamber and up to 3 mm for the Al chamber. The geometrical volume of both chambers is 1.42 cm^3. Generally, the chambers were used air-filled, because of problems with gas accessibility during experiments. However, in the case of neutron and negative pion beams, propane was also used as flushing gas. Its flow rate was more than 100 mm^3.s^{-1} (0.36 l.h^{-1}).

Other details concerning the measurement procedures with ionisation chambers have been published elsewhere[8], the calibration was performed in the primary ^{60}Co photon beam of the IRD. Home-made current readers were used for the measurement. The reproducibility of their readings was better than ±0.1% (2σ). Responses of the chambers were always recalculated to constant temperature and pressure.

Thermoluminescent detectors

Thermoluminescent detectors (TLD) used and tested in our studies are described in Table 2. They were irradiated in sets of at least six detectors. Each set was covered on both sides by 2 mm of Teflon. They were also calibrated in the primary ^{60}Co photon beam of the IRD. TLDs were evaluated using a Pitman 654-Toledo reader.

Table 2. TLDs used and studied.

TLD	Manufacturer	Form, dimensions (mm)*
LiF	PTL 710, 717	Pressed pellet, 4.5×0.85
^{7}LiF	CEC, France	
Al-P glass	CSFR, semi-industrial	Disc, 8 × 1
CaSO$_4$:Dy	Teledyne	5% of TLD by mass in Teflon, disc 12×0.4
CaSO$_4$:Dy	Laboratory, IRD, CSFR	50% of TLD by mass in Si-rubber, disc 8×0.32

* Diameter × thickness

Solid state nuclear track detectors

Solid state nuclear track detectors (SSNTD) used and tested during our studies are listed in Table 3 including etching conditions adopted. They were etched so as to remove a layer from one side as close as possible to Δh ~ 6 μm. Evaluation of the SSNTDs was carried out using an optical microscope. Background track and spot densities were evaluated under the same conditions as the particle-induced events.

Polyethylene activation detectors

Polyethylene activation detectors were used in these studies. The radionuclides produced (^{11}C from ^{12}C) was measured using standard procedures developed in the JINR. All passive detectors were irradiated perpendicularly to the beam.

Results of the evaluation of all detectors were statistically treated and the results are presented with the standard deviation on the reliability level of 95%.

Table 3. SSNTDs used and studied.

SSNTD	Etching conditions
CR-39 (Pershore)	A – 30% KOH, 343K, 25 kV.cm^{-1}, 11 kHz, 2h
	B – 30% KOH, 343K, 1 h; afterwards still 1 h in conditions A
	C – 30% KOH, 343 K, 2 h; afterwards 30% KOH, 298 K, 4 h, 25 kV.cm^{-1}, 10 kHz
Kodak CN-85	A – 2.5N NaOH, 333 K, 2 h
	B – 5.5 N NaOH, 323 K, 4 h
	C – 5.5 N NaOH, 323 K, 4 h, 15kV.cm^{-1}, 5 kHz
Kodak LR-115	2.5 N NaOH, 313 K
Melinex 0	A – 30% KOH, 343 K; 40 kV.cm^{-1}, 5 kHz*
	B – 30% KOH, 343 K; 11 kV.cm^{-1}, 5 kHz**

* For recoiled particles (He, C, O).
** For fission fragments.

RESULTS AND DISCUSSION
Proton beams
Dosimetry with ionisation chambers

An A-150 ionisation chamber was used to determine proton tissue absorbed dose in beams of protons with primary energies 130, 160, 180 and 250 MeV (typical energy spread ±15 MeV). The chamber was positioned free-in-air behind a 5 mm thick Plexiglas plate. In some cases, an additional water shielding in front of chamber was used, simulating in this way positions along the Bragg curve. Energies of protons in these cases were 90 MeV, 55 MeV and from 0 to 80 MeV (Bragg peak position). The characteristic dose rates corresponding to a primary proton current of 1 μA varied from 5 to 25 mGy.s^{-1} (30 to 150) rad.min.$^{-1}$ depending on the proton energy and depth-dose curve modifications. The Bragg–Gray cavity theory allows the prediction of the dose delivered to a medium such as tissue from a measurement of collected charge in the air-filled ionisation chamber. In a ^{60}Co photon beam, the dose in tissue can be calculated from the equation[9]:

$$D_T^G = S_A^W \frac{(\mu_{en}/\rho)_T}{(\mu_{en}/\rho)_W} \overline{W}_A^G J_A^G \quad (1)$$

where S_A^W is the mass stopping power ratio for chamber wall material and air;
$(\mu_{en}/\rho)_T$ and $(\mu_{en}/\rho)_W$ are mass energy absorption coefficients for tissue and wall material respectively;
$\overline{W}_A^G$ is the average energy required to produce an ion pair in air for photons; and
J_A^G represents the number of ion pairs formed per unit mass of air in the sensitive volume of the chamber.

The photon chamber calibration constant is therefore equal to:

$$R^G = \frac{D_T^G}{J_A^G M} = S_A^W \frac{(\mu_{en}/\rho)_T}{(\mu_{en}/\rho)_W} \frac{\overline{W}_A^G}{M} \quad (2)$$

where M is the mass of air in the chamber

For protons, two limiting cases can be defined. In the case of a 'large' cavity, the dose to air in the chamber originates primarily from energy deposition in the air, not in the walls. Therefore, by a modified Bragg–Gray cavity theory,

$$D_T^P(L) = \frac{(dE/\rho dx)_T^P}{(dE/\rho dx)_A^P} \overline{W}_A^P J_A^P \quad (3)$$

where $(dE/\rho dx)_T^P$ and $(dE/\rho dx)_A^P$ are mass stopping powers of, respectively, tissue and air for protons; and $\overline{W}_A^P$ and J_A^P are the proton equivalent quantities to $\overline{W}_A^G$ and J_A^G.

Therefore the calibration constant of the proton chamber is, in this case equal to

$$R^P(L) = \frac{(dE/\rho dx)_T^P}{(dE/\rho dx)_A^P} \frac{\overline{W}_A^P}{M} \quad (4)$$

In the case of a 'small' cavity, the dose to air in the chamber originates primarily from proton energy transfer in the walls, whereas the air contribution is taken as negligible. From Bragg–Gray it therefore follows that in this case

$$D_T^P(S) = S_A^W \frac{(dE/\rho dx)_T^P}{(dE/\rho dx)_W^P} \overline{W}_A^P J_A^P \quad (5)$$

The calibration constant is equal to

$$R^P(S) = S_A^W \frac{(dE/\rho dx)_T^P}{(dE/\rho dx)_W^P} \frac{\overline{W}_A^P}{M} \quad (6)$$

The wall composition of both chambers is given in Table 4. The dimensions of the chambers may correspond to small, intermediate or large cavities, depending on the proton energy. From this point of view it is very important that the ratios of calibration constants depend only slightly on the limiting case assumed. One can see in Table 5 that an A-150 ionisation chamber with a ^{60}Co-based calibration constant would underestimate the proton tissue dose roughly by 1.5%. This correction was always taken into account.

None of data used to calculate the ratios is known exactly. The errors in the energy absorption coefficients, as well as the stopping powers, are estimated to be about ±1–2%, the error in the $\overline{W}$ ratio to about 4%[11,12]. Total uncertainty of the ratios presented in Table 5 is therefore about ±5.1%.

The chamber readings were also corrected for

Table 4. Composition of some materials (by mass per cent).

	K-253[16]	A-150	Tissue
H	3.2	10.13	10.2
C	52.2	77.55	12.3
N	–	3.51	2.8
O	12.0	5.23	72.6
F	25.6	1.74	–
Na	7.0	–	0.08
Ca	–	1.84	–
Others	–	–	1.02

Table 5. Ratios of photon to proton chamber calibration constants.

Chamber	$\dfrac{R^G}{R^P(L)}$	$\dfrac{R^G}{R^P(S)}$
A-150	0.986	0.985
K-253	0.949	0.942

the influence of saturation and wall thickness. It was observed that the signal for a 180 MeV proton beam increases with wall thickness, being about 2% higher for 3.5 mm wall as compared to 0.5 mm. The uncertainty of these corrections is estimated to be 2%, the reproducibility of chamber readings relative to the monitors was typically ±0.2%. The total uncertainty of the measured proton doses can therefore be estimated to be ±5.8%.

As mentioned before, the ionisation chamber K-253 is used for proton dose measurements by the local staff. It was stated that it measures tissue proton absorbed dose with an accuracy of better than ±5%[3,13]. One can see in Table 5, that this approach is not exactly correct. However, when the value of the calibration constant for exposure is used, electrometer readings will correspond with a precision better than ±1% to proton tissue absorbed dose. In this way the approach of the local staff is acceptable. The agreement between the K-253 and TE chambers was, with this approach, seen to be satisfactory (mean ratio 0.982 ± 0.017).

Passive detectors irradiation

TLDs were irradiated to verify whether their response per unit absorbed dose is the same as for photon irradiation. Results of readings were therefore expressed as the response relative to ^{60}Co radiation. They are presented in Table 6. One can see there that the relative response of all TLDs studied is actually very close to unity. This property allows them to be used for many dosimetric studies in high energy proton beams (beam profile, depth dose measurements etc.).

SSNTDs were also exposed to the proton beams studied. For SSNTDs in contact with ^{232}Th fissionable radiators, it was observed that their response is proportional to the fission cross section, with the constant of proportionality (7.4 ± 0.8) × 10^{-6} tracks per proton for a cross

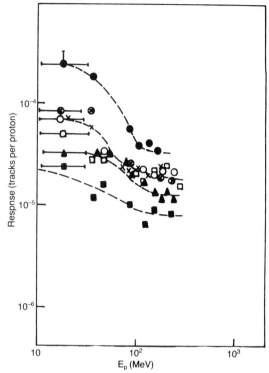

Figure 5. Responses of polymer SSNTDs without fissionable radiators to protons. Key: (●) CR-39, etching conditions A; (○) CR-39, etching conditions B; (▲) Melinex 0; (□) Kodak LR-115, all tracks; (■) Kodak LR-115, through etched tracks; (×) Kodak CN-85, etching conditions B; (⊗) Kodak CN-85, etching conditions C.

section of 10^{-28} m^2. This constant is very close to the value measured for fast neutrons (7.0 ± 0.3) × $10^{-6(14)}$. As far as SSNTDs without fission radiators are concerned, their responses are presented in Figure 5, where it can be seen that:

(a) the responses follow the same tendencies as for fast neutrons, i.e. they are highest for CR-39[14];

(b) the energy dependence of all detectors is qualitatively the same, i.e. the responses decrease with proton energies up to about 100 MeV, and, for higher energies, are more or less constant.

The second property is interesting for proton dosimetry, because the decrease of the responses is correlated with the decrease of kerma–fluence conversion factor. The track density in such a detector is therefore proportional to the proton kerma, the typical error of this correlation is ±15%.

Table 6. Relative responses of TLDs to high energy protons.

Proton energy (MeV)	Relative response* to protons for		
	^{7}LiF	Al-P glass	CaSO$_4$:Dy
179 ± 18	1.01 ± 0.08	0.96 ± 0.06	1.04 ± 0.06
160 ± 18	0.98 ± 0.06	0.94 ± 0.08	0.90 ± 0.10
131 ± 16	–	0.99 ± 0.04	1.06 ± 0.06
90 ± 18	1.11 ± 0.09	1.02 ± 0.05	0.96 ± 0.07
55 ± 16	–	1.01 ± 0.06	0.93 ± 0.07
Bragg peak	1.01 ± 0.10	1.03 ± 0.05	–

*Relative to TE chamber reading.

Table 7. Dosimetric characteristics in high energy neutron beam.

Depth in phantom (g.cm^{-2})	Tissue absorbed dose[a] (mGy)	Relative response[b] of chamber			
		Thimble TE air	Thimble Al air	K-253 No 31038	K-253 No 31076
0.3	11.8 ± 0.6	0.930	0.851[c]	–	–
19.4	21.2 ± 0.7	0.983	0.868	0.895	0.894
34.8	18.5 ± 0.5	0.971	0.874	–	–

[a]Per minute with a proton current on the Be target of 1 μA.
[b]Relative to thimble TE chamber filled with propane, $1\sigma_{rel} \sim \pm 2\%$.
[c]Here the depth was 0.38 g.cm^{-2}.

Neutron beam

Dosimetry with ionisation chambers

All experiments in the neutron beam were performed using a 295 mm depth Plexiglas phantom. The phantom was composed of plates of different thicknesses and an area of 240 × 248 mm^2. Detectors and/or measuring equipment were placed at the front, at a depth of 160 mm and at the back of the phantom. Irradiations were monitored by proton current integration as well as with the K-253 ionisation chamber placed at the exit of last collimator. The reproducibility of monitoring was about ±0.2%. The neutron beams studied were contaminated with photons and charged particles. It was estimated that the contamination in terms of tissue absorbed dose at the front of the phantom was less than 7% for photons and less than 15% for charged particles. In the region of dose maximum the contribution of external charged particles should be lower than 4%[4].

An attempt was made to verify basic dosimetric characteristics with the thimble type TE ionisation chamber, and with propane as the flushing gas.

First, the influence of the thickness of the chamber wall on the signal when the chamber as at the front of the phantom was studied. It was observed that the increase of the signal with wall thickness was very close to the tendency calculated for a pure neutron beam without any contamination from charged particles. Taking into account the precision of our measurements, the contributions of charged particles to tissue absorbed dose at the phantom surface should not be higher than about 7%.

The comparison of signals corresponding to the two filling gases showed that, relative to the ^{60}Co gamma response, the response of propane filling gas was systematically higher. The results obtained are given in Table 7. The readings of the propane filled TE (A-150) ionisation chamber were taken as the best with respect to the tissue absorbed dose. These values measured at three depths are given in Table 7. The errors represent only the reproducibility of measurements. The accuracy of these values was estimated to be not better than ±7%. In any case, our results agree rather well with some previous estimations[4].

Table 7 also presents relative responses for the thimble Al chamber filled with air and for the air equivalent chambers K-253 described above. Their responses are clearly lower than that for the TE chamber.

Passive detectors irradiation

TLDs and SSNTDs were also irradiated in high energy neutron beams. In addition to the monitors mentioned above polyethylene activation detectors (reaction $^{12}C \rightarrow ^{11}C$) were used to establish the fluence of neutrons and hadrons (inside of phantom). TLD irradiations were also monitored with a thimble TE chamber flushed with propane.

Results of the TLD irradiations are presented in Table 8 in terms of tissue kerma measurements of the TE chamber. They agree well with values obtained several years ago (0.58 for ^{7}LiF, 0.59 for Al-P glass[14]. One can see that these responses are clearly lower than 1.00 and that they depend on the position in the phantom. Both these properties are connected with the microdosimetric characteristics of high energy neutron energy transfer to matter. It is known that the dose averaged linear energy transfer (LET) in tissue of the neutron beams studied is, at the front of a phantom, about 68 keV.μm^{-1}[15] and in the region of dose maximum about 36 keV.μm^{-1}. It is also known that the thermoluminscence yield of the TLDs studied in this work decreases with LET[14].

Table 8. TLD readings in the high energy neutron beam relative to TE chamber results.

Depth in phantom (g.cm^{-2})	Relative reading for		
	^{7}LiF	Al-P glass	CaSO$_4$:Dy in Si rubber
0.40	0.54 ± 0.03	0.61 ± 0.03	–
19.4	0.74 ± 0.03	0.80 ± 0.04	0.70 ± 0.05
34.8	0.60 ± 0.04	0.70 ± 0.04	–

From this point of view it is also significant that the relative reading is higher in the region of dose maximum compared with the front of the phantom, because while at the front only 8% of dose is delivered by particles with LET not exceeding 2 keV.μm^{-1}, in the region of the maximum it is about 49%[15]. In any case, results show that one should be careful when using TLDs, even for relative measurements in high energy neutron beams.

SSNTDs irradiation was also monitored with polyethylene activation detectors. It was found that the particle fluence (neutrons and protons with E>20 MeV) decreases continuously inside the phantom. In the region of the dose maximum it is 0.886 ± 0.021 of the value at the front; at the back of the phantom this relative value is equal to 0.673 ± 0.016. Using these results, SSNTD responses could be established, and are presented in Table 9. One can see that the response is practically independent of the position in the phantom. SSNTDs can be therefore used to determine hadron fluences in a phantom irradiated with high energy neutrons.

Finally, the ratio of TE chamber readings and activation detector data established a conversion factor between the high energy neutron fluence and the maximum dose absorbed in the phantom (or human body). The value obtained was (1.33 ± 0.10) × 10^{-10} Gy.cm^2.

Negative pion beams

Ionisation chambers and TLDs were exposed to negative pion beams in the Plexiglas phantom described previously.

TLDs were exposed in the beam with a negative pion energy of 48 MeV (123 MeV/c). Al–P thermoluminescent glasses were exposed in a stack continuing 25 discs, each 1 mm thick. The stack was positioned at the beginning of the 'peak

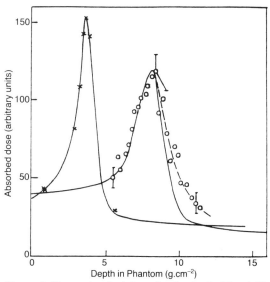

Figure 6. Tissue absorbed dose distribution in 30 and 48 MeV negative pion beams as measured with: (—) Si (Li) semiconductor detector; (o) TLD, Al-P glass; (x) TE chamber filled with air.

region'. TLD readings were compared with the results of measurements performed with a semiconductor (Si-Li) detector calibrated with the air-equivalent ionisation chamber K-253[6]. A comparison of both sets of results is presented in Figure 6. One can see that both sets of data agree well in the front part of the peak region. The TLD response decreases more slowly with the depth in the phantom behind the peak. The reason for this difference is not known, it could follow from different scattering properties of a tissue-equivalent material and the glass in an aluminium holder.

Ionisation chamber studies were performed in a beam with negative pion energy of 30 MeV (95 MeV.c). Relative readings of the TE chamber filled with air are, together with a depth–dose curve obtained with the semiconductor detector are presented in Figure 6. One can see that both sets of data agree very well. The ratio of doses at the maximum and at the 'plateau' is about 3.6, in agreement with previous estimates[5,6].

A determination of the influence of the chamber composition on the value of absorbed dose was attempted. For the thimble TE chamber not only air but also propane was used as a filling gas. The thimble Al chamber filled with air was also tested. The responses of these chambers in three points on the depth curve (plateau, peak, behind peak region) are presented in Table 10. It is evident that the filling gas has a small influence on the TE chamber response at the three points on the depth curve studied. In contrast, the response of the Al

Table 9. Responses of some SSNTDs to high energy neutrons.

SSNTD	Etching conditions	Response, tracks per hadron at the depth in phantom (g.cm^{-2}) of		
		0.30	19.4	34.8
CR-39	C	3.0 × 10^{-5}	2.8 × 10^{-5}	2.6 × 10^{-5}
CN-85	B	4.2 × 10^{-5}	3.9 × 10^{-5}	–
LR-115 all tracks	–	1.8 × 10^{-5}	2.0 × 10^{-5}	2.0 × 10^{-5}
LR-115 through etched	–	4.7 × 10^{-6}	4.5 × 10^{-6}	4.8 × 10^{-6}
MELINEX	A	1.4 × 10^{-6}	1.5 × 10^{-6}	–
MELINEX with Th radiator	B	5.0 × 10^{-6}	5.4 × 10^{-6}	–

Table 10. Responses of thimble-type ionisation chamber in a negative pion beam ($E_{\pi^-} = 30$ MeV).

Depth in phantom (g.cm^{-2})	Relative response* of a chamber	
	TE filled with C_3H_8	Al filled with air
0.80	1.05 ± 0.06	–
3.5 (max)	1.01 ± 0.06	0.72 ± 0.04
4.8	0.97 ± 0.05	0.84 ± 0.05

* Relative to TE air chamber reading.

chamber filled with air is substantially lower, and depends on position, being lower in the peak region. The value corresponding to the region behind the peak is very close to the ratio of calibration constants in photon and charged particle beams[10]. These results show once again that the choice of detector material for high energy particle beams is a problem which should be treated carefully.

Tissue absorbed doses in the negative pion beam were measured. The absorbed dose in the maximum measured with the TE ionisation chamber was equal to about 66 mGy.min^{-1} at a proton current of 1 μA. This is about 25–30% higher than that observed in previous measurements[5,6]. The difference exceeds the error of our results (< ±10%). The reason for this effect will be the subject of future studies.

ACKNOWLEDGEMENT

I am much obliged to V. M. Abazov, V. P. Bamblevski, E. P. Cherevatenko, A. M. Molokanov and V. P. Zorin for their invaluable assistance during experiments, and to O. V. Savchenko for his support of this cooperation.

REFERENCES

1. Raju, M. R. *Pions and Heavy Ions in Radiotherapy: A Brief Review.* LA-UR-74-1530 (Los Alamos) (1974).
2. Komotchkov, M. M. *Radiation Studies in the Department of Radiation Safety of the JINR (1979–89).* Reprint JINR P16-89-539 (in Russian) (Dubna) (1989).
3. Abazov, V. M. et al. *Formation and Research of Therapy Proton Beams at the Reconstructed Phasotron of the LNP JINR.* Reprint JINR P9-86-648 (in Russian) (Dubna) (1986).
4. Abazov, V. M. et al. *Formation and Research of Therapy High Energy Neutron Beam at the Phasotron of the LNP JINR.* Reprint JINR P18-88-392 (in Russian) (Dubna) (1988).
5. Abazov, V. M. et al. *Formation and Research of High Intensity Meson Beams for Medico-Biological and Physical Studies at the Phasotron of the LNP JINR. Composition of Meson Beams.* Reprint JINR P9-90-69 (in Russian) (Dubna) (1990).
6. Abazov, V. M. et al. *Formation and Research of High Intensity Meson Beams for Medico-Biological and Physical Studies at the Phasotron of the LNP JINR. Physico-Dosimetrical Parameters of Beams.* Reprint JINR P9-90-68 (in Russian) (Dubna) (1990).
7. Broerse, J. J., Zoetelief, J., Burger, G., Schraube, H. and Ricourt, J. *A Small Scale Neutron Dosimetry Intercomparison.* EUR 6567 EN (Luxembourg: CEC) (1979).
8. Spurný, F. and Votočková I. *A Contribution to Neutron Beam Dosimetry with Ionization Chambers.* In: Proc. 4th Symp. on Neutron Dosimetry. EUR 7448 EN. (Luxembourg: CEC) Vol. II, p. 327 (1981).
9. Vervey, L. J., Koehler, A. M., McDonald, J. C., Goitein, M., Ma, I-C., Schneider, R. J. and Wagner, M. *The Determination of Absorbed Dose in a Proton Beam for Purposes of Charged-Particle Radiation Therapy.* Radiat. Res. **79**, 34 (1979).
10. IAEA. *Absorbed Dose Determination in Photon and Electron Beams — An International Code of Practice.* Technical Report Series No. 277 (Vienna: IAEA) (1987).
11. Janni, J. *Proton Range–Energy Tables.* At. Data Nucl. Data Tables **27**(2-5) (1982).
12. ICRU. *Average Energy Required to Produce an Ion Pair.* Report No. 31 (Bethesda, MD: ICRU Publications) (1979).
13. Zielczynski, M., Molokanov, A. G. and Cherevatenko, J. P. *Experimental Determination of the Wall Material Influence on the Ionization Chamber Responses in High Energy Particle Beams.* Reprint JINR P16-88-524 (in Russian) (Dubna) (1988).
14. Spurný, F. *Methods of Dosimetry for External Exposure and their Application.* DSc thesis (in Czech) Prague (1984).
15. Serov, A. J., Sychev, B. S. and Cherevatenko, J. P. *Depth Distribution of LET-Spectra in Water Absorber Irradiated with High Energy Nucleons.* Reprint JINR P18-87-670 (in Russian) (Dubna) (1987).
16. Novotný, J., private communications — results of specially performed chemical analysis (1990).

STUDIES RELATING TO 62 MeV PROTON CANCER THERAPY OF THE EYE

V. P. Cosgrove†, A. C. A. Aro†, S. Green†, M. C. Scott†, G. C. Taylor†, D. E. Bonnett‡ and A. Kacperek§
†Medical Physics Group
School of Physics and Space Research, University of Birmingham
Edgbaston, Birmingham B15 2TT, UK
‡Department of Medical Physics
Leicester Royal Infirmary
Leicester LE1 5WW, UK
§MRC Cyclotron Unit, Clatterbridge Hospital
Bebington, Wirral, L63 4JU, UK

Abstract — The 62 MeV proton cyclotron at Clatterbridge Hospital is now being used for the treatment of eye tumours. As part of an on-going study of the Clatterbridge proton beam line, a three-dimensional Monte Carlo program has been developed to model the passage of protons through the various beam modifiers. These include collimators, scattering foils, range shifters and modulators, and are used to provide both beam uniformity and to spread the proton Bragg peak over a tumour volume. The effects of these devices on, for example, the proton energy spectrum and penumbra are calculated and compared with equivalent measurements. A series of microdosimetric measurements has also been made using a planar tissue–equivalent gas filled proportional counter, a redesign of a previous model. Details of the design of this counter, together with measurements on the 62 MeV proton beam are also presented.

INTRODUCTION

Protons and heavy ions are used worldwide in the treatment of benign and malignant tumours[1]. Eye tumours (ocular melanoma) have been treated at Clatterbridge Hospital since June 1989 using a Scanditronix MC60 cyclotron producing a 62 MeV proton beam. The beam layout, illustrated in Figure 1 and described previously[2], begins with a small beam iris, for initial beam shaping, positioned just after the final focusing magnets of the cyclotron. This is followed by a double tungsten scattering foil system together with a brass central stopper, which act as a passive beam spreader[3], the stopper removing the central part of the beam and the foils scattering the protons to provide a uniform distribution at the end of the line. These are all in a vacuum line, beyond which a Perspex (Lucite) range shifter and modulator are located. The range shifter defines the maximum penetration of the beam while the modulator, which is a rotating stepped vane, produces a uniform distribution of dose throughout the target volume. Antiscatter collimators operate as their name suggests, while a final brass patient collimator defines the final target area at the end of the line. This paper reports on the first results from an on-going investigation.

PROGRAM DETAILS

The Monte Carlo program has been designed to model the passage of protons from the beam iris to the isocentre, the point where protons interact with a target volume. Interactions with all the various beam modifiers are considered in the calculations. Brief details of the model and of approximations are as follows:

1. The beam is treated as a circular source of protons with all points and angles chosen using a random number generator. A uniform distribution over this source area is simulated, with a small azimuthal angle introduced into the proton projection, which has a gaussian frequency distribution, to simulate beam broadening.

2. All proton scatter is assumed to follow a gaussian distribution whose width is calculated using the Rossi small-angle multiple scattering relation[4]:

$$\sigma_\theta = \frac{21.2}{pc\beta} \sqrt{\left(\frac{t}{X_0}\right)}$$

where

pc = momentum of incident particle (MeV)
β = v/c of incident particle
X_0 = radiation length of scattering material (g.cm^{-2})
t = scatterer thickness in the same units as X_0

Although this approximation does not account for single or plural scattering, it does account for the majority of scattering events.

3. Energy loss of the protons is calculated assuming the continuous slowing down approximation, with straggling randomly introduced following a gaussian distribution. The width of this distribution, $\sigma_{\Delta E}$, is given by the relation determined by Kellerer[5]:

$$\frac{\sigma^2_{\Delta E}}{\Delta E} = \frac{E_{max}}{2\ln(E_{max}/I)} \qquad E_{max} \gg I$$

where ΔE is the mean energy loss, $E_{max} = 2mc^2 \beta^2/(1-\beta^2)$ is the maximum electron energy possible and I is the mean excitation potential.

4. Since these approximations are only valid for energy losses which are small relative to the initial proton energy, all scattering materials are treated as a series of randomly thin slabs, similar to the method used by Urie et al[6] and Goitein et al[7].

5. The modulator was simulated by randomly choosing a step thickness for each proton to pass through, weighted to the contribution each step makes to the modulator shape.

6. Proton energy and path length relations used in the calculations are from the tables by Janni[8].

BEAM PENUMBRA

Knowledge of the beam penumbra width as a function of thickness of Perspex in the beam line plays an important role in treatment planning. Measurements of penumbra using a silicone diode in a water phantom have been made by Bonnett et al[2] for two beam layouts. The first had the modulator and range shifter placed 600 mm from the final collimator to maximise the length of the beam line in vacuum. The second had the modulator and range shifter placed further upstream, just after the scattering foils. Consequently, the length of beam in vacuum was reduced.

Figure 2 illustrates these measurements for 25 mm diameter collimated beams. The effective thickness of Perspex is defined as the width of the range shifter plus half maximum modulator width. The beam penumbra width is measured between the 90% and 10% isodose lines. This width is seen to be an increasing function of Perspex thickness for the first beam layout, while for the second it remains effectively constant over the same thicknesses.

Monte Carlo calculations were carried out for both these two layouts, and results are also shown in Figure 2. These calculations were made using a two-dimensional version of the code, to reduce computer running time. This is a reasonable approximation since proton scatter is considerably forward peaked.

Again, the increase in penumbra width with effective thickness of Perspex is seen for the first

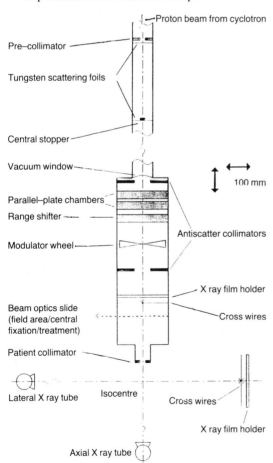

Figure 1. The proton therapy beam line at Clatterbridge.

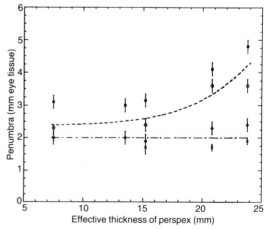

Figure 2. Proton beam penumbra as a function of effective thickness of Perspex. 1st layout: (o) measurement, (□) Monte Carlo. 2nd layout: (x) measurement, (◊) Monte Carlo.

beam layout and a relatively constant penumbra for the second. The calculated penumbras, however, are seen to be higher in magnitude than the equivalent measurements. If the Rossi scattering formula used in calculations was to be reduced by the correction factor experimentally determined by Urie et al[6], then the penumbras are likely to be reduced accordingly. Preliminary calculations using this correction factor, for beams slowed by an effective thickness of 23.8 mm of Perspex, gave an average penumbra width of 3.95 ± 0.2 for the first beam layout and 2.3 ± 0.2 for the second. These compare with measurements of 3.6 ± 0.2 and 1.9 ± 0.1 respectively.

RELATIVE DOSE WITH DEPTH

Measurements of the relative dose with depth in eye tissue are routinely made on the Clatterbridge beam. These measurements are made by exposing a silicon diode stepwise through increasing thicknesses of Perspex. A Monte Carlo program was run to simulate these measurements which are illustrated together in Figure 3. The results are for a 25 mm diameter collimated beam without range shifting.

Although the Monte Carlo calculation predicts a slightly lower relative dose in the plateau region, a difference still being evaluated, good agreement is seen at the overlap of the distal edges. The depth at 90% of the distal edge corresponds to 0.996 of the usually quoted proton range[9]. Measured and calculated ranges using this edge for the two beam layouts are found in Table 1.

A decrease in the proton range is seen for layout 2, and is expected, since the amount of beam transport in vacuum is reduced. This change is due to the extra scatter and energy reduction in air. The Monte Carlo calculations also predict these changes, and compare well with measurements.

Also included in the table are mean proton range values which are directly calculated from tables[8] using the calculated mean proton energy. These values are also consistent with the measurements.

ENERGY SPECTRA

Calculations of the final proton energy spectrum have been compared with measurements made by Kacperek (private communication) using a plastic scintillator detector. Spectra for an unmodulated beam have a large high energy peak with a small low energy tail of events. Figure 4 is a comparison of an experimental and a 3-D calculation for a 10 mm collimated beam. This graph has a logarithmic vertical scale to emphasise the low

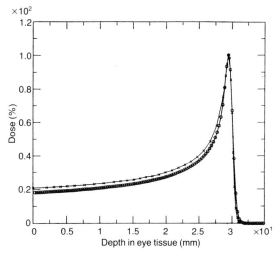

Figure 3. Relative dose with depth in eye tissue. (x) measured with diode, (o) Monte Carlo calculation.

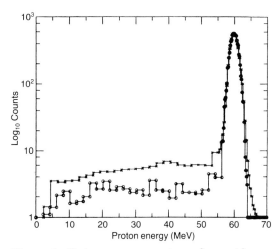

Figure 4. Proton energy spectrum for a 10 mm collimated beam with no range shifting or modulation. (o) Monte Carlo calculation, (x) measurement with plastic scintillator.

Table 1. Range of protons in eye tissue for the two different beam layouts.

Beam layout	1	2
Measurement (using 90% distal edge)	30.4 ± 0.2 mm	29.7 ± 0.2 mm
Monte Carlo calculation (using 90% distal edge)	30.6 ± 0.15 mm	29.85 ± 0.15 mm
Monte Carlo calculation (using mean energy/range relations)	30.5 ± 0.4 mm	29.8 ± 0.4 mm

energy tail, which is itself smoothed into wider bins. The experimental curve has been scaled to the peak height and position of the calculated curve for comparison, as no calibration of the spectrum could be made. A resolution factor was also introduced into the Monte Carlo as an approximation to the resolution of the scintillator detector.

With these approximations, the most useful information that can be gained from the plots is the ratio between peak and plateau area. Figure 5 shows these ratios for various final collimator diameters. Both plots are seen to curve in a similar fashion as collimator size is increased. For small collimator diameters, the low energy events are a significant part of the final spectra. This would be expected if we assume that the low energy events are mostly due to scatter from the final collimator edge.

Although the relative magnitude of the curves differ as collimator size increases, these differences may reflect different modes of cyclotron operation. For example, adjustments were made to the cyclotron to decrease beam current so that the detector count rate could be reduced. More adjustments had to be made for the larger collimator sizes. How this affected the final energy spectrum is unknown.

PROTON MICRODOSIMETRIC DETECTOR

Design details of the planar microdosimetric detector are shown in Figure 6. The detector has a cylindrical cavity 8 cm in diameter and 3 cm deep. To limit energy loss, the entrance window was made of 50 μm thick aluminised Mylar. To simulate 2 μm of tissue, the counter was filled with 7 kPa (70 mbar) of methane-based tissue-equivalent gas. Under this low pressure the Mylar window bowed inwardly. 2 μm sheets of aluminised Mylar were therefore used to define the collection volume to ensure parallel beams of protons had equal path lengths in this volume. Mylar was chosen since it provides a better transparency to the beam than the thinnest metallic grid that could be practically built and it ensured that no slow moving secondary electrons could enter the collection volume, contributing to measured dose.

Although three anodes are illustrated, since it was originally intended to use them to scan wide beams, only the central wire was used in practice. A Perspex window was also included in the design, to enable high energy proton beams to be measured in a water phantom, since it provides a flat entrance window.

MICRODOSIMETRIC MEASUREMENTS

Microdosimetric measurements have been made on the Clatterbridge proton beam using the above counter for three differently range shifted beams[10]. Measurements were limited to a maximum collimator diameter of 10 mm to keep the detector count rate low. Similarly the cylcotron was run at very low beam currents. The y.d(y) curves, illustrated in Figure 7, are all calibrated to the mid-point of the proton edge and normalised to unity. Table 2 shows the corresponding values of $\bar{y}_d$ and $\bar{y}_f$ for each curve. (Note that errors are statistical only).

As expected, the greatest changes in $\bar{y}_d$ occur when the protons are at the end of their range; this value is reduced when such protons are spread over the target volume by the modulator. These results will be linked to radiobiological experiments, made with identical range shifting, at a later date.

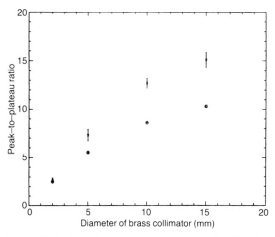

Figure 5. Peak-to-plateau ratios for a series of collimator diameters. (o) plastic scintillator, (x) Monte Carlo.

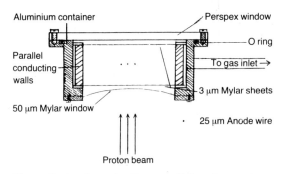

Figure 6. A schematic diagram of the planar microdosimetric detector.

CONCLUSIONS

The Clatterbridge 62 MeV proton beam

transport system has been modelled using Monte Carlo programming techniques. Calculations of beam penumbra, range and energy spectrum have compared favourably with measurements of the same, indicating its usefulness in understanding how each beam modifier will affect the final beam. Further work is under way to improve on the approximations used in the program as well as to increased running efficiency. Monte Carlo predictions of microdosimetric spectra for different beam layouts are also planned.

A planar microdosimetric counter had also been built and tested on the Clatterbridge proton beam. The y.d(y) spectra measured at different depths in Perspex illustrate the wide range of lineal energies that can be provided with this sort of beam. Future work will include measurements of higher energy proton beams as well as experiments to investigate delta ray effects which we have so far been unable to quantify.

ACKNOWLEDGEMENTS

Personal support for V.P.C. has been provided by the Clatterbridge Cancer Research Trust and is acknowledged with gratitude. The support of the staff at both Clatterbridge and the Birmingham Medical Physics Group throughout the research project is gratefully acknowledged.

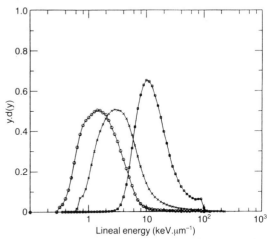

Figure 7. y.d(y) distributions at three depths in Perspex measured with the planar microdosimetric detector. (x) modulated, (□) peak, (o) plateau.

Table 2. Dose parameters for the microdosimetric spectra.

Effective thickness of Perspex (mm)	Position	$\bar{y}_f$ (keV.μm^{-1})	$\bar{y}_d$ (keV.μm^{-1})
3.1	Plateau	1.27 ± 0.002	2.33 ± 0.02
28.5 (14 + $^1/_2$ modulator width)	Modulated	2.55 ± 0.002	5.15 ± 0.02
27	Peak	9.07 ± 0.001	14.52 ± 0.04

REFERENCES

1. Sisterson, J. M. *Clinical Use of Protons and Ion Beams from a World-wide Perspective.* Nucl. Intrum. Methods Phys. Res. **B40/41**, 1350-1353 (1989).
2. Bonnett, D. E., Kacperek, A. and Sheen, M. A. *Characteristics of a 62 MeV Proton Therapy Beam.* In: Proc. 2nd European Particle Accelerator Conf. Vol 2, pp. S12-S14 (1990).
3. Gottschalk, B. *Double-Scattering System with Optimum Dose Uniformity in Proton Radiotherapy.* Harvard Cyclotron Laboratory Technical Note PTA8/1/86 (1986).
4. Rossi, B. *High Energy Particles.* 2nd Ed (New York: Prentice-Hall) pp. 64-67 (1956).
5. Kellerer, A. M. *Mikrodosimetrie. Grundlagen einer Theorie der Strahlenqualität.* Report GSF B-1 (Institut für Biologie der GSF, Munich) (1968). (Also found in ICRU Report 36, *Microdosimetry*, p. 18 (1983).)
6. Urie, M. M., Sisterson, J. M., Koehler, A. M., Goitein, M. and Zoesman, J. *Proton Beam Penumbra: Effects of Separation Between Patient and Beam Modifying Devices.* Med. Phys. **13**(5), 734-741 (1986).
7. Goitein, M. and Sisterson, J. M. *The Influence of Thick Inhomogeneties on Charged Particle Beams.* Radiat. Res. **74**, 217-230 (1978).
8. Janni, J. F. *Proton Range-Energy Tables, 1 keV-10 GeV.* Atomic Data and Nuclear Data Tables **27**(2/3) (1982).
9. Goitein, M., Gentry, R. and Koehler, A. *Energy of Proton Accelerator Necessary for Treatment of Choroidal Melanomas.* Int. J. Radiat. Oncol. Biol. Phys. **9**, 259-260 (1983).
10. Aro, A. C. A., Bonnett, D. E., Cosgrove, V. P., Green, S., Kacperek, A., Scott, M. C. and Taylor, G. C. *Microdosimetric Measurements on the Clatterbridge Proton Therapy Beam.* In: Proc. 2nd European Particle Accelerator Conf. Vol. 2, pp. S17-S19 (1990).

THE PRIMARY ATTENUATION COEFFICIENT OF A p(66)+Be(40) NEUTRON THERAPY BEAM

A. N. Schreuder†, D. T. L. Jones†, S. Pistorius‡ and W. A. Groenewald‡
†Division of Medical Radiation, National Accelerator Centre
PO Box 72, Faure, 7131 South Africa
‡Department of Medical Physics, Tygerberg Hospital
Tygerberg, 7505 South Africa

Abstract — The concept of primary and scattered dose components of a radiotherapy beam is commonly used in radiotherapy planning. Four different methods which have been used for photon beams were applied to determine the primary dose linear attenuation coefficient in water (μ_0) in a p(66)+Be(40) neutron therapy beam. They were: (a) the extrapolation of measured tissue-maximum ratios to zero field size, (b) linear attenuation measurements, (c) dose measurements in phantom with and without a central axis attenuator, and (d) fitting a central axis kerma model, based on convolution techniques, to measured percentage depth dose data. The μ_0 values obtained agree well with each other and vary between 0.071 cm^{-1} and 0.075 cm^{-1}.

INTRODUCTION

Dose calculations often make use of the concept of separating the dose at a point in a medium into primary and scattered components[1]. This partitioning of dose facilitates, for example, dose calculations in irregular and blocked fields[2]. The primary dose, $D_p(d)$, at a depth d may be defined as the energy deposited per unit mass by charged particles generated by the incident uncharged radiation interacting with the medium for the first time. The primary dose is therefore determined by uncharged radiation which interacts within a cross sectional area of radius equal to the mean lateral range, λ, of the liberated charged particles in the phantom material[3]. By definition the primary dose is therefore independent of field size and is only a function of depth. The scattered dose, $D_s(d,A)$, at depth d in a beam with cross sectional area A, is due to charged particles originating from uncharged radiation that has interacted with the medium more than once. The scattered dose may be ascribed to interactions in the medium at a distance greater than λ from the point of interest and is therefore a function of field size and depth. The total absorbed dose, $D_t(d,A)$, anywhere in a beam, may be expressed as:

$$D_t(d,A) = D_p(d) + D_s(d,A) \qquad (1)$$

Since neutrons are indirectly ionising radiations the transfer of energy from neutrons to a medium is a two stage process as it is with photons. The National Accelerator Centre's (NAC) p(66)+Be(40) neutron therapy beam has characteristics similar to those of a 8 MV X ray beam[4]. It may therefore be possible to use the concepts developed for photons to separate neutron doses into primary and scattered components.

Only two quantities are needed to define the primary dose component, viz. the primary attenuation coefficient (μ_0) and a normalising factor which determines the magnitude of the primary dose relative to the total dose for a reference depth and field size[5]. The aim of this study was to determine μ_0 for the p(66)+Be(40) neutron beam, using four different methods which were initially proposed for photon beams. These methods were: (1) extrapolation of measured tissue maximum ratios (TMRs) to zero field size[6,7]; (2) linear attenuation measurements[8]; (3) measurements in phantom with and without a central axis (CAX) attenuator[3,9], and (4) fitting an analytical CAX kerma model, based on convolution techniques, to measured depth dose data[10].

THEORETICAL ASPECTS

TMR extrapolation method

By expressing TMRs measured at different depths in phantom as a function of the geometrical parameter z = rd/(r+d) (r = radius of equivalent circular field, d = depth in phantom)[6,7], the zero-area TMR can be found by linear extrapolation as follows:

$$TMR(z,d) = TMR(0,d) + N(d)z \qquad (2)$$

The TMR(0,d) data, which are proportional to $D_p(d)$, were then fitted with an exponential to obtain μ_0.

Linear attenuation method

Attenuation measurements are generally performed under narrow beam conditions to minimise the amount of scattered radiation reaching the detector[8]. Since it is not possible to obtain an infinitesimally small beam, μ was measured as a function of field size and the data were extrapolated to a zero-area beam[8,11] to obtain μ_0.

The CAX attenuator method

Nizin et al[3] proposed and tested in a ^{60}Co beam[9] a method to measure the primary dose component based on modifying the primary radiation by means of a small diameter attenuator between the source and the point of interest. The main requirements are that the radius of the attenuator should exceed the mean lateral range, λ, of the secondary charged particles and that the attenuator should alter the primary radiation significantly with negligible perturbation of the scattered radiation. It was shown that under these conditions the primary dose at depth d, $D_p(d)$, can be determined from:

$$D_p(d) = \frac{D_t(d,A) - D_t'(d,A)}{\left[1 - \frac{1}{C(d)}\right]} \quad (3)$$

where $D_t(d,A)$ and $D_t'(d,A)$ are total doses at the point of interest without and with the attenuator respectively. The parameter $C(d) = D_p(d)/D_p'(d)$ is the ratio of unattenuated to attenuated primary doses at depth d and is independent of field size. $C(d)$ compensates for a possible hardening of the beam in the attenuator and has to be measured separately under narrow beam conditions[3,9] because it is not possible to measure $D_p(d)$ and $D_p'(d)$ directly.

The CAX kerma model

Pistorius[10] has developed a semi-empirical central axis kerma model which employs convolution techniques. In this model it is assumed that for SSD = ∞ the primary kerma is exponential while the scattered component may be separated into forward and backscattered terms. In a parallel beam where the effective attenuation coefficient remains constant with depth, the backscattered component is proportional to the total fluence moving in the forward direction at that depth. The forward scattered contribution is obtained by the convolution of the primary kerma with a function which describes the spread of the scattered neutrons. For a beam where scattering is predominantly in the forward direction it is assumed that the spread function is separable and that an exponential function may be used to describe the spread of the forward scattered neutrons on the central axis. Summing the primary, forward and backscattered terms gives the following equation for the total kerma[10]:

$$K(s,d,SSD=\infty) = N_1(s)\exp[-\mu(s)d]\{1-N_2(s)\exp[-\Gamma(s)d]\} \quad (4)$$

with

$$N_1(s) = N_p\, BSF(s)\left[\frac{N_s(s)\eta(s)}{\Gamma(s)}+1\right]$$

$$N_2(s) = \frac{N_s(s)\eta(s)}{N_s(s)\eta(s)+\Gamma(s)}$$

and

$$\Gamma(s) = \eta(s) - \mu(s)$$

where N_p = primary kerma normalisation, $N_s(s)$ = integral normalisation of the scatter kernel, $BSF(s)$ = backscatter factor, $\mu(s)$ = effective attenuation of primary component, $\eta(s)$ = effective attenuation of scattered component, s = equivalent square field size, d = depth in phantom.

It can be shown[10] that this equation may be expressed in the form:

$$K(s,d,SSD=\infty) = N.\exp[-\mu(s,d)d] \quad (5)$$

where

$$\mu(s,d) = \mu_0(1-N_2(s)) + \Gamma(s) - d^{-1}$$
$$(\ln\{[c(s)\exp(\Gamma(s)d)-1]/(c(s)-1)\}),$$
$$c(s) = 1/N_2(s)$$
and $\mu_0 = 2\, N_2(s)\,\eta(s)$

EXPERIMENTAL DETAILS

TMRs were measured in a water phantom for field sizes ranging from 2.5×2.5 cm^2 to 25×25 cm^2 and depths of 2 cm to 25 cm. The 2.5×2.5 cm^2 equivalent square field was achieved by inserting a 2 cm diameter steel collimator (field size of 2.7 cm diameter at the depth of maximum dose (d_{max}) in water phantom) into the collimator of the isocentric therapy unit. All the measurements were made with a 0.5 cm^3 tissue-equivalent (TE) ionisation

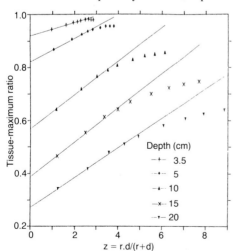

Figure 1. Measured tissue-maximum ratios as a function of z. The solid curves are linear fits (Equation 2) to the data up to a field size of 10×10 cm^2 and r is the radius of the equivalent circular field and d is the depth in phantom.

chamber, fitted with a 4 mm thick A-150 build-up cap and flushed with methane-based TE gas. Typical TMRs as a function of the z geometrical parameter are illustrated in Figure 1. TMR(z,d) is only linear in z for field sizes smaller than 10×10 cm^2 and the data were therefore fitted with Equation 2 up to a maximum field size of 10×10 cm^2.

Attenuation of the neutron beam in water was measured for field sizes ranging from 2.5×2.5 cm^2 to 20×20 cm^2. The measurements were made with the gantry angle set at 180° so that the beam passed vertically upwards through the water tank. Different thicknesses of water were obtained by changing the water level in the tank. All the measurements were made with a 1 cm^3 TE ionisation chamber flushed with TE gas. Measurements showed that by placing the detector 80 cm from the bottom of the tank, which was 140 cm from the source, the number of neutrons scattered into the detector was minimised. To provide the necessary build up, while minimising backscatter, a 1.5 cm thick A-150 disc (2×2 cm^2) was placed in front of the detector. Each set of attenuation data was fitted with an exponential function to obtain μ. Figure 2 shows the values obtained for different field sizes. The error bars represent fitting errors only (1 standard deviation). In order to extrapolate to zero field size, these data were fitted with a third order polynomial[11]. Although there is no physical basis for this it provides an excellent fit to the data ($\chi^2 = 1.5 \times 10^{-6}$).

For the CAX attenuator method measurements were made in a water phantom for various field sizes (5.5×5.5, 10×10 and 20×20 cm^2), using detectors of different sizes and with two different attenuating materials (Fe, Pb). D_t and D'_t were measured at different depths using both a 0.5 cm^3 and a 0.05 cm^3 TE ionisation chamber. Both the Fe and Pb CAX attenuators were 5 cm long by 1 cm radius to satisfy the lateral range requirement[3]. The lateral range, λ, of the charged particles in the neutron beam was measured in a 5.5×5.5 cm^2 field with a 0.05 cm^3 TE ionisation chamber in open geometry. The chamber was coaxial with the central axis of the beam and nylon cylinders of different diameters were successively mounted on the detector. The effective lateral range was found to be 5.4 mm which corresponds to a lateral range of 5.5 mm in water. C(d) values were measured in narrow beam geometry (2 cm diameter collimator) with the same experimental set-up as described above for the conventional attenuation measurements. Illustrative $D_t(d,A)$, $D'_t(d,A)$, $D_p(d)$ and C(d) measurements are shown in Figure 3.

CAX percentage depth doses (CAPDDs) were measured earlier with a 0.3 cm^3 TE ionisation chamber at SSD = 150 cm. A non-linear least squares algorithm was used to fit the CAX kerma model to CAPDDs for depths greater than d_{max} where transient charged particle equilibrium exists and dose is assumed to be equal to kerma. CAPDDs for 12 field sizes ranging from 5.5×5.5 to 29×29 cm^2 up to a maximum depth of 35 cm were used in the fit (828 data points).

Due to the polyenergetic nature of the neutron beam, beam hardening in the water medium causes the attenuation of the primary beam to be non-exponential. A single exponential was therefore fitted to the first 15 cm of attenuation data only for methods 1, 2 and 3.

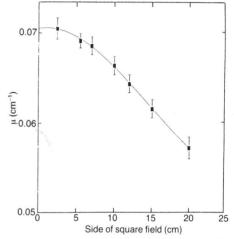

Figure 2. Linear attenuation coefficient as a function of field size. The curve is a 3rd order polynomial fit to the data.

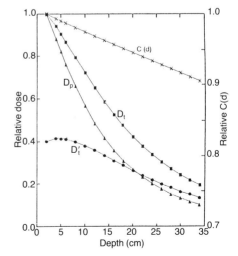

Figure 3. Measured $D_t(d,A)$, $D'_t(d,A)$, $D_p(d)$ and C(d) for a 10×10 cm^2 field with a Fe CAX attenuator. D'_t is normalised to $D_p = D_t = 1$ at a depth of 2 cm. The symbols are defined in the text.

RESULTS AND DISCUSSIONS

The results obtained for μ_0 by the four different methods are given in Table 1. The errors shown represent statistical fitting errors only. The statistical fitting error is an estimate of one standard deviation on the fitted parameter (μ_0). The reproducibility of individual dose measurements was better than 1%. The value of μ_0 varied between 0.071 cm^{-1} and 0.075 cm^{-1}. A value of μ_0 = 0.078 cm^{-1} was calculated from a kerma spectrum[12] and the total cross sections of hydrogen and oxygen[13]. Although this value is somewhat higher than the value measured in this work, the difference can be ascribed to uncertainties in the spectrum in the low energy region and in the oxygen total cross sections in the high energy region. Different techniques of extrapolating TMRs to zero field size have been questioned[14]. However, expressing TMR as a function of the z parameter allows more accurate extrapolation of TMRs. The attenuation measurements (Figure 2) show that narrow beam conditions are achieved with equivalent square fields less than 2.5 × 2.5 cm^2 in size and there is therefore confidence in the μ_0 value obtained with this method. The consistency of the μ_0 value obtained with the CAX attenuator method for different field sizes and attenuating materials lends credence to this technique. Within the experimental errors no significant field size dependence was observed. The CAX kerma model gave an excellent fit to the percentage depth dose data as shown in Figure 4. The standard deviation of the residuals (of all the data fitted) shown in the figure was 0.6 % which is well within the experimental uncertainties of the measurements.

The excellent agreement between the different methods supports the assumption that the CAX kerma model is applicable to the p(66)+Be(40) neutron beam and this in turn allows the calculation of depth doses at any depth and for any field size. In addition, the μ_0 values determined with the CAX kerma model for other filter and wedge combinations in the NAC neutron beam may therefore be used with confidence without the need for independent measurements.

Although the general characteristics of the neutron beam are similar to those of a 8 MV X ray beam (μ_0 = 0.041 cm^{-1} for a typical beam[10]), the μ_0 values obtained here are much larger and are even larger than for a ^{60}Co beam (μ_0 = 0.066 cm^{-1} [15]). The scatter contribution is therefore much larger in the p(66)+Be(40) neutron beam than in high energy photon beams. This complicates neutron therapy planning and emphasises the need for the scattered component to be taken into account in a proper manner.

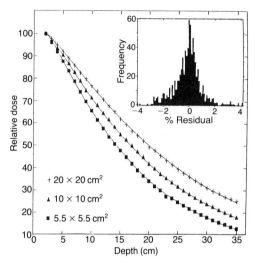

Figure 4. Measured CAPDDs for flattening filter 1. The solid curves are non-linear least square fits of the CAX kerma model to the measured CAPDD data. The residuals are indicated as a frequency distribution of the percentage difference between the measured and calculated data for all the data fitted (12 CAPDD curves, 828 data points).

Table 1. Primary linear attenuation coefficients for a p(66)+Be(40) neutron beam in water.

Method	μ_0 (cm^{-1})	1 standard deviation (%)
Calculated from kerma spectrum	0.078	–
(1) TMR extrapolation	0.075	0.6
(2) Attenuation	0.071	0.5
(4) CAX kerma model	0.072	1.0
(3) CAX attenuator		

Attenuator	Field size (cm × cm)	Ion chamber volume (cm^3)		
Fe	5.5 × 5.5	0.5	0.074	1.3
Fe	10 × 10	0.5	0.073	1.3
Fe	20 × 20	0.5	0.072	1.8
Fe	5.5 × 5.5	0.05	0.070	1.4
Fe	10 × 10	0.05	0.072	1.5
Fe	20 × 20	0.05	0.074	1.7
Pb	10 × 10	0.05	0.073	1.3
Average (CAX attenuator)			0.073	1.7

REFERENCES

1. Clarkson, J. R. *A Note on Depth Doses in Fields of Irregular Shape.* Br. J. Radiol. **14**, 265-268 (1941).

2. Day, M. J. *The Normalised Peak Scatter Factor and Normalised Scatter Functions for High Energy Photon Beams.* Br. J. Radiol. Suppl. **17**, 131-136 (1983).
3. Nizin, P. and Kase, K. *A Method of Measuring the Primary Dose Component in High-Energy Photon Beams.* Med. Phys. **15**(5), 683-685 (1988).
4. Jones, D. T. L., Yudelev, M. and Hendrikse, W. L. J. *Physical Characteristics of the South African High Energy Neutron Therapy Facility.* Radiat. Prot. Dosim. **23**(1-4), 365-368 (1988).
5. Rice, K. R. and Lee, M. C. *Monte Carlo Calculations of Scatter to Primary Ratios for Normalisation of Primary and Scatter Dose.* Phys. Med. Biol. **35**(3), 333-338 (1990).
6. Bjärngard, B. E. and Petti, P. L. *Description of the Scatter Component in Photon-Beam Data.* Phys. Med. Biol. **33**(1), 21-32 (1988).
7. Nizin, P. S. *Geometrical Aspects of Scatter-to-Primary Ratio and Primary Dose.* Med. Phys. **18**(2), 153-160 (1991).
8. Van Dyk, J. *Broad Beam Attenuation of Cobalt-60 Gamma Rays and 6-, 18- and 25-MV X-Rays by Lead.* Med. Phys. **13**(1), 105-110 (1986).
9. Nizin, P. S. and Kase, K. R. *Determination of Primary Dose in ^{60}Co Gamma Beam Using a Small Attenuator.* Med. Phys. **17**(1), 92-94 (1990).
10. Pistorius, S. *PhD Thesis* University of Stellenbosch, South Africa (1991).
11. Robinson, D. M. and Scrimger, J. W. *Monoenergetic Approximation of a Polyenergetic Beam: A Theoretical Approach.* Br. J. Radiol. **64**, 452-454 (1991).
12. Jones, D. T. L., Brooks, F. D., Symons, J. E., Nchodu, M. R., Allie, M. S., Fulcher, T. J., Buffler, A. and Oliver, M. J. *Neutron Fluence and Kerma Spectra of a p(66)/Be(40) Clinical Source.* Med. Phys. (In press).
13. Garber, D. I. and Kinsey, R. R. *Neutron Cross Sections: Volume II, Curves.* BNL 325, Third Ed (Brookhaven National Laboratory) (1976).
14. Mohan, R. and Chui, C. *Validity of the Concept of Separating Primary and Scatter Dose.* Med. Phys. **12**(6), 726-730 (1985).
15. Godden, T. *Gamma Radiations from Cobalt 60 Teletherapy Units.* Br. J. Radiol. Suppl. **17**, 45-49 (1983).

TISSUE-MAXIMUM RATIOS FOR A p(66)+Be(40) NEUTRON THERAPY BEAM

M. Yudelev*, A. N. Schreuder and D. T. L. Jones
Division of Medical Radiation, National Accelerator Centre
PO Box 72, Faure, 7131 South Africa

Abstract — Tissue-maximum ratio (TMR) is a useful concept in isocentric treatment planning as it is independent of SSD. To obviate the necessity for tedious additional measurements a formula for calculating TMRs from central axis depth dose and scatter factor data has been derived for megavoltage photon beams. The applicability of this formula has been tested in the National Accelerator Centre's p(66)+Be(40) neutron therapy beam which has similar physical characteristics to those of 8 MV X rays. It was found that by substituting a collimator scatter correction term for the phantom scatter correction term in the standard formula and scaling the field size appropriately somewhat better overall agreement between measured and calculated TMRs was obtained. For clinical beams less than 16×16 cm^2 all TMRs calculated using both formulae are within 3% of the measured values.

INTRODUCTION

The concept of tissue-air ratio (TAR) was originally devised[1–3] to facilitate isocentric treatment planning with ^{60}Co beams. The definition of TAR precludes its use in megavoltage photon beams where thick build-up caps are required for the measurements in air. The scatter contribution from the build-up cap itself renders the interpretation of the measurements problematical. Furthermore, for small fields, the size of the build-up cap may exceed the field dimensions. To overcome these problems the concept of tissue-phantom ratio (TPR), in which measurements are only made in phantom, was devised[4,5]. Holt et al[6] later proposed the use of tissue-maximum ratio (TMR) (a special case of TPR) for use in megavoltage beams. Another special case of TPR, viz. tissue-standard ratio (TSR) is also used in some applications[7]. The TMR is defined as the ratio of the dose in phantom at point P to that dose at P when P is at the position of maximum build-up. Because all the ratios mentioned above are independent of source-to-surface distance (SSD) they are used in isocentric treatment planning, but are also useful for irregular field[7,8] and asymmetrically collimated field dosimetry[9]. Furthermore, use of these ratios obviates the necessity of separating the dose into primary and scatter components.

To avoid tedious measurements, methods of deriving TMRs using scatter factors and available central axis percentage depth dose (CAPDD) data measured at constant SSD have been formulated[6,10,11]. The concept devised by Khan et al[5] is a general one, applicable to photon beams of any energy. In very high energy photon beams the depth of maximum dose depends on field size as well as on SSD. To account for this and to derive machine-independent functions, Khan et al[11] have chosen the reference depth as the maximum depth of maximum dose. Since the CAPDDs for the National Accelerator Centre's (NAC) p(66)+Be(40) neutron therapy beam are very similar to those of 8 MV X rays[12] it may be possible to calculate TMRs from CAPDDs using similar relationships. To date all patients at NAC have been treated at a fixed SSD. Isocentric treatments are being contemplated for head and neck cases and the use of the TMR formalism will facilitate treatment planning. TMRs have been measured directly and have also been derived from measured scatter factors and existing CAPDD data obtained at the standard SSD = SAD (source-to-axis distance) of 150 cm using the formalism of Khan et al[11]. The measurements were made for a range of filter and field size combinations.

EXPERIMENTAL METHODS

TMR is defined as[6,11]:

$$\mathrm{TMR}(d,W_d) = \frac{D(d,W_d,\mathrm{SSD}+d)}{D(t,W_d,\mathrm{SSD}+d)}$$

where

D = dose at specified depth, field size and distance from source,
d = depth of measurement in phantom,
t = maximum depth of maximum dose
W_d = field size at depth d and distance SSD+d from source.

TMR can also be derived from CAPDD measurements using[11]:

---

* Present address: Gershenson Radiation Oncology Center, Harper Hospital, Detroit, MI 48201, USA

$$\text{TMR}(d, W_d) = \frac{P(d, W, \text{SSD})}{100} \left(\frac{\text{SSD}+d}{\text{SSD}+t} \right)^2 \frac{S_p(W_t)}{S_p(W_d)} \quad (1)$$

where

- P = CAPDD at depth d for field size W at distance SSD from source,
- W, W_t = field sizes for W_d projected to surface and maximum depth of maximum dose respectively
  $W = W_d [\text{SSD}/(\text{SSD}+d)]$
  $W_t = W_d [(\text{SSD}+t)/(\text{SSD}+d)]$

$S_p(W_t)$, $S_p(W_d)$ = phantom scatter correction factors for field sizes W_t and W_d.

The phantom scatter correction factor S_p is defined as the ratio of the dose measured at the reference depth t in phantom for a given field size (W_i) to that for a reference field size (W_o) when the same collimator opening is used[11]. Since the direct measurement of S_p is difficult[11,13] it can be calculated from[11]:

$$S_p(W_i) = \frac{S_{cp}(W_i)}{S_c(W_i)}$$

where S_{cp} and S_c are respectively the total scatter correction factor and the collimator scatter correction factor, both of which are readily measurable. This definition of phantom scatter correction factor is identical to that of normalised peak scatter factor (NPSF)[14].

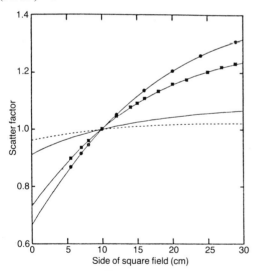

Figure 1. Measured collimator (S_c) and total scatter (S_{cp}) correction factors for square fields using filter F1 together with the derived phantom scatter correction factors (S_p). Also shown are the phantom scatter correction factors for a typical 8 MV X ray beam[14]. The S_c and S_{cp} data have been fitted with 3rd order polynomials. Key: (■) collimator; (●) total; (——) phantom; (- - -) phantom (8 MV).

The total scatter correction factor S_{cp} is defined as the ratio of the dose D at the reference depth t (maximum depth of maximum dose) in phantom for a given field size (W_i) to the dose at the same point and depth for the reference field size (W_o)[11]:

$$S_{cp}(W_i) = \frac{D(t, W_i)}{D(t, W_0)}$$

The measurements are made at SDD = SAD (SDD = source-to-detector distance). The output factor[15] is essentially the same as the total scatter correction factor, except that the former is measured at SDD = SAD+t.

The collimator scatter correction factor S_c is defined as the ratio of the dose D in air at SDD = SAD for a given field size (W_i) to that of a reference field size (W_o)[11]:

$$S_c(W_i) = \frac{D_{air}(W_i)}{D_{air}(W_0)}$$

All TMR and scatter factor measurements were made with a 0.5 cm^3 tissue-equivalent (TE) ionisation chamber, flushed with methane-based TE gas at SDD = 150 cm. For measurements in air a 15 mm thick A-150 build-up cap was used, while for measurements in the water phantom a 4 mm thick build-up cap was used. Recently collimator scatter correction factors were measured in a narrow coaxial cylindrical phantom[16], but no differences were observed between these measurements and those made with appropriate build-up caps. CAPDDs were measured earlier[12] using 0.3 cm^3 TE ionisation chambers.

TMRs (at several depths) as well as both collimator and total scatter correction factors were measured for a range of field sizes and a few filter combinations. The dependence of depth of maximum dose on field size is small[12] and the reference depth for all measurements was 2 cm (i.e. t = 2 cm). The reference field size was 10×10 cm^2 (i.e. W_o = 10).

RESULTS

Measured collimator and total scatter correction factors are illustrated in Figure 1. These measurements were made using flattening filter F1 and have been fitted with 3rd order polynomials. The resultant phantom scatter correction factors (normalised peak scatter factors) are also shown in Figure 1, together with phantom scatter correction factors for a typical 8 MV X ray beam[14]. These curves illustrate the fact that the scatter component in the p(66)+Be neutron beam is greater than that in a 8 MV X ray beam.

Measured TMRs were compared with values calculated using Formula 1. It was found empirically that if the phantom scatter term in Formula 1 was replaced with the corresponding collimator scatter term and the field size for determining the relevant CAPDD was increased to W_d, better agreement between measured and calculated values was obtained.

Calculations were therefore also performed in all cases using the following formula:

$$\text{TMR}(d, W_d) = \frac{P(d, W_d, \text{SSD})}{100} \left(\frac{\text{SSD} + d}{\text{SSD} + t} \right)^2 \cdot \frac{S_c(W_t)}{S_c(W_d)} \quad (2)$$

All measured TMRs have been compared with TMRs derived using both Formula 1 and Formula 2. Illustrative examples of measured and derived (from both Formula 1 and Formula 2) TMRs for square fields are given in Figures 2, 3 and 4 which show some of the data obtained for filter F1, and the F12 and F1W3 filter combinations. F1 is an iron flattening filter (used on its own for small fields) which is always in the beam; F2 is an additional flattening filter which is inserted in the clinical beam for fields with sides greater than 16 cm (the combination of F1 and F2 is designated F12); W3 is a tungsten wedge filter (used for small fields only) which rotates the beam profiles by about 45° at a depth of 10 cm in water. Measurements at different distances confirmed that for the p(66)+Be(40) beam TMRs are independent of SSD over the clinical range of interest.

DISCUSSION

The largest differences between measured and derived TMR data are for filter F12 for 15×15 cm² and 20×20 cm² fields at depths of 10 and 20 cm where differences of up to 7% for both formulae were encountered. For all cases using filter F1 the agreement between direct measurements and data derived with both formulae is less than 3%. It was established that the standard equivalent-square field formula[17] can be used to calculate TMRs for

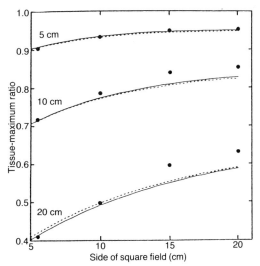

Figure 3. Measured (●) tissue-maximum ratios for the F12 filter combination at different depths in water for different field sizes compared with calculations using Formula 1 (——)[11] and Formula 2 (----) (this work). In the clinical situation this filter combination is only used for fields with sides greater than 16 cm.

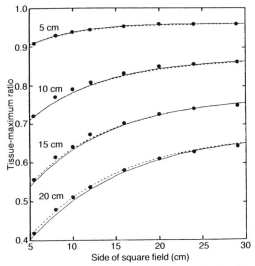

Figure 2. Measured (●) tissue-maximum ratios for filter F1 at different depths in water for different field sizes compared with calculations using Formula 1 (——)[11] and Formula 2 (----) (this work).

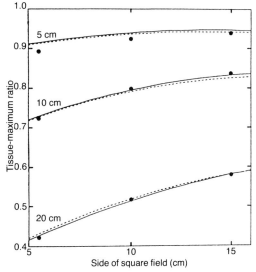

Figure 4. Measured (●) tissue-maximum ratios for the F1W3 filter combination at different depths in water for different field sizes compared with calculations using Formula 1 (——)[11] and Formula 2 (----) (this work).

rectangular fields with both formulae.

Generally the discrepancy between measured and derived TMRs increases with increasing field size and increasing depth in phantom. The data derived using Formula 2 agree overall somewhat more closely with the measurements ($\chi^2 = 0.0131$) than do the data derived using Formula 1 ($\chi^2 = 0.0167$). Because of space limitations in the NAC treatment unit (the distance between the end of the collimator cladding and the isocentre is 34 cm) it is only really practical to consider isocentric treatments for head and neck cases. These cases are usually treated with relatively small fields ($\leq 16 \times 16$ cm^2) for which only filter F1 on its own or in combination with wedge filters is used. If the standard formula[11] for deriving TMRs from CAPDD data in high energy photon beams is modified by replacing the phantom scatter term with the corresponding collimator scatter term and the field size is appropriately scaled, the TMRs obtained can be used for isocentric treatment planning with small fields on the NAC's p(66)+Be(40) neutron therapy facility. Only relatively few and simple in-air measurements of collimator scatter factors are required in addition to existing central axis depth dose data.

REFERENCES

1. Johns, H. E., Whitmore, G. F., Watson, T. A. and Umberg, F. H. *A System of Dosimetry for Rotation Therapy with Typical Rotation Distributions.* J. Can. Assoc. Radiol. **4**, 1 (1953).
2. Johns, H. E. *Physical Aspects of Rotation Therapy.* Am. J. Roentgenol. **79**, 373-381 (1958).
3. Gupta, S. K. and Cunningham, J. R. *Measurement of Tissue-Air Ratios and Scatter Functions for Large Field Sizes, for Cobalt-60 Gamma Radiation.* Br. J. Radiol. **39**, 7-11 (1966).
4. Karzmark, C. J., Deubert, A. and Loevinger, R. *Tissue-Phantom Ratios — An Aid to Treatment Planning.* Br. J. Radiol. **38**, 158-159 (1965).
5. Saunders, J. E., Price, R. H. and Horsley, R. J. *Central Axis Depth Doses for a Constant Source-Tumour Distance.* Br. J. Radiol. **41**, 464-467 (1968).
6. Holt, J. G., Laughlin, J. S. and Moroney, J. P. *The Extension of the Concept of Tissue-Air Ratios (TAR) to High-Energy X-Ray Beams.* Radiology **96**, 437-446 (1970).
7. Hounsell, A. R. and Wilkinson, J. M. *Tissue Standard Ratios for Irregularly Shaped Radiotherapy Fields.* Br. J. Radiol. **63**, 629-634 (1990).
8. Cunningham, J. R., Shrivastava, P. N. and Wilkinson, J. M. *Program Irreg — Calculation of Dose from Irregularly Shaped Radiation Beams.* Comput. Programs Biomed. **2**, 192-199 (1972).
9. Loshek, D. D. *Analysis of Tissue-Maximum Ratio/Scatter-Maximum Ratio Model Relative to the Prediction of Tissue-Maximum Ratio in Asymmetrically Collimated Fields.* Med. Phys. **15**(5), 672-682 (1988).
10. Purdy, J. A. *Relationship between Tissue-Phantom Ratio and Percentage Depth Dose.* Med. Phys. **4**(1), 66-67 (1977).
11. Khan, F. M., Sewchand, W., Lee, J. and Williamson, J. F. *Revision of Tissue-Maximum Ratio and Scatter-Maximum Ratio Concepts for Cobalt 60 and Higher Energy X-Ray Beams.* Med. Phys. **7**(3), 230-237 (1980).
12. Jones, D. T. L., Yudelev, M. and Hendrikse, W. L. J. *Physical Characteristics of the South African High Energy Neutron Therapy Facility.* Radiat. Prot. Dosim. **23**(1-4), 365-368 (1988).
13. Krithivas, G. and Rao, S. N. *Dosimetry of 24 MV X-Rays from a Linear Accelerator.* Med. Phys. **14**(2), 274-281 (1987).
14. Day, M. J. *The Normalised Peak Scatter Factor and Normalised Scatter Functions for High Energy Photon Beams.* Br. J. Radiol. Suppl. **17**, 131-136 (1983).
15. Bewley, D. K., Bradshaw, A. L., Burns, J. E., Cohen, M., Day, M. J., Godden, T. J. Greene, D., Jennings, W. A., Lillicrap, S. C., Smith, C. W. and Williams, P. C. *Central Axis Depth Dose Data for Use in Radiotherapy. Glossary of Terms.* Br. J. Radiol. Suppl. **17**, 143-147 (1983).
16. van Gasteren, J. J. M., Heukelom, S., van Kleffens, H. J., van der Laarse, R., Venselaar, J. L. M. and Westermann, C. F. *The Determination of Phantom and Collimator Scatter Components of the Output of Megavoltage Photons Beams: Measurement of the Collimator Scatter Part with a Beam-Coaxial Narrow Cylindrical Phantom.* Radiother. Oncol. **20**, 250-257 (1991).
17. Day, M. J. and Aird, E. G. A. *The Equivalent-Field Method for Dose Determinations in Rectangular Fields.* Br. J. Radiol. Suppl. **17**, 105-114 (1983).

MONTE CARLO CALCULATIONS OF THE EFFECT OF AIR CAVITIES ON THE DOSE DISTRIBUTION OF d(14)+Be NEUTRONS

P. Meissner
Institut für Medizinische Strahlenphysik
Universitätsklinikum Essen, Hufelandstrasse 55
D-4300 Essen 1, Germany

Abstract — A Monte Carlo program based on a simplified physical model was applied to calculate kerma distributions of pencil beams in a homogeneous water phantom and in an inhomogeneous phantom containing air-filled cylinders with their axis along the pencil beam axis. The front side of the cylinders was at 4 cm depth, the radius and the length of the cylinder were varied. The air cavity causes a shift of the first collision depth dose curve behind it which corresponds to the length of the cylinder. For the scattered neutron dose the situation is more complex. Here the difference between the pencil beam distribution in the homogeneous and in the inhomogeneous phantom was calculated and taken as the scattered neutron dose distribution of a fictive pencil beam of the air-filled cylinder. Like a normal pencil beam component this distribution can be fitted by a function which depends on the radius and on three parameters which are functions of phantom depth. The observed functions are briefly described and compared with those of the pencil beam in the homogeneous phantom. The fictive pencil beam of the air-filled cylinder can be taken as an additional dose component. The convolution integral gives the dose distribution in an extended field. In this distribution the dose increases in a narrow region around the cavity from zero to a peak inside the cavity. Behind the cavity the distribution passes a negative minimum and then returns slowly to zero. However, this distribution is negative.

INTRODUCTION

Air-filled cavities are a serious problem in treatment planning for neutron therapy, especially in the head and neck region. Ignoring an air cavity like the paranasal sinus or the mouth can result in a neutron absorbed dose at the spinal cord, which exceeds the tolerance dose and is dangerous for the patient.

At first one expects that an air cavity changes only the attenuation of the neutrons traversing it. But there should also be an influence on the scattered neutrons, as nearly no collisions occur in the cavity; thus, a cavity which has limited extensions should modify the dose distribution everywhere in its surroundings and not only behind it. Until now, however, only the modified attenuation of the neutrons passing through an air cavity has been considered[1].

To investigate the influence of an air-filled cavity on the neutron dose distribution simplified Monte Carlo calculations were applied to a water phantom containing an air-filled cylinder.

THE METHOD

Normally, treatment planning for fast neutrons is based on dose distributions measured in a homogeneous water phantom. In a patient, however, the dose distribution is influenced by the locally varying atomic composition of the human body. The atomic structure is available only as a mean value in voxels given by nuclear magnetic resonance or X ray computer tomograms. On the basis of these data Monte Carlo calculations can be used to determine dose distributions[2]. However, this is a task for large computers. In a clinical environment the method normally cannot be used, because large computers are not available and calculation times are too long for routine treatment planning. Inhomogeneities are therefore often ignored in treatment planning, or empirical shift factors are applied to approximate the dose distribution in a patient by shifting the isodoses which were calculated for a homogeneous situation.

At a slightly reduced degree of accuracy, which corresponds to that of the dose measurements, one can simplify the physical model and develop a Monte Carlo code that fits into a small computer: the Monte Carlo method is then available in the clinic. However, because of long calculation times the application is still limited to basic investigations.

Following this line for the d(14)+Be neutron beam of the therapy facility in Essen a Monte Carlo computer program was developed. There are five simplifications of the physical model used in this approach:

(1) The calculations are made for a water phantom, as treatment planning is based on dose distributions measured in water.
(2) The Monte Carlo calculations give pencil beam distributions. The integration to the dose distribution in extended fields is a separate task[3].

(3) The radiation source is assumed to be a point source. This is applied not only to the radiation coming from the target but also to the radiation which is produced or scattered in the collimator. This assumption can be justified because of the distance of 125 cm from the target to the isocentre in Essen.
(4) Isotropic elastic scattering is applied for all collisions of neutrons in water. According to quantum theory of scattering this assumption is correct for collisions of neutrons with protons at energies below about 17.5 MeV. The d(14)+Be neutron spectrum fits into that range. For collisions of neutrons with oxygen, however, the model is a serious simplification which can be justified only by the relatively small contribution of these collisions to the neutron absorbed dose in water. The necessary cross sections were digitised from graphs in BNL 325[4] using a graphic tablet.
(5) The Monte Carlo calculations give kerma distributions. The energy deposition in the first collision of the incident radiation, for which charged particle equilibrium does not exist, is scored in the first collision kerma distribution. It is a function of phantom depth, which is stored in a one-dimensional array. The energy depositions in all following collisions, for which charged particle equilibrium exists and the kerma and the dose are equal, are scored in the scattered dose distribution. It is a function of phantom depth and distance from the beam axis, which is stored in a two-dimensional array. The energy depositions of the photons produced by the $^1H(n,\gamma)^2D$ reaction inside the phantom are accumulated in an additional two-dimensional array.

The resulting Monte Carlo program has the advantage that it runs on a personal computer. In the range up to 20 MeV calculated neutron kerma factors deviate on average by -1.6% from the values calculated by Caswell et al[5]. The calculated dose distributions agree quite well with those measured in a water phantom.

The total absorbed dose is a sum of five components. In an extended field each component ($k=1$ to 5) is calculated from the pencil beam distribution $P_k(x,y,z)$ by the evaluation of a convolution integral.

At the reference point in Essen at 5 cm phantom depth on the axis of the 10 cm × 10 cm field with 125 cm target-to-surface distance the components of the total absorbed dose are:

Neutrons:		first collision	47.8%
		scattered	45.5%
Photons from outside			
the phantom:		first collision	1.36%
		scattered	0.23%
Photons from the $^1H(n,\gamma)^2D$			
reaction in the phantom:			5.11%

It has to be emphasised that for neutrons both dose components are nearly equal whereas for the photons coming from outside the phantom the first collision kerma is dominant.

THE INFLUENCE OF AN AIR-FILLED CYLINDER ON THE SCATTERED NEUTRON DOSE DISTRIBUTION OF A PENCIL BEAM

The Monte Carlo program described above was applied to investigate the influence of air-filled cavities on the neutron dose distribution of a pencil beam of d(14)+Be neutrons. The calculations were performed for a homogeneous water phantom and for a water phantom containing an air-filled cylinder. The cylinder had its axis along the pencil beam axis, the z axis of the coordinate system. The front side of the cylinder was at a depth $z_f=4$ cm. The calculations were repeated for different combinations of the radius R_i and the length L_i of the cylinder. For $L_i=2$ cm, R_i was increased from 1 cm to 10 cm. For $R_i=1$ cm and $R_i=2$ cm, L_i was varied from 2 cm to 10 cm resulting in a depth z_b of the back between 6 cm to 14 cm.

As expected, the first collision kerma at any depth z^* behind the air-filled cavity agrees with the value found in the homogeneous phantom at a depth z^*-L_i. Thus, the first collision depth dose curve behind the cavity can be calculated by shifting the curve of the homogeneous phantom by the length of the cavity. The scattered neutron dose on the axis is zero at the back of the air-filled cylinder. Then, within a few centimetres, it increases to a maximum before it decays exponentially with depth. This is similar to the behaviour near the surface, but it cannot be obtained by shifting the depth dose curve of the homogeneous phantom: therefore, the scattered neutron dose distribution of the neutrons (k=2) has to be analysed in detail. The calculated scattered neutron dose distributions of the phantom containing the cavity were subtracted from the dose distribution in the homogeneous water phantom. The differences were taken as separate pencil beam distributions, named pencil beam dose distributions of the air-filled cylinder here. Figure 1 gives an example. The scattered neutron dose distribution of this fictive pencil beam can be evaluated in the same way as the other pencil beam components.

Fitting a pencil beam dose distribution by analytical functions

For the calculation of a dose distribution it is convenient to have analytical functions which describe the pencil beam distributions $P_k(x,y,z)$. The results of the Monte Carlo calculations, however, are mean values of the distributions $P_k(x,y,z)$ in voxels, which are separated by concentric cylinders with equidistant radii r_i and by planes parallel to the surface at equidistant depths z_j. The best way to determine the analytical functions here is a three step procedure. At first at any depth z_j the radial integrals $I_k(r_i, z_j)$ are calculated for all values r_i. Then the radial dependence of the values $I_k(r_i, z_j)$ is fitted using the equation

$$I_k(r_i, z_j) = \int_0^{r_i} P_k(r', z_j)\, dr'$$

$$= A_k(z_j)\, (1 - \exp[-r_i/B_k(z_j)]C_k(Z_j))$$

Finally, functions $A_k(z)$, $B_k(z)$ and $C_k(z)$ are determined which fit the values found at the depths z_j. Here $A_k(z)$ is the value of the integral for $r \to \infty$, it is called the amplitude. $B_k(z)$ is the profile width and $C_k(z)$ is the exponent.

Analytical functions describing the scattered neutron dose distribution of the pencil beam

In a homogeneous phantom the parameters of the scattered neutron dose of the pencil beam are denoted by $A_2(z)$, $B_2(z)$ and $C_2(z)$. The amplitude $A_2(z)$ is the sum of two exponential functions, one positive and slowly decreasing the other negative and more rapidly decaying. Both the profile width $B_2(z)$ and the exponent $C_2(z)$ increase linearly with phantom depth, after a negative deviation at the surface has decayed exponentially within a few centimetres.

For the scattered neutron dose of the fictive pencil beam dose distribution of the air-filled cylinder the parameters are denoted by $A'_2(z)$, $B'_2(z)$ and $C'_2(z)$. They depend in general on the dimensions of the cavity. As an example, Figure 2 gives the amplitude $A'_2(z)$ for a cylinder with a length of 10 cm and a radius of 2 cm which is compared with the amplitude $A_2(z)$ of a pencil beam in a homogeneous phantom. The dependence of the parameters on phantom depth is fitted separately for the range $z_f \leq z \leq z_b$ of the cavity and for the range $z \geq z_b$ behind the cavity.

In the range $z_f \leq z \leq z_b$ the amplitude $A'_2(z)$ decreases with depth like the amplitude in the homogeneous phantom. At a given depth it increases with the radius R_i to the value of the homogeneous phantom. The profile width $B'_2(z)$ depends only on the radius R_i. It increases from lower values with the radius to the value of the homogeneous phantom. The exponent of the profile function $C'_2(z)$ is greater than in the homogeneous phantom, it depends only on the depth behind the front side of the cavity $z-z_f$ and decays exponentially.

In the range $z \geq z_b$ the amplitude $A'_2(z)$ becomes negative and then approaches zero. The deviation from the amplitude of the pencil beam in the homogeneous phantom $A_2(z)$ can be described by

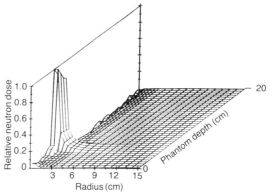

Figure 1. Scattered neutron dose distribution of the fictive pencil beam of the air-filled cavity for d(14)+Be neutrons in a water phantom with an air-filled cylinder calculated as the difference of the scattered neutron dose distribution of a pencil beam in a homogeneous water phantom and the scattered neutron dose distributions of the pencil beam in a water phantom containing an air-filled cylinder. (The cylinder has a radius of 10 cm and a length of 2 cm, the front side is at 4 cm depth.)

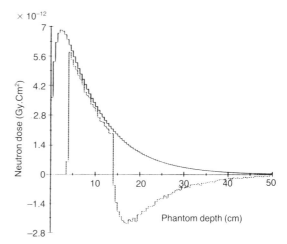

Figure 2. Amplitude of the scattered neutron dose of a d(14)+Be neutron pencil beam as a function of depth in a water phantom. (Solid line, pencil beam in a homogeneous water phantom; broken line, fictive pencil beam of the air-filled cylinder. The cylinder has a radius of 2 cm and a length of 10 cm, the front side is at 4 cm depth.)

a superposition of a slowly decaying negative and a more rapidly decreasing positive exponential function. All parameters of these two exponential functions change linearly with the length of the cavity. Both the width $B'_2(z)$ and the exponent $C'_2(z)$ of the profile function pass a minimum and then increase to the values of the homogeneous phantom. The increase is faster for the exponent than for the profile width. The deviation from the values in the homogeneous phantom can be fitted by superposition of two exponential functions, which depend on the variable $z-z_b$, the distance from the back side of the cavity.

Using the functions described so far, one can calculate the scattered neutron dose distribution for the fictive pencil beam of the air-filled cylinder. The procedure is identical to that used for the other dose components. The resulting dose distribution has to be added to the sum of the first collision and scattered neutron dose components. Inside the cavity it is negative and reduces the dose to zero. As the profile width of the fictive dose distribution is greater than zero, there is also a dose reduction beside the air-filled cylinder, but the effect is relatively small. Behind the cavity, however, the additional term increases within a few centimetres from negative values to a positive maximum and then returns to zero. The neutron depth dose curve is thus modified and decays more slowly than in a homogeneous phantom.

CONCLUSION

The first collision depth dose curve of the neutron pencil beam in a water phantom containing an air-filled cavity can be calculated by shifting the depth dose curve of a homogeneous phantom by the length of the cavity.

The changes in the scattered neutron dose distribution produced by an air-filled cavity in a water phantom can be described by a fictive pencil beam dose distribution of the air-filled cavity, which is the difference between the scattered dose distributions in the homogeneous phantom and in a phantom containing the air-filled cylinder. This distribution can be calculated using a set of analytical functions which correspond to that found for the pencil beam in a homogeneous water phantom.

Consequently one can apply the method, which is already used to calculate the neutron and photon dose components from convolution integrals. As a result one has an additional dose distribution which has to be added as a third term to the first collision and scattered neutron dose. This term modifies the scattered neutron dose distribution of the homogeneous phantom to that of the phantom containing the air cavity.

The introduction of the fictive dose distribution of the cavity describes the effect of the cavity on the scattered neutron dose in a better way than the method of isodose shift, which is valid only for the first collision kerma.

REFERENCES

1. Pfister, G., Prillinger, G., Hehn, G., Krass, C. and Stiller, P. *Absorbed Dose and Recoil Spectra at Critical Tissue Boundaries Characterised by the Absence of Recoil Equilibrium..* In Proc. Fourth Symp. on Neutron Dosimetry, Munich 1981, Vol II, EUR 7448. Eds G. Burger and H. G. Ebert. (Luxembourg: Commission of the European Communities) pp. 91-101 (1981).
2. Emmett, M. B. *The MORSE Monte Carlo Radiation Transport Code System.* ONRL-4972 (1975).
3. ICRU. *Use of Computers in External Beam Radiotherapy Procedures with High-energy Photons and Electrons.* ICRU Report 42 (Bethesda, MD: International Commission on Radiation Units and Measurements) (1987).
4. Garber, D. I. and Kinsey, R. R. (eds) *Neutron Cross Sections, Vol. 1, Resonance Parameters.* BNL 325 (3rd edn), Physics – TID-4500, 18th edn. (Brookhaven National Laboratory Associated Universities Inc. Springfield, IL: Associated Universities) (1973).
5. Caswell, R. S., Coyne, J. J. and Randolph, M. L. *Kerma Factors of Elements and Compounds for Neutron Energies Below 30 MeV.* Int. J. Appl. Radiat. Isot. **33**, 1227-1262 (1982).

NEUTRON SPECTROMETRY AND DOSIMETRY FOR BORON NEUTRON CAPTURE THERAPY

C. A. Perks and J. A. B. Gibson
AEA Environment and Energy, B. 364
Harwell Laboratory, Oxfordshire, OX11 0RA, UK

Abstract — Techniques are described that are used for neutron energy spectrometry of beams of intermediate energy neutrons being set up for boron neutron capture therapy (BNCT) research. These include high resolution proton and alpha recoil spectrometry (in the energy range from 10 keV to 15 MeV), activation detectors and spectrum modification using a variety of scattering materials. As an example, results of measurements made recently on the HB11 filtered beam in the JRC, Petten, High Flux Reactor (HFR) are given. The total fluence rate in this beam (determined from activation detector measurements) is 4.95×10^8 cm^{-1}.s^{-1}, which corresponds to a neutron kerma rate in tissue of 1.59 Gy.h^{-1}. In the energy range above 10 keV, the fluence rate is only 0.912×10^8 cm^{-1}.s^{-1} (about a fifth of the total), but corresponding to over 92% of the total neutron kerma rate in tissue. Consequently, these measurements highlight the importance of making high resolution spectrum measurements using recoil counters for BNCT beams.

INTRODUCTION

Boron neutron capture therapy (BNCT) is a technique, currently in the research stage, for the treatment of cancer[1]. In this technique, ^{10}B is preferentially introduced into the tumour cells. Irradiation with thermal neutrons causes the ^{10}B to split, producing ^{7}Li ions and alpha particles, both of which are short range and highly damaging. Hence, if the partition ratio of ^{10}B concentration in the tumour compared with normal cells is sufficiently large, the tumour cells are killed, whilst the healthy tissue is relatively undamaged. Beams of intermediate energy neutrons (less than 30 keV) are favoured as they are more penetrating than thermal neutrons and are less damaging to the surface tissue than more energetic neutrons.

European research using beams from nuclear reactors, filtered with a combination of appropriate materials, started at Harwell Laboratory. Two beams were developed: one employing a filter of iron, aluminium and sulphur, which produced a highly monoenergetic, 24 keV, beam[2]; and one employing a liquid argon, aluminium and sulphur filter, producing a broad energy spectrum from about 30 eV to 30 keV[3]. These beams were of small diameter and used for radiobiology experiments. More recently, a beam (HB11) has been set up at the High Flux Reactor (HFR) at the JRC, Petten, for experimental work leading to clinical trials of BNCT[4]. This beam is also designed to have a broad energy range similar to the second Harwell beam. Preliminary measurements to check the calculational procedures were made on a mock-up in the HB7 beam at the HFR, Petten[5]. The first spectrometry measurements were undertaken on the HB11 filtered beam in June 1990, but, unfortunately, subsequent to these measurements, it was found that the argon filter was not completely filled and these results have been discounted. Therefore, a second set of spectrometry measurements was made on the HB11 filtered beam in June 1991, as part of the continuing European Concerted Action Programme on BNCT to establish suitable conditions for the first clinical trials.

This paper will review the variety of techniques that are required for characterisation of BNCT beams. The basic data needed for research and, eventually, for therapy, are the neutron energy spectrum, from which the surface and depth doses can be derived. Unfortunately, a high proportion of the neutrons emerging from the beams have intermediate energies which make this measurement difficult. Therefore, a number of techniques have been used including high resolution recoil counter spectrometry, activation foil measurements and a spectrum modification technique specially developed for BNCT beams. As an example of their application, some results will be given for the recent neutron spectrometry measurements made in the HB11 filtered beam at Petten.

NEUTRON SPECTROMETRY METHODS

Most beams developed for BNCT are designed to have a broad neutron energy spectrum ranging from a few eV up to about 30 keV. However, the additional neutrons with energies greater than 30 keV contribute most of the neutron dose to tissue. Consequently, a number of techniques have been adapted to measure the neutron energy spectra of BNCT beams from thermal to 15 MeV.

Proton and alpha recoil counter spectrometry

High resolution proton and alpha recoil counters were developed for field measurements of neutron spectra for radiological protection purposes. Consequently, they are more sensitive than is desirable for BNCT beams and it is necessary to operate them at low reactor power levels (typically a few kW). The Harwell recoil counter spectrometry system consists of three spherical (SP2) proportional counters filled with hydrogen to pressures of approximately 100, 300 and 1000 kPa (1, 3 and 10 atm), covering the energy range from 10 keV to 1.5 MeV. To extend this energy range up to 15 MeV, a ^{4}He (alpha recoil) counter is used. The low energy limit of the alpha recoil counter is nominally 2 MeV. This limit is determined by the gamma ray intensity of the field being measured and by the energy deposited by argon recoils. Consequently, it is necessary to interpolate the neutron energy spectrum between the low energy limit of the ^{4}He counter and the upper energy limit of the proton recoil counters. However, the neutron fluence rate in this energy range in BNCT beams is several orders of magnitude less than the total and, therefore, the uncertainties introduced by this approximation are very small. Since, for neutrons with energies in the range for the ^{4}He counter, the fluence rate relative to those in the range for the other counters is small, the counting statistics obtained using this counter in simultaneous measurements with the other counters are poor. Therefore, unfolding the spectra in the conventional way is not possible. Consequently, it has been necessary to develop a method for estimating the fluence rate in broad energy bins by comparing the areas under the alpha recoil distribution with those obtained for monoenergetic neutrons used for calibration. An improvement to this method would be to use this counter at higher power, although this is not always practicable. Full details of the spectrometry system, its calibration and the method of unfolding the measured pulse height distributions are given elsewhere[6-8].

Activation detectors

Activation detectors can be used to determine the neutron energy spectrum over a wide range of energies. Thermal neutrons are detected using materials which have a significant cross-section to thermal neutrons (e.g. ^{197}Au(n,γ)^{198}Au, ^{45}Sc(n,γ)^{46}Sc and ^{235}U(n,f)). Epithermal neutrons are detected using materials with single, dominant, resonance reactions (e.g.^{115}In(n,γ)^{116}In^m (1.5 eV), ^{197}Au(n,γ)^{198}Au (5 eV, with the gold in cadmium covers to reduce their response to thermal neutrons), ^{186}W(n,γ)^{187}W (18.8 eV), ^{59}Co(n,γ)^{60}Co (132 eV), ^{55}Mn(n,γ)^{56}Mn (337 eV), and ^{63}Cu(n,γ)^{64}Cu (580 eV)). Fast neutrons are detected using detectors with threshold reactions (e.g. ^{115}In(n,n') ^{115}In^m (1.5 MeV), ^{46}Ti(n,p)^{46}Sc (2.4 MeV), ^{56}Fe(n,p)^{56}Mn (5 MeV) and ^{48}Ti(n,p)^{48}Sc (7.1 MeV)).

A range of activation foils are irradiated in the beam and then their induced activities are measured. The neutron energy spectrum can then be determined, with low resolution, using an unfolding program. Alternatively, a calculated input spectrum may be varied so that the measured detector responses are consistent, within the experimental uncertainties, with the adjusted spectrum. At INEL, for example, this is done using a least squares adjustment analysis using the FERRET program[9].

Spectrum modification

In this technique[10], absorbers or scatterers (e.g. discs of ^{10}B of various thicknesses and titanium) are interposed in the beam. The effect of increasing the thickness of the ^{10}B absorber is to attenuate the low and then progressively the higher energy neutron fluence, while the titanium strongly scatters neutrons penetrating the window in aluminium at about 25 keV. The response of various detectors is measured for a range of absorbers, for example: a cadmium covered BF$_3$ counter and vanadyl sulphate bath for the Harwell Al/S/Ar beam; a cadmium covered BF$_3$ counter and the 38.1 mm radius sphere for the Petten HB7 measurements; and the 38.1, 51.0 and 63.5 mm radii spheres for the first HB11 measurements. The changes in the response of the detectors as a function of energy are calculated using a Monte Carlo neutron transport program (e.g. MCNP[11]). The responses of the detectors, together with their response functions as modified by the interposed absorbers, are then used as input to an unfolding program (SENSAK[12]) to adjust a calculated input spectrum.

Other methods

It is proposed to use a small organic (NE102A) scintillator spectrometer to check that the high energy part of the spectrum does not change from the low power used for the recoil counter measurements (about 6 kW, thermal) to the full reactor power of the HFR (45 MW, thermal). This type of device is based on coincident detection of proton recoil induced scintillations. A similar detector has recently been used to make a precision measurement of the charged-to-neutral pion mass difference[13] and the design and operating characteristics of this type of spectrometer are described in more detail by Wishart et al[14]. The NE102A scintillator contains 5.28×10^{22} protons.cm^{-3} and 4.78×10^{22} carbon atoms.cm^{-3}, the neutron scattering length is a few centimetres (depending on energy) and the corresponding scattering length for gamma rays is tens of centimetres. Further, the proton recoiling after a neutron impact is much more likely to be stopped in a small scintillator than are the electrons liberated by gamma rays. In sum, the device can be expected to be an order of magnitude more sensitive to neutrons than to gamma rays. Light from the scintillator is detected in two separate photomultiplier tubes. It is believed that with mm sized scintillators and fast electronics it will be able to operate the spectrometer both at a few kW and at 45 MW to allow direct comparison of the neutron energy spectra at these power levels. A test of this type of device at the low flux reactor (LFR), Petten, will be described in more detail by Crawford et al[15].

PETTEN HB11 NEUTRON SPECTROMETRY MEASUREMENTS

Measurements of the neutron energy spectrum of the HB11 filtered beam were made in June 1991. This filter consists of 1500 mm liquid argon, 150 mm aluminium, 50 mm sulphur, 10 mm titanium and 0.1 mm of cadmium. The spectrum was measured in the neutron energy range from 10 keV to 15 MeV at low reactor power (~6 kW) using proton and alpha recoil counters (Figure 1)[16]. The peak at 50–70 keV can be identified as a window in the liquid argon cross section and the peaks at 20–30 keV and 130–140 keV with windows in the aluminium cross section. Further spectrometry measurements were made at full power using a range of activation detectors by both ECN, Petten and INEL, Idaho Falls. For comparison, the spectrum in terms of neutron fluence rate, normalised to full power, determined from the recoil counter measurements and the INEL activation detectors are presented in Figure 2[17]. These spectra were folded with conversion factors given in ICRU 26[18] to give the spectrum of kerma rate in tissue (Figure 3). Over the neutron energy range from 10 keV to 15 MeV, the total fluence rates obtained were 0.912×10^8 cm^{-2}.s^{-1} from the recoil counters and 0.925×10^8 cm^{-2}.s^{-1} from the INEL activation detectors; a difference of just 1.4%. In terms of the neutron kerma, the values obtained were 1.59 Gy.h^{-1} for the recoil counter measurements compared with 1.47 Gy.h^{-1} for the activation detectors; a difference of 8%. The total neutron fluence rate and kerma rates given by the activation detectors was 4.95×10^8 cm^{-2}.s^{-1} and 1.59 Gy.h^{-1}. Consequently, although the recoil counter spectrometer is only sensitive to about a fifth of the total neutron fluence rate, this corresponds to over 92% of the kerma rate in the beam, highlighting the importance of determining the neutron spectrum with high resolution in this

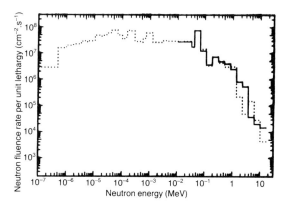

Figure 2. Neutron fluence rate spectrum of the Petten HB11 filtered beam (normalised to a reactor power of 45 MW, thermal). (—) Harwell recoil counters, (·····) INEL activation detectors.

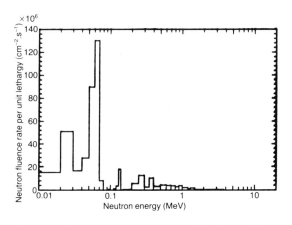

Figure 1. Neutron energy spectrum of the Petten HB11 filtered beam in the energy range from 10 keV to 15 MeV; measured with recoil counters, normalised to a reactor power of 45 MW (thermal).

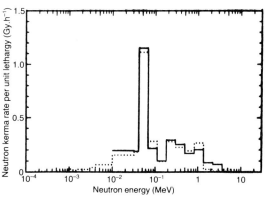

Figure 3. Neutron kerma rate spectrum of the Petten HB11 filtered beam (normalised to a reactor power of 45 MW, thermal). Key as Figure 2.

region. The measurements described refer to the free-in-air condition, further measurements will be required to determine the thermal neutron fluence rate as a function of position in phantoms in order to derive the dose arising from the $^{10}B(n,\alpha)^7Li$ reaction.

CONCLUSIONS

A range of neutron energy spectrometry techniques adapted for the characterisation of intermediate energy neutron beams has been described. The recent measurements made on the HB11 filtered beam, installed in the Petten HFR, have been taken as an example of their application. In particular, these measurements have clearly shown the need for high resolution neutron spectrometry in the energy range above 10 keV, since, although the fluence rate in this energy range is only about a fifth of the total, the neutron kerma rate arising from these neutrons is over 92% of the total.

ACKNOWLEDGEMENTS

This work was funded by the UK Department of Health and is part of the European Concerted Action Programme on BNCT. We would like to thank staff at JRC Petten, ECN Petten, INEL Idaho Falls and the Paul Scherrer Institute for many helpful discussions in preparing this paper and to Mr H. J. Delafield for assistance with the measurements made by Harwell.

REFERENCES

1. Perks, C. A., Mill, A. J., Constantine, G., Harrison, K. G. and Gibson, J. A. B. *A Review of Boron Neutron Capture Therapy (BNCT) and the Design and Dosimetry of a High-intensity, 24 keV, Neutron Beam for BNCT Research.* Br. J. Radiol., **61**, 1115 - 1126, (1988).
2. Perks, C. A., Harrison, K. G., Birch, R. and Delafield, H. J. *The Characteristics of a High Intensity 24 keV Iron-filtered Neutron Beam.* Radiat. Prot. Dosim. **15**, 31 - 40 (1986).
3. Perks, C. A., Constantine, G. and Birch, R. *The Design and Dosimetry of an Al/S/Ar Filtered Neutron Beam.* Radiat. Prot. Dosim. **23**, 329 - 332 (1988).
4. Moss, R. L *Progress Towards Boron Neutron Capture Therapy at the High Flux Reactor Petten.* In: Proc. Workshop on Neutron Beam Design, Development, and Performance for Neutron Capture Therapy, Cambridge, Massachusetts, USA, March 1989, Eds. O. K. Harling, J. A. Bernard, and R. G. Zamenhof. (New York: Plenum Press) (1990).
5. Constantine, G., Perks, C. A., Delafield, H. J., Ross, D. and Watkins, P. R. D. *Neutron Spectrum Characterisation of the HB7 Beam in the HFR, Petten.* (London: HMSO) Report AEA-EE-0024 (1991).
6. Birch, R., Peaple, L. H. J. and Delafield, H. J. *Measurement of Neutron Spectra with Hydrogen Counters. Part I. Spectrometry System, and Calibration* (London: HMSO) Report AERE-R 11397 (1984).
7. Birch, R., Marshall, M. and Peaple, L. H. J. *Measurement of Neutron Spectra with Hydrogen Counters. Part II. Analysis of Proton Recoil Distributions.* (London: HMSO) Report AERE-R 11398 (1984).
8. Birch, R. *An Alpha-recoil Counter to Measure Neutron Energy Spectra Between 2 MeV and 15 MeV.* (London: HMSO) Report AERE-R 13002 (1988).
9. Harker, Y. D., Becker, G. K., Anderl, R. A. and Miller, L. G. *Spectral Characterisation of the Epithermal-neutron Beam at the Brookhaven Medical Research Reactor.* Report to the American Nuclear Society Annual Meeting, Knoxville, Tennessee, USA, June 1990. Submitted to Nuclear Science and Engineering.
10. Constantine, G., Brenen, A., Moore, P. G. F. and Perks, C. A. *Spectrum Measurements on Filtered Neutron Beams for Medical Applications.* In: Reactor Dosimetry: Methods, Applications and Standardisation, ASTM STP 1001. Eds. H. Farrar IV and E. P. Lippincott. (American Society for Testing Materials, Philadelphia) pp. 699 - 709 (1989).
11. Breismeister, J. F. (ed.) *MCNP — A General Monte Carlo Code for Neutron and Photon Transport, Version 3A.* (Los Alamos National Laboratory) Report No. LA-7396-M Rev. 2 (1986).
12. McKracken, A. K. and Packwood, A. *The Spectrum Unfolding Program SENSAK.* AEEW ANSWERS (SENSAK) (1984).
13. Crawford, J. F., Daum, M., Frosch, R., Jost, B. and Kettle, P.-R. *Precision Measurement of the Pion Mass Difference m_{π^-}-m_{π^0}.* Phys. Rev. D, **43**, 46 - 58 (1991).
14. Wishart, L. P., Plattner, R. and Cranberg, L. *Detector for Neutron Time-of-flight Spectrometry with Improved Response to Low Energy Neutrons.* Nucl. Instrum. Meth. **57**, 237 - 244 (1967).
15. Crawford, J. F., Konijnenberg, M., Perks, C. A., Stecher-Rasmussen, F. and Watkins, P. R. D. *A Scintillation Spectrometer for Direct Comparison of Neutron Energy Spectra at High and Low Rates.* Accepted for publication in Proc. Workshop on Dose Components and Biological Effects in Conventional Radiotherapy and Neutron Capture Therapy Essen, Germany 27-29 Feb. 1992, to be published.
16. Perks, C. A., Delafield, H. J. *Neutron Spectrometry Measurements of the Petten HFR, HB11 Neutron Beam.* Accepted for publication in Proc. Int. Workshop and Plenary meeting 'Towards Clinical Trials of Glioma with BNCT' 18-20 Sept. 1991, Petten, Netherlands.
17. Wheeler, F. Private communication.
18. ICRU. *Neutron Dosimetry for Biology and Medicine.* Report 26 (Washington, DC: ICRU Publications) (1977).

DETERMINATION OF DOSE ENHANCEMENT BY NEUTRON CAPTURE OF ^{10}B IN A d(14)+Be NEUTRON BEAM

F. Pöller†, W. Sauerwein‡ and J. Rassow†
Departments of †Medical Radiation Physics and ‡Radiooncology
University of Essen, Hufelandstrasse 55, D-4300 Essen 1, Germany

Abstract — Fast neutron therapy of deep seated tumours using a d(14)+Be neutron beam is still limited because of the steep decrease of the depth–dose distribution. The interaction of fast neutrons in tissue leads to a thermal neutron distribution. Using these thermalised neutrons to produce the neutron capture reaction ^{10}B (n,α) ^{7}Li a modification of the dose distribution can be obtained. The slowing down of the d(14)+Be neutrons, resulting in a thermal neutron distribution in a phantom, has been computed using a Monte Carlo model. This model was experimentally verified by measurements of the thermal neutron fluence rate with the help of gold foil activation. The influence of the ^{10}B concentration and target volume on the distribution of thermal neutron fluence rate and the boron neutron capture dose, is demonstrated. A 15% enhancement of the absorbed dose in a small boronated target volume (100 ppm ^{10}B) at depth of 6 cm in water can be achieved.

INTRODUCTION

Boron neutron capture therapy (BNCT) is based on the interaction of thermal neutrons with ^{10}B localised in tumour cells. The high linear energy transfer (LET) and short range (~ 10 μm) of the products of the reaction ^{10}B (n,α) ^{7}Li gives the possibility of localising the dose imparted to the tumour by selective incorporation of ^{10}B and having a minimal effect on the surrounding healthy tissue [1].

It was the aim of this study to investigate whether a combination of fast neutron therapy with BNCT could be used to improve effectively the depth–dose distribution [2]. Clinical advantages of neutron capture by ^{10}B in the d(14)+Be neutron beam were demonstrated by experiments with cell cultures[3]. The spatial distribution of the thermal neutron fluence rate in a phantom depends strongly on field size and ^{10}B concentration. In the present work, the influence of boron concentration on the distribution of the thermal neutron fluence rate has been studied. The absorbed dose from boron neutron capture in a water phantom, which includes a boronated deep seated target volume, was calculated using a Monte Carlo model.

METHODS

The thermal neutron fluence rate was measured at several depths in a water phantom by activation of gold foils. The use of gold as an activation material allows the thermal neutron fluence rate to be measured separately from the total neutron fluence rate[4]. A Monte Carlo code[5] was adapted to calculate the dose distribution in phantoms. This code, based on isotropic elastic neutron scattering in the centre of mass system, was further developed to calculate the thermal neutron fluence rate distribution and to determine the dose enhancement by neutron capture of ^{10}B in water. The free water molecule has discrete energy states with quantum energies of 0.06 eV, 0.2 eV and 0.48 eV in the energy range of thermal neutrons (i.e. below 1 eV). A numerical calculation of the scattering of thermal neutrons in water was carried out by M. Nelkin[6]. The results were

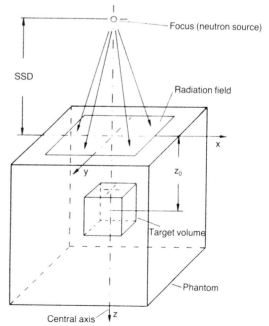

Figure 1. Schematic representation of the neutron irradiation arrangement used, including a water phantom (22 cm × 22 cm × 20 cm) and a deep seated target volume (2 cm × 2 cm × 2 cm) at 6 cm depth.

used in the Monte Carlo simulation with the restriction of one quantum energy transition between thermal neutron and water molecule. Neutron capture reactions of hydrogen at all neutron energies and neutron captures by ^{10}B in the energy range of epithermal neutrons (above 0.54 eV) were neglected. The energy depositions of recoil nuclei were assigned to their starting points. All cross section data were taken from the BNL-325[7]. A schematic representation of the neutron irradiation arrangement used with a cubic phantom (size: 22 cm × 22 cm × 20 cm), which includes a deep seated target volume at a depth of 6 cm on the central beam axis (z axis), is shown in Figure 1. This phantom was modelled in a rectangular mesh lattice with a mesh interval width of 0.5 cm in the x-y-z direction. The thermal neutron fluence rates (integrated over the energy range from 0.001 eV to 0.54 eV) and absorbed doses were calculated in every cubic lattice cell and related to the centre of each cell.

RESULTS AND DISCUSSION

The depth distributions of the thermal neutron fluence rate, Φ_{th}, in water at the central beam axis measured by gold activation and calculated with the Monte Carlo program are shown in Figure 2. The calculation agrees with the measurements to within about 10%. The incident fluence rate per deuteron charge Q of fast neutrons from the Be + d(14.3 MeV) reaction at a distance of 125 cm from the target on the central axis (free-in-air) is about 1.6×10^{12} cm^{-2} C^{-1} [4,8].

Figure 3 shows the depth distribution of the thermal neutron fluence rates at the central axis of a water phantom containing a homogeneous distribution of ^{10}B of different concentrations, obtained

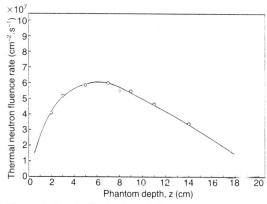

Figure 2. Depth distribution of thermal neutron fluence rate at the central axis of the water phantom (field size: 10 cm × 10 cm at 125 cm). (o) measured values, (—) values calculated by Monte Carlo.

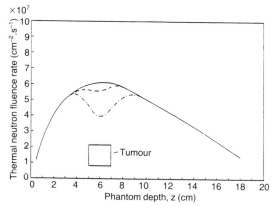

Figure 4. Depth distribution of thermal neutron fluence rate at the central axis in water with a boronated target volume. ^{10}B concentration inside the tumour: (—) 0 ppm, (- - -) 100 ppm, (-·-) 500 ppm.

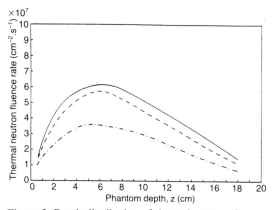

Figure 3. Depth distribution of thermal neutron fluence rate at the central axis in homogeneous boronated water. ^{10}B concentrations: (—) 0 ppm, (- - -) 10 ppm, (-·-) 50 ppm.

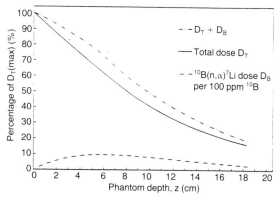

Figure 5. Depth dose distribution of D_B (for 100 ppm ^{10}B), D_T and $D_T + D_B$ on the central beam axis in water relative to D_T at $z = 0$ for incident neutrons from d(14) + Be.

by Monte Carlo calculation. The thermal neutron fluence rate at all depths decreases as the ^{10}B concentration increases. The fluence rate at the depth of 6 cm decreases by about 13% at 10 ppm and 45% at 50 ppm of ^{10}B, compared to the Φ_{th} in the non-boronated phantom.

Figure 4 shows the fluence rates Φ_{th} in a water phantom with a selectively boronated cubic target volume at a depth of 6 cm on the central axis. The attenuation of Φ_{th} in the target volume increases as the boron concentration increases. A reduction in Φ_{th} of about 44% at 500 ppm is observed, while the attenuation of Φ_{th} at 100 ppm is only 16%.

On the central axis the depth distribution of the total dose D_T in the absence of boron and of the calculated boron neutron capture dose D_B per 100 ppm ^{10}B is shown in Figure 5. No RBE factor is applied. A dose enhancement, defined as the ratio of the neutron capture dose D_B to the total dose D_T, was derived and was found in the boronated target volume to be 15 % at a depth of 6 cm.

CONCLUSION

The results presented in this work indicate that the developed 3D Monte Carlo model is a suitable tool for calculating the dose components relevant for BNCT. The calculations demonstrated the modification of the dose distribution by neutron capture with a selective ^{10}B concentration in a target volume or a non-selective uniform distribution within a phantom, produces a local dose enhancement or an improved depth–dose distribution.

ACKNOWLEDGEMENT

This work was supported by the Deutsche Forschungsgemeinschaft.

REFERENCES

1. Fairchild, R. G., Saraf, S. K., Kalef-Ezra, J. and Laster, B. H. *Comparison of Measured Parameters from a 24-KeV and a Broad Spectrum Epithermal Neutron Beam for Neutron Capture Therapy: An Identification of Consequental Parameters.* Med. Phys. **17**(6) 1045-1052 (1990).
2. Pfister, G., Hehn, G. and El-Husseini, F. *Optimization of Fast Spectra Available for Neutron Capture Therapy.* Strahlenther. Onkol. **165**(2/3), 107-109 (1989).
3. Sauerwein, W., Ziegler, W., Szypniewsky, H. and Streffer, C. *Boron Neutron Capture Therapy (BNCT) Using Fast Neutrons: Effects in Two Human Tumour Cell Lines.* Strahlenther. Onkol. **166**(1), 26-29 (1990).
4. Pöller, F., Sauerwein W., Rau, D., Wagner, F. M., Olthoff, K., Rassow, J. and Sack, C. *Neutronenfluenzmessungen im d(14)+Be-Neutronenstrahlungsfeld des Zyklotrons in Essen.* Strahlenther. Onkol. **166**(6), 426-429 (1990).
5. Füg, E. *Monte-Carlo Simulationen von Dosisverteilungen schneller Neutronen in der Umgebung von Knocheninhomogenitäten und experimentelle Überprüfung mit Thermolumineszenzdetektoren.* Diplomarbeit, Universität Essen (1988).
6. Nelkin, M. *Scattering of Slow Neutrons by Water.* Phys. Rev. **119**(2), 741-746 (1960).
7. Garber, D. I. and Kinsey, R. R. *Neutron Cross Sections,* 3rd edn. (Brookhaven National Laboratory), BNL-325, Third Edition, (1976).
8. Brede, H. J., Dietze, G., Schlegel-Bickmann, D. and Kudo, K. *Spectral Neutron Fluence and Tissue Kerma in Colliminated Neutron Beams from Be + d.* In: Proc. Fifth Symp. on Neutron Dosimetry, Munich/Neuherberg 17-21 September 1984, Eds H. Schraube and G. Burger. EUR 9762 EN, (Luxembourg: CEC) pp. 907-916 (1985).

IN-PHANTOM ^{10}B CAPTURE RATES FOR MEDICAL APPLICATIONS AT A REACTOR THERAPY FACILITY

H. Schraube†, F. M. Wagner‡, V. Mares† and G. Pfister§
†GSF - Forschungszentrum für Umwelt und Gesundheit GmbH
München, D(W)8042 Neuherberg Germany
‡FRM - Reaktorstation, Technische Universität
München D(W)8046 Garching, Germany
§Institut für Kernenergetik und Energiesysteme, Universität Stuttgart, D(W)7000 Stuttgart 80, Germany

Abstract — Methods and results are described for the determination of ^{10}B reaction rates in a simple homogeneous phantom of head size by experiment and Monte Carlo calculation. A depth-rate matrix for monoenergetic neutrons is established. Neutron spectra are folded into the matrix in order to simulate experimental results from an existing reactor neutron therapy facility and to predict the characteristics of a facility which is under design.

INTRODUCTION

The application of boron neutron capture therapy (BNCT) for the treatment of tumours in the brain has received renewed attention in the framework of a worldwide effort to develop less toxic chemical boron compounds and sufficiently powerful low energetic neutron sources[1]. BNCT makes use of the effect of high LET particles released from the ^{10}B(n,α)^{7}Li reaction with thermal neutrons in tumour tissue where boron is concentrated, previously administered by chemical compounds. The effect is concentrated in the microscopic tissue environment of boron-containing molecules. The dose distribution in tissue, therefore, depends on the thermal fluence distribution and the boron concentration pattern.

Fast neutron therapy[2], on the other hand, is based on the ionising effect of the recoil protons from fast neutron scattering or of charged particles released by nuclear reactions. The depth–dose distribution depends essentially on the energy of the fast neutrons, i.e. the depth penetration ability increases with increasing neutron energy.

The reactor neutron therapy (RENT) facility[3] at the Forschungsreaktor München, Garching (FRM) was initially designed to provide a fission neutron beam for percutaneous therapy of superficial tumours which are gamma resistent. Due to the low energies of the fission neutrons (mean energy is approximately 2 MeV), the penetration is poor, however the healthy tissue behind the treatment volume may remain relatively unaffected. Because of the high LET of the recoil particles, the biological effect in the treatment region is high compared with neutrons from other therapy sources[4].

It has been considered possible[5] to enhance the clinical effect of fast neutron therapy by the neutron capture reaction. This may be especially suitable for the fission beam because of its comparably low neutron energy, mainly for the treatment of gamma-resistant head and neck carcinoma.

It is the aim of this work to provide a data set which allows calculation of the number of ^{10}B reactions in a homogeneous phantom of head size by a simple scalar multiplication, for any desired spectral distribution of incident neutrons. The data were checked by experimental results obtained at the RENT I beam. Finally, recently calculated epithermal neutron spectra are used to predict the therapy relevant boron rate data for the planned therapy facility at a new reactor (FRM II) under design.

MONTE CARLO CALCULATION OF IN-PHANTOM ^{10}B REACTION RATES

The Monte Carlo program used in the present calculations of ^{10}B capture rates was the MCNP code version 3B[6].

The geometrical alignment used in the Monte Carlo modelling was as follows: A 16 cm cubic phantom was uniformly irradiated from one side by a broad parallel beam of monoenergetic neutrons, collimated to a square profile of 8×8 cm^2 so that phantom and beam axes were coincident. Neutron histories were started from the surface of a plane source, perpendicular to the source-to-phantom axis. A source of parallel flight paths was chosen in order to simulate geometrically the large distance of 5.45 m between source and phantom at the therapy facility. The space between source and phantom was assumed to be vacuum, as the air gap between the therapy beam port and the phantom is only 40 cm. All other scattering effects in the beam tube are assumed to be included into the neutron spectrum incident on the phantom, as used for the experimental verification later on.

The complete representation of thermal neutron scattering by molecules inside the phantom, taking into account the effects of chemical binding and crystal structure, for neutron energies below 4 eV at room temperature was realised by the application of the S(alpha,beta) tables for hydrogen bound in polyethylene.

The number of boron reactions at the points of interest in the phantom centred on its axis was determined in two steps: firstly, the spectral neutron fluences at these points were calculated and averaged over the scoring regions of a cylindrical slab volume (20 mm radius, 2 mm thickness) centred on and perpendicular to the phantom axis. Secondly, the neutron fluences were multiplied by the cross section data of the boron capture reaction and corrected to the actual boron atom density that was on the boron converter used in the experiments.

In order to get sufficiently accurate spatial resolution, the neutron fluence was calculated at 17 points of 10 mm steps, covering the whole cross section of the phantom in the beam direction. Additionally, five points between 0 and 10 mm and five points between 150 and 160 mm were selected to observe in more detail the slope just below the entrance surface and the rear surface of the phantom respectively. At these points, the cylindrical slab volume had to be reduced due to the smaller distances between the points of interest, i.e the thickness of the slab was varied between 1 to 2 mm, whereas the radius remained 20 mm.

CALCULATION RESULTS FOR MONOENERGETIC NEUTRONS

The depth dependence of ^{10}B capture reactions was calculated for 10 logarithmic equidistant energies covering the energy range from 0.02 eV to 20 MeV for a PMMA (polymethylmethacrylate) phantom with density of 1.18 g/cm^2 and an elemental composition by weight of 8% hydrogen, 60% carbon and 32% oxygen.

The results are shown in Figure 1, where each data point represents a detector element which contains 4.8×10^{15} ^{10}B atoms. This corresponds to the actual number of boron atoms as used in the experiment later on. It can be seen that with increasing neutron energy the maximum of the ^{10}B reactions shifts with depth in the phantom.

The use of the relatively large scoring region (20 mm radius) in comparison with the size of the etched track detector was necessary to reduce computing time and to obtain results of sufficient precision throughout the whole depth of the phantom. In order to investigate the effect of the larger scoring region, the fluences and the respective boron reaction rates were calculated for lateral distances, i.e. perpendicular to the main axis, for various depths in the phantom for neutron energies 2 eV – 2 MeV. Sufficient homogeneity of the thermal neutron distribution was found over the area of the 20 mm scoring region.

Additionally, for the neutron energies 200 eV, 2 keV, and 2 MeV, the boron reaction rate was also calculated for a cylindrical scoring region of 10 mm radius. The differences in the number of boron reactions derived from neutron fluence averaged over the cylindrical slab with radius 20 mm and 10 mm, respectively, were less than 1%.

EXPERIMENTAL METHOD

The experiments were performed at the RENT I beam, which is the ^{235}U converter fission beam at the FRM, with a 2.5 cm lead filter. Details of the facility and its physical beam characteristics may be found elsewhere[3].

A multi-layer PMMA phantom was placed at 40 cm distance from the collimator port. One of the layers had a small pit which was just large enough to hold the 0.9 × 3 cm^2 neutron detector foils. Each detector foil was irradiated in an individual run in order to avoid thermal neutron shielding effects. Etched track detectors were used to detect the ^{10}B reaction products.

The base material was LR–115 from Kodak

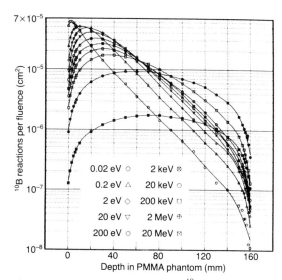

Figure 1. Calculated number of ^{10}B reactions per incident neutron fluence inside the 16 cm cubic PMMA phantom, for monoenergetic neutron beams of 8×8 cm^2 field size (the maximum of the characteristic curves decreases steadily with increasing energy).

which consists of a 100 μm thick polyester carrier and a 6 μm thick red dye cellulose nitrate layer. On the surface, a 1 μl drop of a special solution with natural boron acid was deposited, which fixed — when drying — the boron reproducibly and homogeneously without crystallisation[7]. Each drop contained 80 μg/cm^{-3} ^{10}B, i.e. 4.8×10^{15} atoms, and gave an active detector area of 0.019 cm^2. The detectors were etched after irradiation for 90 min in 2.5 N NaOH at 60°C. After rinsing and drying, the etched pits were counted in a semi-automatic image analysing system.

EXPERIMENTAL VERIFICATION

The spectral fluence distribution of the neutron beam, taken from Koester et al[8], and extrapolated below 10 keV, is shown in Figure 2.

This spectrum was subdivided into 10 log-equidistant energy intervals with the log-mean energies 0.02, 0.2 eV...20 MeV and with energy boundaries at 0.0063, 0.063 eV...63 MeV. The spectral distribution obtained in this way was folded with the ^{10}B reaction matrix which depends on the incident neutron energy and the depth in the phantom (see Figure 1). The result is shown in Figure 3 together with the experimentally obtained number of ^{10}B reactions. In spite of the relatively high statistical uncertainties of the etched track results (5-20%) and the relatively crude energy binning, a generally good agreement is observed.

DEPTH CHARACTERISTICS FOR FRM II CONVERTER SPECTRA

A comprehensive design study has been made[9] for the construction of a new research reactor FRM II with a light-water cooled and heavy-water reflected compact core which can generate a substantially higher thermal neutron fluence. The projected neutron beam from a ^{235}U converter near the core was calculated for two operational conditions, namely the unfiltered and filtered beam. Figure 4(a) shows the fluence for the fast neutron beam without filter, and Figure 4(b) the fluence for the neutron beam using a filter combination of Al and TiD$_2$.

Obviously, this filter combination has the effect that the neutron component below 100 keV is considerably increased and the fast spectral component reduced.

In Figure 5, the resulting depth characteristics of the ^{10}B reaction number are shown, normalised to the fluence at the beam port. For the sake of comparison, the respective depth curve for an uncollided fission neutron beam of Cranberg shape[10] is also shown. It can be seen that the filtered beam generates a higher maximum of the reaction number at a somewhat lower depth in the phantom, compared to the unfiltered beams. The final judgement of the applicability of the filtered beam for an exclusive BNCT treatment, however, has to be performed taking into account the directly induced fast neutron dose which may disturb the specific treatment purpose. The fast neutron dose is clearly reduced, but the quantitative evaluation is beyond the scope of this paper.

CONCLUSION

The task of calculating ^{10}B depth reaction characteristics for BNCT is essentially facilitated by establishing first a matrix of reaction numbers relative to depth in phantom for monoenergetic neutrons. Any specific spectral distribution may then be folded into the matrix to obtain the desired depth characteristics. Comparison of an experi-

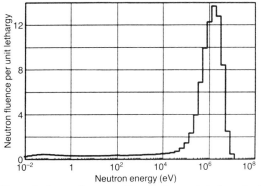

Figure 2. Spectral neutron fluence (per unit lethargy) of the RENT I beam (2.5 cm lead filtered fission beam) at the Forschungsreaktor München FRM (Reference 8).

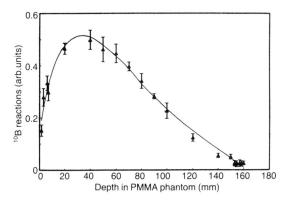

Figure 3. Comparison of calculated (solid line) and experimentally determined number of ^{10}B reactions (single points with error bars) plotted against the depth in the PMMA phantom for the spectral conditions as shown in Figure 2. The error bars represent the statistical uncertainty (1σ) of the track counting.

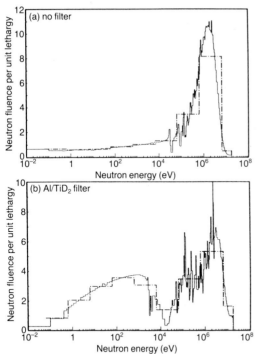

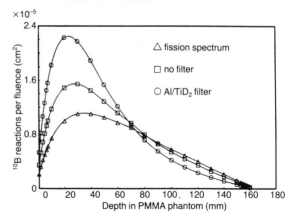

Figure 5. Normalised number of ^{10}B reactions plotted against depth in phantom for an uncollided (Cranberg) fission beam, and the two operational conditions as given in Figure 4.

Figure 4. Neutron spectra of the therapy beam at the planned FRM II reactor: (a) unfiltered beam and (b) beam filtered with Al/TiD$_2$ filter. The dashed lines indicate the energy binning. The mean energies of the intervals are those shown in Figure 1.

mentally determined reaction depth curve with the calculated one showed good agreement, even when a relatively crude energy binning of the spectra is used.

ACKNOWLEDGEMENT

We appreciate the support of Ms C. Hugo, FRM Garching, and Mr E. Weitzenegger, GSF, during the experiments and of Ms G. Schraube, GSF, during the calculations.

REFERENCES

1. Allen,B.(ed.) *Proc. Fourth Int. Symp. on Neutron Capture Therapy, 4–7 December 1990, Sydney, Australia* (in press).
2. Catterall, M. and Bewley, D. *Fast Neutrons in the Treatment of Cancer.* (London: Academic Press) (1979).
3. Wagner, F. M., Koester, L., Auberger, Th., Reuschel, W., Mayr, M., Kneschaurek, P., Breit, A. and Schraube, H. *Fast Neutrons for the Treatment of Superficial Carcinoma.* J. Nucl. Sci. Eng. **110**, 32-37 (1992).
4. Kummermehr, J., Schraube, H., Ries, G., Köster, L., Höver, K.-H., Blattmann, H., Stas, P. and Burger, G. *Biological Effectiveness of Neutrons and Pi-mesons in Gut, Bone, and Transplantable Tumours.* Strahlenther. Onkol. **165**(4) 302-305 (1989)
5. Waterman, F. M., Kuchnir, F. T., Skaggs, L. S., Bewley, D. K., Page, B. C. and Attix, F. M. *The Use of ^{10}B to Enhance the Tumour Dose in Fast-neutron Therapy.* Phys. Med. Biol. **23**, 592-602 (1978).
6. Briesmeister J. F. (ed) *MCNP — A General Monte Carlo Code for Neutron and Photon Transport — Version 3A,* Report LA-7396-M, Rev. 2 (Los Alamos) (1986).
7. Hugo, C. *Diplomarbeit,* Fakultät für Physik, Technische Universität München (1990).
8. Koester, L., Wagner, F. M., Fitzek, Th., Rau, G., Salehi, M. and Schraube, H. *Neutron Fluence and Kerma Measurements Free in Air in a Fission Beam.* Proc. Fifth Symp. on Neutron Dosimetry, EUR 9762 (Luxembourg: CEC) Vol.II, pp. 949-957 (1985).
9. Böning, K. and von der Hardt, P. *Physics and Safety of Advanced Research Reactors.* Nucl. Instr. Methods. **A260**, 239-246 (1987).
10. Ing, H. and Makra, S. *Compendium of Neutron Spectra in Criticality Accident Dosimetry.* IAEA Technical Reports series No. 180 (Vienna: IAEA) (1978).

A FERROUS SULPHATE GEL DOSIMETRY SYSTEM FOR NCT STUDIES: RESPONSE TO SLOW NEUTRONS

M. C. Cantone†, C. Canzi†, U. Cerchiari‡, D. de Bartolo†, L. Facchielli‡, G. Gambarini§,
N. Molho†, L. Pirola† and A. E. Sichirollo‡
†Dipartimento di Fisica dell'Universtitá di Milano, Italy
‡Istituto Nazionale per lo Studio e la Cura dei Tumori, Milano, Italy
§Dipartimento di Fisica del Politecnico di Milano, Italy

Abstract — Neutron capture therapy (NCT) is based on the localised release of energy in tumours from appropriate nuclei bombarded by slow neutrons. It is proposed to measure the spatial dose distribution for some selected nuclei having very large interaction cross sections with neutrons, for different concentrations and concentration gradients, for different geometries and for different neutron beams. To this purpose, it is planned to use a three-dimensional and tissue-equivalent phantom acting as a continuous dosemeter. The dosemeter is a gel solution of ferrous sulphate: the concentration of ferric ions produced by irradiation and correlated to the dose is deduced from the measurement of the proton relaxation times by means of a clinical NMR imaging system. The dosemeter can be used in a mixed field of γ rays, neutrons and heavy particles. As a first step, the response of the gel system to γ rays and slow neutrons was measured.

INTRODUCTION

The method developed and presented here aims to determine the spatial distribution of absorbed doses in a phantom exposed to a mixed field of γ rays, slow neutrons and heavy particles, and discriminate among the contributions of each radiation.

This dosemeter could be of interest for boron neutron capture therapy (BNCT), which essentially takes advantage of the energy released in the region of a tumour by the heavy particles ^{4}He and ^{7}Li ejected in the ^{10}B(n,α)^{7}Li reaction from the boron nuclei previously accumulated in the cancerous tissue. The total energy of these released particles is ~ 2.4 MeV and their maximum range in tissue is of the order of 10 μm, which is approximately the cellular dimension. Hence a selective destruction of cancerous cells with limited damage to surrounding tissue is possible. Generally, this method might be also suitable for studying the dose deposition in selected regions due to the energy released by isotopes which have a high cross section for thermal neutron capture reactions.

To evaluate fully the modalities and the effects of therapy, it is necessary to achieve a full knowledge of the three-dimensional distribution of the absorbed doses due to heavy particles, neutrons and γ rays of the mixed reactor field and also those emitted in capture reactions induced in other nuclei of the tissue.

Due to the variety and the complexity of the dose components, the theoretical determination of three-dimensional dose distributions requires very sophisticated analytical methods based on Monte Carlo approaches with limited simplification possibilities[1] or on deterministic radiation transport analysis techniques[2].

For the purpose of an experimental approach, a suitable dosimetric system is being set up and an appropriate modality of measurement and dose determination.

DOSE MEASURING SYSTEM

The proposed method for determining the spatial distribution of absorbed dose is based on nuclear magnetic resonance (NMR) measurements on a phantom containing ferrous sulphate gel.

In an aqueous solution of ferrous sulphate (which constitutes the conventional Fricke dosemeter) ionising radiation produces a conversion of ferrous ion Fe^{2+} to ferric ion Fe^{3+} whose concentration has a linear dependence on absorbed dose[3]. The spin relaxation times of hydrogen nuclei in the aqueous solution are drastically reduced by both paramagnetic ions Fe^{2+} and Fe^{3+}, but in different amounts. In fact, the longitudinal relation rate increase due to ferric ions is more significant than that due to ferrous ions in the same concentration, and the result is a net correlation between absorbed dose and proton relaxation times T_1 and T_2. Linear relationships between ferric ion concentrations and relaxation rates were found using a 25 MeV X ray beam, in the 2.5–41 Gy range[4]. The correlation is even more evident if the ferrous sulphate solution is incorporated in a gelatin. This peculiarity is of great interest because it gives the possibility of making a phantom of tissue-equivalent material in which the effects of the absorbed dose remain fixed in space, allowing the determination of the spatial distribution. A ferrous sulphate gel dosemeter in the photon-electron mixed field of a linear accelerator has already been tested with NMR measurements[5,6], and a linear dependence of relaxation rate $1/T_1$ on absorbed dose in the 0–40 Gy interval was likewise verified.

This work presents a ferrous sulphate gel as a dosemeter for mixed fields of slow neutrons and γ rays. To calibrate the system, the contribution to the relaxation time variation due to each radiation field must be evaluated.

The energy deposition of slow neutrons in an aqueous solution also containing some nitrogen compound (as in the Fricke solution) is mainly due to nuclear reactions $^1H(n,\gamma)^2H$ and $^{14}N(n,p)^{14}C$. In small volumes the contribution of energy deposition from hydrogen neutron capture is strongly reduced. The indirectly produced ionisation is responsible for the conversion of Fe^{2+} to Fe^{3+}.

The effect of neutrons and γ rays on relaxation times was tested and a calibration of the gel dosemeter in such fields was pursued by correlating the NMR results obtained from the irradiated gel to the results obtained from thermoluminescence dosemeters (TLD) suitably calibrated to measure neutron and gamma absorbed dose in the region of the gelatin.

EXPERIMENTAL METHODS
Ferrous sulphate gel preparation

Ferrous sulphate solutions were prepared with the following composition: 1 mM $Fe(NH_4)_2(SO_4)_2 \cdot 6H_2O$, 1 mM NaCl and 50 mM H_2SO_4.

Powdered gelatin from porcine skin of analytical grade (from FLUKA) in a quantity of 4% of the final total weight is added to triple distilled water in the amount of 75% of the final volume, and mixed at 42°C for 15 min to obtain a good absorption. The preparation is then heated, with a continuous and regular stirring, up to 45°C, in order to have a complete melting of gelatin. At this point, the ferrous sulphate solution is carefully incorporated(25% of the final volume). Care is taken so that the fall in temperature during this operation does not exceed 5°C, and that no air bubble remains in the mixture. The gel is finally poured into cylindrical polyethylene vials 2.5 cm in diameter and accurately sealed. The vials are cooled in a refrigerator at 5°C for 2 h and then maintained at 20°C.

To avoid a different absorption of oxygen by the various samples, all gel preparations are made rigorously under the same conditions regarding temperatures, time and rates of mixing, and modality of gel pouring in vials. The amount of oxygen in the ferrous sulphate gel is significant for the measurement results, because the oxygen concentration affects the rates of oxidation of ferrous iron by ionising radiation, as it is predicted on the basis of the free radical theory applied to the study of the radiolysis of ferrous sulphate solutions.

Irradiation and measurement facilities

The irradiations with thermal neutrons were performed in the mixed neutron and gamma field of the TRIGA MARK II reactor (LENA, Universitá di Pavia). Up to now neutron irradiations were made at the light water swimming-pool-type facility, because a thermal column irradiation port was not available. The transfer of the experiment to the thermal column is scheduled as soon as possible. The reactor operates at 250 kW thermal power. Irradiations were made at a thermal neutron fluence rate of 1.4×10^8 n.cm^{-2}.s^{-1}, uniform in a region of 8×8 cm^2, and with a cadmium ratio of about 91 at the location of the samples.

Gamma irradiations for calibration purpose were made with the ≈ 9000 Ci ^{60}Co source of the Theratron 780C Unit and the 1400 Ci ^{137}Cs source of the irradiator of the Istituto Nazionale dei Tumori in Milan.

The relaxation time measurements were performed with a Somatom Siemens NMR Analyser, operating at 63 MHz, whose magnetic field strength is 1.5 T. The transverse relaxation time T_2 was determined by using a multi-echo sequence with 16 echoes and assuming a monoexponential process. The characteristic times of the sequence were: echo times $T_E = (28 + n\ 92)$ ms where n = 0,...15, and repetition time $T_R = 2.5$ s. The longitudinal relaxation time T_1 was determined using a spin-echo sequence with $T_E = 23$ ms, $T_R = 200, 500, 1000, 1500$ ms with an exponential function fit to 4 points. The samples were placed inside cylindrical plastic containers 8 cm high and 2.5 cm diameter; up to 18 samples could be measured simultaneously, placed in a Perspex container (which produces no NMR signal). A preliminary measurement of the uniformity in the various sample locations was done, and an uncertainty in T_1 and T_2 values of no more than 1% was obtained.

Thermoluminescence dosemeters

To determine the radiation doses of thermal neutrons and gamma rays at the location of the vials containing the ferrous sulphate gel, two pairs of ^{6}LiF–^{7}LiF thermoluminescence dosemeters enveloped in polyethylene foils were placed before and behind each gel container. The dosemeters were TLD-600 with 95.62% ^{6}Li and TLD-700 with 99.93% ^{7}Li, obtained from the

Harshaw Chemical Co., all in the form of chips (3.1 × 3.1 × 0.9 mm^3). The TL reader system was a Model 2000A TL Detector interfaced to a Model 2080 TL Picoprocessor from the Harshaw/Filtrol Co. The system provides both numeric and graphic glow curve analysis. All TLD dosemeters were treated with the same conventional annealing procedure, i.e. were annealed at 400°C for 1 h, quenched with gradual cooling down to room temperature and annealed at 80°C for 24 h. No post-irradiation low temperature annealing was used, but all the measurements were made 48 h after irradiation to allow the decay of low temperature peaks. In the readout system phosphors were heated using a ramp heating rate of 8°C.s^{-1}. Dose information was obtained from the glow curves (TL intensity against temperature) by evaluating the area subtended by the main glow peak (at about 220°C). A control of the background was periodically performed through a second readout; the background in the temperature range of interest does not significantly affect the evaluated area.

The ^{6}Li high capture cross section makes the TLDs that contain ^{6}Li (such as TLD-600) very sensitive to thermal neutrons, and suitable for neutron dosimetric purposes. The ^{6}Li-depleted LiF dosemeters (such as TLD-700) have a low sensitivity to thermal neutrons, and are usually utilised for the determination of gamma ray dose in mixed fields. In a high thermal neutron environment like that of the thermal column of a reactor, the sensitivity of these phosphors to neutrons also has to be considered. The response of thermoluminescence dosemeters to thermal neutrons and gamma rays in mixed fields has been extensively but not exhaustively studied. The properties of TLDs in such mixed fields are still a research object, and the influence on dosimetric response of various factors, like the doping with different activators or the shape and geometry of the phosphor[7-10] is under investigation. The response of the chosen thermoluminescence dosemeters in the mixed field of the TRIGA MARK II reactor prior to their application in the calibration of the ferrous sulphate gel dosemeter has therefore been estimated. The response of both TLDs to gamma rays was previously calibrated in the dose range 2-15 Gy utilising the ^{60}Co and ^{137}Cs sources. In order to separate the contribution of thermal neutrons and of γ rays in the TLD dosemeter response, we have irradiated, with the same neutron flux, some phosphors placed inside and some outside a little box whose walls were made of ^{6}LiF 0.7 g.cm^{-2} thick. This container could be considered a complete shield for thermal neutrons[11] and gave a γ ray attenuation less than 1%. For TLD-600 dosemeters the ratio of glow peak areas of the phosphors outside and inside the ^{6}LiF capsule was about 5 × 10^2; this result was a validation of the good efficiency of the ^{6}LiF container shielding. Therefore the γ ray component in the reactor flux was estimated from the readout of the TLD-700 dosemeters irradiated in the ^{6}LiF shielding box. Finally, the fraction of the luminescence produced in TLD-700 dosemeters by thermal neutrons was deduced from the ratio of the glow peak areas of the phosphors outside and inside the shield. The neutron contribution to such TLD responses was found to be 55% of the overall response. This value was obtained as an average value on some dosemeters. All dosemeters were read the same day, to minimise the effects of reader instability and of phosphor fading.

RESULTS

Relaxation times of non-irradiated gels were measured at different times after preparation. From the data on relaxation rate $1/T_2$ against time shown in Figure 1 it can be seen that the time dependence decreases with time; therefore a waiting time of the order of 15 days was observed before utilising the gel and moreover the results were properly normalised.

To calibrate the dosemeter for gamma radiation, the dose response of the ferrous sulphate gel was determined by irradiating gel samples with ^{60}Co at different doses, in the interval 0-40 Gy. The expected linearity between relaxation rate and absorbed dose in gamma radiation fields has been confirmed (Figure 2). An analogous measurement

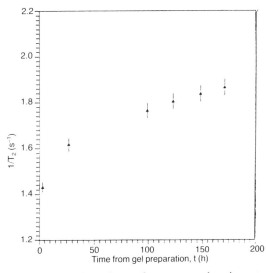

Figure 1. Time dependence of transverse relaxation rate $1/T_2$ of non-irradiated ferrous sulphate gel.

was done with ferrous sulphate solution without gel. From the results it is verified that the addition of gel increases the sensitivity of the ferrous sulphate dosemeter.

The response of the gel dosemeter to the thermal neutron and γ ray mixed field of the reactor is shown in Figure 3, where the relaxation rate $1/T_2$ is given as a function of the total neutron–gamma absorbed dose expressed in terms of the irradiation time, which is proportional to the absorbed dose, since all irradiations were made with the same dose rate.

From the readout retrieved from the TLD-600 and TLD-700 dosemeters coupled to the samples, the contribution to the absorbed dose of thermal neutrons and γ rays was determined. In Figure 4 the TLD-700 responses corrected for the neutron contribution and those of the TLD-600 corrected for the gamma radiation contribution are shown.

Knowing the gamma absorbed dose corresponding to each irradiation, through the gel calibration curve shown in Figure 2, one can subtract the gamma contribution from the relaxation rate values to the n + γ field. It is therefore possible to evaluate the responses of the gel dosemeter to thermal neutrons (Figure 5).

CONCLUSIONS

These preliminary results of gel sensitivity to thermal neutrons and its response for the dose range investigated suggest that more measurements should be done with the aim of obtaining higher sensitivity and more information

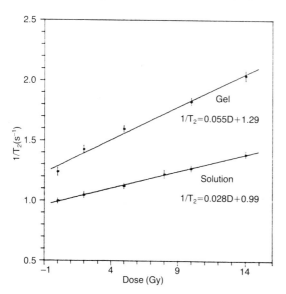

Figure 2. Transverse relaxation rate $1/T_2$ of ferrous sulphate gel and of ferrous sulphate solution plotted against γ ray absorbed dose.

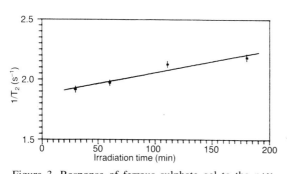

Figure 3. Response of ferrous sulphate gel to the n+γ field of the TRIGA MK II reactor as a function of irradiation time.

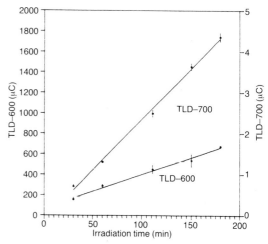

Figure 4. Integral glow values of TLD-600 and TLD-700 corrected for cross-contribution from gamma and neutron radiation respectively.

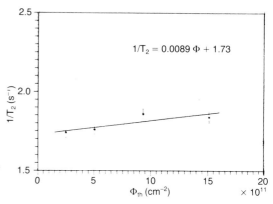

Figure 5. Ferrous sulphate gel relaxation rate response to thermal neutrons as a function of neutron fluence Φ_{th}.

on the reproducibility and accuracy of gel response, the effect of temperature and the decay with time of the information contained in the dosemeter.

Furthermore, by comparing the response of the ferrous sulphate gel with that of a ferrous sulphate gel added with boron, both irradiated in the same neutron–gamma mixed field with the same modality, the α particle contribution to the absorbed dose can be determined. As a preliminary, a calibration of the gel dosemeter for α radiation is essential.

ACKNOWLEDGEMENTS

This work was supported in part by Consiglio Nazionale delle Ricerche and by Ministero dell'Universitá e della Ricerca Scientifica e Tecnologica of Italy.

REFERENCES

1. Zamenhoff, R. G., Clement, S., Lin, K., Lui, C., Ziegelmiller, D. and Harling, O. K. *Monte Carlo Treatment Planning and High-resolution Alpha-track Autoradiography for Neutron Capture Therapy.* Strahlenther. Onkol. **165**, 188-192 (1989).
2. Nigg, D. W., Randolph, P. D. and Wheeler, F. J. *Demonstration of Three-dimensional Deterministic Radiation Transport Theory Dose Distribution Analysis for Boron Neutron Capture Therapy.* Med. Phys. **18**(1), 43-53 (1991).
3. Fricke, H. and Hart, E. J. In: *Radiation Dosimetry,* Vol. 2, Eds F. H. Attix and W. C. Roesch. (New York: Academic Press) (1968).
4. Gore, J. C., Kang, Y. S. and Schulz, R. J. *Measurement of Radiation Dose Distribution by Nuclear Magnetic Resonance (NMR) Imaging.* Phys. Med. Biol. **29**(10), 1189-1197 (1984).
5. Olsson, L. E., Petersson, S., Ahlgren, L. and Mattsson, S. *Ferrous Sulphate Gels for Determination of Absorbed Dose Distributions Using MRI Technique: Basic Studies.* Phys. Med. Biol. **34**, 43-52 (1989).
6. Olsson, L. E., Fransson, A., Ericsson, A. and Mattsson, S. *MR Imaging of Absorbed Dose Distributions for Radiotherapy Using Ferrous Sulphate Gels.* Phys. Med. Biol. **35**, 1623-1631 (1990).
7. Pradhan, A. S. and Bhatt, R. C. *Thermoluminescence Response of LiF: Mg,Cu,P and LiF TLD-100 to Thermal Neutrons,* ^{241}*Am Alphas and Gamma Rays.* Radiat. Prot. Dosim. **27**, 185-188 (1989).
8. Wang, S. S., Cai, G. G., Zhou, K. Q. and Zhou, R. X. *Thermoluminescent Response of* ^{6}*LiF(Mg,Cu,P) and* ^{7}*LiF(Mg,Cu,P) Chips in Neutron and Gamma Ray Mixed Fields.* Radiat. Prot. Dosim. **33**, 247-250 (1990).
9. Cai, G. G., Wang, S. S., Wu, F., Zha, Z. Y., Zhou, K. Q. and Li, P. *Thermal Neutron and Gamma Response of* ^{6}*LiF:Mg,Cu,P and* ^{7}*LiF:Mg,Cu,P TL Films.* Radiat. Prot. Dosim. **35**, 51-53 (1991).
10. Matsumoto, T. *Thermoluminescence Dosemeters for Application to Biomedical Dosimetry.* Nucl. Instrum. Methods. **A301**, 552-557 (1991).
11. Wingate, C. L., Tochilin, E. and Goldstein, N. *Response of Lithium Fluoride to Neutrons and Charged Particles.* In: Proc. Int. Conf. on Luminescence Dosimetry, Stanford CA 21-23 June 1965, (Springfield, VA: NTIS) (1967).

TREATMENT PLANNING OF BORON NEUTRON CAPTURE THERAPY: MEASUREMENTS AND CALCULATIONS

M. W. Konijnenberg†, C. P. J. Raaijmakers†, L. Dewit†, B. J. Mijnheer†, R. L. Moss‡,
F. Stecher-Rasmussen§ and P. R. D. Watkins‡
†The Netherlands Cancer Institute, Antoni van Leeuwenhoek Huis
Plesmanlaan 121, 1066 CX Amsterdam, The Netherlands
‡Institute for Advanced Materials
Joint Research Centre, PO Box 2, 1755 ZG Petten, The Netherlands
§Netherlands Energy Research Foundation ECN
PO Box 1, 1755 ZG Petten, The Netherlands

Abstract — A clinical boron neutron capture therapy (BNCT) facility, employing a high intensity epithermal neutron beam, has recently become available at the High Flux Reactor (HFR) in Petten. Another epithermal beam has been constructed at the Low Flux Reactor (LFR) to test the behaviour of a mixed epithermal neutron and photon field in water phantoms. The neutron and photon fluence distributions have been measured with a BF_3 counter and a ^{235}U fission counter and with a GM counter and a Mg(Ar) ionisation chamber, respectively. Absolute determination of the neutron fluence has been performed with foil stacks and of the gamma ray dose with TLD chips. For the design of a treatment planning system for BNCT, the use of a Monte Carlo technique has been investigated. The outcome of these calculations fits the measurements within 1%.

INTRODUCTION

BNCT is based on the release of high LET radiation in the vicinity of a tumour cell via the thermal neutron capture reaction $^{10}B(n,\alpha)^7Li$. The primary aim of the European collaboration on BNCT is to perform clinical trials, starting with treatment of glioma patients, within a few years[1].

Calculations using either Monte Carlo or deterministic codes have been proposed for BNCT treatment planning[2,3]. Such a procedure not only requires an adequate description of the thermal neutron fluence distribution, but also of the other dose components resulting from epithermal neutrons, fast neutrons and gamma rays. Treatment planning for BNCT based on Monte Carlo calculations has the great advantage that the dose components caused by the induced gamma rays and by ^{10}B capture can, in principle, be included in the calculations of the neutron dose. Such a procedure is not yet available for daily use in radiotherapy departments[4]. External photon beam treatment planning is currently based on empirical knowledge of the dose distribution under reference circumstances together with deterministic calculations. For a better understanding of a mixed epithermal neutron and photon field the reference measurements of the dose distribution have been performed in two types of water phantoms, one rectangular the other cylindrical. It is the aim of this work to investigate the possibilities and limitations of Monte Carlo calculations for treatment planning of BNCT. In a separate paper the preliminary results of a semi-empirical method of BNCT treatment planning will be described[5].

MATERIALS AND METHODS

A filter assembly containing elements of several thicknesses of aluminium, sulphur, titanium, cadmium and liquid argon has recently been installed at the HB11 beam in the HFR to produce an epithermal neutron beam suitable for BNCT[6]. A more versatile testing facility is available at the HN beam in the Low Flux Reactor (LFR) in Petten. The LFR is a 30 kW nuclear reactor of the 'Argonaut' type. The core of this reactor consists of 1.8 kg ^{235}U, surrounded by graphite reflectors, and the neutrons are moderated and attenuated by water. The beam has a field size of 10×10 cm^2 and a neutron fluence rate of 2.74×10^7 cm^{-2}.s^{-1} (measured with Mn foils at the beam exit). Thermal neutrons have been filtered by a set of 1 mm thick Cd sheets, separated by 1.30 m. Two 10 cm long borated polyethylene blocks with a circular opening of 8 cm diameter directly behind the Cd sheets produce a neutron beam with a field size similar to the latest design of the HB11 beam. At the beam exit a 5 mm thick B_4C plate has been placed to capture thermalised neutrons.

No filtering of the photon field is performed. Lead blocks, 5.0 cm thick and with equivalent openings, have been placed behind the neutron collimators to reduce the induced capture gamma rays. The photon field comes mainly from the reactor core and is considerably larger than the dose due to neutron capture gamma rays originating within the phantom. The opposite will apply with the HFR beam, due to the presence of an argon filter. In future experiments a bismuth filter may be installed to obtain some filtering of the gamma rays.

A start has been made with determining the relation between the fluence distribution and the phantom size in water phantoms positioned at the beam exit. A rectangular phantom made of 8 mm Perspex with overall dimensions of $60 \times 50 \times 40$ cm^3 was used for the measurements. A cylindrical phantom, with a diameter of 15 cm and height of 30 cm, made of 1.5 mm thick Perspex, was also applied, bearing a closer resemblance to the shape of a human head. Both phantoms were filled with normal water. In the rectangular phantom, the first point of measurement was situated at 9.5 mm depth within the phantom due to the wall and the thickness of the ^{235}U and the GM counter. In the cylindrical phantom the smallest measurable depth was 3 mm.

RESULTS OF FLUENCE MEASUREMENTS

For relative neutron fluence measurements, a BF$_3$ counter with an effective area of 3 cm^2 was initially used. The spatial resolution of this detector was too large to measure accurately the thermal fluence distribution in the build-up area. A ^{235}U fission counter with an effective area of 0.8 cm^2 (Centronics type FC4A/1000/U235) showed a great improvement of the spatial resolution. The epithermal neutron fluence was determined by covering the detector with a 1 mm thick Cd sheet. The measured values were corrected for dead-time losses of the detector.

Photon dose has been measured with a miniature GM counter (Philips type 18529)

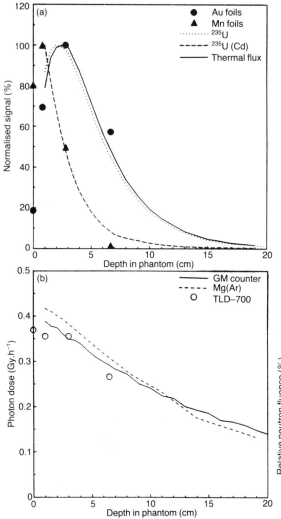

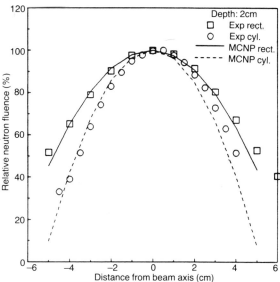

Figure 1. Measured thermal and epithermal neutron fluence rate (a) as well as photon dose rate (b) as a function of depth in a cubical phantom. Neutrons have been measured with a ^{235}U fission counter (with and without Cd) and with Au/Mn foils at four different depths. The thermal fluence rate has been determined by subtraction of the ^{235}U counter (with cadmium) signal from the ^{235}U (without) signal. The photon dose has been determined with a GM counter, a Mg ionisation chamber flushed with Ar, and with TLD chips.

Figure 2. Thermal neutron beam profiles in the rectangular (measurement □; calculation, full line) and in the cylindrical (measurement o; calculation, dashed line) phantom at 2 cm depth, as measured with a ^{235}U fission counter (with and without Cd) or calculated with the MCNP code respectively.

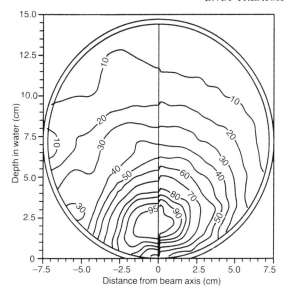

Figure 3. Comparison of the isofluence charts for the cylindrical phantom. A linear extrapolation of the measured ^{235}U beam profiles as input, is shown on the left. The right-hand side of the picture depicts the calculated neutron distribution folded over the ^{235}U fission spectrum, as performed with MCNP. All values are given in the central beam-axis plane.

encapsulated in a 0.5 mm Perspex tube and covered with a 3 mm thick ^{6}LiF cover to reduce the thermal neutron response of the counter. In addition to the GM counter, a neutron–insensitive 1.0 cm^3 cylindrical Mg ionisation chamber (Exradin) flushed with Ar gas was used. Both phantom detectors were calibrated in the reference ECN calibration set-up.

Measured relative depth fluence curves for (epi)thermal neutrons and dose rate curves for photons are shown in Figure 1, for the rectangular phantom. The thermal and epithermal fluence rates, as determined with diluted AuAl and MnNi foils, are also indicated in this figure and given in

Table 1. Both photon detectors indicate similar depth profiles. The GM counter is, however, slightly less sensitive to thermal neutron capture. TLD chips have been placed at four different depths, indicating γ ray dose values in reasonable agreement with those obtained with the other two methods.

Fluence beam profiles for thermal and epithermal neutrons at several depths have been measured in both phantoms. An increase in penumbra for thermal neutrons with depth is apparent, being caused by the increased scattering and capture of the thermalised neutrons. The measured thermal neutron fluence beam profiles at approximately their maxima are shown in Figure 2. At 2 cm depth in the phantom the beam width (at FWHM) was 7.5 cm for the cylindrical phantom and 10.0 cm for the rectangular phantom.

MONTE CARLO CALCULATIONS

The Monte Carlo code MCNP (version 3A)[7] was used for all calculations. This code was run on a SUN Sparc2 workstation at the Joint Research Centre[8]. A full model of the LFR core was used to generate a neutron spectrum integrated over the surface before the first Cd filter in the beam. For this simulation 5×10^5 neutrons and photons were followed in a coupled (n,γ) calculation. The calculated neutron intensity spectrum over this surface was then used to determine the spectrum coming through the filter components and the dose distribution in the two water phantoms, again tracking down the paths of 5×10^5 neutrons.

In the rectangular phantom, the doses have been calculated in a $10 \times 10 \times 10$ cm^3 cube around the beam axis at the entrance end of the phantom. This cube has been divided into a 1.6×10^4 element mesh of $0.25 \times 0.25 \times 1$ cm^3 cells. The remaining part of the phantom has been modelled separately to reduce computing time, but its scattering contribution to the tally cube has been included. For the cylindrical phantom, equivalent sized cells have been used to generate a grid over the beam axis plane. The calculated isofluence distribution for the cylindrical phantom is depicted in Figure 3, together with an interpolation of the measured ^{235}U fission counter response curves over the central surface. The outcome of the MCNP calculations fit the measurements within 1%. The difference in the shape of the calculated and the measured isofluence curves is due to the used interpolation procedure and the detectors's spatial resolution.

CONCLUSIONS

The preliminary results of MCNP calculations

Table 1. Thermal and epithermal neutron fluence rate (Φ_{th} and Φ_{epi}) and γ ray dose (D_γ) at 30 kW power in the cubicle phantom determined from activation of a standard AuAl and MnNi two-foil set and with TLD-700 chips. The photon dose measurements were performed at 10 kW reactor power, extrapolated to 30 kW and corrected for thermal neutron capture.

Depth in phantom (mm)	Spectrum index Au/Mn	Φ_{th} (cm^{-2}.s^{-1})	Φ_{epi} (cm^{-2}.s^{-1})	D_γ (Gy.h^{-1})
0	35.38	2.17E+6	7.64E+5	0.369
8	18.74	8.13E+6	9.51E+5	0.355
28	11.62	1.17E+7	4.73E+5	0.355
67	7.17	6.71E+6	≤ 1.50E+4	0.266

of thermal fluence distributions in phantoms irradiated with a beam of epithermal neutrons, shows that such an approach seems useful for BNCT treatment planning. Several phenomena can be predicted quite well with MCNP. For instance, the neutron fluence rate relative to radial distance in the cylinder shows an increased peaking compared with the results in the cube. Further investigation is required to quantify the influence of field size variation on the dose distribution. All experiments will be repeated at the HB11 beam in the HFR, to check our findings.

ACKNOWLEDGEMENTS

We would like to thank A. Paardekooper and W. Voorbraak of ECN for performing the foil measurements and H. Verhagen of ECN for his support during the photon dose measurements. A detailed MCNP model of the LFR reactor core was provided by W. Freudenreich of ECN.

This work was financially supported by the Netherlands Cancer Foundation (NKB Grant NKI 90-03).

REFERENCES

1. Dewit, L., Moss, R. L. and Gabel, D. *A Proposal for Clinical Pilot Studies for Boron Neutron Capture Therapy.* Proc. Fourth Int. Symp. on Neutron Capture Therapy, Sydney, Australia, 1991 (in press).
2. Nigg, D. W., Randolph, P. D. and Wheeler, F. J. *Demonstration of Three-dimensional Deterministic Radiation Transport Theory Distribution Analysis for BNCT.* Med. Phys. **18**(1), 45-53 (1991).
3. Yanch, J. C., Zhou, X. L. and Brownell, G. L. *A Monte Carlo Investigation of the Dosimetric Properties of Monoenergetic Neutron Beams for NCT.* Radiat. Res. **126**, 1-20 (1991).
4. ICRU. *Use of Computers in External Radiotherapy Procedures with High Energy Photons and Electrons.* Report **42**, (Bethesda, MD: ICRU Publications) (1987).
5. Raaijmakers, C. P. J., Dewit, L., Konijnenberg, M. W., Mijnheer, B. J., Moss, R. L. and Stecher-Rasmussen, F. *A Semi-empirical Method of Treatment Planning of Boron Neutron Capture Therapy.* In: Proc. European Collaboration Workshop 'Towards Clinical Trials of Glioma with BNCT', Petten, the Netherlands (1991) (in press).
6. Moss, R. L., Stecher-Rasmussen, F., Ravensberg, K., Constantine, G. and Watkins, P. R. D. *Design, Construction and Installation of an Epithermal Neutron Beam for BNCT at the High Flux Reactor Petten.* In: Proc. Fourth Int. Symp. on Neutron Capture Therapy, Sydney, Australia (1991) (in press).
7. Briesmeister, J. F. (ed). *MCNP — A General Monte Carlo Code for Neutron and Photon Transport.* **LA-7396-M**, revision 2 (1987).
8. Watkins, P. R. D. *Report of Calculations Performed at Harwell for the Petten HB11 BNCT Facility.* Reactor Physics Note **174**, Harwell Laboratory (1990).

REVIEW ON THE PHYSICAL AND TECHNICAL STATUS OF FAST NEUTRON THERAPY IN GERMANY

J. Rassow[1], U. Haverkamp[2], A. Hess[3], K. H. Höver[4], U. Jahn[5], H. Kronholz[2], P. Meissner[1], K. Regel[5], and R. Schmidt[3]
[1]Institut für Medizinische Strahlenphysik
Universitätsklinikum, D-4300 Essen 1, Germany
[2]Klinik und Poliklinik für Strahlentherapie/Radioonkologie
Universität, D-4400 Münster, Germany
[3]Abteilung für Strahlentherapie, Radiologische Klinik
Universitätskrankenhaus, D-2000 Hamburg-Eppendorf 20, Germany
[4]Institut für Nuklearmedizin, Deutsches Krebsforschungszentrum Heidelberg
D-6900 Heidelberg, Germany
[5]Robert-Roessle-Klinik, Zentralinstitut für Krebsforschung
D-O-1115 Berlin-Buch, und Zentralinstitut für Kernforschung
D-O-8051 Dresden-Rossendorf, Germany

Abstract — All five fast neutron therapy centres in Germany use low energy cyclotrons or neutron generators and are, therefore, at the low energy end of the 21 neutron therapy facilities presently in use worldwide. The depth dose characteristics are all worse than for ^{60}Co gamma rays, the absorbed dose rate is too low and treatment is technically restricted because of lack of modern features like multileaf collimators and full gantry rotation and the capabilities of neutron treatment planning systems cannot be fully utilised. A survey is presented of the statistical and methodological data on neutron treatment in Germany. To avoid masking the potential biological benefits of high LET neutron irradiation by the use of suboptimal equipment and to utilise the real therapeutic benefit for specific tumour types, the German neutron therapy centres urgently need modernisation of their outdated facilities. Specific recommendations of how to meet the requirements of modern neutron therapy are given.

PHYSICAL DATA

The physical data pertaining to the equipment at the five fast neutron therapy centres in Germany are listed in Table 1 and can be characterised by the following features:-

(a) Only low energy cyclotrons and neutron generators, developed in the early 70s, are used.
(b) The half-value depth Z_{50} does not even reach that of ^{60}Co gamma rays (11.6 cm). This results in serious methodological restrictions in the application of fast neutrons and very complicated, time-consuming techniques have to be used to compensate for the poor depth and lateral dose distributions.
(c) The mean nominal absorbed dose rate (excluding Essen) is significantly lower than the desirable one of 0.5 Gy.min^{-1}.
(d) Features of modern electron accelerator therapy units such as multileaf collimators, full gantry rotation and facilities which satisfy the requirements of commercial treatment planning systems are lacking.

TREATMENT STATISTICS AND METHODOLOGICAL DATA

Patient treatment statistics and related methodological data are given in Table 2. They can be summarised in the following statements:-

(a) 2949 patients have been treated with fast neutrons up to the end of 1990 in Germany,
  (i) 50% of them with pure neutron therapy (most frequently 5 to 20 fractions),
  (ii) 33% of them with neutron boost therapy (most frequently 5 to 10 fractions),
  (iii) 17% of them with mixed beam therapy (most frequently 5 to 10 neutron fractions);
(b) 69,350 single neutron fields approximately have been applied in Germany corresponding to 20,000 hours of treatment time including preparation of the patients.

The considerably high number of neutron fields per treatment was necessary to obtain a sufficient dose distribution. In spite of suboptimal technical conditions, a successful treatment was possible by the optimal use of the available units and technical accessories combined with complicated treatment techniques for most patients.

INDICATIONS FOR FAST NEUTRON THERAPY

For the 2949 patients treated in Germany with fast neutrons up to 31 Dec. 1990, 41% of the cases were soft tissue sarcomas, and 16% were cases of head and neck carcinomas including adenoid-cystic carcinomas and other salivary gland tumours.

Resulting from the worldwide experience, often obtained with 'suboptimal technical conditions', the selection of patients for fast neutron therapy of macroscopic residual tumours should in future be based on the following[1–3]:

(1) *Superior advantage of neutrons:* salivary gland tumours (locally extended, well differentiated),
(2) *Significant advantage of neutrons:* osteosarcomas, soft tissue sarcomas, chondrosarcomas, pretreated recurrences (hypoxic, slowly growing), metastatic adenopathies, adenocarcinomas, adenoid-cystic carcinomas.
(3) *With possible advantage:* some head and neck tumours (locally extended), rectum carcinoma recurrences, prostate adenocarcinomas, melanomas (inoperable, recurrent),
(4) *Without therapeutic advantage:* glioblastomas, oesophageal carcinomas, bronchial carcinomas, gastric carcinomas, bladder carcinomas.

REQUIREMENTS AND PHYSICAL CHARACTERISTICS OF MODERN FAST NEUTRON THERAPY FACILITIES

As a principle it can be stated[4] that in cases where fast neutrons (as high LET radiation) are biologically superior to low LET radiations, no methodological disadvantage compared to X rays

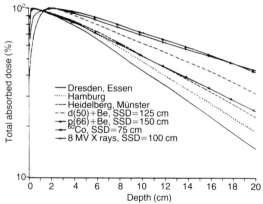

Figure 1. Depth dose distributions in the five German fast neutron therapy centres for cyclotrons (Berlin/Dresden, Essen) and neutron generators (Hamburg, Heidelberg, Münster) compared to photon and modern neutron therapy facilities.

from medical electron accelerators should partially or totally neutralise the potential biological benefit. This requires physical spatial dose distributions comparable to about 8 MV X rays (see Figure 1), sufficiently short irradiation times and equivalent beam delivery systems and treatment conditions:

(i) mean neutron energies $E_n > 20$ MeV resulting from primary particle energies for neutron production of $E_p \geq 60$ MeV or $E_d \geq 50$ MeV in the (p+Be) and the (d+Be) reactions respectively;

Table 1. Characteristics of the five German fast neutron therapy facilities.

Characteristics	Berlin/Dresden	Essen	Hamburg	Heidelberg	Münster
Accelerator	Cyclotron	Cyclotron	Generator	Generator	Generator
Neutron generating reaction	d(14.0) + Be	d(14.3) + Be	d(0.5) + T	d(0.25) + T	d(0.25) + T
Mean neutron energy	5.9 MeV	6.0 MeV	14.1 MeV	14.1 MeV	14.1 MeV
Normal treatment distance	100 cm	125 cm	80 cm	100 cm	100 cm
Half-value depth Z_{50}	7.9 cm	8.1 cm	8.8 cm	10.6 cm	10.5 cm
Penumbra width $P_{80/20}$	2.3 cm	2.4 cm	2.4 cm	2.7 cm	2.7 cm
Mean nominal dose rate D_{T0} (SSD=100 cm, A_0=10 cm×10 cm)	0.15 Gy.min^{-1}	0.5 Gy.min^{-1}	0.08 Gy.min^{-1}	0.10 Gy.min^{-1}	0.12 Gy.min^{-1}
Maximum field size A_{max}	14 cm × 24 cm	20 cm × 20 cm	15 cm × 25 cm	16 cm × 16 cm	16 cm × 16 cm
Multileaf collimator	no	no	no	no	no
Gantry angular range	Fixed at 90°	240°–0°–120°	250°–0°–110°	253°–0°–107°	253°–0°–107°
Full rotation possible	no	no	no	no	no
Isocentric patient table	no	yes	yes	yes	yes
Applied treatment techniques:					
single field therapy	yes	yes	yes	yes	yes
isocentric multiple field therapy	no	yes	yes	yes	yes
rotational therapy	no	yes	no	yes	yes
wedge filter fields	no	yes	no	yes	yes
moulage technique	no	yes	yes	yes	yes
Treatment planning system	Own developed system	Own developed system	Modified commercial system	Own developed system	Modified commercial system

(ii) half-value depth Z_{50} comparable to that of 8 MV X rays (17.1 cm);
(iii) total absorbed dose rate $\dot{D}_T \geq 0.3$ Gy.min^{-1} at normal treatment distance giving about 3 to 5 min irradiation time per single field treatment;
(iv) multileaf collimator to adjust the beam to irregularly shaped target volumes;
(v) capability of full gantry rotation to treat isocentrically with multiple or rotating fields without repositioning of the patient;
(vi) full availability of the facility for medical applications only;
(vii) location within a hospital with access to its infrastructure and personnel.

In the light of these modern requirements, the actual physical characteristics according to Table 3 of the 21 actual fast neutron therapy centres in the world (state: January 1992) demonstrate that only five facilities with $Z_{50} \geq 15$ cm meet the requirements, seven facilities have Z_{50} values between 11.6 cm (^{60}Co gamma rays) and 15 cm and nine facilities, including all five German ones, have half-value depths less than that of ^{60}Co gamma rays. In all German and most of the other centres only the last two general requirements above and sometimes the dose rate requirement are met.

REQUIREMENTS FOR MODERN NEUTRON TREATMENT PLANNING SYSTEMS

As a principle it can be stated that for fast neutron therapy planning, commercial systems (which may be modified by a team of medical physicists in neutron therapy centres) should include all the features for 3D treatment planning of photon and electron therapy, but have the capability of meeting the additional specific requirements of fast neutron therapy:

(1) separate calculation of neutron and photon absorbed dose distributions;
(2) superposition of dose distributions resulting from neutron, X ray, electron or brachy treatments with individually chosen weighting factors for each component (for neutron boost or mixed beam therapy);
(3) direct transfer of CT and MRI data with

Table 2. Patient treatment numbers and frequency of applied treatment modes and fractionation (state: 31 Dec. 1990) (*rotational therapy calculated as therapy with 5 single fields).

Characteristics	Berlin/Dresden	Essen	Hamburg	Heidelberg	Münster	Total
First patient treatments	May 1972	Jan. 1978	Feb. 1976	Oct. 1977	Nov. 1985	
Total number of patients	990	546	822	441	150	2949
Treatment modes:						
pure neutron therapy	516	232	465	165	93	1496
with fractions:						
minimum	4	8	6	6	3	
maximum	15	12	30	22	14	
most frequent	5	10	12	20	9	
neutron boost therapy	12	280	357	276	57	982
with fractions:						
minimum	4	4	1	4	3	
maximum	10	12	18	12	6	
most frequent	5	8	6	10	6	
alternating mixed beam therapy (neutrons+photons/electrons)	462	34	0	0	0	496
with neutron fractions:						
minimum	4	8	0	0	0	
maximum	15	12	0	0	0	
most frequent	5	10	0	0	0	
Number of single fields per fraction:						
minimum	1	1	1	1	1	
maximum	3	6	10	6	6	
most frequent	1–2	3	2	2	3	
Approximate total number of single fields* for:	10600	18700	23250	12700	4100	69350
pure neutron therapy	4600	11000	17400	7600	2600	43200
neutron boost therapy	100	6700	5850	5100	1500	19250
mixed beam therapy	5900	1000	0	0	0	6900
Mean elapsed time per single field irradiation (inclusive of preparation)	15 min	15 min	20 min	15–20 min	20 min	

Table 3. Characteristics of all the fast neutron therapy facilities in the world (state: January 1992).

Place	Country (National car mark)	Neutron generating reaction	Mean neutron energy, E_n (MeV)	Source surface distance (cm)	Half-value depth (cm)	Penumbra width $P_{80/20}$ (cm)
Chicago	USA	d(8.0) + D	7.0	100	9.8	2.2
Hamburg	D	d(0.5) + T	14.1	80	8.8	2.4
Heidelberg	D	d(0.25) + T	14.1	100	10.8	2.7
Münster	D	d(0.25) + T	14.1	100	10.5	2.7
Berlin/Dresden	D	d(13.5) + Be	5.7	100	7.9	2.3
Essen	D	d(14.3) + Be	6.0	125	8.1	2.4
Krakow	PL	d(12.5) + Be	5.2	91	7.7	2.0
Tokyo	J	d(14.0) + Be	5.9	150	8.3	2.3
Riyadh	SA	p(26) + Be	10.0	125	10.3	2.4
Chiba-shi	J	d(30) + Be	12.6	175	11.7	1.9
Orleans	F	p(34) + Be	13.7	169	12.8	1.9
Houston	USA	p(42) + Be	17.5	125	14.0	2.0
Cleveland	USA	p(43) + Be	18.0	125	13.5	2.2
Los Angeles	USA	p(46) + Be	19.4	150	13.1	1.7
Seattle	USA	p(50) + Be	21.3	150	14.8	1.4
Seoul	KOR	p(50) + Be(26)	21.3	150	14.0	(0.5/air)
Detroit	USA	d(50) + Be	21.0	183	15.0	1.5
Clatterbridge	GB	p(62) + Be	26.9	150	16.2	1.6
Louvain-la-N.	B	p(65) + Be	28.3	162	17.6	1.7
Batavia/Ill.	USA	p(66) + Be	28.8	190	16.6	2.0
Faure	ZA	p(66) + Be	28.8	150	16.8	1.3
For comparison		^{60}Co gamma		80	11.6	1.6
		8 MV X rays		100	17.1	0.8

*E_n calculated according formulas[5,6] recommended in ICRU Report 45 (1988)[7]
d+Be: E_n/MeV = 0.42E_d/MeV
p+Be: E_n/MeV = 0.47E_d/MeV − 2.2

Table 4. Commercially available cyclotrons suitable for fast neutron therapy (N) (state: April 1991) (−, partially or +, completely suitable).

Manufacturer	Type	[Particle energies (MeV)]				Magnet mass (t)	Applicability
		p	d	^{3}He	^{4}He		
Scanditronix	MC40	10–40	5–20	13–53	10–40	69	N−
Uppsala, Sweden	MC50	13–50	7–25	17–66	13–50	95	N−
	MC60PF	63	−	−	−	120	N+
Sumitomo Heavy	680	40	25	−	−	100	N−
Industries, Tokyo, Japan	750	70	−	−	−	120	N+
National Superconducting Cyclotron Laboratory, East Lansing, USA, + Ion Beam Applications (IBA), Louvain-la-N., Belgium	?	−	50	−	−	25	N+

differentiation for density and/or hydrogen content pixel by pixel (for fatty and bone tissues and air cavities).

COMMERCIAL CYCLOTRONS FOR MODERN FAST NEUTRON THERAPY

The combined use of cyclotrons for both fast neutron therapy and either nuclear research or radionuclide production cannot be recommended any longer now that the pilot clinical trial phase is past.

Both combinations have the essential disadvantage of time scheduling problems for optimal application. Additionally, the combination with nuclear research has the handicap of the lack of medical infrastructure in a physics institute and the combination with radionuclide production has different requirements for particle energy ranges (low energies: E_p, E_d < 30 MeV) for radionuclide production[8] and high energies for fast neutron therapy.

Multipurpose machines, even with variable energy, are much more complicated and expensive, causing extreme technical problems, compared with two special fixed-energy cyclotrons for one or two particle types, designed specifically for each purpose. Therefore, of the available 24 models of cyclotrons from seven manufacturers, only three are completely suitable and three partially suitable for fast neutron therapy as indicated in Table 4.

CONCLUSION

The German neutron therapy centres, where no essential investment has been undertaken for more than a decade, urgently need modernisation of their outdated facilities:

(a) to be comparable or superior to the physically and technically advanced neutron therapy facilities in the world outside Germany;
(b) to avoid masking the potential biological benefit of high LET neutrons due to lack of technical features which are used in X ray and electron therapy with medical electron accelerators;
(c) to utilise the real therapeutic benefit of neutrons for specific tumour types.

REFERENCES

1. Budach, V. and Sack, H. *Grundlagen und Indikationen der Neutronentherapie.* Dt. Ärztebl. **88**, 2376-2386 (1991).
2. Budach, V. *The Role of Fast Neutrons in Radioonkology - a Critical Appraisal.* Strahlenther. Onkol. **167**, 677-692 (1991).
3. Schmitt, G. and Wambersie, A. *Review of the Clinical Results of Fast Neutron Therapy.* Radiother. Oncol. **17**, 47-56 (1990).
4. Catterall, M. *Radiology Now. Fast Neutrons — Clinical Requirements.* Br. J. Radiol. **49**, 203-205 (1976).
5. Parnell, C. J. *A Fast Neutron Spectrometer and its Use in Determining the Energy Spectra of Some Cyclotron Produced Fast Neutron Beams.* Br. J. Radiol. **45**, 452-460 (1972).
6. Lone, M. A., Ferguson, A. J. and Robertson, B.C. *Characteristics of Neutrons from Be Targets Bombarded with Protons, Deuterons and Alpha Particles.* Nucl. Instrum. Methods **189**, 515-520 (1981).
7. ICRU. *Clinical Neutron Dosimetry Part 1: Determination of Absorbed Dose in a Patient Treated by External Beams of Fast Neutrons.* Report 45 (Bethesda, MD: ICRU Publications) pp. 10 and 12 (1989).
8. Rassow, J. *Beschleuniger für die Radionuklidproduktion.* In: Reiners, C., Harder, D., and Messerschmidt, O. (eds) Strahlenschutz im medizinischen Bereich und an Beschleunigern. Strahlenschutz in Forschung und Praxis Bd. **32**, 183-210 (1992).

NEUTRON CAPTURE THERAPY BEAM ON THE LVR-15 REACTOR

M. Marek†, J. Burian†, Z. Prouza‡, J. Rataj† and F. Spurný§
†Nuclear Research Institute, Rez near Prague, Czechoslovakia
‡Institute of Hygiene and Epidemiology, Prague, Czechoslovakia
§Institute of Radiation Dosimetry, Prague, Czechoslovakia

Abstract — Several configurations of moderating and shielding materials have been designed and measured on the LVR-15 reactor for boron neutron capture therapy (BNCT) purposes. To determine the neutron and gamma ray space–energy distributions in the cylindrical geometry the two-dimensional code DOT with the coupled neutron–gamma data library DLC-36 was used. The experimental verification of the beam parameters was performed in the LVR-15 reactor thermal column empty space with layers of graphite, aluminium, alumina, lead and bismuth. Attention was paid to establishing techniques and instrumentation for monitoring the neutron and gamma ray dose and beam quality. The thermal and epithermal (above the Cd resonance) flux densities were measured by activation foils, the neutron spectrum was determined with a Bonner spectrometer and gamma ray background with a scintillation spectrometer. The distribution of thermal neutrons in the human head phantom was mapped with a small semiconductor detector (Si^{6}Li). Thermoluminescence dosemeters (TLD) were used to determine the gamma ray doses in the free beam and in the phantom geometry. The results demonstrate that the maximal thermal flux of $1.02 \times 10^{13} \mathrm{m}^{-2}.\mathrm{s}^{-1}$ may be obtained at a depth of 3 cm in the polyethylene phantom at reactor power of 1 MW. The therapeutic dose of 20 Gy can be obtained in 31.5 min at the reactor power 5 MW.

INTRODUCTION

Several configurations of moderating and shielding materials have been designed to produce a neutron source with parameters suitable for boron neutron capture therapy (BNCT)[1]. Designs were achieved on the LVR-15 reactor and experimental data were compared with calculated results. The beam of epithermal neutrons described gives the requisite therapeutic dose of 20 Gy.

DESCRIPTION OF FACILITY

A detailed description of the final cylindrical configuration can be seen in Figure 1. It is composed of cylindrical layers (diam. 1000 mm): 150 mm of C, 250 mm of Al_2O_3, 600 mm of Al and 150 mm of S. To improve the neutron–gamma ratio 50+100 mm of Pb and 100 mm Bi are included. A thermal neutron filter (6 mm of B_4C and 1 mm of Cd) reduces the final thermal flux incident upon the phantom so that the production of gamma rays in Pb is also decreased. The diameter of the beam was 18.6 cm.

CALCULATIONS

An analysis of a two-dimensional model was performed by the DOT 4.2 transport code in cylindrical (r–z) geometry. The SAILOR cross-section library was condensed to 33 neutron groups and 13 photon groups for this purpose. As far as the condensation was concerned, neutron and gamma ray spectra resulting from one-dimensional calculations in S_8P_3 approximation by the ANISN transport code with SAILOR-DLC 76 cross-section library were used[2]. All the results were normalised to 1 MW reactor power.

The free-in-air neutron and gamma ray spectra are shown in Figures 2 and 3, respectively. The mean neutron energy is 8.63 keV.

The dose rate distribution in the polyethylene

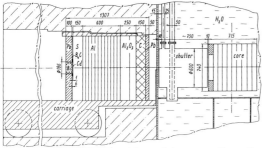

Figure 1. The LVR-15 epithermal beam configuration (sizes in mm).

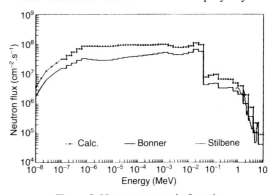

Figure 2. Neutron spectra in free air.

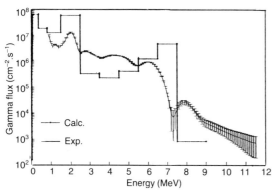

Figure 3. Gamma spectra free-in-air.

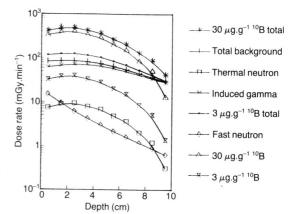

Figure 4. Dose rates plotted against depth in a polyethylene phantom.

phantom of 21 cm diameter and 21 cm height was calculated assuming that in a tumour there would be 30 μg of ^{10}B per g of tumour (30 μg.g^{-1} ^{10}B) and in normal tissue 3 μg of ^{10}B per g of the tissue (3 μg.g^{-1} ^{10}B). The RBE-weighted dose was calculated using RBE values of 1 for all gamma rays, 1.6 for fast neutrons and ^{14}N(n,p) recoil protons and 2.3 for the ^{10}B(n,α)^{7}Li reaction products[3]. The dose distributions are shown in Figure 4. Three criteria are used to describe the BNCT beam[3].

(i) *Advantage depth* (AD) defined as the depth for which the total dose or total therapeutic dose to tumour, from all components of the radiation field would equal the maximum dose to healthy tissue.
(ii) *Advantage ratio* (AR) provides a measure of the ratio of total therapeutic dose to the total background dose integrated over a given depth.
(iii) *Advantage depth dose rate* (ADDR) is the total therapeutic dose rate at the AD.

The advantage depths and ratios estimated from calculations are given in Table 1 for doses with RBE weighting and without.

MEASUREMENTS

Both neutron and gamma ray fluence rates were measured both in the free beam geometry and inside the phantom.

The free-in-air neutron spectrum was obtained using Bonner spheres and a scintillation spectrometer. The Bonner spheres detector with spheres of 2", 3", 5", 8" was calibrated with a ^{252}Cf source. The neutron spectrum from count rates was unfolded using the BASACF program based on Bayes theorem[4] and Distenfeld's response functions[5]. The scintillation spectrometer with a stilbene crystal of 10×10 mm^2 size was used for the estimation of the fast neutron spectrum above 1 MeV. The exclusion of gamma ray background was accomplished by a pulse shape discrimination system. The unfolding was performed by a program described by Marek[6]. The gamma ray spectrum was measured with the same spectrometer but with a NE 213 scintillator (5.08 cm × 5.08 cm in diameter) and raw spectra were unfolded using the FORIST program[7].

All the spectra were measured in the beam centre at 16 cm distance from the shield face. Resulting spectra are shown in Figures 2 and 3 respectively and tabulated in Table 2.

Measurements of fast neutron dose and gamma ray dose as well as the thermal and epithermal neutron fluence rates were performed in the polyethylene phantom described above. Thermoluminescent detectors and Si diodes were used to

Table 1. The advantage depths (AD), the advantage ratios (AR) and the advantage dose rates (ADDR) for the cylindrical polyethylene phantom of 21 cm diameter and 21 cm height.

		RBE weighted
AD (cm) min/max	7.4/7.0	8.6/7.9
AR	2.74	5
ADDR (mGy.min^{-1})	101	127

Table 2. Beam parameters free-in-air at 1 MW reactor power.

		calc.	exp.
Total neutron flux, φ	(10^{12}m^{-2}.s^{-1})	15.9	5.4
Fast neutron flux >30 keV	(10^{12}m^{-2}.s^{-1})	0.41	0.14
Epithermal neutron flux 0.5 eV < φ$_{epi}$ <30 keV	(10^{12}m^{-2}.s^{-1})	9.45	3.97
Thermal neutron flux, φ$_{th}$	(10^{12}m^{-2}.s^{-1})	5.12	1.29
Fast neutron dose rate, $\dot{D}_f$	(mGy.min^{-1})	20.3	9.6
Gamma dose rate, $\dot{D}\gamma$	(mGy.min^{-1})	22	7.0
$\dot{D}_f / \phi$	(10^{-15}mGy.m^2)	21.3	29.7
$\dot{D}\gamma / \phi$	(10^{-15}mGy.m^2)	23.1	21.6

evaluate the gamma dose and the fast neutron dose respectively.

The thermal and resonance neutron distribution in the phantom were obtained by a set of activation detectors with ^{197}Au, ^{55}Mn, ^{186}W and ^{23}Na. The neutron fluence rate distributions were measured with a small semiconductor detector (Si(Li)) both with and without a Cd cover. Results are given in Figure 5 and Table 3. The measurements were taken with the reactor power level at approximately 60 W and 170 kW respectively and all the results were extrapolated for 1 MW.

DISCUSSION

Experimental and calculated spectra of neutrons are in a very good agreement with respect to the relative distribution, but their integral values are rather different. The calculated total neutron flux is 2.9 times higher and the other quantities differ from one another in the same way. This is probably due to some problems with the estimation of the reactor power level especially the uncertainty of extrapolation from the very low level of 60 W to 1 MW power level.

The calculated gamma dose in both the phantom and free-in-air seems to be too high in comparison with the 30 µg.g^{-1} ^{10}B dose (see Tables 2 and 3, respectively). It is probably caused by cross section data for gamma ray production while the influence of gamma rays from the shielding on the results is very small at this depth.

The calculated depth–dose distribution as the main figure-of-merit is shown in Figure 4. The AD$_{min}$ of the beam is 7.0 cm while AD$_{max}$ is 7.4 cm

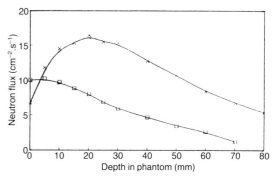

Figure 5. Neutron fluence rate distribution in PE phantom. (□) detector with Cd × 10, (x) thermal.

in polyethylene assuming 30 µg.g^{-1} ^{10}B in a tumour for both, and an effective ten-to-one tumour-to-blood ratio for the former. The AR of the beam was calculated to be 2.74 (with RBE 5.0) and the RBE weighted dose rate at the AD$_{min}$ was calculated to be 127 mGy.min^{-1} at power 1 MW.

CONCLUSIONS

The described filter modelling the beam of 18.6 cm in diameter has been found to be fairly acceptable for BNCT purposes in NRI Rez. The filter is able to produce an RBE weighted dose rate of 127 mGy.min^{-1} at AD of 7 cm. This is enough to deliver the therapeutic dose of 20 Gy in approximately 31.5 min at the reactor power level of 5 MW.

Table 3. Parameters of the cylindrical polyethylene phantom of 21 cm diameter and 21 cm height (1 MW power).

Depth in phantom		(cm)	0.2	3	8
Thermal neutron	calc.		5.48	10.2	2.23
flux, ϕ_{th}	exp.	$(10^{12}m^{-2}.s^{-1})$	3.14	5.0	1.0
Epithermal neutron	calc.		7.36	1.54	0.05
flux, ϕ_{epi}	exp.	$(10^{12}m^{-2}.s^{-1})$	2.30	0.70	0.03
Fast neutron	calc.		15.7	3.8	0.74
dose rate $\dot{D}_f$	exp.	(mGy.min^{-1})	11.3	2.8	1.6
Gamma dose	calc.		36.9	70.9	41.3
rate $\dot{D}\gamma$	exp.	(mGy.min^{-1})	< 20	< 20	< 20
Total therapeutic dose rate in tumour	calc.	(mGy.min^{-1})	98.7	171	36.4

REFERENCES

1. Burian, J., Marek, M. and Rataj, J. *Neutron Beam Parameters on LVR-15 Reactor for Neutron Capture Therapy.* Presented at 4th Int. Symp. on BNCT, Sydney, 3–7 Dec. 1990

2. Burian, J. and Rataj, J. *Neutron Beam Design and Performance for BNCT in Czechoslovakia.* In: Proc. Int. Workshop on Neutron Beam Design Development, and Performance for Neutron Capture Therapy. Eds O. K. Harling, J. A. Barnard and R. G. Zamenhof (New York: Plenum Press) (1990).
3. Clement, S. D., Choi, J. R., Zamenhof, R. G., Yanch, J. C. and Harling, O. K. *Monte Carlo Methods of Neutron Beam Design for Neutron Capture Therapy at the MIT Research Reactor (MITR-II).* In: Proc. Int. Workshop on Neutron Beam Design Development, and Performance for Neutron Capture Therapy. Eds O. K. Harling, J. A. Barnard and R. G. Zamenhof (New York: Plenum Press) (1990).
4. Tichy, M. *The Program BASACF.* (Institute of Radiation Dosimetry). UDZ-254/88 (1990).
5. Diestenfeld, C. H. *Improvements and Tests of the Bonner Ball Spectrometer.* (Brookhaven National Laboratory) BNL-21293 (1973).
6. Marek, M. *Fast Neutron Scintillation Spectrometer with Stilbene Crystal.* PhD Thesis, Czechoslovak Academy of Science (1990).
7. Johnson, R. H. *A User's Manual for COOLC and FORIST.* RSIC Computer Code Collection (Oak Ridge).

EVALUATION OF THE UNDESIRED NEUTRON DOSE EQUIVALENT TO CRITICAL ORGANS IN PATIENTS TREATED BY LINEAR ACCELERATOR GAMMA RAY THERAPY

C. Manfredotti†, U. Nastasi‡, E. Ornato§ and A. Zanini§
†Dipartimento di Fisica Sperimentale, Universita' di Torino, Italy
‡Ospedale "S. Giovanni Battista A.S.", Torino, Italy
§INFN, Sezione di Torino, Italy

Abstract — A careful theoretical and experimental evaluation has been made in order to estimate the neutron dose equivalent for critical organs of patients treated by gamma ray therapy. The machine employed in these studies is the linear electron accelerator MD Class Mevatron, Siemens, installed at the S. Giovanni Battista A.S. Hospital in Torino (Italy) and used for tumour therapy. A calculational method was developed, using two suitable 3D Monte Carlo computer codes: the EGS4 code (Electron Gamma Shower), which evaluates bremsstrahlung gamma rays and photo–neutron production in the accelerator head and the MCNP code (Monte Carlo Neutron and Photon Transport), which simulates the neutron transport and evaluates the organ dose equivalents. In addition, two sets of measurements were made during treatment sessions with 2 Gy therapeutic gamma dose: (a) by means of thermoluminescence dosemeters TLD-600 and TLD-700 placed at various depths inside an anthropomorphic phantom in positions corresponding to the organs; (b) *in vivo* by albedo neutron personal dosemeters UD-802, National Panasonic, positioned on the patients according to the organs. The agreement between experimental and theoretical data is good. The neutron organ dose equivalents are by no means negligible; for instance H_T varies from 2.65 mSv on the thyroid to 42.02 mSv on testes, if normalised to one Gy of maximum absorbed dose in the pelvis. of the primary gamma beam of 15 MeV maximum energy (10 cm × 10 cm field size and source-skin distance of 1 m).

INTRODUCTION

The international organisations responsible for radiation protection, such as NCRP (National Council on Radiation Protection and Measurements), ICRU (International Commission on Radiation Units and Measurements), ICRP (International Commission on Radiological Protection), show increasing interest in the problem of optimising the exposure to ionising radiation in medical applications for diagnosis and therapy, in order to reduce the absorbed dose to a level as low as reasonably achievable. It is particularly important not to underestimate the hazards concerning the undesired exposure to neutrons: the ICRP in publication 60[1] on the basis of a review of the biological information concerning the damage to human tissue due to ionising radiation, introduced 'radiation weighting factors' w_r in addition to 'quality factors'[2]; in the case of intermediate energy neutrons, an increase of dose equivalent values by approximately a factor two is proposed.

A considerable production of undesired neutrons is associated with the electron linear accelerators used for tumour therapy with electron or gamma beams. Neutrons are produced by giant resonance reactions inside the high Z materials of the accelerator head structures, both directly by electrons in (e,e'n) reactions and by bremsstrahlung gamma rays in (γ,n) reactions, when the incident particle energy is higher than the reaction threshold energy Q.

In general, the measurements of neutron dose equivalent in presence of a strong gamma field are not easy to carry out. In the literature, dose equivalent values H due to neutron leakage have been measured at various distances from the machine isocentre, for different therapy accelerators[3–5]; in these cases H varies from 5×10^{-4} Sv to 5×10^{-3} Sv per Gy of primary gamma rays, with gamma ray maximum energy varying from 10 to 25 MeV respectively. On the other hand, it is not easy to evaluate the effective neutron dose equivalent to the patient during a gamma ray treatment: even using computer simulation, it is rather complex to determine the energy spectrum of the neutrons hitting the patient. Usually, the neutron dose equivalent to the organs is estimated theoretically, using broad parallel monochromatic neutron beams[6–8].

In the present experiment a careful evaluation was made of the undesired neutron dose equivalent to critical organs in patients treated by radiotherapy by means of MD Class Mevatron Siemens electron linear accelerator in use at the S. Giovanni Battista A.S. Hospital in Torino (Italy), using both calculations and measurements. Calculations were performed, using two suitable computer codes EGS4 and MCNP, in order to simulate the whole geometry of the system and the physical processes involved. In addition, two sets of measurements were carried out: (a) in an anthropomorphic phantom, with thermoluminescence dosemeters; (b) *in vivo* on three different patients, with personal neutron albedo dosemeters. In each case the irradiation condition was referred to a 2 Gy gamma absorbed dose, produced by a gamma ray beam of maximum energy 15 MeV, centred on the pelvis with 10 cm × 10 cm field size and 100 cm SSD (source-skin distance). The data obtained by the computer simulation are confirmed by the results of the two sets of measurements. The values of the neutron organ dose equivalents, evaluated following ICRP 21 vary from 2.65 ± 0.5 mSv on the thyroid to 42.02 ± 2.32 mSv on the testes per Gy of primary gamma rays; following ICRP 60 the corresponding values of equivalent doses H_T vary from 4.97 ± 0.95 mSv to 72.05 ± 3.61 mSv. In view of the complexity of the numerical work involved in obtaining organ dose equivalents in mathematical phantoms, the consistency between theoretical and experimental results seems to support the confidence in the calculation method.

THE MEDICAL ACCELERATOR FACILITY

The MD Class Mevatron Siemens electron linear accelerator, can produce e⁻ beams with energy of 5, 7, 9, 11, 14 MeV or gamma ray beams of 6 and 15 MeV maximum energy, produced by bremsstrahlung in a ^{197}Au target 0.5 mm thick.

The machine was studied in the high energy configuration, in which electrons of 15 MeV energy produce bremsstrahlung gamma rays of 15 MeV maximum energy (Figure 1).

The system of collimators, all made of ^{186}W, consists of a primary collimator in a fixed location and four secondary collimators of variable position, in order to obtain treatment fields of suitable area. The gamma beam leaving the target is intercepted by a flattening filter in the shape of a cone, radius 2 cm, height 4.8 cm, its function being to flatten the spatial and energetic photon distribution, in order to obtain a beam of intensity as uniform as possible. The angular distribution of photons leaving the flattening filter is strongly peaked in the forward direction and it is limited to an angle of approximately 2 deg[9].

NEUTRON PRODUCTION IN A LINEAR ELECTRON ACCELERATOR

Neutrons are produced by giant resonance reactions, both directly by electrons in (e, e' n) reactions and by bremsstrahlung gamma rays in (γ,n) reactions, when the incident particle energy is higher than the reaction threshold energy Q. In the ^{197}Au target (Q_{197Au} = 8.084 MeV) both reactions occur. It has been proved, with the approximation σ(e,e'n)/σ(γ,n) = 1/137 that the ratio between the neutron yield produced by bremsstrahlung gamma rays Y_b and the neutron yield produced by electrodistintegration Y_e is a function of the electron energy E_e and is directly proportional to the thickness of the target Δx[3]. In the present case, for E_e = 15 MeV and Δx = 0.5 mm, Y_b/Y_e = 0.40. Electrons leaving the target without interaction are eliminated by means of a 0.75 mm thick ^{12}C absorber (Q_{12C} = 18.719 MeV) where neutrons are not produced, owing to the high value of the (γ,n) threshold energy.

As far as the photon interaction with the collimators and the flattening filter in ^{186}W (Q_{186W} = 7.418 MeV) is concerned, it can be assumed that only the gamma rays hitting the flattening filter and the secondary collimators give a considerable contribution to photo-neutron production. In fact, because of the angular and energetic distribution of photons leaving the target, only a small fraction of photons of low energy intercept the primary collimator. The angular distribution of photo-neutrons can be considered to be isotropic. Subsequently, neutrons are uniformly scattered inside the entire structure of the machine, affecting the whole body of the patient and possibly damaging those organs which are

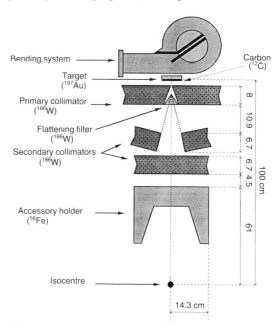

Figure 1. Layout of accelerator head MD Class Mevatron Siemens, in the configuration in which bremsstrahlung gamma rays are produced by a 15 MeV electron beam (high energy configuration).

particularly radiosensitive, such as thyroid, lungs, testes and breast.

CALCULATIONAL METHODS

A calculational method was developed, using two suitable 3D Monte Carlo computer codes, to simulate the entire process, consisting of a first stage of photo-neutron production in the high Z head structure and of a second stage of their transport through the accelerator head and subsequent interaction with an anthropomorphic tissue-equivalent phantom, inside which the critical organs were placed.

EGS4 calculation: photon and neutron production

The accelerator head was simulated by using the Monte Carlo computer code EGS4[10], in the high energy configuration, with a photonic field size of 10 cm × 10 cm at the isocentre plane.

The energy spectrum of bremsstrahlung gamma rays produced by e^- of 15 MeV energy in the ^{197}Au target was evaluated; the code subsequently follows the photons through the accelerator head and, finally, the attenuated energy spectrum of photons leaving the flattening filter was also calculated (Figure 2).

The code was modified in order to consider not only the photoelectric effect, the Compton effect and the pair production, but also the photo-neutron production by giant resonance reactions, in order to evaluate the energy spectra of photo-neutrons produced in the target by (e, e'n) and (γ,n) reactions and in the flattening filter and secondary collimators by (γ,n) reactions only. The results are presented in Figure 3, which shows separately the contribution to the neutron energy spectrum of (γ,n) reactions occurring in the ^{197}Au target and in the ^{186}W flattening filter.

MCNP calculation: neutron transport

The neutron yield evaluated by means of EGS4 was used as the neutron source in MCNP[11].

The code simulates the transport of neutrons through the accelerator head, represented in its actual geometry, and their subsequent interaction

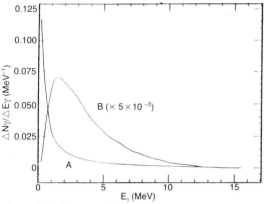

Figure 2. A. Bremsstrahlung gamma rays spectrum as a function of photon energy, produced by 15 MeV e^- on the ^{197}Au 0.5 mm target, per unit electron fluence. B. γ ray spectrum as a function of photon energy leaving the flattening filter, per unit electron fluence ($\times 5 \times 10^{-3}$).

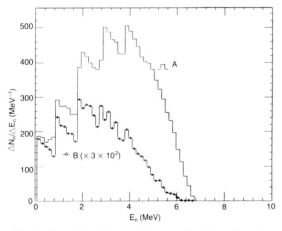

Figure 3. A. Neutron spectrum as a function of neutron energy, produced by (γ,n) reaction in the ^{197}Au target per unit electron fluence. B. Neutron spectrum as a function of neutron energy, produced by (γ,n) reaction in the ^{186}W flattening filter, per unit electron fluence ($\times 3 \times 10^2$).

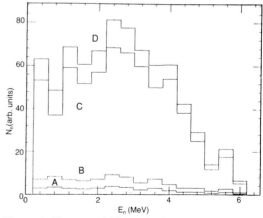

Figure 4. Neutron yields at the isocentre plane, after transport in the accelerator head, per unit neutron fluence. A. Neutron yield produced by photo-disintegration in the ^{197}Au target. B. Neutron yield produced by electrodisintegration in the ^{197}Au target. C. Neutron yield produced by photodisintegration in the ^{186}W flattening filter and secondary collimators. D. Total neutron yield.

with an anthropomorphic phantom, of averaged male and female sex, inside which critical organs were placed, following ICRP 23 data[12].

Neutrons are followed through the accelerator head structure and the attenuated energy spectrum was calculated ($\bar{E}_n$ = 2.49 MeV). Figure 4 shows the total neutron yield incident on the patient at the isocentre plane.

The dose equivalent on the organs (H_T) and the effective dose equivalent on the whole body (H_E) were evaluated using Q values (ICRP 21) and tissue weighting values w_T of ICRP 26[13]; in this case the relationship $H_E = \Sigma H_T w_T$ led to H_E = 15.69 ± 0.8 mSv per Gy of primary gamma rays.

Also, the equivalent doses on the organs and the effective dose (E) on the whole body according to ICRP 60 were evaluated, by using the appropriate w_R and w_T values. From the relationship $E = \Sigma_{T,R} w_T w_R D_{T,R}$, one has E = 29.29 ± 1.52 mSv per Gy of primary gamma rays.

The complete sets of results for the various organs are shown in Table 1.

Table 1. Neutron dose equivalent values (ICRP 21) and equivalent dose values (ICRP 60) for critical organs obtained with MCNP calculation. Irradiation conditions: gamma rays maximum energy 15 MeV; gamma absorbed dose 2 Gy; gamma field size 10 cm × 10 cm; SSD 100 cm; irradiation zone: pelvis.

(a) Using Q values (ICRP 21)

	w_T*	H_T (mSv)
Testes	0.25	84.02 ± 4.65
Breast	0.15	28.40 ± 0.69
Bone marrow (red)	0.12	13.36 ± 2.24
Lung	0.12	13.04 ± 0.60
Thyroid	0.03	5.60 ± 1.01
Bone surface	0.03	1.98 ± 0.12
Remainder	0.30	10.50 ± 0.24

$H_E = \Sigma w_T H_T$ = 15.91 ± 0.85 mSv per Gy of primary gamma rays.
*w_T, tissue weighting factors from ICRP 26.

(b) Using w_R values (ICRP 60)

	w_T**	H_T (mSv)
Testes	0.20	144.31 ± 7.81
Breast	0.05	46.40 ± 1.12
Bone marrow (red)	0.12	22.70 ± 3.69
Lung	0.12	21.42 ± 0.97
Thyroid	0.05	9.94 ± 1.87
Bone surface	0.01	3.21 ± 0.19
Stomach	0.12	15.23 ± 1.08
Bladder	0.05	112.03 ± 6.25
Liver	0.05	22.42 ± 1.21
Oesophagus	0.05	17.96 ± 2.00
Skin	0.01	127.62 ± 3.37
Remainder	0.05	108.57 ± 3.02

$E = \Sigma w_R w_T D_{T,R}$ = 29.29 ± 1.52 mSv per Gy of primary gamma rays.
**w_T, tissue weighting factors from ICRP 60.

MEASUREMENTS ON PHANTOM WITH THERMOLUMINESCENCE DOSEMETERS

Pairs ^{6}LiF TLD-600 (sensitive to n + γ radiation) and ^{7}LiF TLD-700 (sensitive to γ radiation only) were placed inside an anthropomorphic Alderson Rando phantom in synthetic material of density 0.985 g.cm^{-3}, at various depths in positions corresponding to those of the critical organs. The exposure condition was the same as considered in the computer simulation.

An accurate calibration was made with respect to dose equivalent, both for gamma rays and for neutrons. In the first case, TLD-600 and TLD-700 were exposed to a ^{137}Cs source (Amersham, calibrated at the factory within 4% with respect to exposure rate) at the Experimental Physics Department of Torino University (Italy), following a standard procedure. In the second case, TLD-600 were exposed to a suitable irradiation facility designed and built at the JRC Center in Ispra (Varese, Italy). It consists of an Am-Be source (Amersham, calibrated at the factory within 5% with respect to neutron fluence rate) disposed at

Table 2. Comparison between values of neutron dose equivalent on the main organs evaluated by MCNP code, by measurement on anthropomorphic phantom by TLD dosemeters, and measured on three patients by albedo personal dosemeters. Irradiation conditions: gamma rays maximum energy 15 MeV; gamma absorbed dose 2 Gy; gamma field size 10 cm × 10 cm; SSD 100 cm; irradiation zone: pelvis.

| | MCNP | TLD | Albedo dosemeters | | |
			Pat. A	Pat. B	Pat. C
Thyroid	5.60±1.01	5.31±1.59	8.8±2.64	5.28±1.58	5.40±1.62
Lung dx	13.04±0.60	11.48±3.44	13.6±4.08	11.28±3.38	14.04±4.21
Lung sx	13.04±0.60	11.35±3.41	10.5±3.15	10.30±3.09	12.01±3.60
Breast	28.04±0.69	9.99±3.00	27.4±8.22	–	–
*Testes	84.02±4.65	–	–	–	–

*As the gamma beam is centred on the pelvic zone, both albedo and TLD dosemeters give no reliable readings owing to the high gamma fluence in comparison with leakage neutron fluence.

the centre of a 60 cm side cubic graphite moderator, shielded with a cadmium foil 1 mm thick, which produces a neutron energy spectrum similar to the leakage neutron spectrum produced in the accelerator head[14]; the neutron dose equivalent was measured with a Studvsik model 2002 rem counter, giving a response dose equivalent fluence roughly consistent with the ICRU response from thermal energies to 17 MeV. A check of the values measured by the rem counter was made by simulating the entire apparatus by the computer code MORSE: the agreement between the experimental data and the H values obtained from the calculation starting from the calibrated neutron fluence rate and from the Am-Be spectrum quoted in literature was good (within 1%). During the irradiation, the thermoluminescence dosemeters were disposed inside a polythene phantom, so as to reproduce a situation which was similar to the experimental one.

MEASUREMENTS ON PATIENTS WITH PERSONAL ALBEDO DOSEMETERS

Measurements were carried out on three different patients subjected to a radiotherapy treatment by a gamma ray beam of 15 MeV maximum energy, by means of albedo neutron personal dosemeters UD-802 National Panasonic, positioned according to the organs. These detectors are routinely in use at the Health Physics Department of JRC in Ispra (Varese, Italy), and were calibrated with respect to dose equivalent in three energy ranges (thermal, epithermal and fast neutrons) by means of the irradiation apparatus described by Manfredotti et al[14].

For the three patients the dose equivalent values obtained from the reading of albedo dosemeters was normalised to standard exposure conditions: 2 Gy gamma dose, 10 cm × 10 cm field size, 100 cm SSD.

CONCLUSIONS

Table 2 shows the comparison between dose equivalent values on the main organs obtained from MCNP simulation and the two sets of experimental measurements. The agreement between experimental and theoretical data can be considered as good.

Neither the albedo dosemeters placed in relation to the testes nor the TLD dosemeters in the corresponding location in the phantom produced reliable readings. This can be explained by the fact that in this position the dosemeters are directly exposed to the gamma ray beam (which hits the pelvic zone), consequently the gamma dose exceeds the neutron dose equivalent by a factor $10^3 - 10^4$.

It can be observed that the neutron effective dose equivalent to the whole body during a complete radiotherapy treatment of 15–20 sessions, each one of 2 Gy gamma absorbed dose, is significant.

In fact, considering the previous calculated value H_E = 15.69 mSv per Gy of primary gamma rays, the average effective dose equivalent in a complete treatment is about 450 mSv according to ICRP 21. Following the ICRP 60, an E value of 29.29 mSv per Gy of primary gamma rays is expected, corresponding to about 900 mSv in a complete treatment.

It is advisable, therefore, both by experimental field measurements and by means of the computational method developed, to study the possibility of shielding the patient during treatment with suitable screens made of hydrogenated material together with thermal neutron absorbers, so as to reduce the undesired neutron dose equivalent significantly, without altering the therapeutic gamma ray dose.

REFERENCES

1. ICRP. *The 1990–1991 Recommendation of the International Commission on Radiological Protection.* Publication 60. Ann. ICRP **21**(1-3) (Oxford: Pergamon) (1991).
2. ICRP. *Data for Protection against Ionizing Radiation from External Sources,* Publication 21 (Oxford: Pergamon) (1971).
3. NCRP. *Neutron Contamination from Medical Accelerators.* NCRP Report 79 (Bethesda, MD: NCRP Publications) (1984).
4. Sherwin, A. G., Pearson, A. J., Richards, D. J. and O'Hagan, J. B. *Measurement of Neutrons from High Energy Electron Linear Accelerators.* Radiat. Prot. Dosim. **23**, 333-336 (1987).
5. Tosi, G. and Torresin, A. *Neutron Measurements around Medical Electron Accelerators by Active and Passive Detection Techniques.* Med. Phys. **18**, 54-60 (1991).
6. Wittmann, A., Morhart, A. and Burger, G. *Organ Doses and Effective Dose Equivalent.* Radiat. Prot. Dosim. **12**, 101-106 (1985).
7. Hollnagel, R. A. *Effective Dose Equivalent and Organ Doses for Neutrons from Thermal to 14 MeV.* Radiat. Prot. Dosim. **30**, 149-159 (1990).

8. Austerlitz, C., Kahn, B., Eicholz, G. G., Zankl, M. and Drexler, G. *Calculation of the Effective Male Dose Equivalent Relative to the Personal Dose at Nine Locations with a Free Arm Model.* Radiat. Prot. Dosim. **36**, 13-21 (1991).
9. Mohan, R. and Chui, C. *Energy and Angular Distribution of Photons from Medical Linear Accelerators.* Med. Phys. **12**, 592-597 (1985).
10. Nelson, W. R., Hirayama, A. and Rogers, D. W. O. *The EGS4 Code System* (Stanford Linear Accelerator Center) Report 265 (1985).
11. Briesmeister, J. F. (ed.) *MCNP — A General Monte Carlo Code for Neutron and Photon Transport, Version 3A.* (Los Alamos National Laboratory, Los Alamos, New Mexico) (1986).
12. ICRP. *Reference Man: Anatomical, Physiological and Metabolic Characteristics.* Publication 23 (Oxford: Pergamon) (1975).
13. ICRP. *Recommendation of the International Commission on Radiological Protection.* Publication 26 (Oxford: Pergamon) Ann. ICRP **1**(3) (1977).
14. Manfredotti, C., Zanini, A., Rollet, S. and Arman, G. *Simulation and Test of a New Albedo Personal Dosemeter for Neutrons.* Nucl. Instrum. Methods **A284**, 465-475 (1989).

SEVENTH SYMPOSIUM ON NEUTRON DOSIMETRY Berlin, 14-18 October 1991

List of Participants

Alberts W. G., Physikalisch-Technische Bundesanstalt Bundesallee 100, W-3300 BRAUNSCHWEIG GERMANY
Alevra A. V., Physikalisch-Technische Bundesanstalt Bundesallee 100, W-3300 BRAUNSCHWEIG GERMANY
Alves de Aro Antonio C., University of Birmingham, School of Physics and Space Research P.O. Box 363, EDGBASTON, BIRMINGHAM B15 2TT UK
Antolkovic Branka, Ruder Boskovic Institute Bijenicka 54, 41001 ZAGREB CROATIA
Apfel R., Yale University Apfel Enterprises Inc. 25 Science Park, NEW HAVEN CT 06511 USA
Arend E., Universität des Saarlandes, Zentrum f. Umweltforschung Am Markt 5, Zeile 4, W-6601 DUDWEILER GERMANY
Aroua A., Institut de Radiophysique Appliquée Centre Universitaire, CH-1015 LAUSANNE SWITZERLAND
Azimi-Garakani D., Paul Scherrer Institute (PSI), CH-5232 VILLIGEN PSI SWITZERLAND
Barbry Francis, Commissariat à l'Energie Atomique (CEA) Centre d'Etudes de Valduc B. P. no. 21, F-21120 IS-SUR-TILLE FRANCE
Barschall H. H., Department of Physics The University of Wisconsin 1150 University Avenue, MADISON, WISCONSIN 53706 USA
Barthe Jean, CEA - IPSN/DPHD-SDOS B. P. no. 6, F-92260 FONTENAY-AUX-ROSES FRANCE
Bartlett David T., National Radiological Protection Board, CHILTON, DIDCOT OXON OX11 0RQ UK
Becret Claude, Etablissement Technique Central de l'Armement 16 bis, Avenue Prieur de la Cote d'Or, F-94114 ARCUEIL CEDEX FRANCE
Benedetti A., CRESAM Min. of Defence S. Pino A Grado, I-56010 PISA ITALY
Bilski Pawel, Henryk Niewodniczanski Institute of Nuclear Physics ul. Radzikowskiego 152, PL-31-342 KRAKOW POLAND
Binns P., National Accelerator Centre P. O. Box 72, FAURE 7131 SOUTH AFRICA
Blanc Daniel, Centre de Physique Atomique Université Paul Sabatier 118 Route de Narbonne, F-31062 TOULOUSE CEDEX FRANCE
Boltwood, Defence Radiological Protection Service Inst. of Naval Medicine Crescent Road, ALVERSTOKE GOSPORT HANTS PO12 2DL UK
Bos A. J. J., Interfaculty Reactor Institute Mekelweg 15, NL-2629 JB DELFT THE NETHERLANDS
Boschung M., Paul Scherrer Institute (PSI), CH-5232 VILLIGEN PSI SWITZERLAND
Brede H. J., Physikalisch-Technische Bundesanstalt Bundesallee 100, W-3300 BRAUNSCHWEIG GERMANY
Broerse J. J., Institute of Applied Radiobiology and Immunology TNO 151 Lange Kleiweg, P.O. Box 5815, NL-2280 HV RIJSWIJK THE NETHERLANDS
Burgkhardt Bertram, Kernforschungszentrum Karlsruhe Postfach 3640, W-7500 KARLSRUHE GERMANY
Caswell R. S., National Institute of Standards and Technology, GAITHERSBURG, MARYLAND 20899 USA
Chartier Jean-Louis, CEA - IPSN/DPHD-SDOS B. P. No. 6, F-92265 FONTENAY-AUX-ROSES CEDEX FRANCE
Chauvenet Bruno, CEA/LPRI CEN Saclay B.P. No. 52, F-91193 GIF-SUR-YVETTE CEDEX FRANCE
Colautti Paolo, Laboratorio Nazionali INFN Via Romea 4, I-35020 LEGNARO (PD) ITALY
Coppola Mario, ENEA, CRE Casaccia Via Anguillarese 301, I-00060 ROMA ITALY
Cosgrove V.P., Univ. of Birmingham, School of Physics and Space Research P. O. Box 363, EDGBASTON, BIRMINGHAM B15 2TT UK
Crawford J. F., Paul Scherrer Institute (PSI), CH-5232 VILLIGEN PSI SWITZERLAND
Cross W. G., AECL Research Chalk River Labs., CHALK RIVER ONTARIO K0J J10 CANADA
da Cruz Marilia T., Instituto de Fisica Universidade de Sao Paulo C.P. 20516, CEP 01498 SAO PAULO BRAZIL
Curl Ian, UKAEA, Winfrith Technology Center DORCHESTER, DORSET DT2 8DH UK
Curzio G., DCMN Universita degli Studi di Pisa Via Diotisalvi 2, I-56126 PISA ITALY
Dangendorf V., Physikalisch-Technische Bundesanstalt Bundesallee 100, W-3300 BRAUNSCHWEIG GERMANY
DeLuca, Jr. P.M., University of Wisconsin Department of Medical Physics 1300 University Avenue, MADISON, WISCONSIN 53706-1532 USA
Deboodt Pascal, Nuclear Center of Mol (Safety Dpt.) 200 Boeretang, B-2400 MOL BELGIUM
Decossas J. L., LEPOFI Faculté des Sciences 123, rue Albert Thomas, F-87060 LIMOGES FRANCE
Delafield H. J., AEA Environment and Energy B 364 Harwell Laboratory, HARWELL OXFORDSHIRE OX11 0RA UK
D'Errico F., DCMN - Universita degli Studi di Pisa, Via Diotisalvi 2, I-56126 PISA ITALY
Devine R. T., Los Alamos Nat. Laboratory 1222 Big Rock Loop, LOS ALAMOS, NM 87545 USA

LIST OF PARTICIPANTS

Dhermain Joel, DGA/ETCA/CEB/DPN/ER 16 bis Avenue Prieur de la Cote d'Or, F-94114 ARCUEIL CEDEX FRANCE
Dicello John F., Clarkson University 8 Clarkson Avenue, POTSDAM, NY 13699 USA
Dietze G., Physikalisch-Technische Bundesanstalt Bundesallee 100, W-3300 BRAUNSCWEIG GERMANY
Dörschel B., Technische Universität Dresden Institute of Radiation Protection Physics Mommsenstr. 13, O-8027 DRESDEN GERMANY
Draaisma Folkert, IRI afd. Cursorisch Onderwys, Mekelweg 15, NL-2629 JB DELFT THE NETHERLANDS
Drake P., Ringhals Nuclear Power Plant, S-420 22 VÄRÖBACKA SWEDEN
Dubarry Béatrice, LEPOFI Faculté des Sciences 123, avenue Albert Thomas, F-87060 LIMOGES FRANCE
Facius R., DLR, Inst. für Flugmedizin Linder Höhe, W-5000 KÖLN 90 GERMANY
Fajardo Patricia W., Nuclear Engeneering Institute Nuclear Energy National Commission CP 2186, 20001 RIO DE JANEIRO BRAZIL
Fasso Alberto, CERN, CH-2111 GENEVE 23 SWITZERLAND
Fazileabasse Javaraly, Merlin Gerin Provence - Radioprotection BP 1, F-13113 LAMANON FRANCE
Fernandez F., Universitat Autonoma de Barcelona, E-08193 BELLATERRA-BARCELONA SPAIN
Festag Johannes G., Gesellschaft für Schwerionenforschung mbH Planckstr. 1, W-6100 DARMSTADT GERMANY
Fiedler Winfried, Wellhöfer Dosimetrie Bahnhofstr. 5, W-8501 SCHWARZENBRUCK GERMANY
Francis T. M., National Radiological Protection Board, CHILTON, DIDCOT, OXON OX11 0RQ UK
Gaal S., University of Veszprem Egyetem U. 10 PF. 158, H-8201 VESZPREM HUNGARY
Gabel D., Universität Bremen Fachbereich Chemie Postfach 330440, W-2800 BREMEN 33 GERMANY
Gabris Frantisek, Czecho-Slovak Institute of Metrology Tr. L. Novomeskeho IV/487, CS-842 55 BRATISLAVA CSFR
Gambarini Grazia, Dipartimento di Fisica Dell'Università Via Celoria 16, I-20131 MILANO ITALY
Gerdung S., Universität des Saarlandes Abt. Strahlenbiophysik, ZFU Am Markt 5, Zeile 4, W-6602 DUDWEILER GERMANY
German U., Nuclear Research Center Negev POB 9001, 84190 BEER SHEVA ISRAEL
Gibson J. A. B., AEA Environment and Energy Harwell Laboratory B 364, HARWELL DIDCOT OXON OX11 ORA UK
Golnik N., Institute of Atomic Energy Swierk, PL-05-400 OTWOCK POLAND
Grecescu M., Insitut de Radiophysique Appliquée Centre Universitaire, CH-1004 LAUSANNE SWITZERLAND
Green St., School of Physics and Space Research University of Birmingham, EDGBASTON BIRMINGHAM B15 2TT UK
Griffith Richard V., Dept. of Nuclear Energy and Safety IAEA P. O. Box 200, A-1400 WIEN AUSTRIA
Grindborg J. E., Swedish Radiation Protection Institute Box 60204, S-10401 STOCKHOLM SWEDEN
Gubatova Diana, Radiation Safety Laboratory Latvian Medical Academy 226012 Tallin Str. 49, RIGA LATVIA
Guldbakke S., Physikalisch-Technische Bundesanstalt Bundesallee 100, W-3300 BRAUNSCHWEIG GERMANY
Hahn T., Technische Universität Dresden Inst. of Radiation Protection Physics Mommsenstr. 13, O-8027 DRESDEN GERMANY
Hajnal Ferenc, US DOE Environmental Measurement Lab. 376 Hudson Street, NEW YORK, N.Y. 10014 - 3621 USA
Hall Eric J., Columbia University College of Physicians and Surgeons 630 West 168th Street, NEW YORK, N.Y. 10032 USA
Harvey J. R., Nuclear Electric Berkeley Nuclear Laboratories, BERKELEY, CLOUCESTERSHIRE GL13 9PB UK
Hecker O., Physikalisch-Technische Bundesanstalt Bundesallee 100, W-3300 BRAUNSCHWEIG GERMANY
Hess A., Radiologische Klinik und Strahleninstitut Universitätskrankenhaus Eppendorf Martinistr. 52, W-2000 HAMBURG 20 GERMANY
Hoffmann W., Bergische Universität Fachbereich Physik Gaußstr. 20, W-5600 WUPPERTAL 1 GERMANY
Hollnagel Rudolf, Physikalisch-Technische Bundesanstalt Bundesallee 100, W-3300 BRAUNSCHWEIG GERMANY
Huynh V. D., Bureau International des Poids et Mesures Pavillon de Breteuil, F-92312 SEVRES CEDEX FRANCE
Ipe N., Stanford Linear Accel. Center P. O. Box 4349 SLAC Mail Bin 48, STANFORD, CA, 94309 USA
Jahn Ulrich, Robert-Rössle-Klinik Lindenberger Weg 80, O-1115 BERLIN-BUCH GERMANY
Jahr R. Physikalisch-Technische Bundesanstalt Bundesallee 100, W-3300 BRAUNSCHWEIG GERMANY
Jetzke S., Physikalisch-Technische Bundesanstalt Bundesallee 100, W-3300 BRAUNSCHWEIG GERMANY
Jones D., National Accelerator Centre P. O. Box 72, FAURE 7131 SOUTH AFRICA
Jozefowicz Krystyna, Institute of Atomic Energy Radiation Protection Department, PL-05-400 OTWOCK-SWIERK POLAND
Kearsley E., Armed Forces Radiobiology Research Institute Radiation Biophysics Department, SILVER SPRING, MD 20889-5145 USA
Khoshnoodi M., National Radiation Protection Department Atomic Energy Org. of Iran P.O. Box 14155-4494, TEHRAN IR IRAN
Klein H., Physikalisch-Technische Bundesanstalt Bundesallee 100, W-3300 BRAUNSCHWEIG GERMANY
Kluge H., Physikalisch-Technische Bundesanstalt Bundesallee 100, W-3300 BRAUNSCHWEIG GERMANY
Konijnenberg Mark, The Netherlands Cancer Institute Plesmanlaan 121, NL-1066 CX AMSTERDAM THE NETHERLANDS

LIST OF PARTICIPANTS

Koohi-Fayegh R., The University of Birmingham School of Physics and Space Research P. O. Box 363, BIRMINGHAM B15 2TT UK
Kraus W., Bundesamt für Strahlenschutz Waldowallee 117, O-1157 BERLIN GERMANY
Kriens Micheal, Radiologische Klinik Universität Hamburg Abteilung Strahlentherapie Martinistr. 52, W-2000 HAMBURG 20 GERMANY
Krüger U., Schering AG Pharma Forschung Abt. Kernresonanz und Röntgen Postfach 650311, W-1000 BERLIN 65 GERMANY
Kunz A., Universität des Saarlandes, Zentrum f. Umweltforschung, AG Strahlenbiophysik Am Markt 5, Zeile 4, W-6602 DUDWEILER GERMANY
Lawaczeck R., Schering AG Kernresonanz/Röntgen Postfach 650311, W-1000 BERLIN 65 GERMANY
Le Thanh Phung, Commissariat à l'Energie Atomique CEN SACLAY B. P. No. 2, F-91191 GIF-SUR-YVETTE CEDEX FRANCE
Leonowich John, Battelle Pacific Northwest Laboratory MSIN K3-70 P. O. Box 999, MIS K3-70, RICHLAND, WA 99352 USA
Leuthold Gerhard, Institut für Strahlenschutz GSF-Forschungszentrum f. Umwelt und Gesundheit Ingolstädter Landstr. 1, W-8042 NEUHERBERG GERMANY
Lim Taeho, Universität des Saarlandes, Zentrum für Umweltforschung, Abt. Strahlenbiophysik Am Markt 5, Zeile 4, W-6602 DUDWEILER GERMANY
Lommler B., I. Physikalisches Institut der Universität Heinrich-Buff-Ring 16, W-6300 GIESSEN GERMANY
Lounis-Mokrani Zohra, Laboratoire de Dosimetrie / Centre de Radioproction et Surête 02, Bd Frantz Fanon B.P. 1017, 16000 ALGER ALGERIA
Luszik-Bhadra Marlies, Physikalisch-Technische Bundesanstalt Bundesallee 100, W-3300 BRAUNSCHWEIG GERMANY
Makovicka Libor, LEPOFI Faculté des Sciences 123 Rue Albert Thomas, F-87060 LIMOGES FRANCE
Manfredotti Claudio, Dip. Fisica Sperimemtale Università di Torino Via Pietro Giuria n. 1, I-10125 TORINO ITALY
Marchetto A., Centre d'Etudes Nucléaires de Grenoble B. P. No. 85 X, av. des martyrs, F-38041 GRENOBLE CEDEX FRANCE
Marek M., Nuclear Research Institute Rez, CS-250 68 REZ / PRAGUE CSFR
Mares Vladimir, GSF-Forschungszentrum Inst. für Strahlenschutz Ingolstädter Landstr. 1, W-8042 NEUHERBERG GERMANY
Marino Stephen A., College of Physicians & Surgeons of Columbia Univ., Center for Radiol. Res. RARAF-Nevis P. O. Box 21, IRVINGTON, N.Y. 10533 USA
Matzke M., Physikalisch-Technische Bundesanstalt Bundesallee 100, W-3300 BRAUNSCHWEIG GERMANY
Maughan Richard, Gershenson Radiation Oncology Center Harper Hospital 3990 John R., DETROIT, MICHIGAN 48201 USA
Maurício Cláudia L.P., Inst. de Engenharia Nuclear Cidade Universitaria - Ilha de Fundao 2186, RIO DE JANEIRO BRAZIL
Meinhold Charles B., Radiological Sciences Division Brookhaven National Lab., LONG ISLAND, UPTON, NY 11937 USA
Meißner Peter, Univ.-Klinikum Essen, Inst. für Med. Strahlenphysik Hufelandstr. 55, W-4300 ESSEN 1 GERMANY
Menzel H. G., Commission of the European Communities DG XII-D-3, ARTS 3161 rue de la Loi 200, B-1049 BRÜSSEL BELGIUM
Molho Niky, Università degli Studi di Milano, Dipartim. di Fisica Via Celoria 16, I-20133 MILANO ITALY
Morstin Krzysztof, Inst. of Nuclear Physics ul. Radzikowskiego 152, PL-31342 KRAKOW POLAND
Mukherjee Bhaskar, Occupational Health & Safety/Cyclotron Project Australian Nucl. Science and Technology Org. ANSTO / Building 55 PMB-1, MENAI NSW 2234 AUSTRALIA
Nagel G., Deutsche Lufthansa AG Hamburg TW Weg beim Jäger, W-2000 HAMBURG 63 GERMANY
Nikodemova D., Institute of Preventive and Clinical Medicine Limbova 14, CS-833 01 BRATISLAVA CSFR
Nolte R., Physikalisch-Technische Bundesanstalt Bundesallee 100, W-3300 BRAUNSCHWEIG GERMANY
Novotny Tomas, Institute of Radiation Dosimetry Czech Acad. Sci. Na Truhlarce 39/64, CS-180 86 PRAGUE 8 CSFR
Nurdin Guy, Etablissement Technique Central de l'Armement (ETCA) 16 bis Avenue Prieur de la Cote d'Or, F-94 114 ARCUEIL CEDEX FRANCE
de Oliveira W.A., Instituto de Estudos Avancados - IEAV Centro Técnico Aerospacial P.O. Box 6044, SAO JOSE DOS CAMPOS - SP - CEP 12231 BRAZIL
Olko P., Institute of Nuclear Physics Health Physics Lab. ul. Radzikoswkiego 152, PL-31-342 KRAKOW POLAND
Paul D., Faculté des Sciences, Laborat. d'Electronique des Polymères sous Faisceaux Ionique 123 Avenue Albert Thomas, F-870 60 LIMOGES CEDEX FRANCE
Pellicioni Maurizio, Laboratori Nazionali di Frascati INFN Via Enrico Fermi, I-00044 FRASCATI ITALY
Perks C. A., AEA Environmental and Energy B.364 Harwell Laboratory, HARWELL OXFORDSHIRE OX11 0RA UK
Piesch Ernst K.A., Kernforschungszentrum Karlsruhe Postfach 3640, W-7500 KARLSRUHE GERMANY
Pihet P., Universität des Saarlandes, Zentrum f. Umweltforschung, AG Strahlenbiophysik Am Markt 5, Zeile 4, W-6602 DUDWEILER GERMANY

LIST OF PARTICIPANTS

Pitt Eberhard, I. Physikalisches Institut Justus-Liebig-Universität Heinrich-Buff-Ring 16, W-6300 GIESSEN GERMANY
Pöller Fred, Universitätsklinikum Essen Institut für Medizinische Strahlenphysik Hufelandstr. 55, W-4300 ESSEN 1 GERMANY
Portal Guy, CEA - IPSN/DPHD-SDOS B. P. No. 6, F-92265 FONTENAY-AUX-ROSES FRANCE
Posny F., CEA - IPSN/DPHD-SDOS B. P. No. 6, F-92265 FONTENAY-AUX-ROSES CEDEX FRANCE
Pszona S., Institute of Nuclear Studies, PL- 05-400 SWIERK POLAND
Qureshi M.A., QMW College, Dept. of Nucl. Eng. (Nuclear Group) Mile End Road, LONDON E1 4NS UK
Rassow J., Universitätsklinikum Essen Institut für Medizinische Strahlenphysik Hufelandstr. 55, W-4300 ESSEN 1 GERMANY
Regel Kurt, Zentralinstitut für Krebsforschung Berlin-Buch Lindenberger Weg 80, O-1115 BERLIN-BUCH GERMANY
Rimpler A., Bundesamt für Strahlenschutz (BfS) Berlin Waldowallee 117, O-1157 BERLIN GERMANY
Rong Chaofan, Physikalisch-Technische Bundesanstalt Bundesallee 100, W-3300 BRAUNSCHWEIG GERMANY / China Inst. of Atomic Energy P.O. Box 275 (20), 102413 BEIJING P.R. CHINA
Roos Hartmut, Strahlenbiologisches Institut Universität München Schillerstr. 42, W-8000 MÜNCHEN 2 GERMANY
Rosenstock Wolfgang, Fraunhofer-Institut, INT Appelsgarten 2, Postfach 1491, W-5350 EUSKIRCHEN GERMANY
Sannikov A., Institute for High Energy Physics Druzhba Street 14, 121 142284 PROTVINO, MOSCOW REGION RUSSIA
Schmitz Thomas Institut für Medizin Forschungszentrum Jülich Postfach 1913, W-5170 JÜLICH 1 GERMANY
Schraube H., GSF-Forschungszentrum für Umwelt und Gesundheit Ingolstädter Landstr. 1, W-8042 NEUHERBERG GERMANY
Schrewe U. J., Physikalisch-Technische Bundesanstalt Bundesallee 100, W-3300 BRAUNSCHWEIG GERMANY
Schröder Oliver, Inst. für Medizin, Forschungszentrum Jülich Postfach 1913, W-5170 JÜLICH 1 GERMANY
Schuhmacher H., Physikalisch-Technische Bundesanstalt Bundesallee 100, W-3300 BRAUNSCHWEIG GERMANY
Scott Malcolm C., School of Physics and Space Research University of Birmingham, EDGBASTON BIRMINGHAM B15 2TT UK
Seifert R., Strahlenmeßstelle Berlin Soorstr. 84, W-1000 BERLIN 19 GERMANY
Serbat Annick, DGA/ETCA/CEB/DPN/ER 16 bis Avenue Prieur de la Cote d'Or, F-94114 ARCUEIL CEDEX FRANCE
Shaw Peter, National Radiological Protection Board Hospital Lane, COOKRIDGE LEEDS LS16 6RW UK
Siebert B. R. L., Physikalisch-Technische Bundesanstalt Bundesallee 100, W-3300 BRAUNSCHWEIG GERMANY
Silari M., Consiglio Nationale delle Ricerche - ITBA Via Ampere No. 56, I-20131 MILANO ITALY
Simoen Jean-Pierre, CEA/LPRI CE-Saclay BP No. 52, F-91193 GIF-SUR-YVETTE FRANCE
Smirnova Nina, Kurchatov Institute of Atomic Energy Plasma Department Kurchatov sq. 46, MOSCOW 123182 RUSSIA
Sohrabi M., National Radiation Protection Dept. Atomic Energy Organization of Iran P.O. Box 14155-4494, TEHRAN IR IRAN
Spurny F., Institute of Radiation Dosimetry Czech Academy of Sciences Na Truhlarce 39/64, CS-180 86 PRAHA 8 CSFR
Szlavik F., Triumf Meson Facility 4004 Wesbrook Mall, VANCOUVER BC CANADA
Tanner Jennifer, Battelle Pacific Northwest Laboratory MSIN K3-70 Battelle Boulevard, P. O. Box 999, RICHLAND, WA 99352 USA
Tanner Richard, NRPB, CHILTON, DIDCOT OXON OX11 0RQ UK
Taylor G. C., University of Birmingham P. O. Box 363, BIRMINGHAM B15 2TT UK
Thomas D. J., National Physical Laboratory, Division of Radiation Science and Acoustics, Bld. 47 Queens Road, TEDDINGTON, MIDDLESEX TW11 0LW UK
Tommasino Luigi, ENEA / DISP Via Vitaliano Brancati 48, I-00144 ROMA ITALY
Trousil Jaroslav, Institute for Research and Application of Radioisotopes Radiova 1, CS-102 27 PRAHA 10 CSFR
Truter Ernie J., Cape Technikon, Life Sciences Engineering Building P. O. Box 652, KAAPSTAD 8000 SOUTH AFRICA
Varma M., ER-74 E-217 / GTN, Office of Health Physics & Env. Research, US DOE, WASHINGTON, DC 20545 USA
Verhagen Hans W., Netherlands Energy Research Foundation ECN P.O. Box 1, NL-1755 ZG PETTEN THE NETHERLANDS
Vogl K., Institut für Strahlenschutz GSF Forschungszentrum für Umwelt und Gesundheit Ingolstädter Landstraße 1, W-8042 NEUHERBERG GERMANY
Wagner S., Bergiusstr. 2 d, W-3300 BRAUNSCHWEIG GERMANY
Wambersie André, Unité de Radiothérapie Cliniques Univ. St. Luc Avenue Hippocrate 10, B-1200 BRUXELLES BELGIUM
Watt D. E., University of St. Andrews Bute Medical 8 DGS. Annexe Westburn Lane, ST. ANDREWS, FIFE KY16 9TS UK

LIST OF PARTICIPANTS

Weyrauch M., Physikalisch-Technische Bundesanstalt Bundesallee 100, W-3300 BRAUNSCHWEIG GERMANY
White Roger M., Lawrence Livermore National Laboratory L-298 P. O. Box 808, LIVERMORE, CA 94550 USA
Wolber Guy, Electricité de France, Comité de Rad. 3, rue de Messine, F-75384 PARIS CEDEX 08 FRANCE
Yamaguchi Yasuhiro, JAERI 23 Allée de la Perspective, F-91210 DRAVEIL FRANCE
Zamani-Valassiadou Maria, Nucl. Phys. Dept. Aristotle University of Thessaloniki, THESSALONIKI 540 06 GREECE
Zanini Alba, Istituto Nazionale di Fisica Nucleare Via Pietro Giuzia n. 1, I-10125 TORINO ITALY
Zielczynski M., Institute of Atomic Energy Swierk, PL-05-400 OTWOCK POLAND
Zoetelief J., Institute of Applied Radiobiology and Immunology - TNO P.O. Box 5815, NL-2280 RIJSWIJK THE NETHERLANDS

THERMOLUMINESCENCE IN SOLIDS AND ITS APPLICATIONS

K. Mahesh, P. S. Weng and C. Furetta

Contents
Chapter 1. Historical Background of Luminescence and Current Trends in Thermoluminescence
Chapter 2. General Features of Luminescence
Chapter 3. Principles and Methods of Thermoluminescence
Chapter 4. Thermoluminescence Phosphors
Chapter 5. Thermoluminescence Instrumentation
Chapter 6. Models and Theories of Thermoluminescence
Chapter 7. Thermoluminescence-related Phenomena
Chapter 8. Thermoluminescence Applications
Chapter 9. Recent Developments in Thermoluminescence Dosimetry
Appendix   I   Phosphor Terminology
Appendix   II Radiation Quantities and Units
Appendix III Scalar Vector Potentials
Appendix IV Laser Heating Phosphors

Readership – Students, researchers, lecturers and scientists concerned with and applying thermoluminescence techniques.

Scope – This well presented book of 320 pages covers the history, theory and application of thermoluminescence in solids. Very extensive reference lists and bibliographics are included with each chapter.

Price – 50 UKL (Hardback – ISBN 1 870965 00 0). UKL 35 (Softback – ISBN 1 870965 01 9).

ORDER FORM
   To: Nuclear Technology Publishing
        PO Box No 7
        Ashford, Kent, England

       Please supply .... copies of THERMOLUMINESCENCE IN SOLIDS AND ITS APPLICATIONS (*Softback*/Hardback*) at a price of 35*/50* UKL. Remittance of ...... UKL is enclosed*/please invoice me.

Delivery Address   Invoice Address

....

....

....

....

....

Date ....   Signature ....

*Delete as applicable

NTP.J.F.1

7th SYMPOSIUM ON NEUTRON DOSIMETRY
AUTHOR INDEX

Al-Jarallah, M. I. ....(179)
Alberts, W. G. ......(xiv)(313)(323)
Alevra, A. V. ....(223)
Allab, M. ....(329)
Antolkovic, B. ....(31)
Apfel, R. E. ....(343)
Arend, E. ....(213)
Aro, A. C. A. ....(77)(405)
Aroua, A. ....(183)(287)

Baixeras, C. ....(337)
Bardell, A. G. ....(219)(233)
Barelaud, B. ....(363)(367)(371)
Bartlett, D. T. ....(233)(273)
Beauduin, M. ....(41)
Binns, P. J. ....(67)
Birattari, C. ....(193)
Bonnett, D. E. ....(405)
Bos, A. J. J. ....(305)
Boschung, M. ....(183)(243)
Brede, H. J. ....(1)(101)
Brenner, D. J. ....(1)(45)
Britcher, A. R. ....(233)
Broerse, J. J. ....(11)
Buddemeier, B. R. ....(317)
Burghardt, B. ....(179)(267)
Burian, J. ....(453)
Buxerolle, M. ....(125)(239)

Cartone, M. C. ....(437)
Canzi, C. ....(437)
Cartier, F. ....(183)
Caswell, R. S. ....(45)(105)
Cerchiari, U. ....(437)
Chartier, J. L. ....(125)(239)
Coppola, M. ....(35)
Cosgrove, V. P. ....(405)
Coutrakon, G. ....(247)
Covelli, V. ....(35)
Coyne, J. J. ....(49)
Cruz, M. T. ....(143)

De Bartolo, D. ....(437)
Decossas, J. L. ....(363)(367)(371)
Delafield, H. J. ....(227)
Deluca, P. M. Jr. ......(11)(25)(247)
Dewit, L. ....(443)
Di Majo, V. ....(35)
Dicello, J. F. ....(253)
Dietz, E. ....(89)(213)(313)
Dietze, G. ....(11)(31)(165)
Djeffal, S. ....(329)
Domingo, C. ....(337)
Dörschel, B. ....(355)
Dubarry, B. ....(363)(367)(371)

Esposito, A. ....(175)(193)

Facchielli, L. ....(437)
Fargado, P. W. 293
Faust, L. G. ....(171)
Fellinger, J. ....(297)
Fernández, F. ....(337)
Ferrari, A. ....(193)
Francis, T. M. ....(147)(273)
Fratin, L. ....(143)

Gambarini, G. ....(437)
Geard, C. R. ....(45)
Gerdung, S. ....(21)(115)(213)
Gibson, J. A. B. ....(425)
Gmür, K. ....(183)
Golnik, N. ....(139)
Golnik, N. ....(57)
Grecescu, M. ....(183)(287)
Green, S. ....(77)(405)
Grégoire, V. ....(41)
Griffith, R. V. ....(259)
Grillmaier, R. E. ....(115)(213)
Groenewald, W. A. ....(411)
Gueulette, J. ....(41)
Guldbakke, S. ....(313)

Hahn, T. ....(297)
Haight, R. C. ....(11)
Hall, E. J. ....(1)(45)
Hartmann, C. L. ....(25)
Harvey, J. R. ....(325)
Haverkamp, U. ....(447)
Hehn, G. ....(85)
Henniger, J. ....(297)
Herbaut, Y. ....(207)
Hess, A. ....(309)(447)
Hoffmann, W. ....(301)
Hollnagel, R. A. ....(131)(155)
Hough, J. H. ....(67)
Höver, K. H. ....(447)
Hrabovcová, A. ....(291)
Hübner, K. ....(297)
Hudson, I. F. ....(233)
Hunt, J. B. ....(219)
Huynh, V. D. ....(111)

Ipe, N. E. ....(317)

Jahn, U. ....(447)
Jahr, R. ....(xiv)
Jetzke, S. ....(85)(89)(131)
Jones, D. T. L. ....(411)(417)
Józefowicz, K. ....(139)

Kaclík, S. ....(291)
Kacperek, A. ....(405)
Katrrouzi, M. ....(281)
Kawashima, K. ....(11)

Kearsley, E. ....(61)
Khoshnoodi, M. ....(121)
Klein, H. ....(31)(223)
Kluge, H. ....(131)(313)
Knauf, K. ....(97)(223)
Konijnenberg, M. W. ....(443)
Koohi-Fayegh, R. ....(77)
Kopec, M. ....(159)
Kriens, M. ....(309)
Kronholz, H. L. ....(353)(447)
Kunz, A. ....(115)(213)(243)

Laublin, G. ....(41)
Leonowich, J. A. ....(171)
Lerch, P. ....(287)
Lewis, V. E. ....(105)
Lim, T. ....(213)
Liu, J. C. ....(317)
Lommier, B. ....(375)
Lounis, Z. ....(329)
Luguera, E. ....(337)
Luszik-Bhadra, M. ....(313)

Makovicka, L. .....(363)(367)(371)
Manfredotti, C. ....(175)(457)
Marchetto, A. ....(207)
Marek, M. ....(453)
Mares, V. ....(433)
Marino, S. A. ....(45)
Martinuk, Y. ....(347)
Maurício, C. L. P. ....(293)
Meinhold, C. B. ....(151)
Meissner, P. ....(421)(447)
Menzel, H. G. ....(xiv)(11)(115)
Mijnheer, B. J. ....(443)
Miles, C. J. ....(317)
Miller, R. C. ....(45)
Molho, N. ....(437)
More, B. R. ....(219)
Morstin, K. ....(73)(159)
Moss, R. L. ....(443)

Nastasi, U. ....(457)
Nikodemová, D. ....(291)
Nolte, R. ....(21)(101)
Novotný, T. ....(93)

O'Brede, H. J. ....(21)
Olko, P. ....(73)
Olsson, N. ....(11)
Ongaro, C. ....(175)
Ornato, E. ....(457)

Paul, D. ....(363)(367)(371)
Pearson, D. W. ....(25)(247)
Pelliccioni, M. ....(175)(193)
Perks, C. A. ....(85)(227)(425)

AUTHOR INDEX

Pfister, G. ....(433)
Piesch, E. ....(179)(267)
Pihet, P. ....(21)(115)(213)
Piper, R. K. ....(171)
Pirola, L. ....(437)
Pistorius, S. ....(411)
Piters, T. M. ....(305)
Pitt, E. ....(333)(375)
Pöller, F. ....(429)
Portal, G. ....(165)
Posny, F. ....(125)(239)
Prêtre, S. ....(183)(287)
Prouza, Z. ....(453)
Pszona, S. ....(65)

Raaijmakers, C. P. J. ....(443)
Randers-Pehrson, G. ....(45)
Rassow, J. ....(429)(447)
Rataj, J. ....(453)
Rebessi, S. ....(35)
Regel, K. ....(447)
Rimpler, A. ....(189)
Rosenstock, W. ....(351)

Sampsonidis, D. ....(341)
Sannikov, A. V. ....(277)
Sauerwein, W. ....(429)
Savvidis, E. ....(341)

Scharmann, A. ....(333)(375)
Schmelzbach, P. ....(21)
Schmidt, P. ....(297)
Schmidt, R. ....(309)(447)
Schmitz, T. ....(73)
Schmitz, Th. ....(159)
Schraube, H. ..(xiv)(85)(135)(433)
Schreuder, A. N. ....(411)(417)
Schrewe, U. J. ....(27)(101)
Schuhmacher, H. ...(21)(101)(199)
Schulze, J. ....(351)
Scobel, W. ....(309)
Scott, M. C. ....(53)(77)(405)
Seifert, H. ....(355)
Semaschko, N. ....(347)
Shahid, T. ....(77)
Sichirollo, A. E. ....(437)
Siebers, J. V. ....(247)
Siebert, B. R. L. (85)(89)(131)(135)
Silari, M. ....(193)
Simmer, R. ....(333)
Simpson, B. R. S. ....(67)
Smirnova, N. ....(347)
Sohrabi, M. ....(121)(281)
Songsiriritthigul, P. ....(301)
Spurný, F. ....(397)(453)
Stecher-Rasmussen, F. ....(443)
Steele, J. D. ....(273)

Streubel, G. ....(355)

Tanner, J. E. ....(171)(273)
Taylor, G. C. ....(53)(77)(405)
Thomas, D. J. .(85)(135)(219)(233)

Valley, J. -F. ....(183)(287)
Van Dam, J. ....(41)
Vareille, J. C. ....(363)(367)(371)
Vičanová, M. ....(291)
Vilgis, M. ....(267)

Wagner, F. M. ....(433)
Waker, A. J. ....(219)
Wambersie, A. ....(11)(41)(379)
Watkins, P. R. D. ....(443)
Wernli, C. ....(183)(243)
Weyrauch, M. ....(97)
White, R. M. ....(11)
Wittstock, J. ....(223)

Yoder, R. C. ....(317)
Yudelev, M. ....(417)

Zamani, M. ....(341)
Zanini, A. ....(175)(457)
Zielczyński, M. ....(57)(139)
Zoetelief, J. ....(305)

GUIDE TO RADIATION AND RADIOACTIVITY LEVELS AROUND HIGH ENERGY PARTICLE ACCELERATORS

A. H. Sullivan
European Laboratory for Particle Physics, CERN

Contents

Chapter 1   High energy Particle Interactions

Chapter 2   Shielding for High Energy Particle Accelerators

Chapter 3   High Energy Electron Machines

Chapter 4   Induced Radioactivity

Scope The purpose of this guide is to bring together basic data and methods that have been found useful in assessing radiation situations around accelerators and to provide straightforward means of arriving at radiation and induced radioactivity levels that can occur under a wide range of situations, particularly where the basic physics is too complicated to make meaningful absolute calculations.

Readership Researchers, lecturers and scientists in the field of high energy physics; researchers, designers and operators of high energy particle accelerators.

ISBN 1 870965 18 3 (hardback) approx 200 pages. Price £27.00

Nuclear Technology Publishing
P.O. Box 7
Ashford
Kent TN23 1YW
England

APPLIED HEALTH PHYSICS ABSTRACTS AND NOTES

ISBN 0305-7615

– an International abstracts journal published since 1974 with cumulative total of more than 34,000 abstracts covering the world published literature, conferences and reports in all areas of applied health physics (currently about 2,500 abstracts per year). The subject coverage is divided into the following main groups, each subdivided into topic areas

    (i) Radiation Protection
    (ii) Radiation Dosimetry
    (iii) Radiation Effects
    (iv) Measurement Techniques
    (v) Accidents
    (vi) Transport of Radioactive Materials

APPLIED HEALTH PHYSICS ABSTRACTS AND NOTES is a vital tool for all practicing health physicists in hospitals, universities, research establishments and the nuclear industry.

– Published quarterly in January, April, July and October.

– Price 1992 £135 (UK), US$ 295 (outside UK) (AIRMAIL US$ 30 extra if required)

– Please send your subscription order or request a free sample copy:

    Subscription Department
    Nuclear Technology Publishing
    PO Box 7
    ASHFORD,
    Kent TN23 1YW, England